10th South African Conference on Computational and Applied Mechanics (SACAM 2016)

Potchefstroom, South Africa
3 – 5 October 2016

Editor:

Jan-Hendrik Kruger

ISBN: 978-1-5108-3669-3

Printed from e-media with permission by:

Curran Associates, Inc.
57 Morehouse Lane
Red Hook, NY 12571

Some format issues inherent in the e-media version may also appear in this print version.

Printed by Curran Associates, Inc. (2017)

South African Association for Theoretical and Applied Mechanics (SAAM)
c/o Schalk Kok
9-21 Engineering I
Department of Mechanical Engineering
University of Pretoria
South Africa

Phone: 27 21 808 3554
Fax: 27 86 615 5206

Schalk.Kok@uj.ac.za

Additional copies of this publication are available from:

Curran Associates, Inc.
57 Morehouse Lane
Red Hook, NY 12571 USA
Phone: 845-758-0400
Fax: 845-758-2633
Email: curran@proceedings.com
Web: www.proceedings.com

TABLE OF CONTENTS

Foreword

The 10th South African Conference on Computational and Applied Mechanics was held at the Faculty of Engineering, North-West University, Potchefstroom, South Africa from 3 – 5 October 2016. SACAM is held every two years as a forum for presenting current developments in mechanics.

SACAM2016 built on the rich tradition of previous SACAM conferences and received the attention of both local and international participants and served as a platform to share, engage and network on state of the art developments in applied mechanics. The main function of SACAM is to bring together engineers, scientists and applied mathematicians from academia, research and industry to encourage scientific engagement and exchange of ideas in applied mechanics. All participants are thanked for their contributions.

All the papers included in the proceedings were subjected to a peer review process where each paper was reviewed by at least two independent members of the scientific committee to ensure that they met the required standards and fell within the scope of the conference.

The conference was sponsored by Aerotherm CD, Bluetek, the CSIR Centre for High Performance Computing (CHPC), Qfinsoft, M-Tech Industrial and ESTEQ. The organising committee would like thank the sponsors for their support.

Prof CG du Toit
Chairman

Dr JH Kruger
Editor

Under the auspices of the
South African Association for
Computational and Applied Mechanics (SAAM)

NOORDWES-UNIVERSITEIT
NORTH-WEST UNIVERSITY
YUNIBESITI YA BOKONE-BOPHIRIMA ®

SACAM 2016 Scientific Committee and Review Panel

Dr.	Tamer	Abdalrahman	University of Cape Town
Dr.	Alfred Edward Jules	Bogaers	CSIR
Dr.	Johan	Bosman	North-West University
Dr.	Tiri	Chinyoka	University of Cape Town
Dr.	Corne	Coetzee	Stellenbosch University
Dr.	Charles	Crosby	CSIR - CHPC
Dr.	Danie	De Kock	QfinSoft
Prof.	Jaco	Dirker	University of Pretoria
Prof.	Charl Gabriel	Du Toit	North-West University
Dr.	Andrew	Gill	CSIR - CHPC
Prof.	Igle	Gledhill	CSIR
Dr.	Reuben	Govender	University of Cape Town
Dr.	Johan	Heyns	CSIR
Prof.	P Stephan	Heyns	University of Pretoria
Dr.	Jaap	Hoffmann	Stellenbosch University
Dr.	Helen	Inglis	University of Pretoria
Prof.	Attie	Jonker	North-West University
Dr.	Jan-Hendrik	Kruger	North-West University
Dr.	Sunita	Kruger	University of Johannesburg
Prof.	RF	Laubscher	University of Johannesburg
Prof.	Craig	Law	Wits University
Dr.	Philip Wayne	Loveday	CSIR
Mr.	Johan (Jr)	Markgraaff	NWU
Dr.	Andrew	McBride	University of Glasgow
Dr.	Velaphi	Msomi	Cape Peninsula University of Technology
Dr.	Vishnu Visvanathan	Naicker	North-West University
Prof.	Andre Leon	Nel	University of Johannesburg
Dr.	Carl Benjamin	Nel	North-West University
Prof.	Hein	Neomagus	North-West University
Prof.	Graeme John	Oliver	Cape Peninsula University of Technology
Dr.	Oliver	Oxtoby	CSIR
Mr.	Sa-aadat	Parker	University of Cape Town
Prof.	B. Daya	Reddy	University of Cape Town
Dr.	Quinn Gareth	Reynolds	Mintek
Prof.	Pieter	Rousseau	University of Cape Town
Dr.	Mazin	Sirry	University of Cape Town
Dr.	Sebastian	Skatulla	University of Cape Town
Dr.	Willie	Smit	University of Stellenbosch

Under the auspices of the South African Association for Computational and Applied Mechanics (SAAM)

NOORDWES-UNIVERSITEIT
NORTH-WEST UNIVERSITY
YUNIBESITI YA BOKONE-BOPHIRIMA

Dr.	JAvB	Strasheim	University of Stellenbosch
Dr.	Clarence	Suhyah	WITS Reproductive Health & HIV Institute
Dr.	Lagouge Kwanda	Tartibu	University of Johannesburg
Prof.	Thomas	Tengen	Vaal University of Technology
Dr.	Onno	Ubbink	CSIR
Prof.	Kenny	Uren	North-West University
Dr.	Herman	van Antwerpen	M-Tech Industrial
Prof.	Johan	van der Spuy	University of Stellenbosch
Prof.	Martin	van Eldik	North-West University
Dr.	Jan Jacobus	van Rensburg	North-West University
Prof.	Peet	van Schalkwyk	North-West University
Prof.	GPAG	van Zijl	University of Stellenbosch
Prof.	Gerhard	Venter	University of Stellenbosch
Dr.	Martin Philip	Venter	University of Stellenbosch
Mr.	Philip van Zyl	Venter	North-West University

SURFACE ACCURACY AND POINTING ERROR PREDICTION OF A 32 M CLASS RADIO ASTRONOMY TELESCOPE

S. Azankpo*, G. Venter†
Department of Mechanical and Mechatronics Engineering, University of Stellenbosch, Private Bag X1, Matieland 7602, South Africa
*sazankpo@yahoo.com, †gventer@sun.ac.za

ABSTRACT

The African continent witnessed a number of installed telecommunication antennas with 25 m to a 32 m diameters by the Intelsat program during the 1980s. Some of these dishes such as the Ghana 32 m antenna became idle due to the advent of fibre optic communication proving to be faster and more reliable than the telecommunication antennas. The African Very-Long-Baseline Interferometry Network (AVN) aims to convert four of these 32 m class antennas into radio astronomy telescopes. The allowable surface accuracy for a radio telescope is much more stringent than that of a telecommunication antenna. This requires tweaking and testing the telescope's structural integrity and pointing performance to fall within some required tolerances. Generally, the mechanical[1] pointing accuracy of such radio telescopes are influenced by several factors that can be classified as mechanical alignment, structural deformation, and servo drive train errors.

This paper presents the numerical simulation of the surface/pointing accuracy of the Ghana 32 m radio telescope owing to the effect of its structural deformation.

KEYWORDS

Surface Accuracy, Pointing Accuracy, Numerical Simulation, and Structural Deformation.

1 INTRODUCTION

South Africa (SA) is currently the only participating African country in the global Very-Long-Baseline Interferometry (VLBI) science experiments with two telescopes, the 26 m and the 15 m radio telescopes, located at the Hartebeesthoek Radio Astronomy Observatory (HartRAO), in the province of Gauteng. VLBI is a technique whereby radio telescopes with a large distance between them are used to observe the same astronomical radio source, providing a much better resolution than a single dish observation. Countries like Ghana, Mozambique, Kenya, Zambia, Namibia, Madagascar, Botswana, and Mauritius joined the Square Kilometre Array (SKA) - SA in the bid to establish the AVN. The SKA plans to build the world's most sensitive array of radio telescopes to probe the universe in greater details than before [1; 2].

The Ghana 32 m antenna situated at Kuntunse, in the Greater Accra region as shown in figure 1, is earmarked for a conversion into a radio astronomy telescope by the AVN. Its position coordinate is $05°45'01.5''$ N and $00°18'18.4''$ W [3]. The available structural document is a hand drafted manufacturing drawing data pack compiled by TIW Systems Inc. [4] in 1978. A Computer-Aided Drawing (CAD) and Finite Element (FE) model of the antenna were generated from the drawing data pack and site measurements for the numerical analysis as discussed in section 2.

The antenna structure weights about 240,000 kg and sits on top of a concrete building that houses offices, antenna control systems, and other digital backend equipment. It has a Cassegrain focus with paraboloidal primary reflector and a fixed hyperboloidal secondary reflector. The antenna is piloted to point at cosmic sources by means of its elevation and azimuth drive mount systems. Radio wave signals bouncing off the reflectors are transmitted through a beam waveguide (BWG) to the control and processing room and vice versa.

The acceptable surface accuracy of the telescope at an observing frequency of 6.7 GHz, according to Ruze [5], is estimated to be 2.8 mm. The requirement for pointing deviation is less than 10% of the telescope's half-power beam width (HPBW) [6], which is equivalent to 9.6 millidegrees for the Ghana 32 m radio telescope.

[1]There exist other non-mechanical factors such as atmospheric radio frequency (RF) refraction and control system errors that affect pointing accuracy.

Figure 1: The Ghana 32 m Antenna (Image courtesy SKA-SA)

The model was analysed at different elevation angles considering the effect of gravity, the wind and thermal loading with the MSC Patran/Nastran software package. An external program, written in Python, then determines the resulting RMS surface and pointing errors for the Ghana 32 m radio telescope.

The objective of this study is to quantify the RMS surface and pointing errors of the Ghana telescope owing to its structural deformations and to validate the analyses by means of photogrammetry or similar modern surface measurement techniques. Once validated, the FE model may be used to predict the telescope's pointing and tracking performance, and also to optimise the structure for improving pointing performance.

2 THE MECHANICAL MODEL

The mechanical model consist of the CAD and the FE models. The purpose of the CAD model is to generate the component masses particularly for the non-structural components with their centre of mass positions to be used in the FE model.

2.1 THE CAD MODEL

The CAD model was created from the original manufacturer's hand-drafted drawing data pack and consists of about 7,800 structural members. Figure 2 shows the CAD model with the elevation system and the alidade structure assembly. The primary reflector consists of 240 aluminium panels aligned to form a paraboloid surface.

2.2 SIMPLIFICATIONS AND ASSUMPTIONS

To simplify the boundary conditions in the FE model, the following assumptions were made:

- The deformation of the concrete structure upon which the telescope is mounted shall be approximated as a rigid body.
- The azimuth track of the Ghana 32 m telescope is assumed flat and rigid.

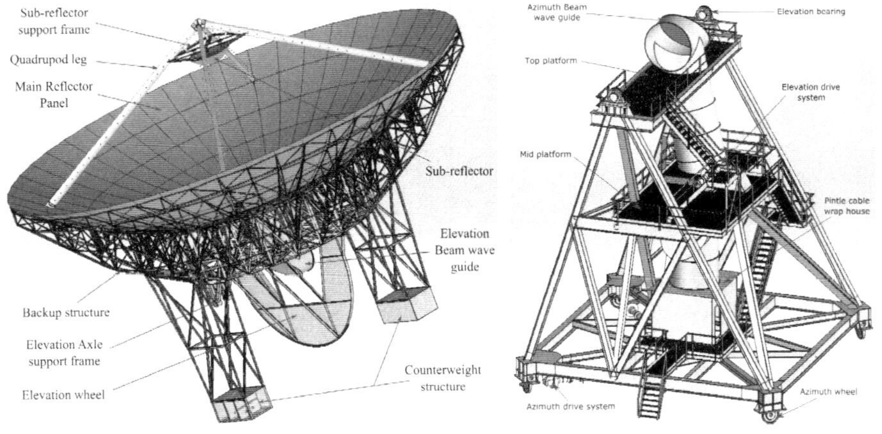

(a) Elevation System (b) Alidade Structure

Figure 2: Ghana 32 m Telescope CAD Model

Structural components with negligible influence on the overall telescope's stiffness were modelled as lumped mass elements. Figure 3 shows the elevation system backup structure trusses simplified as T-section and Z-section.

Figure 3: Simplified Radial Truss Sections

2.3 EXCLUSIONS AND LIMITATIONS

The following are excluded from this research;

- The research does not account for the contribution of degenerated and damaged structural members on the Ghana 32 m telescope since it is not the main focus of the study.
- The displacements of the dish surface nodes tangent to the parabolic surface will not affect the RMS surface and pointing error and therefore are excluded from the pointing error calculation.

2.4 THE FE MODEL

The numerical model consists of 61 different sectional beam elements, mainly angle irons, channels, T-sections, Z-section, I-beams, circular hollow section (CHS), square hollow section (SHS), bars, and elliptical hollow sections. The primary reflector panels, the elevation wheel, and the BWG were created with shell elements. The azimuth drive support structure (figure 4) and the elevation bearing support structure were also modelled with shell elements.

The counterweight, ladder, platforms, sub-reflector and the drive motor assemblies are simplified as non-structural masses at their centre of gravity. RBE2 elements connect the quadrupod legs and the panels to the backup structure and also the elevation/azimuth drive supports to the alidade structure. Figure 5 shows the panels supports created with RBE2 elements linking the primary reflector panels to the elevation backup structure.

Figure 4: Azimuth Drive Support Structure

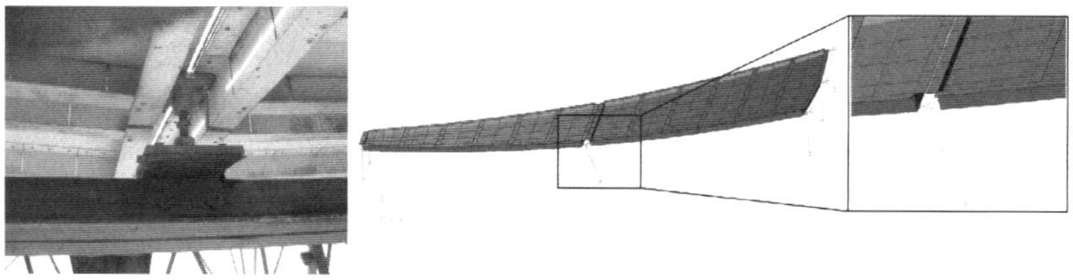

Figure 5: RBE2 Element Panel Support

The complete FE model is shown in figure 6 with 109,952 non-rigid elements, 607,770 degrees of freedom (DOF) and 476 RBE2 elements. Table 1 summarises the physical and mechanical properties of the numerical model.

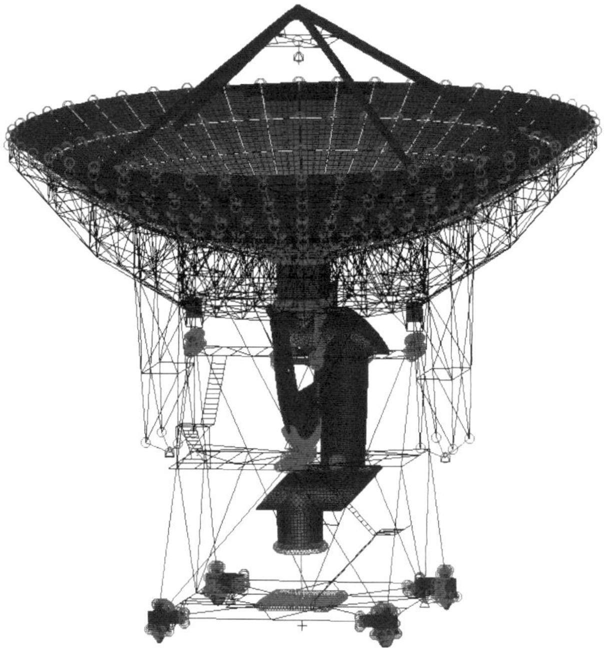

Figure 6: Ghana 32 m Telescope's FE Model

Table 1: Material Properties of Ghana 32 m Telescope

Material	Density (Kg/m^2)	Poisson ratio	Young's modulus (GN/m^2)	Thermal expansivity (m/(m K))	Application region
Al 6061 T6	2700	0.33	68.9	22.2 x 10^{-6}	Main reflector panels
ASTM A36	7850	0.26	200	12 x 10^{-6}	Alidade, backup structure, etc.
AISI 1022	7868	0.29	205	12 x 10^{-6}	Elevation axle
AISI 1040	7845	0.29	200	12 x 10^{-6}	Azimuth axle

Reference: *www.matweb.com*

3 LOAD CASES AND ERROR ANALYSIS

The effect of gravity, thermal and wind loading on the Ghana 32 m telescope was investigated for five different orientations at 0°, 60°, 90°, 120°, 180° elevation angles. The load cases at 120° and 180° elevation angles represent the orientation of the telescope at 60° and 0° respectively rotated through 180° azimuth angle. These selected load cases are to enable comparison with published wind load data and also check the model's consistency.

The gravity loading was implemented in the model by activating 1 g downward translational inertia load. The thermal load, assumed to be a function of the position of the sun, was achieved by applying a 10 K linear thermal gradient vertically across the entire telescope from an ambient temperature of 296.15 K as illustrated in figure 7. This assumption is reasonable as compared to site measured data from the Karoo Array Telescope (KAT-7) and similar analyses used in the thermal analysis of the MeerKAT Telescope [7].

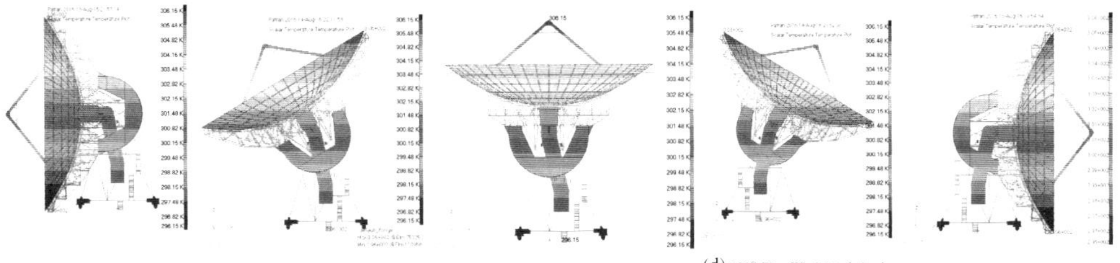

(a) 0° Elev. Wind Attack Angle (b) 60° Elev. Wind Attack Angle (c) 90° Elev. Wind Attack Angle (d) 120° Elev. Wind Attack Angle (e) 180° Elev. Wind Attack Angle

Figure 7: Thermal Loading at Different Elevation Angles

Kron [8] presented how existing wind tunnel pressure coefficient data can be extended to other antennas when simulating wind loading. The wind tunnel test data was obtained from a 26 m antenna with a focal-to-diameter ratio of $\frac{f}{D}$=0.33 which closely compares with that of the Ghana 32 m radio telescope ($\frac{f}{D}$=0.32). A Python program interpolates the wind tunnel pressure coefficient data associated with each of the 11,496 reflector surface nodes on the Ghana 32 m telescope for similar dish orientations. The individual load case nodal wind forces on the dish are calculated from equation 1.

$$F_{sw}^{i} = q\, C_{d}^{i}\, A_{pda}^{i} \tag{1}$$

where

$$q = \frac{1}{2}\rho\, v^2$$

and q is the dynamic pressure, C_d^i is the nodal drag coefficient, A_{pda}^i is the projected dish surface area per grid point normal to the wind direction, ρ is the density of air, and v is the wind speed. The air density, ρ of 1.1684 kg/m^3 is assumed at an average site temperature of 25 °C, 101 kPa pressure, and 22 °C dew point.

Twenty-three different load cases were simulated to investigate the effect of only gravity, thermal, and wind loading. Another set of load cases considered the combined effect of the three loading conditions and the effect of varying wind speed on the Ghana 32 m radio telescope. The dish surface displacements obtained from the FE analysis is corrupted with the reflectors rigid body motions, which are eliminated by applying a series of Zernike Fourier polynomials, Z [9] given by equation 2.

$$Z(x,y) = \sum_{j=1}^{m} C_j \phi_j \tag{2}$$

where ϕ, C, and m represent the Zernike term, Zernike term's coefficient, and the number of polynomial terms respectively. The coefficients of the first three terms quantify the bias, and the X- and Y - tilts which constitute the rigid body motion in the FE analysis results. The subsequent coefficients consider aberration popularly used in optics but not required in the present analysis. The RMS surface and pointing errors are then calculated by fitting an idealised paraboloid through the displaced reflector nodes after the noise in the data cloud are eliminated.

The surface error, σ_i associated with each node on the dish surface is the difference between the undeformed nodal focal distance, uf_d^i and the best fit paraboloid nodal focal distance, ff_d^i as illustrated in figure 8 and expressed by equation 3.

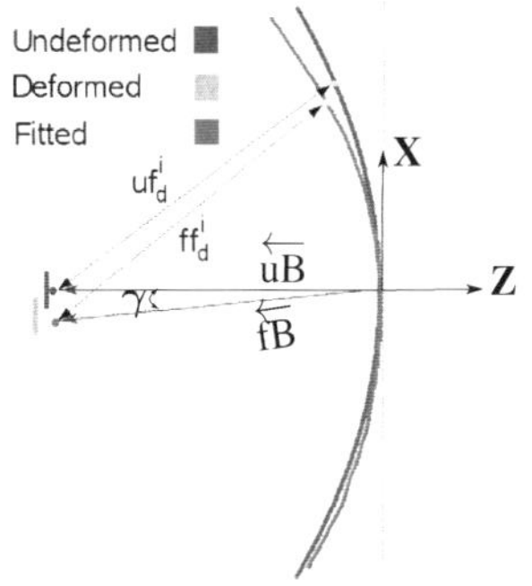

Figure 8: Dish Surface Errors (Image adapted from Joubert [10])

$$\sigma_i = uf_d^i - ff_d^i \tag{3}$$

The reflector surface accuracy, σ_{rms} is the root-mean-square value of the surface errors defined in equation 3, and can be written as equation 4.

$$\sigma_{rms} = \sqrt{\frac{\sum_{i=1}^{n} \sigma_i^2 A_i}{\sum_{i=1}^{n} A_i}} \tag{4}$$

where i, n and A_i are the grid node, the total number of grid nodes, and the surface area per grid node.

The pointing error, γ owing to the telescope's structural deformation is the quantitative measure of the angular offset between the undeformed beam axis and the fitted paraboloid beam axis as also illustrated in figure 8. It is also known as "tracking error" during tracking of a celestial source(s) and is calculated using equation 5.

$$\gamma = \cos^{-1}\left(\frac{\overline{uB}.\overline{fB}}{||\overline{uB}||.||\overline{fB}||}\right) \tag{5}$$

The pointing accuracy, γ_{rms} is the root-mean-square value of the pointing errors of a given number of load cases and is given by equation 6.

$$\gamma_{rms} = \sqrt{\frac{\sum_{l=1}^{t} \gamma_i^2}{t}} \tag{6}$$

where l and t are the load case and the total number of load cases respectively under investigation.

4 PRELIMINARY RESULTS

The surface accuracy due to wind loading on the Ghana 32 m dish based on site median wind speed of 5.56 m/s [6] was compared to published data by Kron [8] on the Jet Propulsion Laboratory (JPL) 26 m antenna for similar wind attack angles as illustrated in figure 9. The RMS surface errors from both dishes show similar trends and order of magnitude. The worse case error occurred at the 60° wind attack angle position with 0.2946 mm and 0.2778 mm for the JPL 26 m and the Ghana 32 m dishes respectively. Both dishes recorded their lowest RMS error at 90° elevation angle.

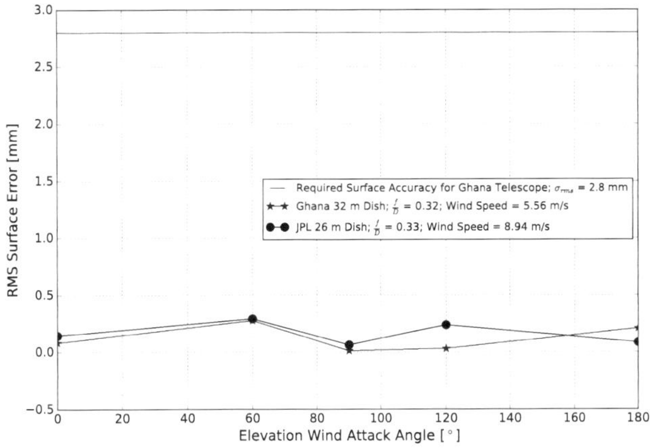

Figure 9: JPL 26 m and Ghana 32 m Dish Surface Accuracy Comparison

The wind loading on the Ghana 32 m radio telescope was varied from 3 m/s to 13.41 m/s with the dish at the 60° wind attack angle position. Figure 10 shows the RMS surface and pointing errors plotted against wind speeds. An incremental value in both the RMS surface and pointing error was registered with 1.619 mm and 0.0011° values at the 13.41 m/s wind speed respectively. This fall within the required tolerance value for science observations as discussed in paragraph 4, section 1. The average maximum wind speed at the site is about 8 m/s.

Figure 11 and figure 12 show the surface accuracy and the pointing errors for the wind, gravity, thermal and combined load cases for different dish orientations of the Ghana 32 m telescope. The surface accuracy values for the load cases fall within the required tolerances except for the dish at the 0° and 180° positions where observations rarely happen [6]. The pointing errors also show compliance for the individual load cases except for the combined load cases at the 60°, 90°, and 120° positions. These errors will be integrated into the radio telescope's pointing model which also includes other measured errors used by software control systems for correcting pointing deviation of the telescope.

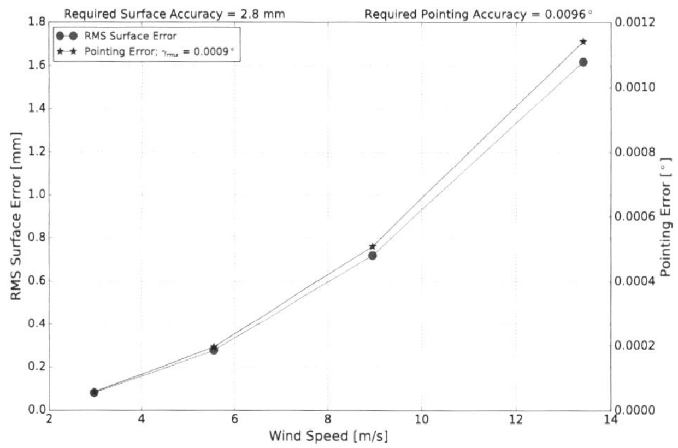

Figure 10: Telescope at 60° subject to Varying Wind Load

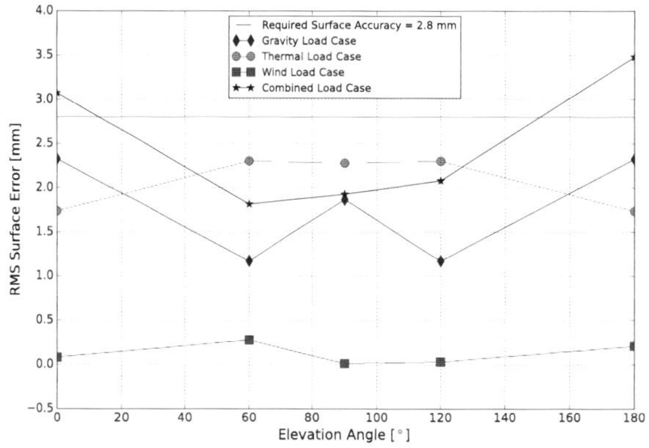

Figure 11: Plot of Ghana 32 m Telescope Surface Accuracy

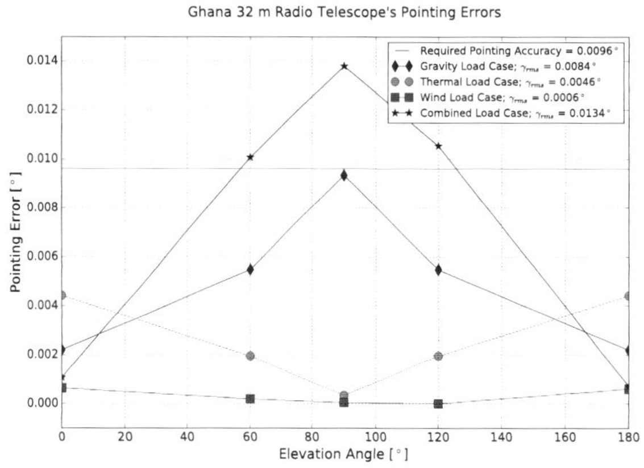

Figure 12: Plot of Ghana 32 m Telescope's Pointing Errors induced by structural deformation

5 CONCLUDING REMARKS

The FE model for the Ghana 32 m radio telescope was created that can serve as a foundation for future analysis aimed at improving the pointing performance of the radio telescope. The preliminary surface accuracy results show compliance with the tolerance requirement of 2.8 mm for science observations.

Additionally, photogrammetry or similar experimental surface measurement techniques will be useful for further comparison and validation. The model can be used to study areas that can be stiffened to obtained increased accuracy and the dynamics analysis of the drive systems to improve the telescope's performance.

ACKNOWLEDGEMENTS

The financial assistance of the National Research Foundation towards this research is acknowledged. Opinions expressed in this paper and the conclusions arrived at, are those of the authors, and are not necessarily to be attributed to the National Research Foundation.

References

[1] Gaylard, M.J., Bietenholz, M.F., Combrinck, L., Booth, R.S., Buchner, S.J., Fanaroff, B.L., MacLeod, G.C., Nicolson, G.D., Quick, J.F.H., Stronkhorst, P. and Venkatasubramani, T.L.: An African VLBI Network of radio telescopes. In: *Proceedings of SAIP2011, the 56th Annual Conference of the South African Institute of Physics,Edited by I. Basson and A.E. Botha*, pp. 473–478. 2011.

[2] Gaylard, M.J.: Expanding radio astronomy in africa. *IOP Conference Series: Materials Science and Engineering*, vol. 44, no. 1, p. 012020, 2013.

[3] Asabere, B.D., Gaylard, M.J., Horellou, C., Winkler, H. and Jarrett, T.: Radio astronomy in Africa: the case of Ghana. In: Engelbrecht, C. and Karataglidis, S. (eds.), *Proceedings of SAIP2014, the 59th Annual Conference of the South African Institute of Physics ISBN: 978-0-620-65391-6*, pp. 296–301. University of Johannesburg, 2014.

[4] TIW: *Ghana Kuntunse 32 m Antenna Assembly Drawing*. TIW SYSTEMS INC., 2460 Embarcadero Way, Palo Alto, CA 94303, rev. 1 edn, 1978.

[5] Ruze, J.: Antenna tolerance theory no.8212;a review. *Proceedings of the IEEE*, vol. 54, no. 4, pp. 633–640, April 1966. ISSN 0018-9219.

[6] Schietekat, S.: Kuntunse VLBI Station Conversion Project: User Requirements Specification(URS). Report Document number:A0200-0001-000, Rev.1, Square Kilometer Array - Africa, 2014 (SKA Internal Document).

[7] Mutzberg, U.: Technical Note on Structural Analyses - MeerKAT Dual Offset Antenna. Report TN-1012033-21300-01-01, R3.0, Square Kilometer Array South Africa/Vertex Antennentechnik GmbH, 2014.

[8] Kron, M.: Load distribution on the surface of paraboloidal reflector antennas. *moment*, vol. 10, p. 2, 1971.

[9] Genberg, V.L.: Optical surface evaluation. In: Cohen, L.M. (ed.), *Proceedings of the SPIE Symposium on Structural Mechanics of Optical Systems*, vol. 0450, pp. 81–87. 1984.

[10] Joubert, N.J.D.: *Numerical design optimisation for the Karoo Array Telescope*. MScEng, Mechanical Engineering, University of Stellenbosch, Stellenbosch, South Africa, 2009.

A design and analysis of a multifunctional composite structure for nano-satellites

Primary Author:

Mr Jeffrey C. Ball (Cape Peninsula University of Technology - PLMCC/F'SATI) – jcball@worldonline.co.za

Supervisor:

Prof. Graeme Oliver (Cape Peninsula University of Technology) – oliverg@cput.ac.za

Co-Supervisor:

Mr Micheal Petersen (Cape Peninsula University of Technology-PLMCC) – petersenmi@cput.ac.za

Abstract

This paper covers the methodology and consideration used to create a concept multifunctional composite structure for a Cube Satellite (CubeSat). A brief background into technology and the design considerations needed for this research has also been provided. A micro-mechanics model was used to provide the predicted material properties for the twill weave carbon fibre fabric lamina properties. A numerical finite element analysis was done using the lamina data from the micro-mechanics equations and the macro-mechanics input parameters for a mixed model based composite laminate specification in ABAQUS. The addition of a progressive failure analysis tool for ABAQUS was used to provide accurate material failure characteristics when the various lamina ultimate tensile strengths are reached. Empirical tests in the form of tensile and bending/flexure tests of the concept material has also been done to verify the validity of the numerical results from the FEA simulations based on the predicted data of the mathematical models. Good agreement was seen between the numerical FEA results and the empirical material tests that were done, this provides a sound confirmation and validation that the methodology and approach used for the numerical analysis is acceptable.

Keywords

Finite element analysis, micro-mechanics, CubeSat, nano-satellite, structural design, multifunctional composite, progressive failure analysis.

Introduction:

In the aerospace industry the requirement for optimising each and every subsystem concurrently is a fundamental challenge. It is important that all they subsystems of an aerospace system are optimised together to allow the best overall performance possible (Fortescue et al., [1]). There is no point over optimising a single subsystem, when other subsystems are not and cannot achieve the same level of performance. In essence, the system can only really be made as good as the worst performing subsystem. Within the nano-satellite community, research efforts are now turning to multifunctional (also called SMART) composite structures. The improvement of the functionality of the structural component allows some otherwise basic activities from other subsystems to be handled now by the structure, thus opening up room for further capabilities to be added to those relieved subsystems.

Cube Satellites or CubeSat's for short, are a class of satellite that typically ranges in mass from 1 to 10 kg. CubeSats are becoming very popular with small organizations and educational institutions across the world as a means to do space exploration at a fraction of the cost. Generally when major institutions or companies launch large satellites, there is a portion of the payload of the launch vehicle that is not utilised. CubeSats address this opportunity by allowing multiple smaller institutions to "purchase" an area inside launch pods that are placed in this available space. The company that began the CubeSat program created a set of standards that governed the dimensional, weight and other geometric specifications in order to allow organisations to develop geometrically similar CubeSats to be easily placed in the launch pods. Due to the size and weight of CubeSats, they fall into a category known as nano-satellites. Inevitably due to the strict restrictions in space and weight available, CubeSat designs have to continuously optimise utilisation of space to allow more electronics and sub-systems to be built into the satellite, but still not to exceed the weight restriction assigned to its volumetric size.

CubeSat design requirements:

CubeSats have standardized structural dimensions every manufacturer and satellite producer must conform to. The outer structural dimensions specifications are strictly controlled by the standards to ensure that the satellite will interface successfully with the launch pod placed in the launch vehicle. Multiple satellites produced or created by various institutions or companies across the world can then be integrated into a single launch pod on a particular launch vehicle. The standardisation, for obvious reasons reduces the launch cost each CubeSat unit in comparison to uniquely designed satellites. A full detailed document of the CubeSat design standards is available (Cal-Poly-SLO, [2]), however a list of the key design considerations from the document is provided below:

- Deployables must me contained by the CubeSat.
- Rails must have a minimum width of 8.5 mm.
- Rails must have a surface roughness of less than 1.6 μm
- The ends of the rails must have at least a 6.5 mm x 6.5 mm contact area.
- Maximum mass of the 1U CubeSat shall be 1.33kg.
- The rails and stand-off, which contact the P-POD and adjacent CubeSats must be hard anodized aluminium to prevent cold welding within the P-POD.

Space environment considerations:

The space environment differs to that of the environment we know on earth and it differs quite a lot. The earth's atmosphere provides a great deal of protection and calmness to the environment we experience on the surface of the earth. Although is space there is no rain or weather in terms of how we know it to be, other factors that are not experienced on the ground are now introduced. The satellite deals with the earth's environment during its creation, assembly and launch. Once in orbit the effects of radiation, debris and thermal environment can cause components to perform in different ways that in some cases can be implicit and if missed will show up as an emergent property of the system. The environments in which the satellite will be exposed to needs to be considered for the structural design of a spacecraft.

The pre-launch phase of the satellites life can be many month or years in length and exposure to the atmosphere can create issues such as corrosion or distortion of the part depending on how the components are stored with in the construction and development phase. The launch environment is the most dangerous time of a satellites lifecycle (Fortescue et al., [1]), this is due to the large gravity loads, vibrations and shocks experienced by the satellite as the payload of the launch vehicle. The space environment has large amounts of radiation and energized particles which are carried to earth by solar winds. The high energy and radioactive neutrons, protons, electrons and ions mainly affect the integrated circuit components which can create single event upsets in the electronic subsystems (Fortescue et al., [1]). Thermal radiation from the sun creates heating issues and thermal management in the satellite needs to be accounted for. Due to the lack of atmosphere, convection to the environment is not available and thus the thermal heat dissipation can only be conducted between parts of the satellite and radiated back into space. All of the key environmental factors has been considered in the concept design creation for the Multi-functional composite (MFC) structure for the CubeSat.

Multifunctional Composite Technology:

Composite materials, namely plastic reinforced composites are structurally made up of a combination of two materials, a fibre and a matrix material (Campbell, [3]). The fibre that will be the focus in this research area is mainly glass and carbon fibres. Composites are created by layering multiple fibre sheets and impregnating it with a matrix, usually an epoxy or polyester resin system. Depending on the method of construction, the curing process may require an atmospheric condition with a different temperature and/or pressure compared to ambient.

The purpose of Multi-functional composite (MFC) structures is to create a composite which performs multiple structural functions as well as multiple non-structural functions (Gibson, [4]). As taught in engineering, material structural functions consist primarily on strength, fatigue strength, ductility, stiffness, energy absorption, damping, and thermal stability. Non-structural function could consist of electrical conductivity, thermal conductivity, material condition sensing and deformation, energy harvesting, energy storage, self-healing capabilities, electromagnetic interference (EMI) shielding, biodegradability and multiple other functions (Gibson, [4]).

A review of previous academic research into composite material structures for CubeSats reveals that a CubeSat project (NUTS) based at the Norwegian University of Science and Technology, has been in a research and development phase of a carbon fibre structure since 2011 (Haland, [5]). The NUTS project however only considered the use of composite materials and no integration of other materials to create secondary functions or overall MFC structure. Previous development in MFC structural technology for spacecraft has been conducted for use in larger satellite platforms.

For this research effort, the primary focus areas for the first prototype MFC structure for a CubeSat configuration, considers the integration of electrical conducting tracks and basic circuitry components onto a carbon composite structural material layup. The main focus is on the mechanical properties and mechanics of the laminate construction. Composite micro-mechanics formulations and the material properties of the copper foil layer is analysed and used to create the numerical model of the overall material laminate configuration. Numerical modelling software has been used to perform the overall macro-mechanics modelling of the laminate configuration. The software package used for this research is *ABAQUS FEA*. The addition of a composite progressive failure analysis (PFA) plug-in tool for *ABAQUS*, *Autodesk Helius PFA* has been used to produce a more accurate composite material characterisation.

Material and structural concept design:

A material laminate concept design was chosen from various different design configurations. Consideration of using a core material, the location and placement of the copper foil layer, and the method of manufacture were all driving factors in the selection process. Mechanical properties of the complete laminate configuration were also considered. A review of standard printed circuit boards (PCB's) that are used on current CubeSat's was conducted and used as a benchmark or baseline for the first iteration of the material configuration design. The main information used from the review was to create an initial iteration that had similar dimensional characteristics to the standard PCB panels currently used, but to have improved mechanical properties based on the known information of the carbon composite material that was chosen. One of the primary reasons for the specific type of carbon fibre chosen for the design is the availability of the material in the research lab, material cost consideration for a prototype and manufacturability of the composite panel based on the capabilities of the equipment available at the time. Figure 1 illustrates the composite layup chosen for the CubeSat panels.

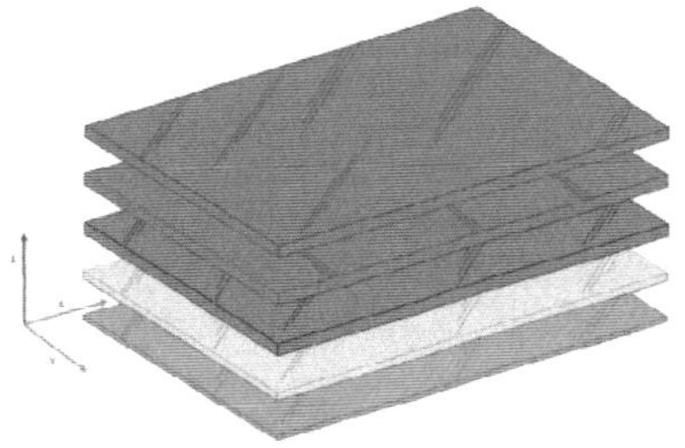

Figure 1: Composite laminate design (Carbon fibre 0/45/0 degree, adhesive/backing layer, copper foil layer.

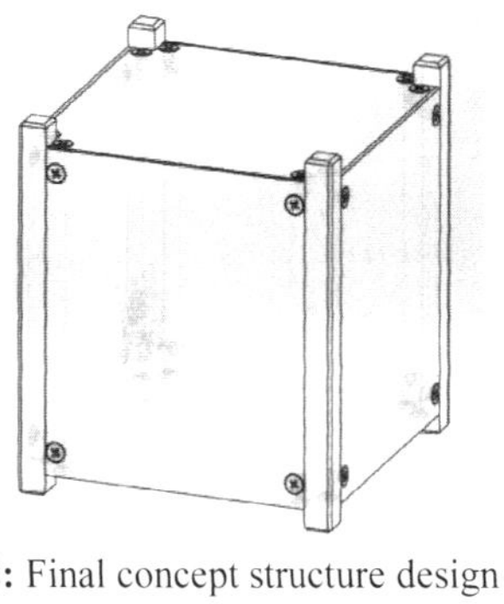

Figure 2: Final concept structure design chosen.

The composite panels are integrated in the overall structural design for the prototype CubeSat. The CubeSat standards and the environmental conditions that the structural will be exposed to were all considered. Once again multiple design concepts were created and the selected concept was chosen based on manufacturability, practicality, modularity of the design, space optimisation and access to internal subsystems during assembly of the CubeSat. The selected design uses fasteners to attach the composite panels to aluminium support posts in the corners of the CubeSat structure. The selected structural design conforms the CubeSat standards document for the structural design. Each panel can be removed or replaced by a custom or purpose designed panel to provide deployable functions for equipment such as solar panels, sensors or antennas. Figure 2 shows the concept design that was finally chosen. The concept material layup and structural design will be the concepts modelled in the numerical and empirical analysis phases.

Micro-mechanics formulation of the composite material:

The numerical analysis in ABAQUS FEA will perform the macro-mechanics formulation and simulation of the complete laminate construction. Micro-mechanics needs to be done to find the material properties for each carbon lamina. The orientation of the lamina, adhesive and copper foil layers will be considered in the composite material laminate specification in ABAQUS. After reviewing various literature information and numerical formulations for the micro-mechanics of woven fabric composites, the implementation of the model presented in a composite handbook by (Barbero, [6]). The application of this theory is further explained in a paper by (Adumitroaie and Barbero, [7]), for use with twill weave composite fabric materials. A twill weave fabric is the construction type used in this research. During the paper a review of a repetitive unit cell (RUC) is used to consider only a complete section of the woven fabric profile. The RUC is the smallest area of the 2D fabric that has a full representation of the woven geometry as depicted in Figure 3. The dotted rectangular section indicates the smallest area of the fabric that is needed in order to be repeated to form the completed fabric profile. Figure 3 also shows a geometrical representation of the 3D profile of the fabric and the variable used to define the profile (fibre width, gap between tows, thickness, etc.).

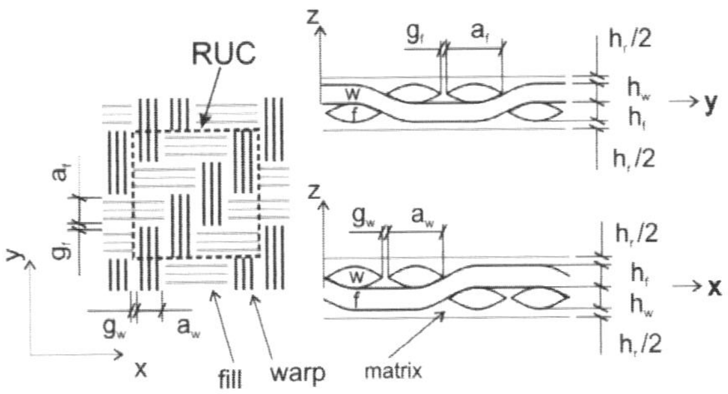

Figure 3: RUC of a 2/2 twill weave fabric geometric parameters (Adumitroaie and Barbero, [7])

The models presented by both (Adumitroaie and Barbero, [7]) and (Barbero, [6]), are an attempt to create a set of generalised equations that can be applied to any woven fabric composite material implementing classical laminate theory (CLT), and by characterising and parameterising the fabric construction. In this case however, focus is only placed on the twill weave fabric equations to stay within the research scope.

Simplified CLT and rule of mixtures (ROM) equations are initially applied to find the tow composite fibre properties with respect to its own material coordinate system (infused with the resin), the next stage is an to evaluate the tows within the fabric construction in terms of the effect created by the undulations of the fill and wrap tows of the fabric. The fill and wrap cords are individual termed as a tow. A tow is made up of much finer hair like carbon filaments. For the twill weave fabric used in this case, the tow structure used for the fill and wrap cords of the fabric are identical in physical structure. Firstly, the transversely isotropic compliance matrix of the fabric tow needs to found using equation 1.

$$S = \begin{pmatrix} 1/E_{11} & -\nu_{21}/E_{22} & -\nu_{31}/E_{33} & 0 & 0 & 0 \\ -\nu_{12}/E_{11} & 1/E_{22} & -\nu_{32}/E_{33} & 0 & 0 & 0 \\ -\nu_{13}/E_{11} & -\nu_{23}/E_{22} & 1/E_{33} & 0 & 0 & 0 \\ 0 & 0 & 0 & 1/G_{23} & 0 & 0 \\ 0 & 0 & 0 & 0 & 1/G_{13} & 0 \\ 0 & 0 & 0 & 0 & 0 & 1/G_{12} \end{pmatrix} \quad (1)$$

E_{11} : Modulus of elasticity in the fibre direction.
E_{22} : Modulus of elasticity in the direction transverse to the fibres.
G_{12} : In-plane shear modulus.
G_{23} : Out of plane shear modulus.
ν_{12} : In-plane Poisson's ratio.

Then a rotation matrix for the fill and wrap needs to be constructed, where the fill undergoes a clockwise rotation due the undulation of the tow through the fabric and the wrap undergoes firstly a 90 degree rotation about the z-axis in Figure 3, relative to the fill direction and then undulates as it crosses the fabric (Adumitroaie and Barbero, [7]). The transformation matrix for the fill and wrap tows are given by equations 2 and 3 respectively.

$$T_f = \begin{bmatrix} c_f^2 & 0 & s_f^2 & 0 & 2c_f s_f & 0 \\ 0 & 1 & 0 & 0 & 0 & 0 \\ s_f^2 & 0 & c_f^2 & 0 & -2c_f s_f & 0 \\ 0 & 0 & 0 & c_f & 0 & -s_f \\ -c_f s_f & 0 & c_f s_f & 0 & c_f^2 - s_f^2 & 0 \\ 0 & 0 & 0 & s_f & 0 & c_f \end{bmatrix} \quad (2)$$

$$T_w = \begin{bmatrix} 0 & c_w^2 & s_w^2 & -2c_w s_w & 0 & 0 \\ 1 & 0 & 0 & 0 & 0 & 0 \\ 0 & s_w^2 & c_w^2 & 2c_w s_w & 0 & 0 \\ 0 & 0 & 0 & 0 & -c_w & -s_w \\ 0 & c_w s_w & -c_w s_w & c_w^2 - s_w^2 & 0 & 0 \\ 0 & 0 & 0 & 0 & s_w & -c_w \end{bmatrix} \quad (3)$$

Where:

$$s_f(x) = sin\theta_f(x)$$
$$c_f(x) = cos\theta_f(x)$$
$$s_w(y) = sin\theta_w(y)$$
$$c_w(y) = cos\theta_w(y)$$

Thereafter, the transformed compliance matrix can be found using equations 4 and 5. This is now with respect to the material coordinate system (x,y,z).

$$\bar{S}_f(x) = (T_f(x))^T \cdot S \cdot T_f(x) \quad (4)$$

$$\bar{S}_w(x) = (T_w(x))^T \cdot S \cdot T_w(x) \quad (5)$$

For each transformed compliance matrix, a 6x6 matrix would have resulted for the fill and wrap. Applying the plane stress assumption to the compliance matrix we will only consider the terms of the matrix that correspond with i,j = 1,2,6. The reduced stiffness matrix in the laminate co-ordinate system for the fill and wrap will be:

$$Q^f = \begin{bmatrix} \bar{Q}_{11}^f & \bar{Q}_{12}^f & 0 \\ \bar{Q}_{21}^f & \bar{Q}_{22}^f & 0 \\ 0 & 0 & \bar{Q}_{66}^f \end{bmatrix} = \begin{bmatrix} \bar{S}_{11}^f & \bar{S}_{12}^f & 0 \\ \bar{S}_{21}^f & \bar{S}_{22}^f & 0 \\ 0 & 0 & \bar{S}_{66}^f \end{bmatrix}^{-1} \quad (6)$$

$$\bar{Q}_w = \begin{bmatrix} \bar{Q}_{11}^w & \bar{Q}_{12}^w & 0 \\ \bar{Q}_{21}^w & \bar{Q}_{22}^w & 0 \\ 0 & 0 & \bar{Q}_{66}^w \end{bmatrix} = \begin{bmatrix} \bar{S}_{11}^w & \bar{S}_{12}^w & 0 \\ \bar{S}_{21}^w & \bar{S}_{22}^w & 0 \\ 0 & 0 & \bar{S}_{66}^w \end{bmatrix}^{-1} \quad (7)$$

The reduced stiffness matrix is the inverse of the reduced compliance matrix in terms of the laminate material coordinates. Figure 4 provided in the book by (Barbero, [6]) shows the material coordinate system vs. the laminate coordinate system (x,y,z) and the angle of undulation of the fill and wrap tows as needed in equations 2 and 3.

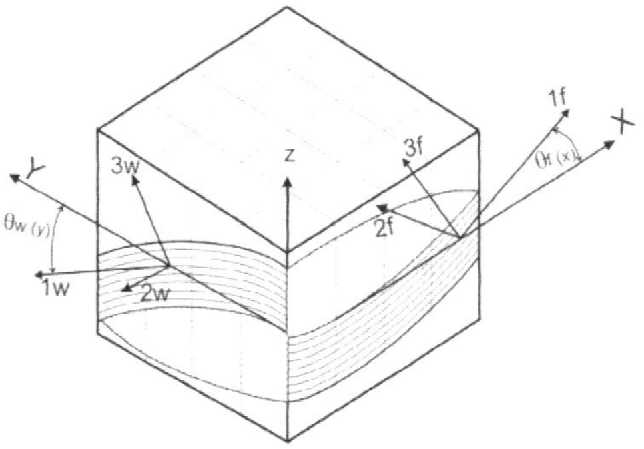

Figure 4: The material coordinate system vs. the laminate coordinate system (x,y,z). (Barbero, [6])

Taking the theory presented and the formulations provided by (Barbero, [6]) and (Roylance, [8]), a combined and simplified formula can be created to find the extensional stiffness matrix [A] as seen in equation 8.

$$[A] = \frac{4}{(a_f + g_f)(a_w + g_w)} \sum_{k=1}^{N} \left[\bar{Q}_{ij}^{f,w}\right] \times \left(\frac{Z}{2}\right) \tag{8}$$

Where:

- N is the in-plane traction.
- Z is the thickness of the final lamina.
- $\alpha_{f,w}$ is the width of the tow for the fill and wrap.
- $g_{f,w}$ is the gap between the fill and wrap tows.

The final matrix that is needed is the laminate compliance matrix which is found by the inverse of the extensional stiffness matrix [A]. Once the laminate compliance matrix is known, the lamina properties of the fabric composite can then be found from the primary values of the matrix and the lamina height. A program was written in MATLAB using the known material inputs and the equations of this micro-mechanics model to produce the lamina data that would be inputted into the FEA software.

Numerical FEA modelling of the material and structure:

A virtual tensile and bending test model has been created in ABAQUS, along with a structural model on which a static load analysis, Natural frequency and dynamic load response analysis has been conducted. A small impact test has also been done to view the stress and possible interaction characteristics if struck with debris. Note it is assumed that the reader has a background with FEA and a basic understanding of ABAQUS. The specific loads and analysis steps for each simulation will be expanded on in the relevant sections that follow.

Overall for the FEA model of the composite material, only each individual lamina is being considered in determining the mechanical properties of the plies. Although this is not a true representation of the ply stacking thickness as there can be an in and out of phase shift between the fabric layers, where the fabric tows clash or interlock between fabric layers when stacked. This formulation however is sufficient enough to provide an indication that is lower than the real world value by a certain margin due to the in accuracy of the overall volume fibre fraction. This can be used logically as an offset for manufacturing inaccuracies or small impurities that may be present in the material and create inconsistencies. Due to the high strength to weight ratio requirement, the most optimal setup needs to be obtained for the construction. The implications of damage mechanics are considered only on a very light level by implementing the PFA plugin for ABAQUS. The key in the aerospace industry is not to design too much redundancy into the components to make them overly strong and heavy, but also to allow enough margin in the design for slightly more loading than what is expected. The working condition of nano-satellites once in space are very low stress in terms of direct loading, but rather focus needs to be more of the effect of the environmental conditions that may damage or deteriorate the material strength.

A mixed modelling or lamina level technique was adopted for the creation of the FEA model. Each lamina will be modelled separately and stacked in a ply layering sequence to represent the material configuration that is chosen. Continuum shell elements will be used to create the element structure of the FEA model. Continuum shell elements require the explicit geometry of the model and the laminate. Generally 3D solid elements would be used, however continuum shell elements enforce the FSDT (First-order Shear Deformation Theory).

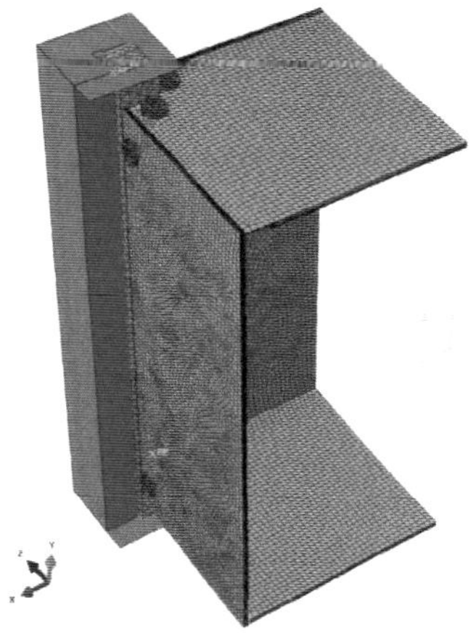

Figure 5: Quarter FEA model of the structural assembly.

Continuum shell elements only have displacement degrees of freedom (Barbero, [9]) and can therefore be stacked as well as connected to solid elements. Continuum shell elements discretize the 3D geometry in the same manner as solid elements would.

For the multifunctional composite FEA model is first created by extruding a profile sketch with the required dimensions to the required thickness to create a 3D solid body. The body is then partitioned through the thickness to create the various layers at their relative thicknesses. The overall thickness is very close to 1.4 mm, where each carbon fabric layer (3 layers) is 0.3 mm thick, the adhesive layer is 0.2 mm thick, the FR4 material backing on the copper foil is 0.18 mm and the copper layer is 20 micron. The mesh used for the model was created by using a global seeding of 0.0025 (model drawn in metres to maintain SI units) and the vertical edges (that run through the thickness) were meshed to 5 nodes per partitioned layer in the part. This provides a fine enough resolution to view the shear stress transition through each lamina ply. After the meshing has been done, it is important that the stacking orientation should be assigned to the part regions in this module before proceeding to the composite layup. A structural or sweep mesh can be done to provide the best mesh results.

The material allocations and section assignment were done using a composite material layup in ABAQUS. Firstly before the composite layup can be done, the material properties of each layer needs to be created. The adhesive and copper material properties can be entered into ABAQUS using the general isotropic material type input. The FR4 material and the carbon composite material ply properties will use a lamina material type. The carbon lamina properties are the outputs of the MATLAB program written to determine the lamina properties using the micro-mechanics models. All the data has been kept to standard SI units to eliminate conversion errors. It must be noted that the isotropic material layers are also included in the composite layup, however the materials assigned to these layers are also isotropic thus it will not be an issue. The composite lamina need to their rotation angles specified relative to the selected co-ordinate system. For the woven fabric composite lamina, the properties of $E11 = E22$, thus the angle can be either 0 or 90 degrees. In this case the alignment of the upper and lower carbon composite layers is 0 degrees and the middle layer at 45 degrees to those two layers. The rotation is about an axis normal to the ply surface.

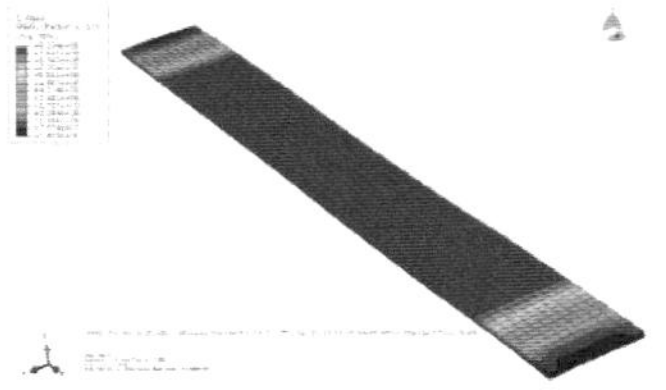

Figure 6: FEA tensile test stress plot.

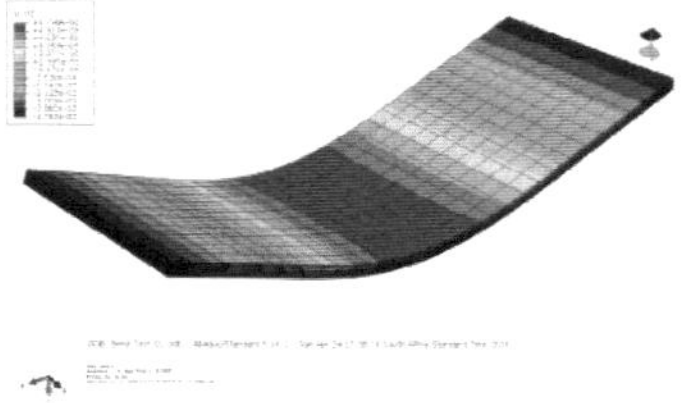

Figure 7: FEA bend deflection.

The structural analysis consisted of a static gravity load to simulate the launch loading of the satellite. The natural frequency analysis of the structure is a relatively simple operation is ABAQUS, the static load test previously mentioned is suppressed and a frequency step is created for the simulation. The first 10 eigenvalues are solved for to compute the first 10 natural frequencies of the structure. The frequency range of interest is 0 to 100 Hz, as the spacecraft test data in the various literature presented does not cover frequencies beyond 100 Hz. The natural frequencies will be evaluated and considered in a design sense.

A steady-state-dynamic modal step is created to measure the dynamic response of the structure under loading. Once again the frequency range of interest is 0 to 100 Hz. In this case the range has been extended to 120 Hz to capture any response behaviour near the 100 Hz frequency that may be present. For each natural frequency found in the previous analysis, the dynamic response of the structural model will be measured. In this case a gravity load is applied once again to the overall structure to simulate the frequency response will in launch conditions. The vibrations and loading during launch would be the most likely time of the satellite system life cycle where a modal response to a loading under a vibration pattern may create interfacing issue to the launcher or damage internal subsystems or even the structure itself if the response deflection is too high. No damping factors have been applied to the model, this has been done to produce a direct output of the frequencies of the CubeSat structure. The deflection and stress response to the dynamic loading condition will be measured and taken into account for the data output request from the analysis.

A simulation model of an impact on the composite material was also done to show the characteristics of the material during impact. According to the literature reviewed in the theory section the average impact velocity of debris in space is approximately 4km/s. As an initial value, the ``projectile" or debris will be set to impact the composite panel at 400 m/s which is at a similar velocity to that of a rifle. Once the model has been solved, the velocity will be increased to 1km/s and then to 4km/s to see the progression of the effect to the material. The debris in this test is a sphere of aluminium 1 mm in diameter and represents the majority of the debris field in space.

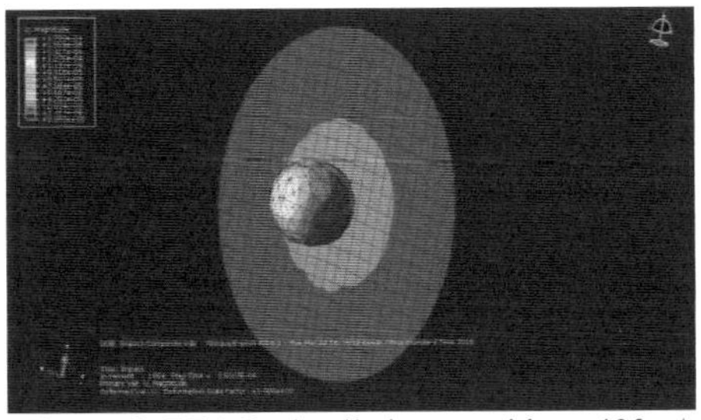

Figure 8: FEA Projectile impact side at 400m/s.

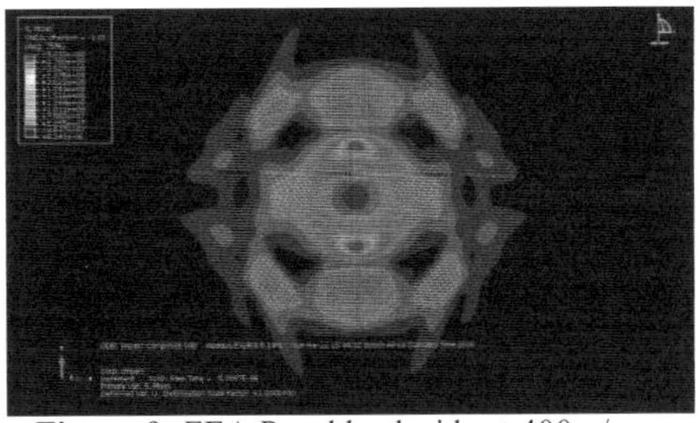

Figure 9: FEA Panel back side at 400m/s.

Figure 8 and 9 above illustrate the results of the impact test of a 2mm diameter aluminium particle at 400m/s. In this test, a maximum deflection of is 1.7 mm and a maximum tensile von Mises stress of 303 MPa is seen near the centre of the panel, on the opposite side to impact. Tests at 1km/s and 4km/s created a large simulation error due to the number of deleted elements and the level of distortion. It is safe to assume that an impact at those velocities would destroy the panel.

As a general note all FEA models the boundary conditions are applied to the corner support posts of the satellite via an encastred surface that represents the housing that the satellite is launched in. The representative surfaces can be seen in Figure 5 (the blue planes). For the impact analysis a smaller section of the assembly was considered to reduce computational time and expense, as a global seeding of approximately 0.002 to capture the fine stress. The full assembly models and material sample simulations used a large global seeding of approximately 0.01.

Figure 10 and 11 shows the results of the steady state analysis of the structural response to a 20g gravity load applies to the centre of gravity in the vertical axis of the satellite (vertical axis in the image). Two concept structures have been tested, one with full surface panels and another with cut-outs in the side panels. The results of the full surface side panels have been illustrated below.

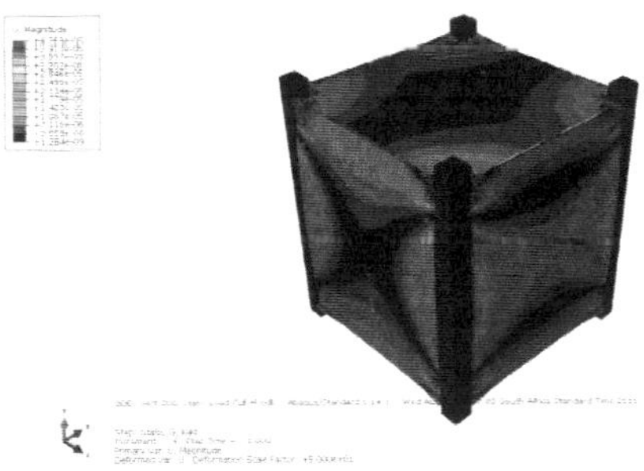

Figure 10: Overall deflection magnitude (U).

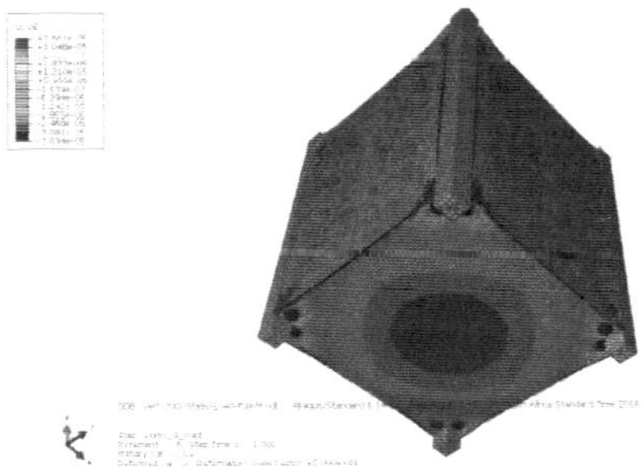

Figure 11: Deflection in the vertical (U2) axis, bottom view.

For the solid panel concept structure, a maximum deflection of 0.037 mm is seen at the centre of the bottom surface of the panel. This result indicates that the structure is principally very rigid, even when subjected to load well beyond the upper limits of the gravity loads that are expected in its working environment. Although no internal dummy loads to represent the internal payload was used in this simulation, the mounting positions of the internal payload would distribute the load directly into the lower sections of the aluminium posts and not have a great deal of impact on the deformation of the aluminium corner rails.

Empirical test results:

Physical samples of the composite material were tested using both a tensile and bend methods as defined by the ASTM D3039/D3039M and D790-02 standard and represented the setup of the numerical model test simulations. Table 1 summarises the results of the tensile test done on 12 material samples.

Sample Number	Maximum Load [KN]	Tensile Extension [%]	Corrected Gauge length [mm]	Extension [mm]	Strain:	Maximum Tensile Stress [MPa]	Modulus of Elasticity [GPa]
1	11.383	1.6	179.989	2.88	0.016	325.229	20.327
2	10.639	1.6	179.981	2.88	0.016	303.971	18.998
3	10.937	1.7	179.994	3.06	0.017	312.486	18.382
4	9.101	1.4	179.995	2.52	0.014	260.029	18.573
5	11.464	1.6	179.993	2.88	0.016	327.543	20.471
6	10.972	1.5	179.994	2.70	0.015	313.486	20.899
7	11.509	1.6	179.997	2.88	0.016	328.829	20.552
8	11.408	1.7	179.993	3.06	0.017	325.943	19.173
9	5.678	0.9	179.992	1.62	0.009	162.229	18.025
10	11.11	1.5	179.995	2.70	0.015	317.429	21.162
11	12.315	1.7	179.993	3.06	0.017	351.857	20.697
12	11.554	1.6	179.994	2.88	0.016	330.114	20.632

Maximum Load [KN]	Tensile Extension [%]	Corrected Gauge length [mm]	Extension [mm]	Strain:	Maximum Tensile Stress [MPa]	Modulus of Elasticity [GPa]
10.673	1.533	179.993	2.760	0.015	304.929	19.824

Table 1. Tensile test data (average values in the lower block)

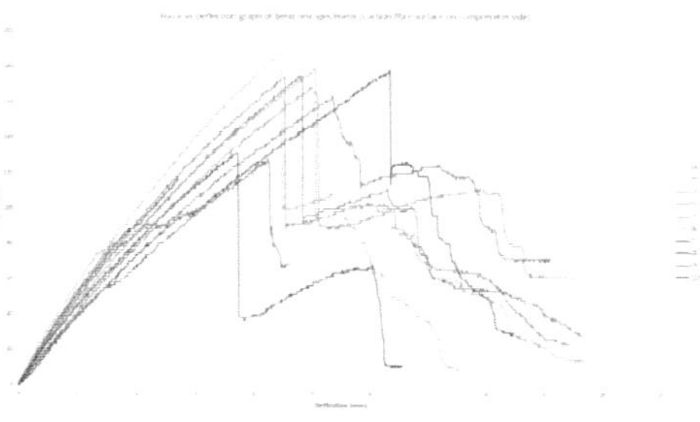

Graph 1: illustrate the results from the copper foil on compression side and tensile side during bend test respectively.

Comparison of the numerical and empirical data:

The two graphs below provide a comparison between the numerical FEA simulations using a PFA tool verses the data from the similar empirical tests done on the material for tensile and bend loading.

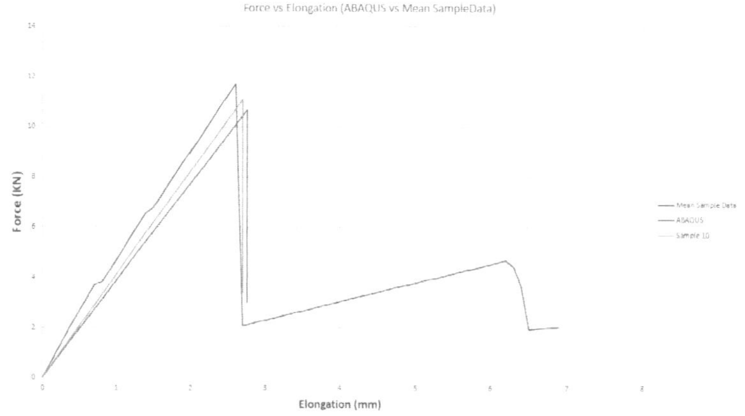

Graph 2: Tensile test comparison between numerical and empirical results.

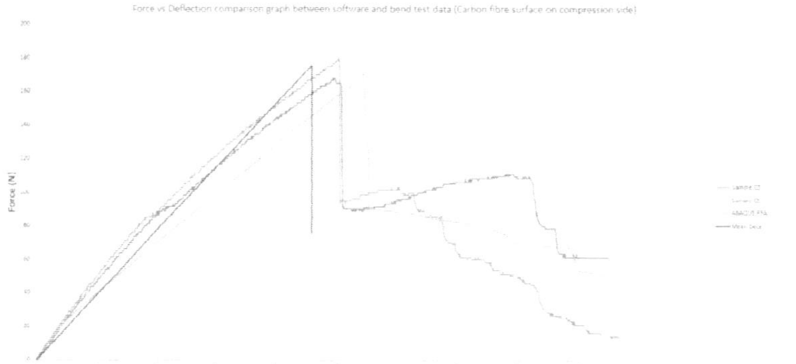

Graph 3: Material bend test comparison between numerical and empirical results.

Conclusion:

In conclusion a concept design for the multifunctional composite material and a CubeSat structure that made use of this material configuration provided positive results. The selection of the material and the design of the CubeSat structure considers the operational environment and conforms to the satellite standards that are already established. An application of systems think had to be taken to view the structural subsystem holistically and consider the interfacing between other subsystem and the subsystem requirements of the structural assembly. The concept structure would not need to undergo any special weaver process other than having the composite material outgas approved, if approved carbon composite material is not used for the panel construction of the flight model structure. The use of the mathematical micro-mechanics model presented by (Adumitroaie and Barbero, [7]) and (Barbero, [6]), along with the implementation of the macro-mechanics modelling capabilities of ABAQUS FEA using a PFA plug in tool provided good numerical data correlation to that of the empirical data. The material tests conformed to the ASTM standards for composite and in each case the test was replicated as accurately as possible in a physical and virtual testing setup. The comparison of the data provides a means of verification of the numerical predictions and shows that the methodology used for the prediction of the multifunctional concept material properties and characteristics is acceptable. Although the data did not correspond exactly with one another, the margin of error seen between the numerical and empirical data is within tolerance, when considering the vast number of possible variables when dealing with composite materials and their numerical prediction models. Accumulatively, small defects in the material during manufacture, the integration of secondary material to the composite, the test equipment itself, all add up to create this variation seen in the test data. The impact test that was done on the material showed no penetration of the composite panel with the debris striking it at 400m/s, however when the impact velocity was increased to the next stage of 1 km/s, model was unable to converge due to the distortion of the model components. Logically analysing the impact velocity of debris on the satellite structure, it is quite clear that any debris striking the satellite at a velocity of 4 km/s would create catastrophic damage to the satellite system. For this very reason, the space environment needs to have space debris control measures. All in all considering the number of variables, the error between the comparisons of the results is relatively small and has a good material characteristic representation during loading until the failure stages.

References:

1. Fortescue. P, Swinerd. G, and Stark J. "Spacecraft Systems Engineering". Wiley, 2011. ISBN 9781119978367.
2. Cal-Poly-SLO. "CubeSat Design Specification". The CubeSat Program. California Polytechnic State University, Rev. 13.
3. Campbell, F.C. 2003. "Manufacturing Processes for Advanced Composites". Elsevier Science. ISBN 9780080510989.
4. Gibson, R.F. "A review of recent research on mechanics of multifunctional composite materials and structures". Elsevier Ltd., 2010.
5. Haland, A. "Development of mechanical structure for NUTS CubeSat - Semester Report". Norwegian University of Science and Technology, 2013.
6. Barbero E.J. "Introduction to Composite Materials Design", Second Edition. Composite Materials. Taylor & Francis, 2010. ISBN 9781420079159.
7. Adumitroaie A. and Barbero E.J. "Beyond plain weave fabrics ii. mechanical properties". Composite Structures, 93(5):1449 – 1462, 2011. ISSN 0263-8223.
8. Roylance. D. "Laminated Composite Plates". Lecture Notes. Massachusetts Institute of Technology (MIT), 2000.
9. Barbero E.J. "Finite Element Analysis of Composite Materials using AbaqusTM". Composite Materials. Taylor & Francis, 2013. ISBN 9781466516618.

Modelling the structural behaviour of a round nozzle to flat plate interface in pressure vessels

Wilhelm Beyers[1*], Albert Zapke[2], Gerhard Venter[3]

[1,3] Stellenbosch University

[2] Enexio Management GmbH

15682447@sun.ac.za

Abstract

In the field of pressure vessel design, much research has been done to predict the structural behaviour of round nozzles connected to spherical and cylindrical plates. This has been motivated by the fact that the majority of pressure vessels in industry have round bodies, as these are better suited to withstanding pressure. A large demand however also exists for flat plate pressure vessels with round nozzles, but there is a lack of research and guidelines that accurately predict how these structures behave. This paper investigates the categorisation of the structural behaviour arising from both internal pressure and nozzle loading in both the nozzle and the flat plate.

Keywords

FEM, Pressure Vessel, Meta-Modelling

Introduction

The purpose of this research is to build a meta-model that will accurately predict the structural behaviour at the interface region between various round nozzles and a flat plates. A meta-model is a mathematical model that predicts the behaviour of another model, based on a set of test or simulation results.

The motivation for this research comes from the field of air-cooled heat exchanger header box design, in particular plug type header boxes. Heat exchanger header boxes, as seen in Figure 1, are most often rectangular flat-sided pressure vessels. They are constructed from metal plates that are cut to size and welded together to form the box. Their purpose is to act as a manifold, channelling process fluid between a piping system and the finned tubes of an air-cooled heat exchanger bundle. In the case of the heat exchanger configuration shown in Figure 2, an inlet header receives process fluid from a piping system and directs the flow into a set of finned tubes. A second, third and fourth header box redirect the flow of process fluid from one set of finned tubes to another and finally an outlet header directs the flow from the finned tubes into an outlet piping system.

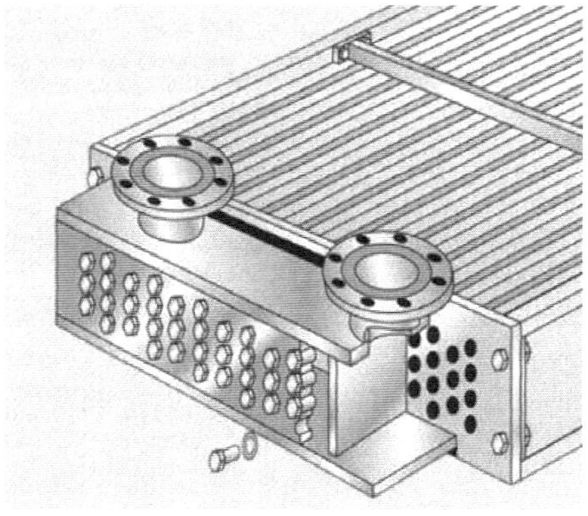

Figure 1: Typical Plug Type Heat Exchanger Header Box and Finned Tube Bundle [1]

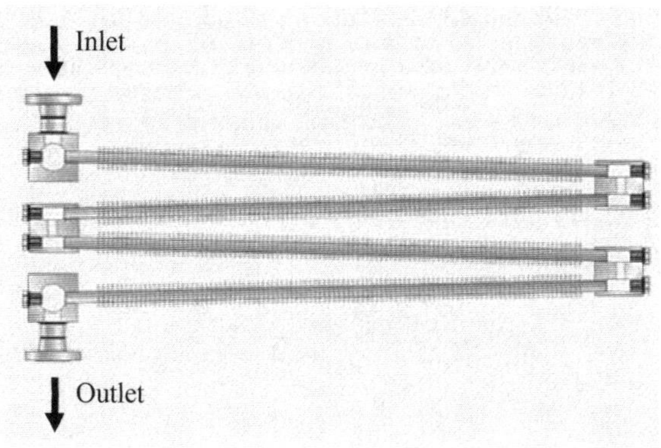

Figure 2: Multiple Pass Air-cooled Heat Exchanger [1]

According to the Pressurised Equipment Regulations (Department of Labour RSA, 2015) and South African National Standard 347 [2], new pressure vessels that are designed to operate at internal pressure at or above 50 kPa(g) have to be categorised according to their Fluid and Pressure-times-Volume risk. All categories other than sound engineering practice (SEP) shall be designed and manufactured to an accepted Health and Safety Standard design code. The most widely used design code is the ASME Boiler & Pressure Vessel Code Section VIII (ASME, 2015) [3].

Division 1 of the ASME code provides a set of rules and guidelines that dictate how a wide variety of pressure vessels must be designed. This is known as "Design by Rule". In cases where the code does not provide the guidelines for the design of a certain pressure vessel, or part thereof, the designer is permitted to use other design methodologies, provided they are "as safe as" the rules in Division 1. In the case of a plug type header boxes, this rule is most regularly applied to the design of the header box nozzles and the area where they connect to the box.

The method that is used in this study to fill this gap left by Division 1 is to employ the rules given in Part 5 of Division 2 of the ASME code. This part of the code allows for "Design by Analysis" by means of using finite element (FE) analysis. This allows for an accurate analysis of components whose design requirements are not specified in Division 1 and also allows the designer to see how these components will interface structurally with the components that can be designed according to the rule of Division 1.

Procedure

The goal of this study is to build a meta-model, based on FE analysis results, which will predict the structural behaviour of a flat plate to nozzle interface. The planned procedure to achieve this goal will be to first generate the data necessary for constructing a meta-model. This will require setting up and testing a large number of finite element (FE) models of nozzle to flat plate intersections. Each model will have a different combination dimensions and together these models will span the design space and provide discrete data at a multitude of points. By means of support vector regression, a numerical model will be fitted to the data in order to provide a continuous prediction of the structural behaviour of the nozzle to flat plate intersection for any point in the design space.

Model setup

Two separate models were set up to be tested. The first, shown in Figure 3 and Figure 4, consists of a large flat plate with a nozzle in the centre. This model was set up as a simplified test case, with only a limited number of dimensions to be varied. Its purpose was to test the process and toolset used to set up a meta-model and to provide a platform where any errors could be corrected, before moving on to more complex models.

Figure 3: Large Flat Plate and Nozzle

Figure 4: Nozzle Close Up

As we are interested in the results around the base of the nozzle and where is connects to the flat plate, these areas have been meshed in fine detail, while the surrounding mesh has been made more course to reduce the computation expense of the model. Specific attention was paid to the mesh around the base of the nozzle, seen in Figure 5 and Figure 6, to ensure that there were enough nodes in the areas where results are to be extracted. The red lines in Figure 6 show exactly where the stress results will be extracted. The details regarding stress extraction and why these lines are necessary are discussed in more detail later. The figures also show how the model has been segmented to allow for stress extraction at 8 equally spaced points around the junction. The boundary condition applied to this model is a simply supported constraint around the outer edge of the plate.

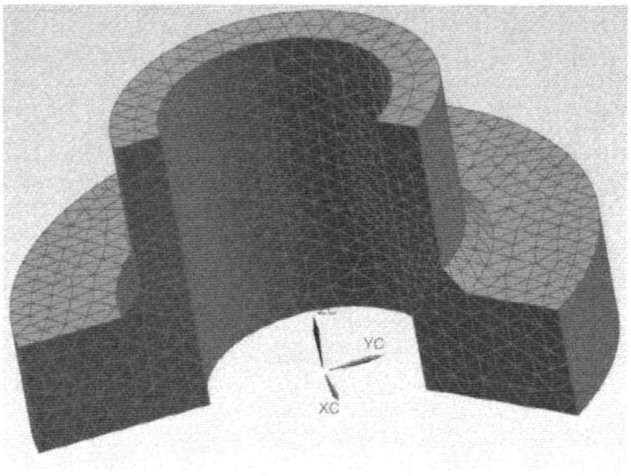

Figure 5: Mesh Detail

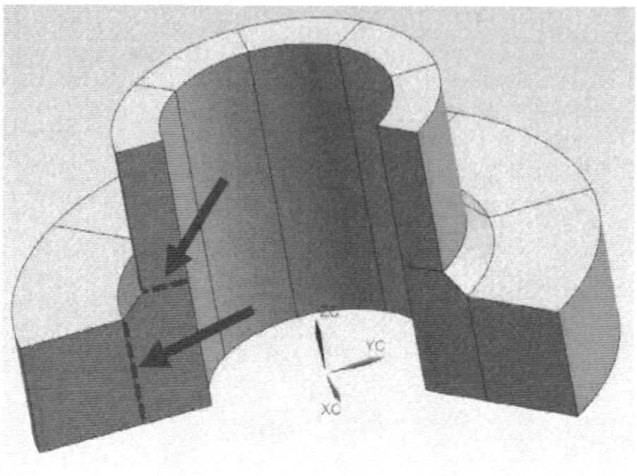

Figure 6: Model Detail

The second model, seen in Figure 7 and Figure 8, was created to more accurately simulate the boundary conditions that a nozzle would experience as part of a plug type header box. It has seven dimensions that can be varied. These are the nozzle diameter, nozzle length, box height, box width, box length and the thicknesses of the two sets of opposing box side. (Each set of parallel sides on the box have the same thickness) In this model the ends of the box are held with fixed constraints and enough room is left between the constraint and the nozzle so as not to affect the area of interest. The same detailed and segmented meshing pattern described previously was also applied to this model.

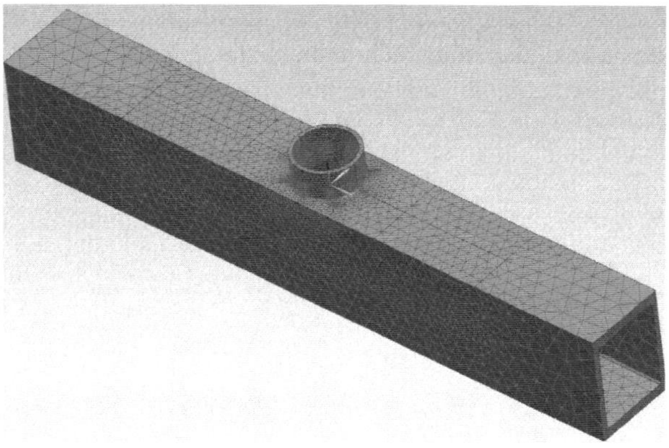

Figure 7: Box Shell and Nozzle

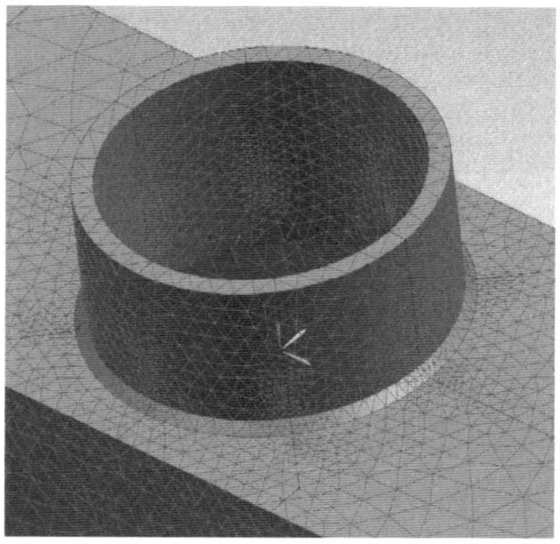

Figure 8: Nozzle Close Up

Two sets of loads are applied to these models, pressure loads and nozzle loads. Pressure loads are applied to the areas that would form part of the inside of a pressure vessel. This load is also accompanied by an equivalent force which is applied to the top edge of the nozzle, in order to account for the effect of having the nozzle connected to a pressurised piping system. This force is applied in line with the axis of the nozzle and acts to pull the nozzle away from the flat plate. The magnitude of this force is calculated as per Equation 1.

$$Equivalent\ Force = \frac{\pi \times (Nozzle\ Diameter)^2}{4} \times Internal\ \text{Pressure} \quad (1)$$

Nozzle loads are applied as shown in Figure 9, where, F_x, F_y and F_z represent forces applied to the top edge of the nozzle and M_x, M_y and M_z are moments applied to the top of the nozzle. In the models, each of these loads are applied as separate load cases so that the results can be analysed separately and combined as necessary during post processing.

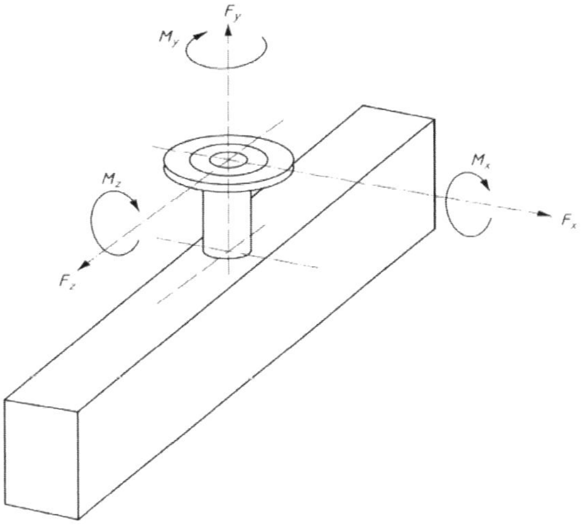

Figure 9: API Nozzle loading [4]

Results Processing and Stress Extraction

ASME Section VIII Division 2 Part 5 [3], which deals with FE analysis of pressure vessel, gives specific guidelines on how stress results from structural analyses should be evaluated. According to the code, stress results need to be linearised through the thickness of the pressure vessel at the point of interest. This is done in order for stresses to be separated into three components, namely the membrane, bending and peak stresses. These individual components are then used for comparison with the failure criteria of a material. Figure 10 shows a visual illustration of what each of these components represent. The positions at which stresses are linearised are called stress classification lines (SCL).

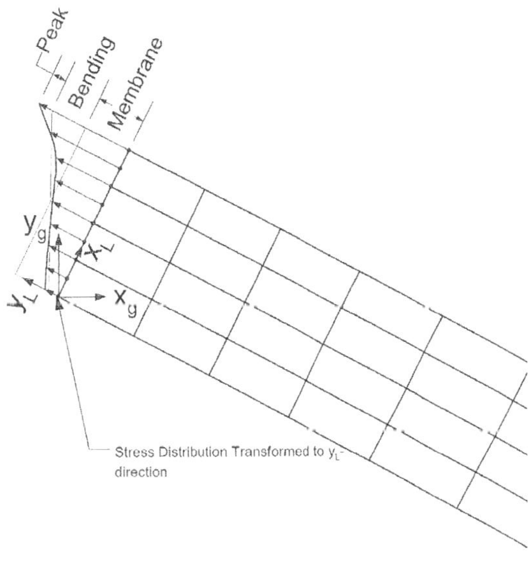

Figure 10: Stress classification [2]

Membrane stress is the average of the component stresses along the SCL and is calculated using Equation 2. σ_{ij} represents the component stress values at the nodes along the SCL. $\sigma_{ij,m}$ is calculated for each of the six nodal stress components which are then combined using the von Mesis formulations to yield an equivalent membrane stress for the SCL position.

$$\sigma_{ij,m} = \frac{1}{t}\int_0^t \sigma_{ij}dx \qquad (2)$$

Bending stress represents the linearly varying portion of the component stresses along the SCL. The same procedure is followed as before, using Equation 3 instead, to calculate the bending stress values. A minor difference however is that while the membrane stress calculations are performed for all six nodal stress components, the bending stress calculations only consider three of these components. According to the ASME code, only stress components perpendicular to the SCL should be considered.

$$\sigma_{ij,b} = \frac{6}{t^2}\int_0^t \sigma_{ij}\left(\frac{t}{2} - x\right)dx \quad (3)$$

Peak stress is the component that exceeds the membrane plus bending stress. Peak stress is however not relevant to this study as it is only used for fatigue failure assessment, which is not part of what this study is attempting to address.

Finally, the ASME code also provides guidance as to where SCL's should be placed. Figure 11 and Figure 12 are examples of this, showing where an SCL should be positioned to analyse the stress at the toe of the weld between the nozzle and the pressure vessel wall.

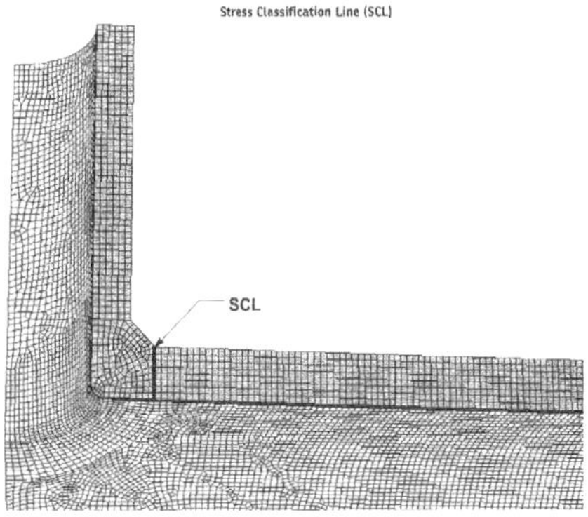

Figure 11: SCL positioning [3]

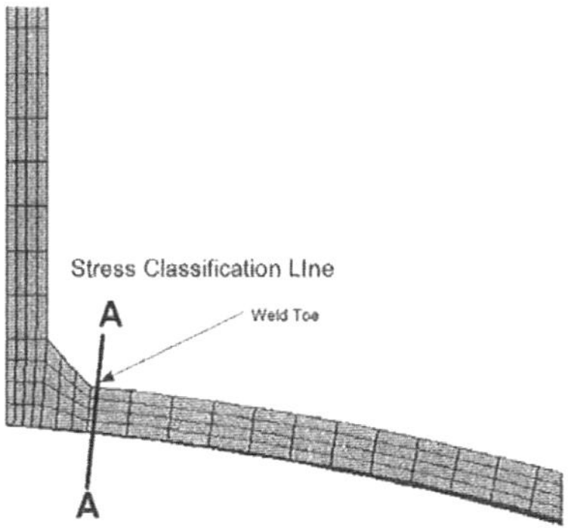

Figure 12: SCL at weld toe [3]

Automation

Due to large number of FE models that need to be set up, simulated and analysed in order to generate a meta-model, it is not feasible to construct and analyse them each individually by hand. Instead, a programming toolset has been developed to automate the process of setting up, testing and analysing the results of the FE analyses necessary for generating the data. Figure 13 shows a schematic of how this toolset works.

The process starts by setting up a parametric CAD model of the structure to be simulated using Siemens NX. In this model, each parameter or dimension is linked to a variable in the program that can be altered in order to modify the size and properties of the structure. NX then allows for this CAD model to be directly linked to its FE analysis software where a detailed FE model is set up to incorporate all the detail mentioned above.

In parallel to this first step, a script written in Python, generates the required set of design of experiment (DOE) sample points that will be simulated to create the data point from which to construct the eventual meta-model. The DOE is based on Latin Hyper Cube sampling and each sample point takes the form of a set of dimensions which correspond the parametric variables in the CAD model.

Next a Python script, which is called from within the NX software, draws in the set of dimensional combinations that need to be sampled. The script loops through the set and for each sample point takes the dimensions and applies them to the corresponding variables in the parametric model. Once the CAD model is updated, the FE model, which is linked to it, is also updated. This update re-meshes the structure according to its new dimensions while retaining the original loading and boundary conditions. It also retains any meshing preferences from the initial model, such as mesh seeds and mating faces. Once each model update is complete, the script exports a Nastran solver deck for that model, with a unique file name, to be analysed at a later stage.

Based on the information in the solver decks and the dimensional inputs linked to each one, another Python script determines exactly which nodal results will be required to calculate the necessary stress results at the end of the process. The solver decks are then modified and the correct syntax is added to ensure that only the necessary results are printed to the output file when they are run. This step significantly reduces the time required for the subsequent post-processing of the results.

Due to the large number of models being considered, it is also not practical to analyse the solver decks on a standard computer. Instead, the solver decks are uploaded to a high performance computer (HPC) where they are analysed in parallel.

Finally, a last Python script reads each set of results files and calculates the relevant stresses for each load case at all of the points of interest. This process involves first sorting the nodal results into groups, depending on which SCL they are associated with. The distance between each node along a SCL is then calculated and the nodes are sorted so as to arrange them one after the other through the thickness of the plate or nozzle. Once this step is complete the stress results are linearised according to the earlier description and the relevant membrane and bending stresses calculated and exported for subsequent use.

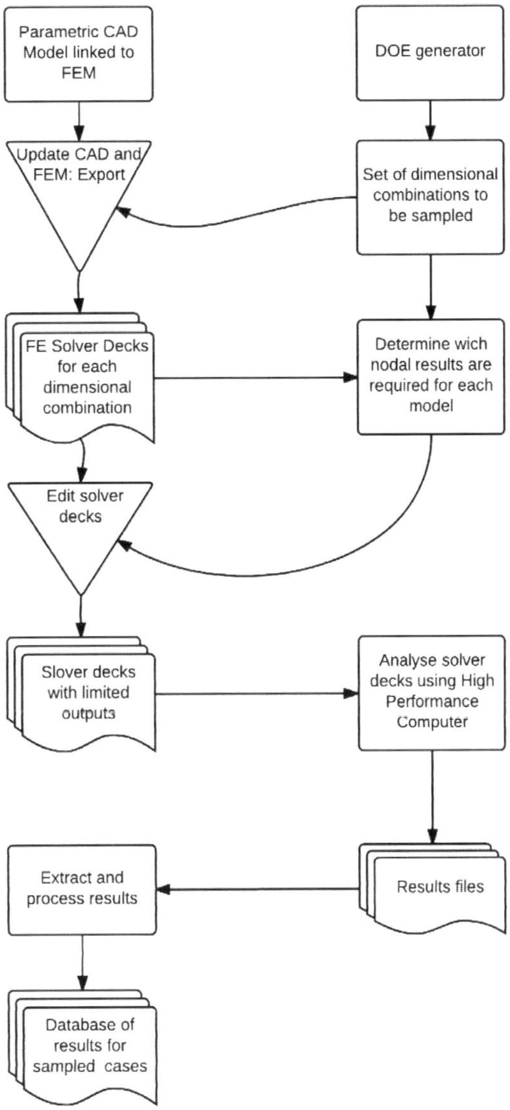

Figure 13: Flow chart of automation process

Mathematical Model

With the automation process in place, a set of 2500 sample cases were simulated. With all the data collected and processed, a mathematical model had to be fit to it in order to be able to predict the stress values for new design cases.

Two such models were tested. The first was a polynomial response surface model. The result showed that a 4th order response surface made a fair approximation of the results, but was unable to account for the majority of the non-linearity in the results. A support vector regression model was also tested and was found to be very well suited and capable of accounting for the non-linearity in the model. Figure 14 shows a plot which compares the results predicted by the support vector model to those of the actual sampled values. In this plot, the closer the points lie to the 45° line in the middle of the graph, the more accurate the predictions are. In the plot there are 240000 points, each being displayed as a 90% transparent dot, in order that it may be clearly seen which areas are more densely populated. This large number comes from the fact that there are 2500 sample cases, which each yield the 6 linearised stress components values for each of the 16 SCLs in the model.

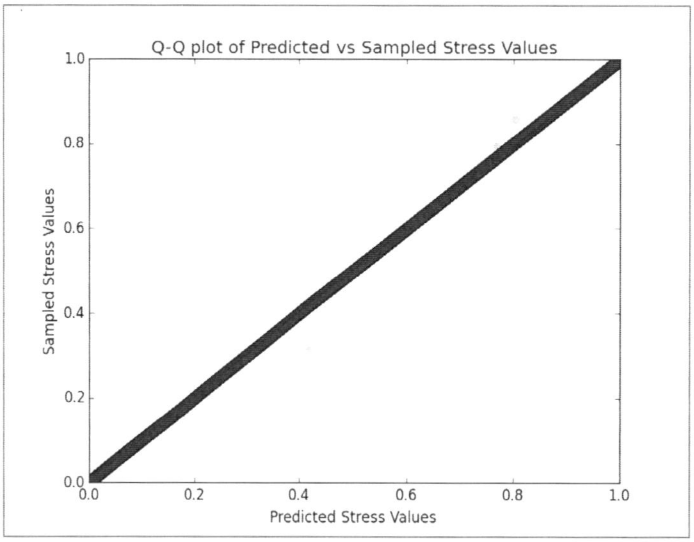

Figure 14: Predicted vs Sampled Stress values

Graphical User Interface

In order to make the results of the meta-model more accessible a graphical user interface (GUI) was created with which to select model inputs and request result outputs. The layout of this GUI is show in Figure 15. The top half of the interface focusses on allowing the user to extract results for a specific design case with fixed dimensions and loads. On the left hand side of this part, each slider controls a dimension, which can be varied in the meta-model. On the right hand side, the components of the load case and the allowable stress can be specified. From here a graph can also be generated to show the predicted stresses at each of the stress extraction points around the nozzle.

The lower half of the interface allows the user to go further than this and study what the overall effect would be of changing specific dimensions. Using the sliders here, the users can set up a new combination of dimensions which are different from the set above and then request a graph of how the stress values change, as the dimensions morph from the first set to the second. Examples of result plots mentioned here are shown in the following section.

Figure 15: Meta-model Post Processing GUI

Results

Although the first model was only intended to be used to test the process of constructing a meta-model, it did produce some interesting results. By adding a few additional stress extraction points at intervals further away from the nozzle junction, the plot in Figure 16 was produced. The results are for the load case when only the force F_x was applied. The figure shows a plot of the bending stress in the flat plate at several points moving away from the nozzle junction. Each red bar represents the results from one of the eight segments around the perimeter of the nozzle. The left hand side of each red bar show the bending stress at the base of the nozzle junction and moving from left to right, they show how the stress values decay with increasing distance from the nozzle. The blue lines are sine curves that were fit to the results with which it is possible to interpolate between the segments.

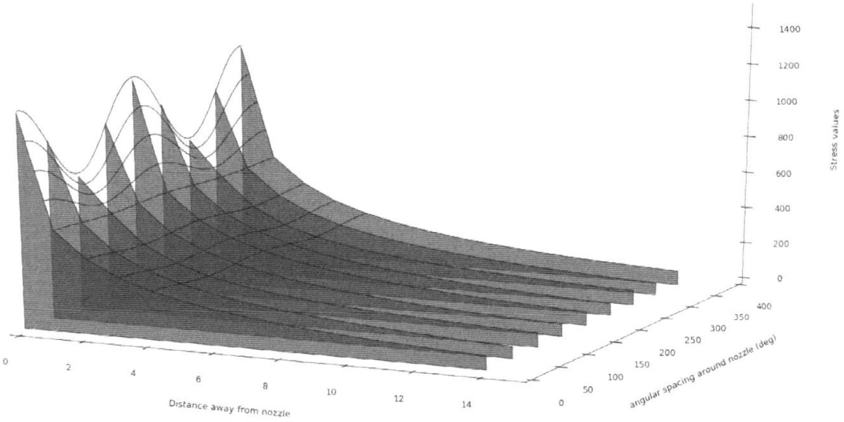

Figure 16: Bending Stress Distribution in Flat Plate

Looking at the second model, Figure 17 shows an example of a stress prediction plot, generated by the meta-model created in this study. In this plot, the eight groupings shown along the x-axis correspond to the eight points around the nozzle where stresses are predicted. In each of these groupings, four coloured bars are shown. The red bar shows the predicted membrane stress that will be present in the flat plate and the blue bar predicts the membrane plus bending stress that will be present in the flat plate. In a similar manner, the yellow and green bars respectively predict the membrane and membrane plus bending stresses that will occur in the in the nozzle at the specified point. (Refer to Figure 6 for visual illustration) The red line in the graph shows the allowable limit for membrane stress and the blue line shows the allowable limit for membrane plus bending stress.

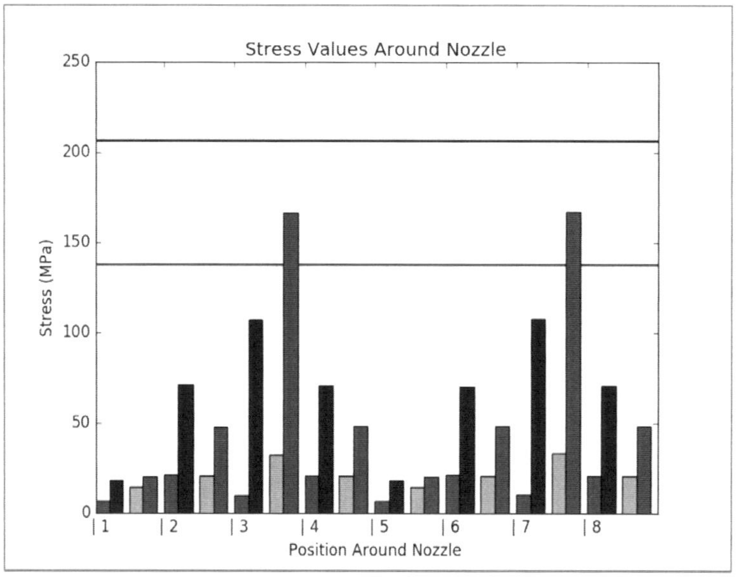

Figure 17: Sample Result Output

Figure 18 and Figure 19 show the results from two studies that were performed using the second part of the GUI. In the case of Figure 18, a design case was entered where all of the dimensions of the model were kept constant, except for the length of the nozzle which was varied. The same procedure was followed for Figure 19, only here the height of the box between the nozzle and the bottom was varied. These graphs show all the results described above plotted on a single graph with the only distinction being between membrane stress in red and membrane plus bending stress in blue. This can be done as the focus here is not to find specific answers at specified points, but rather to analyse the trend of the results and find the best design region.

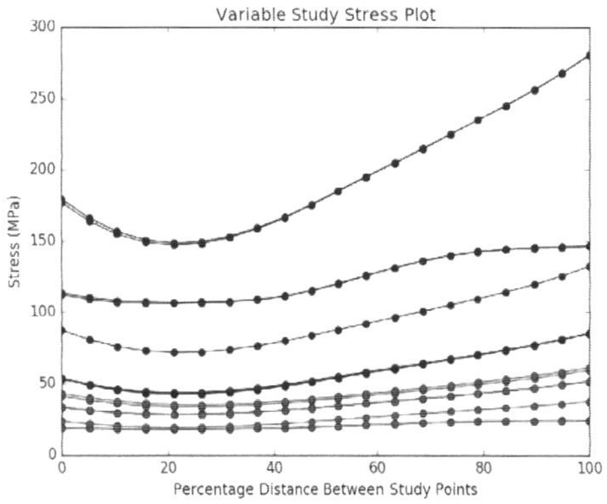

Figure 18: Stress Variation with Nozzle Length Change

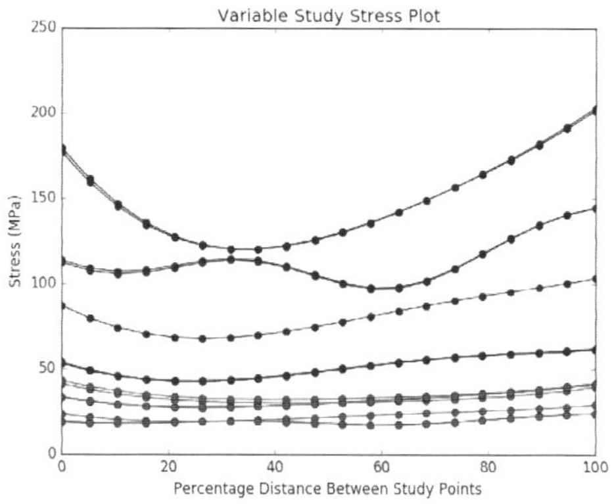

Figure 19: Stress Variation with Box Height Change

Conclusion

Thus far this study has reached its main objective, which was to build a meta-model that will predict the structural behaviour at the interface region between various round nozzles and flat plates. Further work still remains in validating this model, but preliminary results have shown the model to be accurate. Additional goals have also been achieved in the process, of which the most notable was the development the toolset which automates the process of setting up, simulating and analysing the results of FE models in order to generate data from which to construct a meta-model. As this is a generalised toolset, which is not bound to a single type of structure, it now has the ability to analyse a wide variety of new structures with relative ease in order to build predictive meta-models for their structural behaviour. The use of this tool in subsequent similar studies will also significantly reduce the time required to complete them.

References

[1] Kelvion Thermal Solutions (Pty) Ltd, "Air Fin Coolers Headers In Detail," 2016. [Online]. Available: http://www.kelvion.com/za/products/finned-tube-heat-exchangers/air-fin-coolers/headers/. [Accessed 3 May 2016].

[2] SABS, "SANS 347 : 2012 SOUTH AFRICAN NATIONAL STANDARD Categorization and conformity assessment criteria for all pressure equipment," 2012.

[3] ASME, "Rules for Construction of Pressure Vessels," in *ASME Boiler & Pressure Vesssel Code*, New York, The American Society of Mechanical Engineers, 2015.

[4] API, Air-Cooled Heat Exchangers for General Refinery Service petrochemical and natural gas industries, 6th ed., 2006.

[5] R. Myers, D. Montgomery and C. Anderson-Cook, Response Surface Methodology, 3rd ed., Hoboken: Wiley & Sons, 2009.

[6] M. Kassner, "Insight: OpenSource Gaze Tracking," 2011. [Online]. Available: http://fab.cba.mit.edu/classes/MIT/864.11/people/Moritz_Kassner/insight.html. [Accessed 30 June 2015].

Interface information transfer between non-matching, nonconforming interfaces using radial basis function interpolation

A.E.J. Bogaers[1,*], S. Kok[2], B.D. Reddy[3,4], T. Franz[5,6]

[1]Advanced Mathematical Modelling, Modelling and Digital Sciences, CSIR, South Africa
[2]Department of Mechanical and Aeronautical Engineering, University of Pretoria, South Africa
[3]Department of Mathematics and Applied Mathematics, University of Cape Town, South Africa
[4]Centre for Research in Computational and Applied Mechanics, University of Cape Town, South Africa
[5]Division of Biomedical Engineering, Department of Human Biology, University of Cape Town, South Africa
[6]Research Office, University of Cape Town, South Africa
*abogaers@csir.co.za

ABSTRACT

In this paper we outline the use of radial basis function interpolation (RBF) to transfer information across non-matching and nonconforming interface meshes, with particular focus to partitioned fluid-structure interactions (FSI). In general, transferring information across a non-matching interface presents itself as a nontrivial problem. RBF interpolation, which requires no global connectivity information, provides an elegant means by which to negate any geometric discrepancies along the interface. The aim is to investigate the feasibility of RBF interpolation, with a strong focus on a comparison between a conservative and consistent formulation.

KEYWORDS

radial basis function interpolation, non-matching interface mesh, FVM-FEM coupling, consistent, conservative

INTRODUCTION

Fluid-structure interactions is the two-way coupled analysis of deformable structures and the corresponding interactions with fluid flow. A few common examples include flutter analysis of aero-elastic structures (Farhat et al. [1], Rifai et al. [2]) or blood flow through the cardiovascular system (Torii et al. [3], Wolters et al. [4]). To couple the two domains requires the transfer of information across the shared interface. The information to be transferred typically include the interface displacements transferred from the solid domain to the fluid domain, and interface stresses to be transferred from the fluid domain in the form of interface pressures and wall shear stresses.

The numerical properties between the fluid domain and the solid domain differ sufficiently that they are naturally solved with very different geometric discretisation requirements. For example, the fluid solver typically requires far more degrees of freedom than the shared structural problem. Or conversely, the structural domain, either around sharp edges or in vicinities of high stresses, may require localised mesh refinement. Consider for example Figure 1, illustrating the potential mismatch along a curved interface. The interface meshes are both mismatched and nonconforming, where due to the curvature there are both gaps and overlaps. The problem may further be complicated by the choice of using different numerical schemes on each of the sub-problems. The solid domain is typically solved using the finite element method (FEM) where interface information is located at element nodal coordinates with well defined internal interpolation shape functions. The fluid field solver on the other hand, typically based on the finite volume method (FVM), defines quantities at face centres where the internal interpolation function can at best be described as a face-constant step function.

Accurate information transfer is critical to the accuracy and stability of partitioned FSI solution schemes. There are a large number of interface information transfer schemes available in literature; many are based either on physical arguments (Cebral and Lohner [5], Farhat et al. [6], Jiao and Heath [7]) or approaches based on mathematical arguments (Beckert and Wendland [8], Lombardi et al. [9], Quaranta et al. [10], Smith et al. [11]).

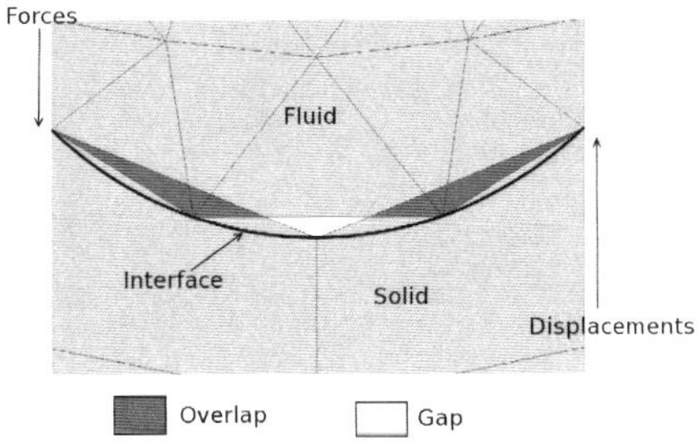

Figure 1: Illustrative example of a non-matching mesh along a curved interface.

In this paper we aim to outline the use of radial basis function (RBF) interpolation to transfer information. Multivariate interface transfer methods have become very popular, primarily because they require no mesh connectivity information and is therefore well suited to coupling FVM-FEM discretisation schemes, and further negates any potential geometric discrepancy. In general when considering RBF interpolation, there is a trade-off between transferring the information in a conservative or in a consistent sense, with no way of satisfying both conditions. By conservative transfer, we imply that the sum of forces are transferred exactly, or that equal work is done along the interface. For consistent information transfer, we adopt the definition to imply that information is transferred in such a manner that a constant stress state can be transferred exactly (or patch test is satisfied).

Many researchers stipulate that conservative information transfer is crucial to the overall stability and accuracy of FSI simulations (Lombardi et al. [9], Piperno et al. [12]). To enforce conservatism using multivariate transfer, the interpolation/projection functions are often constructed on the basis of satisfying virtual work along the interface. In this paper, we aim to demonstrate that this approach in essence results in a zero-order scheme, with the possibility of transferring unrealistic and oscillatory stress states (similar observations have also been noted in de Boer et al. [13]). Furthermore, the conservative formulation can not be used to transfer information between discretisation schemes of different orders. We further aim to show, that while a consistent formulation is not provably conservative, it is convergent, and hence conservative, within the limit of mesh refinement.

The outline for the remainder of the paper is as follows. We start by outlining the mathematical notation and formulation for consistent and conservative interface information transfer. We provide a brief overview of RBF interface information transfer and aim to compare conservative and consistent interface transfer via a patch test and transfer of an analytical function across a curved interface.

INTERFACE CONDITIONS

The coupled FSI problem is a two-field problem, with a fluid domain Ω_f and a solid domain Ω_s which share a common interface Γ_{FSI}. By allowing each of the two domains to be discretised independently, each of the two domains have distinct interfaces, namely Γ_f and Γ_s, where $\Gamma_f \neq \Gamma_s$. The partitioned FSI problem can be viewed as a two-field problem with jump conditions along the interface which need to be satisfied in the form of the kinematic and dynamic continuity conditions. Each FSI cycle requires satisfying the equilibrium of interface tractions $\boldsymbol{t}_f$ and $\boldsymbol{t}_s$, and compatibility of interface velocities, i.e.

$$\boldsymbol{t}_s = \boldsymbol{t}_f, \quad \frac{\partial \boldsymbol{d}_s}{\partial t} = \boldsymbol{u}_f \quad \text{along}\, \Gamma_{\text{FSI}},$$

where $\boldsymbol{t}_f = p_f \boldsymbol{n}_f - \boldsymbol{\sigma}_f \cdot \boldsymbol{n}_f$ and $\boldsymbol{t}_s = \boldsymbol{\sigma}_s \cdot \boldsymbol{n}_s$. p_f denotes the fluid pressure along the interface, $\boldsymbol{\sigma}_f$ the fluid viscous stress tensor and $\boldsymbol{\sigma}_s$ the solid stress tensor; the outward pointing normals along Γ_s and Γ_f are $\boldsymbol{n}_s$ and $\boldsymbol{n}_f$ respectively.

In the event that the meshes are both matching, and the numerical schemes and order of internal shape functions are exactly the same, the discrete nodal quantities can simply be transferred, i.e. $\boldsymbol{U}_f = \boldsymbol{U}_s$ and $\boldsymbol{T}_s = \boldsymbol{T}_f$. For non-matching meshes, an intermediate projection or interpolation step is required:

$$\boldsymbol{U}_f = H_{fs}\boldsymbol{U}_s, \text{ and } \boldsymbol{T}_s = H_{sf}\boldsymbol{T}_f. \tag{1}$$

Here H_{AB} represents the transformation matrix to transfer information from mesh B to mesh A. $\boldsymbol{U}$ and $\boldsymbol{T}$ are the vectors of discrete values at the interface points and can be defined by the approximations

$$\boldsymbol{u}(\boldsymbol{x}) \approx \sum_{i=1}^{n^u} N_u^i(\boldsymbol{x})\boldsymbol{U}^i, \quad \text{and} \quad \boldsymbol{t}(\boldsymbol{x}) \approx \sum_{j=1}^{n^t} N_t(\boldsymbol{x})_t^j \boldsymbol{T}^j. \tag{2}$$

$N_{u,t}$ here represents the spatial interpolation function used for the displacement and tractions respectively. Typically for the FEM, $N(\boldsymbol{x})$ is the internal basis/shape functions and a step function for the FVM method; n^u and n^t are the number of degrees of freedom (DOF) along the interface where the discrete displacement and traction quantities are known.

CONSERVATIVE INFORMATION TRANSFER

The general consensus in literature is that the information transfer should be conservative (see for example Beckert and Wendland [8], Farhat et al. [6], Jaiman et al. [14, 15], Lombardi et al. [9], Quaranta et al. [10]). By the strictest of definitions this would imply that the integrated quantities should be equal on both Γ_f and Γ_s. The concentrated loads, typically located at the nodal coordinates for the FEM and at face centres for the FVM method, can be defined as

$$\boldsymbol{F}_f = \int_{\Gamma_f} \boldsymbol{t}_f d\Gamma, \text{ and } \boldsymbol{F}_s = \int_{\Gamma_s} \boldsymbol{t}_s d\Gamma.$$

Conservation of forces can then be expressed as

$$\boldsymbol{F}_s = \sum_{i=1}^{n_s} \boldsymbol{F}_s^i = \sum_{j=1}^{n_f} \boldsymbol{F}_f^j = \boldsymbol{F}_f. \tag{3}$$

Along a non-matching interface, there are an infinite number of nodal load vectors which will satisfy conservation. A convenient argument often used to enforce conservation is the definition of virtual work. For steady state (or for very small time steps), energy can be stated to be globally conserved over the interface if

$$\int_{\Gamma_f} \boldsymbol{u}_f \cdot \boldsymbol{t}_f \boldsymbol{n}_f d\Gamma_f = \int_{\Gamma_s} \boldsymbol{u}_s \cdot \boldsymbol{t}_s \boldsymbol{n}_s d\Gamma_s. \tag{4}$$

Using the approximations in equation (2), the semi-discrete form of equation (4) becomes

$$[M_{ff}\boldsymbol{U}_f]^T \boldsymbol{T}_f = [M_{ss}\boldsymbol{U}_s]^T \boldsymbol{T}_s, \tag{5}$$

where matrices M_{ff} and M_{ss} are defined as

$$M_{ff}^{ij} = \int_{\Gamma_f} N_f^i N_f^j d\Gamma, \quad \text{and} \quad M_{ss}^{ij} = \int_{\Gamma_s} N_s^i N_s^j d\Gamma.$$

While not strictly correct, matrices M_{ss} and M_{ff} are often referred to in this context as mass matrices (Jiao and Heath [7]).

Given some displacement transformation matrix H_{fs} such that $\boldsymbol{U}_f = H_{fs}\boldsymbol{U}_s$, and substituting this into equation (5), it is possible to construct a global traction transformation matrix,

$$\boldsymbol{T}_s = \left[M_{ff}H_{fs}M_{ss}^{-1}\right]^T \boldsymbol{T}_f. \tag{6}$$

Choosing $H_{sf} = \left[M_{ff} H_{fs} M_{ss}^{-1}\right]^T$, will result in global conservation of interface stress states. If we further recognise that the discrete concentrated forces are defined by $\boldsymbol{F}_s = M_{ss}^T \boldsymbol{T}_s$ and $\boldsymbol{F}_f = M_{ff}^T \boldsymbol{T}_f$, then equation (6) can be rewritten as

$$\boldsymbol{F}_s = H_{fs}^T \boldsymbol{F}_f. \tag{7}$$

Energy along the interface will therefore provably be conserved if the transpose of the displacement interpolation matrix is used to project the concentrated nodal forces. It is perhaps important to note, that the force interpolation presented in equation (7) contains no information regarding the solid domain's internal interpolation and integration schemes. This immediately imposes a limitation, as it restricts passing information between equal order schemes only (i.e. linear FVM to linear FEM).

CONSISTENT INFORMATION TRANSFER

In this paper the definition of consistent transfer is adopted from de Boer et al. [13], which states that a constant stress state should be transferred exactly . The terminology of consistency stems from the requirement that the interpolation and integration of quantities along the interface should be consistent with the sub-domain schemes. In order to transfer a constant quantity requires that the row-sum of both H_{sf} and H_{fs} be equal to 1. Consistent information transfer therefore requires that the displacement and force transformation matrices be constructed independently of each other. In other words

$$\boldsymbol{U}_f = H_{fs} \boldsymbol{U}_s, \quad \boldsymbol{T}_s = H_{sf} \boldsymbol{T}_f, \quad H_{sf} = H_{fs}^T. \tag{8}$$

The consistent approach, when using RBF interpolation, translates to transferring the field quantities in the form of pressures and shear stresses directly. These field quantities are then integrated along the solid interface using integration rules consistent with the solid domain. For arbitrary nonlinear interface fields, with large gaps between the two sub-domain interfaces, there is no guarantee that such a consistent scheme will be energy conserving. The fact that the consistent scheme is not provably conservative, does not, as mentioned in Farhat et al. [1], imply that the transfer of information will lead to inaccurate or unstable FSI simulations. Transient partitioned FSI solvers are by construction non-conservative due to the time lag between the solid and flow solvers within a given time step. The primary concern should then rather be whether the accuracy and errors introduced by the information transfer are less than those already present. In the limit case of a constant stress state (patch test), provided the constant is exactly transferred, the system will be in static equilibrium, and hence energy conserving. We will further demonstrate that the error in energy, introduced for nonlinear stress states, disappears in the limit of mesh refinement.

RADIAL BASIS FUNCTION INTERPOLATION

The class of multivariate interpolation is based on the idea of using a global interpolation function to transfer information. Multivariate interpolation requires no connectivity information and is therefore well suited to the transfer of information between interface meshes with arbitrary geometric mismatches.

Radial basis functions (RBF) have in particular gained large popularity in the field of multivariate approximation theory. RBF is comparatively simple to implement, the underlying mathematical properties are well understood and provides good interpolation properties (for a numerical comparison to Kriging and the moving least-squares method, see Krishnamurthy [16]). RBF interpolation is based on fitting a series of splines, or basis functions to interpolate information from one point cloud to another.

Let us assume we wish to transfer information, $\boldsymbol{s}(\boldsymbol{x})$, from mesh A to mesh B, where $\boldsymbol{s}(\boldsymbol{x}) = \{\boldsymbol{u}, \boldsymbol{t}\}$. At the centres, $x_1, x_2, ..., x_{n_A}$ (either nodal coordinates or integration points) of mesh A, we know the discrete values of $\boldsymbol{s}(\boldsymbol{x})$, which we denote here as $g_1, g_2, ..., g_{n_A}$, where n_A is the number of DOFs at which information is known along the interface of mesh A. We now wish to construct a continuous function which allows us to interpolate the known values, at known locations, from A to the centres of B. The interpolation function, using RBF with an additional linear polynomial, has the following form:

$$\boldsymbol{s}\left(\boldsymbol{x}\right)=\sum_{i=1}^{n_A}\alpha_i\phi\left(||\boldsymbol{x}-\boldsymbol{x}_{A_i}||\right)+\sum_{j=1}^{m}p_j\left(\boldsymbol{x}\right)\beta_j, \tag{9}$$

where ϕ is the chosen basis function, $||\cdot||$ refers to the Euclidean distance (in 3 dimensions, $||\boldsymbol{x}-\boldsymbol{x}_{A_i}|| = \sqrt{\left(x-x_{A_i}\right)^2+\left(y-y_{A_i}\right)^2+\left(z-z_{A_i}\right)^2}$) and the coefficients α_i are to be solved so that the condition

$$\boldsymbol{s}\left(x_{A_i}\right)=g_i,\quad \text{for } i=1,...,n_A \tag{10}$$

is satisfied. $p_j\left(\boldsymbol{x}\right)$ are the monomial terms of a polynomial of degree m, and β_j are m additional constants introduced due to the additional polynomial terms. The m additional constants are obtained by including an additional m constraints in the form

$$\sum_{i=1}^{n_A}p_i\left(\boldsymbol{x}\right)\alpha_i=0. \tag{11}$$

The inclusion of the polynomial has an important consequence. If the function $\boldsymbol{s}\left(\boldsymbol{x}\right)$ can be described exactly by a linear polynomial, and the included polynomial $p\left(\boldsymbol{x}\right)$ is linear, then the polynomial will be exactly reproduced. This follows from, given that ϕ is a conditionally positive definite function, there provably exists a unique function $\boldsymbol{s}\left(\boldsymbol{x}\right)$ which satisfies both equations (10) and (11) (see Wendland [17] for a proof of this, and Beckert and Wendland [8] for a more detailed discussion). In this paper a linear polynomial of the form $p\left(x\right)=\beta_0+\beta_1x+\beta_2y+\beta_3z$ is used. By using a linear polynomial we can provably transfer constant information as well as rigid body motion (when transferring displacement). The inclusion of the linear polynomial does place some mild restrictions on our choice of information centres. While no additional sampling points are necessary, it does mean that at least 4 points must not fall along a plane. If Γ_{FSI} is in fact a flat face, then a linear polynomial cannot be used.

Using equations (10) and (11) the following matrix problem can be defined to solve for the RBF coefficients:

$$\begin{bmatrix}\boldsymbol{g}_A\\ \boldsymbol{0}\end{bmatrix}=\begin{bmatrix}\boldsymbol{M}_{AA} & \boldsymbol{P}_A\\ \boldsymbol{P}_A^T & \boldsymbol{0}\end{bmatrix}\begin{bmatrix}\boldsymbol{\alpha}\\ \boldsymbol{\beta}\end{bmatrix}. \tag{12}$$

Here $\boldsymbol{\alpha}$ is the vector containing the coefficient sets $\boldsymbol{\alpha}_i$, and $\boldsymbol{\beta}$ is the vector containing the polynomial constants, both to be solved for. For a linear polynomial $p\left(\boldsymbol{x}\right)$, $\boldsymbol{P}_A$ is an $n_A\times 4$ matrix where each row i is given by $\left\{1,x_{A_i},y_{A_i},z_{A_i}\right\}$, for $i=1,2,...,n_A$. Finally, $\boldsymbol{g}_A$ refers to the matrix of known values to be interpolated to mesh B. $\boldsymbol{M}_{AA}$ is an $n_A\times n_A$ matrix containing the evaluations of the RBF basis functions

$$\boldsymbol{M}_{AA}=\begin{bmatrix}\phi_{A_1A_1} & \phi_{A_1A_2} & \cdots & \phi_{A_1A_{n_A}}\\ \vdots & \vdots & \ddots & \vdots\\ \phi_{A_{n_A}A_1} & \phi_{A_{n_A}A_2} & \cdots & \phi_{A_{n_A}A_{n_A}}\end{bmatrix}, \tag{13}$$

where $\phi_{A_1A_2}=\phi\left(||\boldsymbol{x}_{A_1}-\boldsymbol{x}_{A_2}||\right)$. Once the coefficients $\boldsymbol{\alpha}$ and $\boldsymbol{\beta}$ have been solved, the interpolated quantities on mesh B, $\boldsymbol{g}_B$ can then be found by

$$\left[\boldsymbol{g}_B\right]=\begin{bmatrix}\phi_{BA} & \boldsymbol{P}_B\end{bmatrix}\begin{bmatrix}\boldsymbol{\alpha}\\ \boldsymbol{\beta}\end{bmatrix}. \tag{14}$$

In other words,

$$\boldsymbol{g}_B=\begin{bmatrix}\phi_{BA} & \boldsymbol{P}_B\end{bmatrix}\begin{bmatrix}\boldsymbol{M}_{AA} & \boldsymbol{P}_A\\ \boldsymbol{P}_A^T & \boldsymbol{0}\end{bmatrix}^{-1}\begin{bmatrix}\boldsymbol{g}_A\\ \boldsymbol{0}\end{bmatrix}. \tag{15}$$

NAME:	DEFINITION
C^0 compactly supported piecewise polynomial (C^0):	$(1-(\|\|x\|\|/r))_+^2$
C^2 compactly supported piecewise polynomial (C^2):	$(1-(\|\|x\|\|/r))_+^4(4(\|\|x\|\|/r)+1)$
Thin-plate spline (TPS):	$\|\|x\|\|^2 \ln \|\|x\|\|$
Multi-quadratic biharmonic (MQ):	$\sqrt{\|\|x\|\|^2+a}$
Cubic:	$\|\|x\|\|^3$
Quintic:	$\|\|x\|\|^5$

Table 1: List of RBF basis functions used in this study. Basis functions obtained from [8, 13, 19, 20].

The transformation matrix H_{BA} is therefore the first n_B rows and n_A columns of the matrix

$$\begin{bmatrix} \phi_{BA} & \boldsymbol{P}_B \end{bmatrix} \begin{bmatrix} \boldsymbol{M}_{AA} & \boldsymbol{P}_A \\ \boldsymbol{P}_A^T & \boldsymbol{0} \end{bmatrix}^{-1}. \tag{16}$$

The matrix inverse in equation (15) is usually not computed explicitly. We are only interested in $\boldsymbol{g}_B$, which can be found by solving equation (12) and performing the dot product in equation (14).

The list of RBF basis functions employed in this study are shown in Table 1. r in the C^0 and C^2 basis functions refer to the choice of support radius and the subscript $+$ indicates that only positive quantities are taken into account. How to choose r is important to the overall behaviour and interpolation quality of compactly supported functions. A larger value of r typically leads to very good interpolation results. Choosing r too large however, leads to ill-conditioned systems. Equally, smaller values of r leads to a sparsely populated banded matrix which is beneficial for efficient linear system solutions. For good interpolation results, r is typically recommended to be set to $r = 2r_{\max}$, where $r_{\max}$ is the radius which includes all points. For the MQ RBF function, the shape of the spline is controlled via the parameter a. Small choices of a lead to sharp cone-like splines which flatten out as a is increased. De Boer *et al.* [13, 18], suggest values of a in the range $10^{-5} - 10^{-3}$ for a domain of unit length.

NUMERICAL ANALYSIS

PATCH TEST

The purpose of the patch test is to determine whether the transfer scheme can exactly represent a constant stress state. The patch test geometry used in this analysis is shown in Figure 2(a), where the discretised meshes along the curved interface are incompatible, and contain gaps and overlapping regions (see Bathe and Ledezma [21] for an alternative proposal of a patch test benchmark problem). We apply a constant, normal pressure along the top of the fluid domain of $p = 100$Pa. The fluid domain is prescribed with a density and viscosity of $\rho = 1$kg/m^3 and $\mu = 10$kg/ms respectively. For the solid domain, Young's modulus is set to E $= 200$GPa with a Poisson's ratio of $\nu = 0.49999$ to represent an incompressible material. The material and problem descriptions are chosen such that the solid domain displacements are very small. The resulting displacements are small enough to be near negligible, but because of the coupled nature of an FSI problem, inaccuracies, even at such small levels remain fundamentally important to the overall solution of the coupled system.

In Figure 2, we show the pressure state for both the conservative and consistent information transfer. The consistent approach exactly transfers the pressure state, and the FSI system is in perfect equilibrium (to illustrate this we summarise the norm of the transferred forces in Table 2). By contrast, the interface stress state arising from the conservative approach is highly oscillatory. The conservative approach transfers the magnitude of forces in each distinct direction exactly without accounting for the slight differences in surface areas and norms and therefore, despite being exactly conservative, the results deriving from the conservative scheme are incorrect. The inaccuracy of the stress state further manifests itself in an oscillatory displacement field (not shown here).

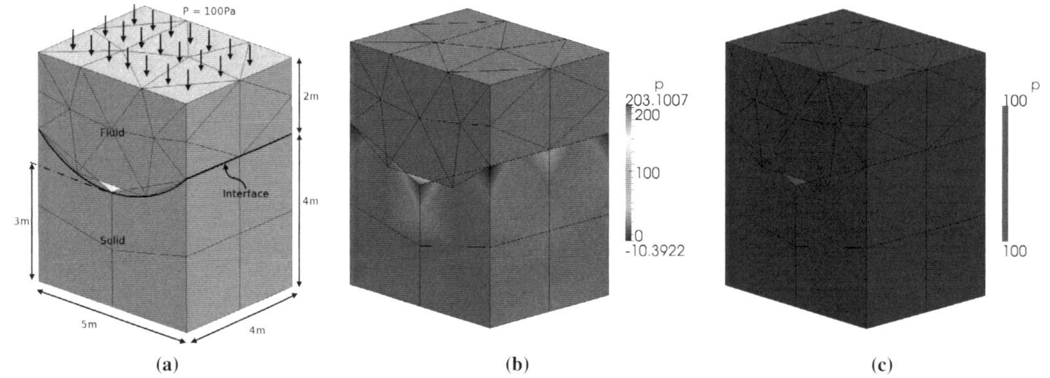

Figure 2: (a) Constant pressure patch test problem description solved using (b) conservative information transfer and (c) consistent information transfer.

	Conservative		Consistent	
	Fluid	Solid	Fluid	Solid
$\sum F_x$	-1.34e-07	-1.34e-07	-1.07e-10	-8.84e-11
$\sum F_y$	-2000.00	-2000.00	-2000.00	-2000.0 0
$\sum F_z$	-0.332	-0.332	-0.332	1.016e-17
$\lvert\sum F\rvert$	2000.00	2000.00	2000.00	2000.00

Table 2: Comparison of the sum of the transferred forces for the constant pressure patch test.

ANALYTICAL TEST FUNCTION

In this section, we compare the accuracy of the consistent and conservative approaches by transferring an analytical function across a curved interface, for different levels of grid refinement. We make use of the smooth, nonlinear function

$$\boldsymbol{s}(x, z) = \sqrt{\cos\left(x^2 + z^2\right)}. \tag{17}$$

To measure the accuracy of the transfer schemes we make use of the relative L^2 error defined by

$$\boldsymbol{s}_{\text{error}} = \sqrt{\frac{\sum_{i=1}^{n}\left(\boldsymbol{s}_{\text{exact}}^{i} - \boldsymbol{s}_{\text{interpolated}}^{i}\right)^2}{\sum_{i=1}^{n}\left(\boldsymbol{s}_{\text{exact}}\right)^2}}. \tag{18}$$

The interface mesh used for the analysis (depicted in Figure 3) is fairly typical of FSI simulations. The fluid interface is described with 2D triangular elements coming from a FVM domain, and the solid interface consists of quadrilateral elements with 4 quadrature points. It is important to mention that we limit our analysis to linear FE meshes for the structural domain. We already mentioned that the conservative scheme cannot be used to transfer information between linear to quadratic fields. Therefore, rather than biasing the results to the order incompatibility of the internal interpolation and quadrature rules, we wish to demonstrate that the conservative approach is a zero-order method even when transferring between two linear discretisation schemes.

In Figure 4(a) we report the error in displacement as a function of the choice of RBF basis functions for the simultaneous refinement of both the solid and fluid interfaces. These results are further compared to the error incurred using a linear FVM method. All the basis functions results in displacement errors and convergence rates below those already present within the fluid domain solver.

In Figure 4(b) we show the relative L^2 norm pressure error. The conservative and consistent schemes are indicated with dashed (- · -) and solid lines (–) respectively. For comparison, we include the pressure error for both linear

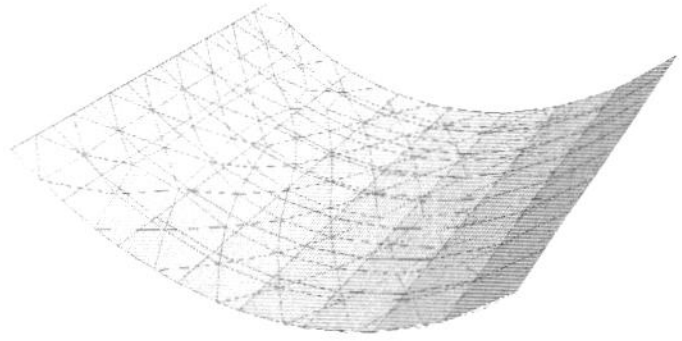

Figure 3: Illustrative example of the typical surface mesh mismatch using triangular fluid and four-noded solid surface elements (red: fluid mesh, black: solid mesh).

and quadratic FEM interpolation. The analysis most importantly shows that the conservative approach results in a non-convergent pressure transfer. This implies that, while the conservative formulation provably transfers the correct magnitude of forces, the spatial distribution and orientation of these forces are incorrect. The same trend has been observed in de Boer et al. [13]. This is a fairly concerning result: it implies that no matter how fine the fluid and solid mesh along a curved interface, the interface stress states will never be correct. As a side note, the conservative pressures are computed from the transferred concentrated forces by $P_s = [M_{ss}]^{-1} H_{fs}^T F_f$.

Since the consistent approach cannot provably conserve energy, we show the difference in work done along the interface using the consistent formulation in Figure 4(c). For a scheme to be globally conservative, this error should be exactly 0. This is indeed the case when using the conservative approach (by construction), and therefore is not shown. All the basis functions result in an energy error convergence rate of approximately 3.5, almost an order higher than the displacement and pressure convergence rates. Therefore, while the consistent approach is not provably conservative, the incurred energy errors decrease consistently with mesh refinement. Finally in Figure 4(d), we show the pressure error incurred by the consistent formulation as a function of mesh mismatch, illustrated by modifying the ratio of fluid to solid elements along the interface. The error in pressure transfer reaches a minimum at roughly twice as many fluid to solid elements with the error remaining fairly constant as the number of fluid elements is further increased.

CONCLUSION

In this paper, we outlined the use of radial basis function interpolation for interface information transfer. RBF interpolation requires no global connectivity information, and is therefore an elegant means by which to transfer information across a non-matching interface between FVM-FEM schemes. We focused, in particular, on a comparison between conservative and consistent formulations. We demonstrated that the conservative formulation is a zero-order method, where the error made in the spatial distribution of the transferred stress state does not reduce with the simultaneous refinement of the interface meshes. We further demonstrated that the consistent formulation, while not provably conservative, provides consistently decreasing errors in work done along the interface with mesh refinement. Furthermore, the rate of convergence of transfer errors (for both displacement and pressure) are higher than the associated FVM errors. The quintic function provided the highest rates of convergence but should be used with caution, as it rapidly leads to poorly conditioned matrices. Overall, the cubic function appears to provide a good compromise between accuracy and stability, with the C^2 and C^0 basis functions also providing promising results.

REFERENCES

1 Charbel Farhat, Kristoffer G Van der Zee, and Philippe Geuzaine. Provably second-order time-accurate loosely-coupled solution algorithms for transient nonlinear computational aeroelasticity. *Computer methods in applied mechanics and engineering*, 195(17):1973–2001, 2006.

2 Steven M Rifai, Zdeněk Johan, Wen-Ping Wang, Jean-Pierre Grisval, Thomas JR Hughes, and Robert M Fer-

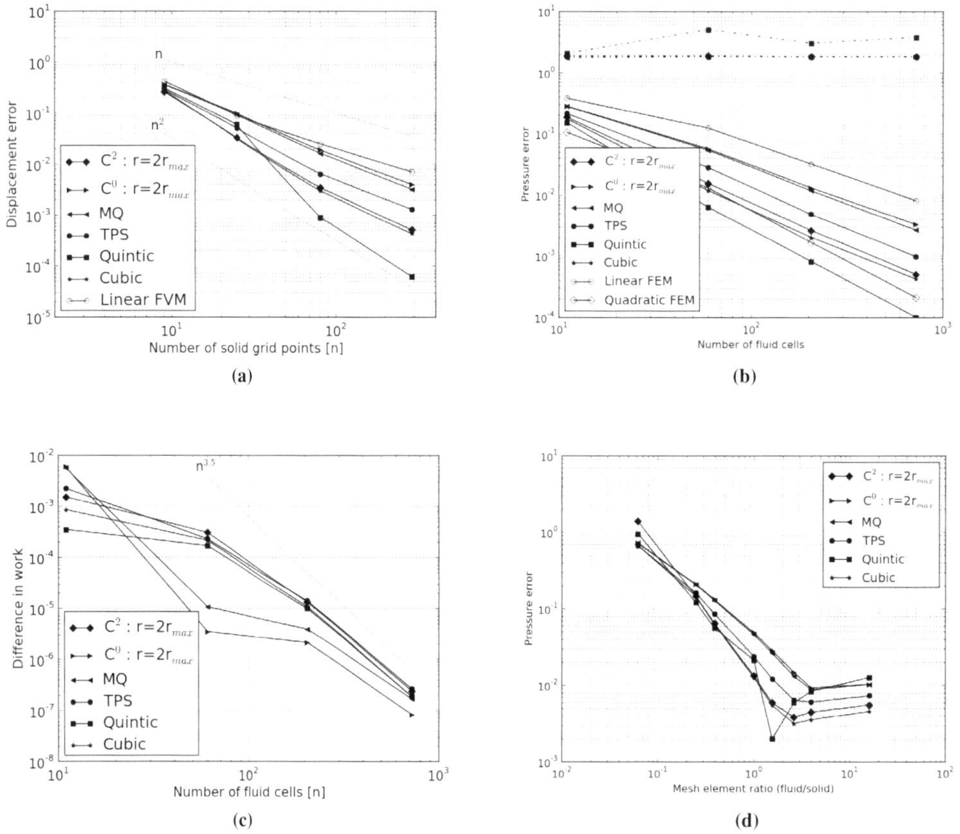

Figure 4: Results obtained from the transfer of the analytical test function. (a) Plot of displacement error for simultaneous refinement of the fluid and solid meshes (same for both conservative and consistent formulations). (b) Comparison of pressure error for the simultaneous refinement of fluid and solid interface discretisation. Solid lines (—) represent consistent transfer and (- · -) dashed lines represent conservative transfer. (c) Difference in work done across the interface. Only consistent transfer is shown, as the conservative transfer by construction is energy preserving. (d) Plot of L^2 relative pressure error using consistent transfer as a function of the ratio of the number of fluid to solid elements.

encz. Multiphysics simulation of flow-induced vibrations and aeroelasticity on parallel computing platforms. *Computer methods in applied mechanics and engineering*, 174(3):393–417, 1999.

3 Ryo Torii, Marie Oshima, Toshio Kobayashi, Kiyoshi Takagi, and Tayfun E Tezduyar. Computer modeling of cardiovascular fluid–structure interactions with the deforming-spatial-domain/stabilized space–time formulation. *Computer Methods in Applied Mechanics and Engineering*, 195(13):1885–1895, 2006.

4 BJBM Wolters, MCM Rutten, GWH Schurink, Ursula Kose, J De Hart, and FN Van De Vosse. A patient-specific computational model of fluid–structure interaction in abdominal aortic aneurysms. *Medical engineering and physics*, 27(10):871–883, 2005.

5 Juan Raul Cebral and Rainald Lohner. Conservative load projection and tracking for fluid-structure problems. *AIAA journal*, 35(4):687–692, 1997.

6 Charbel Farhat, Michael Lesoinne, and P Le Tallec. Load and motion transfer algorithms for fluid/structure interaction problems with non-matching discrete interfaces: Momentum and energy conservation, optimal discretization and application to aeroelasticity. *Computer methods in applied mechanics and engineering*, 157(1):95–114, 1998.

7 Xiangmin Jiao and Michael T Heath. Common-refinement-based data transfer between non-matching meshes in multiphysics simulations. *International Journal for Numerical Methods in Engineering*, 61(14): 2402–2427, 2004.

8 Armin Beckert and Holger Wendland. Multivariate interpolation for fluid-structure-interaction problems using radial basis functions. *Aerospace Science and Technology*, 5(2):125–134, 2001.

9 M Lombardi, N Parolini, and A Quarteroni. Radial basis functions for inter-grid interpolation and mesh motion in fsi problems. *Computer Methods in Applied Mechanics and Engineering*, 256:117–131, 2013.

10 Giuseppe Quaranta, Pierangelo Masarati, and Paolo Mantegazza. A conservative mesh-free approach for fluid-structure interface problems. In *International Conference for Coupled Problems in Science and Engineering, Greece*. Citeseer, 2005.

11 Marilyn J Smith, Dewey H Hodges, and Carlos E S. Cesnik. Evaluation of computational algorithms suitable for fluid-structure interactions. *Journal of Aircraft*, 37(2):282–294, 2000.

12 Serge Piperno, Charbel Farhat, and Bernard Larrouturou. Partitioned procedures for the transient solution of coupled aroelastic problems part i: Model problem, theory and two-dimensional application. *Computer methods in applied mechanics and engineering*, 124(1):79–112, 1995.

13 Aukje de Boer, Alexander H van Zuijlen, and Hester Bijl. Comparison of conservative and consistent approaches for the coupling of non-matching meshes. *Computer Methods in Applied Mechanics and Engineering*, 197(49):4284–4297, 2008.

14 Rajeev K Jaiman, Xiangmin Jiao, Philippe H Geubelle, and Eric Loth. Conservative load transfer along curved fluid–solid interface with non-matching meshes. *Journal of Computational Physics*, 218(1):372–397, 2006.

15 RK Jaiman, X Jiao, PH Geubelle, and E Loth. Assessment of conservative load transfer for fluid–solid interface with non-matching meshes. *International journal for numerical methods in engineering*, 64(15): 2014–2038, 2005.

16 Thiagarajan Krishnamurthy. Comparison of response surface construction methods for derivative estimation using moving least squares, kriging and radial basis functions. In *Proceedings of the 46th AIAA/ASME/ASCE/AHS/ASC structures, structural dynamics and materials conference, AIAA-2005-1821*, pages 18–21, 2005.

17 Holger Wendland. Konstruktion und untersuchung radialer basisfunktionen mit kompaktem träger. *Göttingen, Georg-August-Universität zu Göttingen, Diss*, 1996.

18 A De Boer, MS Van der Schoot, and H Bijl. Mesh deformation based on radial basis function interpolation. *Computers and Structures*, 85(11):784–795, 2007.

19 Holger Wendland. Piecewise polynomial, positive definite and compactly supported radial functions of minimal degree. *Advances in computational Mathematics*, 4(1):389–396, 1995.

20 B Fornberg, TA Driscoll, G Wright, and R Charles. Observations on the behavior of radial basis function approximations near boundaries. *Computers and Mathematics with Applications*, 43(3):473–490, 2002.

21 Klaus-Jürgen Bathe and Gustavo A Ledezma. Benchmark problems for incompressible fluid flows with structural interactions. *Computers and structures*, 85(11):628–644, 2007.

10th South African Conference on Computational and Applied Mechanics
Potchefstroom
3-5 October 2016

Energy usage optimisation of heavy haul freight trains

A. Bogaers* and N. Botha

Modelling and Digital Science, CSIR

abogaers@csir.co.za

ABSTRACT

The aim of this paper is to outline a preliminary investigation into an energy optimisation model with the aim of eventually being incorporated into a real-time driver assist program. A significant portion of the South African economy is sustained by a large and extensive mining industry. Integral to sustaining the mining economy is the transport of mined raw material via freight rail over large distances, typically from the mines to central distribution or processing centres. Due to the heavy tonnage and long distances an enormous amount of energy is required. Using classical mechanics, an energy usage model for typical freight trains based on available tractive power and typical rolling stock resistances (curvature, friction, gravity etc.) is considered. The aim is to find the optimal velocity profile over the full distance of the intended track, for which the total locomotive energy usage is reduced. We will illustrate the robustness of the proposed model to predict optimal operational velocity profiles for a number of interesting scenarios, many supported by similarly related analytical studies.

KEYWORDS

Energy, Optimisation, Freight rail

INTRODUCTION

The large amount of energy used in the operation of heavy haul freight trains has led to much research into energy efficient operation [1, 2, 3, 4, 5, 6]. Some of the earliest literature on optimal train control dates back to the 1960's when Ishikawa [3] determined the optimal control for a train traveling along a flat straight track. It was found that by maximising the Hamiltonian of the control problem (assuming that the control variables follow linear relationships with simple bounds) it can be shown analytically that the optimal energy control for a train on a flat track starts with a period of maximum power (maximum acceleration) followed by a period of partial power (cruising), a period of no power or braking (coasting) and finally a period of maximum braking, as illustrated in Figure 1(a). As the trip time is reduced, the cruising phase will decrease until it disappears, leaving only the full power phase, the coasting phase followed by maximum braking, as illustrated in Figure 1(b).

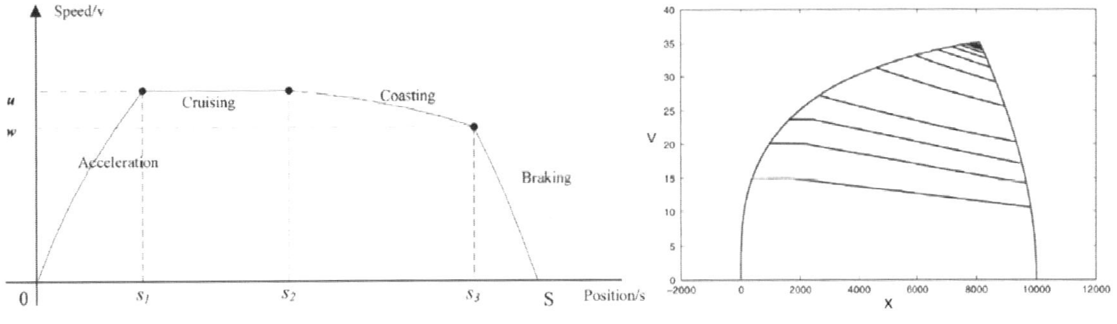

(a) (b)

Figure 1. (a) An illustration of the various phases, or switching points between the different driving phases for optimal control on a flat track. (b) The analytical profile for various control phases on a flat track for varying trip times. Images taken from Su et. al. [7] and Howlett et. al. [2] respectively.

In general, minimising the total energy for a trip within a given time frame can be fairly complex. The terrain typically consists of a number of inclines, descents and curves. Furthermore, for safety reasons, either in regions of sharp corners, or steep track gradients, speed limits are imposed. To further complicate matters, locomotive technology does not produce constant tractive power or braking force; both traction and braking efficiencies are typically functions of velocity, illustrated in Figure 2 for an E14 locomotive.

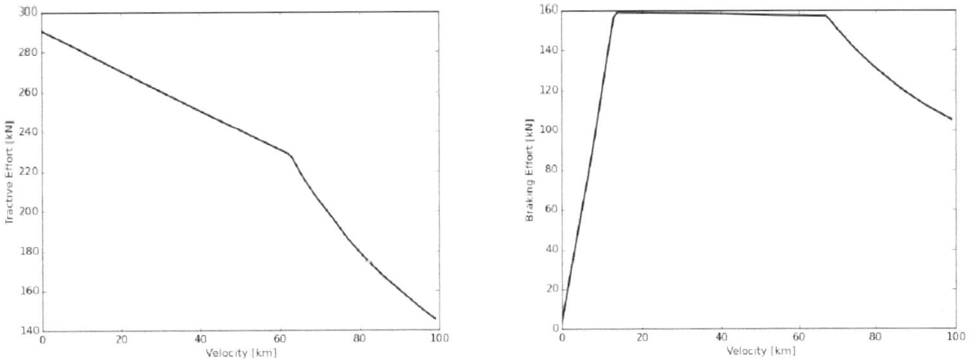

(a) (b)

Figure 2. Maximum available (a) tractive and (b) braking effort for an E14 locomotive as a function of velocity, images produced from Lombard [8].

To enable application of optimal train control in practice, the analytical studies have been extended to include speed limits, variable traction efficiencies and variable gradients [2]. These analytical studies are limited to lumped mass models. While evidence tends to suggest that by minimising energy one naturally improves the overall train coupling forces, to date there has been no formal attempt to optimise both in-train coupling forces and energy usage. In an attempt to improve the cost effectiveness and meet increasing scheduling constraints heavy haul operators are continuously increasing the length of trains. This leads to higher inter-train coupling forces, increased wear and an increased risk of derailment. Longer trains further pose the possibility of a single train simultaneously experiencing a multitude of different track conditions. For example, the lead portion of a train may be traveling downhill while the trailing end of the train is still traveling uphill, see Figure 3.

Minimising energy usage, while simultaneously including coupling forces, implies that analytical methods are no longer strictly suitable. To adopt such a multi-parameter optimisation problem would require the use of non-linear optimisation methods. Some early attempts have focused on using evolutionary optimisation methods. Consider for example the work presented by McClanachan et al. [9] who used genetic algorithms (GA) to optimise the speed profile. The study focused on a preliminary investigation illustrating that GA can provide speed profiles closely matching those predicted by analytical methods. The results appear sufficiently promising to warrant further investigation by including coupling forces.

GA algorithms are general purpose directed random optimisation algorithms which employ no special knowledge of the problem, to determine optimal solutions. They typically require several tens of thousands of function evaluations, similar to other evolutionary methods and related methods, such as ant colony optimisation, particle swarm optimisation or cuckoo search. More information on these methods can be found in a review by Wahab et. al. [10]. These evolutionary methods can be extremely expensive (orders of several hundreds to thousands of CPU hours depending on the computational complexity of the function evaluations, and the profile of the design space). Due to these computational constraints it is currently infeasible to converge to machine precision, often resulting in highly sporadic, unrealistic results.

In this paper we aim to pose an argument for gradient based optimisation methods. Gradient based methods are classified as non-linear optimisation methods, and utilise information pertaining to the gradients and gradient landscape of a cost function. They can therefore be considered as directed search algorithms and if the gradients associated with a cost function are available, typically require significantly fewer iterations to reach convergence than evolutionary methods.

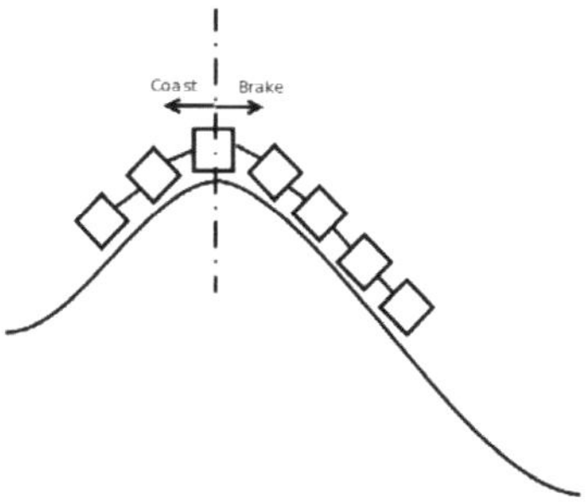

Figure 3. Long train on a hill experiencing different forces.

PROBLEM FORMULATION

The general non-linear optimisation problem can be posed as [minimise $f(x)$], subject to a number of equalities and inequality constraints, namely $b(x) \geq 0$ and $c(x) = 0$.

The optimisation problem can be formulated as finding the sequence and magnitude of locomotive power which completes a given trip, over a given terrain, within the specified time limit, while minimising the overall energy used. We define energy as the integral of locomotive traction, $f(x)$

$$E(x) = \int f(x)dx, \tag{1}$$

where x is the train location. The problem is subject to a number of constraints which can be summarised as

- Total travel time,

- speed limits,
- maximum available power (traction and braking), which varies with velocity,
- maximum rail-wheel adhesion (prevent slipping),
- coupling forces.

The locomotive traction forces can be defined by

$$F_{\text{Loc}} + F_r + F_{cv} + m_{\text{tot}} g_a = m_{\text{tot}} a, \tag{2}$$

where F_{Loc} is the force exerted by the locomotives, m_{tot} represents the total mass of all the locomotives and wagons within the train [11]. a represents the acceleration, F_r the lumping of all the rolling resistances, g_a the force due to gradient changes and F_{cv} represents the resistances due to curvature. The rolling resistance term F_r includes the resistance due to airflow friction losses and wheel rail frictional losses. Typically F_r takes the following form

$$F_r = m(c_0 + c_1 v + c_2 v^2), \tag{3}$$

where the constants $c_{0,1,2}$ are typically empirically/experimentally determined based on the given locomotive and wagon properties as well as the track configuration. Slope and curvature are illustratively shown in Figure 4 and the associated losses/forces due to inclines or curvature can be approximated as $g_a = g\sin(\theta)$ where θ is the angle of the slope and curve resistance as $F_{cv} = 0.004mD$ where $D = 0.5d_{wheelbase}/R$. R here represents the radius of curvature of the track curve.

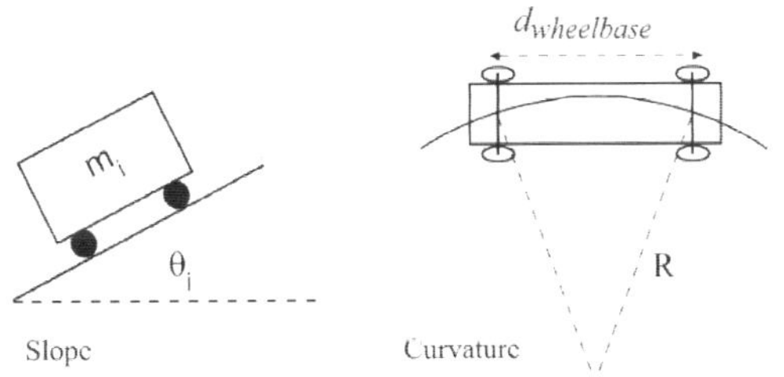

Figure 4. Simplistic illustration of losses due to gradients (slopes) and curvature [14].

Most gradient based methods require smooth, twice differentiable and continuous function definitions. Problem formulation is therefore of critical importance. Our design variable is therefore chosen to be velocity. While counter intuitive, this results in a relatively well behaved problem formulation. The track is discretised into a number of n segments. At each discrete location x_i, we assume we have available the track height h_i, and for each segment between two respective positions we know the track radius of curvature R_j . At x_0 and x_n the starting and final velocities v_0 and v_n are known. The unknown velocities $\{v_1 : v_{n-1}\}$ are therefore the velocities to be solved for, or the so called design variables, ***dV***, which minimises the total energy usage. Therefore, given a series of design variables ***dV*** = $\{dV_1, dV_2, ..., dV_{n-1}\}$, our known velocities within a given iteration are given by

$$\boldsymbol{v} = \{v_0, dV_1, dV_2, ..., dV_n, v_n\}. \tag{4}$$

The cost function can be computed for each section of the discredited track, Δx_j,

$$
\begin{aligned}
\Delta x_j &= x_{i+1} - x_i \text{ for } i = 1, 2, ..., n & v_{average,j} &= \tfrac{v_{i+1}+v_i}{2}, \\
\Delta t_j &= \frac{x_{i+1} - x_i}{v_{average,j}} & a_j &= \frac{v_{i+1} - v_i}{\Delta t_j}, \\
\Delta h_j &= \frac{h_{i+1} + h_i}{2} & g_{a,j} &= \sin(\arctan(\frac{\Delta h_j}{\Delta x_j}))9.81, \\
F_{r,j} &= m(c_0 + c_1 v_{i+1} + c_2 v_{i+1}^2) & F_{loc,j} &= m_{total}(a_j + F_{r,j} + g_{a,j} + F_{cv,j}).
\end{aligned} \tag{5}
$$

Given a initial guess for velocities at each section of the track we can compute the locomotive traction and breaking forces $\boldsymbol{F}_{loc}$ needed to achieve the given velocities $\boldsymbol{dV}$.

$\boldsymbol{F}_{loc}$ would be positive if traction is applied and negative if braking is present. The convergence rates of gradient based methods reduce when handling highly skewed landscapes. We therefore normalise the forcing function to be between $\{0, 1\}$. The force function and braking function **b** is then defined as

$$
f_i = \begin{cases} F_{loc,j}/P & \text{if} \quad F_{loc,j} > 0, \\ 0 & \text{if} \quad F_{loc,j} \leq 0 \end{cases} \tag{6}
$$

and

$$
b_i = \begin{cases} F_{loc,j}/B & \text{if} \quad F_{loc,j} \leq 0, \\ 0 & \text{if} \quad F_{loc,j} > 0 \end{cases} \tag{7}
$$

where P is the maximum available traction force to the locomotive and B is the maximum available braking force. It is perhaps important to note that in typical trains all wagons also apply braking forces. For long trains when using pneumatic braking, the braking signal may take a long time to propagate along the length of the train, which has been omitted for the current study. The cost function to be optimised then becomes,

$$
E(x) = \sum_{j=1}^{n-1} f_j \Delta x_j. \tag{8}
$$

Speed limits are handled as upper and lower bounds on the set of design variables $\boldsymbol{dV}$ and therefore can be handled by most optimisation algorithms.

Speed limits on tracks are typically imposed for safety reasons either in regions of sharp corners or other environmental factors which requires due diligence. In the absence of speed limits, it is possible to obtain estimates of maximum cornering speeds via simple summation of centrifugal forces and gravitational forces acting on a wagon running on a superelevated track with cant angle θ, see Figure 5 [12]. The critical, or maximum velocity for a given wagon can be found when the torque due to centrifugal force equals the torque due to the gravitational force, and is described as

$$
v_{\max} = \sqrt{\frac{gRl\sin^2\theta}{\cos\theta\,[h\cos\theta + l\sin\theta]}}. \tag{9}
$$

g represents the gravitational acceleration, R the radius of curvature of the section of track, l is half the gauge of the track and h the height of the center of gravity of the wagon/locomotive above the rail.

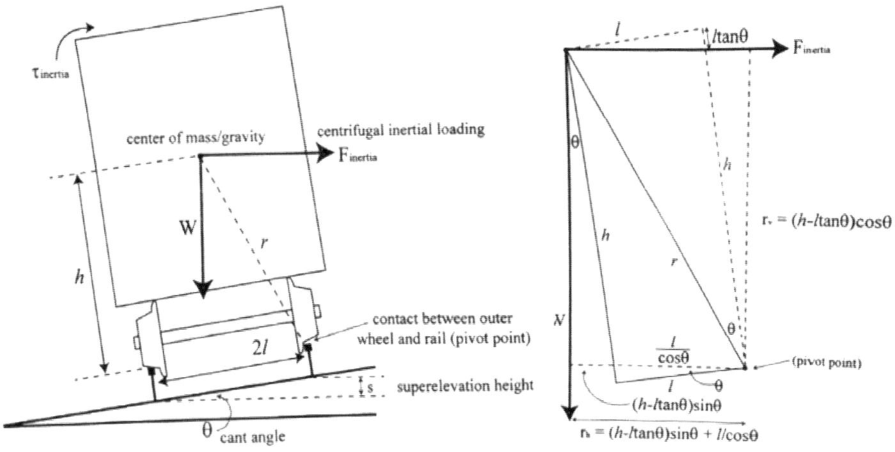

(a) (b)

Figure 5. (a) Forces acting on a wagon on a super-elevated track used to compute the critical cornering speeds and (b) the schematic representation [12].

The remaining constraints on the system, namely the maximum allowable time constraint, maximum force and braking constraint are mathematically posed as

$$
\begin{aligned}
&t_{\text{total}} \leq t_{\max}, \\
&f_j \leq f_{\max}(v), \qquad b_j \leq b_{\max}(v), \qquad \text{for } j = 1, 2, ..., n-1, \\
&f_j > 0, \qquad b_j > 0.
\end{aligned} \tag{10}
$$

The traction and braking constraints are posed as functions of velocities (i.e. $f_{\max}(v)$) since typical locomotive traction efficiencies are dependent on velocity and are not constant (as is often assumed in literature). An alternative cost function can be defined to optimise for minimum trip time for which the cost function then becomes

$$\text{minimise } T = \sum_{j=1}^{n-1} t_j. \tag{11}$$

It is further possible to define a multi-object optimisation problem to minimise for a combination of trip time and energy usage for which the cost function becomes

$$\text{minimise } ET = \omega\left(\int f(x)dx\right) + (1-\omega)\sum_{j=1}^{n-1} t_j, \tag{12}$$

where ω is a weighting factor to give preference to either energy or time. Typically, because energy used and travel time length scales differ, it would be advisable to include an upfront scaling factor to provide some equality between energy and time. Optimising for minimum trip time and the multi-parameter energy and time optimisation are both potentially useful for trip scheduling.

Due to the large number of constraints, we made use of sequential quadratic programming (SQP). SQP is useful for solving non-linear optimisation problems, where the problem is transformed into a succession of quadratic optimisation problems, with a quadratic objective and the linear constraint system consisting of equalities and inequalities. A detailed description of SQP can be found in the work by Bonnans et. al. [13].

RESULTS

To highlight the ability of the proposed energy optimisation model, a number of scenarios are tested. It has previously been shown [3] that the optimal energy profile can be summarised (following Figure 1) as a period of acceleration at maximum power, followed by a period of partial power with no braking (cruising), a period of no power and no braking (coasting) and finally maximum braking. As the trip time is decreased the period of partial power disappears, thereby only having three phases of maximum power, coasting and finally maximum braking. And lastly, for the shortest trip time there is only a maximum power phase, followed by maximum braking.

To test whether our non-linear optimisation formulation can reproduce these optimal energy curves, we analyse a train with 1 locomotive towing 20 wagons loaded with 100 tonnes each. The locomotive has a constant power output of 400kN and constant braking tractive effort of 500 kN. The total trip is 10 km and is discretised into 100 segments. The optimisation algorithm reproduced the expected analytical results, shown in shown in Figure 6 for different trip times. These results verify that our model applied to train on a flat track compares well to previous work. Some of the constants and values used are given in Table 1. [14]

Table 1: Values for constants.

Constant	Assigned value
c_0	$7.6558\text{x}10^{-3}$
c_1	$1.08\text{x}10^{-4}$
c_2	$1.4915\text{x}10^{-}5.$

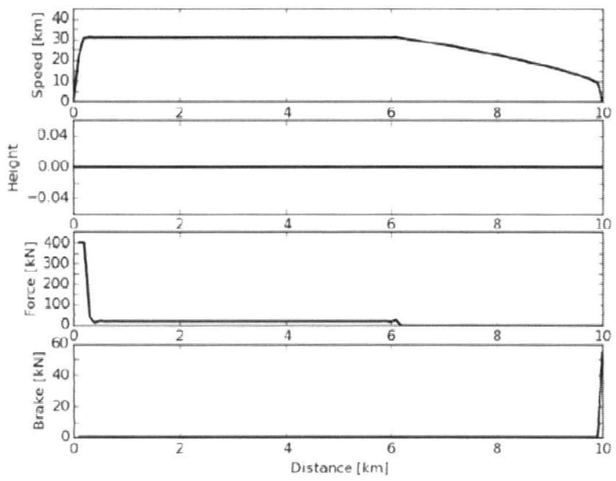

(a)

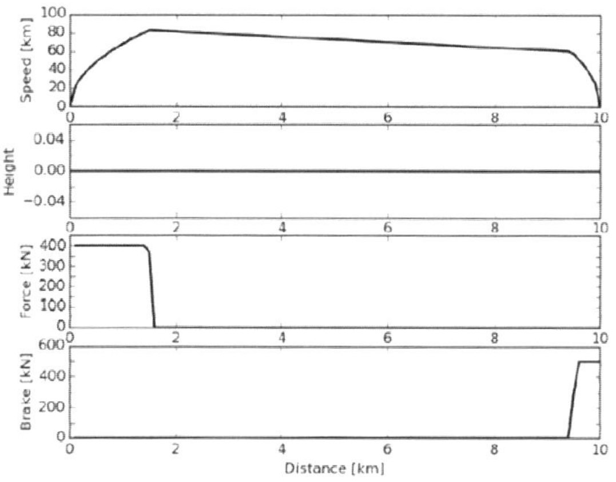

(b)

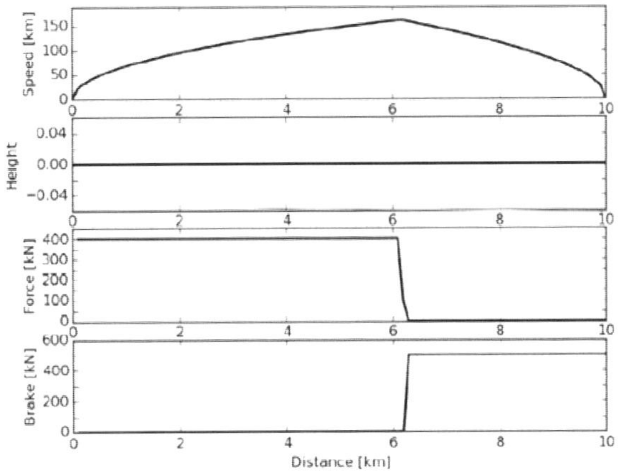

(c)

Figure 6. The optimised trip with a single locomotive pulling 2000 tonnes on a flat terrain for a time limit of (a) 25 min, (b) 10 min and (c) minimum trip time (7.3 min).

In Figures 6-8 the top subplot shows the train speed versus distance (the position on the track) with the track elevation versus distance in the second subplot. The third and fourth subplot shows the traction and braking force respectively, versus distance.

An important distinction is that real locomotives do not have constant tractive and braking efficiencies. For an E14 locomotive (using the maximum traction and braking curves as a function of velocity) we obtain a velocity and tractive effort profile, as shown in Figure 7, for minimum trip time, using the same track and number of wagons as used to generate the results in Figure 6. Following the expected profile of maximum power followed by maximum breaking, the optimisation algorithm successfully limits the maximum available tractive effort as a function of velocity. When comparing the results using the constant (Figure 6(c)) and velocity dependent (Figure 7) efficiencies there is a clear difference between the application of power and braking to get the same velocity profile.

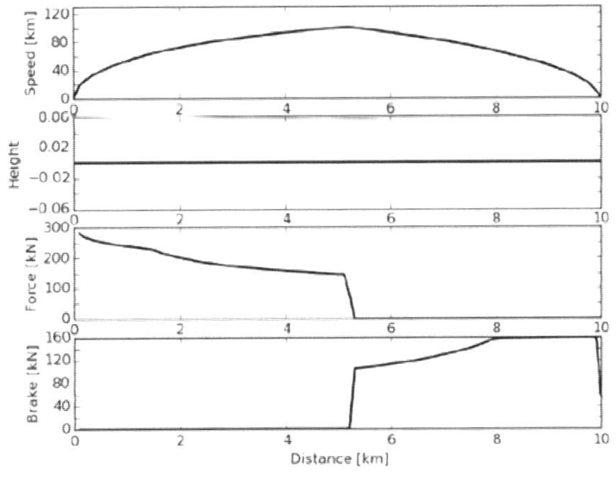

Figure 7. The optimised trip with a single locomotive pulling 2000 tonnes on a flat terrain for a E14 locomotive with variable maximum available traction and braking.

Finally a test track with steep gradients is considered, since typical tracks have numerous hills and descents. Following the work by Howlett et. al. [1, 2], it is expected that the optimal energy profile across a small hill would be to maintain cruising velocity by applying partial power to get over the hill. The gradient based optimisation can accurately predict this behavior. Consider here a 16km track with a 20m height change between 8 to 10km. Here we analyse a train with a locomotive with maximum constant power of 400kN, towing 2000 tonnes, where the optimal velocity profile for a 30 minute trip is shown in Figure 8(a). Notice how the cruising velocity is maintained by the application of partial power.

A really interesting problem occurs the moment an incline is too steep given the available power. In order to overcome the incline gradient the train requires application of power prior to the steep gradient, allowing momentum to carry the mass of the train over the incline. Even while full power is applied on the hill the train decelerates. To illustrate the capability of the gradient based optimisation algorithm to cope with this complex situation we use the same track with the locomotive now towing 6000 tonnes. The results for this simulation is shown in Figure 8(b).

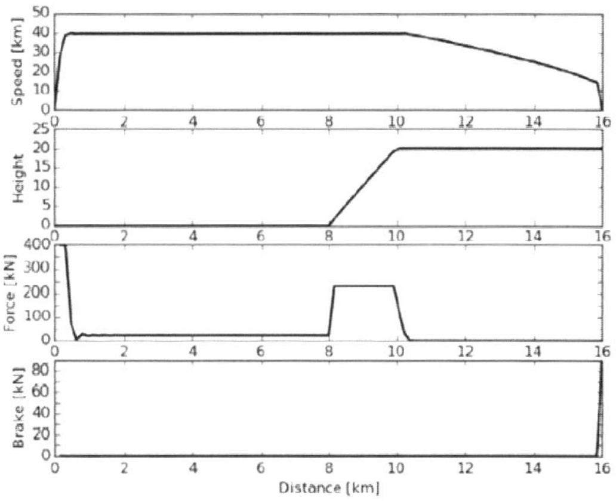

(a)

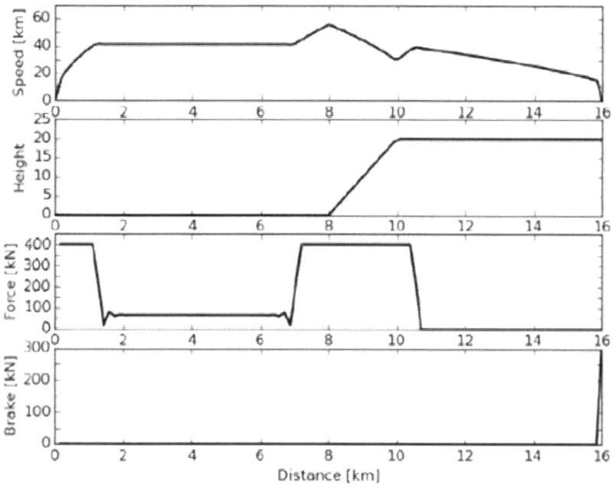

(b)

Figure 8. (a) Inclined track with train pulling 2000 ton and (b) pulling 6000 ton.

CONCLUSION

In this paper we outlined a preliminary investigation into a gradient based energy optimisation model with the aim of eventually being incorporated into a real-time driver assist program.

It was established that for a flat terrain the optimal control strategy starts with maximum acceleration at maximum power, followed by a period of cruising, a period of coasting and finally maximum braking. The cruising period disappears with decreasing total trip time, so that there is only three phases: maximum power, coasting and maximum braking. For the shortest trip time there is only a maximum power phase, followed by maximum braking. When the terrain has steep gradients the application of additional power is necessary, while for very steep gradients the application of additional power may be necessary before the gradient. The numerical model mirrors the behaviour from literature for each scenario, which verifies that our model applied to a train on a flat track as well as steep gradients, correlates well with the work of previous authors.

In this study the train has been described mathematically as a lump mass. This lump mass model has laid the groundwork for further investigation into integration of the coupling forces between the different locomotives and wagons by using a longitudinal dynamics model.

REFERENCES

1 Howlett, P, "Optimal strategies for the control of a train", Automatica, Vol. 32. No. 4, pp. 519-532, 1996.

2 Howlett, P.G. et. al., "Local energy minimization in optimal train control", Automatica, Vol. 45. No. 11, pp. 2692-2698, 2009.

3 Ichikawa, K., "Application of optimization theory for bounded state variable problems to the operation of train", Bulletin of JSME, Vol. 11. No. 47, pp. 857-865, 1968.

4 Jiaxin, C. and Howlett, P., "A note on the calculation of optimal strategies for the minimization of fuel consumption in the control of trains", Automatic Control, IEEE Transactions on, Vol. 38.(11), pp. 1730-1734, 1993.

5 Khmelnitsky, E., "On an optimal control problem of train operation", Automatic Control, IEEE Transactions on, Vol. 45. No. 7, pp. 1257-1266, 2000.

6 Liu, R.R. and Golovitcher, I.M., "Energy-efficient operation of rail vehicles", Transportation Research Part A: Policy and Practice, Vol. 37. No. 10, pp 917-932, 2003.

7 Su, S. et. al., "A Subway Train Timetable Optimization Approach Based on Energy-Efficient Operation Strategy", IEEE Transactions on Intellegent Transportation Systems, Vol. 14. No. 2, 2013.

8 Lombard,P.J., "Lecture notes in train service planning. Introductory course on multi-disciplinary concepts in railway engineering", University of Pretoria, July 2015.

9 McClanachan, M. and Cole, C., "Optimising the operation of a heavy haul train", In CORE 2012: Global Perspectives; Conference on railway engineering, 10-12 September 2012, Brisbane, Australia, page 569. Engineers Australia, 2012.

10 Wahab, A.b. et. al., "A Comprehensive Review of Swarm Optimization Algorithms", PLoS One, Vol. 10. No. 5, e0122827.

11 Cole, C, "Chapter 9 Longitudinal Train Dynamics", Editor Iwnicki, S., Handbook of Railway Vehicle Dynamics (pp. 239-277), CRC Press, 2006.

12 Moody, J.C., "Critical Speed Analysis of Railcars and Wheelsets on Curved and Straight tracks", Bates Collage, Honors Thesis, Paper 107, 2014.

13 Bonnans, J.F., et al., "Numerical optimization: Theoretical and practical aspects", Springer Science & Business Media, 2006.

14 Chou, M. and Xia, X., "Optimal cruise control of heavy-haul trains equipped with electronically controlled pneumatic brake systems", Control Engineering Practice, Vol. 15. No. 5., pp. 511-519, 2007.

Bulk Air Cooler Energy Optimisation through Simulation Coupled with an Optimisation Platform

Authors: Waldo Bornman[1*], Jaco Dirker[1], Deon C. Arndt[2], Josua P. Meyer[1]**

[1] *Department of Mechanical and Aeronautical Engineering, University of Pretoria, Pretoria, Private Bag X20, Hatfield 0028, South Africa.*

[2] *Enoveer Engineering Innovation, Pretoria, PO Box 60111, Pierre van Ryneveld 0045, South Africa*

**Corresponding Author:*

Email address: waldo.bornman@gmail.com

Phone: +27 (0)82 318 1847

***Alternative Corresponding Author:*

Email address: jaco.dirker@up.ac.za

Phone: +27 (0)12 420 2465

ABSTRACT

Mine cooling systems account for around 25% of the total energy consumption of a deep level mine. Bulk Air Coolers (BACs) contribute significantly towards this consumption due to the extensive cooling load introduced in order to sufficiently cool mine ventilation air to achieve acceptable underground working conditions. In this study an explicit modelling method is developed to predict the outlet conditions of a mechanical draft, direct-contact BAC as the inlet parameters are varied. This allows for reduced computational cost required for performance predictions which are useful in first order energy optimisation applications. The model was developed by combining forward and first principle methods, including underlying thermodynamic and heat transfer principles, and was used together with a non-linear optimisation algorithm to minimise the BAC energy consumption. The BAC model's characteristic system parameter was determined from and validated against measured data obtained from a deep-level mine. Based on the optimisation results, an average energy saving opportunity of 13% was identified on the consumption for a period of high energy demand by altering the operating conditions. This study highlights that such a modelling method coupled with an optimisation platform could be a useful tool to reduce energy costs of multi-component cooling systems.

KEYWORDS

Thermal analysis, Simulation, Bulk Air Cooling, Optimisation

INTRODUCTION

Bulk Air Coolers (BACs) are used in deep level mines to cool ambient air for underground ventilation and cooling use. They operate on the same principle as direct contact cooling towers and rely on mass and energy transfer to exchange heat between liquid water and air. Due to this, BACs have significant chilled water requirements and are responsible for a substantial portion of the mine's energy consumption as they form part of the mine's refrigeration system, typically accounting for around 25 % of the mine's total energy consumption [1]. In order to reduce the energy consumption of the mine's refrigeration system, BACs offer obvious potential due to their significant magnitude of total cooling demand. To minimise the BAC's energy consumption, an explicit mathematical model is required from which the outlet thermal conditions of a BAC can be predicted with acceptable accuracy to facilitate the coupling with an optimisation algorithm allowing real time energy minimisation.

Modelling of direct contact cooling towers was first formally documented by Merkel in the early 19th century [2]. Klimanek and Bialecki [3] and Kloppers and Kröger [4] indicated that the *Merkel*, *e-NTU* and *Poppe* methods are among the most popular models used for predicting the performance of cooling towers and are extensively used in cooling tower design. Although they are widely used during design stages, these models are not ideally suited to optimisation as they require iterative computation.

Khan *et al.* [5] considered a sensitivity analysis and reported that the two most important parameters influencing the performance of counter-flow cooling towers are the air inlet wet bulb temperature and the water outlet temperature. These models were based on the steady-state energy and mass balance of incremental volume sections, with the solution obtained through iterative computation. These are also iterative in nature and are not ideal for optimisation purposes.

Part-load cooling tower operation has been considered by various studies. Wang [6] indicates that the water-to-air mass flow ratio, $\dot{m}_w/\dot{m}_a$, has a significant effect on cooling tower performance. Söylemez [7] performed optimisation in terms of $\dot{m}_w/\dot{m}_a$ and found that the optimum ratio increased as the water inlet temperature increased. The models used here were based on the *e-NTU* method which requires iterative computation.

According to Wang *et al.* [8], the complex models constructed to model the various components of the cooling tower in refined detail are not always practical. Therefore, they suggested that actual measured process data could reflect the actual process parameters and performance with more ease and simplicity. The objective of their study was to develop a model with which the outlet water temperature can be calculated in terms of variables that include the water inlet temperature, the inlet air psychometric state and the water mass flow rate. The model was therefore simplified and limited by neglecting air flow as a variable, since there was no air flow variance. Similarly, Pan *et al.* [9] also used this model that was based on the multi-model modelling method described by Xue and Li [10] and also requires iterative computation to obtain the solution.

Serna-González *et al.* [11] suggested using the *Merkel* method while Rubio-Castro *et al.* [12] considered the *Poppe* model, both coupling their models with a mixed-integer non-linear programming algorithm to minimize the total annual cost. Although yielding satisfactory results, these methods are also limited for real time optimisation due to the iterative computation requirement of the models. It contains valuable insight for cooling tower design considerations. Serna-González *et al.* [11] achieved operational cost savings of 19.65% on the cooling tower considered.

In a summary of the three most common cooling tower models (the *Merkel*, *e-NTU* and *Poppe* methods), Klimanek and Bialecki [3] confirmed the iterative requirement of these models to determine the solutions. The initiatives mentioned above have, to date, only been applied to conventional cooling towers, where the objective of the cooling tower is to cool the water flowing through it and not to cool the air as is the case in BAC's. Also, little to no information is available for modelling of thermodynamic components for real time optimization. This is needed to reduce the energy requirements of a large cooling system by adapting the operating point depending on altered operation requirements and surrounding atmospheric conditions. Arndt [13] presented an explicit method, without the requirement for iterative computation, to model cooling towers based on the heat balance heat transfer rate principles. The advantage of this method has not been fully explored and will be used as the proposed modelling method to simulate the BAC in conjunction with an optimisation platform. The novelty of this study thus lies in the operational cost minimisation that has not yet been applied to integrated BAC tower systems by coupling a bulk air cooling tower model with an optimisation platform.

PROPOSED MODELLING METHOD

In this section, the proposed cooling tower modelling method is briefly presented.

Consider a counter flow direct contact cooling tower as shown in Figure 1. Air is mechanically drawn upwards through the tower by means of an electric fan bank while water is sprayed into the air-stream at the top and falls downward under the influence of gravity. Known quantities include the air inlet enthalpy ($h_{a,i}$), air inlet dry bulb temperature ($T_{a,i}$), air mass flow rate ($\dot{m}_a$), water inlet temperature ($T_{w,i}$), and the water mass flow rate ($\dot{m}_w$). From these parameters the air outlet enthalpy ($h_{a,o}$) and the water outlet temperature ($T_{w,o}$) need to be determined explicitly. In addition to this, for optimisation purposes, the equations used to construct the prediction model should ideally be continuous functions without step changes.

The following simplifications are applied:

- The reduction of the water flow rate (mass transfer) by evaporation is neglected since it has an insignificant impact on the thermal behaviour of the tower;
- The heat transfer surface area remains constant and;
- In the general operating range of a cooling tower the saturation enthalpy curve is approximately linear between 7 °C and 20 °C [14] and is modelled as such.

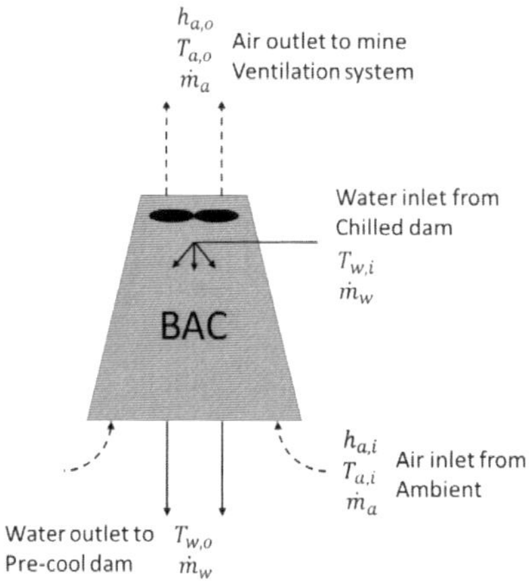

Figure 1 BAC schematic diagram

Based on the heat balance principle and heat transfer rate fundamentals, it is possible to express the steady state air and water outlet conditions of such a cooling tower in terms of the air and water inlet conditions by using Eqs. (1) and (2) [13].

$$h_{a,o} = \frac{(1-r)}{(\Gamma-r)}h_{a,i} + \frac{(\Gamma-1)}{(\Gamma-r)}\phi T_{w,i} \quad \textbf{(1)}$$

$$T_{w,o} = \frac{\frac{(\Gamma-1)r}{(\Gamma-r)}h_{a,i}}{\phi} + \frac{(1-r)}{(\Gamma-r)}T_{w,i} \quad \textbf{(2)}$$

Where:

$$\Gamma = \exp\left(-\beta\left(\frac{1}{C_w} - \frac{1}{C_a}\right)\right);\ r = \frac{C_a}{C_w};\ \phi = \frac{h_s}{T_s}$$

Here, C_a and C_w are mass transforms where $C_a = \dot{m}_a$ and C_w is defined as: $C_w = \frac{C_{p,w}\dot{m}_w}{\phi}$, with $C_{p,w}$ being the specific heat of the water. The symbol h_s refers to the enthalpy of the saturated air at the inside wetted surface temperature (T_s), measured in °C. β [W/K] is the characteristic system parameter that represents the inverse of the effective overall thermal resistance.

The air outlet temperature ($T_{a,o}$) can subsequently be calculated from the air outlet enthalpy obtained in Eq. (1), by using Eq. (3) [13].

$$T_{a,o} = \frac{(T_{a,i} - T_{w,i})}{(h_{a,i} - \phi T_{w,i})}h_{a,o} + T_{a,i} - \frac{(T_{a,i} - T_{w,i})}{(h_{a,i} - \phi T_{w,i})}h_{a,i} \quad \textbf{(3)}$$

From Eq. (1), β can be determined by minimising the prediction output errors in terms of the measured output values (in this case the exit water temperature) or alternatively from manufacturer catalogue data. This is implemented in the study by minimising the sum of the squares of the prediction errors that is represented in Eq. (4) for a data set consisting of N data points:

$$\beta_{min} = MIN \sum_{k=1}^{N} \left(T_{w,o,calc,k} - T_{w,o,real,k}\right)^2 \tag{4}$$

$T_{w,o,calc,k}$ represents the calculated water outlet temperature from Eq. (2) based on β, while $T_{w,o,real,k}$ is the measured cooling tower water outlet temperature for each data point, k, respectively.

SYSTEM PARAMETER DETERMINATION

The proposed modelling method was used to predict the thermal performance of a BAC at the Kopanang deep level gold mine in South Africa, which is located in a moderately cold eastern plateau interior at an altitude of 1336 m. A schematic flow diagram of the cooling system configuration is given in Figure 2. Atmospheric air is drawn into the BAC and cooled before it is passed through the mine. Chilled water is supplied to the BAC from a chilled dam which is cooled by a surface refrigeration plant at a relatively constant (3 ± 1 °C) water set point temperature which is on average equal to the BAC design water inlet temperature. Water that passed through the BAC is collected into a pre-cool dam before being pumped through the refrigeration plant where it is cooled to the set point temperature before it enters the chilled dam. The other main components shown in the figure (hot dam and pre-cooling tower) are used to pre-cool the water obtained from the mine, before it is also passes through the refrigeration plant.

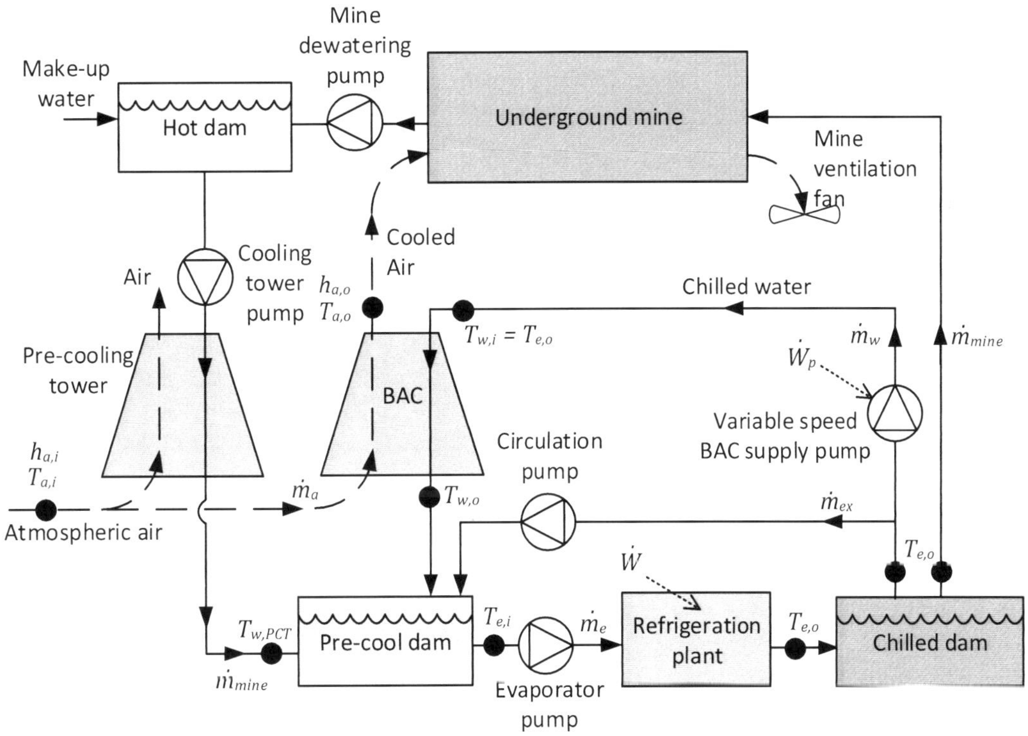

Figure 2: Mine cooling system schematic

A variable speed drive (VSD) on the BAC supply water pump is used to vary the water flow through the BAC while the air flow is controlled by the mine ventilation fans according to mine ventilation requirements. The design data of the bulk air cooling tower is presented in Table 1.

Table 1: BAC design operational parameters

Description	Value
Water inlet temperature, $T_{w,i}$	3 °C
Water outlet temperature, $T_{w,o}$	9 °C
Water flow, $\dot{m}_w$	750 kg/s
Air flow, $\dot{m}_a$	1250 kg/s
Air inlet wet bulb temperature, $T_{a,i,wb}$	18 °C
Air outlet wet bulb temperature, $T_{a,o,wb}$	7 °C
Air inlet relative humidity	50 %
Pump motor rating	225 kW

Daily average values of data logged by the mine for November and December 2013 were used to capture the daily climatic trends as the climate changes, instead of the variance throughout each day. The accuracy of the measuring equipment is summarised in Table 2 [15]–[17].

Table 2: Measurement instrumentation accuracy

Parameter	Instrument	Unit	Accuracy
Water temperature	Wika model TC40 type T thermocouple	°C	± 0.5 °C
Water flow rate	Krohne UFM 3030 in-line ultrasonic flow meter	l/s	± 0.5 %
Air temperature	Testo 6682 with 6616 probe	°C	± 0.15 °C
Air relative humidity	Testo 6682 with 6616 probe	%	± 1 %

All required BAC parameters were available as logged data on the mine's data system, except for the air mass flow rate ($\dot{m}_a$) and water outlet temperature ($T_{w,o}$). These two parameters were determined via a mass and energy balance across the BAC and pre-cooling dam with Eqs. (5) and (6). $T_{w,o}$ can be calculated by the assumption that the specific heat capacity of the water ($C_{p,w}$) remains constant for the temperature range considered. Since the pre-cooling dam is covered, evaporation from the dam is neglected. In addition to this, all pipes, pumps and valves are assumed to be adiabatic:

$$T_{w,o} = \frac{\dot{m}_e \cdot T_{e,i} - \dot{m}_{mine} \cdot T_{w,PCT} - (\dot{m}_e - \dot{m}_{mine} - \dot{m}_w) \cdot T_{e,o}}{\dot{m}_w} \quad \textbf{(5)}$$

After determining the BAC water outlet temperature, the energy balance principle in the BAC can be used to obtain the air mass flow rate, $\dot{m}_a$, via Eq. (6).

$$\dot{m}_a = \frac{C_{p,w} \cdot \dot{m}_w (T_{w,o} - T_{e,o})}{h_{a,i} - h_{a,o}} \quad \textbf{(6)}$$

From the measured data and Eq. (4) the characteristic model parameter resulting in the minimum error was determined to be $\beta = 844.47$ W/°C. Based on this parameter, the water outlet temperatures of the dataset can be predicted. A Root Mean Square of Relative Error (RMSRE) value of 0.107 and a correlation coefficient of R = 92.3 % are obtained. The predicted water outlet temperature remained within a ±1.5 °C error for 91.9 % of the time as can be seen in Figure 3.

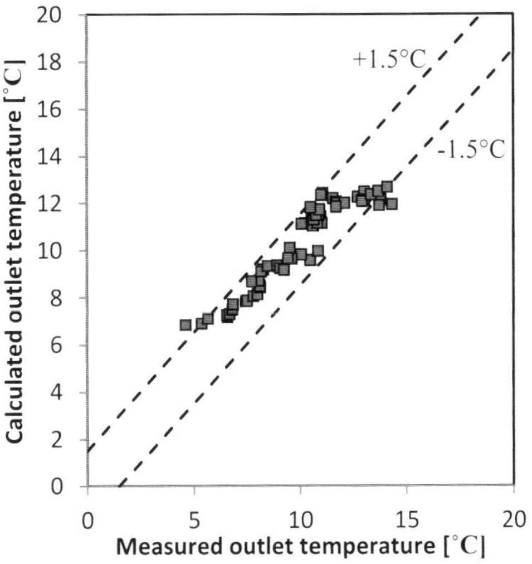

Figure 3: Parameter determination correlation

ENERGY CONSUMPTION MINIMISATION

BAC water flow control was previously considered by Du Plessis *et al.* [18] who reported significant energy savings achieved on *in situ* experimental results where BAC flow reduction was applied amongst other cooling system flow control. A saving of 33.3 % was achieved on the combined system, although the contribution due to the BAC control on its own was not quantified. Coupling the developed BAC model to an optimisation platform facilitates the determination of the energy savings potential of the BAC. Furthermore, the energy savings potential can be maximised by controlling the water flow rate as the ambient conditions and air flow varies. The water inlet temperature is assumed constant as the chilled dam temperature is controlled to a fixed set point delivered by the refrigeration plant as mentioned earlier.

System data from the mine for January to March 2013 was considered for the optimisation. The average BAC air outlet enthalpy of 35 kJ/kg recorded during the summer was used as an air outlet condition constraint while controlling the water flow rate. This value corresponds well with typical acceptable underground conditions [19]. The BAC simulation model was coupled to a commercial optimisation platform, called the Analytic Solver Platform developed by Frontline Systems Inc. This platform uses a generalised reduced gradient (GRG) optimisation method to optimise non-linear optimisation problems. This method chooses a basis and thereafter it determines a search direction. Subsequently, a line search is performed on each major iteration in order to solve systems of nonlinear equations at each step to maintain feasibility. This platform offers a multi-start method coupled to the GRG engine in order to yield global optimal solutions within the boundary conditions specified [20].

In order to model the integrated BAC power consumption, the supply pump power and chilled water power requirements are used in conjunction with the proposed BAC model. The fan power was not part of the energy optimisation control variables as the air flow is controlled by the mine ventilation requirements. Eq. (7) was used to determine the pump power [21].

$$\dot{W}_p = \frac{\Delta P \cdot \dot{m}_w}{\rho \eta_{m,p}} \qquad \textbf{(7)}$$

where the pressure differential over the pump is expressed as described in Eq. (8):

$$\Delta P = C\left(\frac{\dot{m}_p{}^2}{\rho}\right) \qquad \textbf{(8)}$$

The power requirement to chill the water for the BAC was determined from the water mass flow rate through the BAC, the increase in the water temperature in the BAC and the thermodynamic coefficient of performance of the refrigeration plant:

$$\dot{W}_{BAC} = \frac{\dot{m}_w \cdot C_{p,w}(T_{w,o} - T_{w,i})}{COP} \tag{9}$$

The coefficient of performance (COP) was obtained from the mine's data system and updated for each 24-hour interval. By combining the pump power and the refrigeration power to service the BAC yields the total power consumption by the BAC and can be computed with Eq. (10).

$$\dot{W}_{BAC,tot} = \dot{W}_p + \dot{W}_{BAC} \tag{10}$$

The optimised results are shown in Figure 4. The baseline power requirement is shown by the dashed blue line at the top, while the reduced power requirement is shown by the solid black line. An average reduction in electrical energy consumption of 337 kW or 517 632 kWh for the period was realised. This resulted in a reduction of 13 % of the average energy consumption of the BAC. From Figure 4, it can also be seen that the BAC over-cooled the air during the first part of January, the first half of February, towards the end of February and around the end March. This can be seen from the BAC power difference line that reaches peaks during those timeframes as well as the outlet enthalpy obtained being lower than the target outlet BAC enthalpy. It is also evident that the optimised power consumption decreases substantially when the ambient air enthalpy decreases as can be seen around the first part of January, as well as around the first and last ends of February along with the last half of March. This is due to the reduced chilled water flow requirement resulting in less chilled water consumption and reduced pumping power. Furthermore, it can also be seen that the ambient air enthalpy increased at the end of January as well as in the middle- and end- of February, resulting in increased power consumption where the optimised power reaches the baseline power. This is due to the BAC running at full load at these ambient conditions in an attempt to achieve the desired air outlet enthalpy of 35 kJ/kg. At these conditions, it should also be noted that the BAC air outlet enthalpy increased above the targeted 35 kJ/kg for both the baseline and optimised scenarios. This is due to the BAC being incapable of reducing the air enthalpy further as it was running at full cooling capacity.

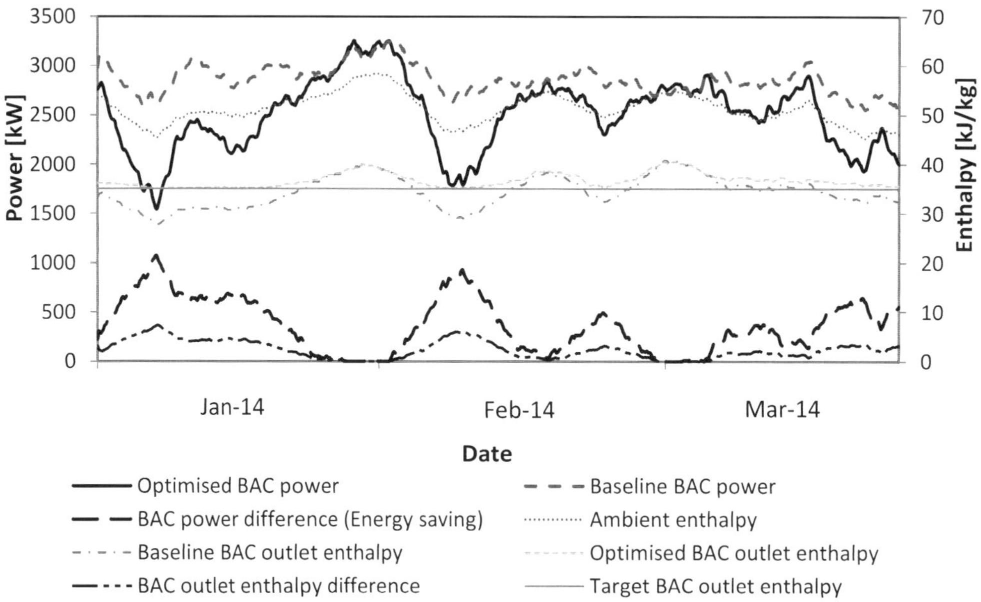

Figure 4: BAC energy optimisation results for January to March 2014

OPERATIONAL PARAMETERS

As a result of the optimisation, the opportunity arises to utilise the simulation model coupled to the optimisation platform to determine the ideal chilled water supply flow rate for all external uncontrolled system parameter configurations; being the air flow rate as well as ambient air conditions. As mentioned earlier, it should be noted that a constant BAC water inlet temperature of 3 °C was considered here, similar to the energy minimisation. It should also be noted that other water inlet temperatures could be used along with the simulation model for application to other systems with their unique or varying water inlet temperatures. For this determination, ambient air enthalpies ranging between 35 kJ/kg and 65 kJ/kg were used. The lower limit was chosen as the BAC would not be utilised for ambient air below 35 kJ/kg, while the maximum was chosen as it represents the maximum typical summer daytime air enthalpy. Air flow rates between 750 kg/s and 2000 kg/s were considered while water flow rates were varied between 150 kg/s and 750 kg/s as these were the range of measured air and water mass flow rate data by the mine.

In Figure 5, the horizontal axis represents the air mass flow rate, while the vertical axis indicates the optimised water flow rate at which the BAC should be operated in order to yield the desired air outlet enthalpy of 35 kJ/kg. The various curves represent the ambient enthalpy in increments of 1 kJ/kg. From the graph it is evident that the minimum water flow rate of 150 kg/s satisfies the cooling demand for all variations in air flow rate for ambient enthalpies below 37.6 kJ/kg. It is also evident that the BAC could only cool 788.5 kg/s of air at the highest ambient enthalpy of 65 kJ/kg, as the maximum chilled water flow rate of 750 kg/s would not be able to yield outlet air at an enthalpy below 35 kJ/kg for higher air flow rates at this inlet condition.

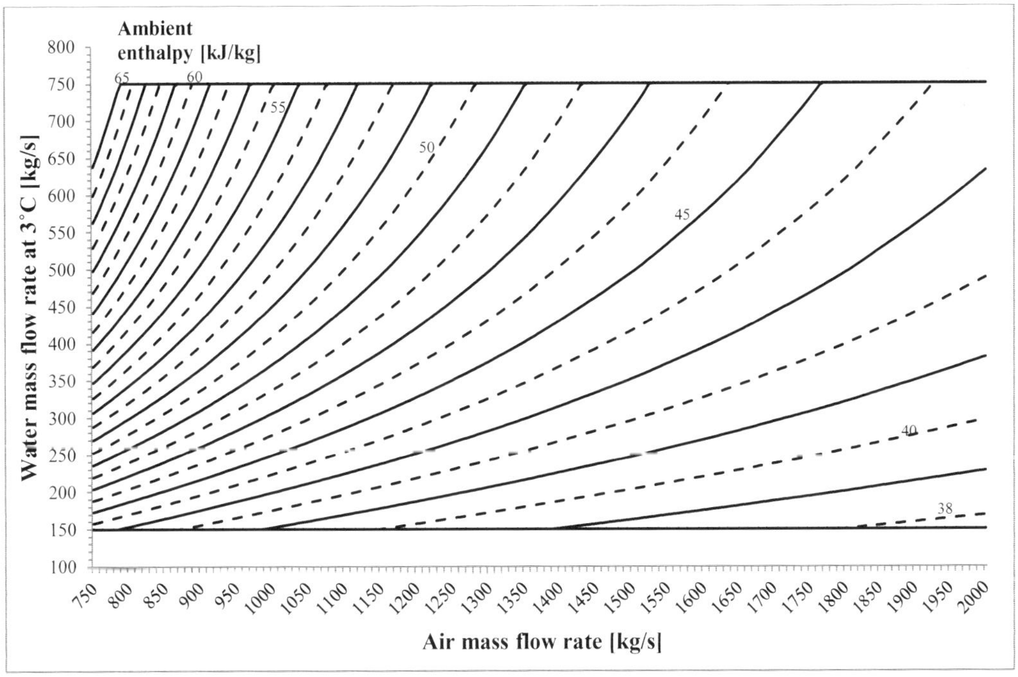

Figure 5: Optimised BAC water flow at 3°C for variations in air flow and ambient enthalpies

For comparison, two alternative water inlet temperatures were considered. For this purpose 2 °C and 4 °C were examined. Due to space limitations, the results were not included in the form of graphs. It was found that for a water inlet temperature of 2 °C the minimum water flow rate of 150 kg/s would be able to cool ambient air at all air flow rates for air inlet enthalpies below 38.2 kJ/kg, which results in a 1.3 % increase in enthalpy compared to the 3 °C water inlet temperature case. At the maximum water flow rate of 750 kg/s, a 2 °C water inlet temperature would be able to cool up to 869.4 kg/s of air at the maximum inlet enthalpy of 65 kJ/kg. This results in a 10.3 % increase in air flow rate compared to the 3 °C water inlet temperature case. Similarly for a water inlet temperature of 4 °C, the results decreased by 1.5 % and 11.8 % respectively. It is therefore evident that variations in the water inlet temperature is more sensitive to air volumes that the BAC would be able to cool than the air inlet conditions.

The graph in Figure 5 is valuable to determine the optimised water flow rate of a cooling tower or BAC in operations where the flow is controlled manually by an operator instead of automatic flow control. In such events, the operator could incrementally vary the supply water flow rate based on current air flow requirements as well as the varying ambient conditions.

It should be noted that the advantage of utilising real time optimisation for automated applications would be the possibility to re-evaluate the optimum more accurately on a continuous basis as well as allowing for frequent re-calibration where the model parameter could be refined as the performance of the BAC or cooling tower changes.

PRACTICAL APPLICATION AND FURTHER OPPORTUNITY

The contribution considered in this study included only the BAC as a single component of the total cooling system due to its significant cooling demand. The energy optimisation simulated in this study could be implemented on real BAC's by utilising the suggested simulation model and optimisation platform as part of the system control typically programmed onto programmable logic controllers (PLC's) controlling the cooling system. The water and air inlet and outlet conditions would be measured and used to calibrate the simulation model. The calibration could be re-evaluated periodically, as mentioned earlier, in order to ensure that simulation result remain accurate. The resulted optimised water flow rate would then be communicated to the BAC supply pump VSD which controls the BAC water flow rate.

The successful implementation of this system would result in the opportunity to extend the cooling system energy optimisation by coupling the BAC simulation model to an integrated model representing more components of the cooling system.

CONCLUSION

An explicit BAC modelling method is presented in this paper. The model is based on energy balance and heat transfer rate principles with the objective of reducing computational intensity while maintaining acceptable accuracy to be used for system optimisation. The parameter identification can be obtained either by using the equipment manufacturer's catalogue data or measured data from installed BACs, together with a suitable multivariable non-linear optimisation algorithm. The parameter identification achieved a correlation coefficient of $R = 92.3$ % and RMSRE value of 0.107.

The modelling method was applied to an optimisation case study to determine the energy savings potential of controlling the water flow rate through the mine's BAC in order to obtain a targeted constant air outlet enthalpy of 35 kJ/kg. Energy savings of 13 % on the BAC energy consumption are proposed by the optimisation model for the summer months.

The proposed simulation model coupled with the optimisation platform were used to determine pre-set, optimised water flow rates at the full operational range of air flow and ambient conditions of the BAC. In doing so, various curves were populated for incremental changes in ambient conditions while the optimised water flow rate was given for variations in air flow rates. These values could be used as operational parameters on manually adjusted systems.

ACKNOWLEDGEMENTS

Kopanang mine for the cooling system data

Sunél Bornman for proof reading the paper

REFERENCES

1 W. Bornman, "Improved mine cooling system performance through the control of auxiliary systems," M-Eng Dissertation, North-West University, Potchefstroom, South Africa, 2012.

2 J. A. Queiroz, V. M. S. Rodrigues, H. A. Matos, and F. G. Martins, "Modeling of existing cooling towers in ASPEN PLUS using an equilibrium stage method," *ENERGY Convers. Manag.*, vol. 64, pp. 473–81, 2012.

3 A. Klimanek and R. A. Bia, "Solution of heat and mass transfer in counter flow wet-cooling tower fills," *Int. Commun. Heat Mass Transf.*, vol. 36, pp. 547–53, 2009.

4 J. C. Kloppers and D. G. Kröger, "A critical investigation into the heat and mass transfer analysis of counterflow wet-cooling towers," *Int. J. Heat Mass Transf.*, vol. 48, pp. 765–77, 2005.

5 J. Khan, B. A. Qureshi, and S. M. Zubair, "A comprehensive design and performance evaluation study of counter flow wet cooling towers," *Int. J. Refrig.*, vol. 27, pp. 914–23, 2004.

6 S. K. Wang, *Handbook of air conditioning and refrigeration*, 2nd ed. New York: McGraw-Hill, 2000.

7 M. S. Söylemez, "On the optimum performance of forced draft counter flow cooling towers," *Energy Convers. Manag.*, vol. 45, no. 15–16, pp. 2335–41, 2004.

8 J. Wang, S. Shieh, S. Jang, and C. Wu, "Discrete model-based operation of cooling tower based on statistical analysis," *Energy Convers. Manag.*, vol. 73, pp. 226–33, 2013.

9 T. Pan, S. Shieh, S. Jang, W. Tseng, C. Wu, and J. Ou, "Statistical multi-model approach for performance assessment of cooling tower," *Energy Convers. Manag.*, vol. 52, no. 2, pp. 1377–85, 2011.

10 Z. K. Xue and S. Y. Li, "Multi-model modelling and predictive control based on local model networks," *Control Intell. Syst.*, vol. 34, no. 2, pp. 105–12, 2006.

11 M. Serna-González, J. M. Ponce-Ortega, and A. Jiménez-Gutiérrez, "MINLP optimization of mechanical draft counter flow wet-cooling towers," *Chem. Eng. Res. Des.*, vol. 88, pp. 614–25, 2010.

12 E. Rubio-castro, M. Serna-gonzález, J. M. Ponce-ortega, and M. A. Morales-cabrera, "Optimization of mechanical draft counter fl ow wet-cooling towers using a rigorous model," *Appl. Therm. Eng.*, vol. 31, no. 16, pp. 3615–28, 2011.

13 D. Arndt, "Integrated dynamic simulation of large thermal systems," PhD Thesis, University of Pretoria, Pretoria, South Africa., 2000.

14 C. Lombard, "Two-Port Simulation of HVAC Systems and Object Approach," PhD Thesis, University of Pretoria, Pretoria, South Africa, 1996.

15 "Testo 6682 Instruction manual." [Online]. Available: https://www.testo.org/resources/media/global_media/produkte/testo_6682/Testo_6682_Ex_-_Manual.pdf. [Accessed: 09-Jul-2014].

16 "WIKA cable thermocouple Model TC40 Data sheet." [Online]. Available: http://www.wika.co.za/upload/DS_TE6540_en_co_2045.pdf. [Accessed: 17-Jul-2014].

17 "Krohne UFM 3030 Techinical Datasheet." [Online]. Available: http://cdn.krohne.com/dlc/TD_UFM3030_en_060424.pdf. [Accessed: 07-Sep-2014].

18 G. E. Du Plessis, L. Liebenberg, E. H. Mathews, and J. N. Du Plessis, "A versatile energy management system for large integrated cooling systems," *Energy Convers. Manag.*, vol. 66, pp. 312–25, 2013.

19 G. Pingye and C. Chen, "Field experimental study on the cooling effect of mine cooling system acquiring cold source from return air," *Int. J. Min. Sci. Technol.*, vol. 23, no. 3, pp. 453–6, 2013.

20 "Frontline Solver User Guide." [Online]. Available: http://www.solver.com/files/_document/FrontlineSolvers_UserGuideV11.pdf. [Accessed: 20 May 2014].

21 G. E. Du Plessis, D. C. Arndt, and E. H. Mathews, "The development and integrated simulation of a variable water flow energy saving strategy for deep-mine cooling systems," *Sustain. Energy Technol. Assessments*, vol. 10, pp. 71–8, 2015.

The use of digital image correlation in tyre research

Botha Theunis Richard*, Guthrie Alan Glenn, Stallmann Martin Joachim, Els Pieter Schalk

University of Pretoria

Theunis.Botha@up.ac.za, glenn.guth@gmail.com, joachim@stallmann.co.za, Schalk.Els@up.ac.za

ABSTRACT

Advances in camera technology have brought about the proliferation of inexpensive, high resolution and high frame rate cameras. At the same time the increase in computational power of computers has allowed algorithms which determine the depth of a scene and enabled the tracking of features in an image to be run in real time. These technological advances have resulted in the use of Digital Image Correlation (DIC) in many applications. In this paper a summary of DIC techniques used in modern tyre research is provided. These techniques allow for tyre model parameterisation such as the measurement of the tyre slip-angle and longitudinal slip measurement. Other algorithms enable the improvement of our understanding of the mechanisms behind the forces generated by a tyre. These techniques enable the deformation and strain measurement on the outside as well as the inside of a tyre in a non-contact manner and therefore do not influence the measurements and are performed in post processing on the raw camera images. This not only provides a visual representation of the test but also allows multiple algorithms to be implemented on the same raw images each proving different measurements.

KEYWORDS

digital image correlation, tyre research, slip angle, deformation

INTRODUCTION

In vehicle dynamics the majority of forces which excite the vehicle are generated by the tyres, especially at low vehicle speeds where aerodynamic forces can be neglected. Therefore, the tyre is a fundamentally important aspect in vehicle dynamics. For this reason multiple institutes and universities have on-going research in the field of tyre research. The advent of inexpensive digital cameras and the increase in computational power of computers have enabled multiple non-contact measurements such as displacement, velocity and strain to be determined using inexpensive cameras.

In tyre research there are multiple measurements which are of interest. The lateral force produced by a tyre is generated by the sliding and deformation of the tread over the terrain at the contact patch. This deformation and sliding motion varies over the contact patch. It can however be shown that the total lateral force generated can sufficiently be captured by a single parameter which is the average side-slip angle of a tyre [1], [2]. Commercial side-slip angle measurement sensors are available such as Kistler's Correvit S-HR [3] which uses both Doppler Effect and an absolute measuring method. However, these sensors are expensive and provide unsatisfactory results at speeds below 15km/h and have a limited measuring range of ±20deg.

The longitudinal force generated by a tyre is similarly generated by both the sliding and deformation of the tread in the longitudinal direction. The total generated force is therefore dependent on the amount of slip between tyre and terrain. The sliding motion of the contact patch is also non-uniform [4] with the tyre adhering to the terrain in certain regions and other region sliding over the terrain. It can however also be shown that the total longitudinal force can be described by the average sliding velocity between the tread and terrain. This parameter is termed the tyre slip ratio [5]. The slip ratio definition often used in many studies is given as [6]:

$$S = -\frac{(V_t - V_x)}{V_t} \qquad (1)$$

where S is the slip ratio as a fraction, V_t is the theoretical speed (at zero slip) of the center of the wheel and V_x is the actual tyre speed in the contact path. The theoretical tyre speed V_t is often merely represented as the vehicle longitudinal speed V.

The tyre speed in the contact patch is determined as:

$$V_x = \omega R_e \qquad (2)$$

where ω is the tyre angular velocity and R_e is the effective tyre radius.

With in-situ tyre tests, the tyre rotational speed, vehicle velocity and tyre rolling radius needs to be measured to determine the slip ratio. All of these measurements are made using separate measurement devices with the rolling radius often considered constant. There are problems associated with each of these measurements such as noise in wheel speed measurements, delay and accuracy of GPS based vehicle speed measurements and the difficulty in measuring the rolling radius.

In Botha and Els [7] , [8] Digital Image Correlation (DIC) is used to measure the tyre longitudinal slip and slip angle on un-deformable terrain. The developed slip angle measurement system provides better results at low speeds compared to commercial measurement devices. The longitudinal slip ratio measurement technique measures the tyre and terrain velocity at the contact region of the tyre, taking the rolling radius into consideration directly. These techniques and results will be summarised in this paper.

The vertical force generated by a tyre is dependent on the tyre vertical deformation, material properties and tyre pressure. Models used in vehicle simulations vary from basic single point contact where the tyre is modelled as a spring to more complex models using a bulk modelling approach. More complex models are high fidelity finite element models which are used in tyre research rather than vehicle dynamics simulations.

All these tyre models are often characterised by applying a vertical load. Of interest is the loading curve of the tyre which showing applied load with respect to the tyre vertical deformation. The parameters of the tyre model are altered until the measured vertical force characteristics are obtained. However, in complex models multiple tyre properties can be changed to obtain the same effect and each modification needs to be compared to other measurements such as the contact patch size, lateral stiffness and rotational stiffness. This paper proposes an alternative method whereby the deformations of multiple points on the tyre are measured while a vertical force is being applied. These points will consist of points on the sidewall and contact patch. The tyre properties can then be optimised to until the tyre exhibits the same deformation. This should result in a tyre model which has the same force displacement characteristic

VELOCITY MEASUREMENT USING DIGITAL IMAGE CORRELATION

In order to determine the velocity of an object using a camera requires the tracking of points from one frame to another. One of the fastest feature tracking algorithms is the Lucas-Kanade optical flow algorithm [9]. The algorithm first obtains unique features on the images which should be easy to track. These features are typically defined as corner features which exhibit a high gradient in two directions.

The unique features are tracked across image pairs using optical flow which makes three assumptions namely, brightness consistency, small motion and that a small region around feature exhibits the same motion. The 2D pixel coordinates can be transformed into 3D coordinates, with origin relative to the camera, using two cameras which capture the same scene from different viewpoints. The method is very similar to one of the ways a human captures depth though the use of two eyes. This method is often termed stereovision and works by determining the coordinate difference $(x_l - x_r)$ of an object in two images (see Figure 1) commonly referred to as disparity. Thus, to obtain 3D coordinates using cameras requires solving the problem of correspondence matching between the two images. The obtained disparity can then be used to determine the 3D coordinate of a pixel. In order to perform this coordinate transformation requires that the cameras are calibrated to remove lens distortion and to obtain the relative translation and rotation between the cameras. From the calibration process the epipolar geometry is also obtained. The epipolar geometry maps a feature projected onto one image as a line on the other image. This can reduce the problem of global correspondence searching to a more simplistic line correspondence search between the two images.

There are various algorithms used to determine the disparity from two images. One algorithm of note is the Semi Global Block Matching Algorithm (SGBM) developed by Hirschmuller and Box [10]

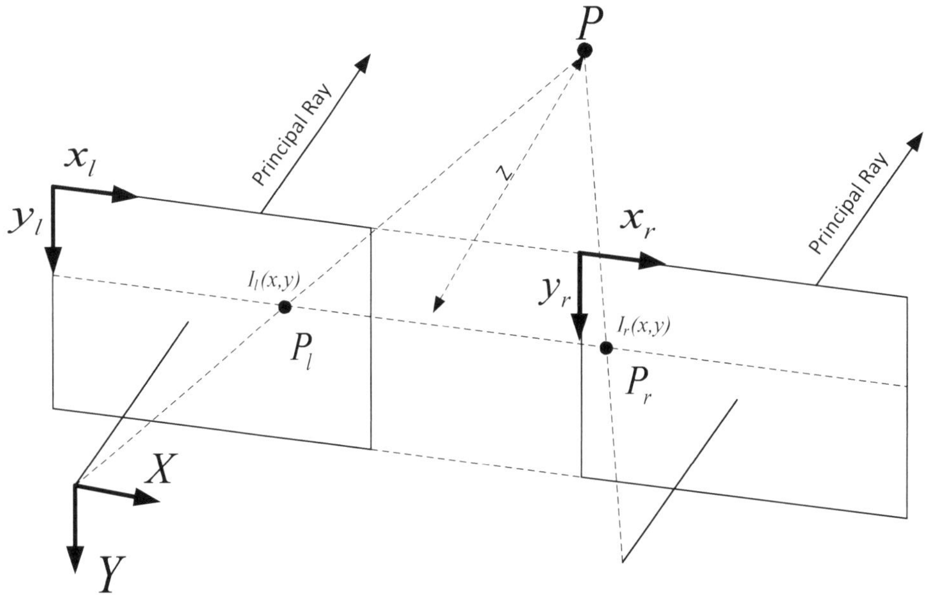

Figure 1. Stereo vision methodology

SIDE SLIP ANGLE MEASUREMENT

The side-slip angle can be measured by measuring the lateral and longitudinal velocity of the tyre. In order to accomplish this, using cameras, a calibrated stereo rig is mounted such that it has the same heading direction of the tyre. Images are recorded which overlap as the tyre moves over the terrain. Features are tracked using the algorithms mentioned earlier to determine the velocity of these features. The motion of the camera can be obtained by fitting a motion model to the features velocities in a least squares mean. The problem solved is the rotation matrix $\boldsymbol{R}$ and translation vector $\boldsymbol{t}$ can be determined [11]:

$$\boldsymbol{P_2} = \boldsymbol{R}\boldsymbol{P_1} + \boldsymbol{t} \tag{3}$$

Where the N 3D points set $\boldsymbol{P_1}$ undergoes rigid motion to yield the points set $\boldsymbol{P_2}$. The method starts of by finding the centres of $\boldsymbol{P_1}$ and $\boldsymbol{P_2}$:

$$\overline{\boldsymbol{P_1}} = \frac{1}{N}\sum_{i=1}^{N} \boldsymbol{P_1^i} \tag{4}$$

where $\boldsymbol{P_1^i}$ is the ith 3D point correspondence. Next the covariance matrix, $\boldsymbol{\Sigma}$, between points set $\boldsymbol{P_1}$ and $\boldsymbol{P_2}$ are determined as:

$$\boldsymbol{\Sigma} = \sum_{i=1}^{N} (\boldsymbol{P}_1^i - \overline{\boldsymbol{P}_1})(\boldsymbol{P}_2^i - \overline{\boldsymbol{P}_2})^T \tag{5}$$

The rotation matrix $\boldsymbol{R}$ is obtained from the Singular Value Decomposition (SVD) of the covariance matrix:

$$\boldsymbol{\Sigma} = \boldsymbol{U}\boldsymbol{S}\boldsymbol{V}^T \tag{6}$$

$$\boldsymbol{R} = \boldsymbol{V}\boldsymbol{U}^T \tag{7}$$

If the determinant of $\boldsymbol{R}$ is negative the solution corresponds to an inflection, the rotation matrix is then recalculated as:

$$\boldsymbol{R} = \boldsymbol{V}\begin{bmatrix} 1 & 0 & 0 \\ 0 & 1 & 0 \\ 0 & 0 & -1 \end{bmatrix}\boldsymbol{U}^T \tag{8}$$

The translation vector is the obtained from:

$$\boldsymbol{t} = \overline{\boldsymbol{P}_2} - \boldsymbol{R}\overline{\boldsymbol{P}_1} \tag{9}$$

A RANdom SAmple Consesus (RANSAC) [12] algorithm is used to fit the motion model in (3) to all tracked features in the presence of outliers.

Testing was conducted by mounting the camera at a fixed angle relative to the longitudinal travel direction of the vehicle. Therefore, if the vehicle moves in a straight line the system experiences the slip angle which the cameras are set at. The camera algorithm is compared to a commercial slip angle measurement system in Figure 2 for a slip angle of 10deg. This figure shows lower noise, especially at low speeds for the camera based system compared to the commercial system. The system has also been successfully tested on off-road conditions such as rough terrain, snow, ice, gravel and mud terrain conditions.

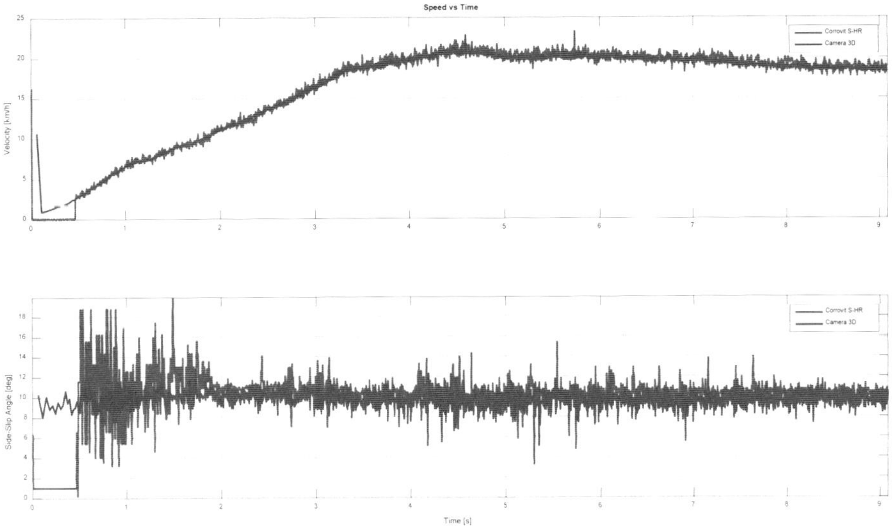

Figure 2. Vehicle speed and side-slip angle comparison

LONGITUDINAL SLIP RATIO MEASUREMENT

The longitudinal slip of the tyre is obtained by measuring the velocity of features around the contact patch region. These features are then classified into features lying on the tyre and the terrain respectively by identifying the contact line between the tyre and terrain. The contact line is obtained by identifying the region furthest away from the cameras. As the contact line is where the depth from the cameras is a maximum. A quadratic curve is fitted between all the points using a RANSAC algorithm. Features are classified as lying on the tyre if they are above the contact line but are still close enough to the contact line in order to determine the velocity of the tread and not of the sidewall. The terrain features are taken as all features on the terrain side of the contact line. Figure 3a shows the identified contact line and classified points. After classification a RANSAC algorithm is used on all the tyre and terrain features to determine the respective velocities in the presence of outliers. This process determines the tyre tread velocity as well as the terrain velocity from which the longitudinal tyre slip can be determined.

This methodology has been used to measure the longitudinal slip under heavy braking on rough hard terrains. In such a test the tyre rolling radius changes continuously due to the vertical excitation. The tyre longitudinal slip also changes as a result of the carrying vertical load. In these conditions other methods may not be able to capture the longitudinal slip accurately as the rolling resistance changes continually. Figures 3a shows the contact interface line as well as classified features (blue lines are tracked terrian features and green are tracked tyre features) and the tyre and vehicle velocities respectively. Figure 3b shows the velocity of both the vehicle and the tyre. In this test the lock-up of the tyre are evident from the sudden decrease in tyre speeds.

Figure 3a. Tyre contact interface and classified features

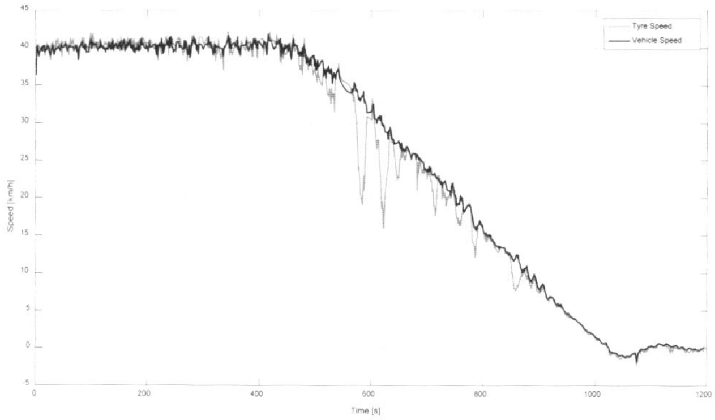

Figure 3b. Tyre and vehicle speed

TYRE DEFORMATION MEASUREMENT

In this study the tyre deformation measurement is primarily performed from the inside of the tyre. This is because the tyre contact patch of the tyre is obscured by the terrain when viewed from the outside. Figures 4a and 4b show the test setup in laboratory conditions where the tyre is compressed with a longitudinal cleat. The actuator pushes the cleat into the tyre until a certain deflection is obtained while both force and displacement of the actuator is measured. Two cameras are used and mounted such that they capture the whole contact patch longitudinally and at least half the contact patch laterally. In this test discrete points were placed on the inside of the tyre as shown in Figures 5a. These points can then be used in a finite element model. The displacement of nodal points from the model and measurements can then be used to validate or optimise the material properties of the model. A obtained displacement field where the tyre is compressed by a lateral cleat in shown in Figure 5b. The plot in Figure 5b shows the displacement of the contact patch overlayed over the geometry of the finite element model. The model can now be excited using the same loading conditions and the tyre parameters can be updated or validated by using the displacements of discrete points.

Figure 4a: Tyre contact interface and classified features **Figure 4b:** Tyre and vehicle speed

Figure 5a: Tyre contact interface and classified features

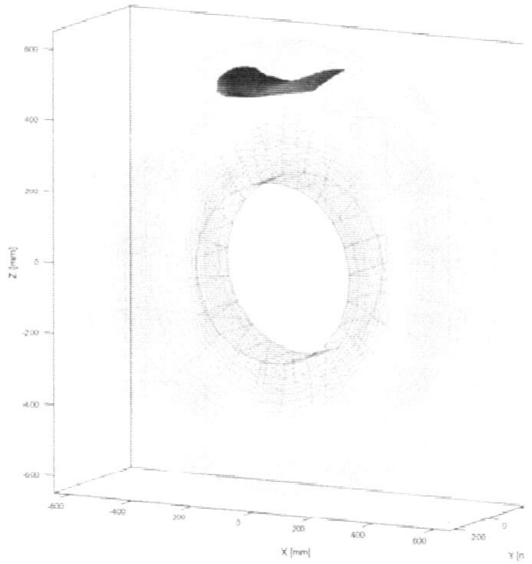

Figure 5b: Tyre and vehicle speed

The test setup shown in Figures 4a and 4b, is a static test setup. This means that the tyre is not allowed to rotate and means this method cannot be used on a rotating wheel on a vehicle. To address this problem a set of cameras were mounted on a rail on the inside of a tyre. The rail allows the cameras to rotate freely relative to the wheel. A stabilization system is used to keep the cameras fixed such that the cameras always view the contact patch. The camera stabilization concept is the same one that was developed in [13] however the implementations has been refined as shown in Figure 6. The system consists of two large rings gears with 174 teeth each that are mounted on rollers which allow the rings to rotate about the rim's main axis of rotation. One of the large gears is mounted on the inside the rim and the other is mounted on the outside of the rim. These gears each mesh with a 15 tooth pinion gear. The pinion gears are rigidly mounted to a shaft which passes through the wall of rim. The pinion shaft is mounted in a bearing housing with radial seals which stop air from escaping the inflated wheel. Having two gear sets with identical ratios connected together with a common shaft creates a planetary gear system that constrains the two rings to rotate at the same angular velocity. Since the two large gears must always rotate at the same angular velocity it is possible to keep the inner ring stationary by simply holding the outer ring stationary.

Some further improvements to the system were that the rim has been completely custom machined with no bead seats and designed to be able to split in half to allow for easy assembly and disassembly of the delicate camera system. This inner profile of the rim was also modified to increase the available area inside the tyre and allow the cameras to be placed further from the tread. The system is also integrated with a six component wheel force transducer which measures all of the forces and moments acting on the wheel. This complete system has been dubbed T2CAM.

This system allows the deformation of the inside of the tyre, close to the contact patch, to be measured. Instead of using discrete points, since the tyre is rotating, a speckle pattern is sprayed on the inside of the tyre and full field deformation is performed on the region in view. Figure 7a shows a test where the wheel was driven over a trapezoidal bump with the tyre slightly off the edge of the bump. The profiles obtained over the bump are superimposed on to the bump in Figure 7b. In this figure the deformation looks representative of what one would expect. These deformation measurements are obtained along with all force and moment components produced by the tyre from the wheel force transducer. This process can be used to validate a tyre model in both deformation and generated forces. This system has additional possibilities of measuring the deformation of the tyre relative to a reference profile from which the lateral and longitudinal deformation can be measured as well. This also means that the velocity profile of the tyre can be measured. This system opens up new possibilities in the field on tyre research as a large area of the contact can be measured using a non-contact based approach.

I all the studies represented in this paper Point Grey Flea3 USB 3 cameras were used. The specifications of the cameras are shown in Table 1.

Table 1: Point Grey Flea3 specifications

Parameter	Value
Resolution	1280 x 1024
Frame Rate	150 FPS
Megapixels	1.3 MP
Price	600USD

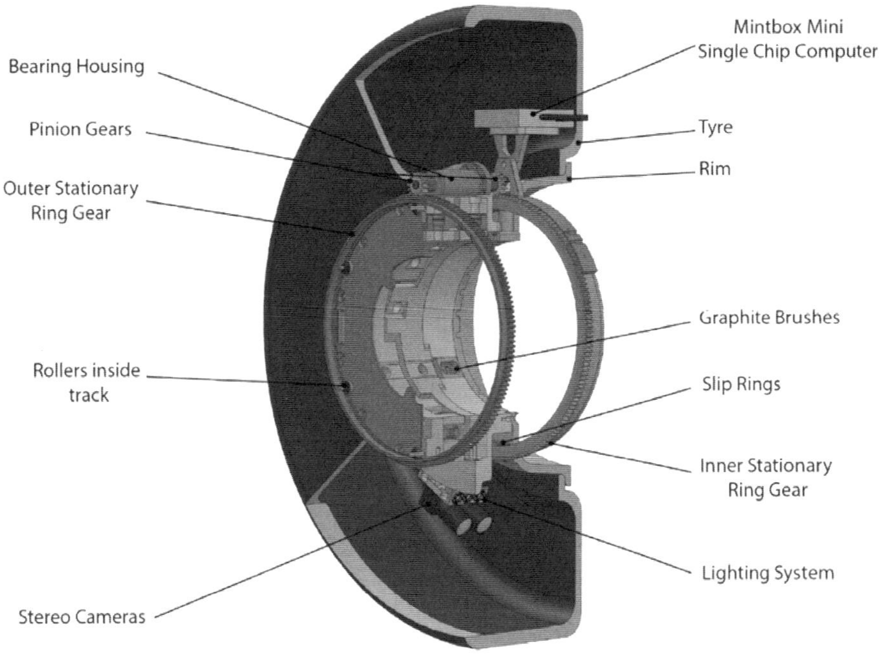

Figure 6: Tyre contact interface and classified features

Figure 7a: Tyre contact interface and classified features

Figure 7b: Tyre and vehicle speed

CONCLUSION

This paper summarises the use of digital image correlation in tyre research. Tyre research is an on-going research field and the use of cameras to perform non-contact measurements which can be used for parameterisation and modelling of tyres opens up new means of tyre testing. In this paper the side-slip angle of a tyre can be measured directly as it moves over a terrain. The side slip angle measurement shows better accuracy than commercial sensors especially at the low speeds at which tyre testing is performed.

The longitudinal slip measurement technique allows the velocity of the vehicle and tyre to be measured at the contact patch using a single sensor. The method addresses some problems associated with current methods used in in-situ tests. The system takes the rolling radius directly into consideration which means that it is ideal technique to be used when the tyre rolling radius changes continually such as over rough terrain.

The vertical deformation of the contact patch can also be measured using camera systems which are mounted on the inside. This will allow tyre parameterisation to be conducted of physics based models which model the actual structure of the tyre. Instead of parameterisation the force-deflection curve the deformation of the tyre carcass can be optimised for and the force-deflection curve can be used as validation. A system has also been developed which allows these measurement to be performed on the vehicle in in-situ test. These techniques opens up new possibilities to tyre reseach.

REFERENCES

1. Fiala, E. 1954 Seitenkrafte am rollenden Luftreifen, VDI, Bd. 96, Nr.29 , 11

2. Bakker, E., Nyborg, L. and Pacejka, H.B. 1987. Tyre modelling for use in vehicle dynamics studies. Conference: Society of Automotive Engineers international congress and expo, Detroit, MI, USA, 23 Feb 1987; Other Information: 870421

3. Kistler. Correvit S-HR Sensor, 2014, [Viewed 09,09,14 from http://www.kistler.com/us/en/product/velocity/CSHRA22111]

4. Deur , J., Ivanović, V., Troulis, M., Miano, C., Hrovat, D., and Asgari, J., 2005 Extensions of the LuGre tyre friction model related to variable slip speed along the contact patch length, Vehicle System Dynamics: International Journal of Vehicle Mechanics and Mobility, 43:sup1, 508-524, DOI: 10.1080/00423110500229808

5. Sakai, H., 1981, Theoretical and Experimental Studies on the Dynamic Properties of Tyres Part 1: Review of theories of rubber friction. International Journal of Vehicle Design 1981(2) pp. 78-110.

6. Wong, J.Y., 2001, Theory of Ground Vehicles, Third Edition, John Wiley & Sons

7. Botha, T.R. and Els, P.S., 2015a, Digital image correlation techniques for measuring tyre-road interface parameters: Part 1–Side-slip angle measurement on rough terrain. *Journal of Terramechanics*, *61*, pp.87-100.

8. Botha, T.R. and Els, P.S., 2015b, Digital image correlation techniques for measuring tyre–road interface parameters: Part 2–Longitudinal tyre slip ratio measurement. *Journal of Terramechanics*, *61*, pp.101-112.

9. Lucas, B.D. and Kanade, T., 1981, An iterative image registration technique with an application to stereo vision. *Proceedings of the 1981 DARPA Imaging Understanding Workshop* 1981: pp. 121–130.

10. Hirschmüller, H., 2005, Accurate and efficient stereo processing by semi-global matching and mutual information. In *Computer Vision and Pattern Recognition, 2005. CVPR 2005. IEEE Computer Society Conference on* (Vol. 2, pp. 807-814). IEEE.

11. Arun, K. S., Huang, T. S. and Blostein S. D., 1987, Least-Squares Fitting Of Two 3-D Point Sets. *IEEE Transactions On Pattern Analysis And Machine Intelligence*, Vol. Pami-9, No. 5, September 1987.

12. Fischler, M.A. and Bolles, R.C., 1981, Random sample consensus: A paradigm for model fitting with applications to image analysis and automated cartography. *Communications of the ACM*, 1981;24:6, pp. 381–395.

13. Guthrie, A.G., Botha, T.R., Els, P.S., 2015. 13th European Conference of the International Society for terrain Vehicle Systems. Rome, pp. 157–165.

10th South African Conference on Computational and Applied Mechanics
Potchefstroom
3 – 5 October 2016

Fatigue Crack Life Estimation of a Notched Blade-Like Component using Finite Element Modelling

Jacob CP Brits *[1], Stephan P Heyns[1], and Helen M Inglis[1]

[1] Centre for Asset Integrity Management, Department of Mechanical Engineering, University of Pretoria, Lynnwood Road, Pretoria, South Africa

BritsJacques@tuks.co.za

ABSTRACT

Damage and failure in turbomachinery blades are primarily caused by fatigue. Fatigue crack life (FCL) estimation makes proper budgeting and planning for maintenance possible. Fatigue crack life is defined as the number of load cycles required to propagate a crack once initiation has occurred. In this paper, the FCL of a blade-like component is estimated from a physics-based model in the commercial FEA software MSC. MARC/Mentat and is then compared to experimental measurements.

A finite element estimate of the remaining life was obtained by computing stress intensity factors at increasing crack lengths and substituting these into the Paris law. In the experimental setup, a steel cantilever beam with a machined initial notch was loaded cyclically. Crack growth was monitored periodically using a travelling microscope. The experimental results were in good agreement with the FEM result.

KEYWORDS

Crack propagation, Finite element modelling, Fatigue crack life

NOMENCLATURE

a	Crack length
C	Crack growth model constant
E	Elasticity modulus
K	Stress intensity factor
m	Crack growth model exponent
N	Number of load cycles
Y	Geometry and loading factor
ρ	Density
σ	Stress

INTRODUCTION

A primary cause of damage and failure in turbomachinery blades is fatigue crack propagation [1]. Failure of a blade can cause enormous damage, which may lead to unplanned downtime, repair, maintenance, and inspection costs. Thus, the prediction of the remaining useful life (RUL) is important. A machine or system's remaining useful life is defined from the current state until the life has reached a point where it is no longer useful [2]. In this paper, fatigue crack life (FCL) will be investigated and is defined as the number of load cycles required to propagate a crack once initiation has occurred on a component or machine member [3], [4]. Better maintenance planning and budgeting is made possible with FCL estimation, as it can be used to estimate the overall RUL of a machine or system [5].

The FCL of turbomachinery blades is largely influenced by crossing over critical speeds during start-ups and shut-downs [6]. These transient conditions result in a combination of large centrifugal and bending stresses on the blades [7]. Vibration induced stress on the blades increases as they pass through their natural frequency, resulting in resonance [7], [8]. A strong correlation exists between blade failure and the level of stress acting on the blades [9].

In numerous investigations of turbine and compressor blade failures, it has been found that a common failure location is on or near the root of the blade [9]–[14]. The contact surfaces between blades and their attachment to the rotor shaft was the origin of the cracks [9], [15], whereas the cracks near the root originated on the leading edge due to high bending stresses present at the blades' first order resonance frequency [11], [12].

Crack propagation is considered as any increase in crack size or length once initiated and can be sorted into three stages: short cracks, long cracks, and final fracture [16]. In this paper, the second stage (or long cracks) is considered. These cracks propagate in a direction normal to the applied load [17] where localized stresses and material structure primarily controls the progression rate of these cracks [4].

The stress intensity factor K is a measure of the intensity of the stress field at the crack tip [18] and is dependent on the method at which the load is applied to the cracked specimen and the configuration of the crack,

$$K = Y\sigma\sqrt{\pi a}, \tag{1}$$

where σ is the far-field stress, a is the crack length, and Y is a geometric factor which accounts for different geometry and loading configurations. If the stress intensity factor at a crack tip exceeds the fracture toughness of the material crack growth will occur.

FCL growth models describe the relationship between the crack growth rate, material constants, stress intensity factors, and the size of a crack [19], [20]. These models are developed using experimental crack growth data, based on fracture mechanics [21]. Paris's law is the most common among the growth models,

$$\frac{da}{dN} = C(\Delta K)^m \tag{2}$$

where $\frac{da}{dN}$ is the crack growth rate, ΔK is the cyclic stress intensity range, and C and m are material constants. When equation 2 is integrated, the number of load cycles until failure can be estimated,

$$N_f = \int_{a_i}^{a_f} \frac{da}{C(\Delta K)^m}. \tag{3}$$

This estimation corresponds to a specific failure crack size [17], [19], [20].

A specific failure mode of a system or component can be described with a physics-based model [22]. A physics-based model is simulated under specific operating conditions and solved by either deterministic equations or equations derived from experimental data to estimate FCL [23]. To make this model useful, complete knowledge of the physical process is required [22]. An intuitive prediction of FCL is then the result of the cause and effect relationship described within this model [2] .

Physics-based models or Finite Element Models (FEM) have been shown to be a useful tool in the estimation of FCL in previous work. The FCL of compressor blades, with one and two cracks, were compared by Poursaeidi and Bakhtiari [24] using Franc3D, fracture mechanics software developed at Cornell University. The estimated FCL results from both the Paris and Forman-Newman-de Koning (FNK) growth models indicated that the life of the blade with two cracks is shorter opposed to the blade with only one crack, as the two cracks grow individually until they are large enough to merge.

The stress in a compressor blade, with a specific tip displacement, was calculated using FEM by Witek [25]. This stress was used along with analytically calculated stress intensity factors to estimate the FCL using the Paris growth model and showed promising results. Citarella and Lepore [26]simulated a turbine blade in a Dual Boundary Element Model (DBEM) to obtained stress intensity factors at specific crack sizes. This was then used in a growth model for FCL estimation.

In this study the fatigue crack life of a steel cantilever beam with an initial machined notch vibrated at constant amplitude is measured experimentally. MSC.MARC/Mentat, commercial Finite Element Analysis (FEA) software, is used to estimate the FCL. The numerical results are compared with experimental measurements.

EXPERIMENTAL SETUP

The test specimen for this experiment is a steel cantilever beam with geometric and material properties shown in Table 1. The beam is fixed on one end with a Zonic servo-hydraulic shaker attached near the middle of the beam. A 2.5 mm deep v-notch was machined into the test specimen, 50 mm from the base, as an initial crack. A constant sinusoidal displacement, at 8 Hz, was applied using the shaker to propagate the crack. Displacement was used as control, since the stiffness of the beam reduces as the crack propagates, decreasing the load measured by the load cell. The crack size was measured after a specific number of load cycles, the interval reducing as the crack increases in size. By performing a frequency sweep modal analysis on the test specimens, the frequency response and mode shapes were obtained. These results can then be used to validate the numerical model's parameters.

Table 1. Test specimen properties

Property	**Value**
Geometric	
Length	1000 mm
Width	50 mm
Thickness	12 mm
Material	
Elasticity modulus, E	207 GPa [4]
Density, ρ	7860 kg/m^3 [4]

Figure 1 shows a schematic of the experimental setup that was used to propagate the crack in the test specimens. The load is applied 425 mm from the fixed end, the location chosen to be away from any nodes, peaks, or valleys of the first six cantilever bending mode shapes. The Polytec PSV400 scanning laser vibrometer was used to measure the velocity at specific points on the test specimen as well as the mode shapes and frequency response functions during the modal analysis. The Polytec system was also used to control the shaker and to acquire data. A travelling microscope was used to measure the crack size visually. Figure 2 shows photos of the experimental setup used to propagate the crack in the test specimen. A sturdy base was used to mount all the components of the setup. Three different stress cases, 196.2 MPa, 226.3 MPa, and 232.9 MPa, were considered in the experiments to determine the Paris' growth model constants.

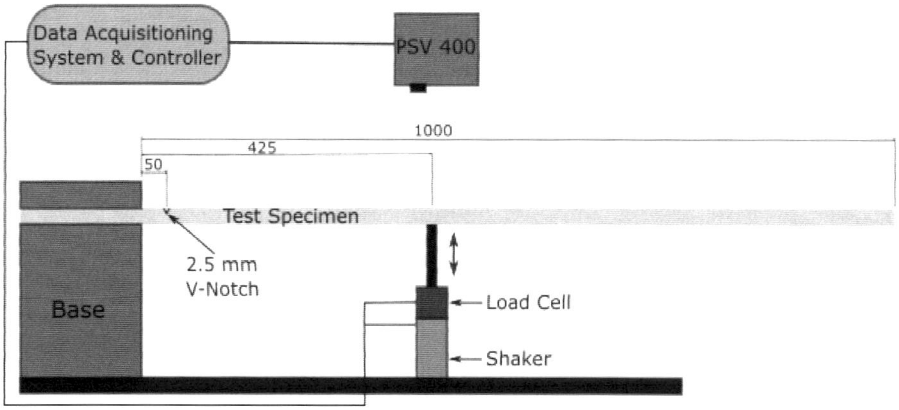

Figure 1. Schematic of experimental design

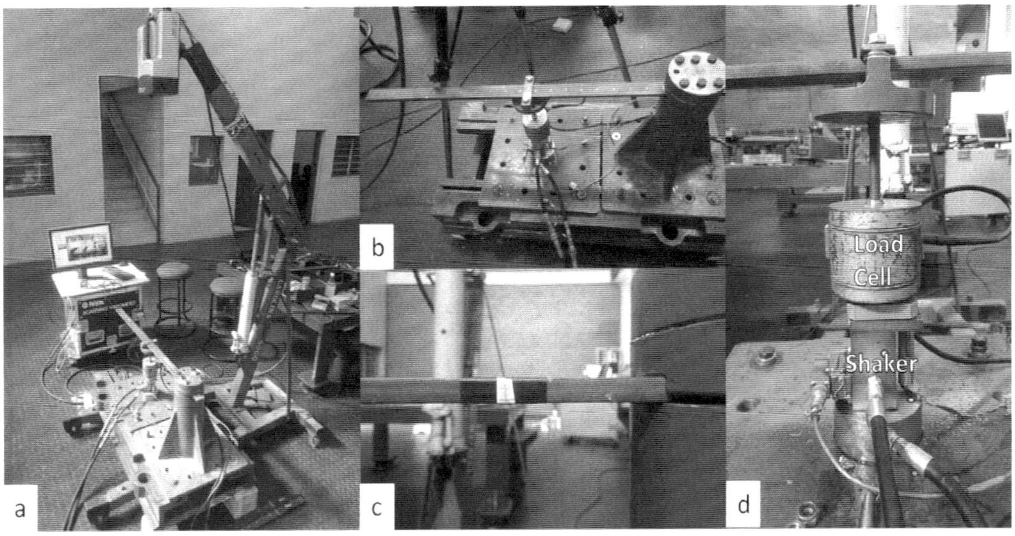

Figure 2. Experimental setup: (a) overview, (b) fixed test specimen with shaker attached, (c) machined notch and (d) load cell and Zonic shaker attached to the test specimen

NUMERICAL APPROACH

This section covers the finite element analysis procedure. To validate the simulation parameters used in the numerical model, a modal analysis was performed and results were compared with experimental results, reported in the results section. This allowed us to check that the simulation parameters were as close as possible to the experimental setup. Having done this initial check, crack growth analyses were performed. The software used to do all the simulations was MSC.Marc/Mentat 2014.2 along with Matlab R2013b.

Figure 3 shows the 2D beam with the boundary conditions used in the simulations. The same geometric and material properties were used as shown in Table 1. The bottom left corner is fully constrained for all directions and rotations where the top left corner is only constrained in the global X-direction to allow slight deformation caused by the Poisson's effect. 427.5 mm from the fixed side, 6 mm from the bottom, the beam is further constrained, which acts as the connection from the shaker. A constraint in the global Y-direction is used for the modal analysis, where a fixed sinusoidal displacement in the Y-direction is used in the crack growth analysis. The model consisted out of 40 025 nodes and 38 400 Q4 assumed strain elements. Assumed strain elements were used, as they perform well under bending without adding large computational expenses.

The crack was modelled using MSC.Marc/Mentat's 2D VCCT remeshing method along with the advancing front quad global mesh adaptivity. This method uses the maximum hoop stress theorem to determine the direction and size of the crack growth and remeshes automatically as the crack propagates. The element edge length remeshing parameter was set to 0.5 mm and the crack growth increment size set to 0.1 mm. All three load cases were modelled in a transient environment at 8 Hz. An initial analytical crack of 2.5 mm was inserted 50 mm from the fixed side perpendicular to the top surface. The crack was inserted by selecting element edges in the meshed model, MSC.Marc/Mentat then separates the elements adjacent to the selected edges. This separation then creates the crack tip or crack front [27]. This method of crack insertion allows the user to add cracks in components during the analysis in MSC.Marc/Mentat, therefore it is not necessary to have a crack created before the analysis as part of the geometry [27]. However, this obviously means that the shape of the initial crack is not the same as the machined notch. A fixed crack growth increment was chosen to ensure propagation at 0.125 s, thus 12 s were needed to crack through the beam. The solution parameters were chosen to converge at a displacement tolerance and force residual of 0.001 using 2400 steps of 0.005 s long.

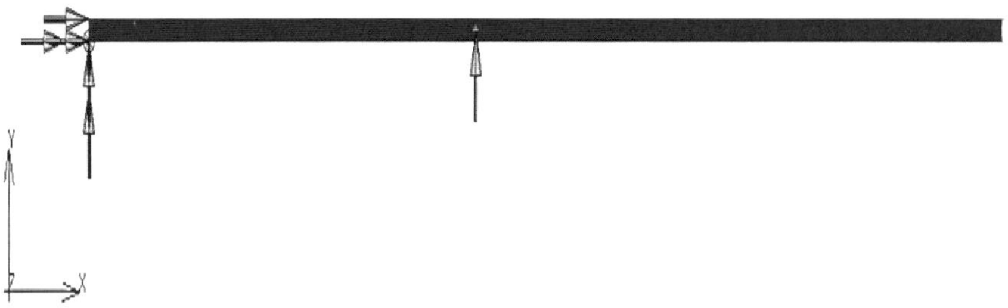

Figure 3. 2D FEM with boundary conditions

RESULTS

This section presents and discusses the experimental and numerical modal analysis and crack propagation results. The numerical crack growth results are compared to experimental measurements for validation.

Modal Analysis Results

The experimental and FEM modal analysis results are shown in Figure 4. It is evident that the obtained mode shapes are similar. Slight variations in the magnitude of the mode shapes and natural frequencies are present with a maximum error of 1.2 %, indicating that the system was modelled with the correct material properties and boundary conditions. These results also show that the analytically inserted crack is a valid representation of the notch machined in to the test specimen.

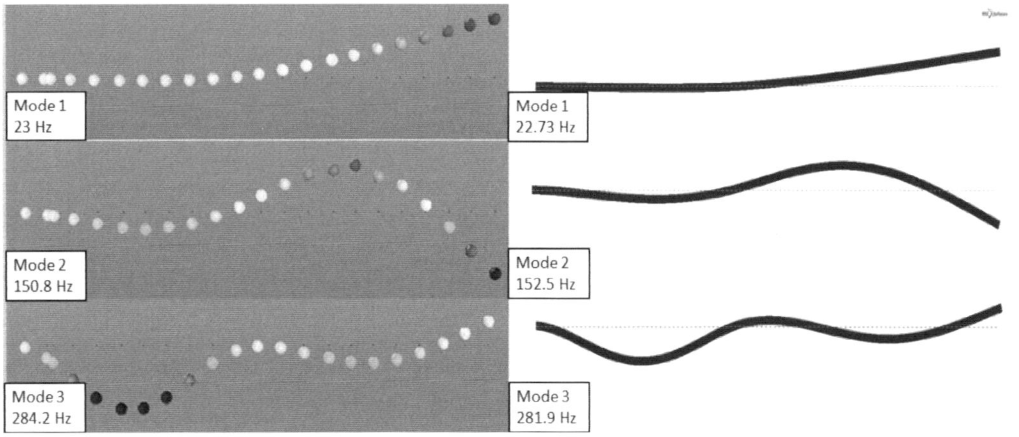

Figure 4. Experimental (left) and numerical (right) mode shapes

Paris Growth Model Parameter Estimation

The Paris growth model parameters (as divined in equation 2) are derived from the experimental lifetime results, making use of an analytical formulation for the stress intensity factors. The stress intensity factors for the experimental data were calculated using equation 1, with the geometric and loading factor for a beam under pure bending [28] given by equation 4.

$$Y = \sqrt{\frac{2}{\pi\alpha}\tan\left(\frac{\pi\alpha}{2}\right)}\ \frac{0.923 + 0.199\left(1 - \sin\left(\frac{\pi\alpha}{2}\right)\right)^4}{\cos\left(\frac{\pi\alpha}{2}\right)}\ ;\ \text{with } \alpha = \frac{a}{b}. \tag{4}$$

where a is the crack length and b the thickness of the beam, with assumption that the crack length is uniform through the width of the test specimen.

The experimental growth rate and associated stress intensity factors are then plotted against one another on a log-log scale, see Figure 5. By fitting a straight line to the scattered points

$$\log\left(\frac{da}{dN}\right) = m\log(\Delta K) + \log(C), \tag{5}$$

it is possible to determine the Paris growth model parameters. According to Budynas and Nisbett [4], m usually ranges between 2 and 4, with a value of 3 for steel, while C ranges between 5.61×10^{-12} and 1.36×10^{-10}, with a value of 6.89×10^{-12} for steel. The experimentally obtained values are therefore well in line with the expectations and shown in Table 2.

Table 2. Fitted Paris Growth Model Constants

Paris' law constant, C	6.96×10^{-12} MPa$\sqrt{m}$
Paris' law exponent, m	3.085

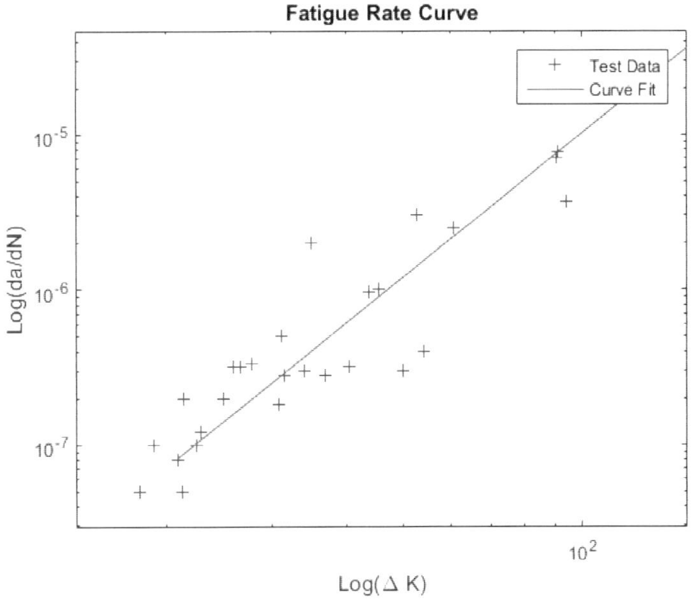

Figure 5. Fatigue rate results with experimental and Paris growth model results

Crack Propagation Results

The stress intensity factors at specific crack lengths were obtained from numerical simulations, and were used as inputs to the Paris growth model to determine the crack growth curves of the notched beam. Figure 6 compares the crack growth curves from the numerical and experimental results. The FEM results obtained compare well with the experimental results, with the largest crack growth error of 22 % at the final crack length in the 196.2 MPa load case. This error could be due to the assumption that the crack length is constant throughout the thickness of the beam, or the 2D simplification, or the modelling parameters. The results from both the FEM and experiments from the other load cases compare very well, thus the methodology followed to estimate the FCL using FEM can be validated.

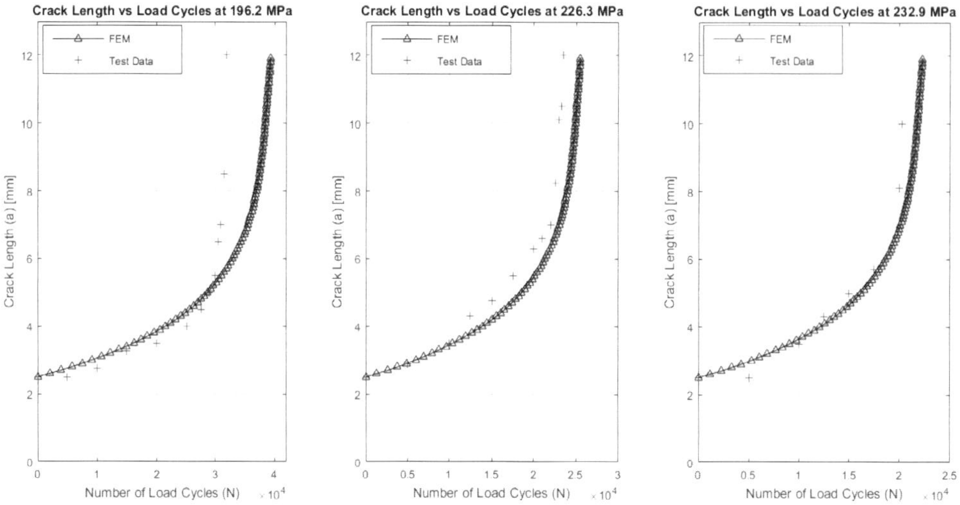

Figure 6. Crack growth vs. number of load cycles

The crack paths from the FEM and experiment are compared in Figure 7. The figure shows the numerical and experimental results are in very good agreement, without any distinct differences. Thus, it can be concluded that the numerical approach is adequate to model the crack propagation for a notched cantilever beam.

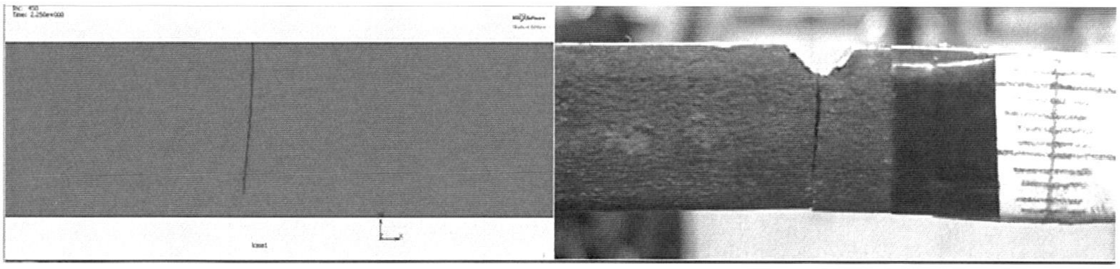

Figure 7. Comparison between FEM (left) and experimental (right) crack paths

CONCLUSION

A physical model was built and simulated to predict the FCL of the notched cantilever beam and as validation; it was compared to the experimental results. With simple geometry, such as a cantilever beam, FEM is not really necessary, as analytical solutions for the stress intensity factors can be found in literature. A numerical model would be a valuable tool to obtain stress intensity factors of complex geometries. The FCL and the crack paths from the experimental and numerical results showed to be in good agreement. Thus, it can be concluded that the FCL of a blade-like component can be estimated adequately using the proposed numerical approach.

FUTURE WORK

No material has exactly the same properties, slight variations in geometry between parts are present, and the loading conditions of said parts may vary slightly, thus, uncertainties or stochasticity are present in every system. Future work includes the effect of uncertainties on the lifetime of a more complex geometry and makes use of FEM and Monte Carlo simulations to determine FCL estimation reliability.

REFERENCES

[1] R. I. Stephens, A. Fatemi, R. R. Stephens, and H. O. Fuchs, *Metal Fatigue in Engineering, 2nd*

Edition, 2nd ed. New York: John Wiley and Sons, 2000.

[2] M. Mishra, J. Saari, D. Galar, and U. Leturiondo, *Hybrid Models for Rotation Machinery Diagnosis and Prognosis*. Lulea University of Technology, 2014.

[3] M. D. Sangid, "The Physics of Fatigue Crack Initiation," *Int. J. Fatigue*, vol. 57, pp. 58–72, 2013.

[4] R. G. Budynas and J. K. Nisbett, *Shigley's Mechanical Engineering Design*, 9th ed. New York: McGraw-Hill, 2009.

[5] H. Sanchez, S. Sankararaman, T. Escobet, S. Frost, and K. Goebel, "Analysis of two modeling approaches for fatigue estimation and remaining useful life predictions of wind turbine blades," pp. 1–11, 2013.

[6] J. S. Rao, S. Suresh, R. Ratnakar, and R. Narayan, "A procedure to predict influence of acceleration and damping of blades passing through critical speeds on fatigue life," *Adv. Vib. Eng.*, vol. 10, no. 2, pp. 89–101, 2011.

[7] X. Wu, "Life Prediction of Gas Turbine Materials," *Gas Turbines*, pp. 215 – 282, 2010.

[8] C. Booysen, P. S. Heyns, M. P. Hindley, and R. Scheepers, "Fatigue life assessment of a low pressure steam turbine blade during transient resonant conditions using a probabilistic approach," *Int. J. Fatigue*, vol. 73, pp. 17–26, 2015.

[9] M. Nurbanasari and Abdurrachim, "Crack of a First Stage Blade in a Steam Turbine," *Case Stud. Eng. Fail. Anal.*, vol. 2, no. 2, pp. 54–60, 2014.

[10] Y. S. Choi and K. H. Lee, "Investigation of blade failure in a gas turbine," *J. Mech. Sci. Technol.*, vol. 24, no. 10, pp. 1969–1974, 2010.

[11] J. S. Kubiak, G. R. Gonzalez, D. R. Juarez, J. G. Nebradt, and F. E. Sierra, "An Investigation on the Failure of an L-0 Steam Turbine Blade," *J. Fail. Anal. Prev.*, vol. 4, no. 3, pp. 47–51, 2004.

[12] C. B. Meher-Homji and G. Gabriles, "Gas Turbine Blade Failures - Causes, Avoidance and Troubleshooting," *Proceedings of the 27th Turbo Machinery Sympoisum*. pp. 129–180, 1998.

[13] L. Witek, "Experimental Crack Propagation and Failure Analysis of the First Stage Compressor Blade subjected to Vibration," *Eng. Fail. Anal.*, vol. 16, no. 7, pp. 2163–2170, 2009.

[14] E. Poursaeidi, A. Babaei, F. Behrouzshad, and M. R. Mohammadi Arhani, "Failure Analysis of an Axial Compressor First Row Rotating Blades," *Eng. Fail. Anal.*, vol. 28, pp. 25–33, 2013.

[15] Z. L. Xu, J. P. Park, and S. J. Ryu, "Failure Analysis and Retrofit Design of Low Pressure 1st Stage Blades for a Steam Turbine," *Eng. Fail. Anal.*, vol. 14, no. 4, pp. 694–701, 2007.

[16] G. E. Totten, "Fatigue Crack Propagation," *Adv. Mater. Process.*, vol. 166, no. 5, pp. 39–41, 2008.

[17] R. W. Hertzberg, *Deformation and Fracture Mechanics of Engineering Materials*, 3rd ed. John Wiley and Sons, 1996.

[18] A. Liu, "Summary of stress Intensity Factors," *ASM International*. pp. 980–1000, 1996.

[19] E. E. Keller, "Real Time Sensing of Fatigue Crack Damage for Information-Based Decision and Control," The Pennsylvania State University, 2001.

[20] M. Blažić, S. Maksimović, Z. Petrović, I. Vasović, and D. Turnić, "Determination of fatigue crack growth trajectory and residual life under mixed modes," *Stroj. Vestnik/Journal Mech. Eng.*, vol. 60, no. 4, pp. 250–254, 2014.

[21] S. Vethe, "Numerical Simulation of Fatigue Crack Growth," Norwegian University of Science and Technology, Trondheim, 2012.

[22] O. F. Eker and F. Camci, "State-Based Prognostics with State Duration," *Eur. Conf. Progn. Heal. Manag. Soc.*, no. April 2012, pp. 1–8, 2013.

[23] B. Wu, Z. Tian, and M. Chen, "Condition-based maintenance optimization using neural network-based health condition prediction," *Qual. Reliab. Eng. Int.*, vol. 29, no. 8, pp. 1151–1163, 2013.

[24] E. Poursaeidi and H. Bakhtiari, "Fatigue crack growth simulation in a first stage of compressor blade," *Eng. Fail. Anal.*, vol. 45, pp. 314–325, 2014.

[25] L. Witek, "Simulation of Crack Growth in the Compressor Blade subjected to Resonant Vibration using Hybrid Method," *Eng. Fail. Anal.*, vol. 49, pp. 57–66, 2015.

[26] R. Citarella and M. Lepore, "Fatigue Crack Growth by FEM-DBEM Approach in a Steam Turbine

Blade," *Ind. Eng. Manag.*, vol. 04, no. 02, pp. 2–6, 2015.

[27] MSC Software, "Volume A: Theory and user information," *Msc. Software, Palo Alto, USA*, 2014.

[28] P. P. Benham, R. J. Crawford, and C. G. Armstrong, *Mechanics of Engineering Materials*, 2nd ed. London: Pearson, 1996.

[29] ASM International, "Fatigue," in *Elements of Metallurgy and Engineering Alloys*, 2008, pp. 243–265.

Calibration of the Discrete Element Method and the Effect of Particle Shape

C.J. Coetzee

Department of Mechanical & Mechatronic Engineering, Stellenbosch University, Stellenbosch, South Africa

ccoetzee@sun.ac.za

ABSTRACT

The Discrete Element Method (DEM) is used across a wide range of applications. However, accurate predictions can only be made if the input parameters are carefully selected. In this work, a large shear box was used to measure the shear resistance of particles up to 40 mm in size which could then be used to determine the inter-particle friction coefficient. Confined compression tests were used to determine the particle stiffness. Two methods were used to create the clump particles: a manual process and an optimised process to create clumps comprising of 2, 4 or 8 spheres. Each clump type was individually calibrated to obtain a unique set of parameter values for each. The angle of repose is often used for the calibration of the friction coefficient and was included in the calibration process for comparison with the shear test results. The calibration process was validated by modelling hopper discharge. The results showed that care should be taken when the angle of repose is used to calibrate the particle-particle friction coefficient. It can result in a friction value which is too low for use in other applications, although the angle of repose is accurately predicted.

KEYWORDS

Discrete element method (DEM), calibration, shear test, angle of repose, hopper discharge

INTRODUCTION

The Discrete Element Method (DEM) has become the method of choice for modellers and engineers to validate and optimise the design of bulk material handling systems [1-3]. It is estimated that up to 40% of the capacity of industrial plants is wasted because of bulk handling problems [4].

Before any DEM modelling can be attempted with confidence, an accurate set of material input parameter values is needed. DEM is still lacking robust calibration procedures that are efficient from both an experimental and numerical point of view. In the literature, parameters are often not measured and the values are assumed without justification [5]. How the parameter values were obtained is often not mentioned and whether they were measured or calibrated is not clear. Together with this, the final simulation is often not validated [6].

The particle size and shape distributions are also considered to be input parameters. In DEM codes, spherical particles are usually preferred due to the efficiency of contact detection. However, when using spherical particles, the bulk friction of the assembly is usually too low when compared to real granular material like crushed rock. Two methods to increase the bulk friction exist to be implemented separately or in combination. The one method is to include contact rolling resistance and the other is to make use of non-spherical particles. Examples of non-spherical particles include ellipsoids, super-quadrics, polygons and clumps (clusters) [7]. Clumps can be formed by adding two or more spherical particles together to form one rigid particle [8].

When laboratory setups are modelled in DEM, it might be possible to accurately model the size of the particles. However, when large scale industrial applications are modelled, it would normally not be possible to accurately model the particle size since it would be computationally too demanding. In order to decrease the total number of particles, it is possible to increase the particle size. It has been shown by Obermayr et al. [9] and Ucgul et al. [10] that particles can be scaled up in size and still the draft force in soil can be accurately predicted. In this study the actual size of the particles was used to build the model particles.

For the discrete element method to be successfully used by industry, a calibration procedure is needed for materials with relatively large particle sizes. Therefore the main objective of this study was to make use of a calibration procedure to determine all the DEM parameter values needed to model crushed rock with particles up to 40 mm in size. For this purpose, a large shear box was used. Different clump types were used and each was individually calibrated. This is important since the performance of different particle shapes cannot be compared if the exact same parameter values are used for all. If the modelled particle shape is changed, the parameter values should be re-calibrated to take this change into account. To validate the calibration process, the calibrated parameter values were used to model hopper discharge. The commercial DEM software package, PFC3D (version 4.0), was used in this study [8].

THE DISCRETE ELEMENT METHOD

Discrete element methods are based on the modelling of the motion of granular material as discrete particles [11]. Using the soft particle approach in its simplest form, each contact is modelled with a linear spring in the contact normal direction (secant stiffness k_{cn}) and a linear spring in the contact tangential direction (tangent stiffness k_{cs}). Frictional slip is allowed in the tangential direction, governed by the friction coefficient μ. The particles are allowed to overlap and the amount of overlap is used in combination with the spring stiffness to calculate the contact force components. Damping can be applied at the contacts through a viscous dashpot in the normal and the tangential direction by specifying the critical damping ratio $0 \leq \xi \leq 1$. Artificial damping α can also be applied to the governing equation at local level. In this study, local damping was only used to achieve a static equilibrium state. During all subsequent modelling α was set to zero and only contact viscous damping ξ was applied. For a detailed description of the DEM code used, see [8].

THE LARGE SHEAR BOX

The direct translational shear test (Jenike shear cell) is commonly used to determine the bulk properties of granular materials [12]. The normal and shear stresses acting on the cross section of the sample are measured and used to determine bulk properties such as the internal friction angle [13].

The shear box consisted of a cylindrical container divided into a top and a bottom half. The container had an inner diameter of 590 mm and a total height of 330 mm. The top half of the container was fixed to a support frame while the bottom half rested on a set of tracks, thus enabling it to be moved horizontally on steel castor wheels, Figure 1(a).

To facilitate the normal load on the sample, a lid was placed on top of the sample, Figure 1(b). The normal load was applied by means of a linkage mechanism that was connected to the lid. A constant shear velocity of 1 mm/s could be applied over a total shear distance of 70 mm. Two commercially available 50 KN load cells (HBM S9) were used to measure the normal load on the lid and the shear force on the bottom half respectively. The horizontal displacement was recorded with an HBM WA L linear variable differential transducer (LVDT). All the sensors were connected to an HBM Spider8 data logger and the data acquisition was performed by means of HBM Catman software at a sampling frequency of 50 Hz.

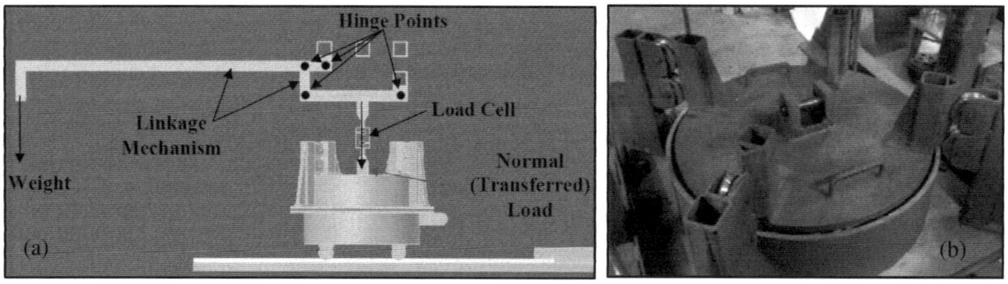

Figure 1. The large shear box showing the layout (a) and the lid support (b).

CALIBRATION OF DEM PARAMETER VALUES

Only non-cohesive frictional material was considered and hence the DEM parameters under investigation included the particles size and shape distribution, particle density, particle-particle friction coefficient, particle contact stiffness, particle-wall friction coefficient, wall contact stiffness and the contact damping. Clumps were used and rolling friction ignored. The calibration process is presented below in the same sequence that it was executed.

Contact Damping and Wall Stiffness

Contact damping is very difficult to measure [14, 15]. A local damping constant $\alpha = 0.7$ was used to obtain an initial static equilibrium state. It was assumed that the material had a fairly high damping coefficient and that all the calibration tests were quasi-static in nature. Therefore the contact viscous damping was set to a critical ratio of $\xi = 0.8$ (with $\alpha = 0$) in all the models after initial static equilibrium was achieved. The wall stiffness was set to a value roughly one order higher than the particle stiffness to ensure rigid steel walls. In this study a wall stiffness of $k_w = 10$ MN/m was used in all the models.

Particle-Wall Friction Coefficient

A basic inclined plane experiment [16] was used to estimate the particle-wall friction coefficient, μ_w, for each of the interface materials in this study. Each test was repeated 25 times and the average friction coefficient with bare mild steel was found to be $\mu_w = 0.54$ and with coated mild steel it was $\mu_w = 0.62$. These values were directly implemented in the DEM model.

Particle Size and Shape

Two methods were used to create clumps: a manual process based on visual inspection and an optimised method. In the manual process, a random sample of roughly 300 rocks was taken and the particles were classified into three distinct particle shapes, Figure 2(a): a pyramid shape (37%), an elongated shape (42%) and a largely spherical shape (21%). The clump representations of the particle shapes can be seen in Figure 2(b). A uniform distribution was used to generate the clumps with the minimum and maximum sphere radii 12.5 mm and 13.4 mm respectively for the spherical particles, 10.0 mm and 12.5 mm for the elongated particles and 6.2 mm and 8.7 mm for the pyramid shaped particles. The factor C (Figure 2b) was used to determine the centre distance between spheres in a clump. A uniform distribution between 0.5 and 1.0 was used to create the elongated particles and a fixed value of 1.0 for the pyramid shaped particles.

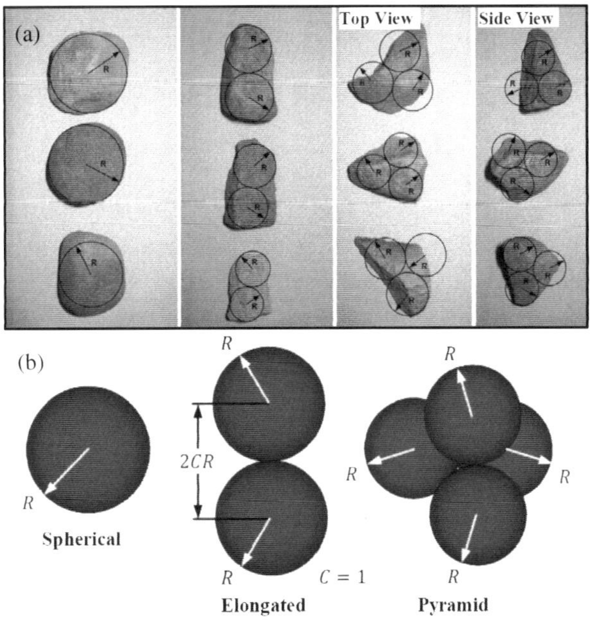

Figure 2. The rock particles (a) and the manual clump representation (b).

In the optimised process, 10 representative rock particles were selected from the sample. Each of these particles was then scanned using a NextEngine3D laser scanner [17], Figure 3. The surface mesh was then imported into ASG3D [18] to generate the equivalent clump for each of the 10 selected rock particles. ASG3D makes use of optimisation techniques together with the specified number of spheres per clump to calculate the diameter and relative position of each sphere within the clump. The algorithm ensures the best fit in terms of particle volume and surface by minimising the distance between mesh vertices and the surface of their nearest sphere.

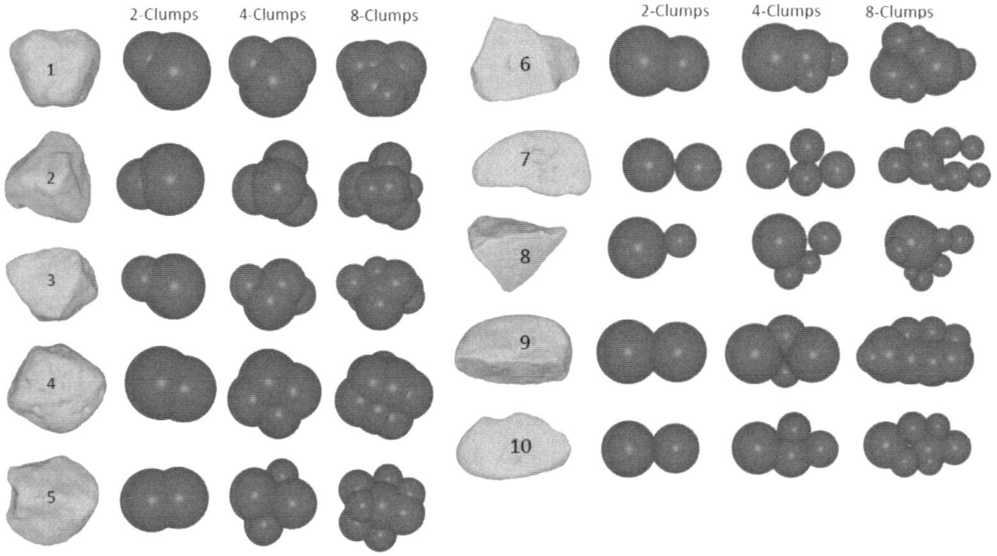

Figure 3. The optimised 2-clumps, 4-clumps and 8-clumps for each of the ten scanned particles.

The number of spheres per clump was investigated to see what effect it had on the calibration and validation processes. Three sets of clumps were created using 2, 4 and 8 spheres respectively as shown in Figure 3 and referred to as 2-clumps, 4-clumps and 8-clumps. As reference, the same optimisation procedure was used to create single spheres. The clump models were then exported in PFC3D format where each of the 10 clumps was used in equal numbers to create a system of particles with the required bulk volume.

Particle Density, Bulk Density and Porosity

The calibration was focused on the bulk density, with the particle density and the porosity simply a result thereof.

Experimental Approach

The bulk density was determined by filling a container of a known volume with a sample of the bulk material [19]. The bulk mass, m, of the material was used together with the container volume, V_b, to calculate the bulk density, $\rho_b = m / V_b$. The shear box was used as the container and the results from three samples were 1412.0 kg/m^3, 1383.6 kg/m^3 and 1397.8 kg/m^3 respectively with an average value of ρ_b = 1397.8 kg/m^3.

The particle density was determined by weighing approximately 1 kg of particles and determining the volume of that same sample by submerging it under water. It was assumed that the particles did not absorb any water during the short time the experiment was conducted. This procedure was repeated for ten samples and resulted in an average particle density ρ_p = 2586.5 kg/m^3. The porosity [19] is defined as the ratio between the void volume, V_v, and the bulk (total) volume, $n = V_v / V_b = 1 - \rho_b / \rho_p$. The measured results from the three samples were 0.45, 0.47 and 0.46 with an average value of n = 0.46.

DEM Approach

The shear box was modelled in DEM using the same geometric dimensions as in the experiment. The clumps under investigation were then generated in a volume above the shear box. The particles were given an initial density of 2500 kg/m^3 and the particle stiffness and particle-particle friction coefficient were set to the values under investigation. The particles were then allowed to rain down under the influence of gravity and allowed to reach a static equilibrium condition by applying local damping $\alpha = 0.7$. The bulk density of the assembly was then determined and the ratio of the bulk density and the target value used to determine the particle density for the next iteration. After each change in particle density, the assembly was allowed to reach a new equilibrium state. The iteration process was stopped when the bulk density was within 1% of the measured value and this could be achieved within 2 to 3 iterations.

In general, the particle stiffness ranged from 0.5 MN/m to 3.0 MN/m while the particle-particle friction coefficient ranged from 0.1 to 1.1. It was shown that for all values of the particle stiffness and friction coefficient and all the clump types, the bulk density could be accurately calibrated within 1%. The porosity varied between 0.38 and 0.48 in all cases. This compared well to the measured values which ranged between 0.45 and 0.47, especially for higher particle-particle friction values.

Particle Stiffness

Confined uniaxial stress loading is often used in geotechnical engineering to measure soil consolidation [20].

Experimental Approach

The shear box was filled with material, the bed carefully levelled and the initial height measured. The lid was then placed on top of the bed and the load on the linkage was increased in increments of 10 kg until a maximum of 40 kg was reached, resulting in a maximum normal stress on the bed of approximately 66 kPa. The load was then again decreased in increments of 10 kg. Three of these loading-unloading cycles were completed per test, Figure 4. The confined bulk stiffness was defined and obtained by calculating the average of the gradients for the second and third loading cycles. The whole test was repeated three times to ensure repeatability and the bulk stiffness ranged between 8.39 MN/m^2 and 9.10 MN/m^2 with an average value of 8.69 MN/m^2.

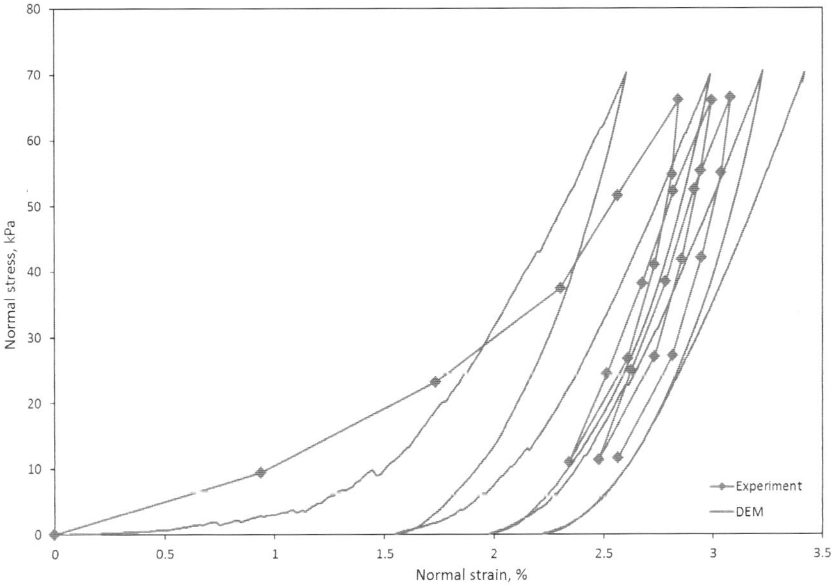

Figure 4. A typical stress-strain curve from a confined compression test showing the experimental result and the DEM result using 2-clumps.

DEM Approach

At this stage in the calibration process, only the particle-particle friction coefficient and particle stiffness had not yet been calibrated. Coetzee and Els [21, 22] showed that the confined compression test is not sensitive to the particle-particle friction and that there is a linear relation between the confined bulk stiffness and the particle stiffness. Therefore, the particle-particle friction coefficient was kept constant at $\mu_p = 0.9$ while the confined compression test was modelled using particle stiffness values between 0.5 MN/m and 3.0 MN/m. The bulk stiffness was taken as the average of the last three loading cycles and is given in Figure 5 for each clump type. It can be seen that a linear relation was found between the bulk stiffness and the particle stiffness, similar to [23, 24]. The linear relation and the average measured bulk stiffness of 8.69 MN/m^2 were used to find the calibrated particle stiffness through interpolation. This resulted in a particle stiffness of 1.08 MN/m for the manual clumps, 1.08 MN/m for the 2-clumps, 1.11 MN/m for the 4 clumps, and 1.02 MN/m for the 8-clumps. These values were used in the subsequent calibration of the particle-particle friction coefficient for each clump, Table 1. There was no definite trend in the data for the different clump types and the values were all within 9% of each other. The only exception was the stiffness of the spheres at 0.75 MN/m. This could be attributed to the number of spherical particles needed to fill the volume as compared to the number of clumps.

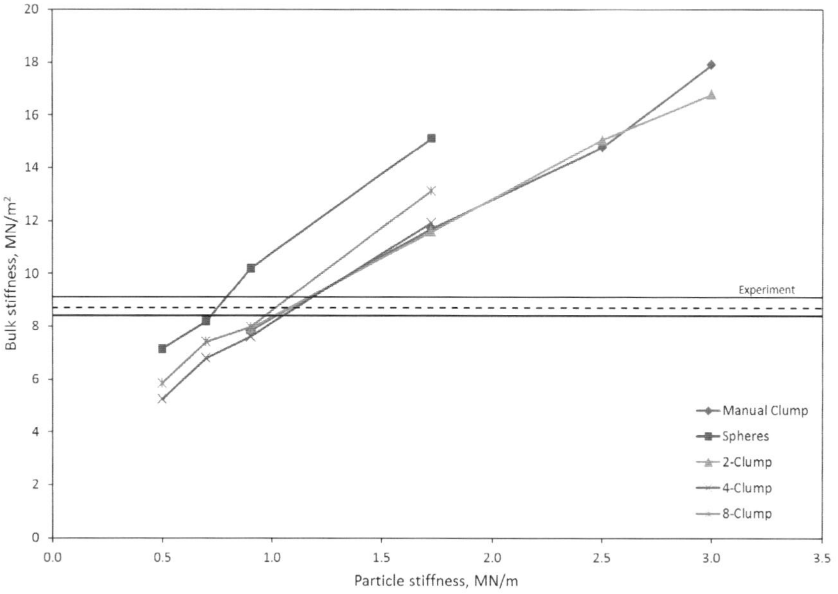

Figure 5. The bulk stiffness as a function of the particle stiffness for each clump type. The horizontal lines indicate the minimum, average and maximum range of experimental measurements.

Particle Friction Coefficient: Shear Test

The next and last step in the calibration process was to determine the particle-particle friction coefficient, μ_p.

Experimental Approach

Figure 6 shows a typical strain softening stress-strain curve from the experiment using normal loads of 10.5 kPa, 24.0 kPa, 38.3 kPa and 52.4 kPa respectively. Figure 7 shows the shear stress plotted against the normal stress. Two values of the shear stress were used, namely the maximum value and the residual value (average between 8% and 12% shear strain). A linear relation was assumed and the slope of the line defined as the internal friction angle. Three tests were conducted and taking the maximum and the minimum of all measurements, the internal friction angle varied between 50.1° and 56.3° with an average of $\phi = 52.8°$.

DEM Approach

It was shown by [21], [23] and [25] that the shear test results are influenced by both the particle-particle friction and the particle stiffness (for low stiffness values). Therefore it was important to first calibrate the particle stiffness before the shear test was modelled. Particle-particle friction coefficients of μ_p = 0.1, 0.3, 0.5, 0.7, 0.9 and 1.1 were used and new material samples were generated. In each case, the particle density was again calibrated as before and the particle stiffness as calibrated in the previous section was used. Four normal loads were applied and this resulted in stress-strain curves similar to the experimental results, Figure 6. Again, the maximum and residual shear stresses were determined and plotted against the normal stress to determine the internal friction angle, Figure 7.

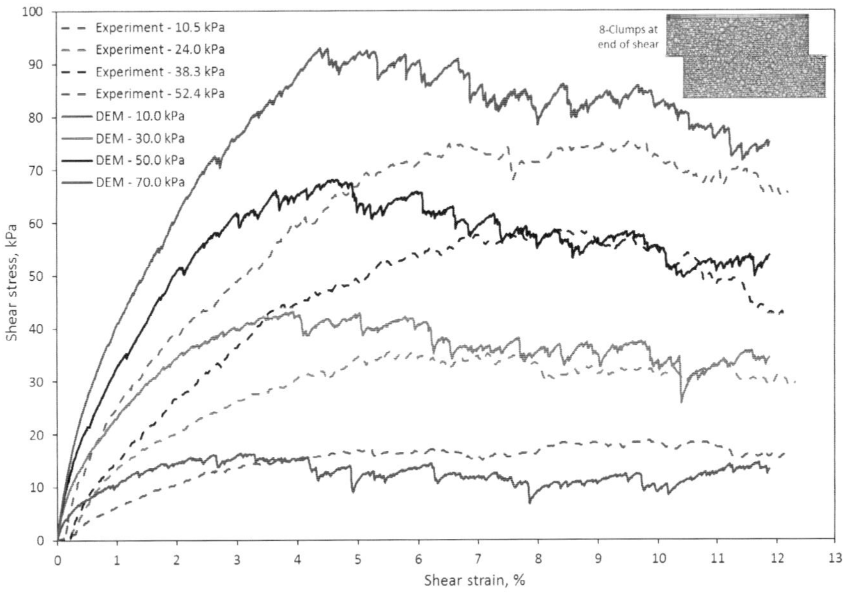

Figure 6. Typical shear stress-strain curves from the direct shear test using different normal loads. The DEM results are from the 2-clumps with a particle-particle friction coefficient of 0.7.

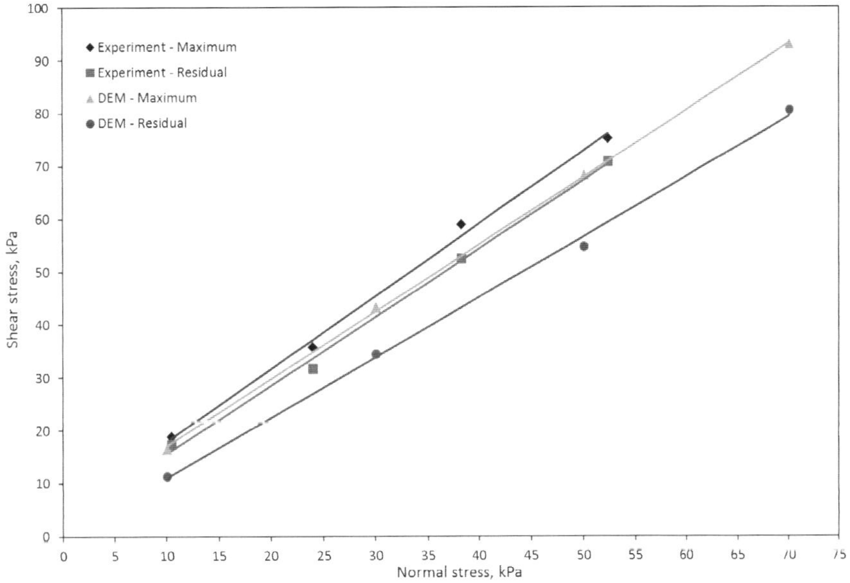

Figure 7. The maximum and residual shear stress plotted against the applied normal stress for the experiment and the DEM model. The slope of the fitted lines was defined as the internal friction angle.

The average internal friction angle for each test is plotted in Figure 8. It is clear that the relation between the particle-particle friction coefficient and the material internal friction angle is not linear, and that it approaches an asymptotic value. A similar trend was found by Simons et al. [25]. Using friction coefficients of 0.1, 0.3 and 0.5 the results from using the manual clumps, 2 clumps, 4-clumps and 8-clumps did not differ much, although the internal friction angle increased slightly in that same order. Using higher friction coefficients (0.7 to 1.1) resulted in bigger differences between the clump types and the same general trend was visible, except for a friction coefficient of 0.7 where the 8-clumps displayed a lower internal friction angle than the 4-clumps. Without rolling resistance, the spheres were not able to reproduce internal friction angles higher than 25°. When compared to the experimental results, a particle-particle friction coefficient between 0.7 and

0.8 for all clumps, except the manual clumps, would result in an acceptable internal friction angle as indicated by the shaded region in Figure 8. For the manual clumps, the particle friction coefficient should be between 1.0 and 1.1. A friction coefficient higher than 1.0 might seem unrealistic, but is often used in DEM to compensate for the lack in accurate shape modelling [27-29].

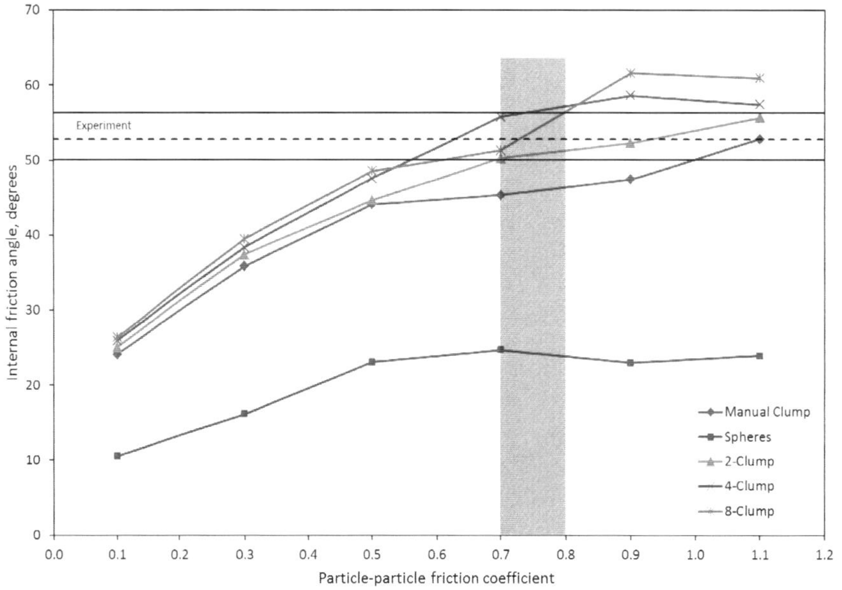

Figure 8. The material internal friction angle as a function of the particle-particle friction coefficient for the different clump types. The horizontal lines indicate the minimum, average and maximum range of experimental measurements.

Particle Friction Coefficient: Angle of Repose

We propose the shear test as a means of calibrating the particle-particle friction coefficient, however, several authors [24, 26] instead made use of the angle of repose. Therefore the angle of repose was also used experimentally and numerically to see which method was more accurate in this case.

Experimental Approach

The angle of repose was measured by filling a bottomless tapered cylindrical container with material. The container had a lower diameter of 260 mm an upper diameter of 290 mm with a height of 370 mm. The container was then slowly lifted vertically to allow the material to exit at the bottom and form a static pile. The angle of the pile was measured at eight different positions and the experiment repeated four times. The angle of repose ranged between 33.6° and 38.5° with an average value of 36.6°.

DEM Approach

For each of the parameter sets used to calibrate the particle-particle friction, the angle of repose was also modelled including a friction coefficient μ_p = 0.2. The experimental setup was followed as closely as possible and the angle of repose measured at 8 different positions around the pile. Using 2-clumps, the effect of the number of particles on the angle of repose was first investigated by using 1500 and 3000 particles respectively. The results showed no significant difference and in all further models 1500 particles were used.

The results are summarised in Figure 9. With an increase in the particle-particle friction coefficient between 0.1 and 0.3 there was an increase in the angle of repose for all clump types. With a further increase in the friction coefficient, the angles of repose remained largely unchanged and within the bounds of the experimental results, except for the 2 clumps where the angle of repose remained slightly lower. When compared to the shear box results (Figure 8), it is clear that the angle of repose reached the asymptotic value at lower particle friction coefficients. If the angle of repose test was used for calibration, any value of the friction coefficient between 0.3 and 1.1 for the manual clumps, 4-clumps and 8-clumps would have resulted in an acceptable value of the angle of repose. Using the 2-clumps, it would have resulted in a friction coefficient roughly between 0.8 and 1.0.

The difference between the angle of repose results and the direct shear results is significant. Although both show the same general asymptotic trend, it is the actual values that are important. If one were to make use of the angle of repose alone to calibrate the particle friction, there is no way to tell which value to use for the friction coefficient since any value between 0.3 and 1.0 could be deemed appropriate for the manual clumps, 4-clumps and the 8-clumps.

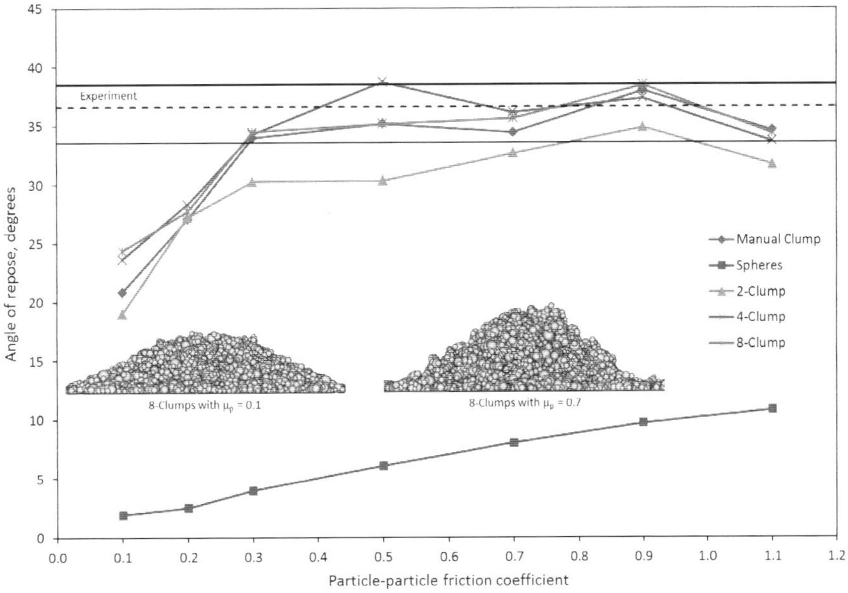

Figure 9. The angle of repose as a function of the particle-particle friction coefficient for the different clump types. The horizontal lines indicate the minimum, average and maximum range of experimental measurements. Two examples using the 8-clumps with different friction coefficients are shown.

Table 1 - The calibrated parameter set for each clump type

Parameter	Spheres	Manual Clump	2-Clumps	4-Clumps	8-Clumps
Contact damping ratio	0.8	0.8	0.8	0.8	0.8
Wall stiffness, MN/m	10	10	10	10	10
Particle-wall friction coefficient*	0.54, 0.62	0.54, 0.62	0.54, 0.62	0.54, 0.62	0.54, 0.62
Particle stiffness, MN/m	0.75	1.08	1.08	1.11	1.02
Particle-particle friction coefficient	1.1	1.1	0.9	0.7	0.7
Particle density, kg/m^3	2358.7	2664.7	2300.0	2081.0	1652.9
Bulk density, kg/m^3	1396.6	1397.0	1396.2	1396.9	1396.7
Porosity	0.40	0.48	0.47	0.46	0.46
Bulk stiffness, MN/m^2	8.69	8.69	8.69	8.69	8.69
Internal friction angle, °	23.9	52.8	52.3	55.8	51.3
Angle of dilatancy, °	-	15.2	16.0	14.3	13.5
Angle of repose, °	10.8	34.6	34.8	36.1	35.6

* Values for uncoated and coated steel respectively

HOPPER DISCHARGE

The discharge of the material from a hopper was investigated to validate the calibration process. The hopper had a rectangular outlet with two horizontal trapdoors that spanned the whole width. The hopper was filled with approximately 132 kg of material and suspended 1 m above the ground using a load cell (HBM S9, 50 kN). The weight of the material inside the hopper was measured as a function of time. The geometry of the setup used in the experiments was used to build an accurate DEM model.

A sensitivity analysis was first conducted to determine the sensitivity of the model to the particle-wall friction coefficient, the particle stiffness and the contact viscous damping. This was done with a hopper opening of 250 mm and the 2-clumps. All the parameters not investigated were kept constant and assumed the calibrated values (Table 1). The particle-wall friction coefficient was changed from the calibrated value of 0.62 to 1.0 and this had no effect on the results. The particle stiffness was then reduced by a factor of 10 to a value of 0.1 MN/m and increased by a factor of 10 to a value of 10 MN/m and this too had no significant effect on the results. The contact viscous damping ratio was varied between 0.4 and 1.0 and this too had no significant effect on the results.

Figure 10 shows the results from the 4-clumps using particle-particle friction coefficients of 5.0, 0.7 (calibrated), 0.5 and 0.1 respectively. The calibrated friction coefficient was the most accurate over the whole discharge with lower friction coefficients predicting flow rates too high and the friction coefficient of 5.0 predicting too low a flow rate. Using the calibrated friction coefficient for each clump type, the results are compared in Figure 11. There was no significant difference in the results from the 4-clumps and the 8-clumps. These two clump types together with the manual clumps compared well to the experimental result although the manual clumps predicted a slightly higher flow rate. The 2-clumps showed an even higher flow rate but for practical reasons still compared well to the experiment. The spheres showed a very high flow rate as expected. All the clump types were able to accurately predict the mass of material (7.5 kg) that remained on top of the trapdoors at the end of the discharge. Note that the initial oscillation observed in the experimental results was due to the elasticity of the hopper suspension system.

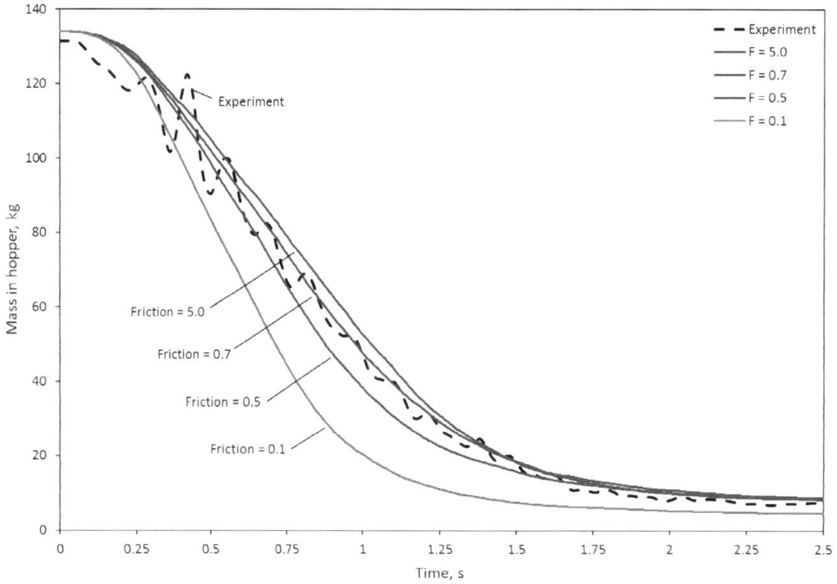

Figure 10. The mass in the hopper as a function of time using the 4-clumps and different particle-particle friction coefficients. The hopper opening was 250 mm.

CONCLUSION

A calibration procedure was proposed to determine the parameter values needed for accurate DEM modelling of non-cohesive frictional granular material with particles up to 40 mm in size. Two methods were used to model the particle shape. In the first, the clumps were created manually and in the second an automated optimisation process was used to create different clump types, comprising of 2, 4 and 8 spheres. The different clump types were all put through the calibration process to determine a unique set of parameter values for each. The calibration process was then validated by modelling hopper discharge and comparing the results from each clump type to the experimental results.

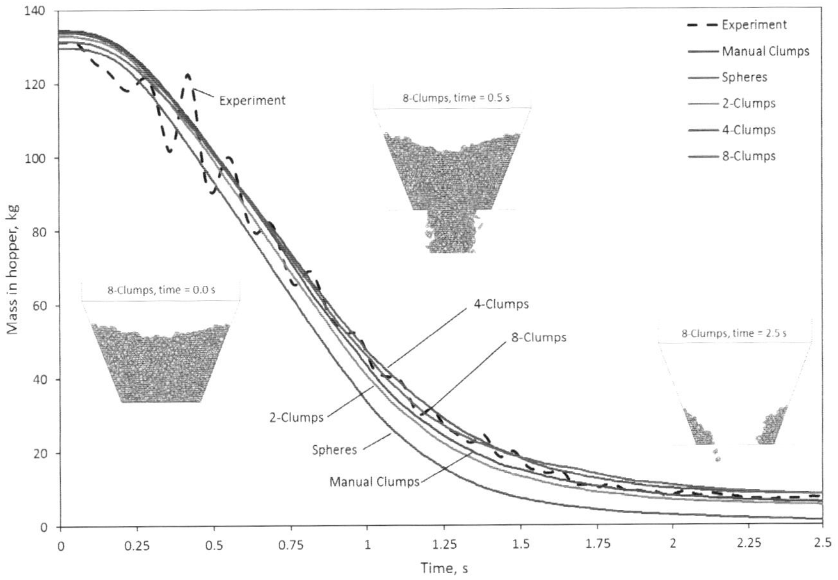

Figure 11. The mass in the hopper as a function of time using the manual clumps, spheres, 2-clumps, 4-clumps and 8-clumps. The insets show snapshots of the hopper discharge using the 8-clumps.

It is concluded that the direct shear test was better than the angle of repose for calibrating the particle-particle friction coefficient in the validation case studied here. The angle of repose tended to reach an asymptotic value at lower friction coefficients than the direct shear test. This can result in the analyst using a particle-particle friction coefficient that is too low when a different application is modelled, although the angle of repose was predicted accurately. Comparing the different clump types, they all performed well in the validation tests with the manual clumps, 4-clumps and 8-clumps slightly more accurate than the 2-clumps. It is concluded that most non-spherical particles, with the parameter values calibrated, could accurately model hopper discharge. The clumps created manually performed as well as the optimised clumps.

ACKNOWLEDMENTS

The work presented in this paper has been partially funded by the National Research Foundation (NRF), grant number 90317 (2014-2016).

REFERENCES

1 Nordell, L.K., "Particle Flow Modelling: Transfer chutes and other applications", International Materials Handling Conference (BELTCON 9), Johannesburg, South Africa, 1997.

2 Hastie, D.P., Grima, A.P., Wypych, P.W., "Validation of particle flow through a conveyor transfer hood via particle velocity analysis", 2nd International Conference and Exhibition on Storage, Handling and Transporting Bulk (Bulk Europe 2008), Prague, Czech Republic, 2008.

3 Gröger, T., Katterfeld, A., "On the numerical calibration of discrete element models for the simulation of bulk solids", Conveying and Handling of Particulate Solids (CHoPS-05), Sorrento, Italy, 2006.

4 Tijskens, B., Van Besin, Vandewalle, Ramon, "Large scale DEM computation - expectations and recent results", Conveying and Handling of Particulate Solids (CHoPS-05), Sorrento, Italy, 2006.

5 Asaf, Z., Rubinstein, D., Shmulevich, I., "Determination of discrete element model parameters required for soil tillage", Soil Tillage Res., Vol. 92, pp. 227–242, 2007.

6 Marigo, M., Stitt, E.H., "Discrete Element Method (DEM) for Industrial Applications: Comments on Calibration and Validation for the Modelling of Cylindrical Pellets", KONA Powder Part. J., Vol. 32, pp. 236–252, 2015.
7 Lu, G., Third, J.R., Müller, C.R., "Discrete element models for non-spherical particle systems: From theoretical developments to applications", Chem. Eng. Sci., Vol. 127, pp. 425–465, 2015.
8 Itasca, PFC3D, Particle Flow Code in 3 Dimensions, User's Guide, 4th edition, (www.itascacg.com), 2008.
9 Obermayr, M., Vrettos, C., Eberhard, P., Dauwel, T., "A discrete element model and its experimental validation for the prediction of draft forces in cohesive soil", Journal of Terramechanics, Vol. 53, pp. 93–104, 2014.
10 Ucgul, M., Fielke, J.M., Saunders, C., "Three-dimensional discrete element modelling of tillage: Determination of a suitable contact model and parameters for a cohesionless soil", Biosystems Engineering, Vol. 121, pp. 105-117, 2014.
11 Cundall, P.A., Strack, O.D.L., "A discrete numerical method for granular assemblies", Geotechnique, Vol. 29, pp. 47–65, 1979.
12 Carson, J.W., Wilms, H., "Development of an international standard for shear testing", Powder Technology, Vol. 167, pp. 1-9, 2006.
13 Härtl, J., Ooi, J.Y., "Experiments and simulations of direct shear tests: porosity, contact friction and bulk friction", Granular Material, Vol. 10, pp. 263-271, 2008.
14 González-Montellano, C., Fuentes, J.M., Ayuga-Téllez, E., Ayuga, F., "Determination of the mechanical properties of maize grains and olives required for use in DEM simulations", J. Food Eng., Vol. 111, pp. 553–562, 2012.
15 Wang, L., Zhou, W., Ding, Z., Li, X., Zhang, C., "Experimental determination of parameter effects on the coefficient of restitution of differently shaped maize in three-dimensions", Powder Technology, Vol. 284, pp. 187–194, 2015.
16 Barrios, G.K.P., de Carvalho, R.M., Kwade, A., Tavares, L.M., "Contact parameter estimation for DEM simulation of iron ore pellet handling", Powder Technology, Vol. 248, pp. 84–93, 2013.
17 NextEngine, 3D Scanner HD, Available at: www.nextengine.com.
18 Cogency, ASG Software. Available at: www.cogency.co.za.
19 Head, K.H., "Soil Technicians' Handbook", London, Pentech Press, 1989.
20 Cabalar, A.F., "Applications of the oedometer, triaxial and resonant column tests to the study of micaceous sands", Engineering Geology, Vol. 112, pp. 21–28, 2010.
21 Coetzee, C.J., Els, D.N.J., "Calibration of discrete element parameters and the modelling of silo discharge and bucket filling", Computers and Electronics in Agriculture, Vol. 65, pp. 198-212, 2009.
22 Coetzee, C.J., Els, D.N.J., "Calibration of granular material parameters for DEM modelling and numerical verification by blade–granular material interaction", Journal of Terramechanics, Vol. 46, pp. 15–26, 2009.
23 Lommen, S., Schott, D., Lodewijks, G., "DEM speedup: Stiffness effects on behavior of bulk material", Particuology, Vol. 12, pp. 107–112, 2014.
24 Paulick, M., Morgeneyer, M., Kwade, A., "Review on the influence of elastic particle properties on DEM simulation results", Powder Technology, Vol. 283, pp. 66–76, 2015.
25 Simons, T.A.H., Weiler, R., Strege, S., Bensmann, S., Schilling, M., Kwade, A., "A Ring Shear Tester as Calibration Experiment for DEM Simulations in Agitated Mixers – A Sensitivity Study", Procedia Eng., Vol. 102, pp. 741–748, 2015.
26 Frankowski, P., Morgeneyer, M., "Calibration and validation of DEM rolling and sliding friction coefficients in angle of repose and shear measurements", AIP Conf. Proc., pp. 851–854, 2013.
27 Li, S., "Parameter estimation approach for particle flow model of rockfill materials using response surface method", In ICCM2014, Cambridge, England, pp. 1–12. 2014.
28 Härtl, J., Ooi, J.Y., "Experiments and simulations of direct shear tests: Porosity, contact friction and bulk friction", Granular Matter, Vol. 10(4), pp. 263–271, 2008
29 Markauskas, D., Kačianauskas, R., "Investigation of rice grain flow by multi-sphere particle model with rolling resistance", Granular Matter, Vol. 13(2), pp.143–148, 2011

An alternative approach to system identification

J.M Crous*, S. Kok, P.S. Heyns
Department of Mechanical and Aeronautical Engineering
University of Pretoria
Jacq.Crous@gmail.com

August 22, 2016

Abstract

The classical system identification approach to solving inverse problems is compared to the following statistical learning techniques: Ridge regression, Regularised Kernel regression methods, Principal Component regression and Thresholded Partial Least Squares regression. It is shown that these big data techniques perform better than the classical techniques on larger data sets in terms of accuracy, efficiency and in capturing the characteristics of the inverse solution.

1 Introduction

System identification is a subfield of control theory where a mathematical model representing a system is constructed given input-output data [1]. This field has developed out of engineering fields where non-autonomous systems, discrete-time models and disturbance modelling are of interest to capture the behaviour of a dynamical system [2]. The motivation behind system identification is two-fold: firstly to understand the physics of the system by constructing models that have physically interpretable variables and secondly to accurately model the dynamic behaviour of the system. System identification can be pursued with both these motivations in mind or with only one of them as the primary objective [3]. In this work the second motivation is adopted as alternative methods are investigated and compared to the classical approach.

The importance of accurate and efficient system identification algorithms have not diminished although virtual experiments have become increasingly common. Rather system identification has become a key interface where the numerical predictions and the experimental investigations are to be reconciled [4].

Introducing machine learning techniques to system identification problems was recently surveyed by Pillonetto et al.[1] Their work primarily focused on applying regularised kernel-based techniques to stable linear systems and showed that they can outperform conventional techniques if applied correctly. Kernel methods is a novel approach to the inverse problem presented by the system identification problem. However the approach is still similar to the conventional techniques by viewing the input and output data as two time series where the output is related to the immediate and some (or all) previous input data points.

In this work we take a further step and view this data as independent variables in high dimensional space, where each data point is considered as a new variable.

2 QUARTER CAR MODEL

2.1 THE QUARTER CAR MODEL

The quarter car model is a simplified suspension model used to model the vertical displacement, velocity and acceleration that a car would experience given an uneven road profile. In this study we consider a linear quarter car model in order to illustrate the ideas presented before.

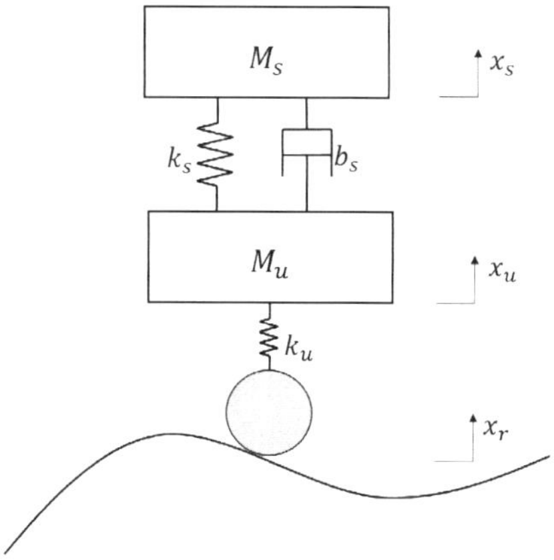

Figure 1: Schematic of quarter car model.

A schematic of the model is shown in Figure 1. The tire is modelled by a spring, k_u. The effective mass of the wheel axis, also called the unsprung mass, is denoted by M_u. Likewise the effective vehicle body mass, also called the sprung mass, is denoted by M_s. The vertical displacements from static equilibrium for M_u and M_s are represented by x_u and x_s respectively. The road profile is represented by x_r. The suspension travel $x_s - x_u$ is measured and compared to the set point ($r = 0$). The model parameters were taken from ISOMatlab, these parameters correspond to that of a car, and is presented in Table 1.

The system is described by the set of second order differential equations [6]:

$$M_s\ddot{x}_s + b_s(\dot{x}_s - \dot{x}_u) + k_s(x_s - x_u) = 0 \tag{1a}$$

$$M_u\ddot{x}_u - b_s(\dot{x}_s - \dot{x}_u) + b_u(\dot{x}_u - \dot{x}_r) - k_s(x_s - x_u) + k_u(x_u - x_r) = 0 \tag{1b}$$

Table 1: Model parameters.

Parameter	Unit	Value
M_s	kg	400
M_u	kg	40
k_s	$\frac{N}{m}$	21000
k_t	$\frac{N}{m}$	150000
b_s	$\frac{Ns}{m}$	1500

This set of equations can be written as a system of first order differential equations. To this end we define the following variables:

$$v_1 = x_s,\ v_2 = \dot{x}_s,\ v_3 = x_u,\ v_4 = \dot{x}_u \tag{2}$$

Transforming the variables as described above allows one to write the set of second order differential equations as a system of first order differential equations, which is commonly referred to as the state space model. For a linear system the state space model has the form:

$$\dot{x} = Ax + Bu, \tag{3a}$$
$$\vec{y} = Cx + Du. \tag{3b}$$

Where x denotes the state of the system, u the control/external input to the system and y the outputs of the system. Equation 3a is usually called the state equation, where as 3b is usually called the output equation. The quarter car model is therefore cast in the form of the state space model:

$$\begin{bmatrix} \dot{v}_1 \\ \dot{v}_2 \\ \dot{v}_3 \\ \dot{v}_4 \end{bmatrix} = \begin{bmatrix} 0 & 1 & 0 & 0 \\ -\frac{k_s}{m_s} & -\frac{b_s}{m_s} & \frac{k_s}{m_s} & \frac{b_s}{m_s} \\ 0 & 0 & 0 & 1 \\ \frac{k_s}{m_u} & \frac{b_s}{m_u} & \frac{k_s+k_u}{m_u} & -\frac{b_s}{m_u} \end{bmatrix} \begin{bmatrix} v_1 \\ v_2 \\ v_3 \\ v_4 \end{bmatrix} + \begin{bmatrix} 0 \\ 0 \\ 0 \\ F_r \end{bmatrix}. \tag{4}$$

Here F_r represents the displacement of the wheel due to the road profile. From equation 1b it is seen that:

$$F_r = \frac{k_u}{m_u} x_r \tag{5}$$

The output equation is:

$$\begin{bmatrix} y_1 \\ y_2 \\ y_3 \\ y_4 \end{bmatrix} = \begin{bmatrix} 1 & 0 & 0 & 0 \\ 0 & 1 & 0 & 0 \\ 0 & 0 & 1 & 0 \\ 0 & 0 & 0 & 1 \end{bmatrix} \begin{bmatrix} v_1 \\ v_2 \\ v_3 \\ v_4 \end{bmatrix} \tag{6}$$

The above equations are used to define the quarter car model as a dynamic system object in MATLAB. These objects are derived from the DynamicsSystem superclass.

2.2 ROAD PROFILE GENERATION

The ISO 8608 standard (Mechanical Vibration - Road surface profiles - Reporting measured data) provides a means to classify road profiles according to different damage levels [7]. Using these guidelines as a starting point it is possible to back-track the specifications given and reconstruct a road profile that adheres to a specified road profile class. The road profile classes specify the lower and upper limits to the values of the Power Spectral Density (PSD) for a specific class of road. To generate the artificial road profile one uses a stochastic representation, which is a function of the PSD of vertical displacements. This PSD is obtained through a Fourier Transform of the auto-correlation function of the stochastic process describing the road profile [5]. The artificial road profile can then be generated from:

$$h(x) = \sum_{i=1}^{N} \sqrt{\Delta n} 2^k 10^{-3} \left(\frac{n_0}{in} \right) \cos\left(2\pi i \Delta n x + \phi_i\right). \tag{7}$$

Here $x \in [0, L]$, where L is the total length of the road. The samples are equally spaced over the length of the road $\Delta n = \frac{1}{L}$. For computing the bandwidth of the PSD let n_{max} denote the maximum theoretical sampling spatial frequency (limited by the size of sampling interval B); then $N = \frac{n_{max}}{\Delta n}$; k is a constant that depends on the class of the road profile. It varies (discretely) between 3 and 9 corresponding to class A to H. Finally ϕ denotes a random phase angle drawn from a uniform distribution having a value between $-\pi$ and π.

Figure 2 shows the road profiles generated for the various classes from A to E.

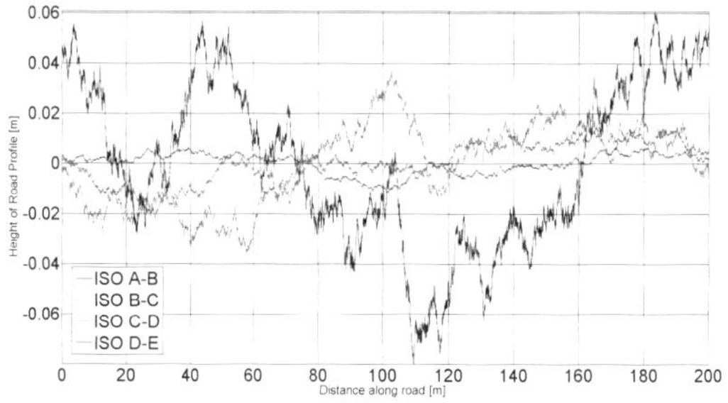

Figure 2: Generated Road Profiles for various classes

2.3 DATA GENERATION

Four data sets were generated each being different in size. A set of N road profiles with a length of 200 meters were generated using equation 7. Each road profile is considered an observation. In order to prevent the same or very similar road profiles from being included into a single data set a way to filter the signals were devised: Taking a single road profile, a comparison is made with all the other profiles and a number is obtained as a measure of its overall similarity to the other profiles. Similarity can be measured by various kernels. For this work a simple linear kernel was used. The overall similarity was then computed as:

$$S = \sum_j \frac{x_i \cdot x_j}{x_i \cdot x_i}, \; i, j = 1, \ldots, N \text{ and } i \neq j, \tag{8}$$

where S is the overall similarity of the road profile. The sum runs over all the other road profiles in the set. A similarity threshold is set so that when a high similarity is encountered the newly generated road profile is discarded from the set and a new profile would be generated. Continuing in this way four sets consisting of 30, 50, 100 and 200 different road profiles and their responses were generated.

3 CONVENTIONAL TECHNIQUES

The general model structure of the conventional linear system identification techniques is [1]:

$$y = G(q,\theta)u + H(q,\theta)e. \tag{9}$$

Here u is the input to the system, y is the system's response and e is the noise present in the data. The transfer function G maps the input data to the output data and H maps the noise to the output additive disturbance. The transfer functions are functions of q, a shift operator, which is used to shift the data forward in time. This is particularly used when working with discrete time systems. The variable θ denotes the parameters of the system. For black-box models, where no physical insight into the system is required, the transfer functions have the following form:

$$G(q,\theta) = \frac{B(q)}{A(q)}, \; H(q,\theta) = \frac{C(q)}{D(q)}. \tag{10}$$

Where $A(q), B(q), C(q)$, and $D(q)$ are polynomials of q^{-1}. Different choices of these functions lead to alternative versions of system identification techniques. In this work ARX (AutoRegressive model with eXternal input), ARMAX (AutoRegressive Moving Average model with eXternal input), BJ (Box-Jenkins) and OE (Output-Error polynomial) were all considered and OE was chosen as it yielded the best results. The structure of the OE model is obtained by setting $H = 1$.

4 STATISTICAL LEARNING TECHNIQUES

The approach to system identification presented by statistical learning techniques is fundamentally different to that of the conventional techniques. The statistical learning techniques attempt to find features in the input and output spaces from which regression can be done. The regression analysis is thus centred on the feature space, where the conventional techniques perform regression with variables.

4.1 Ridge Regression

In problems where the condition number is high the Least-Squares approach yields solutions that exhibit high variance and are poor estimations of the solution [8]. The problem arises due to some regression coefficients become uncontrollably large. In order to deal with this, a penalty term can be added to the regression algorithm, limiting the sizes of the regression coefficients. To this end shrinkage methods incorporate a continuous regularisation to the regression algorithm. Related coefficients are driven towards zero by increasing the added penalty constraint. For ridge regression the following minimization problem is solved:

$$\min_{\beta}\{||X\beta - y||^2 + \lambda\beta^T\beta\}. \tag{11}$$

The solution to equation 11 is:

$$\hat{\beta}^{\text{ridge}} = (X^TX + \lambda I)^{-1}X^Ty. \tag{12}$$

The regularisation constant λ can be found using cross-validation.

4.2 Regularised Kernel Methods

Kernel methods introduce flexibility into the problem of estimating the mapping between the input and output spaces by not imposing a predetermined structure onto the mapping but rather finding the mapping in a possibly infinite dimensional functional space $\mathscr{H}$. However this functional space is requiring to be a complete Hilbert space. The problem can be simplified to the following optimisation problem [1]:

$$\min_{g\in\mathscr{H}}\left(\sum_{i=1}^{N}(y_i - g(x))^2 + \gamma\|g\|^2_{\mathscr{H}}\right). \tag{13}$$

For the representer theorem it then follows that this optimisation problem has the solution[1]:

$$\hat{g}(x) = \hat{C}K(x) \tag{14}$$

where $\hat{C}$ is given by:

$$\hat{C} = (K + \lambda I)^{-1}Y. \tag{15}$$

This estimate is the Kringing estimate of a Gaussian random field in spatial statistics [8].

4.3 Principal Component Regression

Principal Component Regression (PCR) is a regression technique that uses derived inputs. The derived input space is composed of the principal components which are the right singular values of the original input space. These principle components are the directions in the input space that maximises the variance. Finding these directions thus solves the following optimisation problem [8]:

$$\max_{z}\left(z^T X^T X z\right). \tag{16}$$

The optimisation problem is subject to the vectors found being unit length and being conjugate to one another with respect to the sample covariance of the features of X. Least Squares is then used to regress the derived input space onto the output space.

4.4 Partial Least Squares

The Partial Least Squares (PLS) algorithm is used to model relationships between two sets of data. In order to do this it uses a derived input space. When performing regression, inputs are weighted based on their similarity to responses [8]. PLS is a powerful tool due to its ability to analyse data that is noisy and that has many collinear or incomplete variables [9]. A covariant matrix is constructed which quantifies the correlation between input and output variables. A conjugate gradient approach is then used to search for directions of high correlation [8, 10].

The SIMPLS algorithm [10], which is used in this work, casts PLS as the following optimisation problem:

$$\max_{r_a, q_a}(r_a^T \mathrm{Cov} q_a). \tag{17}$$

The so-called weight vectors r_a and q_a are also the left and right singular vectors of the covariant matrix constructed. The optimisation problem is subject to the following constraints: The weight vectors are required to be normalised such that their lengths are 1. The weight vector r_a must be conjugate with respect to the matrix $X^T X$. The input matrix X contains all the input data from the system and its size is $n \times p$ where n refers to observations and p to variables. This implies that the search directions are required to be conjugate to the variance of the input data.

4.5 Thresholded and adapted thresholded strategies for PLS

The data set sizes presented in this work do not allow the application of PLS directly due to its high computational costs. Therefore alternative strategies are needed to reduce the data and would allow for PLS to be employed. Thresholded PLS (T-PLS) has been developed for cases where high dimensional data are encountered, i.e., data sets that have a large number of variables and observations [8]. For T-PLS linear regression is performed between the input and output space. Once this is done the regression coefficients are used to determine which predictors have a strong correlation to the response variables. Setting a threshold, which is determined by cross-validation, the predictors that have a high correlation to the responses are retained. For this particular application this does not provide much benefit since the correlation between the road profile and the corresponding response of the quarter car model is highlighted. Therefore the thresholded strategy is adapted for the time series setting as follows: The predictors are compared with the target input signal (for which the response is to be predicted). This gives a correlation between what is to be predicted and how it correlates to the existing data. Since the correlation between the input and output space is known only the predictions and responses that yield a high correspondence to the target input signal are retained and PLS is then applied to this reduced data set.

5 Comparison of Algorithms

The algorithms presented above are now applied to the data generated from the Quarter Car model. Since the application of the algorithms is with inverse problems in mind, the vertical response is chosen to be the input while the road profile is the output. This presents a further challenge to the algorithm: the vertical response of the quarter car is a smooth curve along the length of the road, but the road profile is coarse and jagged. The results presented in Table 2 are derived from data sets that have 5% white noise added to them. In order to investigate the robustness of the algorithms with regards to noise the largest set of 200 observation were used and 0%, 2% and 5% white noise was added to it, the results are presented in Table 3. The results reported in both Table 2 and 3 are scaled with 10^{-3}.

Table 2: Comparison of various techniques.

Number of observation	OE	Ridge	Kernel	PCA	T-PLS
30	2.40	2.70	2.70	4.40	2.70
50	1.50	1.80	1.90	3.90	1.90
100	1.40	1.10	1.10	1.10	1.10
200	1.50	0.78	0.79	0.75	0.78

Table 3: Comparison of various techniques on a set of 200 observation with various levels of white noise added to the data.

Noise Level	OE	Ridge	Kernel	PCA	T-PLS
0%	1.40	0.31	0.36	0.36	0.39
2%	1.40	0.47	0.47	0.47	0.47
5%	1.50	0.76	0.79	0.75	0.78

Cross validation was used to determine the regularisation parameters for each of the techniques. The Mean Square Errors (MSE) of the algorithms were calculated with respect to the test signal. These results are presented in Table 3. The performance between the Output-Error (OE) model and the best performing statistical leaning techniques are presented graphically in Figures 3 and 4. Figure 3 shows the road profile estimated by the aforementioned techniques and Figure 4 presents the errors made by each one. The performance of OE and the other techniques are similar on smaller data sets. As the data set size increases however, the statistical learning methods perform better. For the last two data sets the statistical learning methods

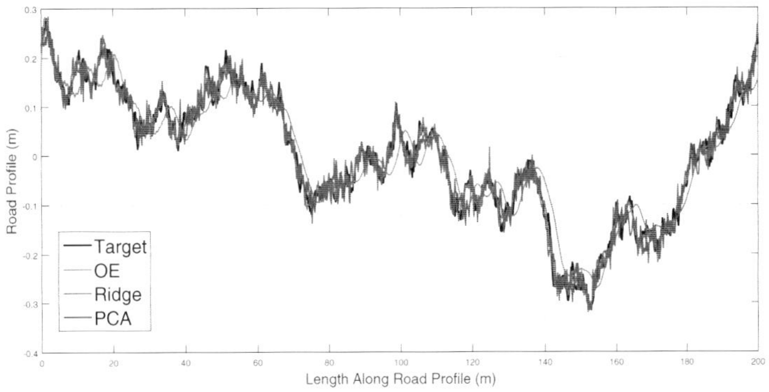

Figure 3: Estimation of Road Profile by best performing statistical learning techniques and the OE model.

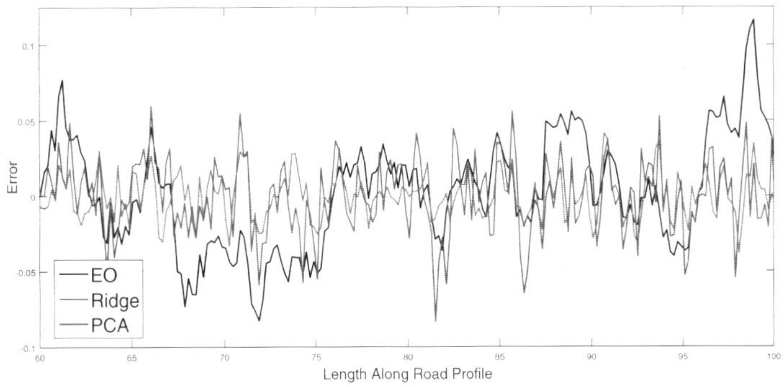

Figure 4: Sample of the road profile and the corresponding errors made by each of the techniques presented in Figure 3

also executed faster than the OE model. The greatest improvement with data set size was made by PCA, while ridge regression performed the best overall when comparing all the data set sizes. The robustness of the OE model is seen to be better that the statistical learning techniques, this suggest that these methods should be used in conjunction with filters, smoothers or similar algorithms that removes noise from the data. Alternatively when considering Table 2 it can be seen that the statistical learning techniques still show improvements in accuracy as the data set size increases even in the presence of high noise levels: thus more data also is a remedy for overcoming the noise level in the data. The statistical learning models were also able to capture the roughness of the road surface where the OE model gave a smooth approximation. The statistical learning methods performed better on the larger data sets in terms of accuracy, efficiency and capturing the characteristics of the road profile.

6 Conclusion

By comparing statistical learning techniques to classical techniques in system identification it has been established that in a data rich environment statistical learning techniques are accurate and efficient. The statistical learning techniques outperformed the Output Error (OE) model (which was the best performing model of the that classical system identification techniques) on larger data sets and was thus shown to be a viable alternative to these classical methods on larger data sets.

References

[1] G. Pillonetto, F. Dinuzzo, T. Chen, G. De Nicolao, and L. Ljung, "Kernel methods in system identification, machine learning and function estimation: A survey," *Automatica*, vol. 50, no. 3, pp. 657–682, 2014.

[2] L. A. Aguirre and C. Letellier, "Modeling nonlinear dynamics and chaos: a review," *Mathematical Problems in Engineering*, vol. 2009, 2009.

[3] S. A. Billings, *Nonlinear system identification: NARMAX methods in the time, frequency, and spatio-temporal domains.* John Wiley & Sons, 2013.

[4] G. Kerschen, K. Worden, A. F. Vakakis, and J.-C. Golinval, "Past, present and future of nonlinear system identification in structural dynamics," *Mechanical systems and signal processing*, vol. 20, no. 3, pp. 505–592, 2006.

[5] M. Agostinacchio, D. Ciampa, and S. Olita, "The vibrations induced by surface irregularities in road pavements–a matlab® approach," *European Transport Research Review*, vol. 6, no. 3, pp. 267–275, 2014.

[6] R. N. Jazar, *Vehicle dynamics: theory and application.* Springer Science & Business Media, 2013.

[7] I. Standard, "8608," *Mechanical vibration – road surface profiles – reporting measured data*, 1995.

[8] T. J. Hastie, R. J. Tibshirani, and J. H. Friedman, *The elements of statistical learning: data mining, inference, and prediction.* Springer, 2009.

[9] S. Wold, M. Sjöström, and L. Eriksson, "Pls-regression: a basic tool of chemometrics," *Chemometrics and intelligent laboratory systems*, vol. 58, no. 2, pp. 109–130, 2001.

[10] S. De Jong, "Simpls: an alternative approach to partial least squares regression," *Chemometrics and intelligent laboratory systems*, vol. 18, no. 3, pp. 251–263, 1993.

Finite Element Analysis of the Tread Quenching of Railway Wheels

J.L. Cuperus[1,2], G. Venter[1] and D.C. Blaine[1]

1) Department of Mechanical and Mechatronic Engineering, Stellenbosch University
2) Transnet Engineering

Louwrens.cuperus@transnet.net

ABSTRACT

The heat treatment process that is used during the manufacturing of monobloc railway wheels is designed to induce a compressive residual hoop stress in the rim of the wheel. This residual stress is often ignored during finite element analysis of railway wheels. This is likely due to the complexity of the analysis needed to predict the residual stress field. A number of non-linear finite element simulations were performed, each one testing a different assumption or simplification. Based on the simulation results, if quantitative residual stress results are required, the important factors to take into account are material creep, the strain hardening model and parameters, latent heat of phase transformation, phase transformation strain, surface temperature dependent boiling heat transfer coefficients, and radiation.

KEYWORDS

Heat treatment simulation, residual stress, finite element method, phase transformation, transformation induced plasticity.

INTRODUCTION

As part of the manufacturing process, substantial effort goes into inducing the desired residual stress pattern to the wheel through a tread quenching process. However, due to the difficulty of including phase transformation effects and the need for material data that are typically not available, standards and specifications usually do not explicitly require that the initial residual stress field be taken into account (BS EN 13979-1 [1], UIC 510-5 [2], Matjeke [3]). Consequently, many analysts ignore the initial residual stress pattern when finite element analyses are performed, even when fatigue and final residual stress fields are investigated (Portesi, Gallo and Lombardo [4], Ha and Kang [5]).

In this paper, tread quenching is simulated on a class 5M2A wheel using MSC Marc. By running various simulations - taking different boundary conditions and metallurgical effects into account - we try to establish which parameters and physical phenomena are important to consider during these simulations and which can be ignored in future.

In order to predict the residual stress state through simulating the heat treatment process, a highly accurate material model, that covers the entire temperature range of the process, is required. As this material data is not available for the wheel in question, and constitutes a prohibitively vast undertaking to obtain experimentally, published material data were used instead. Since the purpose of this study is to determine how sensitive the model is to various parameters and metallurgical phenomena, only a representative material

SIMULATION PARAMETERS

Simulations were performed using the heat treatment process as well as particular material properties from Gordon and Perlman [6], in which a commuter car wheel was analysed. The wheel starts at a uniform initial temperature of 870 °C, at which point the entire wheel is assumed to be fully austenised. A water spray tread quenching operation is then applied for two minutes. This is followed by a four minute dwell at room temperature during which the wheel is handled between operations. The four minute dwell is followed by an isothermal annealing stage at 500 °C for 5 hours, after which the entire microstructure is assumed to be pearlite. Finally, the wheel is left to cool to room temperature, taken as 25°C.

Accounting for all the interactions pertinent to the heat treatment process is a very complex simulation problem that is a current research topic (Ariza *et al* [7], Song *et al* [8]). See Mackerle [9] for an extensive bibliography covering 25 years.

For the purposes of the current discussion, material data are broken up into general material data and data accounting for the effects of phase transformation. Most of the major technical challenges in this work were associated with the effects of phase transformation.

General Material Data

From Gordon and Perlman [6], the thermal properties of the material are shown in Table 1, with the mechanical properties in Table 2. The material model parameters from Gordon and Perlman were used as it was the only source for the relevant parameters, which covered the entire temperature range, known to us at the time. It wasn't possible to establish how similar the material is to the wheel being simulated, other than it also being a railway wheel. As stated before, only a representative or plausible material model is required given the goal of the study.

Table 1: Thermal Material Properties for Railway Wheel from Gordon and Perlman [6]

Temp (°C)	Specific Heat c_p (J/kg.K)	Thermal Conductivity (W/m.K)
0	419.5	59.71
350	629.5	40.88
703	744.5	30.21
704	652.9	30.18
710	653.2	30.0
800	657.7	25.0
950	665.2	27.05
1200	677.3	30.46

Table 2: Mechanical Material Properties for Railway Wheel from Gordon and Perlman [6]

Temp (°C)	Young's Modulus E (GPa)	Poisson's Ratio ν	Initial Yield Stress σ_y (MPa)	Tangent Modulus E_T (GPa)	Tangent CTE α (10^{-6})
24	213	0.295	422.9	21.66	5.3
230	201	0.307	424.7	25.73	8.81
358	193	0.314	366.7	20.29	10.25
452	172	0.320	291.0	14.89	10.95
567	102	0.326	132.3	5.93	11.38
704	50	0.334	39.4	0.92	11.3
900	43	0.345	11.7	0.085	11.25

Note that the specific heat capacity and the coefficient of thermal expansion (CTE) in Table 1 and Table 2 were not used for the nominal case, but only to investigate the results of omitting phase transformation effects. Specific heat capacity and CTE data for the nominal case (which takes all phase transformation effects into account) are discussed in the next subsection.

A simple bilinear strain hardening model is used. The Young's modulus describes the stress-strain relationship for any stress less than the yield stress, while the tangent modulus is applicable for any stress equal to the yield stress (with the yield stress constantly being updated).

Due to the high temperature and the presence of an annealing stage, it is clear that material creep could have a marked effect on the residual stress field. This was investigated and largely confirmed by Kuhlman *et al* [10] by performing finite element analyses on a 40~inch railway wheel. As the creep data for the wheel material in question is not available, the material creep as given by Kuhlman *et al* (also used by Gordon and Perlman) was used. Kuhlman's viscoelastic creep equation was converted to metric units to obtain:

$$\dot{\varepsilon} = 1.5305(10^{-93})\sigma_{eq}^{12.5}e^{\left(\frac{-29840}{T+273}\right)} \quad (1)$$

with the equivalent stress (σ_{eq}) in Pascal and the temperature (T) in degrees Celsius.

Material Property Dependence on Phase Transformation

Modelling the phase transformations is a complex problem that can rarely be performed with the pre-defined options in non-linear finite element packages. Figure 1 shows the general interactions during phase changing heat treatment. Since we are not interested in predicting the microstructure, it is preferable to artificially account for the effects of the microstructure on the temperature and stress fields in order to simplify the simulation. This is simplified in the current study by the fact that the final phase of the steel throughout the wheel is known to be pearlite (PRASA Rail Chair for Maintenance and Engineering Management [11]).

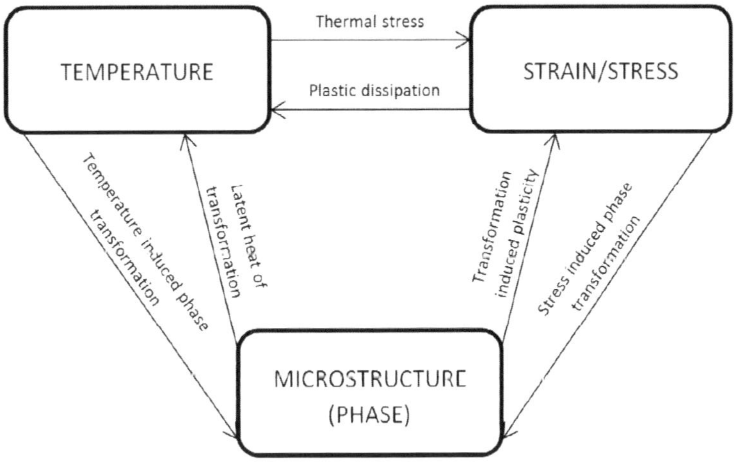

Figure 1: Metallo-Thermo-Mechanical Coupling

The latent heat of transformation can be artificially accounted for in the specific heat capacity of the material. The primary drawback to this approach is that it is not possible to account for the changes in the temperature range over which the transformation takes place as a result of supercooling or the stress in the component. The temperature dependent specific heat capacity that accounts for austenite to pearlite transformation of eutectoid (1078) steel (ASM International [12]), also used by Kuhlman *et al* [10], was used and is shown in Table 3. Here the specific heat capacity is given as the average value over 50 °C increments, ranging from 50 °C to 800 °C. To avoid numerical instabilities, a continuous relationship for latent heat as a function of temperature was estimated by straight line interpolation between values (see Figure 2).

Table 3: Specific Heat of Eutectoid (1078) Steel

Temp (°C)	**Specific Heat c_p (J/kg.K)**
50-100	490
150-200	532
200-250	548
250-300	565
300-350	586
350-400	607
450-500	670
550-600	712
650-700	770
700-750	2081
750-800	615

During phase transformations, the incremental strain $d\varepsilon_{ij}$ can be broken down as:

$$d\varepsilon_{ij} = d\varepsilon_{ij}^{e} + d\varepsilon_{ij}^{p} + d\varepsilon_{ij}^{th} + d\varepsilon_{ij}^{pt} + d\varepsilon_{ij}^{trip} \quad (2)$$

where $d\varepsilon_{ij}^{e}$ is the elastic strain, $d\varepsilon_{ij}^{p}$ the plastic strain, $d\varepsilon_{ij}^{th}$ the thermal strain, $d\varepsilon_{ij}^{pt}$ the phase transformation strain, and $d\varepsilon_{ij}^{trip}$ is the transformation induced plasticity (TRIP).

The first of the transformation induced strains, the phase transformation strain, is driven by a difference in the specific volume of the parent and product phases. Thus it is a purely spherical (isotropic) strain tensor. Similarly to the latent heat of transformation, this transformation induced change in volume (as a result of crystal structure alone) can be included by modifying the coefficient of thermal expansion, as this too is a spherical tensor. According to Lement [13], the change in volume (percent) from the face-centred cubic (FCC) structure of austenite to pearlite, a combination of body-centred cubic (BCC) ferrite and orthorhombic cementite, at 20 °C (based on measured lattice parameters) is:

$$\Delta V\% = (4.64 - 2.21 \times \%C)\% \quad (3)$$

where $\%C$ denotes the weight percentage carbon content.

As Equation 3 is only valid at 20 °C, and the austenite to pearlite transformation takes place at a much higher temperature, the equation is modified in order to take into account the different thermal expansion coefficients of austenite and pearlite. Using the temperature invariant coefficients of thermal expansion for austenite and pearlite, respectively, by Jung *et al* [14], $\propto_A = 24(10^\wedge 6)$ and $\propto_P = 15(10^\wedge 6)$, it is possible to extrapolate the change in volume given by Equation 3 to another desired temperature. For this extrapolation, the same temperature range over which the phase transformation is captured in the specific heat capacity data (as indicated by the spike between 700 and 750 °C in Figure 2a) is used. The transformation induced volume change is calculated for the middle of this range (725 °C) and then converted from a volume change (ΔV) to an equivalent linear dimensional change (ΔL). This relationship between isotropic volume and linear expansion is typically approximated as:

$$\Delta V \approx 3\Delta L \quad (4)$$

This change in length is then superimposed onto the conventional coefficient of thermal expansion. The resultant CTE as a function of temperature is shown in Figure 2b. To avoid numerical instabilities owing to discontinuities at both 700 and 750 °C, linear ramps were used instead of a constant value over the phase transformation temperature range. The temperature invariant coefficients of thermal expansion for austenite and pearlite (by Jung *et al*) are used for all analyses accounting for transformation effects, rather than those given in Table 1, as the data in Table 2 do not distinguish between the different microstructures. Between 700 and 750 °C the conventional CTE value was taken as the average of that of austenite and pearlite.

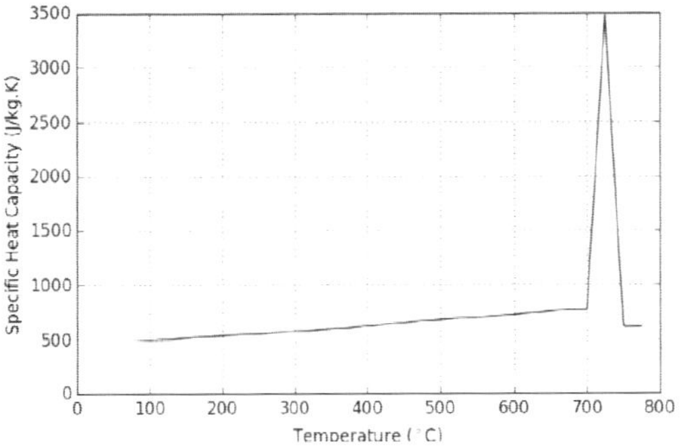

(a) Specific Heat Capacity

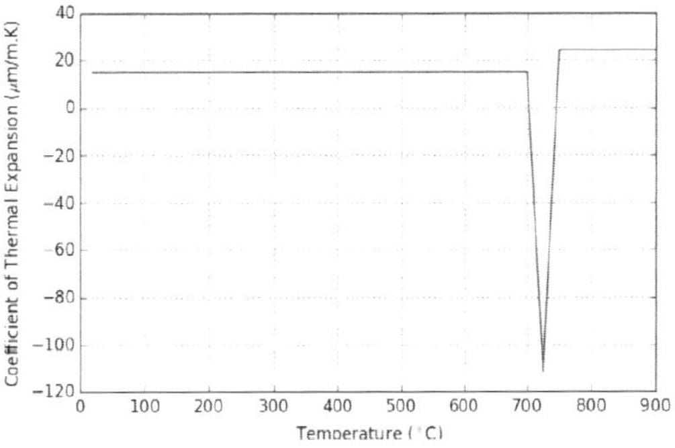

(b) Coefficient of Thermal Expansion

Figure 2: Modifications to Specific Heat and CTE to Account for Transformation Effects

The second of the transformation induced strains, TRIP, results from yielding at a low stress state during the phase transformation and is a traceless tensor. Two mechanisms have been put forward and are both widely accepted. The first is known as the Greenwood-Johnson effect [15]. As already stated, the parent and product phases have different specific volumes and as such there is a change of volume associated with the phase transformation. This volume change can quickly cause the softer material, typically the austenite in the case of steel, to yield in order to accommodate the change in volume. This is particularly true for diffusive transformations where the transformation starts at a nucleation site and propagates outwards, yielding the surrounding austenite matrix. The second mechanism is known as the Magee effect and is only associated with displacive transformations. As such, it is not of interest in this work as the product phase is pearlite. The interested reader is directed to the bibliography by Mackerle [9].

The TRIP model used in this work was developed by Leblond over four publications [16] [17] [18] [19]. It is itself broken up into three terms, with the incremental strains, respectively, proportionate to the stress increment, temperature increment, and product phase fraction increment. The relationship is expressed as rate of change:

$$\dot{\varepsilon}^{trip} = \dot{\varepsilon}(\dot{\sigma}_{eq}) + \dot{\varepsilon}(\dot{T}) + \dot{\varepsilon}(\dot{z}) \tag{5}$$

where σ_{eq} is the equivalent stress, T is the temperature, and z is the phase fraction of the product phase.

Similar to Şimşir and Gür [20], TRIP was calculated in Marc's ANEXP (anisotropic thermal expansion) user subroutine[21], adding the traceless TRIP strain tensor to the isotropic thermal strain tensor. Given the material data available, some approximations had to be made to use LeBlond's models. Firstly, phase transformation was assumed to take place at a constant rate from 750 to 700°C. The global (macroscopic average of all phases) yield stress and strain hardening characteristics are based on the data given in Table 2 at the current temperature and sum of plastic strain and TRIP strain. The yield stress and strain hardening characteristics of austenite were approximated as that of the global material (given in Table 2) at 750 °C throughout the entire transformation period.

During the transformation, all or most of the strain hardening accumulated in the austeninte is lost. For this reason it is acceptable to include TRIP into the thermal strain tensor as most, if not all, of the plasticity is confined to the austenite matrix. As the temperature at an integration point passes through a pre-set temperature, the plastic strain tensor and equivalent plastic strain scalar are reset to zero. Marc's NEWSV (new state variable) user subroutine was used to set the temperature at which plasticity is to be reset. Once the plasticity at a particular integration point is reset, the state variable for that integration point is changed to an unattainably low temperature to ensure that it is only reset once.

Finite Element Mesh and Boundary Conditions

A fine mesh was used in order to obtain a high resolution and not for the purpose of mesh convergence. An axisymmetric mesh (Figure 3) with 5 327 elements and 16 616 nodes is used. Finer meshes were used to establish mesh convergence, but were not used for the bulk of the simulations due to excessive computational cost.

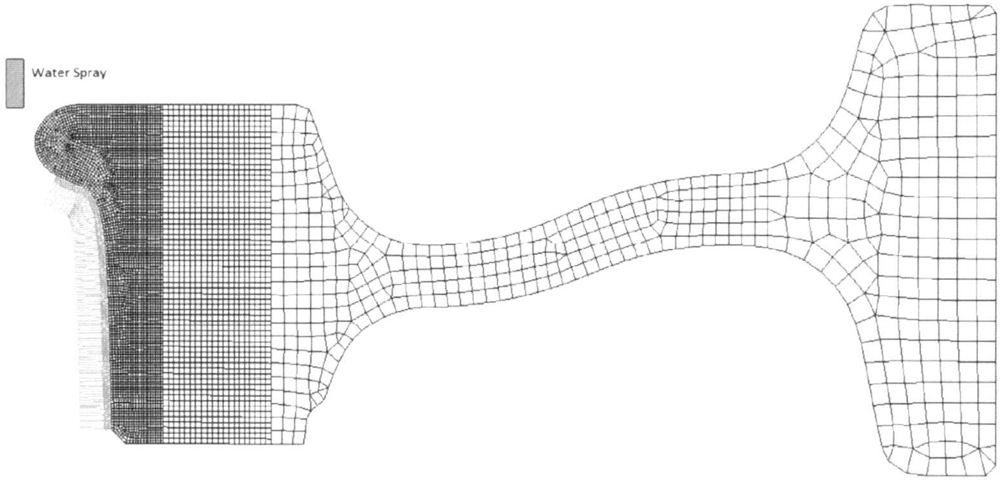

Figure 3: Axisymmetric Mesh

The heat transfer coefficients (HTC) from Kuhlman *et al* [10] were used. Quenching was applied to the tread using a convection coefficient of 3066 $W/m^2.K$ and water temperature of 23 °C (see Figure 3 for tread quenching area). Natural convection was implemented on the remaining wheel surface with a convection coefficient of 28 $W/m^2.K$ and an ambient temperature of 25 °C. Radiation was included using an emissivity of $\varepsilon = 0.81$ (Çengel and Ghajar [22]) to space at the relevant ambient temperature for the stage in the heat treatment process.

The constant heat transfer coefficient of Kuhlman *et al* is thought to potentially be a primitive approximation of the actual heat transfer, warranting more detailed investigation. The boundary condition representing the spray quenching was also solved using the correlations published by Mudawar's group, specifically those reported by Hall *et al* [23]. When a hot surface is quenched, five distinct heat transfer regimes can be observed (see Figure 4). Starting from a sufficiently high temperature, the first regime is that of film boiling. During this regime, the surface is covered by a vapour blanket which acts to restrict the heat transfer. As the temperature drops, the "departure from film boiling" point will be reached, separating the film boiling and film wetting regime (Point A). The vapour blanket is now intermittent and partial liquid contact with the surface is established. Upon further cooling, the heat flux continues to decrease until the minimum heat flux point is reached, which separates the film wetting and transition boiling regions (Point B). During the transition boiling regime, a decrease in the surface temperature acts to increase the heat flux. Due to this positive feedback nature, this regime is typically very short lived. As the temperature continues to decrease, the point of maximum heat flux, or critical heat flux, is reached, which separates the transition and nucleate boiling regimes (Point C). As the surface cools down further, it enters the single-phase cooling regime (Point D).

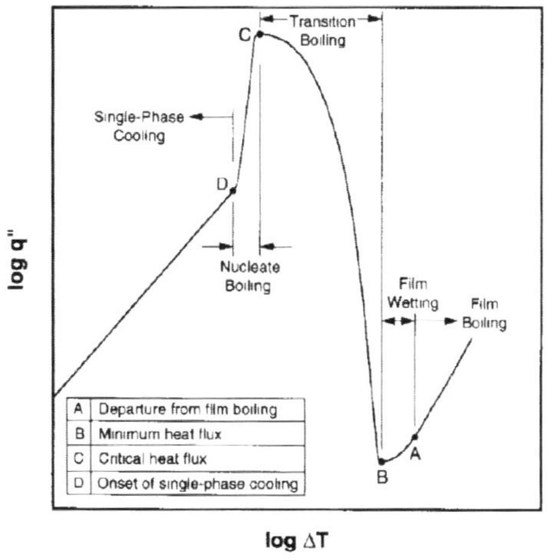

Figure 4: Boiling Curve (Hall *et al* [23])

The spray characteristics used during the rim quenching of the wheel in this study are not known as the information is proprietary. The spray characteristics used for the simulations (see Table 4) were chosen such that they fall within the tested ranges of Mudawar's studies (the middle of the acceptable range was used where possible), while extracting roughly the same amount of thermal energy during the quenching process as the temperature invariant HTC by Kuhlman *et al*. Table 4 also shows the quenchant (water) properties that are used (from Çengel and Ghajar [22]). The resultant temperature dependent heat transfer coefficient is shown in Figure 5. No radiation is applied to the sprayed surface during the quenching stage as radiation was not accounted for during the studies of Mudawar's group or Kuhlman *et al*.

Table 4: Spray and Quenchant Properties

Description	Value
Mean drop velocity U_m	20 m.s^{-1}
Sauter mean diameter d_{32}	7.44(10^{4}) m
Volumetric spray flux Q''	9(10^{3}) m^3/s.m^2
Thermal conductivity k_f	0.679 W/m.K
Dynamic viscosity μ_f	282(10^{-6}) kg/m.s
Fluid density ρ_f	957.9 kg/m^3
Steam density ρ_g	0.5978 kg/m^3
Heat of vaporisation h_{fg}	2 257 kJ/kg
Surface tension σ	58.9(10^{-3}) N/m
Specific heat c_p	4 217 J/kg.K
Subcooling ΔT_{sub}	77 °C

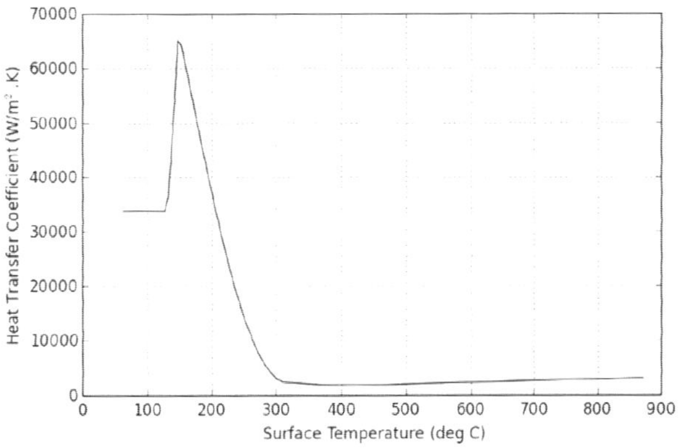

Figure 5: Spray Quenching Heat Transfer Coefficient

SENSITIVITY TO MATERIAL PARAMENTERS AND BOUNDARY CONDITIONS

The axisymmetric model is used to investigate variations in parameters relating to material behaviour, phase transformations and boundary conditions (summarised in Table 5). Since one of the major objectives of the heat treatment is to induce residual compressive hoop stress in the rim of the wheel, it is used as the metric to judge the effect of the various changes.

Table 5: Model Parameter Configuration

Case	Hardening Law	Creep Model	Latent Heat	Transf. Strain	TRIP Strain	Plastic Heat	Radiation	Quench HTC
Isotropic	Isotropic	Incl.	Incl.	Incl.	Incl.	Incl.	Incl.	Mudawar
Kinematic (Nominal)	Kinematic	Incl.	Incl.	Incl.	Incl.	Incl.	Incl.	Mudawar
No Creep	Kinematic	Omtd	Incl.	Incl.	Incl.	Incl.	Incl.	Mudawar
Perfect Plastic	N/A	Incl.	Incl.	Incl.	Incl.	Incl.	Incl.	Mudawar
No Transf. Strain	Kinematic	Incl.	Incl.	Omtd	Omtd	Incl.	Incl.	Mudawar
No Transf. Heat	Kinematic	Incl.	Omtd	Incl.	Incl.	Incl.	Incl.	Mudawar
No TRIP	Kinematic	Incl.	Incl.	Incl.	Omtd	Incl.	Incl.	Mudawar
No Transf. Kinetics	Kinematic	Incl.	Omtd	Omtd	Omtd	Incl.	Incl.	Mudawar
No Radiation	Kinematic	Incl.	Incl.	Incl.	Incl.	Incl.	Omtd	Mudawar
Constant HTC	Kinematic	Incl.	Incl.	Incl.	Incl.	Incl.	Incl.	Kuhlman
No Plastic Heat	Kinematic	Incl.	Incl.	Incl.	Incl.	Omtd	Incl.	Mudawar

The nominal case in this study uses kinematic hardening and takes all transformation effects into account. This include transformation strain, transformation induced plasticity, latent heat of transformation, and the loss of strain hardening during phase transformation. The heat generated by plastic deformation is also taken into account. The quenching heat transfer is calculated using the work by Mudawar's group. Radiation is included on all surfaces being cooled by natural convection. The resultant residual stress field is shown in Figure 6.

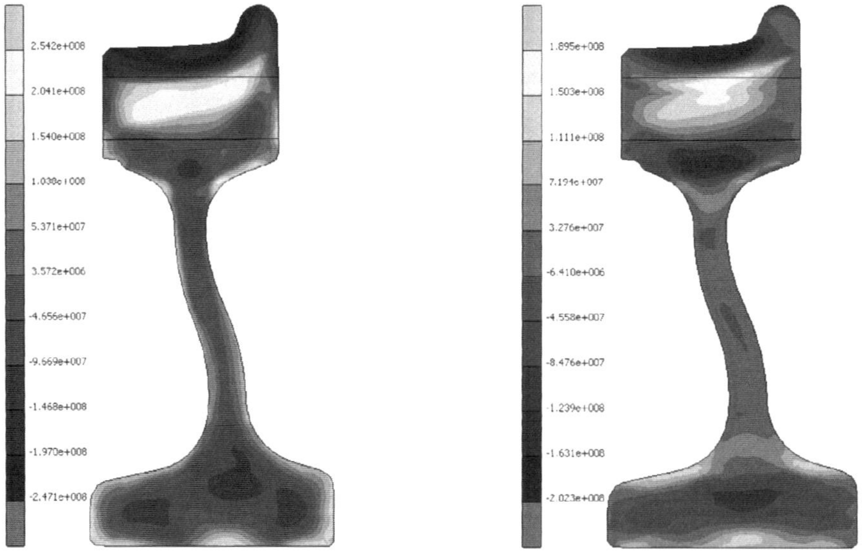

(a) Circumferential stress

(b) Axial stress

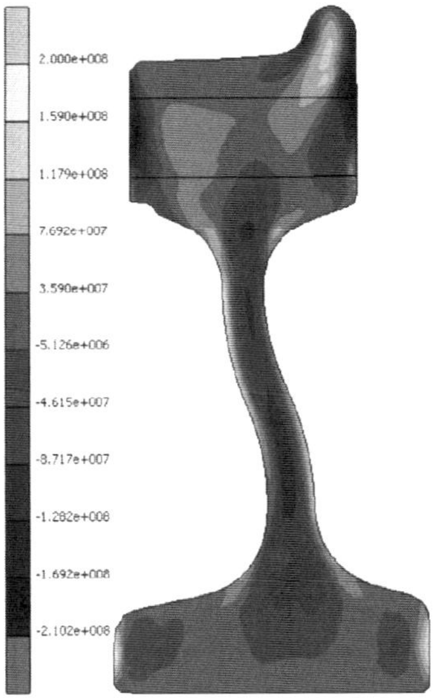

(c) Radial stress

Figure 6: Residual Stress for Nominal Case

On the material side, isotropic and kinematic hardening models are compared. Isotropic hardening is faster to solve, most likely due to the delay in reverse plasticity as the wheel returns to a uniform temperature after the quench. For the remaining simulations, kinematic hardening is used since Kuhlman *et al* and Rammerstorfer *et al* [24] reported it to be more representative. The effect of omitting creep as well as strain hardening is also investigated.

To investigate the various transformation effects, a number of combinations of model parameters are considered, as summarised in Table 5. The transformation strain is ignored (as well as the transformation induced plasticity as it is caused by the transformation strain) by using the temperature dependent thermal expansion coefficient of Gordon and Perlman [6] (given in Table 2) which does not include transformation effects. This temperature dependent CTE is also compared to a constant CTE (taken as 12×10^{-6} rather than 15×10^{-6} as given by Jung *et al* to better match the data by Gordon and Perlman). This is done in order to gauge the importance of the temperature dependence since constant values were used to construct the CTE accounting for phase transformation. The latent heat of transformation is also excluded by using the specific heat capacity from Gordon and Perlman (given in Table 1). Transformation induced plasticity is then omitted while still including the volumetric transformation strain and the latent heat of phase transformation. Finally all transformation effects are ignored simultaneously.

Different boundary conditions are also investigated. Radiative heat transfer is excluded. The effect of approximating the spray quenching process with a constant HTC is investigated by using the value used by Kuhlman *et al*. The heat generated by plastic deformation is also excluded.

Figure 7 shows the hoop stress along a line from the tread surface on the taping line, extending 63 mm radially down into the rim. Figure 7c excludes the curve associated with the exclusion of plastic heat generation since the curve is coincident with the *Reference* curve. Figure 8 through Figure 10 show the hoop stress histories for the different material variations, transformation kinetics and boundary conditions respectively. The hoop stress is extracted at two nodes, one on the tread surface on the taping line and one 63 mm below that. The data are also separated at 360 s elapsed time, with the first time span covering the quenching and dwell periods and the second covering the annealing and cooling to room temperature. The reason for the temporal split being the dramatic difference in the rates of change for the hoop stress.

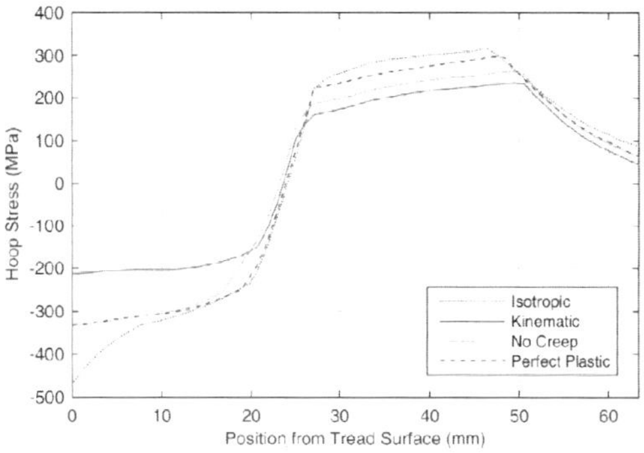

(a) Material model variations

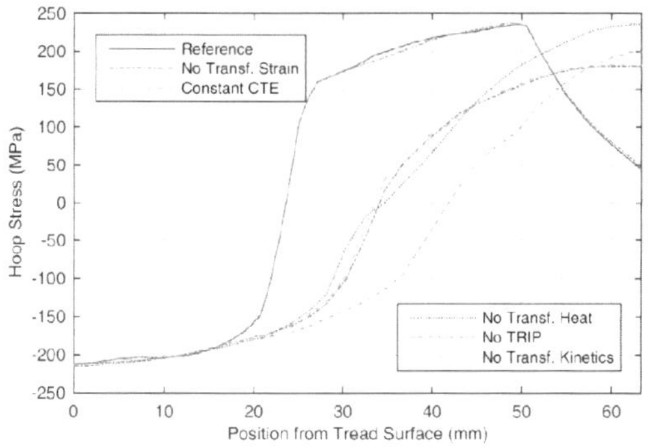

(b) Transformation kinetics variations

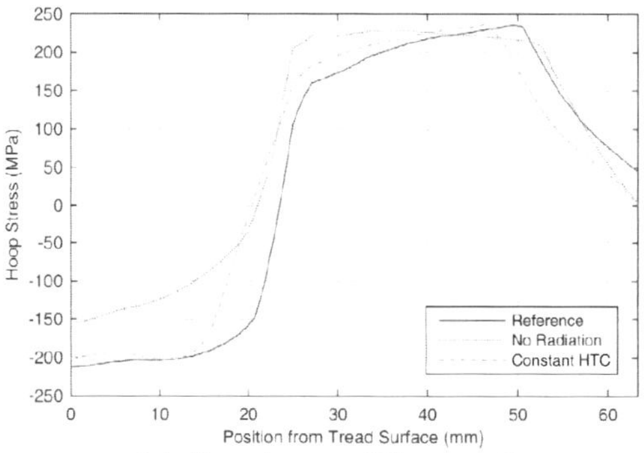

(c) Boundary condition variations

Figure 7: Hoop Stress Comparison at Process End

Material Effects

Figure 8 shows that there is an appreciable difference in the stress fields produced by the isotropic and kinematic hardening laws, particularly at the tread surface where the hoop stress is more than 100% higher for the isotropic model. Creep and strain hardening also seem to have a much more pronounced effect close to the surface than deeper into the wheel, although not to the same degree as the isotropic hardening model. Given the strong influence of the hardening behaviour of the material, a simple bilinear strain model (consisting of Young's and tangent moduli) might be insufficient to make accurate predictions. Figure 8b and Figure 8d show that creep starts to exert a strong influence early during the annealing stage; thus it is imperative to include creep, using Equation 1, if quantitative results are desired.

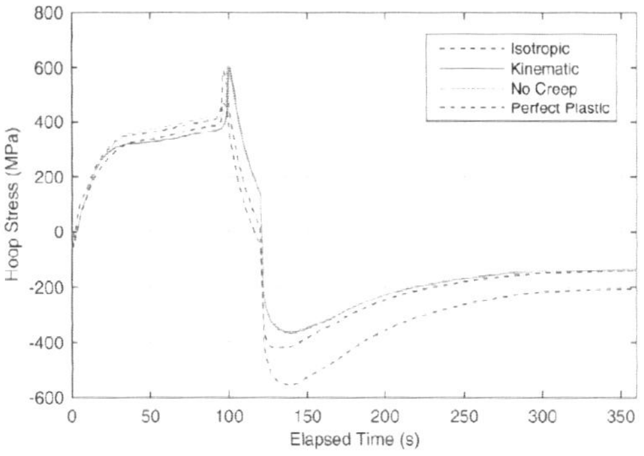

(a) Surface – Quench and Dwell

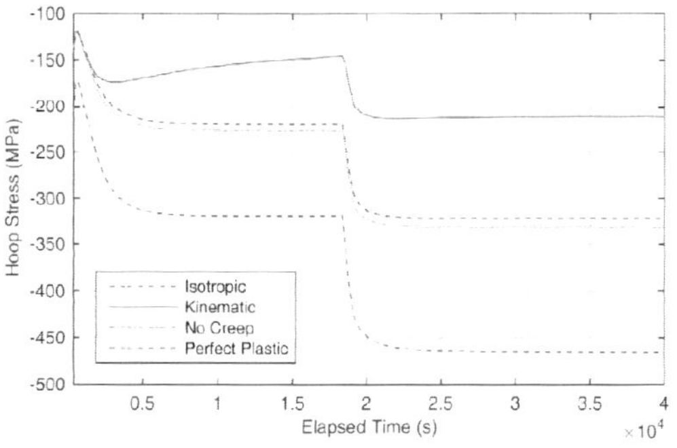

(b) Surface – Anneal and Cooling

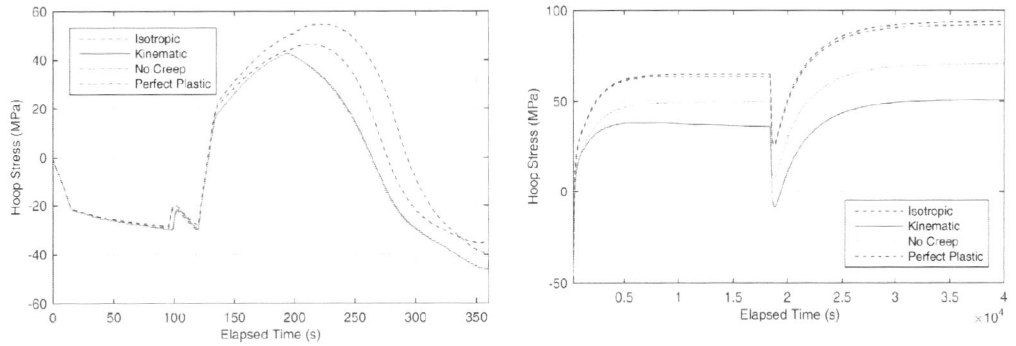

(c) 63 mm – Quench a (d) 63 mm – Anneal and Cooling

Figure 8: Hoop Stress History for Different Material Properties

Phase Transformation Effects

From Figure 7b it can be seen that TRIP has a very limited impact on the residual hoop stress; albeit with a slightly compromised implementation due to the lack of austenite specific material data. Given the complexity of including TRIP, it would not be worthwhile to include it in future analyses, especially when considering the uncertainty already accrued throughout such a complex model. Furthermore, resetting the plastic strain tensor and equivalent strain scalar was also omitted with no discernible difference. Without TRIP and resetting plastic strain, there is no need for user defined subroutines in MSC Marc, providing that explicit phase transformation is not simulated.

Figure 7b also shows that, when excluding phase transformation influences, there is only a small difference between the temperature dependent CTE from Table 2 (annotated as *no transformation strain*) and a constant CTE. Thus it seems acceptable to construct CTE values from temperature invariant CTE data if necessary.

The phase transformation kinetics on a whole plays a crucial role in predicting the residual hoop stress. Figure 7b shows that omitting phase transformation effects almost doubles the depth to which the residual stress remains compressive. The transformation strain and latent heat of transformation can both be seen to have a significant influence on the residual stress. The latent heat of transformation can also be seen from Figure 9a to advance the collapse of the vapour blanket, causing an increase in the heat extracted during the quenching stage.

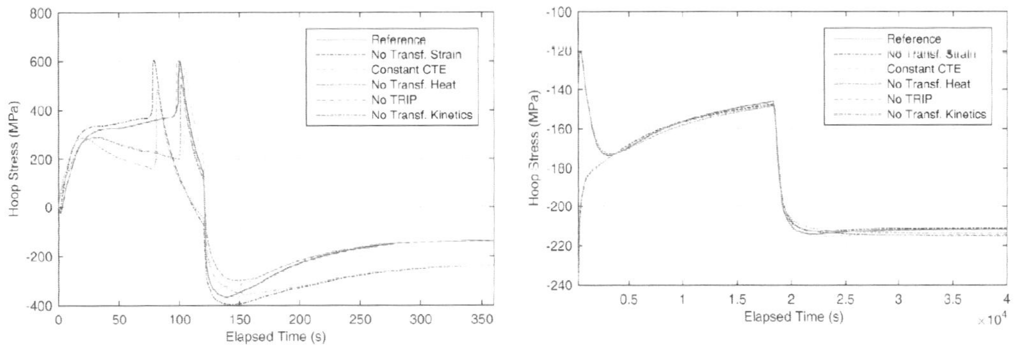

(a) Surface – Quench and Dwell (b) Surface – Anneal and Cooling

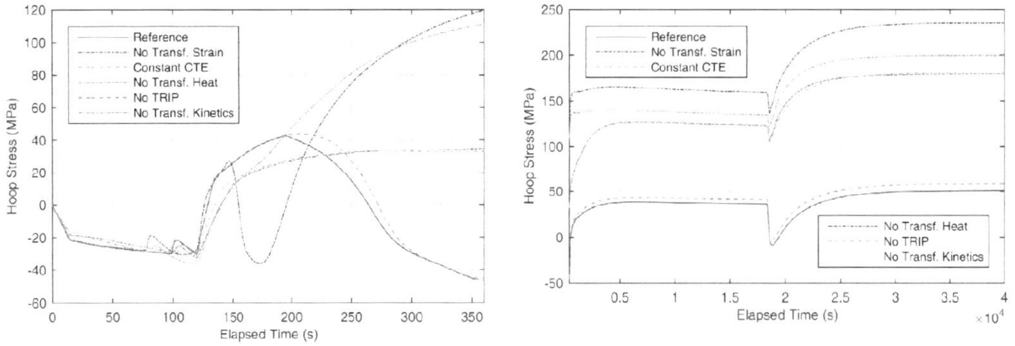

(c) 63 mm – Quench and Dwell (d) 63 mm – Anneal and Cooling

Figure 9: Hoop Stress History for Different Transformation Kinetics Considerations

Boundary Conditions

Insofar as boundary condition variations are concerned, the absence of radiation seems to produce the greatest distortion to the residual stress (see Figure 7c). The use of a constant HTC also has a marked influence and seems to shift the entire stress curve closer to the surface. The greatest deviation can be seen in the range from 15 mm to 25 mm below the tread surface. The heat generation due to plastic deformation has a negligible effect and it would not be necessary to consider this in future simulations, allowing a decoupled analysis.

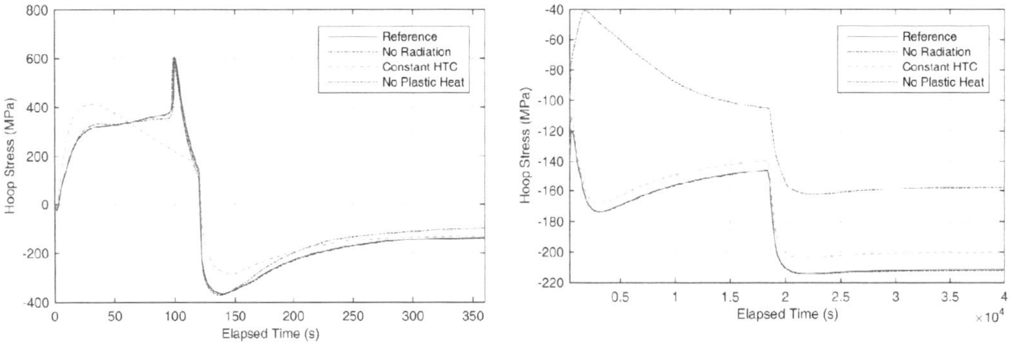

(a) Surface – Quench and Dwell (b) Surface – Anneal and Cooling

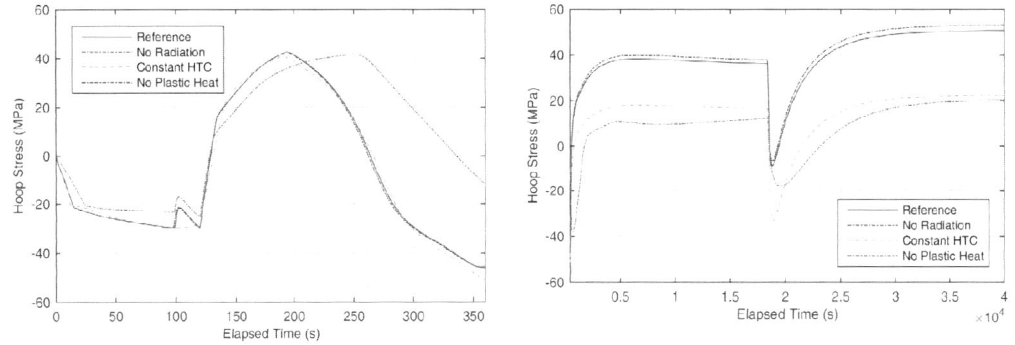

(c) 63 mm – Quench and Dwell (d) 63 mm – Anneal and Cooling

Figure 10: Hoop Stress History for Different Boundary Conditions

Important Parameters

It is clear that certain approximations and simplifications are more detrimental to accurate results than others.

As already observed by Kuhlman *et al* [10], material creep plays an important part in predicting the residual stress level. Creep does not seem to have a great influence over the stress pattern, only the actual stress values. The hardening model, as well as the stress strain curve, exert a great influence, especially close to the tread. Due to the importance of strain hardening, the simple bilinear hardening model used in this work might be inadequate for quantitative predictions if the material exhibit highly non-linear hardening behaviour.

The only transformation parameters that exhibit a strong influence on the residual stress are the latent heat of phase transformation and the volume change associated with phase transformation. It is important to note that these transformation effects are even important for qualitative indications of the stress pattern. Given the challenges of including transformation induced plasticity and its almost indiscernible impact, together with the build-up of uncertainty in complex models like these, it does not seem worthwhile to include transformation induced plasticity when only the residual stress is of interest.

Radiation can be seen to have a significant effect and should be included even for merely qualitative results. Approximating the spray quenching HTC with a constant value is seen to have a marked effect, albeit over a relatively narrow radial zone. Furthermore, due to the wildly varying heat flux and cooling rates close to the tread, it might have a significant effect if the phase transformation is explicitly simulated.

CONCLUSION

A nominal case for simulating the development of the residual stress field due to tread quenching of a railway wheel is presented. Variations in model parameters relating to material behaviour, phase transformations and boundary conditions are thereafter implemented and the results of these simulations are presented. Phase transformation effects appear to exert the highest influence on the residual stress field, with transformation strain and latent heat of transformation being very important. The strain hardening and creep behaviour of the material also have a significant influence on the residual stress field. The hardening models evaluated were only adequate for qualitative simulations and further research is needed in which a more accurate stress-strain relationship is used. The heat transfer boundary conditions exhibit marked influence on the simulation results.

REFERENCES

1. BS EN 13979-1, in Railway applications - Wheelsets and bogies - Monobloc wheels - Technical approval procedure - Part 1: Forged and rolled wheels (2011).

2. UIC 510-5, in Technical Approval of Solid Wheels (2003).

3. Matjeke, V. J. & Mabaso, F. G., in Transnet Freight Rail Specification for the Supply of Cast Wheel for Trailing Stock (2011).

4. Portesi, M., Gallo, R. & Lombardo, F. (2005).

5. Ha, M.-T. & Kang, C.-G., Fatigue Analysis of Railway Wheels according to UIC Standards, presented at 12th International Conference on Control, Automation and Systems, 2012 (unpublished).

6. Gordon, J. & Perlman, A. B., Estimation of Residual Stresses in Railroad Commuter Car Wheels Following Manufacture. Proceedings of the International Mechanical Engineering Congress and Exhibition, ASME RTD 15, 13-18 (1998).

7. Ariza, E. A., Martorano, M. A., de Lima, N. B. & Tschiptschin, A. P., Numerical Simulation with Thorough Experimental Validation to Predict the Build-up of Residual Stresses during Quenching of Carbon and Low-Alloy Steels. ISIJ International 54 (6), 13961405 (2014).

8. Song, K. J. et al., Constitutive model coupled with mechanical effect of volume change and

transformation induced plasticity during solid phase transformation for TA15 alloy welding. Applied Mathematical Modelling (2014).

9. Mackerle, J., Finite element analysis and simulation of quenching and other heat treatment processes - A bibliography (1976-2001). Computational Material Science 27, 313-332 (2003).

10. Kuhlman, C., Schitoglu, H. & Gallagher, M., The Significance of Material Properties on Stresses Developed During Quenching of Railroad Wheels. Proceedings of the Joint ASME IEEE Railroad Conference, 55-63 (1988).

11. Maintenance, & Management}, E., Tech. rep., 2012.

12. ASM International, ASM Handbook, Volume 01 - Properties and Selection: Irons, Steels, and High-Performance Alloys (ASM International, 1990).

13. Lement, B. S., Distortion in tool steels (American Society for Metals, 1959).

14. Jung, M., Kang, M. & Lee, Y.-K., Finite-element simulation of quenching incorporating improved transformation kinetics in a plain medium-carbon steel. Acta Materialia 60, 525-536 (2012).

15. Greenwood, G. W. & Johnson, R. H., The deformation of metals under small stresses during phase transformations. Proceedings of the Royal Society A 283 (1394), 403-422 (1965).

16. LeBlond, J. B., Mottet, G. & Devaux, J. C., A theoretical and numerical approach to the plastic behaviour of steels during phase transformations - I. Derivations of general relations. Journal of the Mechanics and Physics of Solids 34 (4), 395-409 (1986).

17. LeBlond, J. B., Mottet, G. & Devaux, J. C., A theoretical and numerical approach to the plastic behaviour of Steels during phase transformations - II. Study of classical plaplastic for ideal-plastic phases. Journal of the Mechanics and Physics of Solids 34 (4), 411-432 (1986).

18. LeBlond, J. B., Devaux, J. & Devaux, J. C., Mathematical modelling of transformation plasticity in steels I: Case of ideal-plastic phases. International Journal for Plasticity 5, 551-572 (1989).

19. Leblond, J. B., Mathematical modelling of transformation plasticity in steels II: Coupling with strain hardening phenomena. International Journal for Plasticity 5, 573-591 (1989).

20. Şimşir, C. & Gür, C. H.,

21. MSC Software, Marc 2013.1 Volume D: User Subroutines and Special Routines, 2013.

22. Çengel, Y. A. & Ghajar, A. J., Heat and Mass Transfer - Fundamentals and Applications (McGraw-Hill, 1998).

23. Hall, D. D., Mudawar, I., Morgan, R. E. & Ehlers, S. L., Validation of a Systematic Approach to Modeling Spray Quenching of Aluminum Alloy Extrusions, Composites, and Continuous Casting. Journal of Materials Engineering and Performance 6 (1), 77-92 (1997).

24. Rammerstorfer, F. G., Fischer, D. F., Mitter, w., Bathe, K. J. & Snyder, M. D., On thermo-elastic-plastic analysis of heat-treatment processes including creep and phase changes. Computers & Structures 13, 771-779 (1981).

EXPERIMENTAL INVESTIGATION OF THE EFFECTIVE THERMAL CONDUCTIVITY IN THE NEAR-WALL REGION OF A PACKED PEBBLE BED

M De Beer*, CG Du Toit, H Dreyer

North-West University, Potchefstroom, South Africa

*maritz.db@gmail.com

ABSTRACT

Experimental test facilities are used to determine the effective thermal conductivity of packed pebble beds; however the experimental results reported are often case specific and do not necessarily account for wall effects in the near-wall regions. Not all of the experimental data reported in current literature provide associated uncertainties, thus limiting the use of the data to validate new effective thermal conductivity models. Recently the Near-wall Thermal Conductivity Test Facility (NWTCTF) was developed at the North-West University in Potchefstroom, South Africa. The purpose of the NWTCTF is to perform detailed investigations of the conduction and radiation heat transfer phenomena in the near-wall region of a packed pebble bed at higher temperatures. Experimental tests were performed for structured as well as randomly packed pebble beds at temperatures of up to 800°C. This paper describes the test facility as well as the experimental tests performed and reports the experimental results obtained together with associated uncertainties. The results clearly demonstrate the effect of the near-wall region on the effective thermal conductivity.

KEYWORDS

Effective Thermal Conductivity, Packed Pebble Beds, Experimental Investigation, Heat Transfer

INTRODUCTION

Various thermal-fluid industrial applications involve heat transfer processes through packed beds of spherical particles, including catalytic reactors, drying processes and pebble bed gas-cooled reactors (PBRs) [1]. The effective thermal conductivity is an important parameter that is representative of the overall heat transfer through a packed bed of spheres. During upset conditions in a PBR, when convective cooling is absent, radiation and conduction are the two dominant heat transfer mechanisms that contribute to the overall effective thermal conductivity.

The effective thermal conductivity forms a part of the self-acting decay heat removal chain in a PBR [2]. In order to have a fundamental understanding of the heat transfer phenomena and to determine the inherent safety characteristics of the reactor design, one must be able to accurately predict the effective thermal conductivity. In order to achieve this emphasis must be placed on the heat transfer at very high temperatures in the near-wall regions of a packed pebble bed. It is important to consider higher temperatures as the thermal radiation's contribution to the overall heat transfer increases significantly at temperatures above 650 °C [1,4,5,6]. The porous structure of the packed bed also changes significantly in the region near any wall, this variation in packing structure is known as the wall effect and influences the magnitude of the effective thermal conductivity in the wall region.

To ensure reliable results are obtained with new and existing effective thermal conductivity models one must be able to validate and verify the results with the use of experimental data. Van Antwerpen [2] concluded that far more experimental research has been done at low temperatures with small diameter spheres than at high temperatures with larger spheres [7,8,9]. Experimental test facilities including the High Temperature Oven (HTO) [10], the SANA-I test facility [11] and the High Temperature Test Unit (HTTU) [12] were used to conduct experimental tests on randomly packed pebble beds with larger spherical particles (diameters ranging from 30 mm to 60 mm) at very high temperatures of up to 1500°C.

None of the test facilities focused specifically on the heat transfer in the near-wall region of a packed pebble bed. Van Antwerpen [2] also noted that care should be taken when comparing the experimental results with the effective thermal conductivity results of developed models as experimental test facilities used are mostly case specific and cannot necessarily be applied to a general model. Another shortcoming is the fact that not all of the experimental data reported in current literature provide associated uncertainties, thus limiting the use of the data to validate new models. Rousseau *et al.* [12] proposed and applied a comprehensive method to the HTTU experimental data in order to derive the effective thermal conductivity of the pebble bed together with associated uncertainties. Therefore the HTTU results could be used with confidence in the verification and validation of new effective thermal conductivity models.

A detailed investigation of the heat transfer in the near-wall region was not included in the HTTU experiments and it was concluded that further studies must be done to investigate the influence of the wall effects on the bed effective thermal conductivity [2]. This led to the development of a new experimental test facility, the Near-wall Thermal Conductivity Test Facility (NWTCTF), at the North-West University in Potchefstroom, South Africa. The purpose of the NWTCTF is to perform detailed investigations of the conduction and radiation heat transfer phenomena in the near-wall region of a packed pebble bed at higher temperatures.

The aim of this paper is to provide an overview of the NWTCTF test facility and the experimental tests performed. The effective thermal conductivity results obtained for the three separate experimental tests each with a different packing structure namely random, Simple Cubic (SC) and Body Centred Cubic (BCC) will be presented together with associated uncertainties. The purpose of the paper is not to perform a detailed analysis and evaluation of the results.

EXPERIMENTAL TEST FACILITY

The NWTCTF test facility was specifically developed to conduct physical tests to characterise the total thermal conductivity in the near-wall region. Fig. 1 shows a schematic representation of a side view of the NWTCTF test section. The test section has a cubical configuration with side lengths of 420 mm in which the spheres are placed to form the packed bed. Depending on the packing configuration used the packed bed consisted of approximately 350 machined graphite spheres. Each sphere has a diameter of 60 mm and was inherited from the PBMR HTTU test facility [13].

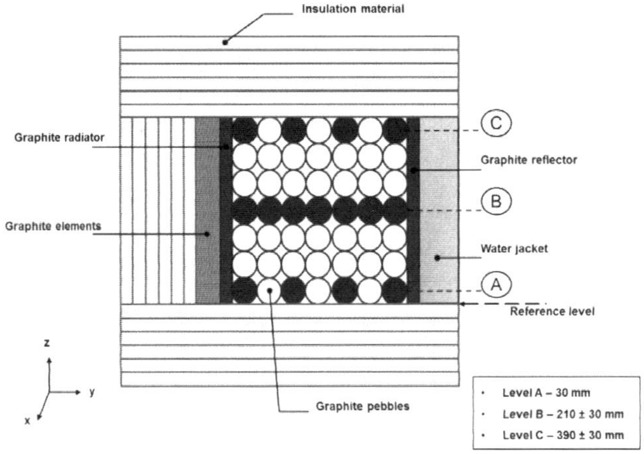

Fig. 1 Representation of a vertical section through the NWTCTF test section [14].

On one side of the bed there is an inner graphite "radiator" wall with nine graphite electrodes (or heater elements) positioned on the outside right next to it. These were designed so that it can heat up the radiator wall to approximately 1600°C. On the opposite side of the bed there is an outer graphite "reflector" wall with a water jacket positioned right up against it. The water jacket was designed so that it can maintain the reflector wall temperature at an almost homogeneous temperature. The rate of heat extracted via the water jacket can be determined by measuring the temperature difference and mass flow rate of the water flowing through the water jacket. The entire test section is enclosed by a layer of insulation material with a thickness of approximately 230 mm in an effort to minimise the heat losses to the surrounding environment. The various components of the NWTCTF test section together with the insulation layers can be seen in Fig. 2.

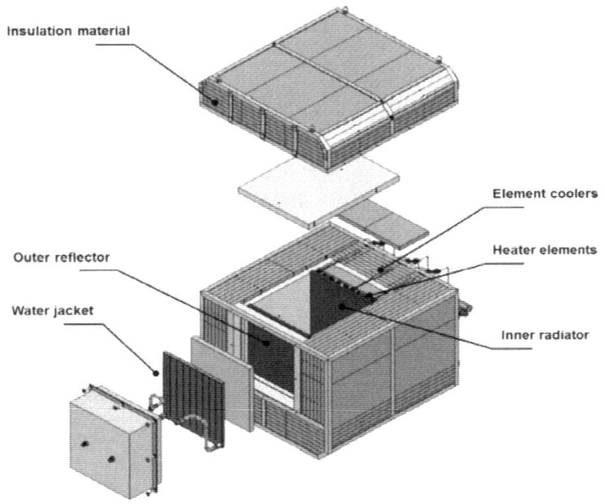

Fig. 2 Exploded view of the NWTCTF test section [14].

The NWTCTF test section fits inside a pressure vessel, as can be seen in Fig. 3, in order for a vacuum to be drawn on the entire assembly. This is necessary for the preparation as well as the execution of the experimental tests. During the preparation stage of the experimental tests the entire test section is subjected to a rigorous bake-out procedure in order to remove as much of the moisture naturally contained inside the porous insulation material. All tests are also completed whilst a near-vacuum nitrogen environment is maintained inside the pressure vessel to eliminate the effects of natural convection inside the pebble bed.

Fig. 3 Photographs of the NWTCTF test section positioned inside (left) and outside (right) the pressure vessel [14].

The various auxiliary systems of the NWTCTF include the vacuum, nitrogen supply, heating and water cooling systems. The water cooling system supplies to both the water jacket heat exchanger and the element coolers. Besides the main instrumentation and control system the NWTCTF also has two separate safety systems to ensure the safety of the operators and to protect the system during testing.

Thermocouples were placed inside the packed bed along the y-coordinate parallel to the primary heat flux direction at three different axial levels as shown in Fig. 1 by the dark coloured spheres. For levels A and C there were two graphite spheres with thermocouples placed at the centre of each sphere. As level B represents the area of primary interest for the heat transfer through the packed bed, 12 spheres instrumented with thermocouples were spaced throughout the bed at this level. Two thermocouples were also embedded in the inner graphite radiator and one in the outer graphite reflector.

The thermocouple wires were threaded through the insulation walls to the middle of the packed bed and inserted into the various spheres taking care to minimise the disruption to the packing structure of the bed. Fig. 4 shows a photograph of a half full randomly packed bed with threaded thermocouple wires. A small hole with a depth of 30 mm and a diameter of 1 mm was drilled into the graphite spheres that were instrumented to ensure a tight fit once the thermocouple wire was inserted into the sphere. Two types of thermocouples were used for the temperature measurements namely type B Platinum-Rhodium and Type K Inconel thermocouples with a maximum operating temperature of 1750°C and 1350°C and an instrument uncertainty of ±5°C and ±3°C respectively.

Fig. 4 Half full randomly packed bed of spheres showing thermocouple wires [14].

It is important to note here that in order to correctly characterise the wall effect, one would ideally only have the packing structure influenced by the walls at the inner radiator and/or outer reflector, which is perpendicular to the heat flux, and not at the top, bottom and sides. However, for the current tests, no special measures were taken to eliminate the wall effects at the top, bottom and sides as the focus was placed on getting the test facility in good working order and obtaining a first set of experimental results.

EXPERIMENTAL METHOD

At the start of each experimental test the insulation surrounding the packed bed was subjected to a bake-out sequence. During the bake-out sequence the insulation material was heated up uniformly to 300°C using three small electrical heater units with elements placed throughout the insulation material at evenly distributed positions. As the moisture was released over time it was removed from the vessel by maintaining a near-vacuum environment inside the vessel and continuously flushing the vessel with nitrogen gas.

After the bake-out sequence was completed the experimental tests were conducted with the surface temperature of the inner radiator maintained at 400, 500, 600, 700 and 800°C respectively. Once the desired surface temperature was reached and the system reached complete steady-state the experimental data was gathered. In total five experimental tests were conducted for three different packing structures. Two separate but identical experimental tests were conducted at different times for each of the temperature cases and packing structures to prove the repeatability of the results.

After the first test the plant was switched off and allowed to cool down completely. The vessel was opened and the pebble bed was unpacked to check for any signs of corrosion of the graphite pebbles as well as the condition of the electrical connections of the heater elements. The test section was packed with graphite spheres for a second time following a similar packing structure to that of the first test. At a much later time the plant was switched on again to conduct the second experimental test. Due to time constraints only one test was performed for the BCC packed bed.

EXPERIMENTAL DATA AND UNCERTAINTY ANALYSIS

The uncertainty associated with the average value of each of the measurements was calculated to determine the confidence interval associated with the measured experimental data. The total measurement uncertainty consisted of two types of uncertainties namely the instrument uncertainty and the statistical variance and was derived using the methodology proposed by Rousseau *et al.* [12]. For a more detailed description of the methodology consult Rousseau *et al.* [12]. The drift uncertainty was not considered in the calculation of the total measurement uncertainty; however it must be included in future tests.

According to Rousseau *et al.* [12] the derivation of the effective thermal conductivity, $k_{eff}(y)$, from the measured temperature distribution is based on a simple Fourier conduction rate equation as shown in Eq. (1):

$$k_{eff}(y) = \frac{\dot{Q}(y)}{A\dfrac{dT}{dy}} \tag{1}$$

with $\dot{Q}$ the heat transfer rate, T the temperature, A the applicable area in the pebble bed through which the heat transfer is taking place and y the coordinate perpendicular to the area. Thus, in order to determine the effective thermal conductivity as a function of position it was necessary to first determine the heat transfer rate as well as the temperature gradient through the packed bed as a function of position.

The calculation of the heat transfer distribution was based on the heat flow paths shown schematically in Fig. 5. The primary measurement is the heat extracted through the outer reflector via the water jacket, $\dot{Q}_{out}$. The bed was discretised into 56 equally spaced increments along the y-coordinate direction. For each of the increments an estimated heat loss value was calculated for the various insulated walls surrounding that increment based on the measured temperature differences across the insulation layers. The heat transfer through each increment was then calculated sequentially as the average of the heat transfer through the cold face and the heat transfer through the hot face, which is equal to the cold face plus all the losses. Once again, all the relevant instrument and measurement uncertainties were taken into account in order to determine the overall measurement uncertainty of the heat transfer rates.

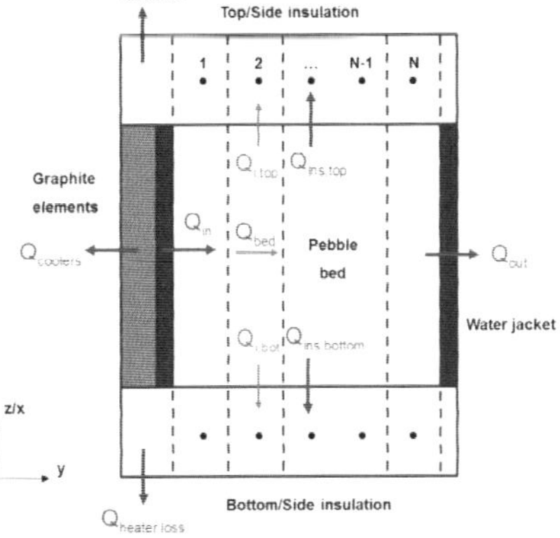

Fig. 5 Schematic representation of the heat flow paths through the test section.

To obtain the uncertainty of the calculated effective thermal conductivity values, the uncertainty of the bed heat transfer as well as the uncertainty of the temperature gradient function had to be known. The uncertainty associated with the effective thermal conductivity values was calculated using Eq. (2):

$$u(k_{eff}) = \sqrt{\left[\frac{\partial(k_{eff})}{\partial Q_{bed}} \cdot u(Q_{bed})\right]^2 + \left[\frac{\partial(k_{eff})}{\partial(dT/dr)} \cdot u(dT/dr)\right]^2} \tag{2}$$

with the partial derivatives shown in Eq. (3) and (4) respectively.

$$\frac{\partial(k_{eff})}{\partial Q_{bed}} = \frac{-1}{A\,dT/dr} \tag{3}$$

$$\frac{\partial(k_{eff})}{\partial(dT/dr)} = \frac{Q_{bed}}{A(dT/dr)^2} \tag{4}$$

RESULTS

The measured temperature and heat transfer distribution results as well as the derived effective thermal conductivity are presented in the following section. Due to space constraints only the results for the 800°C case of each experimental test for the various packing structures will be presented. The measured temperature distributions along the length of the test section from the inner radiator wall to the outer reflector wall are shown in Fig. 6. The inner radiator surface temperature was maintained at 800.1, 809.9 and 824.1°C for the random, SC and BCC packed bed respectively, resulting in an outer reflector wall surface temperature of 162.9, 149.9 and 149°C. The maximum measurement uncertainty for all of the test results is ±5.0°C, thus the plotted error bars are barely visible.

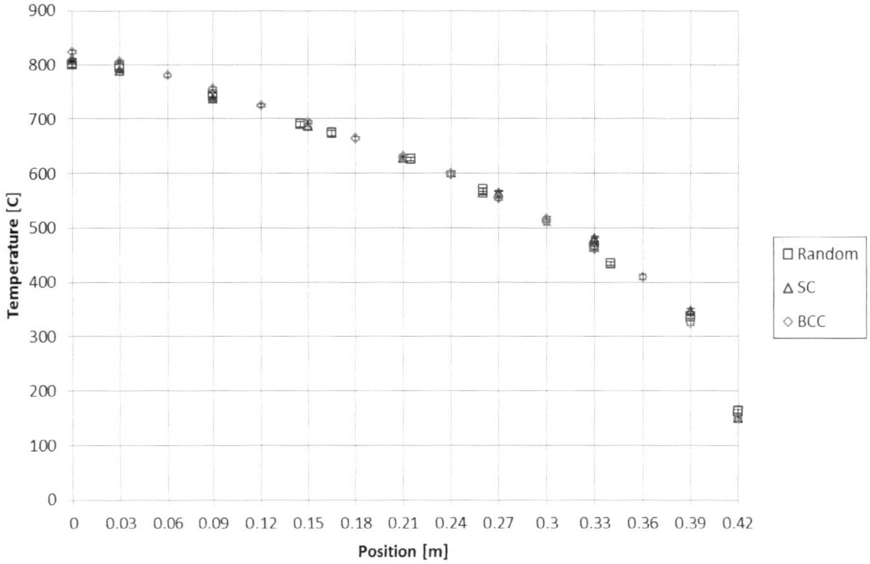

Fig. 6 Measured temperature distributions through the test section for a random, simple cubic and body centred cubic packing structure [14, 15].

The heat transfer distributions through the test section for the three different packing configurations with associated uncertainties are shown in Fig. 7. For the different experimental tests the heat transfer distribution is in the same order of magnitude with the error bands overlapping. As can be seen in Fig. 7 the heat transfer distributions for the structured packed beds are higher than that of the randomly packed bed. The results show the influence of the changes in packing structure on the heat transfer phenomena as the structured packed beds have a higher porosity than the randomly packed bed. The influence of the changes in packing structure is especially prominent in the wall region next to the inner reflector at higher temperatures.

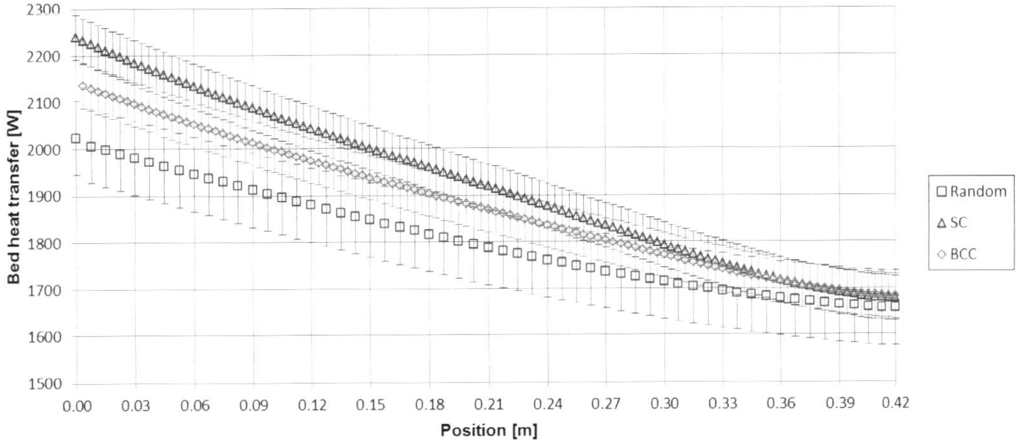

Fig. 7 Heat transfer distributions through the test section for a random, simple cubic and body centred cubic packing structure [14, 15].

The general form of the functions derived to describe the heat transfer distributions is shown in Eq. (5):

$$Q_{bed}(y) = c_0 y^3 + c_1 y^2 + c_2 y + c_3 \tag{5}$$

with $c_0 - c_3$ the polynomial coefficients specific to each test case. Table 1 presents the calculated values for the polynomial coefficients as well as the heat extracted via the water jacket and the heat losses together with uncertainties. The amount of heat extracted via the water jacket is similar for all three tests as the temperature difference between the water inlet and outlet temperatures was controlled at approximately 10°C during all tests. The heat losses through the insulation side walls are higher for the structured packed beds. The structured packed beds had more contact points between the pebbles and the side walls, thus providing more conduction paths through which heat losses could occur.

Table 1 Heat transfer results.

Packing structure	Random	SC	BCC
c_0	1409.152	1740.044	1283.017
c_1	139.039	-162.293	167.455
c_2	-1212.280	-1578.056	-1381.443
c_3	2041.778	2234.781	2132.127
$u(Q_{bed})$	79.6 W	47.9 W	47.6 W
$Q_{water\ jacket}$	1657.6 ± 73.1 W	1677.5 ± 48.9 W	1681.1 ± 48.9 W
Q_{losses}	366.7 ± 31.4 W	445.7 ± 6.2 W	454.0 ± 6.3 W

A fifth order polynomial was fitted through the measured data for each of the experimental tests to derive a function describing the temperature distribution through the bed. The temperature distribution results for each of the test cases are shown in Fig. 8.

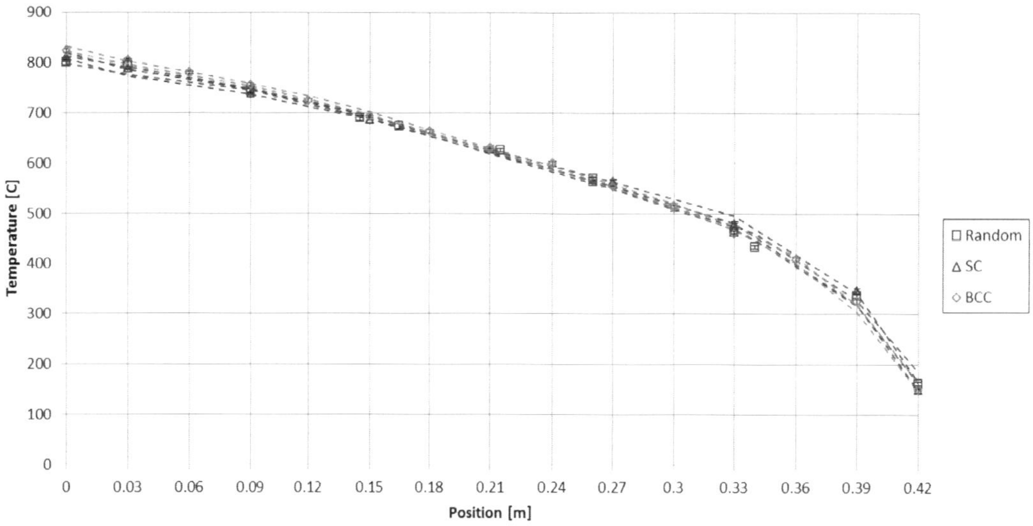

Fig. 8 Fifth order polynomial curve fit through the measured experimental data for each of the packing structures considered [14, 15].

The temperature gradient functions with its uncertainty bands for each of the test cases is shown in Fig. 9. It can be seen that the temperature gradient is very similar for the different packing structures with slight differences that can be observed in the near-wall regions.

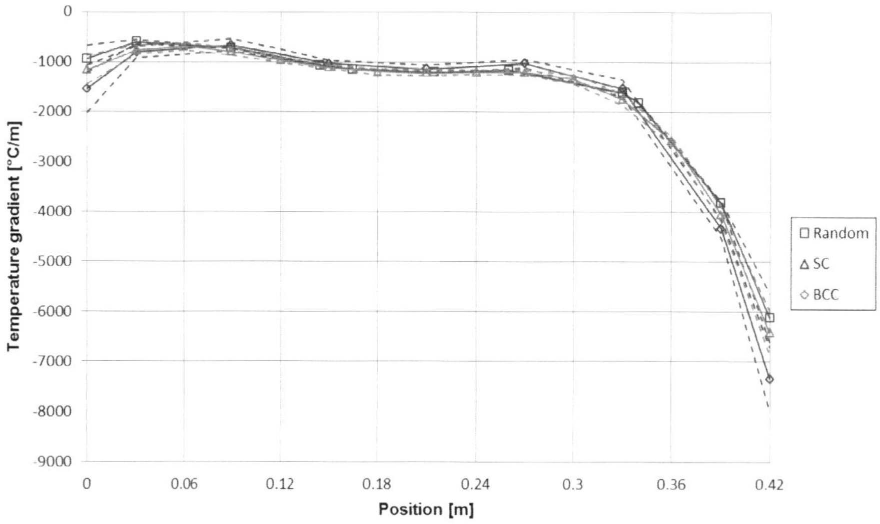

Fig. 9 Temperature gradient functions with uncertainties for each of the test cases [14, 15].

Now, with the temperature gradient functions and the heat transfer distributions known the bed effective thermal conductivity can be derived. The effective thermal conductivity results as a function of position and temperature are shown in Fig. 10 and Fig. 11 respectively.

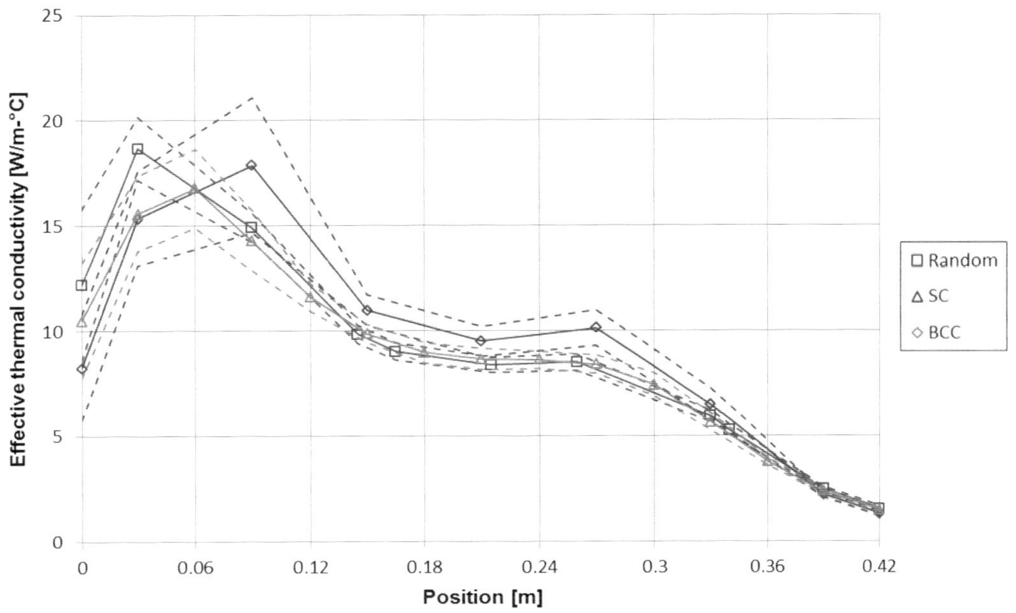

Fig. 10 Effective thermal conductivity results as a function of position for each of the test cases [14, 15].

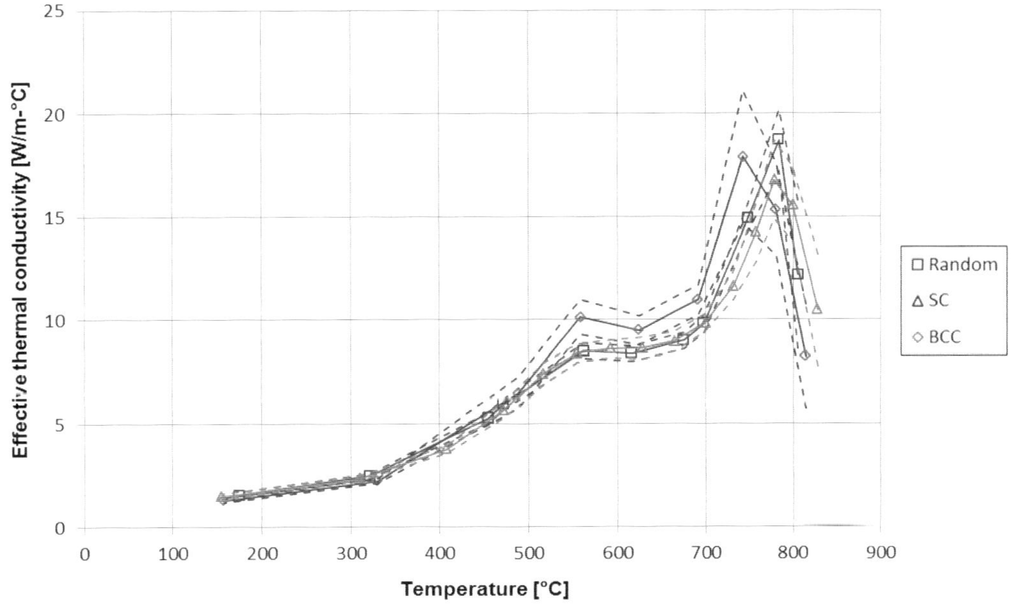

Fig. 11 Effective thermal conductivity results as a function of temperature for each of the test cases [14, 15].

The effective thermal conductivity results for the three different packing structures are very similar for the bulk region in the bed, between 0.09 m and 0.33 m. The structured beds have effective thermal conductivity values that are slightly higher than that of the randomly packed bed. The wall effect present in the near-wall region next to the inner reflector, between 0.0 m and 0.09 m, has a definite effect on all three of the packed beds. This can be seen in the notable difference between the magnitude of the effective thermal conductivity in the bulk and near-wall regions, with a prominent reduction in the total effective thermal conductivity next to the inner reflector wall.

It is interesting to observe in Fig. 10 that there seems to be a two and half sphere region near the hot wall, a two sphere core region and a two and a half sphere region near the cold wall. This is in accordance with the variation in porosity, or porous structure. The relationship between the porous structure, overall effective thermal conductivity, the radiation and conduction contributions and temperature should be studied in detail.

One would expect the effective thermal conductivity curve of the structured packed beds to have a more linear trend, however the wall effect as well as the effect of radiation becomes more significant at higher temperatures. It must be kept in mind that the insulation walls of the test section also had an effect on the results and one would ideally want to eliminate the presence of these effects in future tests. For the SC packed bed it was also noted that irregularities were present in the packed bed, in which gaps existed between some of the adjacent spheres. This will also have a direct effect on the predicted conduction heat transfer as the conduction phenomenon was essentially eliminated between the spheres where the irregularities were present.

CONCLUSION

Detailed experimental data with associated uncertainties are important for the verification and validation of effective thermal conductivity models. A lot of experimental work has been done on the heat transfer in packed pebble beds; however a need exists for experimental tests focusing on high temperatures and the near-wall region.

The newly built NWTCTF test facility was used to determine the effective thermal conductivity of packed pebble beds with three different packing structures namely random, SC and BCC. From the results it is clear that the near-wall effects have a significant influence on the effective thermal conductivity results for all of the packed beds. However, it is important to note that the setup of the test section still included multiple wall effects and some irregularities were present in the packing structure of the structured beds. For future experimental tests one would want to minimise or eliminate the multiple wall effects and ensure that no irregularities exist in the packed bed.

The authors have recently developed a numerical model based on the experimental data to predict the effective thermal conductivity and separate the radiation and conduction components thereof [16] and are currently busy with the refinement of the model.

ACKNOWLEDGEMENTS

This work is based upon research supported by the South African Research Chairs Initiative of the Department of Science and Technology and National Research Foundation (Grant No 61059). Any opinion, finding and conclusion or recommendation expressed in this material is that of the author(s) and the NRF does not accept any liability in this regard.

NOMENCLATURE

A	Area
k_{eff}	Total effective thermal conductivity
Q	Heat transfer rate
T	Temperature
y	y - Coordinate

ABBREVIATIONS

BCC	Body Centred Cubic
HTO	High Temperature Oven
HTTU	High Temperature Test Unit
NWTCTF	Near-Wall Thermal Conductivity Test Facility
PBR	Pebble Bed gas-cooled reactor
SC	Simple Cubic

REFERENCES

1. Zhou, J., Yu, A. & Zhang, Y. 2007. A boundary element method for evaluation of the effective thermal conductivity of packed beds. *Journal of Heat Transfer*, 129:363-371.
2. Van Antwerpen, W. 2009. Modeling the effective thermal conductivity in the near-wall region of a packed pebble bed. Potchefstroom, South Africa: North-West University. (Thesis - PhD).
3. Van Antwerpen, W., Du Toit, C.G. & Rousseau, P.G. 2010. A review of correlations to model the packing structure and effective thermal conductivity in packed beds of mono-sized spherical particles. *Nuclear Engineering and Design*, 240:1803-1818.
4. Breitbach, G. & Barthels, H. 1980. The radiant heat transfer in the high temperature reactor core after failure of the afterheat removal systems. *Nuclear Technology*, 49:392-399.
5. Cheng, G.J. & Yu, A.B. 2013. Particle scale evaluation of the effective thermal conductivity from the structure of a packed bed: Radiation heat transfer. *Industrial & Engineering Chemistry Research*, 52(34):12202-12211.
6. Talukdar, P., Mendes, M.A.A., Parida, R.K., Trimis, D. & Ray, S. 2013. Modeling of conduction-radiation in a porous medium with blocked-off region approach. *International Journal of Thermal Sciences*, 72(2013):102-114.
7. Widenfield, G., Weiss, Y. & Kalman, H. 2003. The effect of compression and preconsolidation on the effective thermal conductivity of particulate beds. *Powder Technology*, 133(2003):15-22.
8. Mandal, D., Sathiyamoorthy, D. & Vinjamur, M. 2012. Experimental measurement of effective thermal conductivity of packed lithium-titanate pebble bed. *Fusion Engineering and Design*, 87(2012):67-76.
9. Béttega, R., Barrozo, M.A.S., Corrêa, R.G. & Freire, J.T. 2013. CFD simulation of heat transfer inside packed beds: Evaluation of effective thermal conductivity. *JP Journal of Heat and Mass Transfer*, 8(2):137-148.
10. Breitbach, G. & Barthels, H. 1980. The radiant heat transfer in the high temperature reactor core after failure of the afterheat removal systems. *Nuclear Technology*, 49:392-399.
11. Stöcker, B. & Niessen, H.F. 1997. The SANA-I-experiments for self-acting removal of the afterheat from a pebble bed.
12. Rousseau, P.G., Du Toit, C.G., Van Antwerpen, W. & Van Antwerpen, H.J. 2014. Separate effects tests to determine the effective thermal conductivity in the PBMR HTTU test facility. *Nuclear Engineering and Design*, 271:444-458.
13. Rousseau, P.G. & Van Staden, M. 2008. Introduction to the PBMR heat transfer test facility. *Nuclear Engineering and Design*, 238:3060-3072.
14. De Beer, M. 2014. Characterization of thermal radiation in the near-wall region of a packed pebble bed. Potchefstroom, South Africa: North-West University. (Dissertation - Masters).
15. Dreyer, H. 2015. Investigation into the conduction and radiation heat transfer in the near-wall region of a packed pebble bed. Potchefstroom: North West University. (Project Report).
16. Rousseau, P.G., Du Toit, C.G. & De Beer, M. 2015. A methodology for the characterization of the effective thermal conductivity in the near-wall region of a packed pebble bed. Paper presented at the International Congress on Advances in Nuclear Power Plants, Nice, France, 3-6 May.

A SYSTEM CFD SIMULATION OF THE EFFECT OF A PRESSURE PULSE ON THE PERFORMANCE OF A REACTOR CAVITY COOLING SYSTEM OF A VERY HIGH TEMPERATURE REACTOR

CG du Toit[1*], PG Rousseau[2], JS Jun[3], MH Kim[3]

[1]North—West University, Potchefstroom, South Africa

[2]University of Cape Town, Cape Town, South Africa

[3]Korea Atomic Energy Research Institute, Daejeon, Korea

[*]jat.dutoit@nwu.ac.za

ABSTRACT

The reactor cavity cooling system (RCCS) of a high temperature reactor must be able to operate naturally or passively safe to avoid accidents due to human error or the failure of components. The RCCS under consideration in this investigation is an air-cooled system with a U-tube-like configuration. It consists of an annular down-comer channel which is connected via a bottom-header to 220 riser tubes. The down-comer and risers are connected to the atmosphere via inlet and outlet manifold systems respectively. In this study the entire system is simulated using two network codes, GAMMA+ and FLOWNEX. The purpose of the study is to investigate the effect that a positive pressure transient acting on the outlet of the RCCS will have on the performance of the cooling system.

The agreement between the results obtained with the two codes is good and serves as a verification of the methods that are used. It has been found that the pressure increase may result in a reversal in the flow. When the pressure returns to its original value the flow does not reverse to the original flow direction. Although the heat removal is not severely affected, it leads to an unacceptable increase in the temperature of the concrete containment. This should be taken into account in the design of the RCCS system as it relates to the reactor safety case.

KEYWORDS

Reactor Cavity Cooling System, System CFD, Very High Temperature Reactor, Radiative heat transfer, Convection heat transfer

INTRODUCTION

Since the accidents at Three Mile Island nuclear power plant and at Fukushima nuclear power plant, much emphasis has been placed on the systems to remove decay heat effectively. The reactor cavity cooling system (RCCS) is an important part of these systems. To avoid accidents due to human error or the failure of components, the RCCS must be able to operate naturally or passively safe, particularly in the case of very high temperature reactors (VHTR). It is therefore understandable that natural processes be considered in this regard. The RCCS can therefore be a buoyancy driven water cooled system [1-2], a buoyancy driven air-cooled system [3-4], or a combined water-air-cooled buoyancy driven system [5-6].

Korea has chosen the 200 MW_{th} PMR (PMR200) General Atomic prismatic core VHTR reactor as the candidate reactor for the Nuclear Hydrogen Development and Demonstration (NHDD) project. The inlet temperature to the core is considered to be 490 ^{o}C and the outlet temperature from the core to be 950 ^{o}C [6]. Apart from removal of the residual and decay heat, it also requires the proper cooling of the reactor cavity under normal operating conditions. The RCCS is the only safety-related system for the residual and decay heat removal and its performance impacts directly on the maximum temperatures of the fuel, reactor pressure vessel (RPV) and concrete during off-normal conditions. The General Atomic air-cooled RCCS design concept, which is a completely passive system, has not yet been deployed in a VHTR. Therefore, experimental programs are under way within the Argonne National Laboratory (ANL) [4] and Korea Atomic Energy Research Institute (KAERI) [5] to demonstrate its viability and performance. Code-to-code benchmarks are also required to prove the system design tools for simulation of the RCCS natural circulation. This should include the cavity convection and the flow inside the RCCS ducts as affected by radiation heat transfer from the vessel.

In previous studies by Du Toit et al. [7] and Rousseau et al. [8] results obtained with the GAMMA+ [9] and Flownex [10] system or network codes of the expected flow and heat transfer in the RCCS were compared. The purpose of the current study is to investigate the effect that a positive pressure transient acting on the outlet of the RCCS will have on the performance of the cooling system using both GAMMA+ and Flownex.

RCCS LAYOUT

The layout of the RCCS that was studied in the GAMMA+ and Flownex simulations is described in the following paragraphs. Only the relevant parts of the geometry of the reactor, RCCS and reactor building are considered.

The layout of the manifold inlet/outlet structure of the RCCS is shown in Fig. 1, whilst the cross section in Fig. 2 shows the radial layout of the reactor vessel and RCCS ducts [11]. Fig. 1(a) shows the duct arrangement for the entire system. Note that in parts of the system the hot (red) outlet ducts are contained inside the cold (blue) inlet ducts which form a concentric annulus around the hot ducts. Also shown in the Fig. 1(a) is the concrete wall (grey) of the containment building. Fig 1(b) shows the arrangement of the hot ducts. Not shown in the figure is the bottom header that connects the down-comer annulus to the riser tubes or ducts. The heated air rises in 220 riser ducts or tubes that are arranged around the reactor cavity, is then collected in an outlet duct, header and manifold structure and finally exits through the top of the outlet chimneys to the atmosphere. The arrangement of the cold ducts is shown in Fig. 1(c). The atmospheric air enters at the bottom of the inlet chimneys (the tops are closed off), flows down the inlet manifold, header, inlet duct and is then distributed in the down-comer annulus. The air then flows through the bottom header to the riser tubes.

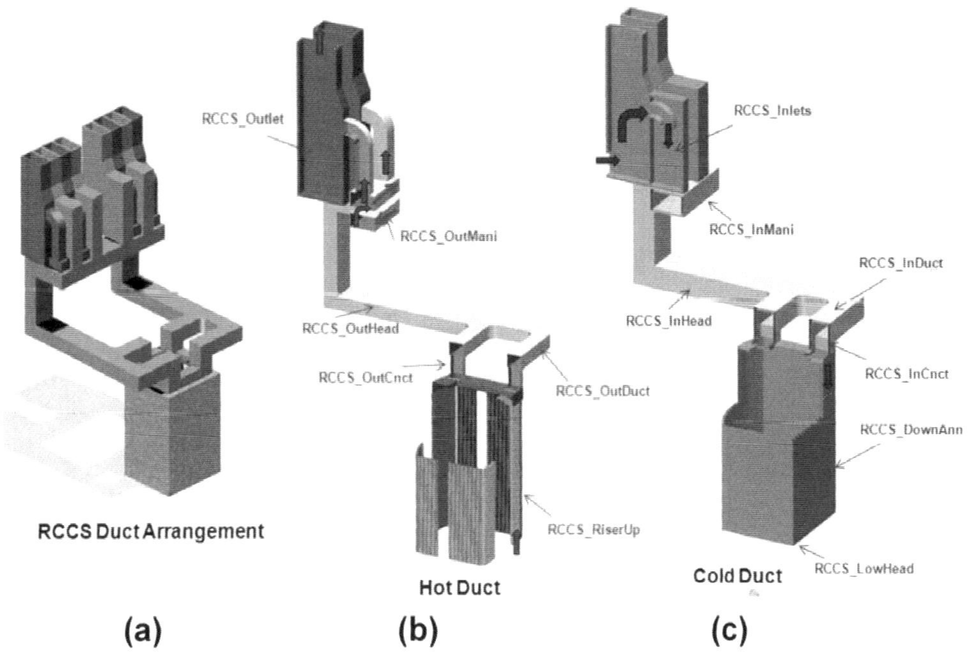

Figure 1. Manifold inlet/outlet structure of the air-cooled RCCS [11].

Fig. 2 shows only a part of the cross section through the RPV, reactor cavity, RCCS and concrete wall and is not to scale. The arrangement shown in Fig. 2 has a vertical length of 18.5 m. The gaps between the riser tubes can be seen where some of the radiant heat from the RPV can pass through to the down-comer wall. For this reason the down-comer wall is protected by an insulation layer. Heat can also be reflected from the insulation layer onto the back of the riser tubes. The riser tubes and the down-comer walls are made of SS304 stainless steel, whilst the RPV is made of SA508 steel.

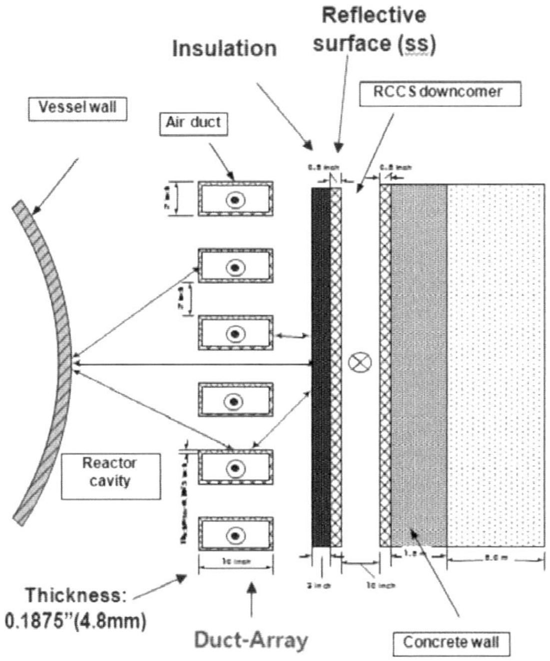

Figure 2. Radial layout of reactor vessel and the air-cooled RCCS [11].

Two benchmark calculations were considered as characterised by the temperature of the inner wall of the RPV. The boundary and calculation conditions for the two cases are summarized in Table 1.

Table 1. Boundary and calculation conditions for two benchmark cases considered

RPV Inner Surface Temperature	250 °C and 350 °C
Ambient Air Inlet Temperature	40 °C
Ambient Air Inlet Pressure	100 kPa
Outer Concrete Boundary	adiabatic
Emissivity of all solid parts	0.8
Emissivity of microtherm insulation	0.1

The pressure transient was simulated by increasing the pressure at the hot outlet from 100 kPa to 100.14 kPa (corresponding to a wind speed of approximately 60 m/s) over a period of 50 s, whilst maintaining the pressure at the cold inlet at 100 kPa. The pressure at the outlet was then maintained at 100.14 kPa for 150 s and then decreased again to 100 kPa over a period of 50 s.

MODELS

Only a brief description will be given of the GAMMA+ and Flownex models. A more detailed discussion of the models can be found in Du Toit et al. [7] and Rousseau et al. [8].

GAMMA+ [9] is a system thermo-fluid code that solves the equations for the multi-dimensional heat conduction, fluid flow, chemical reactions, multi-component molecular diffusion, and radiation heat transfer. For the integration of the partial differential equations a semi-implicit ICE-type (Implicit Continuous-fluid Eulerian [12] staggered grid approach with adaptive time-stepping is used. Together with the multi-dimensional analysis features, GAMMA+ has a one-dimensional analysis capability for a general pipe network flow to simulate the complicated core flow distributions including the coolant channel flow, the axial bypass flow and the cross flow in the VHTR [13]. Flow paths are defined by fluid blocks (control volumes) which are connected by means of junction blocks (pipes/ducts). Pipe walls and solid structures are defined by means of solid blocks (control volumes). Convection heat transfer between the fluid and the solids is achieved by linking the relevant fluid and solid surfaces and then defining the correlation to be used to obtain the Nusselt number. Radiation heat transfer is realized by specifying the relevant surfaces and providing the associated view factor matrix. The flow through the inlet and outlet manifolds, down-comer and riser tubes are modeled using a one-dimensional approach, whilst the flow in the cavity is modeled using a three-dimensional approach. The RPV wall, riser tube walls, down-comer walls and the concrete wall are modeled using a three-dimensional approach. Since the previous studies [7,8] improvements were implemented in GAMMA+ for the calculation of the convection heat transfer associated with the surfaces of the reactor cavity.

Flownex is a systems CFD (pipe network) code that solves the transient one-dimensional equations for the conservation of mass, momentum and energy employing a finite volume based implicit pressure correction method [14], using the θ-method [15] to perform the time integration. A weighting factor between zero (fully explicit) and one (fully implicit) is provided that determines the degree of forward and backward Euler integration, and a value of 0.7 is usually applied since it produces a good balance between accuracy and stability. A Flownex model consists of 1D elements and the associated nodes (control volumes). In the case of the flow path the pipe elements are responsible for the momentum transfer between the flow nodes. Within the nodes and across any cross-section in the pipes the fluid and flow properties are assumed to be uniform. The heat transfer between the solid nodes and the fluid nodes of elements is handled by convection elements. Heat transfer through the solids is accounted for by conduction elements and the radiation heat transfer between interacting surfaces is evaluated using surface and spatial radiation heat transfer elements. The flow through the inlet and outlet manifolds, down-comer, riser tubes and in the cavity, and the heat transfer through the RPV wall, riser tube walls, down-comer walls and the concrete wall are modeled using a one-dimensional approach. Flownex does not have the capability to calculate the heat transfer coefficient associated with the reactor cavity surfaces. In the current study, therefore, the relevant heat transfer coefficients implemented in the Flownex models were extracted from the corresponding GAMMA+ results.

RESULTS

The steady-state results that form the basis of the pressure transient analyses are summarised in Table 3 for a RPV inner wall temperature of 250 °C and for a RPV inner wall temperature of 350 °C in Table 4.

Table 3. Steady-state results for GAMMA+ and Flownex for a RPV inner wall temperature of 250 °C

Parameter	Previous		Current	
	GAMMA+	Flownex	GAMMA+	Flownex
Total heat loss at RPV surface (MW)	0.803	0.826	0.954	0.964
Radiation heat transfer at RPV surface (MW)	0.683	0.699	0.641	0.642
Convection heat transfer at RPV surface (MW)	0.120	0.127	0.313	0.321
Mass flow rate of air in chimney (kg/s)	10.86	11.12	11.33	11.63
Exit air temperature in chimney (°C)	113	113	122.7	121.9
Maximum temperature of riser tube (°C)	157	161	161.0	157.5
Maximum temperature of down-comer wall (°C)	151	135	151.3	129.4
Maximum temperature of concrete wall (°C)	43	43	42.3	42.5
Pressure drop in riser tube (kPa)	0.22	0.24	0.22	0.25
Temperature rise in riser tube (°C)	68	72	77.7	80.5
RPV average outer wall surface temperature (°C)	238	237	235.6	235.4

Both the previous [7,8] and the current results are summarised in Tables 3 and 4. Let us take the averages of the GAMMA+ and the Flownex results for each study and then determine the percentage increase from the previous to the current results for each parameter. We then take average of the percentages obtained for each parameter for 250 °C and 350 °C. We find that the total heat loss at the RPV surface increased by 15.7%. The radiation heat transfer at the RPV surface decreased by 6.3%, whilst the convection heat transfer at the RPV surface increased by 159.6%. The increase in total heat transfer results in an increase of 8.0% in the air exit temperature and an increase of 11.8% in the rise of the air temperature over the riser tubes. The other parameters increased on average by only 1.8%. It is seen that the major change between the previous [7,8] and current results is the convection heat transfer which mainly affects the temperature of air in the riser tubes.

Table 4. Steady-state results for GAMMA+ and Flownex for a RPV inner wall temperature of 350 °C

Parameter	Previous		Current	
	GAMMA+	Flownex	GAMMA+	Flownex
Total heat loss at RPV surface (MW)	1.545	1.588	1.755	1.783
Radiation heat transfer at RPV surface (MW)	1.373	1.416	1.293	1.310
Convection heat transfer at RPV surface (MW)	0.172	0.172	0.462	0.473
Mass flow rate of air in chimney (kg/s)	12.75	13.09	13.03	13.46
Exit air temperature in chimney (°C)	159	160	171.6	170.9
Maximum temperature of riser tube (°C)	228	234	232.7	225.1
Maximum temperature of down-comer wall (°C)	212	195	213.2	181.1
Maximum temperature of concrete wall (°C)	44	44	43.0	43.2
Pressure drop in riser tube (kPa)	0.23	0.26	0.24	0.27
Temperature rise in riser tube (°C)	111	118	123.8	129.1
RPV average outer wall surface temperature (°C)	326	326	322.6	322.9

The variation in the mass flow rate, monitored at the outlet, as a function of time is shown in Fig. 3. The agreement between the GAMMA+ and the Flownex results is good. In the case of a RPV inner wall temperature of 250 °C, the flow reverses in direction during the pressure transient and then remains in the reverse direction once the pressure has returned to its ambient value. This is in accordance with the previous results [7,8]. However, in the case of a RPV inner wall temperature of 350 °C the mass flow rate also decreases, but never reverses in direction, and increases again to reach its original value at the end of the pressure transient. This is contrary to what was found in the previous study [7,8]. It will appear that for the given pressure transient there might be a critical total heat transfer rate beyond which the mass flow rate does not reverse in direction during the pressure transient. This needs to be studied in more detail looking at other possible pressure transients and RPV inner wall temperatures.

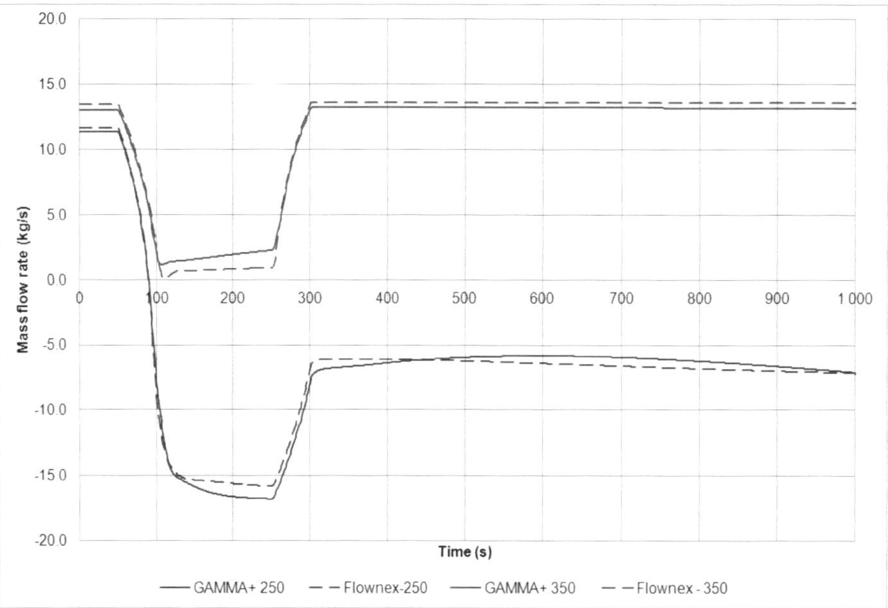

Figure 3. Mass flow rate as function of time as monitored at the outlet.

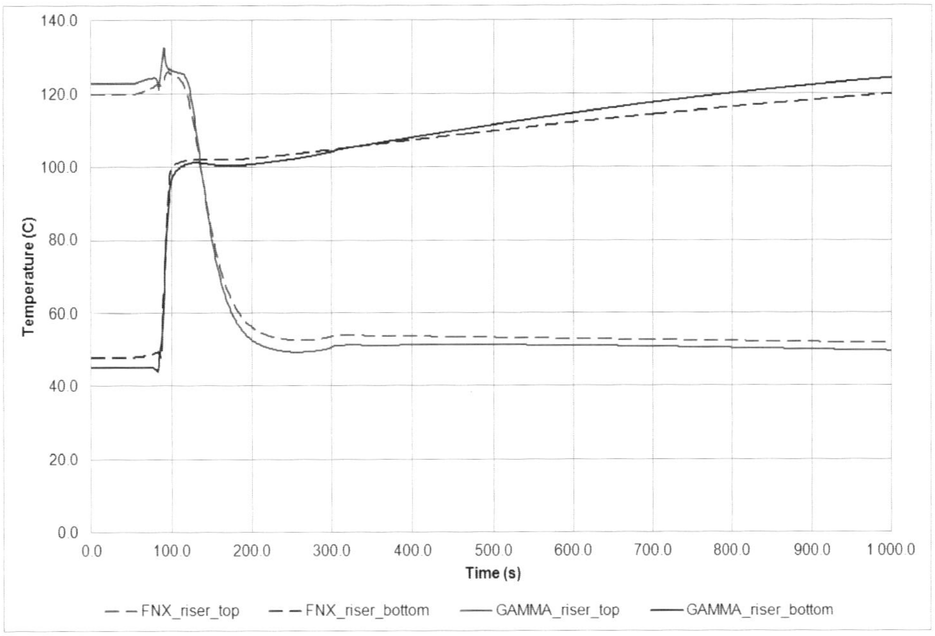

Figure 4. Riser temperatures as function of time for 250 °C.

The variation in the temperature as a function of time of the air inside the riser tubes at the top and at the bottom for a RPV inner wall temperature of 250 °C is shown in Fig. 4. The positions at which the temperatures were monitored are approximately 400 mm from the top of the riser tubes and approximately 400 mm from the bottom of the riser tubes. The agreement between the GAMMA+ and Flownex results is good. Looking at the variation in the temperature at the top of the riser tubes, we observe a slight increase in the temperature as the mass flow rate decreases and then a sharp decrease in temperature when the ambient air, being flushed through the outlet manifold after the reversal of the flow, reaches the top of the riser tubes. The final value of the temperature is a reflection of the heat transferred to the air over the first 400 mm.

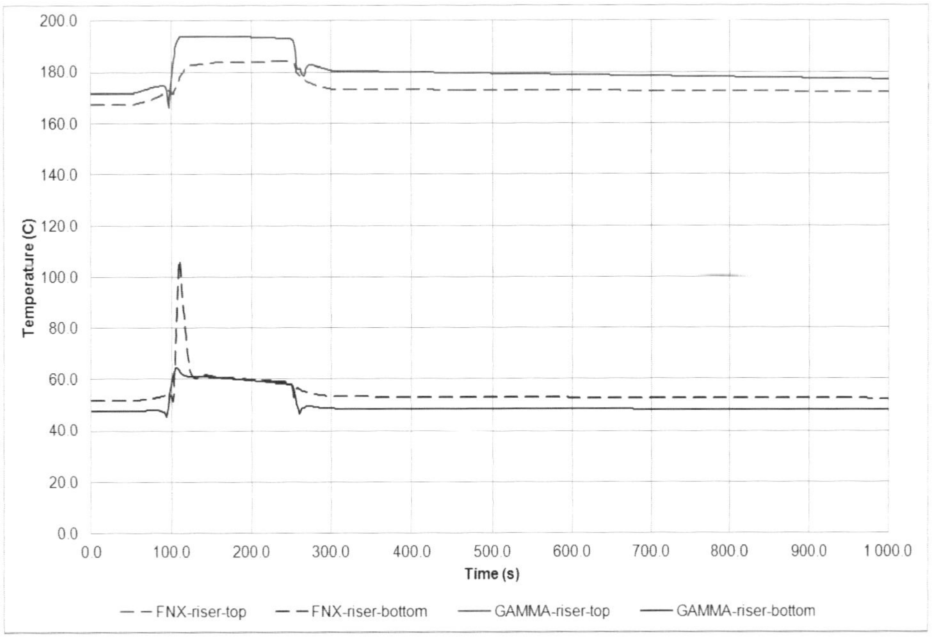

Figure 5. Riser temperatures as function of for 350 °C

The temperature of the air in the riser at the bottom of the riser tubes sharply increases during the reversal of the flow, then remains virtually constant for a while and finally increases steadily during the latter part of the pressure transient and after the transient. The last mentioned effect is probably due to the heating up of the lower part of the riser tubes wall as the hotter air flows through the tubes requiring higher wall temperatures for the heat to be transferred from the reactor cavity through the tube walls to the air inside the tubes. The oscillation present in the GAMMA+ temperature results, particularly at the top of the riser tubes, must be investigated to determine the cause.

The variation in the temperature as a function of time of the air inside the riser tubes at the top of the riser tubes and the bottom of the riser tubes for a RPV inner wall temperature of 350 °C is shown in Fig. 5. The positions at which the temperatures were monitored are the same as mentioned in the previous paragraph. The agreement between the GAMMA+ and Flownex results is again good. The spike in temperature at the bottom of the riser tube in the Flownex results is due to the flow almost coming to a standstill (See Fig. 3). At the low mass flow rate heat is still being transferred to the fluid resulting in a rapid increase in temperature. The temperature spike is dissipated almost immediately when the mass flow rate starts to increase again. The spike is also visible in the GAMMA+ temperature results, but less pronounced and also occurs slightly earlier. In the GAMMA+ results the lowest mass flow rate occurs a little earlier and does not go down as low as in the Flownex results.

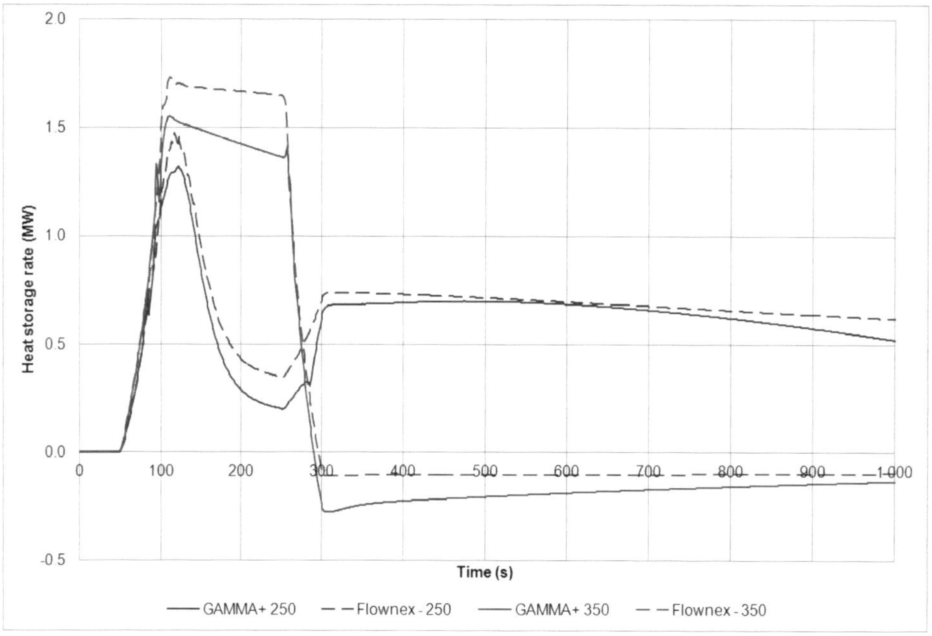

Figure 6. Rate at which heat is stored in system as a function of time.

The variation of the rate at which heat is stored in the system as a function of time is shown in Fig. 6. The agreement between the GAMMA+ and Flownex results is good with some differences in the maximum and minimum values reached. The heat stored Q_T in the system is defined as:

$$Q_T = Q_{T,RPV} + \left(\dot{m} \times h\right)_{in} - \left(\dot{m} \times h\right)_{ex} \quad \ldots (1)$$

where $Q_{T,RPV}$ is the total rate at which heat is added by the RPV surface to the system, $\dot{m}$ the mass flow rate and h the enthalpy. The subscripts *in* and *ex* refer to the inlet and exit (under normal positive flow conditions) respectively. The second term in eq. (1) is thus the rate at which heat is entering the system at the inlet and the third term is the rate at which heat is leaving the system at the exit. When the flow direction reverses heat enters at the exit and leaves at the inlet. In the case of a RPV inner wall temperature of 250 °C we see how the rate at which heat is stored in the system initially increases rapidly, then decreases rapidly and finally increases markedly again during the pressure transient $(50 \text{ s} < t < 300 \text{ s})$. After the pressure transient the rate at which heat is added to the system slowly decreases from the value reached at the end of the transient. In the case of a RPV inner wall temperature of 350 °C the rate at which heat is stored in the system during the pressure transient initially increases rapidly, then decreases slowly and finally decreases rapidly, reaching a stage where heat is being released. After the pressure transient the rate at which heat is released from the system decreases slowly.

In Fig. 7 the rate at which heat is stored as a function of time in the riser tube walls, and the down-comer inner wall (including the insulation), the down-comer outer wall and concrete, as obtained from the Flownex results for a RPV inner wall temperature of 250 °C, are separated. Also included in Fig. 7 is the associated variation in the mass flow rate. During the transient $(50 \text{ s} < t < 300 \text{ s})$ the majority of the heat is absorbed by the stainless steel walls of the riser tubes due to their high volumetric thermal capacity $(\approx 4000 \text{ kJ/m}^3.\text{K})$ and thermal conductivity $(\approx 16 \text{ W/m.K})$. The tubes are also directly exposed to the radiation heat from the RPV wall surface and the convection heat from the hot air in the reactor cavity. The stainless steel down-comer walls and the concrete (volumetric thermal capacity $\approx 250 \text{ kJ/m}^3.\text{K}$; thermal conductivity 2 W/m.K) are shielded from the radiation heat from the RPV wall by the insulation (thermal conductivity 0.022 W/m.K). The major mechanism to transfer heat to the down-comer walls and the concrete is convection from the air and then conduction to the concrete.

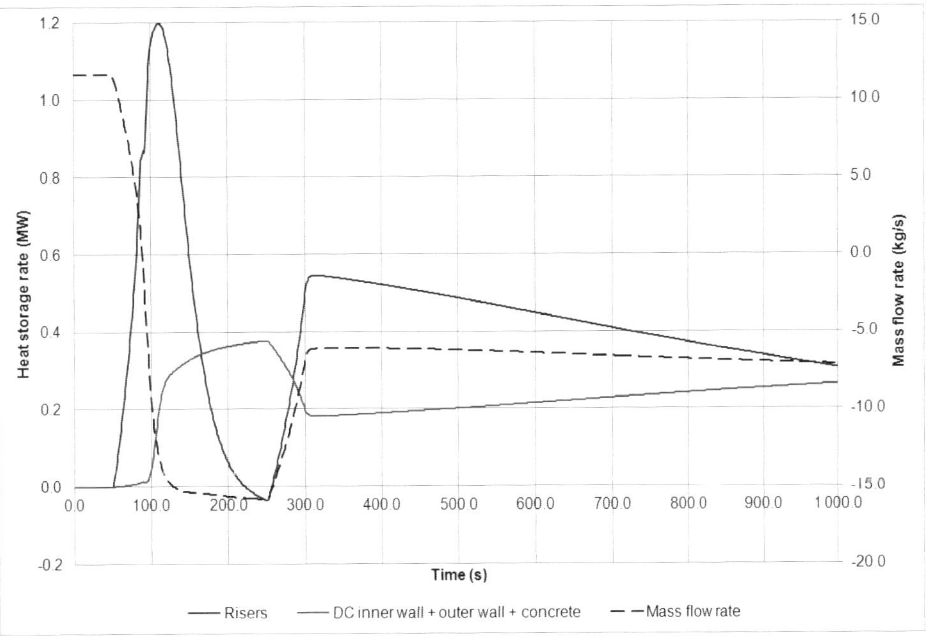

Figure 7. Rate at which heat stored in structures as a function of time for 250 °C.

During the first part of the pressure transient when the flow rate decreases and the flow direction is reversed, there is rapid increase in the rate at which heat is transferred to the riser tube walls. When the mass flow rate stabilises between 120 s and 250 s the rate at which heat is transferred to the riser tubes rapidly decreases and for a short while between 220 s and 250 s the riser tubes releases heat to the air. During the last stage of the pressure transient the rate at which heat is stored in the riser tubes increases rapidly again until it reaches a maximum value at the end of the pressure transient. After the transient the rate at which heat is transferred to the riser tubes decreases slowly. The rate at which heat is transferred to the down-comer walls and the concrete only starts to increase rapidly during the last stage of the flow reversal. The rate then increases at slower rate when mass flow rate stabilises between 120 s and 250 s and then decreases markedly again until the end of the pressure transient. After the pressure transient the rate at which heat is transferred to the down-comer walls and the concrete increases slowly. During the period of 1000s the heat stored in the riser tube walls is 397.55 MJ and the heat stored in the down-comer walls and concrete is 219.13 MJ. The concrete reaches a temperature of 65 °C after approximately 22 h and a final temperature of 127.2 °C after 16 days. In the case of a RPV inner temperature of 350 °C the heat stored in the riser tube walls during the period of 1000 s is 194.17 MJ and the heat stored in the down-comer walls and concrete is 13.26 MJ.

Table 5. Steady-state results for GAMMA+ and Flownex for a RPV inner wall temperature of 250 °C for reverse flow.

Parameter	Previous	Current	
	Flownex	GAMMA+	Flownex
Total heat loss at RPV surface (MW)	0.812	0.914	0.940
Radiation heat transfer at RPV surface (MW)	0.688	0.634	0.630
Convection heat transfer at RPV surface (MW)	0.124	0.280	0.310
Mass flow rate of air in chimney (kg/s)	-10.4	-10.9	-10.7
Exit air temperature in chimney (°C)	117.3	122.3	126.6
Maximum temperature of concrete wall (°C)	117.8	122.4	127.2
RPV average outer wall surface temperature (°C)	237.7	236.4	235.7

The steady-state results for GAMMA+ and Flownex for a RPV inner wall temperature of 250 °C for reverse flow are summarised in Table 5. When the current results are compared with the previous results [7,8] it can be seen that the increase in total heat loss is due to the significant increase in convection heat transfer. When the results for reverse flow are compared with those for positive flow in Table 3 it can be seen that there is only a 2.6% difference in the heat transfer rates. The major difference is the increase in the maximum temperature of the concrete wall from 42.5 °C to 127.2 °C.

It can be concluded that the GAMMA+ and the Flownex results compare very well and that the study serves as a further verification of the two codes. The results from the experimental tests [5] being performed at KAERI will be used to validate the ability GAMMA+ and Flownex to simulate the natural convection in as reactor cavity cooling system.

CONCLUSION

The purpose of the current study was to investigate the effect that a positive pressure transient acting on the outlet of the reactor cavity cooling system of the proposed 200 MW_{th} PMR (PMR200) General Atomic prismatic core VHTR reactor -will have on the performance of the cooling system using both GAMMA+ and Flownex. Since the previous studies [7,8] improvements were implemented in GAMMA+ for the calculation of the convection heat transfer associated with the surfaces of the reactor cavity. The relevant heat transfer coefficients implemented in the Flownex models for the reactor cavity surfaces were extracted from the corresponding GAMMA+ results.

Two cases were considered, namely for RPV inner wall temperatures of 250 °C and 350 °C respectively. For the steady-state results it was found that the total heat loss at the RPV wall increased on average by 15.7% compared to the results obtained by Du Toit et al. [7] and Rousseau et al. [8] due to an increase of 159.6% in the convection heat transfer from the RPV wall surface. As in the previous studies [7,8] it was found that the flow direction reverses during the pressure transient for a PRV inner wall temperature of 250 °C and then remains in the reverse direction after the pressure transient. However, in the case of a RPV inner wall temperature of 350 °C it was found that the flow direction does not reverse as found in the previous studies [7,8]. It will appear that for the given pressure transient there might be a critical total heat transfer rate beyond which the mass flow rate does not reverse in direction during the pressure transient. This needs to be studied in more detail looking also at other possible pressure transients and RPV inner wall temperatures.

The rate at which heat is stored and released in the system was also analysed and the response of the riser tube walls, the down-comer wall and the concrete wall were separated. It was found that the riser tube walls respond much faster and also absorbs the majority of the heat.

It can be concluded that the GAMMA+ and the Flownex results compare very well despite the differences in the modelling approaches and that the study serves as a further verification of the two codes. Formal validation of the codes via comparison with results from the experimental programmes mentioned in the introduction is envisaged for the future. Furthermore, the fact that flow reversal can occur under certain circumstances should be taken into account in the design of the RCCS system as it relates to the reactor safety case.

ACKNOWLEDGMENTS

This work is based on the research supported by the South African Research Chairs Initiative of the Department of Science and Technology and National Research Foundation of South Africa (Grant No 61059), as well as supported by the Nuclear R&D Program of the NRF of Korea grant funded by the Korean government (Grant code : NRF-2012M2A8A2025679).

Any opinion, finding and conclusion or recommendation expressed in this material is that of the authors and the NRF does not accept any liability in this regard.

REFERENCES

1 Lommers, L.J., Mays, B.E., Shahrokhi, F., "Passive heat removal impact on AREVA HTR design," Proc. of HTR 2012., HTR2012-8-010, pp. 1-12, (2012).

2 Tsuji, N., Nakano, M., Takada, E., Tokuhara, K., Ohashi, K., Okamoto, F., Tazawa, Y., Inaba, Y., Tachibana, Y., "Study of the applicability of CFD calculation for HTTR reactor," Proc. of HTR 2012., HTR2012-8-009, pp. 1-9, (2012).

3 Kim, S.H., Tak, N.I, Kim, M.H., "Comparison of axisymmetric and 3D simulations of the heat transfer in RCCS," Trans. of Korean Nuc. Soc Autumn Meeting., pp. 75-76, (2008).

4 Lomperski, S., Pointer, W.D., Tzanos, C.P., Wie, T.Y.C., Krauss, A.R., "Generation iV Nuclear Energy System Initiative. Air-Cooled Option RCCS Studies and NSTF Preparation," ANL-GenIV-179, Argonne National Laboratory, (2011).

5 Bae, Y.Y., Hong, S.D., Kim, Y.W., "Scaling analysis of reactor cavity cooling system for PMR200," Proc. of HTR 2012., HTR2012-7-020, pp. 1-8, (2012).

6 Kim, M.H., Tak, N.I, Lim, H.S., "Thermal-fluid assessment of the design options for reactor vessel cooling in a prismatic core VHTR" Annals of Nuc. Eng., 37, pp. 1774-1782, (2010).

7 Du Toit, C.G., Rousseau, P.G., Jun, J.S., Noh, J.M., "1D and 3D numerical simulation of the reactor cavity cooling system of a very high temperature reactor." Proc. of IHTC-15, IHTC15-8982, pp. 1-13, (2014).

8 Rousseau. P.G., Du Toit, C.G., Jun, J.S., Noh, J.M., "Code-to-code comparison for analysing the steady-state circulation in an air-cooled RCCS using GAMMA+ and Flownex." Nuc. Eng. and Design, vol. 291, pp. 71-89, (2015).

9 Lim, H.S. and No, H.C., "GAMMA Multidimensional Multicomponent Mixture Analysis to Predict Air Ingress Phenomena in an HTGR", Nuc. Sci. and Eng., vol. 152, 87-97, (2006).

10 M-Tech Industrial, Flownex SE Version, 8.5.0.2466, http://www.flownex.com, (2015).

11 Jun, J.S., "Common input data list of KAERI-NWU joint GAMMA+/FLOWNES code-to-code benchmark study of the natural circulation in the air-cooled RCCS," NHDD-RD-CA-12-016, Korea Atomic Energy Research Institute, (2012).

12 Harlow, F.H. "A numerical fluid dynamic calculation method for all flow speeds", J. Comput. Phys., vol. 8, pp. 197-213, (1971).

13 Jun, J.S., Lim, H.S., Jo, C.K., Noh, J.M., Venneri, F., "Thermal-fluid Characteristics of the Transuranics Fuel in a Deep-burn HTR Core", Nuc. Eng. and Design, vol. 241, pp. 3867-3878, (2011).

14 Greyvenstein, G.P., "An implicit method for the analysis of transient flows in pipe networks", Int. J. Numer. Meth. Engng., 53, pp.1127-1143, (2002).

15 Lewis, R.W., Morgan, K., Thomas, H.R., Seetharamu, K.N., "The Finite Element Method in Heat Transfer", John Wiley & Sons, Chichester, (1996).

Methodology for the Mechanical Characterisation of Uncoated Woven Polymer Textiles

D.R. Ellis*, M.P. Venter, G. Venter
Department of Mechanical and Mechatronic Engineering, Stellenbosch University
*16425103@sun.ac.za

Abstract

Uncoated woven textile materials are often used as a reinforcing cover for inflatable structures. In recent times, the numerical modelling using finite element analyses of these structures has become more prevalent as a means of streamlining prototyping of new designs and material changes. This method of simulation requires the use of representative material models that can capture the often complex behaviour that these materials exhibit. Such models need accurate parameters from physical material testing in order to perform well.

The methodology required for reliably determining the material properties of uncoated woven textiles, with an emphasis on that where polypropylene is the base material, along with a new approach for the response modelling thereof, is covered with this research.

The shortcomings of previous test methods and rigs were identified and used as design criteria for a newly developed rig. The developed test rig is capable of biaxial load-control testing of various load ratios with the added functionality that a shear loading can be applied. Further, the two biaxial directions of the rig are self-balancing ensuring that a state of tension can be tested alone, mitigating the necessity of time consuming specimen set up attempting to avoid unwanted material shear due to incorrect placement. The biaxial tensile load-unload response of the material for a single cycle to full-load will be tested for using a selection of load ratios. The material response from these tests are modelled with the proposed method using polynomial regression. The performance of this regression model was evaluated using independent and intermediate load test results and found to be capable of determining the unload response for loads not physically tested for.

Keywords: Biaxial tensile test; Fabrics/Textiles; Mechanical behaviour; Material modelling

Introduction

The increased popularity of woven polymer textile materials in applications where specific strength is required has lead to a surge in numerical modelling of these structures. This surge is driven primarily by economics as a reduction in material usage whilst retaining product quality could potentially lead to a greater profits. Numerical models, constructed using the Finite Element Method (FEM), that can represent the structure sufficiently accurate to base design changes off of need to represent both the structure and loading scenarios as best possible. Research undertaken by Venter [1] showed that a major stumbling block during the development of a numerical model for a inflatable dunnage bag was the lack of a robust and practical means of modelling the complex woven material within the finite element environment. Analysis tools for their efficient design lag behind those for conventional materials, partly because woven fabrics do not behave as a continuum and have intricate interfibre relations [2]. Changes in fabric kinematics occur with loading that lead to several sources of nonlinearity requiring modelling techniques not required for conventional and continuous materials.

At their lowest level, fabrics are constructed of individual fibres of which many grouped together constitute an individual tow which ultimately are woven as one to construct the material at the fabric level. The weave architecture as well as the individual tows that make up the general material used during this research can be seen in Figure 1 as viewed under a microscope. At this level the non-continuous, non-homogeneous nature is evident. Material characterisation can be conducted at any of the three levels depending on the desired application. Investigation at too high a level of detail often leads to impractical results when used to simulate large structures. Parsons et al. [3] found that models incorporating too detailed geometries are computationally prohibitive and too complicated when investigating actual full-scale structures. The sporadic squashing and deformation of tows caused by the weaving process, visible in the material shown in Figure 1, further illustrates the difficulty in modelling at that level. For the intended use of this research a fabric level investigation is required. At this level, material response is effectively homogenized over the area of investigation.

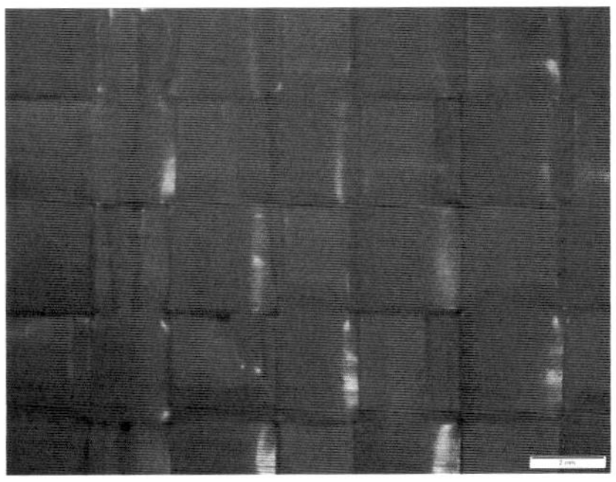

Figure 1: Plain woven material architecture as viewed under a microscope

The material in inflatable structures undergo large out-of-plane displacements from their stowed to deployed shapes, but the material remains primarily loaded in-plane, and therefore a planar response is sufficient for characterisation. Lomov et al. [4] noted the coupled response between tows, and Venter [1] the coupled response between primary (Warp or machine direction(MD)) and secondary (Weft or cross-direction(CD)) material directions under tensile conditions. This coupled response stems from the crimp interchange between tows in which the straightening of a tow in one direction, or decrimping, causes the subsequent apparent shortening of the tows normal thereof, thereby increasing their degree of crimp. This interchange can occur without any actual elongation of tows, and is therefore a geometric feature driven by the fabric architecture. Figure 2 shows a through-thickness representation of the material architecture where the effect of the warp tow being tensioned can be seen as a shift of the weft tows further apart. Investigating this interchange requires a means to vary the load/displacement ratio between the two orthogonal material directions in order to determine the change in response caused thereby. The summarised requirements of a test fixture are thus:

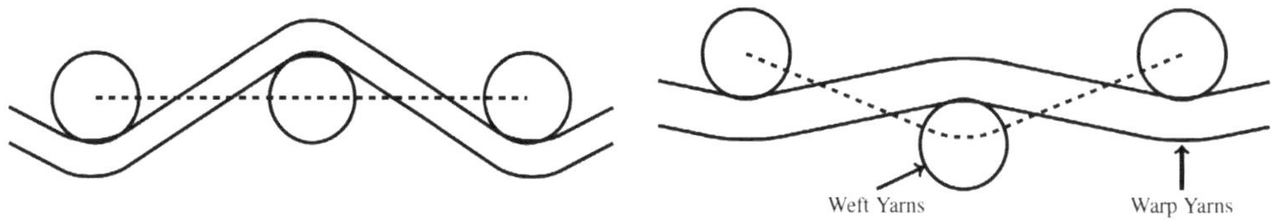

Figure 2: Crimp interchange

1. Fabric-level testing
2. Biaxial tensile loading of various load ratios
3. Ability of shear application
4. Ease of repeatable specimen loading and set up

Mechanical Test Methods

Various test apparatus exist for evaluating tensile and shear behaviour of woven textiles, ranging in complexity and capability. The uniaxial bias tensile and picture frame tests are popular methods used to capture shear response. In both these tests the specimen is loaded at 45° relative to the fibres in the load direction. The bias test forms a local trellis shear mode located in the centre of the specimen. The simplicity of this method allows for numerous tests to be run at varying angles. It does however have a major drawback in that Galliot and Luchsinger [5] found the strain distribution to be non-homogeneous. Further, in most instances it cannot be used for uncoated materials as fibres are not connected on either end of the sample due to cutting the sample at an angle relative to the tows. This results in load being applied to tows that rely only on the negligible friction between adjacent tows to withstand this force. Generally the sample will fall apart without much force due to the lack of a coating keeping it in tact. The picture frame test allows for global trellis shear deformation. The boundary conditions that ensure homogeneity in the sample however have the drawback that undesired fibre tensions result. Galliot and Luchsinger [5] indicate that due to these weaknesses of both test methods the results from both are commonly combined to investigate shear response. Benchmark testing at the fabric-level was undertaken by Cao et al. [6]. Their testing focused on aiding the simulation of forming processes and therefore testing was done solely on shear response utilizing both the bias extension test and picture frame test. Testing performed for such draping simulations aim to measure large out-of-plane responses and therefore are not constructive to the proposed research.

The biaxial test machine developed and tested by Galliot and Luchsinger [5], [7] shares many of the sought qualities. Their proposed method was developed specifically for testing the difficult to determine shear response, but also allows for biaxial testing at different load ratio's. The use of five independently controlled electromechanical actuators for gripping each of the four sides of a cruciform specimen geometry allows for the mitigation of conventional shear testing drawbacks. The shear stress is introduced in a biaxially pre-stressed sample by means of a shear ramp. The complexity and cost of this rig is however prohibitive.

Utilizing a standard tension-torsion machine, Cavallaro et al. [2] designed and patented a multi-axial tension and shear fixture. The shear loading is applied to a biaxially preloaded sample by rotating one of the directions relative to the other, in-plane. A clear insight to the effect of pretension on shear response and the effect of biaxial load ratio was found during their research.

Proposed Experimental Fixture

Design Overview

The design of a custom rig to fulfil the requirements for testing was necessary. The details of the biaxial and shear test rig (BASR) are briefly covered within this section.

Material grips should be able to supply a sufficient clamping load to prevent slipping as well as a boundary condition that can easily be replicated in a finite element analysis (FEA). Wrap type grips were designed such that a loose slotted roller is used around which the material is wrapped. This slotted roller has flat spots machined such that sliding blocks can constrain the rotational degree-of-freedom whilst still allowing the assembly to slide back and forth. The slot jeopardizes the structural integrity of the roller such that any deformation in the roller would be noticeable in the displacement-field of the loaded sample, thus introducing inaccuracies. Therefore, a second fixed pin is used as a stop for the sliding assembly to be loaded against to increase the bending strength of the slotted roller. Figure 3 shows a CAD representation of the design.

Figure 3: CAD model of wrap type grip

Load is applied using pneumatic cylinders controlled with proportional pressure regulators. This setup allows for load-controlled tests to be run only and not displacement-controlled. Each biaxial direction has two cylinders controlled with a communal regulator. Material grips are attached to these cylinders via a fixture that floats on linear bearings and shafts to ensure an accurately guided assembly. Both directions share the same plane of movement to prevent moments from acting on the bearings which could lead to stiction on the linear bearings and subsequent inaccuracies. Both cylinders in every set are connected to each other via compression members. These members allow for the assembly to be mounted on the same guide shafts and linear bearings as the grip fixtures, ensuring that initial material misplacement will not affect results through the introduction of a shear component, but will rather float to, and settle at, a position where only tension can exist. One of the biaxial directions has the ability to rotate 30° in total, or 15° to either side of the neutral position, in-plane, around a stationary shaft fixed to the midway point between the two material grips. The neutral position is perpendicular to the stationary biaxial direction. This rotation replicates the method of in-plane shear application used by Lomov et al. [4]. A CAD representation of the rig is shown in Figure 4.

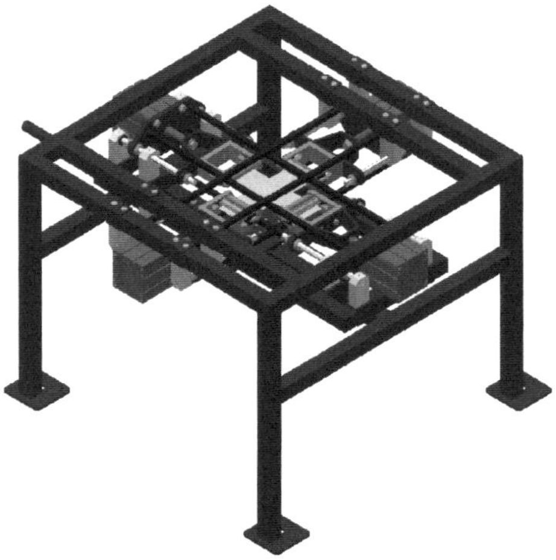

Figure 4: CAD model of BASR

Performance Envelope

The performance of the rig was tested for as well as its ability to determine material properties of conventional metallic materials according to the ASTM E8M standard.

Load Measurement Sensitivity

Applied load is measured as the internal pressure in the cylinders. A theoretically determined scaling factor, calculated as the internal cross-sectional piston area on the pressurised side of the cylinder, is used to convert from a pressure measurement to that of load. The sensitivity of this setup to rate of testing was investigated. The test involved a linear ramp from zero to 4000 N using five different inflation rates. Pressure results were plot using the theoretical factor. To adjust the theoretical factor to that of actual load, a linear regression model was fitted to measurements from the load cell and pressure transmitter. The minimum and maximum difference in gradient is shown in Figure 5. The data shows reasonable correlation to the theoretical factor and indicates very little influence on increased flow rate, within the range tested.

Efficacy using ASTM Metallic Testing

The efficacy of the rig to determine material properties of a metallic material is used to determine it's feasibility for mechanical testing. The ASTM E8M standard and 1000-series aluminium samples are used in conjunction with a conventional displacement-controlled tensile tester and Digital Image Correlation (DIC) to determine the properties of the metallic samples. This data is used as the datum for further testing. The test is repeated using the BASR for both of the biaxial directions. Custom grips were manufactured to clamp the samples in the rig. In total 15 tests were run shared equally amongst the three devices. The engineering stress-strain curve is shown in Figure 6 for each of the devices. Two sets of data per device are shown in the figure as indication to the repeatability of the rig.

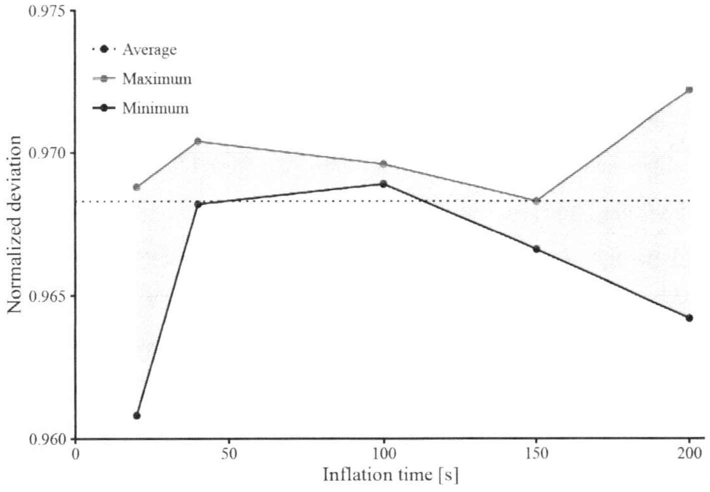

Figure 5: Deviation from theoretical pressure conversion factor for increased flow rates

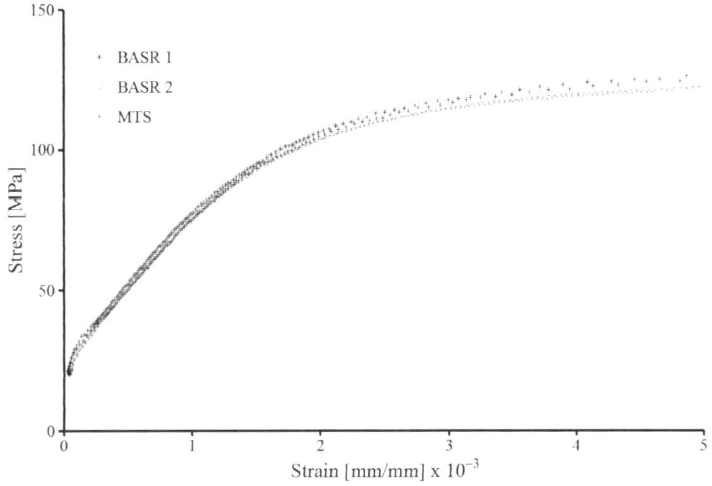

Figure 6: Engineering stress-strain curve as tested for using ASTM E8M

Table 1: Biaxial load ratios tested

Load ratio No.	Biaxial load ratio [CD:MD]
I	3000:3000
II	3000:2500
III	3000:2000
IV	3000:1500

Experimental Method

Sample Preparation

Meticulous sample preparation proved to be paramount by Cao et al. [6] in order to achieve high repeatability between tests. Preparation includes primarily exercising caution while cutting each sample to ensure individual tows are traceable along the length of the specimen, and mechanically preconditioning the cut cruciform to free tows from one another, creating a material more representative of typical loading environments.

Loading Variations

No agreed method exists for testing uncoated woven textiles for biaxial load-unload response at this time. In practice, load cases are designed around the sought response. Physical testing should be done in a manner that attempts to aid the numerical modelling thereof. Here, two approaches are taken. The widely used multi-cycle approach is followed and compared to the proposed method of single cycle testing.

Multi-cycle testing is applied in literature using various load cases, where all attempt to capture the material response at different loads. Chen et al. [8] used a load profile where a pretension is enforced where after the load is increased above and decreased below that of the pretension, in cycles. For this paper the applied load is cycled in increments that increase with 10% in each consecutive loading cycle between no-load and the required increment step until full-load is reached. This multi-cycle approach is often time consuming to conduct and generates a material response that is difficult to utilize within a numerical material model.

The proposed testing method is one where material response be tested for using a single cycle to a maximum load below that required for failure. The response model uses this single-cycle test data to calculate load-unload behaviour for intermediate loads to those tested for. A comparison between single- and multi-cycle tests conducted under similar conditions is shown in Figure 7. The results indicate that, for the material tested, the maximum measured strain is largely independent to the cyclic nature of the preceding load path.

Single-cycle tests were performed at intermediate loads to aid in supporting the proposed response modelling method of the succeeding section. The coupled nature between orthogonal material directions was tested for using the load ratio's in Table 1.

The focus of physical testing was on determining a method that will better improve the numerical application thereof. Extensive testing to gain an exact understanding of the material response under various loading conditions is beyond the scope of this research. These excluded factors include creep measurements, shear response and a large amount of load ratio testing.

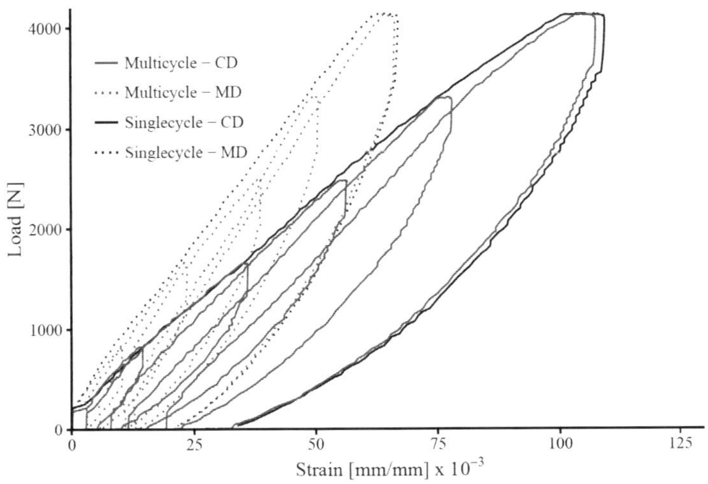

Figure 7: Comparison of single- and multi-cycle tests

Proposed Response Modelling Method

It is proposed to model the load and unload sections of the single-cycle response with two independent third order polynomial functions. The unload material response for intermediate values is then approximated using that of the full-cycle where the starting point of the full-cycle is shifted to that of the value of interest. Figure 8 shows this method compared to intermediate response measurements for the material cross-direction.

Conclusion

The necessity for a means to accurately determine the mechanical response of uncoated woven textiles in a repeatable fashion has been discussed. Factors such as the non-continuum and inhomogeneous characteristics together with the coupled response due to crimp interchange solidify the argument that this response cannot be determined using methods developed for metallic materials. The shortcomings of current biaxial test rigs were identified and formalized as design criteria for the proposed rig. These factors include difficulty in achieving various biaxial load ratio testing and time consuming procedures to ensure proper specimen placement within the grips. The performance of the proposed rig in determining material properties was evaluated using the ASTM E8M standard for metallic materials. These results were compared to those measured using a conventional displacement-controlled tensile tester. It was found that the proposed rig determined the elastic properties accurately, but measured higher strength in the plastic region of material response.

An alternative testing procedure to the widely used multi-cycle approach is given as that requiring only a single-cycle. Tests were performed using both of these methods and compared to one another to determine the plausibility of the approach. The preliminary results indicate that the envelope that

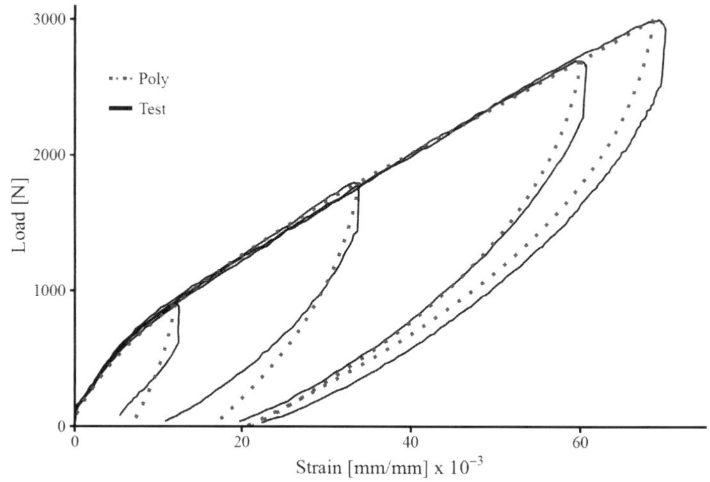

Figure 8: Performance of proposed material response modelling approach

encompasses the results from multi-cycle testing is similar to that of a single-cycle, demonstrated in Figure 7. This single-cycle response for each of the material directions is modelled using two independent polynomials representing the load and unload behaviour. It was found that loads intermediate to that of the full-cycle can be modelled sufficiently accurate using the regression models when the unload response is shifted appropriately along the load path to the required load magnitude. This single-cycle load path reduces the time required for testing and aims at modelling the material response in a computationally more efficient means.

Future Work

It is envisioned to model the material response to an array of load ratios using polynomial regression or Artificial Neural Networks. These models will then be incorporated into finite element analyses as user-defined subroutines to determine the material constitutive relationship during simulations. The accuracy of this model will be compared to that of analytical- and numerical unit cells, and conventional material models available within commercial finite element software packages.

References

[1] Venter, M. P. (2015). *A Methodology for Numerical Prototyping of Inflatable Dunnage Bags*. Phd dissertation, Stellenbosch University.

[2] Cavallaro, P., A. Sadegh, and C. Quigley (2007). Decrimping behavior of uncoated plain-woven fabrics subjected to combined biaxial tension and shear stresses. *Textile Research Journal 77*(6), 403–416.

[3] Parsons, E., T. Weerasooriya, S. Sarva, and S. Socrate (2010). Impact of woven fabric: Experiments and mesostructure-based continuum-level simulations. *Journal of the Mechanics and Physics of Solids 58*(11), 1995–2021.

[4] Lomov, S. V., P. Boisse, E. Deluycker, F. Morestin, K. Vanclooster, D. Vandepitte, I. Verpoest, and A. Willems (2008). Full-field strain measurements in textile deformability studies. *Composites Part A: Applied Science and Manufacturing 39*, 1232–1244.

[5] Galliot, C. and R. Luchsinger (2010). The shear ramp: A new test method for the investigation of coated fabric shear behaviour Part I: Theory. *Composites Part A: Applied Science and Manufacturing 41*(12), 1743–1749.

[6] Cao, J., R. Akkerman, P. Boisse, J. Chen, H. S. Cheng, E. F. de Graaf, J. L. Gorczyca, P. Harrison, G. Hivet, J. Launay, W. Lee, L. Liu, S. Lomov, A. Long, E. de Luycker, F. Morestin, J. Padvoiskis, X. Q. Peng, J. Sherwood, T. Stoilova, X. M. Tao, I. Verpoest, A. Willems, J. Wiggers, T. X. Yu, and B. Zhu (2008). Characterization of mechanical behavior of woven fabrics: Experimental methods and benchmark results. *Composites Part A: Applied Science and Manufacturing 39*, 1037–1053.

[7] Galliot, C. and R. Luchsinger (2010). The shear ramp: A new test method for the investigation of coated fabric shear behaviour Part II: Experimental validation. *Composites Part A: Applied Science and Manufacturing 41*(12), 1750–1759.

[8] Chen, W., C. Gao, D. Zhang, and J. Gong (2015). Shear test method of architectural coated-fabrics subjected to biaxial tensile loading. *International Conference on Textile Composites and Inflatable Structures* (7), 544–554.

Computational multiphase flow modelling and dimensional analysis study of oxygen lancing of pyrometallurgical furnaces

Markus W. Erwee[1,2,*], Quinn G. Reynolds[1], Johan H. Zietsman[2,3].

1 Mintek, 2 University of Pretoria, 3 Ex-Mente

markuse@mintek.co.za

ABSTRACT

Furnaces are high temperature vessels used to beneficiate valuable products from various materials. Molten materials have different chemical and physical properties, and these different properties are exploited to separate valuable materials from waste products at temperatures in excess of 1600°C.

Furnaces are lined with high melting point materials (refractories), which help contain the molten material. An opening, or tap-hole, located in the refractory wall of the furnace is used to remove or "tap" material at regular intervals such that new material can be processed.

During normal operation, a refractory clay is used to keep the furnace tap-hole closed. Once the furnace is ready to be tapped, the clay is removed by drilling and/or oxygen lancing.

Lancing with oxygen is carried out by injecting oxygen through a steel pipe that is inserted into the tap-hole. The tip of the lance is ignited, and oxidation of the iron in the lance releases energy which is used to melt away clay in the tap-hole. During this process, the lance may penetrate the furnace, injecting oxygen into the molten material and subsequently damaging the refractory lining and tap-hole.

The effect of lancing is difficult to quantify experimentally due to the harsh environment inside the furnace. Operational studies have however shown that it affects tap-hole life and process performance. Computational modelling offers an alternative approach of studying this effect. The problem involves fluid flow with multiple gas and liquid phases present. Accurate modelling of these phases and their physical properties requires that dimensional similarity be maintained between the computational model (implemented in OpenFOAM®), the real-world process, and any validation experiments done.

In this paper, a preliminary dimensional analysis is presented together with a computational modelling study addressing the problem of fluid flow resulting from gas injection into multiphase liquids.

KEYWORDS

Furnace tapping, Oxygen Lancing, Tap-hole wear, OpenFOAM, Computational Modelling, Dimensional Analysis

INTRODUCTION

Pyrometallurgical furnaces are used to convert raw materials to molten products at high temperatures. The processes involve melting and subsequent separation (physical and/or chemical) of valuable material from gangue (unwanted oxides). Examples include platinum smelting, ferro-alloy production (e.g. ferrochromium, ferromanganese), iron- and steelmaking, etc. Typically, molten oxide material (slag) floats on top of molten metal due to the difference in density of these materials, as shown schematically in Figure 1.

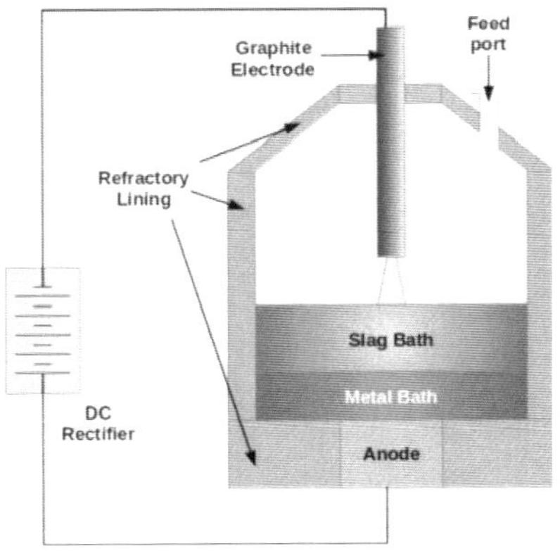

Figure 1: Schematic of a typical direct current (DC) electric arc furnace showing a metal and slag bath (Jones and Reynolds [1])

Furnace tap-holes

Regardless of the process involved, material needs to be tapped from furnaces at regular intervals after processing completes. Tapping of molten material typically occurs through furnace tap-holes. Tap-hole designs vary significantly depending on the nature of the process, but share a common trait: they are opened and closed at regular intervals and are considered high-wear areas in furnaces (Nelson and Hundermark [2]).

Depending on the process, a furnace might have a single tap-hole for metal and slag or separate tap-holes for each of these components. An example of a pilot-scale furnace which has two tap-holes (one for tapping slag and one for tapping metal) is shown in Figure 2.

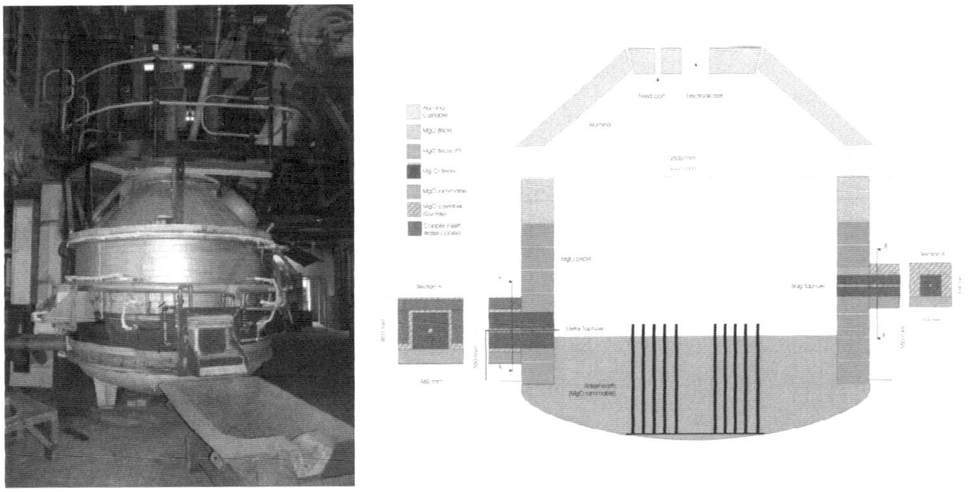

Figure 2: Pilot scale furnace at Mintek which has separate tap-holes for metal and slag (Geldenhuys and Jones [3])

The tap-block assembly has many different designs, depending on the material that it will be in contact with, and the typical heat transfer around the tap-hole during processing as well as during tapping. These systems can be quite complex and rely heavily on computational modelling as well as designer and operator experience to ensure longevity of this part of the furnace. Typically, the tap-hole itself is merely a channel created by an assembly of refractory materials. The channel is closed once tapping is concluded, using refractory clay that expands and hardens upon heating. The clay is removed using a combination of drilling and/or lancing with oxygen, both of which contribute significantly to wear of the tap-hole.

Furnace tap-hole wear and lancing

Aside from wear due to mechanical and chemical erosion by material flowing through the tap-hole channel at regular intervals, furnace tap-holes can wear significantly due to the opening and closing processes involved.

Typically, furnace operators would opt to use a drill to remove the hardened refractory clay plug from the furnace to start tapping. However, this is often unable to complete the opening process and oxygen lancing is employed. Oxygen lancing involves using a consumable steel pipe as a 'cutting torch'. Oxygen flows through the lancing pipe and reacts with iron in the steel, an exothermic reaction that provides intense heat at the tip of the lance. The thermal energy produced is focussed on the tap-hole clay inside the tap-hole which melts away, allowing for metal or slag to start flowing from the furnace interior.

Lancing is frequently an inexact, manual process. The hot tip of the lance hence runs the risk of coming into contact with not only the tap channel, but also the material inside the furnace. This may lead to increased wear of the tap-hole channel and the refractory lining in and around the tap-hole assembly, and also potentially influence the quality of the products due to interaction between the molten material and the oxygen injected by the lance.

Background to this work

In this work, the aim is to identify means of quantifying the effect that the lancing operation has, in particular when the lance penetrates the tap-hole completely and oxygen interacts with molten material inside the furnace. It is not possible to quantify this effect properly experimentally due to extreme temperatures which implies that one cannot see into the furnace, hence the need for a computational model. The problem is quite complex, involving multiphase as well as chemical reaction between molten material and oxygen.

To date, the authors have been progressively building a more complex model for the problem using both experimental work (experimental 'cold' models) and computational work (using the OpenFOAM® platform[4], specifically the interFoam solver). An initial computational model was built in a two-dimensional domain using a single liquid and gas. This model was validated against experimental work using a rectangular Perspex tank to simulate a furnace vessel, with water and air as the two fluids (Erwee et al[5]). Results from the two-dimensional computational model showed some correlation with the experimental model, but many deviations occurred due to the simplification of the computational geometry to 2D. A three-dimensional computational model was subsequently constructed and tested (Erwee, Reynolds and Zietsman [6]), in addition to which some changes to the experimental model were suggested. The detailed results of this work will not be repeated here, but an example of a typical comparative result is given in Figure 3.

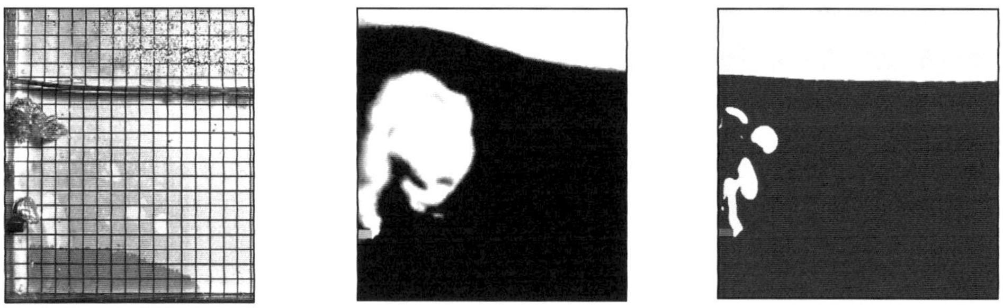

Perpex tank with water (experimental model)

High-speed camera image

2D model (~42k elements)

Black = Water, White = Gas

3D model (~600k elements)

Black = Water, White = Gas

Figure 3: Initial experimental models for one liquid phase with gas injection to represent the lancing problem

The ultimate aim of the work is to construct a multiphase computational flow model of the lancing problem which can be validated experimentally, such that the impact of lancing on the performance of the tap-hole can be quantified. The envisaged model ultimately should be capable of quantifying the effect of lancing on flow as well as also heat transfer in and around the furnace tap-hole. This requires a step-by-step approach, starting with the construction of a well-validated multiphase computational flow model.

Purpose of this paper

This paper comprises two parts. The first deals with dimensional analysis of the flow problem. The second presents preliminary results from the computational model in OpenFOAM® which will be validated against an experimental cold-model rig using materials at room temperature at a later date.

DIMENSIONAL ANALYSIS OF THE MULTIPHASE FLOW PROBLEM

The experimental rig used for validation of the multiphase computational flow model is a rectangular tank (for more detail see Erwee et al [5]). A schematic of the tank is given in Figure 4, while the variables of interest for the multiphase flow problem (applicable for both the tank and a furnace) are given in Table 1.

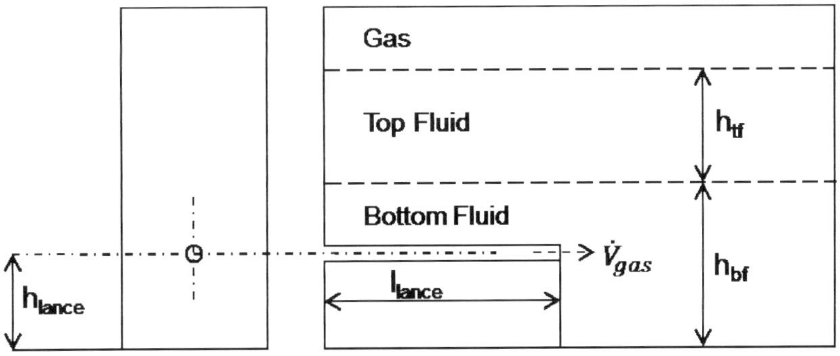

Figure 4: Dimensions of the rectangular tank that is used for experimental validation, showing key dimensions and parameters for the problem.

Table 1: Variables of interest for the lancing problem

Variable	Unit	Comment
h_{lance}		From the bottom of the tank to the centreline of the lance
l_{lance}	m	Depth of penetration of the lance into either fluid[1]
h_{tf}		The variable depth of the top fluid (e.g. slag)
h_{bf}		The variable depth of the bottom fluid (e.g. metal)
$\dot{V}_{gas}$	m^3/s	Gas flow rate from lance
ρ_{tf}, ρ_{bf}	kg/m^3	Density of top fluid (tf) and bottom fluid (bf) respectively[2]
μ_{tf}, μ_{bf}	Pa.s	Viscosity of the respective fluids[3]
$\sigma_{gas\text{-}tf}, \sigma_{gas\text{-}bf}, \sigma_{tf\text{-}bf}$	kg/s^2 or N/m	Surface tension between gas and top fluid, gas and bottom fluid and the top fluid and bottom fluid
g	m/s^2	Gravitational acceleration

Dimensional analysis using the variables in Table 1 yields 10 functional groups, given in Table 2.

[1] It should be noted that the tank width, depth and height is considered infinitely larger than the depth of penetration of the lance (l_{lance}) and that the lance diameter is infinitely smaller than the distance from the lance tip to the walls of the tank (or furnace).

[2] The density of the gas is considered very small in comparison to the density of either the fluids (this is due to the fact that the metal and slag densities are larger than that of the gas by at least three orders of magnitude).

[3] As for density (see footnote 2).

Table 2: Functional groups for dimensional analysis of the lancing problem

$\frac{\sigma_{tf-bf}}{\sigma_{gas-tf}}$	$\frac{\mu_{tf}}{\mu_{bf}}$
$\frac{\sigma_{gas-bf}}{\sigma_{gas-tf}}$	$\frac{\rho_{tf}}{\rho_{bf}}$
$\frac{h_{bf}}{h_{tf}}$	$Mo = \frac{g\mu_{tf}^4}{\rho_{tf}\sigma_{gas-tf}^3}$
$\frac{l_{lance}}{h_{tf}}$	$Re = \frac{\dot{V}_{gas}\rho_{tf}}{(h_{tf} - h_{lance})\mu_{tf}}$
$\frac{h_{lance}}{h_{tf}}$	$Eo = \frac{g(h_{tf} - h_{lance})^2\rho_{tf}}{\sigma_{gas-tf}}$

Familiar groups from the dimensional analysis are the Morton number (ratio of viscous forces to surface tension forces), Reynolds number (ratio of inertial forces to viscous forces) and the Eötvös number (ratio of buoyancy to surface tension forces).

COMPUTATIONAL MULTIPHASE FLUID MODEL

Previous work (Erwee et al [6]) on the lancing problem was performed using the standard interFoam solver in OpenFOAM® (OpenFOAM [4]) to solve a series of cases in which the gas flow rate and depth of penetration of the lance was varied. interFoam is however only useful for fluid flow problems involving two phases. For this work, the multiphaseInterFoam solver was used to account for the additional liquid phase present. This solver is based predominantly on work done by Ubbink (Ubbink [7]) and is not described in full detail here.

The model was constructed in a three-dimensional domain. A tetrahedral mesh was created in Gmsh 2.12 (Geuzaine and Remacle [8]) and then imported into OpenFOAM®. The tetrahedral mesh was then converted to a polyhedral mesh using the polyDualMesh application within OpenFOAM. The resulting mesh contained roughly 420000 elements (the amount of elements increases as the depth of penetration of the lance increases, due to mesh refinement around the tip of the lance).

An example of a polyhedral mesh for this problem is given in Figure 5.

Figure 5: Polyhedral mesh (3D) for studying the lancing problem

As a starting point, only one gas flow rate and depth of penetration was considered. However, the depth of the bottom fluid, e.g., metal was varied.

The rectangular tank (schematically shown in Figure 4) is 0.2 m high, 0.4 m wide and 0.045 m deep in both the computational and experimental model. The cylindrical lance has an internal diameter of 2.91 mm. The centreline of the lance is h_{lance} = 0.045 m above the bottom of the tank, whilst the depth of penetration of the lance from the left wall, l_{lance} = 0.01 m for all the cases considered in this initial model. A volumetric gas volume flow rate of 7.5 litres per minute was used. To test this model, the depth of the bottom fluid was varied while keeping other model parameters constant. The dimensionless values were calculated for the "test" values used for the fluid physical properties (equivalent to oil and water) in order to qualitatively test the functioning of the computational model.

Table 3: Dimensionless numbers for the lancing problem using test fluid properties

h_{bf} (m) →	0.0225	0.045	0.675	h_{bf} (m) →	0.0225	0.045	0.675
$\frac{\sigma_{tf-bf}}{\sigma_{gas-tf}}$		1		$\frac{\mu_{tf}}{\mu_{bf}}$		1	
$\frac{\sigma_{gas-bf}}{\sigma_{gas-tf}}$		1		$\frac{\rho_{tf}}{\rho_{bf}}$		0.5	
$\frac{h_{bf}}{h_{tf}}$	0.176	0.0428	0.818	Mo		2.86×10^{-11}	
$\frac{l_{lance}}{h_{tf}}$	0.078	0.095	0.121	Re	2.13×10^{7}	1.75×10^{7}	1.38×10^{7}
$\frac{h_{lance}}{h_{tf}}$	0.353	0.429	0.545	Eo	954	505	197

In general, both metal and slag properties differ significantly from the test values used in the present simulations. The values are highly dependent on temperature and chemical composition and often poorly defined for each fluid due to a lack of experimental data. By way of example, the dimensionless quantities have been estimated for a slag (40% CaO-40% Al_2O_3-20% SiO_2) in contact with pure iron at 1600°C using data from Mills (Mills [9]) and Guthrie (Guthrie [10]). The results are shown in Table 4. From the table it is evident that the dimensionless numbers are significantly different from the initial model with hypothetical values. Future models will use real numbers for slag and metal phases. The challenge is matching values of slag and metal with fluids at room temperature for the experimental cold model.

Table 4: Dimensionless numbers for the lancing problem using more realistic fluid properties

h_{bf} (m) →	0.0225	0.045	0.675	h_{bf} (m) →	0.0225	0.045	0.675
$\frac{\sigma_{tf-bf}}{\sigma_{gas-tf}}$		1.07		$\frac{\mu_{tf}}{\mu_{bf}}$		0.30	
$\frac{\sigma_{gas-bf}}{\sigma_{gas-tf}}$		0.004		$\frac{\rho_{tf}}{\rho_{bf}}$		0.39	
$\frac{h_{bf}}{h_{tf}}$	0.176	0.0428	0.818	Mo		2.49×10^{-13}	
$\frac{l_{lance}}{h_{tf}}$	0.078	0.095	0.121	Re	1.89×10^{5}	1.55×10^{5}	1.22×10^{5}
$\frac{h_{lance}}{h_{tf}}$	0.353	0.429	0.545	Eo	0.36	0.19	0.07

TYPICAL RESULTS FROM SIMULATIONS

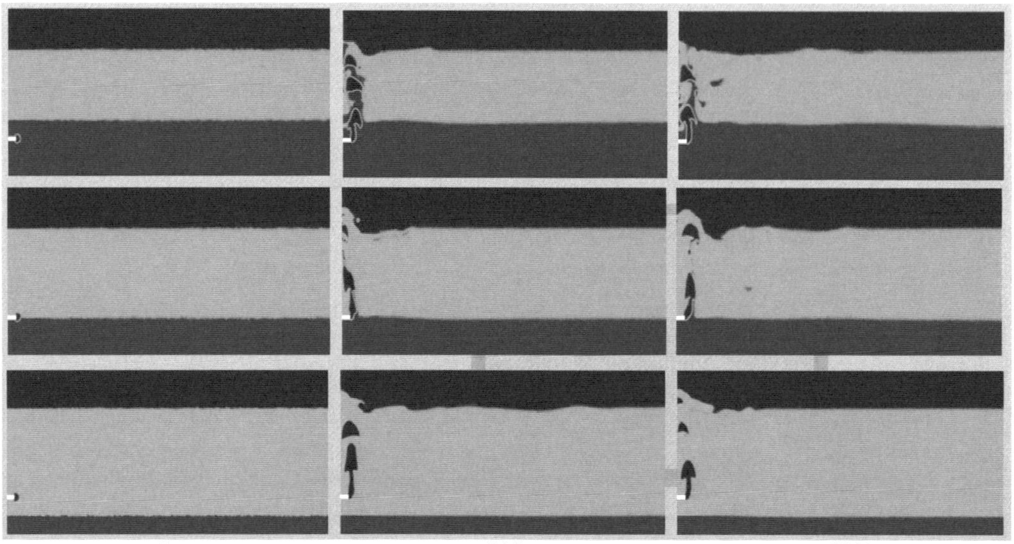

t = 5 ms t = 500 ms t = 2000 ms

Figure 6: A typical set of results from the computational model for the case of lancing with two liquid phases present (red = gas, grey is less dense liquid and blue is the densest liquid). From top to bottom, the level of the bottom fluid is 0.0675 m, 0.045 m and 0.0225 m.

Simulations were run in parallel using facilities at the Centre for High Performance Computing in Cape Town. Each simulation was run on a single computational node using 2 x 12-core Intel Haswell-architecture processors, and took approximately 20 hours to simulate 2.5 s of real time.

Qualitatively, several interesting effects were observed that have some relevance to the problem of tap-hole lancing:

- Interaction of the gas plume with the bottom or top fluid does lead to interaction of material against the wall – this could have a significant effect on possible refractory wear. Refractory wear occurs by a combination of chemical and mechanical erosion in most furnaces. Molten metal has a higher thermal conductivity than slag, hence, when molten metal is carried up the side wall, it is possible that the increased thermal load could make the refractory more susceptible to chemical wear.

- There is significant entrainment of the denser bottom fluid into the less dense top fluid – this can lead to metal losses (aside from metal lost by oxidation). Although lancing is usually a process of relatively short duration in comparison with tapping, it is often performed repeatedly throughout a tap if material is difficult to extract from the furnace. Further investigation into how often this occurs and what amendments should be made to tapping practices is warranted.

CONCLUSION AND CURRENT WORK

A multiphase fluid flow model has successfully been set up and tested for a problem involving lancing into pyrometallurgical furnaces. Dimensional analysis reveals that significant work is required to match future cold model experiments to the computational model due to the differences between slag and metal properties, and typical room-temperature experimental fluids.

The qualitative behaviour of the computational model does allude to what is observed in practice, i.e. that there are significant near-wall effects, and entrainment of fluids into one another. Future work will include validating this multiphase model experimentally using a cold model. Additional physics will also be introduced into the model by including the possibility of thermochemical reaction of oxygen gas with the metal and slag.

ACKNOWLEDGEMENTS

This paper is published by permission of Mintek. The authors also gratefully acknowledge support from the Centre for High Performance Computing (CHPC).

REFERENCES

1 Jones, R.T and Reynolds, Q.G. "Design Methods for DC Arc Furnaces to enhance Furnace Integrity". COM 2015 - The Conference of Metallurgists, Toronto, Canada: Canadian Institute of Mining, Metallurgy and Petroleum, 2015, pp 1-10.

2 Nelson, L.R., Hundermark, R.J. 'The tap-hole' - key to furnace performance. SAIMM, 2014. Furnace Tapping Conference. Vol. S80, pp. 1-32, Muldersdrift, South Africa, 2014

3 Geldenhuys, I. J. and Jones, R.T. "What Scale Should Your Smelting Testwork Be Done At, And What Do You Get For The Money You Spend?" 6th Southern African Base Metals Conference, Phalaborwa, North West Province, South Africa: The Southern African Institute of Mining and Metallurgy, 18-20 July 2011, pp. 477-492.

4 OpenFOAM® Version 3.0.1. Accessed 2016, from http://www.openfoam.org

5 Erwee, M. W., Reynolds, Q. R., Zietsman, J. H., Cromarty, R. D., & Lexmond, A. S. "Towards Computational Modelling of Multiphase Flow in and around Furnace Tap-holes due to Lancing with Oxygen: an Initial Computational and Cold Model Validation Study". pp. 174-183. INFACON XIV, Kyiv, Ukraine, 2015.

6 Erwee, M. W., Reynolds, Q.G. and Zietsman, J.H. "Comparison of 2D and 3D Computational Multiphase Fluid Flow Models of Oxygen Lancing of Pyrometallurgical Furnace Tap-Holes". Journal of Metals, Published Online 17 March 2016. http://dx.doi.org/10.1007/s11837-016-1873-6.

7 Ubbink, O. "Numerical prediction of two fluid systems with sharp interface". Ph.D. Thesis. London, England: Department of Mechanical Engineering, University of London.

8 Geuzaine, C. and Remacle, J.F. "Gmsh: A three-dimensional finite element mesh generator with built-in pre- and post-processing facilities". International Journal for Numerical Methods in Engineering, 79(11), 2009, 1309-1331.

9 Mills, K.C. "The Estimation of Slag Properties". Southern African Pyrometallurgy . Cradle of Humankind, South Africa, 6-9 March 2011: Southern African Institute of Mining and Metallurgy.

10 Guthrie, R.I.L. Engineering in Process Metallurgy. New York: Oxford University Press, 1989, pp. 440-441.

Numerical and Experimental studies of a pulverised fuel vertical spindle mill

Andre Govender[1,*], Walter Schmitz[2], Reshendren Naidoo[2]

[1] Eskom Holdings SOC Ltd., Grootvlei Power Station, Grootvlei, Mpumalanga, 2420, South Africa

[2] University of the Witwatersrand, School of Mechanical, Industrial and Aeronautical Engineering, 1 Jan Smuts Avenue, Johannesburg, 2000, South Africa

Govendad@eskom.co.za

ABSTRACT

This study aimed to address the shortfalls associated with conventional time-based monitoring techniques by developing an online pulveriser condition monitoring philosophy. A CFD model based on the physical geometry of a coal pulveriser was developed to predict the static pressure drop across major internal components as a function of the pulveriser air flow. Validation of the CFD model was assessed by comparing the experimentally determined and numerically calculated static pressure profiles. In this regard, an overall congruity of 95% was achieved. Candidate damage scenarios were simulated to assess the viability of employing the static pressure measurements as a means of detecting changes in mechanical pulveriser condition. Application of the validated CFD model proved to be advantageous in identifying worn pulveriser components through statistical analysis of the static pressure drop. Particle size distribution analyses provided a means for comparison of changes in the particle size distributions resulting from each modification applied. Hence, the interplay between the mechanical condition of the pulveriser and the size of particles discharged by the pulveriser was established. The application of this condition monitoring model can impact mill performance by proactively addressing changes in mill condition, thereby improving mill output, reducing cost and mill downtime.

KEYWORDS

Vertical spindle mill, Condition monitoring, Particle size distribution, Static pressure

1. INTRODUCTION

Coal pulverising systems, the location of which is highlighted in Figure 1, are found in all fossil fuel-fired power plants to finely grind, dry and preheat, classify as well as pneumatically transport the pulverised coal (PF) particles to the furnace for combustion. The pulverisers are generally classified under the boiler plant system.

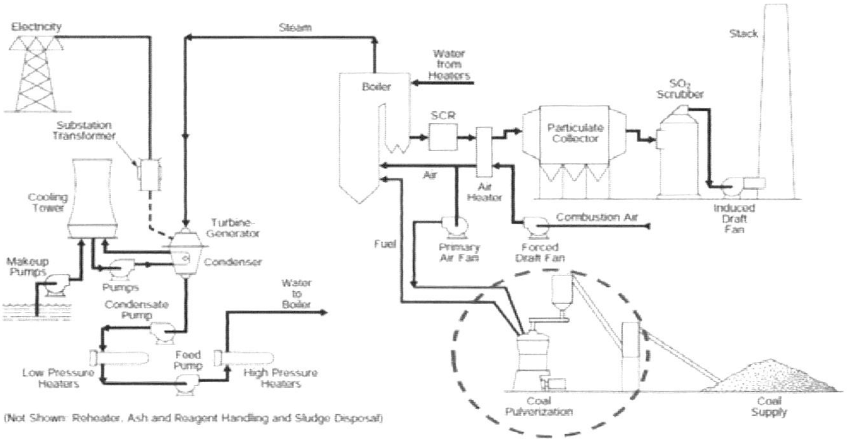

Figure 1. System arrangement of a typical coal-fired power plant (Kitto and Schultz [1])

A pulveriser system typically comprises of the mechanical components as illustrated in Figure 2. Hot primary air, obtained from an air heater, is utilised to dry and transport PF during the grinding process. The pulveriser receives hot primary air from the air preheater. The temperature of the primary air entering the pulveriser is controlled using tempering air in order to maintain a fixed air/coal temperature at the pulveriser outlet. In the pulveriser, raw coal is mechanically reduced from a nominal size of about 40 mm to a fineness in the range of approximately 70 % (by mass) passing a 200 mesh sieve (75 μm). The dried and ground coal is then air-swept to the burners for combustion.

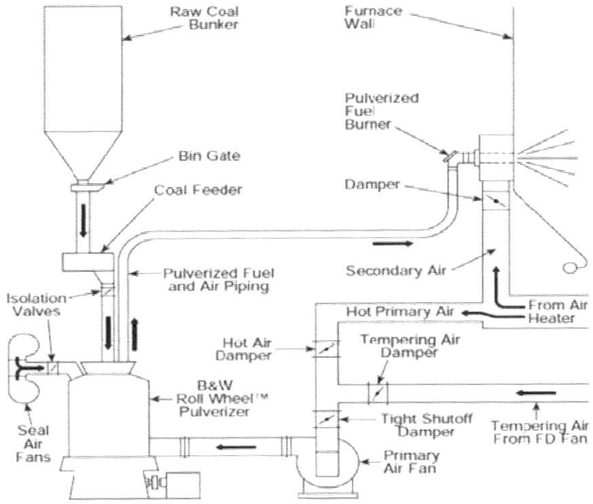

Figure 2. Schematic illustration of a coal pulverisation plant (Kitto and Schultz [1])

The lack of pulveriser condition monitoring techniques is perhaps a primary determinant of poor pulveriser performance leading to unstable and inefficient combustion, high emissions as well as reduced power plant efficiency. An inherent aspect of pulveriser operation is the wearing process of pulveriser components. The mechanical condition of internal components are critical as they aid in achieving the operative purposes of the pulveriser especially with regards to grinding efficiency. Conventional strategies employed to identify worn components are ineffective as they are performed on a time-based mode as opposed to being predictive or condition-based.

1.1. Vertical spindle mill (ball and ring)

Vertical spindle mills (VSM) are generally characterised by three different designs. These are bowl mills, ring roll mills, and ball and ring mills. Since the principle of operation of these designs is similar, only a description of the 8.5E VSM (ball and ring type) is provided. Figure 3 represents a simplified schematic diagram of a typical VSM. The pulveriser consists of nine hollow steel balls (730 mm in diameter) which are located freely between two grinding rings. The bottom grinding ring is connected to a shaft and is designed to rotate at 40 rpm. The top grinding ring remains stationary and is spring loaded during normal pulveriser operation. Coal enters through the central pipe from a volumetric or gravimetric feeder and settles on a rotating table. The volume feed flow is controllable as the feeder speed is variable. Whilst on the table, coal is radially displaced outward onto the grinding path (bottom ring) by means of centrifugal forces. The raw coal is ground as it passes under the grinding balls. Preheated primary air is introduced through primary air ports or throat gaps located around the periphery of the bottom grinding ring. As the primary air emerges through the equally spaced throat gap area it captures ground coal particles and effectively forms a fluidized bed region directly above the throat. On-site tests performed at a South African power plant revealed that a large percentage of moisture in coal is removed in this region during the grinding process (De Wet Krüger, [2]).

1.2. Pressure drop

The static pressure drop in a pressurised coal pulveriser is usually determined by measuring the difference in static gauge pressure between the pulveriser inlet (highest pressure) and the pulveriser outlet (lowest pressure). The pressure drop is dependent on frictional flow resistance, mass flow rates of air/coal as well as the quantity of heat energy lost during the pulverisation process (Flynn, [3]). A significant contributor to frictional flow resistance is the geometrical configuration of the pulveriser hence there is a distinct relationship between pressure drop and the pulveriser geometry (Lindsley, [4]). Miller and Tillman (2008) mention that the pulveriser pressure drop is directly proportional to the pulveriser air/coal throughput. Moreover, the pressure losses increase with a decrease in coal Hardgrove Grindability Index (HGI) and classifier opening percentage (Miller and Tillman, [5]). Another factor which affects the pressure drop is the mechanical condition of pulveriser internal components (Flynn, [3]). Worn components lead to poor grinding capability thereby increasing the coal recirculation load inside the pulveriser which subsequently intensifies the pressure drop. In industry, the pressure drop is a 'general indication' of the quantity of coal present in the pulveriser (recirculating load) and not the coal feed rate or throughput (Flynn, [3]). For safe pulveriser operation, the pressure drop is maintained within a specified range. For the 8.5E VSM (ball and ring), normal operating pressure drop ranges between 3 – 4.5 kPa. However, this threshold is pulveriser load, condition as well as coal quality dependent (Fan and Rees, [6]).

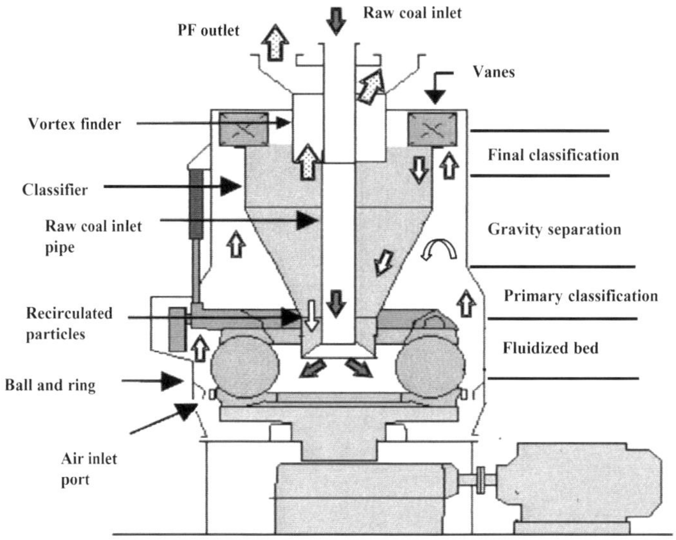

Figure 3. Schematic representation of a vertical spindle mill with static classifier, showing air and coal flow paths (Adapted from Parham et al. [3])

1.3. Literature review

The earliest application of Computational Fluid Dynamics (CFD) in coal pulverisers excluded the grinding rollers and journal assembly from the computational domain (Bhasker, [7]). The model was solved using the finite volume method. A fully three-dimensional model was simulated. The model employed the standard κ-ε turbulence model which satisfied the Reynolds Averaged Navier-Stokes equations of motion for an incompressible fluid flow. Coal particles having a constant particle size of 25 μm were injected into the converged continuous air flow phase. The results of the CFD model were qualitative and produced distribution plots of velocity and static pressure as well as particle trajectories. This author concluded that the presence of solid components inside the pulveriser, as well as particle drag forces, influence the air-coal trajectories resulting in unequal air and coal flow distribution at the pulveriser exit.

A previous study utilised a granular Eulerian-Eulerian approach to generate the parametric behaviour of multiphase flows in a simplified VSM (tyre and table) model (Vuthaluru et al., [8]). The flow domain of the three-dimensional model was restricted to the areas between the pulveriser casing and the classifier cone, hence, the grinding zone, classifier and mill outlet pipes were not captured. Tetrahedral cells were used to discretize the three-dimensional computational domain and the trajectories of two different sized particles were simulated, 100 μm and 500 μm. The CFD results showed that the flow field within the pulveriser was slightly symmetric and demonstrated that the two different particle sizes injected followed different trajectories. The 100 μm particles were carried by the air flow to the top of the pulveriser while the 500 μm remained at the bottom, perhaps due to gravitational separation. No quantitative validation data was provided and the solution domain lacked satisfactory mesh resolution.

In the Eulerian-Eulerian approach employed by (Vuthaluru et al., [8]) the fluid dynamics equations for momentum and continuity were solved separately for the air and particle phases. Best practice guidelines suggest that when using this approach, modelling particle size distributions are complex and the hydrodynamics of each particle cannot be adequately represented (CD-Adapco, [9]). Furthermore, the modelling of turbulence for each phase has a substantial impact on computational costs and the strong coupling present between the individual phases renders it immensely difficult to achieve solution convergence. Hence the approach can be considered to be non-viable for large complex models.

An expansion on the previous model was explored by employing the Eulerian-Lagrangian approach to model the pulveriser (Vuthaluru et al., [10]). The focal point of this study was based on investigating the effect of air flow distribution on the wear pattern inside the pulveriser. Model input variables that were incorporated included the pulveriser static pressure and total pressure, velocity direction and turbulence intensity. Turbulence was modelled using the default κ-ε turbulence model and the segregated numerical solver was employed. The dispersed phase was simulated with coal particles of a single size (250 μm) to simplify the solution. The findings of this research study highlighted the versatility of employing CFD calculations to develop condition monitoring systems.

In the Euler-Lagrangian approach, the model solves the Navier-Stokes equations for the continuous phase using the Euler method (Vuthaluru et al., [10]). The equations of motion for individual particles in the particulate phase are solved under the Lagrangian framework. Best practice guidelines suggest that in utilising this approach comprehensive information can be obtained for individual particles such as:

- Particle-particle interaction, particle impact on wall boundaries in the computational domain.
- Particle size distribution analysis.
- Modelling of heat and mass transfer between particles and the surrounding fluid.

However, the fundamental drawbacks of the Euler-Lagrangian approach are that it requires more computational power to solve complex models and it is limited to smaller concentrations of particles. Up to 60 % of a cell volume can be occupied by particles using this approach. Whilst in the Euler-Euler approach, up to 100 % of a cell volume can be occupied by either the particle or continuous phases (CD-Adapco, [9]).

A full scale multi-outlet CFD model of a bowl type VSM was developed for the evaluation of classifier efficiency and to quantitatively determine the influence of classifier vane angle settings on the pulveriser output (Shah et al., [11]). Five geometrically modified classifier vane scenarios were assessed between 100 % and 45 %. The results of the simulations succinctly illustrated the unprecedented influence of the vane position on the air flow resistance in the classifier region. By evaluating the particle size distribution at the pulveriser exit, this team was able to demonstrate that an optimal PF fineness could be achieved at a classifier opening of 65 %. Additionally, coal mass balances were performed and presented as a percentage deviation from the mean. The simulated results demonstrated good accuracy and correlated well within a 7 % error when compared to physical plant measurements, emphasizing the viability of employing CFD to accurately predict flow phenomena in coal pulverisers. A highlight of this study to be noted was the exclusion of the two-way coupling model, as a result the interaction of the forces present between coal particles and the continuous air-phase was not considered.

Bhambare et al. (2010) expanded on the previously developed pulveriser CFD models by including two-way coupling between air and coal as well as moisture evaporation from coal. The single-phase findings of this study correlated well to the single-phase results presented by previous authors which highlighted the uneven air flow distribution at the throat region. This team aimed at determining whether the non-uniform air flow distribution at the throat had any relation to the non-uniform air and particle size distribution in the pulveriser outlet pipes. A hypothetical case of even air flow distribution was created at the throat region. Coal and air mass flow in the pulveriser outlet pipes were expressed as a percentage deviation from the mean for before and after the hypothetical test scenario simulated. Although the standard deviation of the primary air flow in the pulveriser outlet pipes improved from 3.8 % to 0.3 % with the modification, the non-uniformity of the coal flow only improved from 3.5 % to 2.8 %, emphasizing a poor correlation between the two measured outcomes (Bhambare et al., [12]).

1.4. Objectives of this study

Previous CFD studies were predominantly performed on geometrically simplified tyre and table VSM designs. These studies explored the single and two phase flow dynamic pattern exhibited within the pulveriser with the intent of understanding and improving air and PF flow distribution. However the models were simplified and the impact of the pulveriser mechanical condition on the performance of the pulveriser has not been investigated. The objective of this research was to assess the viability of employing static pressure measurements to detect changes in the mechanical condition of critical pulveriser components predominantly impaired by wear.

2. On-site measurements to establish the single phase VSM static pressure profile

Existing VSM static pressure measurement points (Figure 4a, G and F) are spatially biased hence, they were considered inadequate to establish a comprehensive picture of the flow dynamic pattern along the vertical height of the pulveriser. Therefore additional pressure measurement points (Figure 4a, A, B, C, D, E) were instituted in accordance with best practice guidelines at strategic locations on one VSM (Shaw, [13]). For the static pressure measurement, 9 mm square-edged holes were drilled through the pulveriser walls. The holes drilled were visually inspected to ensure that they were free of any obstructions that could negatively influence the static pressure measurements. Half inch steel nipples were subsequently welded onto the external surface of the pulveriser casing, following which a half inch ball valve was fitted to the free end of the steel nipple. Each of the five static pressure tapping point installations were identical and the architecture followed is schematically presented in Figure 4b. In accordance with best practice guidelines, 6 mm impulse lines were then connected to the quick coupling and linked to Siemens differential pressure transducers and linked to a National Instruments (NI) Data Acquisition System (DAS) USB 6008.

Subsequently, quantification of the static pressure values, as well as the trend exhibited, was successfully accomplished for three air mass flow rates. All air flows tested exhibited a similar static pressure trend, thereby highlighting the accuracy and precision of the measurement technique. All pressure transducers were calibrated by a South African National Accreditation System (SANAS) approved laboratory. The uncertainty of measurement reported for each pressure transducer was ±25 Pa. The experimental results obtained are presented in Figure 8.

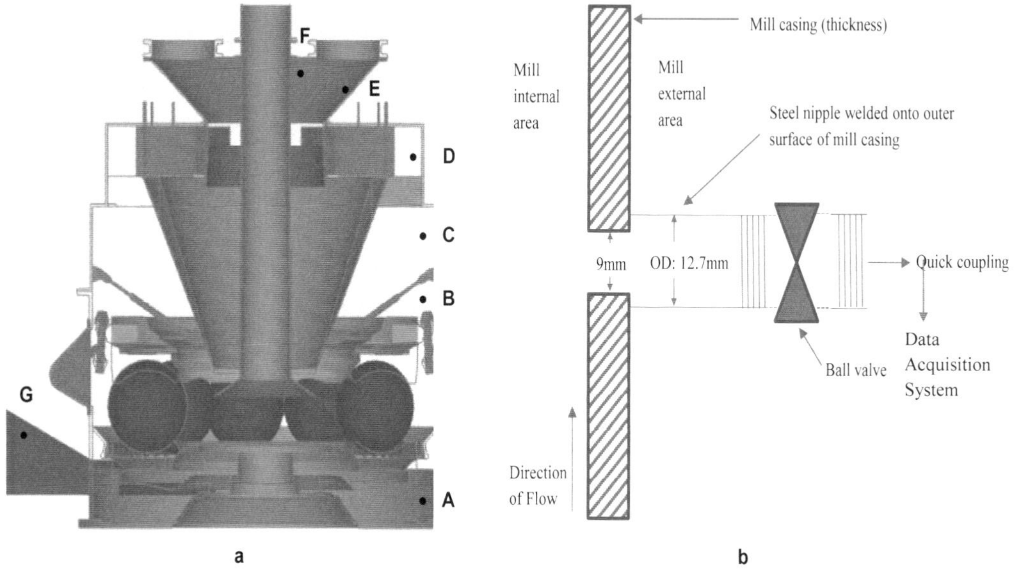

Figure 4. Cross section of pulveriser showing location of static pressure tapping points (a) and (b) cross section of pulveriser wall showing the tapping point installation architecture

3. Development and Validation of actual vertical spindle mill (ball and ring) CFD model

3.1. Geometry creation, boundary conditions and mesh development

Figure 5a illustrates an isometric view of the three dimensional domain considered in the numerical simulation. This model, referred to as the base case scenario, was prepared in accordance to the existing VSM design as constructed on-site. Best practice guidelines suggest selecting boundary locations at areas where the fluid flow is adequately defined or can be reasonably approximated (Ansys, [14]). Therefore, the choice of inlet and outlet boundary locations was limited to the primary air (PA) inlet duct and the PA/PF outlet pipes. A holistic approach was taken to model the pulveriser by extending the domain further upstream of the pulveriser PA inlet to include both the left hand (LH) and right hand (RH) primary air ducts. The inclusion of these components sanctions for a more representative flow profile to be attained at the pulveriser inlet. The inlet and outlet boundaries were defined as mass flow inlets and pressure outlets, respectively. Additionally, to assess the static pressure profile and particle size distribution at the pulveriser outlet, both PF pipe outlet turrets were considered. For the pulveriser simulation, the option of unstructured mesh generation was selected based on the high degree of accuracy achievable with this meshing strategy for complex geometries. The models chosen included the surface remesher, polyhedral, prism layer and extruder and a mesh refinements study culminated in the selection of the most plausible mesh resolution for areas defined with high velocity gradients. This ensured a good compromise between solution accuracy and solution run time.

3.2. Physics models defined and model input parameters

The appropriate selection of physical models is essential to define the primary variables as well as the mathematical formulation that will govern how the physical phenomenon is characterized for a given numerical solution. In actual operation, the bottom grinding ring and throat gap are in a constant state of rotation. However in this simulation the flow is modelled in steady state. The ideal gas model was employed to derive the density of the continuous phase (air) as a function of temperature and pressure. The gravity model was selected to account for gravitational acceleration. When these two models are used in combination, the gas density becomes a function of the absolute pressure. This necessitated appropriate specification of the system operating pressure.

The air flow velocity encountered in the pulveriser domain is not significantly high (<200 m/s), hence it results in a low Mach number (< 0.3) and the flow was categorised as being incompressible (Vuthaluru et al., [10]). Therefore the segregated flow model was selected. In this approach, the flow equations for pressure and velocity are solved sequentially in an uncoupled approach or segregated manner. The viscous flow regime for the pulveriser simulation was considered to be turbulent (Vuthaluru et al., [10]). To model turbulence, the Realizable K-Epsilon Model (RKE) and the Reynolds Stress Turbulent Model (RST) were applicable. Best practice guidelines highlight that the RKE model provides a good compromise between robustness, accuracy and computational cost and is suitable for flow domains prevalent with recirculation.

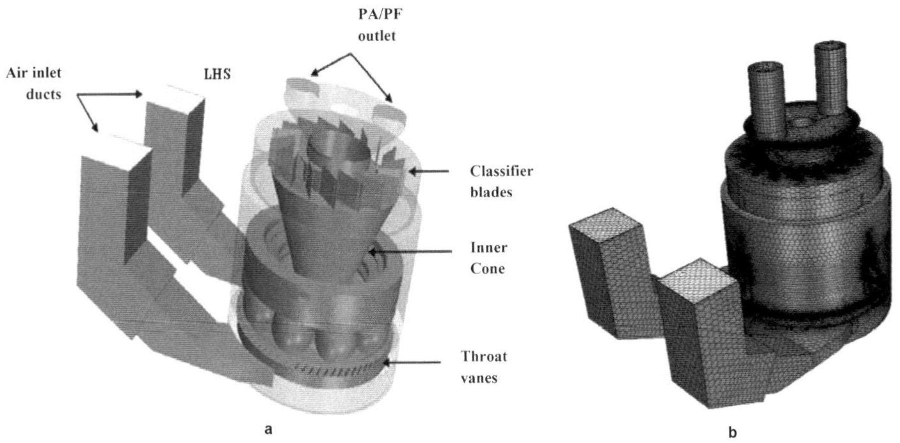

Figure 5. Perspective view illustrating (a) the pulveriser domain and (b) geometrical domain populated with polyhedral cells

Alternatively, the RST Model is better suited for domains where high swirling flows are anticipated; however it is computationally expensive and may require more iterations than the RKE to obtain a fully converged solution (CD-Adapco, [9]). Turbulent flows are significantly affected by the presence of walls, where the viscosity affected regions have large gradients in the solution variables. This necessitates the use of a suitable wall treatment model to effectively capture the flow profile of wall bounded flows. The wall y+ value is a non-dimensional distance used to describe how coarse or fine a mesh is for a particular flow (Salim and Cheah, [15]). In the pulveriser simulation, a combination of coarse and fine meshes has been used. The throat gap and vortex finder have finer meshes whereas their adjacent components have coarser meshes. The all y+ wall treatment is a hybrid treatment that emulates the high y+ wall treatment for coarse meshes and the low wall treatment for fine meshes. Therefore the all y+ wall treatment model was selected and this treatment produced y+ values less than 300.The single-component material model was used to model the air phase. Table 1 presents the input data employed in the simulation. In each simulation, the static gauge pressure measured experimentally at the pulveriser outlet (Figure 4, F), was employed to calculate the absolute pressure at the pressure outlet boundary. Additionally, equal primary air flow rates were applied to each primary air duct entering the pulveriser. Simulations were performed with air at ambient temperature.

3.3. Single-phase (cold) pulveriser simulation results and discussion

The construction of scalar and vector contour plots in StarCCM+ proved particularly useful as they provided a visual representation of the overall flow phenomenon that exists in the simulated domain. Figure 6 shows the contour plots generated to conceptualize the effect of the pulveriser internal components on the magnitude of the air flow velocity along the height (Z direction) of the pulveriser. The velocity profile was discernibly affected by both the throat gap and vortex finder region. The highest air flow velocity is observed across the throat gap region. This is congruent with results from a previous study (Vuthaluru et al. [10]).

Figure 7 shows the scalar contour plots exposing the effect of the pulveriser internal components on the static pressure distribution pattern along the same central plane generated for the velocity analysis. The highest static pressure was observed in the primary air duct, with a slight reduction as the air enters the air plenum chamber region. The contour plot highlights a higher pressure at the far right side of the air plenum chamber. This could be related to the arrangement of the PA ducts, as the primary air enters at the left side, which may cause the flow to be forced towards the far right end, resulting in a higher pressure. A significant reduction in the static pressure was evident across the throat gap region and based on literature, this observation was not entirely unexpected. Notably, the static pressure above the throat gap up to the classifier is relatively consistent, thereby highlighting components in this region to have a smaller effect on the overall static pressure losses in the system. Analogous to the higher velocity gradient evident in the vortex finder, there appeared to be a similar trend with a higher pressure gradient observed in this area.

Table 1. Input physics data for the pulveriser simulation

	Units	Simulation 1	Simulation 2	Simulation 3
Air inlet temperature	^{o}C	25.1	25.1	25.1
Air inlet mass flow	kg/s	9.8	8.0	7.0
Static pressure at pulveriser outlet	Pa	1007.4	628.4	396.5
Atmospheric Pressure	Pa		85000	
Absolute Pressure	Pa	86007.4	85628.4	85396.5

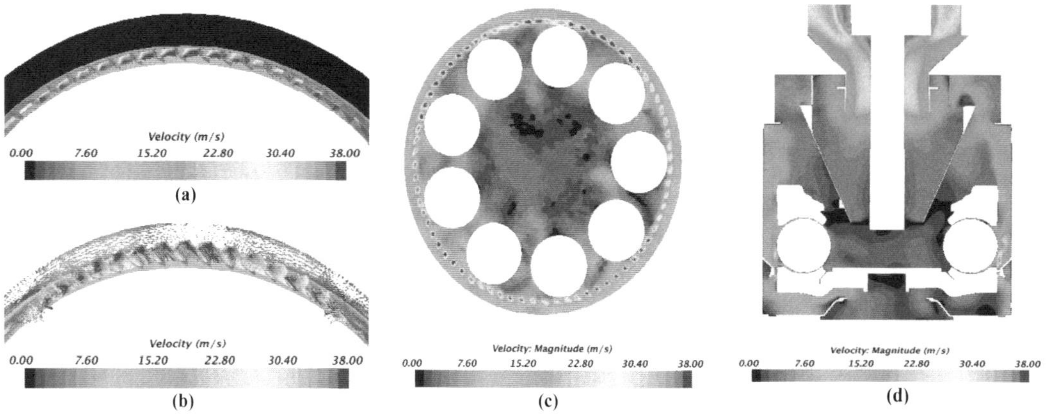

Figure 6. Scalar contour plot (a) highlighting the magnitude of the air velocity in the throat gap, (b) vector contour plot illustrating the effect of the angled throat vane, (c) scalar contour plot depicting even air flow distribution at throat gap and (d) depicting the influence of pulveriser internal components on the velocity profile

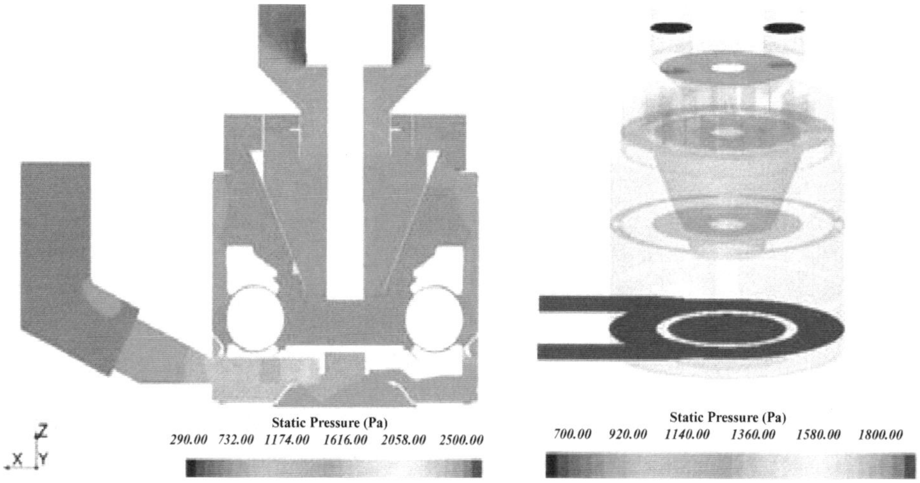

Figure 7. Scalar contour plots highlighting the static pressure profile along the cross sectional area of the pulveriser domain simulated

3.4. Validation of the CFD model through experimental correlation

CFD validation refers to the establishment of a correspondence between fluid flow profiles predicted numerically through computation and actual fluid flows depicted in real life. The conventional approach to CFD validation requires relevant experimental work to be performed in order to establish a standard form of reference (Marvin, [16]). The installation of additional static pressure ports along the vertical height of the pulveriser provided a means for a fundamental validation of the numerical model to be performed. The ability of the numerical model to accurately predict the static pressure profile in the pulveriser can be evaluated by the intensity of the correlation between the experimentally determined and numerically predicted static pressure values. It must be noted that the measurement points created in CFD were positioned at the same location as in the physical experiment (Figure 4) by employing a derived part. Three primary air flow rates were assessed. The excellent correlation demonstrated between the static pressure profiles obtained experimentally and the static pressure profiles predicted through computational modelling, as presented in Figure 8, serves to validate the CFD model developed. The overall incongruity (not shown) between the experimental and numerically predicted static pressure results was calculated to be less than 5 %.

3.5. Introduction of the solid phase

Given the shortcomings posed by the Eulerian framework to provide detailed information about particle trajectories, the Lagrangian framework emerged as the most suitable approach to model the particulate phase for the pulveriser considered in this study.

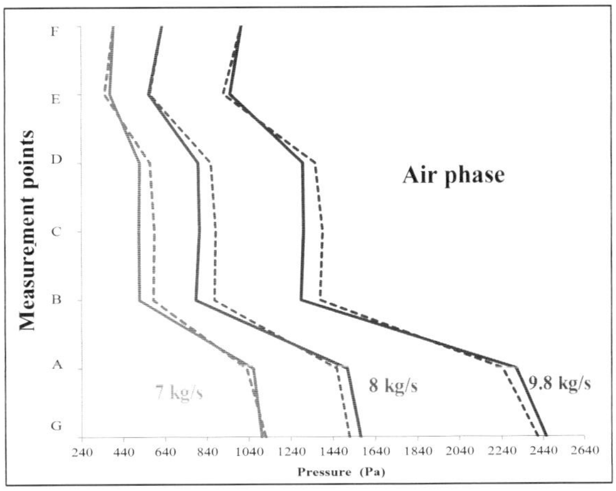

Figure 8. Experimental (solid) and simulated (dashed) static pressure profiles

The introduction of the solid-phase into the computational domain is accomplished through the use of injectors in StarCCM+. Injectors are responsible for defining the location, size, and direction at which solids particles enter the domain (CD-Adapco, [9]). Coal flows from the grinding table outward radially onto the grinding track, therefore it was postulated that ground coal particles emerge from the location highlighted in Figure 9. A part type injector was deemed most suitable for this application. When a part injector is employed, particles are injected from the centroid of every cell defined in the part. 34 constant particle size injectors were manually set up to inject an equal number of coal particles ranging from 1 μm up to 400 μm.

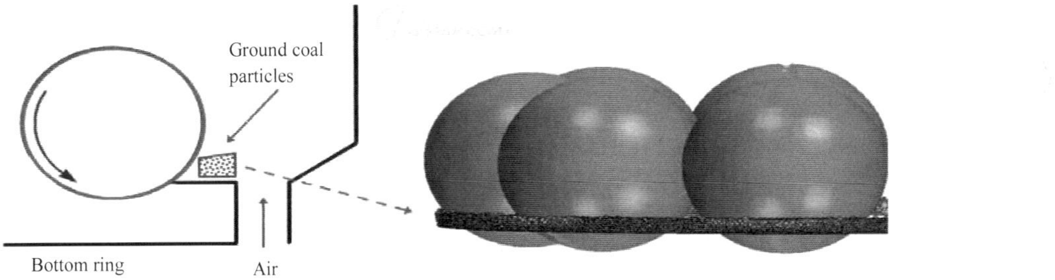

Figure 9. Schematic highlighting the injection location of ground coal particles

The uncoupled approach was selected to define the Lagrangian phase. In this approach, the continuous phase influences the particle paths through terms such as drag in the momentum equation and heat transfer in the energy equation. Therefore, although the effect of the discrete phase on the continuous phase is neglected, the discrete phase pattern can be established and assessed in StarCCM+ based on a fixed continuous phase flow field (CD-Adapco, [9]). Forces exerted on the coal particles due to its velocity relative to the continuous air phase were considered through the selection of the drag force model. Additionally, the turbulent dispersion model was selected to account for turbulence. In a separate study inclusion of the particle loading highlighted poor static pressure prediction using two way coupling model (Govender, [18]).

The determination of the coal particle size distribution at the pulveriser exit was assessed quantitatively by completing a particle size count balance as described by Equation 1. An iterative approach was adopted to delineate the optimum particle residence time for the pulveriser simulation, being 24 seconds. This approach ensured that the simulation was not stopped prematurely by allowing particles enough time to reach the exit if their size was satisfactory enough to do so with the prescribed air flow conditions.

$$Coal\ passing\ (\%) = \frac{(Number\ of\ particles\ at\ exit\ )}{(Number\ of\ particles\ injected)} \times 100 \qquad ...(1)$$

4. Development of a condition monitoring philosophy

Condition monitoring approaches, employing statistical evaluation of measured outcomes over a given period, are extensively practiced in industry to maintain a certain level of performance from processes, as it is an efficient and cost-effective tool that is able to provide early detection of damaged plant components. In order to ensure the successful development of a condition monitoring system for a given process, a sensing system that is able to monitor the system response to changes coalesced with an intelligent and rigorous data analysis procedure are fundamental prerequisites (Farrar and Worden, [17]). Subsequent to the identification of the single-phase flow model demonstrating the best potential for the development of a pulveriser condition monitoring philosophy, it provided a platform to assess the measured outcomes under conditions that mimic wear scenarios, typically those that cannot be adequately tested in real life. Common pulveriser damage scenarios were simulated to assess the viability of employing the static pressure measurements to achieve the aim.

4.1. Damage scenarios

The inner cone directs larger coal particles rejected by the classifier to the grinding zone. Therefore, it performs an important role in coal classification. Figure 10a presents the modified inner cone (scenario 1) replicated in StarCCM+ to mimic the actual mechanical degradation. The vortex finder is attached to the top plate of the classifier and is responsible for directing the ground coal particles from the classifier into the outlet plenum following which they are discharged into the PF pipes. It assists the classifier vanes in providing the centrifugal forces required for PF delivery to the coal burners. The image illustrated in Figure 10b highlights the damaged vortex finder (scenario 2) created where part of the bottom right hand side was removed.

4.2. Simulation results and discussion

The results for static pressure variations, observed for the ensuing scenarios, are presented as 'static pressure drop' across each respective zone. Each zone is identified by the measurement points as outlined in Figure 4. In addition, comparison of the model-predicted outcomes for each candidate wear scenario is highlighted through the percentage variation value calculated in relation to the base case (undamaged scenario).

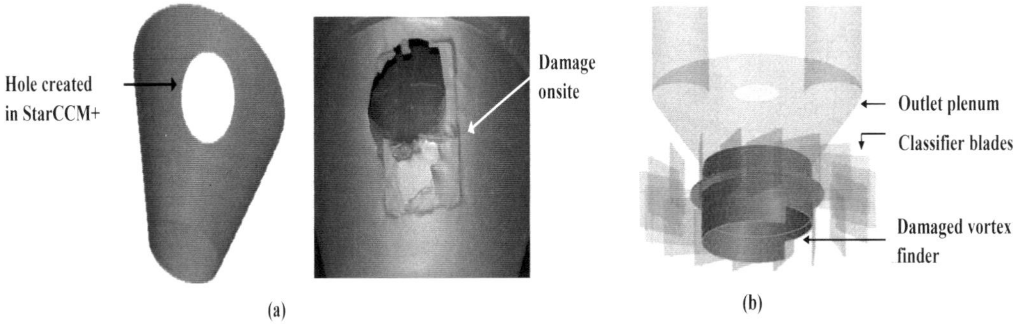

Figure 10. Schematic highlighting (a) damaged inner cone and (b) damaged vortex finder

Following the application of the modified inner cone and damaged vortex finder in StarCCM+, the static pressure drop results are presented in Table 2. For scenario 1, there is a considerable variation from the base case scenario for the classifier zone (BF and CF, Figure 4). Additionally, a notable change in pressure drop is highlighted across BC. As the deficiency was created adjacent to the measurement points BC, it can be concluded that the worn condition simulated contributed to the change in static pressure values sensed in the pulveriser domain. Relatively, the static pressure drop observed in other areas can be considered to be trivial.

Following the assessment of the static pressure drop caused by deficient components highlighted in Figure 10, it was also of particular importance to establish the effects of the aforementioned conditions on the particle size distribution at the pulveriser exit. A particle 'size count' balance proved to be a suitable means to categorically elucidate the quantity of different sized particles that exit the pulveriser subsequent to the application of the worn components in the CFD model.

Table 2. Static pressure drop results

Pressure drop	Base case (Pa)	Scenario 1 (Pa)	Variation (%)	Scenario 2	Variation (%)
ΔP_{GA}	168.41	162.11	3.74	163	3.21
ΔP_{AB}	865.39	881.57	1.87	900.21	4.02
ΔP_{BC}	-12.52	-8.51	32.06	-12.67	1.20
ΔP_{BF}	371.03	247.48	33.30	331.03	10.78
ΔP_{CF}	383.55	255.99	33.26	343.70	10.39
ΔP_{GF}	1404.83	1291.16	8.09	1394.23	0.75

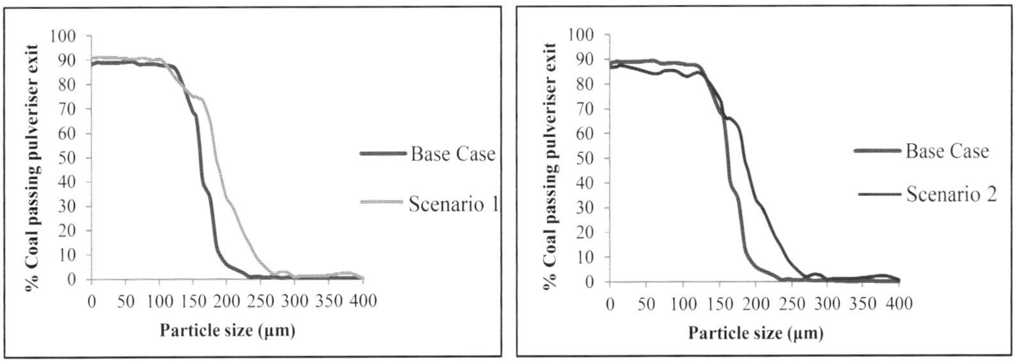

Figure 11. Comparison of the particle size distributions generated to elucidate the effect of the inner cone and vortex finder modifications on the model predicted output

5. Conclusion

The application of geometrically modified pulveriser internal components in CFD highlighted the definitive and statistically significant effect of sub-optimal pulveriser mechanical condition on the associated static pressure drop predicted and the particle size distribution. The assessment of the measured outcomes from the modified scenarios investigated highlights a general trend where a reduction in the static pressure drop is observed as different components deteriorate. Additionally, this study provided evidence to substantiate the general tenet that as the pulveriser condition deteriorates, satisfactorily ground coal particles as required for efficient combustion is circumvented due to larger particles exiting the pulveriser.

6. References

1 Kitto, J. & Schultz, S., 2005. Steam/ Its generation and use. 41st ed. Barberton, Ohio, U.S.A. The Babcock and Wilcox Company.

2 De Wet Krüger, J. I., 1977. A study of the parameters affecting the design of Vertical Spindle Mills, MSc. Dissertation (Eng), Johannesburg, South Africa: University of the Witwatersrand.

3 Flynn, D., 2003. Modelling and control of pulverised fuel coal mills. Thermal Power Plant Simulation and Control. London, U. K. The Institution of Electrical Engineers, pp. 63-99.

4 Lindsley, D., 2000. Power-plant control and instrumentation. London: The institute of Electrical Engineers.

5 Miller, B. & Tillman, D., 2008. Combustion Engineering Issues for Solid Fuel Systems. Academic Press: London, United Kingdom.

6 Fan, G. & Rees, N., 1997. An intelligent expert system (KBOSS) for power plant coal mill supervision and control. Control Engineering Practice, Volume 5, pp. 101-108.

7 Bhasker, C., 2002. Numerical simulation of turbulent flow in complex geometries used in power plants. Advances in Engineering Software, Volume 33, pp. 71-83.

8 Vuthaluru, H., Pareek, V. & Vuthaluru, R., 2005. Multiphase flow simulation of a simplified coal pulveriser. Fuel processing technology, Volume 86, pp. 1195-1205.

9 CD-Adapco, 2014. StarCCM+ Version 9.02.

10 Vuthaluru, R. et al., 2006. Investigation of wear pattern in a complex coal pulveriser using CFD modelling. Fuel Processing Technology, Volume 87, pp. 687-694.

11 Shah, K. V., Vuthaluru, R. & & Vuthaluru, H. B., 2009. CFD based investigations into optimisation of coal pulveriser performance: Effect of classifier vane settings. Fuel Processing Technology, Volume 90, p. 1135–1141.

12 Bhambare, K., Ma, Z. & Lu, P., 2010. CFD modelling of MPS coal mill with moisture evaporation. Fuel Processing Technology, Volume 91, pp. 566-571.

13 Shaw, R., 1960. The influence of hole dimensions on static pressure. Journal of Fluid Mechanics, Volume 7, pp. 550-564.

14 ANSYS, 2012. Ansys Fluent User's Guide, Version 14.5. United States of America: ANSYS, INC.

15 Salim, M. & Cheah, S., 2009. Wall y+ strategy for Dealing with Wall-bounded Turbulent Flows. Hong Kong, Proceedings of the International Multi Conference of Engineers and Computer Scientists 2009 Vol II.

16 Marvin, J., 1995. Perspective on computational fluid dynamics validation. AIAA Journal, Volume 33, pp. 1778-1787.

17 Farrar, C. & Worden, K., 2007. An introduction to structural health monitoring. Philosophical transactions of the royal society, Volume 365, pp. 303-315.

18 Govender, A., 2016. Development of a condition monitoring philosophy for a pulverised fuel vertical spindle mill. MSc. Dissertation (Eng), Johannesburg, South Africa: University of the Witwatersrand.

10th South African Conference on Computational and Applied Mechanics
Potchefstroom
3 – 5 October 2016

Ride comfort sensitivity to speed over different roads for an off-road vehicle

R.P. Gräbe*, C. Kat, P.S Els

Vehicle Dynamics Group, Department of Mechanical and Aeronautical Engineering, University of Pretoria, South Africa

Corresponding author: cor-jacques.kat@up.ac.za

ABSTRACT

Ride comfort is a complex field that incorporates psychological effects, ergonomics, climate, noise and vibration exposure. Vehicle ride is evaluated by means of objective measurements and/or subjective evaluation. Ride comfort evaluations are done by driving vehicles at the same speeds over the same roads and comparing the objective and/or subjective measurements. The question is how precise a vehicle's speed should be regulated during experimental ride comfort evaluations to ensure that the occupant does not perceive any change in vibration caused by a change in speed instead of actual vehicle differences. The paper investigates the sensitivity of ride comfort measurements to a change in speed over different road profiles. A validated non-linear ADAMS model of an off-road vehicle was used. The BS 6841 standard was used to quantify the ride comfort and investigate its sensitivity to changes in speed over different roads. The results showed that the ride comfort measure is dependent on the nominal speed, the road roughness and the difference in the speed between two tests. The precision to which the speed is to be controlled was shown to be dependent on these variables as well as the difference threshold of a human.

KEYWORDS

Ride-comfort, speed sensitivity, rough terrains, off-road vehicle, just noticeable difference

NOMENCLATURE

CRV	Component ride value
DOF	Degree of freedom
DLC	Double lane change
JND	Just noticeable difference
ORV	Overall ride value
APG	Aberdeen Proving Ground
r.m.s.	Root-mean-square
r.m.q.	Root-mean-quad
r.s.s.	Root-sums-of-squares
r.s.q.	Root-sums-of-quads
4S4	Four state semi-active suspension

1. INTRODUCTION

Ride has always been an important aspect in vehicle design. Ride is evaluated by means of objective measurements using standards such as ISO 2631-1 [1] or BS 6841 [2] and/or subjective evaluation. Ride evaluations are done to evaluate different suspension settings on a vehicle or comparisons of ride between different vehicles. Ride evaluations are normally done by driving the vehicle at the same speed over the same road and then comparing objective and/or subjective measurements between the two vehicles or a single vehicle with different suspension settings. In order to have a feasible comparison it is necessary that the road and speed is the same between tests. The road can easily be "fixed" by testing on a test track with a road surface that is non-deformable and therefore does not change with each pass. The speed of the vehicle can be controlled using different methods, the questions is how precise a vehicle's speed should be regulated during the experimental ride evaluations to ensure accurate and repeatable results.

A further important question is, that if a change in speed results in a change in the ride of the vehicle, how big of a change in speed will result in it being perceived by a human. In other words, if a small change in speed over the same road results in a difference that is perceivable by a human, then this may result in actual differences due to a change in suspension settings or differences between the two vehicles being obscured. Mansfield and Griffin [3] investigated the difference in magnitude between two stimuli which is just sufficient for the difference to be detected (i.e. just noticeable difference (JND)). They determined the JND for subjects seated on a vehicle seat excited with road profiles played through an actuator.

This paper investigates the sensitivity of ride comfort measurements to a change in speed over different road profiles ranging from smooth to rough roads using a validated 15 degree of freedom, non-linear MSC.ADAMS/View model of a Land Rover Defender 110. The differences in the objective ride comfort measurements caused by a change in speed are compared to the difference thresholds stated in Mansfield and Griffin [3] as an approximate indication of whether these differences will be large enough to be perceived by a human.

2. SIMULATION MODEL

A non-linear 15 DOF full vehicle MSC.ADAMS/View model of a Land Rover Defender 110 was used to simulate the vehicles response over different roads. The vehicle was modelled as two bodies with a torsional spring between them representing chassis stiffness which was determined experimentally. The vehicle is fitted with a four state semi-active hydro-pneumatic suspension system (4S4), developed by Els [4], that can be set to either a handling or ride mode. For the purpose of this study the vehicle model was run with the suspension setting in the ride mode. The suspension model includes suspension geometry, friction, bump stops, suspension bushings and damper characteristics which were experimentally measured and included in the model. The spring stiffness of the suspension was modelled as ideal adiabatic compression for which Van Der Westhuizen [5] achieved the best correlation to experimental measurements for high frequency inputs. An FTire model of the Michelin LTX A/T^2 235/85R16 tyre was used.

The road was modelled as a 2D road i.e. the road is given to the model as a profile defined by the longitudinal distance and vertical height for the left and right track. Three road profiles were created from the ISO 8608 [6] standard for a Class B, Class D and Class F road. ISO 8608 [6] specifies parameters for a 2D profile and recommends that left and right tracks be modelled to possess the property of isotropy. The 2D road profiles were generated and extended over the width of the road so that the left and right side of the vehicle experience the same excitation. In addition to the three ISO 8608 road profiles, a road was created from the experimentally measured Belgian paving test track at Gerotek [7]. The displacement power spectral density plot of the Belgian paving in Figure 1 shows how it relates to the ISO 8608 [6] road profiles.

Els [4] validated the full vehicle model over a 100 mm high, single discrete obstacle (locally known as an Aberdeen Proving Ground (AGP) bump) for low frequency vertical inputs. Cronjé [8] performed validation on the lateral dynamics by doing a double lane change (DLC) up to 70 km/h. Figure 2 shows the correlation obtained between experimental results and simulation results for weighted high frequency vertical acceleration measured on the right rear seat over Gerotek's [7] Belgian paving. Good correlation was obtained with regards to the trend of the data.

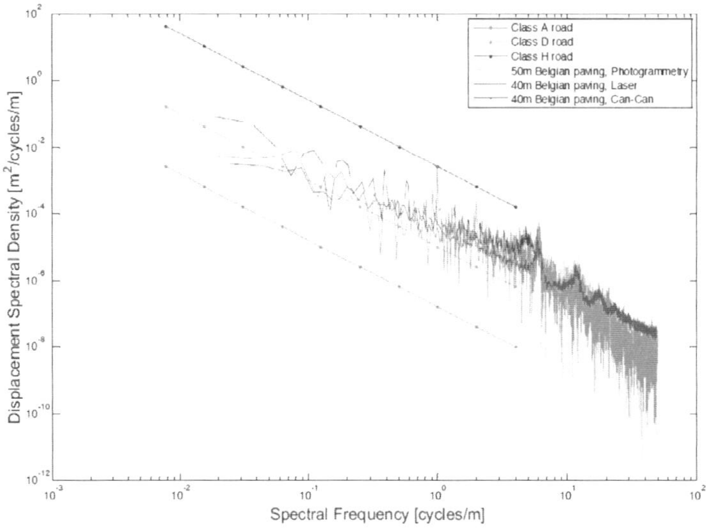

Figure 1: Displacement spectral density of Gerotek's [7] Belgian paving together with a class A, D and H road. Becker [9] used photogrammetry, laser and a Can-Can machine to measure the Belgian paving. The road profile used was measured with the Can-Can machine.

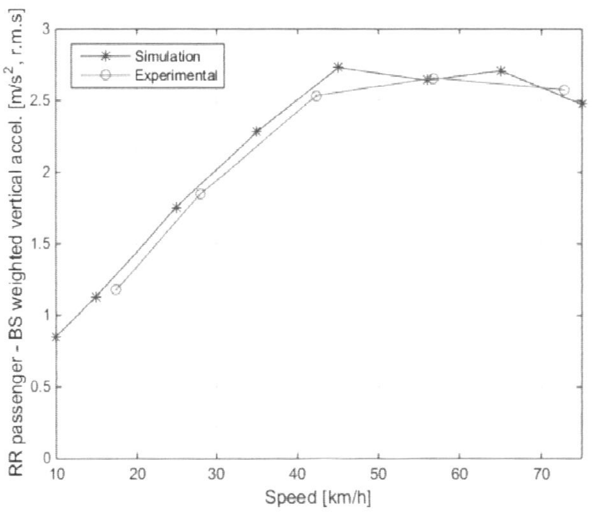

Figure 2: Comparison between experimental results and simulation results of the weighted r.m.s. of the vertical acceleration measured on the right rear seat.

2. METHOD

The MSC.ADAMS/View model of the Land Rover Defender 110 was simulated over the four road profiles (Class B, Class D, Class F and Belgian paving) at a range of speeds (i.e. 10, 15, 20, 25, 35, 45, 56, 65 and 75km/h). Translational as well as rotational acceleration in and about the x-, y-, and z-directions were measured in the ADAMS/View model. The accelerations were recorded at four locations in the vehicle: the driver seat, front passenger seat, left rear seat and right rear seat. The BS 6841 [2] standard was used to calculate the component ride values (CRV) and the overall ride value (ORV). Figure 3 shows the process followed to calculated the CRV, PRV and ORV. Each component of acceleration is multiplied by a certain weighting function and then multiplied by an axis multiplication factor. After checking the crest factor the component ride value (CRV) is calculated by using Equation (1) or (2). The BS 6841 [2] standard states that the CRV can be calculated using Equation (1) for crest factors below 9 while the BS 6841 [2] standard states that the crest factors should be below 6. All simulations conducted had crest factors below 4, allowing the use of Equation (1). The point ride value (PRV) and the ORV is then calculated by using Equations (3) and (4).

$$r.m.s = \sqrt{\frac{1}{N}\sum_{i=1}^{N} a_i^2} \tag{1}$$

$$r.m.q = \sqrt[4]{\frac{1}{N}\sum_{i=1}^{N} a_i^4} \tag{2}$$

$$r.s.s = \sqrt{r.m.s_{a_x}^2 + r.m.s_{a_y}^2 + r.m.s_{a_z}^2} \tag{3}$$

$$r.s.q = \sqrt[4]{r.m.q_{a_x}^4 + r.m.q_{a_y}^4 + r.m.q_{a_z}^4} \tag{4}$$

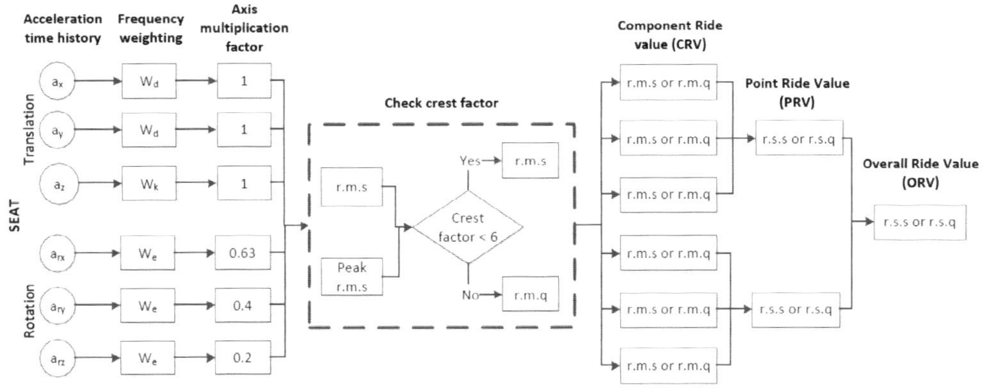

Figure 3: Method used for the weighting of vibration data for the analysis of ride comfort as suggested by Griffin [10].

3. RESULTS

The r.m.s. of vertical accelerations measured at the four locations in the vehicle gave the same trend over the speed range considered. Only the accelerations at the right rear seat are therefore presented. Figure 4 shows the vertical CRV and the ORV over the four roads at the eight speeds. The vertical CRV value and the ORV for the three ISO roads are similar, implying that the vertical acceleration is the dominant component. The ORV value is however slightly higher than the vertical CRV over the Belgian paving indicating that the other components have a slightly higher contribution over this road than for the ISO generated roads. Due to the ORV and the vertical CRV not differing much and considering that the JND in Mansfield and Griffen [3], for seated subjects, considers only vibration in the vertical direction, it was decided to use the vertical CRV for comparison with JND results. At speeds over 15 km/h, on the class F road, the vehicle starts to have significant periods of no contact between the tyres and the road, therefore Figure 4 only shows the weighted r.m.s. up to 15 km/h.

Figure 5 to Figure 7 shows the absolute difference and percentage difference in the vertical CRV at speeds of 1, 3 and 5 km/h higher than the desired speed. The absolute difference is calculated from the data in Figure 4 by subtracting the vertical CRV value at, for example, 15 km/h from the vertical CRV at a speed of 15+1 km/h, 15+3 km/h and 15+5 km/h, respectively. The percentage difference is calculated by dividing the absolute difference at each speed by the r.m.s. at each speed. These plots therefore presents the sensitivity of the ride comfort measure (i.e. the vertical CRV) to changes in speed over the different roads.

Figure 5(a) and Figure 5(b) shows the absolute difference and percentage difference, respectively, in the vertical CRV for a 1 km/h increase from the desired speed over the different roads. Figure 6(a) and Figure 6(b) depicts a 3 km/h increase and Figure 7(a) and Figure 7(b) depicts a 5 km/h increase.

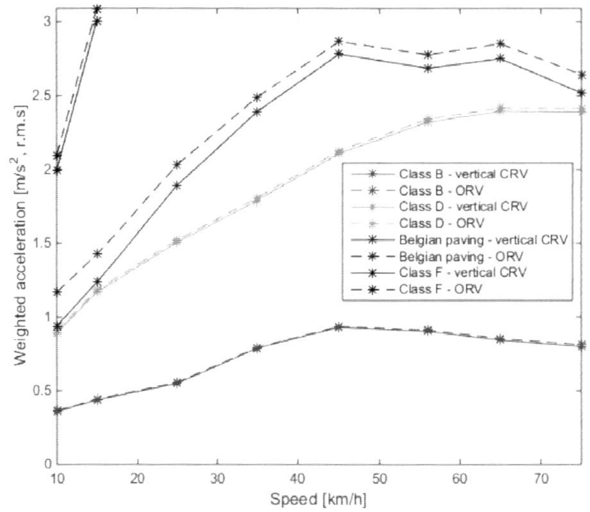

Figure 4: Comparison between the ORV and the vertical CRV for the various road profiles in simulation.

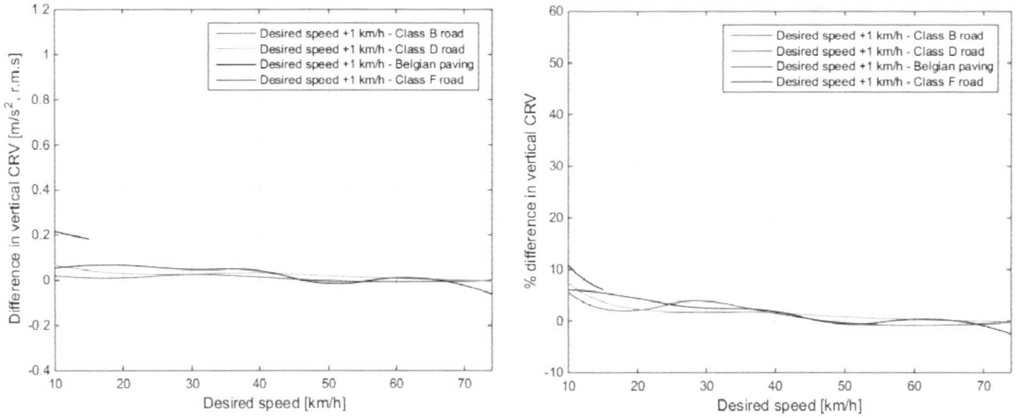

(a) (b)

Figure 5: (a) Absolute and (b) percentage difference in the vertical CRV for an increase of 1 km/h above the desired speed over various roads.

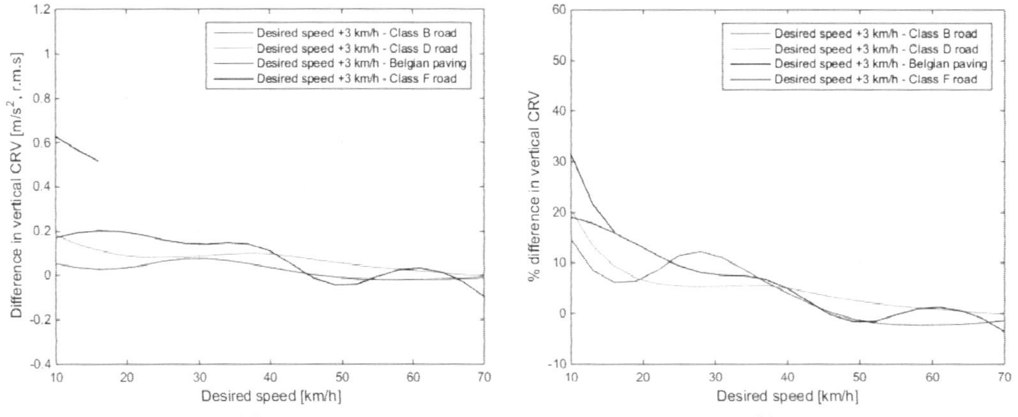

(a) (b)

Figure 6: (a) Absolute and (b) percentage difference in the vertical CRV for an increase of 3 km/h above the desired speed over various roads.

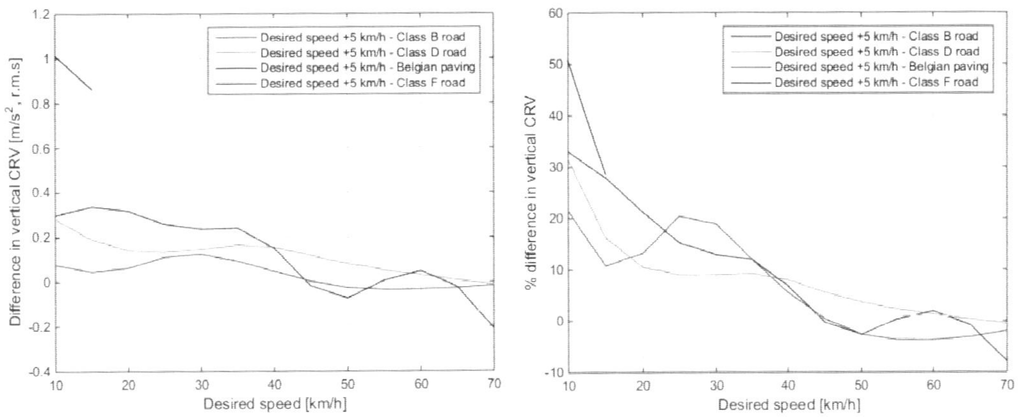

(a) (b)

Figure 7: (a) Absolute and (b) percentage difference in the vertical CRV for an increase of 5 km/h above the desired speed over various roads.

4. DISCUSSION

The ORV and the vertical CRV are very similar for the ISO roads, whereas the Belgian paving has an ORV that is larger than the vertical CRV as shown in Figure 4. This is due to the fact that the generated ISO 8608 [6] road profiles induces less roll, lateral acceleration and yaw than the Belgian paving. This is expected as the ISO 8608 [6] road profiles are constant over their width (i.e. left and right tracks are identical).

From Figure 5 to Figure 7 it is observed that for a certain increase in speed the difference in the vertical CRV tend to increase as the roughness of the road increases. It is also observed that the difference in the vertical CRV decreases as the speed increases. From this observation it could appear as if the best speed to perform suspension tests would be at higher speeds where the vertical CRV is less sensitive to a change in speed. From the results in Figure 4 however it can be seen that if this vehicle is driven over a Class D road above 45 km/h the ride can be classified as very uncomfortable according to the BS 6841 [2] standard. Therefore, it will not be sensible to drive the vehicle over this specific road at the higher speeds where the vehicle is less sensitive to a change in speed.

Also, from these figures we can observe that the difference in the vertical CRV is larger for greater increases in vehicle speed. This implies that if a vehicle is being evaluated over a specific road, a difference in the vertical CRV will be obtained if the speeds between two runs are different. For example, consider the vehicle driving over the Class B road. If the vehicle drives 1 km/h faster than the previous run that was done at 25 km/h it results in a 3.4 % higher vertical CRV. Whether the speed needs to be controlled within 1 km/h or 5 km/h precision will be determined by the difference threshold of a human. Mansfield and Griffin [3] obtained difference thresholds for whole-body vibration on a car seat. They obtained these difference thresholds for male and female subjects with a lab setup of a seat that was excited with a signal that was a reproduction of the vertical accelerations measured at the seat guide in a small car. The absolute difference thresholds from Mansfield and Griffin [3] for the male subjects are shown in Table 1. Note that Mansfield and Griffin [3] showed that there was no significant difference between the difference thresholds of their male and female subjects. The table shows the 25th percentile, median and 75th percentile values for the absolute difference thresholds obtained for two of the stimuli that were used i.e. Tarmac track 0.4 and Tarmac track 0.8. These two stimuli resulted in weighted r.m.s. values of 0.4 m/s^2 and 0.8 m/s^2 at the seat surface in the vertical direction.

The vehicle model driving over the Class B road at 15 km/h and at 35 km/h resulted in a weighted r.m.s. value of 0.44 m/s^2 and 0.79 m/s^2 respectively. The difference thresholds of Tarmac track 0.4 are therefore used as an approximate indication of whether the difference in the vertical CRV value over the Class B road at 15 km/h will be perceivable. Similarly, the difference thresholds of the Tarmac track 0.8 are used as an approximate indication of whether the difference in the vertical CRV value over the Class B road at 35 km/h will be perceivable. Table 2 shows the difference in the vertical CRV value over the Class B road at the two speeds. Only the difference in the vertical CRV value for a 3 km/h and 5 km/h increase from the desired speed of 35 km/h over the class B road falls within the ranges specified in Table 1. This implies that these differences may be large enough to be perceived by a human whereas the other differences may not be large enough. Therefore, for the class B road at a desired speed of 15 km/h it may be sufficient to control the speed with a precision of 5 km/h without a human perceiving any changes in vibration due to a change in speed.. On the other hand the speed will have to be controlled with a precision of 1 km/h over the class B road at a desired speed of 35 km/h in order to avoid a human perceiving any changes in vibration due to a change in speed.

	Tarmac track 0.4 (0.40) [m/s^2, r.m.s.]	Tarmac track 0.8 (0.80) [m/s^2, r.m.s.]
25th percentile	0.0530	0.0590
Median	0.0600	0.0880
75th percentile	0.0830	0.1110

Table 1: Absolute difference thresholds for male subjects for Tarmac track 0.4 and 0.8 from Mansfield and Griffin [3].

	Class B @ 15 km/h (0.44) [m/s^2, r.m.s.]	Class B @ 35 km/h (0.79) [m/s^2, r.m.s.] "*" = above JND threshold
+1 km/h	0.0104	0.0220
+3 km/h	0.0301	0.0610*
+5 km/h	0.0473	0.0940*

Table 2: Difference in vertical CRV for a class B road at speeds of 15 and 35 km/h.

The most suitable JND data available was used, but it is important to remember that the frequencies present in the vibration signal that a human experiences has an effect on the JND threshold. Ideally JND should be determined for the same frequencies and magnitudes to which a human will be exposed to while testing a vehicle. It is also important to keep in mind that the simulation results presented here are strongly dependant on the type of vehicle, vehicle suspension and on the road profile.

5. CONCLUSION

The results showed that the objective ride comfort measures are dependent on the nominal speed, the road roughness and the difference in the speed between two tests. The precision to which the speed is to be controlled was shown to be dependent on these variables, but more specifically on the difference threshold [4] of a human.

REFERENCES

[1] International Organization for Standarization, *ISO 2631-1 Mechanical vibration and shock - evaluation of human exposure to whole-body vibration,* Geneva, 1997.

[2] British Standards Institution, *BS 6841 - British Standard Guide to Measurement and evaluation of human exposure to whole-body mechanical vibration and repeated shock,* London, 1987.

[3] N. Mansfield and M. Griffin, "Difference thresholds for automobile seat vibration," *Journal of Applied Ergonomics,* vol. 3, no. 31, pp. 255-261, 2000.

[4] P. S. Els, "The ride comfort vs. handling compromise for off-road vehicles," Philosophiae Doctor, University of Pretoria, 2006.

[5] S. Van der Westhuizen and P. Els, "Comparison of different gas models to calculate the spring force of a hydropneumatic suspension," *Jounal of Terramechanics,* no. 57, pp. 41-59, 2014.

[6] International Organization for Standarization, *ISO 8608 - Mechanical vibration - Road surface profiles - Reporting of measured data,* Geneva, 1995.

[7] Gerotek Test Facilities, *Vehicle testing facility,* Pretoria: ARMSCOR, 2016.

[8] P. Cronjé, "Impoving off-road vehicle handling using an active anti-roll bar," Masters of Engineering, University of Pretoria, 2008.

[9] C. Becker, "Profiling of rough terrain," Masters of Engineering, University of Pretoria, 2008.

[10] M. Griffin, Hanbook of Human Vibration, London: Academic Press Limited, 1990.

Modeling and Development of a Magneto-Rheological Equipped Hydro-Pneumatic Semi-Active Suspension System

Gerhardus S Heymans

University of Pretoria
Pretoria, South Africa
11030501@tuks.co.za

P Schalk Els

University of Pretoria
Pretoria, South Africasc
schalk.els@up.ac.za

ABSTRACT

The Vehicle Dynamics Group (VDG) at the University of Pretoria have developed a semi-active hydro-pneumatic suspension system for application on an off road vehicle. Els (2006) who developed this 4 State Semi-Active Suspension System ($4S_4$) designed it such that it can switch characteristics between two discrete spring - as well as between two discrete damping characteristics, all incorporated in a single suspension strut. Recently, the $4S_4$ was further developed with the aim of extending its damping characteristics to be continuously variable through the use of Magneto-rheological (MR) technology, a system which was now called the $MR4S_4$. The damping characteristics of each of the magnetic flow valves are controlled independently by the application of electric currents through the coils. These MR valves are also able to block flow completely which achieves switching between the discrete spring characteristics from independent accumulators.

The $MR4S_4$ prototype was successfully modelled through a physics based approach and its characteristics validated against experimental results. However, during the characterisation process it was realized that the $MR4S_4$ suffers from undesirable response time delays. Thus a research initiative was initiated which aimed to improve the response time characteristics of the system and reduce the delays observed during switching of characteristics. The factors influencing the response time of the MR system was studied through a combination of simulation and experimental work as will be highlighted in this paper.

KEYWORDS

Magneto-rheological (MR) fluid, $MR4S_4$, Semi-Active Suspension System, Physics Based Modelling, System Response Time.

INTRODUCTION

In the world of vehicle dynamics, it's a well-known fact that the suspension characteristics required for ride comfort and handling are contradictory. Performance vehicles sacrifice passenger comfort to improve handling while luxury sedans compromise stability and control to ensure better ride comfort over rough roads.

In an attempt to solve this inherent compromise fully active suspension systems are often used in high performance sports cars and allow the vehicle to adapt rapidly to changes in dynamic conditions, although at the cost of considerable power consumption. Perhaps the most elegant solution to this compromise can be found between the two extremes of a fully active suspension system and a passive suspension. Semi-active suspensions systems have been found as a good solution to these extremes. As a result, semi-active suspensions have enjoyed considerable attention of late by both industry and academia and has become widely available for demanding off-road vehicle applications.

In an attempt to solve this compromise, Els (2006) developed the 4-State Semi-active Suspension System or $4S_4$ to switch between suspension characteristics which are tailored for either ride comfort or handling. The $4S_4$ was designed as a low power-consuming semi-active suspension system which achieves various characteristics by switching solenoid valves to allow spring and damping contributions from two independent accumulators and two independent dampers. However, once Els had proved that his system could successfully solve the compromise, be it at a maximum switching rate of 100 milliseconds, further development was launched into replacing its simple orifice dampers with more continuously controllable magneto-rheological fluid based magnetic valve dampers. This concept sparked the birth of the $MR4S_4$, a semi-active suspension system, which could theoretically change its damping characteristics through a continuous range while being able to switch between discrete spring states. A magneto-rheological damper is considered a semi-active damper as it can develop a large range of damping force characteristics with the application of only proportionally small amounts of electric energy (magnetic fields) to the system. A section view of the current $MR4S_4$ as developed and implemented in 2015 is depicted in Figure 1 showing the two respective MR valves and spring accumulators.

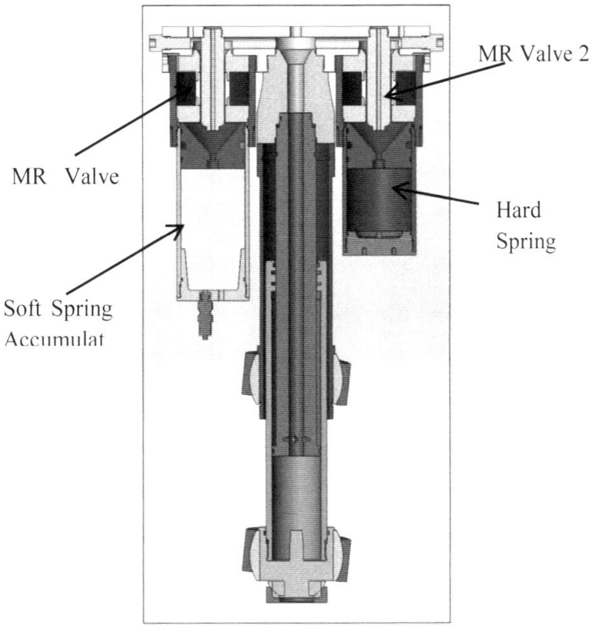

Figure 1: Sectional View of MR4S4

Once manufactured and assembled an extensive experimental characterization was performed on the $MR4S_4$. During the characterization process the $MR4S_4$ strut forces and pressures were measured as well as the individual accumulator and damper pressures so that a thorough picture of the operation of the system could be obtained. A total of 28 channels were recorded during this process. The characterisation aided the modelling and validation of the individual spring and damper elements within the suspension system as well as allowing one to validate the overall characteristics of the unit. Figure 2 shows the $MR4S_4$ installed on the test bench and actuated by a linear hydraulic Schenk actuator while the pressure, force and temperature measurements across a range of points in the system were recorded.

Figure 2: $MR4S_4$ Experimental Characterization

Physics Based Modelling Approach

After experimental characterization had been performed, an appropriate system model was developed based on the thermodynamic - and fluid flow properties of the system. This was done in contrast to much of the literature available on Magneto-Rheological damper models where it is practice to capture the characteristics of MR based dampers through 'curve' fitted models. However, for this application a physics based modelling approach was essential in order to capture the large non-linearity of the system and account for the complex interactions between the elements of the $MR4S_4$. The model calculates the volume of the displaced fluid, that flows from the main strut and then splits along the two flow paths through the independent MR valves and into the independent accumulators from which the model then calculates the overall force characteristics.

The flow split between each of the flow paths ($\dot{q}_1$ and $\dot{q}_2$) is obtained by iteratively solving a complex set of inter-linked non-linear equations which govern the accumulators and the MR dampers.

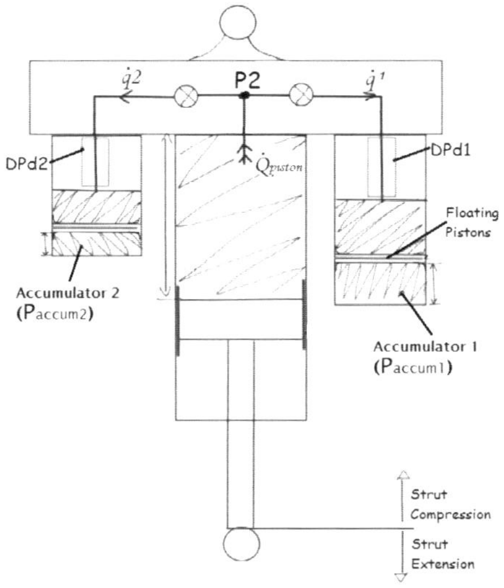

Figure 3: Physics Based Modelling Approach

The flow rate of $\dot{q}_1$ and $\dot{q}_2$ are intricately linked to the conditions at both the MR dampers as well as the accumulator states for each time step. The flow split has to be iteratively estimated at each time step while taking into account the respective pressure drops across the MR dampers and the connected accumulator spring pressures. The pressure drops across the MR dampers (DP_{d1}, DP_{d2}) are a function of both the current through the MR damper coil as well as the fluid flow rate. This pressure drop across the MR damper is also a function of the geometry, orifice size, magnetic field applied and material properties of the MR fluid and the derivation of how this element was modelled will be discussed in this paper. Similarly the spring element is a function of compression history, the initial conditions as well as the gas compression process. The gas compression is modelled by the Benedict-Webb-Rubin (BWR) real gas model (Cooper and Goldfrank, 1967) as was proved appropriate by Els (2006).

The physics based approach solves a set of equations for each time step to obtain a possible solution to the split condition. The flow split, i.e. the ratio between the flow in each of the flow path, is estimated and related to the total measured flow rate from the main piston strut by equation 1 and 2.

$$q_1 = split * Q_{pist} \quad [1]$$

$$Q_{pist} = Velocity_{Piston} * Area_{Piston} = q_1 + q_2 \quad [2]$$

The flowrate is integrated to find the volume change in both the accumulators due to the fluid flow.

$$dV_1 = q_1 * \frac{1}{fs} \quad [3]$$

$$dV_2 = q_2 * \frac{1}{fs} \quad [4]$$

After obtaining the solutions, the pressure drops across the dampers as well as the pressures in the accumulators are calculated using the flowing functions.

$$DP_{d1} = f_{damper}(q_1, Coil\_1\_Current) \quad [5]$$

$$DP_{d2} = f_{damper}(q_2, Coil\_2\_Current) \quad [6]$$

$$P_{accum1} = f_{SpringRate}(P_{01}, V_{01}, dV_1, T_{g1}) \quad [7]$$

$$P_{accume2} = f_{SpringRate}(P_{02}, V_{02}, dV_2, T_{g2}) \quad [8]$$

Once these calculations have been performed the model determines whether the obtained solution, or assumed split condition, results in an equilibrium pressure condition at the centre (P2). At equilibrium both flow paths are give the same P2 value as defined by equation 9. This formulation assumes that there are negligible fluid frictional losses during flow in the system, or in other words that these losses can be incorporated by the damper elements which also oppose fluid flow.

$$P_{accum1} + DP_{d1} = P_2 = P_{accum2} + DP_{d2} \quad [9]$$

If this condition is not met, the model returns to equation 1 and iterates the above processes by changing the split condition assumed. A non-linear Matlab based iterative solver is used on to iteratively solve for the split condition and finds the equilibrium split condition. After this is obtained, the output force of the suspension system can simply be modelled as in equation 10.

$$F_{MR\,4S_4} = P_2 * A_{piston} \quad [10]$$

The MR Damper Element Model

To model the damper element, as defined by the functions in equations 5 and 6, the Philips equation (Phillips, 1969) relates the pressure drop across a MR valve to the valve geometry, MR fluid properties and Magnetic field strength applied across the fluid. The formulation accounts for the pressure drop over the valve as a combination of the MR fluid friction, which is MR viscosity based and proportional to the flow gap, and the magnetic field induced shear stress effects.

It was however noticed that the Phillips formulation did not accurately capture the high velocity pressure drop characteristics observed experimentally across the MR valve. This is possibly explained by the fact that at high velocities the fluid has less opportunity to form magnetic poles and combine into structures which can resist fluid flow and create the desired damping effect. Additionally, MR fluid is a non-Newtonian fluid and exhibits a shear thinning at high velocities and therefore the viscosity becomes a function of the fluid shear rate. A scaling factor was added to reduce the magnetic field effect at high velocities while the viscosity of the fluid was also modelled as a function of shear rate to capture the reduced viscosity effects.

$$\Delta DP_{d(i)} = \Delta P_\eta + \Delta P_\tau = \frac{12\eta(\dot{\gamma})QL}{g^3 w} + \frac{c\tau_{y(field)}L}{g} * \frac{1}{2}(2 - e^{-(q*\dot{\gamma})}) \quad [11]$$

Where, η is the MR fluid Viscosity, $\dot{\gamma}$ is the Fluid Shear Rate, Q is the MR fluid Volume Flow Rate, L is the Length of MR fluid interaction with the magnetic fields while g and w are the MR Valve Flow channel width and circumference length respectively. Finally the damper pressure drop is further defined by the MR fluid Magnetic Field Dependant Fluid Shear Stress, $\boldsymbol{\tau_{y(field)}}$, as well as a Philips equation scaling constant, c.

The magnetic field dependant shear stress of the MR fluid was furthermore modelled using "Dr. Dave's" empirical formulation, equation 12, for the shear stress of MR fluid vs. magnetic field strength, which is a widely used formulation for estimating the shear strength changes in MR fluid. (Carlson, 2005). Although the application of this equation appears trivial, it must be noted that Φ and H are related to one another through the magnetic 'circuit within the MR valve.

$$\tau_{y(field)} = C * 271700 * \Phi^{1.5239} * \tanh(6.33 * 10^{-6} * H) \qquad [12]$$

In this equation the MR fluid Magnetic field dependant shear stress is a function of the Magnetic Flux Density (Φ), through the MR fluid as well as the Magnetic field strength (H) across the fluid with a final scaling function (C).

Pneumatic Spring Element Model

Els (2006) determined that the Benedict-Webb-Rubin (BWR) real gas model gave good correlation for the experimentally observed spring characteristics within the $4S_4$. For this reason the BWR real gas model was again used to model the $MR4S_4$ which operates under similar pressure and temperature conditions as the original suspension unit. Again good correlation was observed with this spring model.

$$P(v) = \frac{(R * T_g)}{v} + \frac{B_0 R T_g - A_0 - \frac{C_0}{T_g^2}}{v^2} + \frac{bRT_g - a}{v^3} + \frac{\left(c\left(1 + \frac{\lambda}{v^2}\right) e^{-\frac{\lambda}{v^2}}\right)}{v^3 T_g^2} \qquad [13]$$

The above BWR model was incorporated in a function, as stated in equation 7 and 8, which calculates the accumulator pressure as a function of the initial conditions and the respective fluid volume change during that time step. A higher accuracy of the model can be obtained by incorporating the gas temperature.

Modelling Validation and Correlation

In order to validate the $MR4S_4$ model it was exposed to identical 'conditions' as was seen during experimental characterisation tests. The same initial conditions and displacement time history as was experimentally recorded was used in the simulation. During this validation process, multiple correlation tests were performed at various velocities and with the MR valves activated by various coil currents. Although not all correlations depicted in Figure 4 and 5 were exact, it was found that the model was able to capture the extreme non-linearity's of the system to a large extent throughout its operational range as Figure 4 and 5 demonstrates.

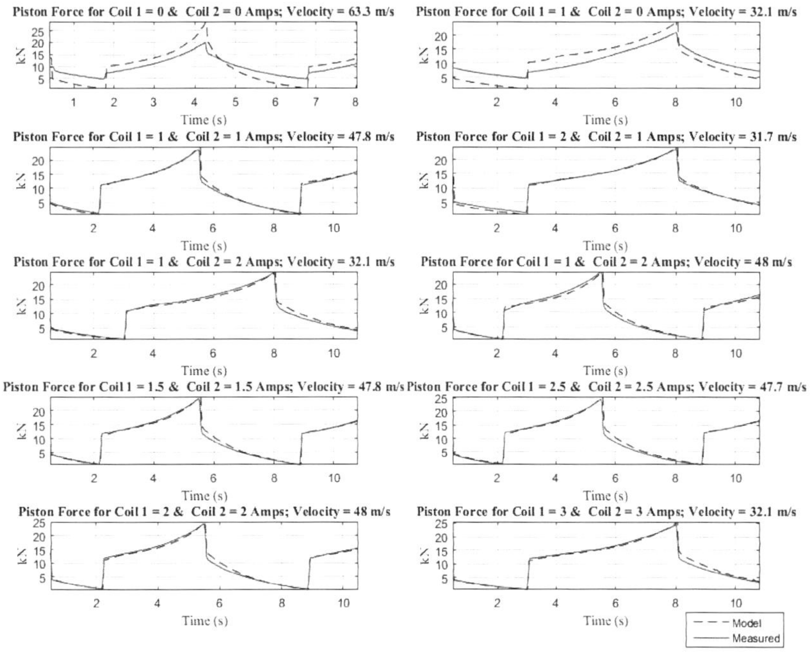

Figure 4: Force vs. Time correlation of the overall $MR4S_4$ model vs. experimental measurements.

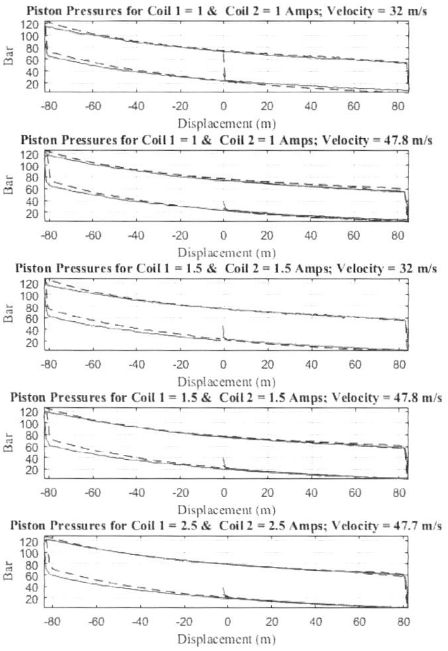

Figure 5: Force vs. Displacement correlation of the overall $MR4S_4$ model.

MR $4S_4$ Modelling Conclusion

Although good correlation was observed during this analysis, it was realized that this correlation deteriorates when there is state switching present. In other words, the correlation diminishes when the MR valve coil currents are changed during operation. The deviation from the model was soon explained by a inherent MR valve response time delay observed during state switching, a behaviour which required further analysis. The remainder of this paper will focus on the research conducted in order to minimize the response time characteristics of the MR valve such that it will align better with the model and allow for faster and more instantaneous switching as is captured by the model.

Reducing MR Valve Response Characteristics

The response time delay behaviour of the MR valve was not only detrimental towards the accuracy of the physics based model but also brought about negative implications for the control of the $MR4S_4$ characteristics in a vehicle application. Due to the impact that the response time has on the accuracy of the model this continued work is also discussed in this paper.

Additional research focused on reducing the time response properties of the MR valve while keeping the off-state damping and full locking ability of the original MR valve. This work entailed detailed magnetic study and optimization of the MR valve.

Finite Element Magnetic Study

As a start to the MR valve magnetic study the original MR valve was imported and setup in a 2D finite magnetic simulation package called Finite Element Method Magnetics (*FEMM4.2*). This package supports 2D planar or 2D axis-symmetric problems, the latter of which was used in the case of the MR valve.

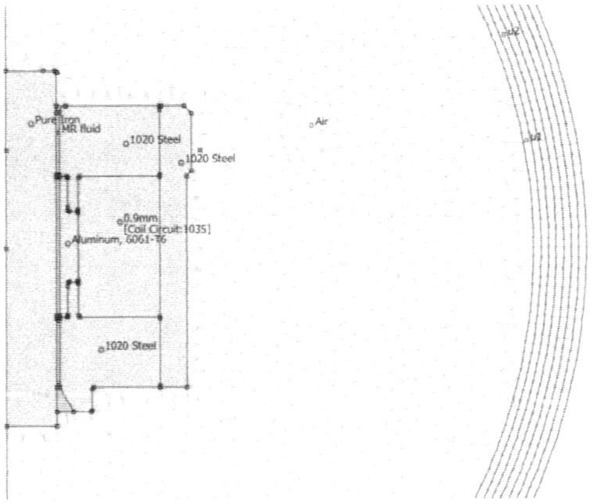

Figure 6: MR Valve FEMM mesh.

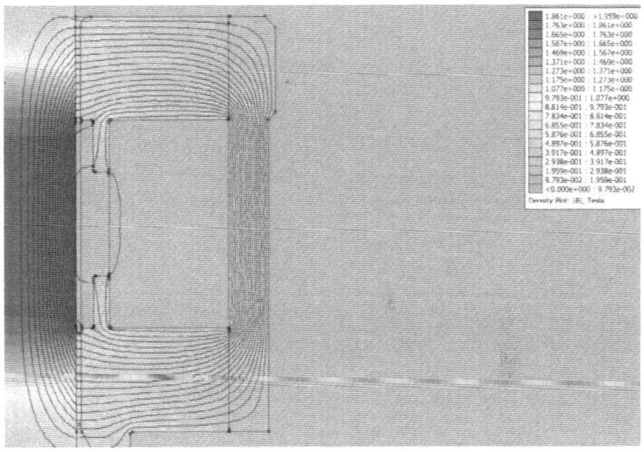

Figure 7: Solution Obtained from FEMM for a coil current of 1 Amp.

In addition to obtaining magnetic field strength and magnetic flux density results from the simulations, the coil circuit properties are calculated based on the coil diameter, number of turns and wire diameter. The coil's resistance, inductance, power dissipation, response time (τ) as well as voltage drop are calculated based on the applied coil current.

Original MR valve Response Properties

In order to verify the validity of the FEMM solutions obtained, the response time properties as well as circuit resistance and voltage drop were experimentally investigated for the same original MR valve. The MR valve was given a step in drive voltage, while the response of the current through the coil was recorded. The coil current was driven with both 6V and 12V battery supplies while the coil currents were respectively recorded. Additionally, tests were conducted with and without the presence of the MR fluid in the flow gap. Each test was repeated 3 times with extremely good repeatability, depicted by Figure 8.

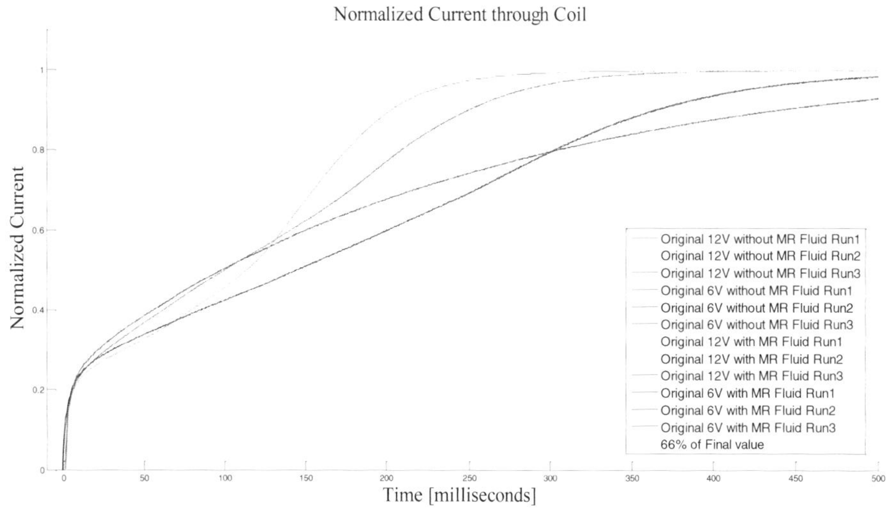

Figure 8: Response Time Characteristics of Original MR valve coil current.

This investigation yielded an extremely good correlation between the results obtained from experimental and simulated studies which are illustrated in Table 1. Experimentally it was determined that the current in the coil reached 66% of its steady state value in 215 milliseconds for the case where MR fluid was present in the valve energized by 6V. Correspondingly the simulation suggested the Tau value for this case to be 227 ms. This is a remarkably close comparison. Furthermore it was noted that the voltage drop and resistance measured was also very close to the simulated value given that the simulation was run at the current measured in the experimental results. Again the same trends were observed in the 12V study as well as the study without MR fluid present in the valve.

Table 1: 6V and 12V comparison of simulation and experimental results for a magnetic valve with and without MR fluid present.

	Simulation With MR Fluid	Experimental With MR Fluid
	6V Simulation	6V Experimental
Tau (miliseconds)	227	215
Voltage Drop (V)	6.71	6.45
Resistance ('R')	6.133	6.3
Current (Amps)	1.095	1.095
	12V Simulation	12V Experimental
Tau (miliseconds)	131	115
Voltage Drop (V)	13.57	12.44
Resistance ('R')	6.133	6.3
Current (Amps)	2.213	2.213

	Simulation Without MR Fluid	Experimental Without MR Fluid
	6V Simulation	6V Experimental
Tau (miliseconds)	212	229
Voltage Drop (V)	6.71	6.335
Resistance ('R')	6.133	6.3
Current (Amps)	1.095	1.095
	12V Simulation	12V Experimental
Tau (miliseconds)	127	125
Voltage Drop (V)	13.6	12.5
Resistance ('R')	6.133	6.3
Current (Amps)	2.28	2.28

* Note these are 66% Tau measurements

This FEMM validation investigation provided the required confidence that the FEMM package can be trusted to give at least the correct trends and approximate solution to a magnetic characteristics of the MR valve and could thus it can be use with some confidence as a tool to develop a more responsive MR valve.

Optimizing MR valve for response time characteristics

An investigation was launched which aimed to keep the same MR valve outer dimensions and fluid flow area, as this is related to the off state characteristic, while decreasing the magnetic flux resistance. The reason for this was that a decrease in the magnetic resistance of the valve could increase the magnetic field strength. Because a higher magnetic field strength was not required the coil size and number of turns could be reduced while simultaneously reducing response times of the system.

During the above process the FEMM package was used in a co-simulation with Matlab. From this co-simulation, geometry, materials, coil turns and diameters etc. could be changed parametrically and the simulation performed and results extracted. A Matlab based non-linear optimization was used in order to obtain a solution which would provide an adequate magnetic field strength of over 180 kA/m while minimizing the response properties of the coil. The design requirement was set at a coil response time of less than 30 milliseconds. However, it was soon realized that the there would be no setup which would satisfy both of these requirements. To overcome this problem it was decided that a double coil setup would need to be implemented in which a small coil can be used to adjust the MR valve characteristics to 40% of the locking requirement, where after a bigger coil will be used to lock the MR valve such as to achieve the flow blocking requirement of at least 180 kA/m, but at a much slower response time.

The surface plots of Figure 9 highlight the investigation to find appropriate magnetic field strength and response time properties for various coil sizes and number of turns while energizing these coils with their respective maximum carrying capacity current. The magnetic field strength plotted in Figure 9 was determined from the solution of the FEMM solver as an average across the MR gap. The first point which satisfies the magnetic field strength requirement of 180 kA/m has a response time of 48 milliseconds which does not meet the second criterion.

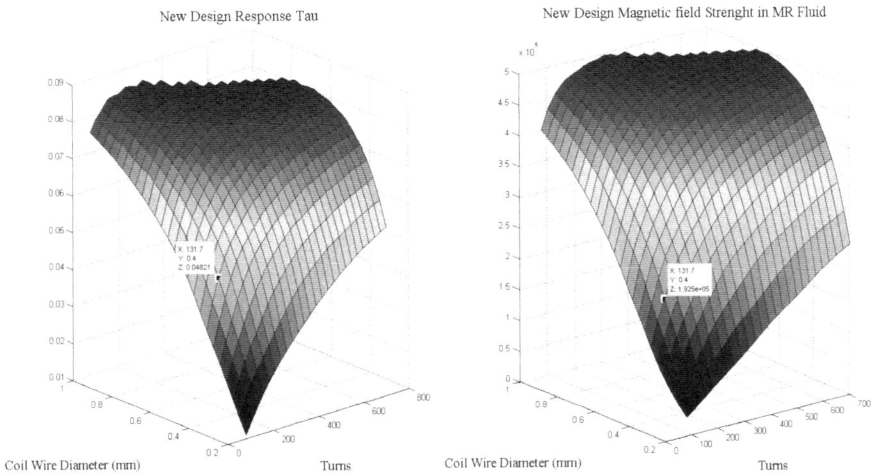

Figure 9: New design Response and Magnetic Field strength Properties.

The above investigation however became instrumental as a design tool from which the coils could be designed. The small coil was selected to have 200 turns of a 0.2 mm diameter wire which would give a resultant response time of 23 milliseconds and 70 kA/m magnetic field strength in the MR gap. Similarly the large locking coil was specified to have 350 turns with a diameter of 0.4 mm which should securely lock the MR valve with a magnetic field strength of 320 kA/m be it with a slower response time of 65 milliseconds.

Final Optimized MR valve Geometry and Properties.

The new optimized double coil MR valve was then designed based on the decisions made from FEMM simulation study. The final geometry was again imported into the 2D software for analysis and an estimate of the characteristics.

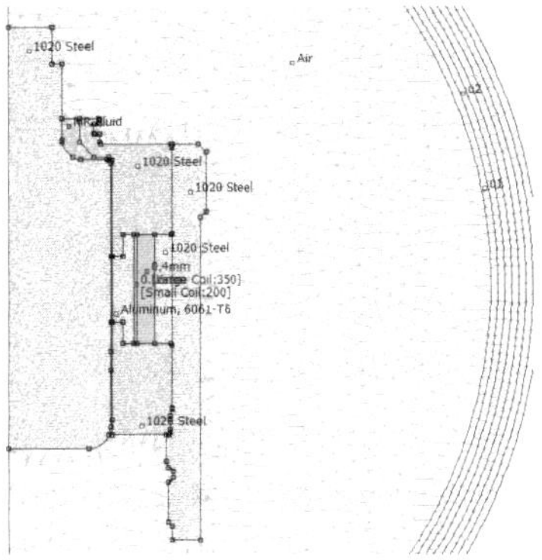

Figure 10: 2D FEMM import of the new MR valve

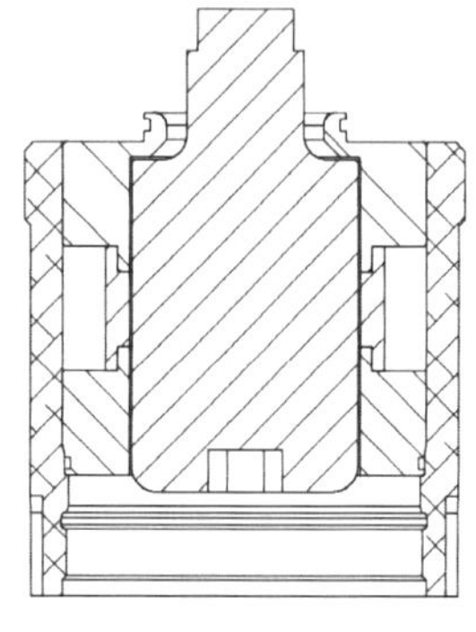

Figure 11: Sectional View of the new MR valve design

The new design's magnetic field strength in the MR gap as well as its coil response properties was investigated through the simulation software. The results are indicated in Figure 12. The small coil allows switching between states within 26.5 milliseconds throughout its current carrying range. At the same time the larger coil has been selected to provide sufficient magnetic field strength to achieve flow blocking at values larger than 180 kA/m which would suggest that if the large coil was energized with only 0.8 A, one can expect flow blocking within about 120 ms, although increasing the current could bring down the blocking to about 60 ms.

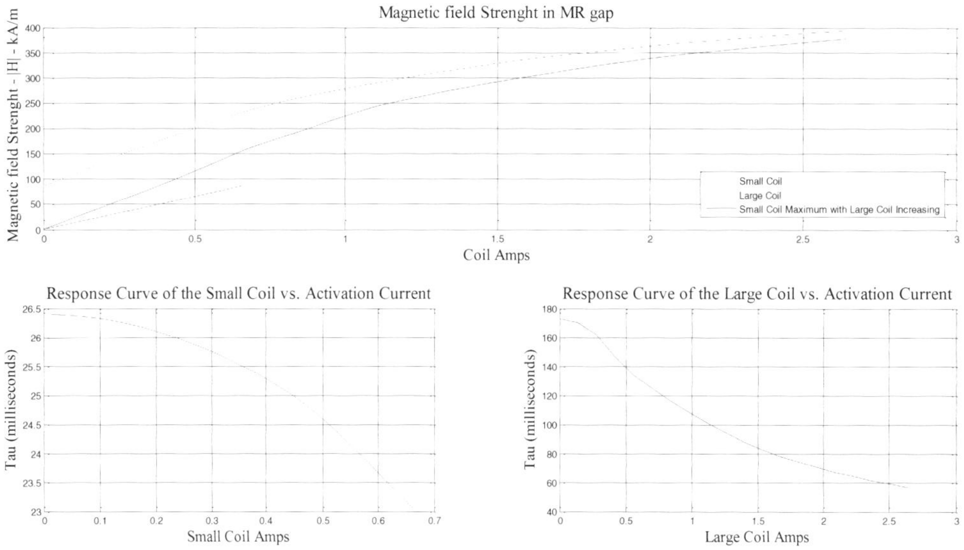

Figure 12: Magnetic field and Response properties of the newly designed MR valve

Experimental Response time Characterization of the new MR valve.

After manufacturing of the designed MR valve, the response times of the respective coils were experimentally tested. Here the experimental results recorded surpassed what was predicted by the FEMM package. Experimentally it was observed that the current though the small coil reaches its 66% steady state level almost instantly after the coil has been activated while it takes 18 ms for the current to rise to 95% of the steady state value. This was the case for both the 6V and 12V experiments conducted on the small coil with respective 0.25 and 0.5A steady state coil currents.

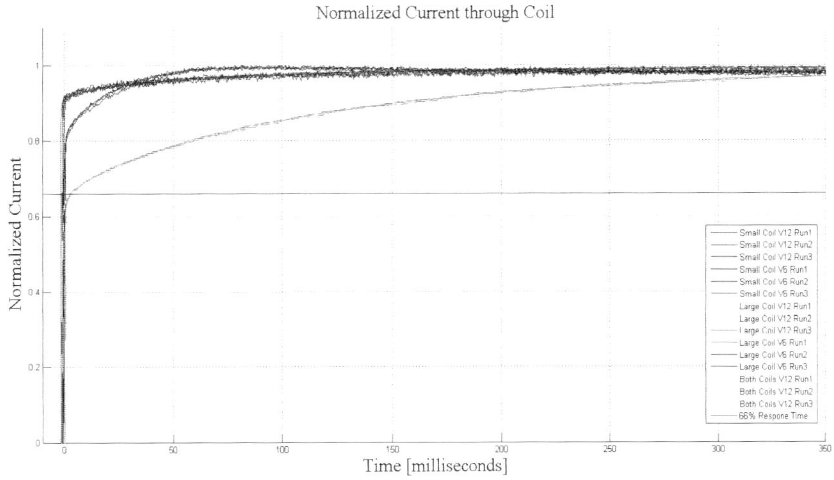

Figure 13: Coil Current Response Curves to Step Voltage Supplies

However the above experimental work seems to suggest that the large coil creates a magnetic field which saturates to some degree when activated by 12V. This conclusion can be drawn because the large coil only uses 30 milliseconds to reach 95% of its steady state value when 6V is applied to it. However, as soon as 12V is applied, the settling time becomes almost 10 times longer. This clearly suggests that the magnetic field has saturated some part of the MR valve causing the inductance to remain high and the response time to be slow.

The results brings some questions as to why the FEMM solution was not reflected in the experimental results or vice versa. This deviation suggests some more validation is required of the FEMM package. However, one can clearly conclude from the above experimental results that the response time of the new MR valve is almost 10 times faster than that of the old design.

Conclusion

This paper focused on the derivation of a physics based model for the $MR4S_4$ as well as discussing the work conducted in order to improve the response time of the $MR4S_4$ system. The initial MR valve was characterised by a typical response time of 230 milliseconds to achieve 95% of its steady state value. In contrast after the development of the new MR valve the system now displays a response characteristic of less than 30 milliseconds to reach 40 percent of the required magnetic field while full flow blocking in the MR valve is reached in 120 milliseconds. Therefore the response characteristics of the full locking system has been halved at least while the new valve now allows for a much more responsive control of the damping properties through the use of the faster switching small coil, an ability which the large coil in the initial MR valve did not possess.

The above MR valve research and development has helped improve the MR $4S_4$ model correlation. Additionally, the application of this faster MR valve will enable control systems to more accurately control the $MR4S_4$ to achieve more desirable characteristics as it enables them to switch between damping and spring states faster.

This work opens up further avenues of research which could use the $MR4S_4$ model to study the possible contributory effect the $MR4S_4$ system, while the newly developed MR valve could serve as a basis for investigating required control strategies in order to improve the dynamics, comfort and safety of applicable vehicles of the future.

REFERENCES

Aly, A.A. and Salem, F.A., 2013, "Vehicle Suspension Systems Control: A Review", *International Journal of Control, Automation and systems,* Vol 2, pp. 46-51.

Carlson, J D., 2005, "MR Fluids and Devices in the Real World." *International Journal of Modern Physicis B*, Vol 19.

Els, P.S., 2006, *The Ride Comfort vs. Handling Compromise for Off-Road Vehicles Ph.D. Thesis.* Pretoria, Mechanical Engineering, University of Pretoria, http://repository.up.ac.za/handle/2263/43431.

Goldfrank, H.W. and Cooper, J. C., 1967, "B-W-R constants and new correlations." *Hydrocarbon Processing, Vol. 46.*

Phillips, R.W., 1969 "Engineering applications of fluids with variable yield stress". *Ph. D Thesis.* Berkeley: University of California.

Spencer, D. S. (1996). Phenomenological Model of a Magnetorheolgical Damper. To Appear in: *ASCE Journal of Engineering Mechanics, currently obtained at http://citeseerx.ist.psu.edu/viewdoc/download?doi=10.1.1.119.4990&rep=rep1&type=pdf.*

Developing a methodology to fit a constitutive creep model to 1Cr-1Mo-0.25V cast alloy steel

Gareth J. Howard*[1], Helen M. Inglis[1], Francesco Pietra[1], Schalk Kok[1]

[1]Department of Mechanical and Aeronautical Engineering, University of Pretoria

g-j-howard@hotmail.com

Abstract: Creep data is rarely available in a form that allows for easy implementation into commercial FEA software. This paper presents a method for effectively creating a creep material model in the commercial FEA software ANSYS using experimental creep data from the Japanese National Institute for Materials Science (NIMS). An optimisation methodology is proposed that solves for ANSYS creep model coefficients in order to approximate the response of 1Cr-1Mo-0.25V cast alloy steel. Through the use of the downhill simplex method and a form of Tikhonov regularisation, a family of creep curves are developed for the material. By implementing these curves into commercial FEA software, remaining useful life estimations can be made for components of interest in thermal power stations.

Keywords: Creep, constitutive model, finite element analysis, optimisation, NIMS experimental data

INTRODUCTION

Due to the increasing demands on power stations worldwide there is an on-going effort to increase component operational temperatures, thus increasing plant efficiencies. As a consequence the structural loads on components which are exposed to these increased temperatures also increase. Many of these power stations have deteriorated markedly with age and are now operating under conditions other than those for which they were designed. For this reason the industry needs more accurate methods of modelling materials and predicting remaining useful life [1]. The research which is reported in this paper arises from challenges encountered in modelling the creep life of a 1Cr-1Mo-0.25V cast alloy steel component in service on a South African power plant.

Creep is defined as time-dependent irreversible deformation at elevated temperatures [2, 3]. Most metals experience creep at elevated temperatures of approximately 40% of their melting point temperature. Creep is generally considered to occur in three stages, namely primary creep, secondary or steady-state creep, and tertiary creep, as depicted in Figure 1. Primary creep is characterised by a transient response in which strain rate $\dot{\varepsilon} = d\varepsilon/dt$ decreases with time. This decreasing strain rate is believed to be due to a changing microstructure, causing increasing resistance to dislocation movement. Secondary creep, which comprises most of the part's life, is characterised by approximately linearly increasing strain and is hence also referred to as steady-state creep. During secondary creep there is an approximate balance between dislocation increase due to strain hardening and dislocation annihilation due to recrystallization, hence the approximately steady creep strain rate. Lastly, tertiary creep is characterised by exponentially increasing strain rate until eventual rupture occurs. The onset of tertiary creep is usually macroscopically characterized by the onset of necking and damage, and hence tertiary creep is short lived. Additionally, an instantaneous strain ε_0 is usually present due to the initial loading conditions. This strain is mostly elastic, however it can contain plastic strain depending on the magnitude of the applied loading.

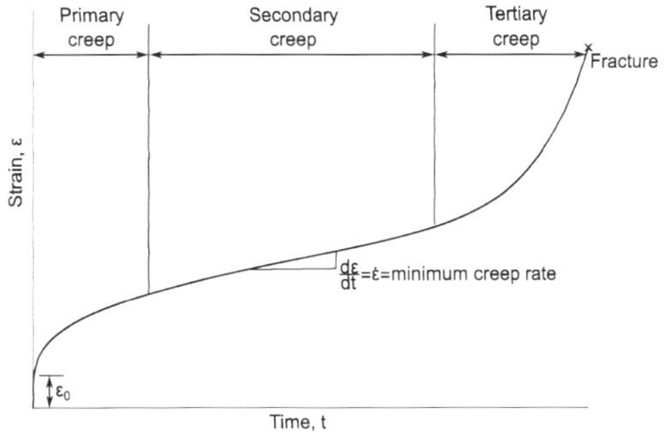

Figure 1: Typical creep curve showing three stages of creep

This paper focuses on methods of fitting creep material models to experimental data, specifically experimental creep data for 1Cr-1Mo-0.25V cast alloy steel operating between 450 and 550°C. This data is published by the Japanese National Institute for Materials Science (NIMS) [4] as part of an international effort to increase the size of creep databases.

As creep is a function of time, temperature and stress, many data points are needed to fully characterise the standard creep models. Even the most extensive data sets available, such as those from NIMS, are insufficient to calculate model coefficients directly. Optimisation techniques are used in conjunction with the published experimental creep data to determine optimal creep model coefficients. The following paper summarises the techniques used to model the experimental creep data.

CREEP MODELS

Constitutive equations approximate the response of materials to applied loading, material state, and environmental conditions. These material models are necessary inputs to analysis, including Finite Element Analysis (FEA). In simulation of the long term creep behaviour of structures operating at elevated temperatures, we make use of constitutive models for both the creep strain and the creep strain rate. Creep strain constitutive models relate the creep strain ε to time t, operating temperature T, and local stress σ in the form $\varepsilon = f(t, T, \sigma)$. Creep strain rate models also include the creep strain as a variable, $\dot{\varepsilon} = f(\varepsilon, t, T, \sigma)$ [5]. A wide range of constitutive models are available to represent the creep behaviour of engineering materials at elevated temperatures, in which each regime (primary, secondary and tertiary) is usually modelled separately. The purpose of this section is not to give an exhaustive breakdown of all creep constitutive models that are available, but rather to touch on the more common models and demonstrate the basics behind the formulations of such models.

A summary of some of the most common primary creep models is given in Table 1. All of these models have the capability of representing the decreasing creep strain rate that occurs in the primary creep regime over time t. The rate of this decrease and magnitude of the accumulated creep strain is adjusted by the constants A, B and c. An interesting fact regarding the primary creep models is that they all, barring the exponential model, have the ability to model secondary creep as well. In the special case in which the deformation mechanism does not change between the primary and secondary creep regimes this could prove to be useful.

Table 1: Constitutive models for primary creep [5]

Model	Constitutive equation
Logarithmic	$\varepsilon = A\log(1 + Bt)$
Power	$\varepsilon = At^c$
Exponential	$\varepsilon = A[1 - \exp(-Bt)]$
Hyperbolic sine	$\varepsilon = A\sinh(Bt^c)$

In a similar fashion the steady state creep strain rate $\dot{\varepsilon}_{min}$ in the secondary regime can also be represented by power, exponential or hyperbolic sine functions of stress. The classical form of these models is tabulated in Table 2.

Table 2: Constitutive models for secondary creep [5]

Model	Constitutive equation
Power	$\dot{\varepsilon}_{min} = A\sigma^n$
Exponential	$\dot{\varepsilon}_{min} = A\exp(B\sigma)$
Hyperbolic sine	$\dot{\varepsilon}_{min} = A\sinh(B\sigma)$

With the above models it is possible to model primary creep and secondary creep separately, or both primary and secondary regimes with an additive combination of the above laws. Additionally, there are various tertiary creep constitutive models available. These models usually take the form of power or exponential functions, which often include a damage accumulation parameter [6]. However, tertiary creep is often not modelled in FEA packages. This is due to the fact that the onset of tertiary is commonly seen as component failure.

It is worth noting that there is not one single creep model that is best suited for all creep deformation problems, rather a model is chosen that best fits the available data. Many data sheets available neglect both the primary and tertiary creep regimes, and report only secondary creep data. In these cases it becomes increasingly difficult to model the entire creep response, and the analyst will need to do much work with the use of justified assumptions.

By now it is clear that there are various constitutive models available for modelling the different creep regimes. With this in mind, consider the commercial FEA software ANSYS, and its available creep models. The deformation of a specimen under load at elevated temperatures comprises of an elastic, plastic and creep deformation. ANSYS therefore makes use of additive decomposition of the creep strain rates, which automatically implies that all three strain rates are independent of one another [7].

ANSYS has available 13 independent creep equations that can be used to model either primary, secondary or both primary and secondary creep regimes. These 13 creep models, which show significant similarities to those constitutive models tabulated above, can be found in the ANSYS user guide [7]. For clarity, consider the combined time hardening model given by equation (1),

$$\varepsilon = \frac{10^{x_1}\sigma^{x_2}t^{x_3+1}\exp\left(-\frac{x_4}{T}\right)}{x_3+1} + 10^{x_5}\sigma^{x_6}t\exp\left(-\frac{x_7}{T}\right), \tag{1}$$

where the coefficients x_i ($i = 1, 2, ..., 7$) are model coefficients that are to be determined by the user based on the experimental data .

NIMS EXPERIMENTAL CREEP DATA

As mentioned above, the NIMS creep data set for 1Cr-1Mo-0.25V cast steel [4] will be used throughout this investigation. It is a well-known fact that different casting processes result in a variation of material properties. This can lead to different material properties for "paper equivalent" materials ordered from different manufacturers. Additionally, when casting large parts, different areas of the part are exposed to different cooling and tempering rates, leading to a natural material property variation through the part. In order to take this into account the experimental data from NIMS is presented for nine different castings, all having undergone different heat treatments.

In order to fit constitutive creep models ($\varepsilon = f(t, T, \sigma)$, $\dot{\varepsilon} = f(\varepsilon, t, T, \sigma)$), it is necessary to arrange the experimental data in a similar form. For this reason, creep rupture data is of no use when attempting to generate creep constitutive models. An in depth discussion into the data presented in the NIMS data sheets is out of the scope of this paper. It is however worth noting that for the purpose of this work only the reported creep data tables will be used. An example of the creep data presented for a single casting is shown in Table 3.

Table 3: Extract for casting VbA from NIMS creep data sheet [4]

T (°C)	σ (%)	ε_0 (%)	Time to total strain, $t_{\%}$ (hrs)				Time to tertiary creep, t_{ter} (hrs)	$\dot{\varepsilon}_{min}$ (%/hr)	Time to rupture, t_f (hrs)
			0.50%	1.00%	2.00%	5.00%			
450	412	0.257	5890	15600	19400	NA	15100	2.59E-05	19841.6
	373	0.238	24400	36500	40500	NA	30000	6.12E-06	40712.2
	333	0.199	46300	68900	80400	85300	53000	2.85E-06	85835.9
500	294	0.193	1400	2120	2680	3180	1580	1.13E-04	3373.0
	265	0.175	1870	2890	3870	4850	1780	5.93E-05	5283.2
	235	0.145	6700	28600	NA	NA	43800	2.09 E-05	44004.8
	196	0.114	6980	16700	33100	59100	37200	5.35 E-05	60844.8
550	235	0.150	106	201	324	457	168	2.17 E-03	524.0
	196	0.122	500	1720	4250	NA	5040	3.76 E-04	5425.0
	196	0.138	373	1510	NA	NA	NA	NA	NA

As was noted in the section on constitutive models for creep, the different creep regimes are modelled with different equations, with tertiary creep often not modelled in FEA packages. It is therefore necessary to determine which of the experimental data points are in the tertiary creep regime, and omit these points from the experimental data which will be used to fit the models. From the data it can be seen that the time to 5% strain $t_{5\%}$ is approximately equal to the time to fracture t_f for all cases. Additionally, there are data points for which the time to total strain is greater than the time to tertiary creep ($t_{\%} > t_{ter}$). With all of these data points omitted from the usable set, data points falling into either the primary or secondary creep regime are left. These are shown highlighted in Table 3.

It is important to note that the creep strain at zero time is exactly zero, and any instantaneous strain is due to the initial application of the load. The data supplied by NIMS does not make any distinction between the different strains in the specimen (e.g. elastic, plastic and creep), and thus the reported strains are total strains. Constitutive creep models normally do not include any contributions for this initial instantaneous creep, and model only creep strain. Therefore the total strains reported in the NIMS data have to be adjusted by instantaneous strains for each specific temperature and stress combination. An example of processed data is tabulated in Table 4 for casting VbA at 450°C. It can be seen that the data is now in the desired form $\varepsilon = f(t, T, \sigma)$.

Table 4: Pre-processed extract for casting VbA at 450°C

ε (%)	t (hrs)	T (°C)	σ (MPa)
0.243	5890	450	412
0.262	24400	450	373
0.301	46300	450	333

OPTIMISATION OF MODEL COEFFICIENTS

As mentioned above, the combined time hardening model has been chosen to attempt to model the available creep data. This choice is based solely on the form of the experimental data, and is by no means claimed to be an optimal model which should be used for all creep modelling purposes. Consider a scatter plot of the experimental data for multiple castings at 550°C and 196MPa shown in Figure 2. Close inspection of the experimental data reveals that the data has a linear trend with an offset on the y-axis. Additionally, it can be seen that the data for different castings can vary by up to an order of magnitude. This discourages the use of a single fit to model the entire data set due to the poor confidence intervals that will result.

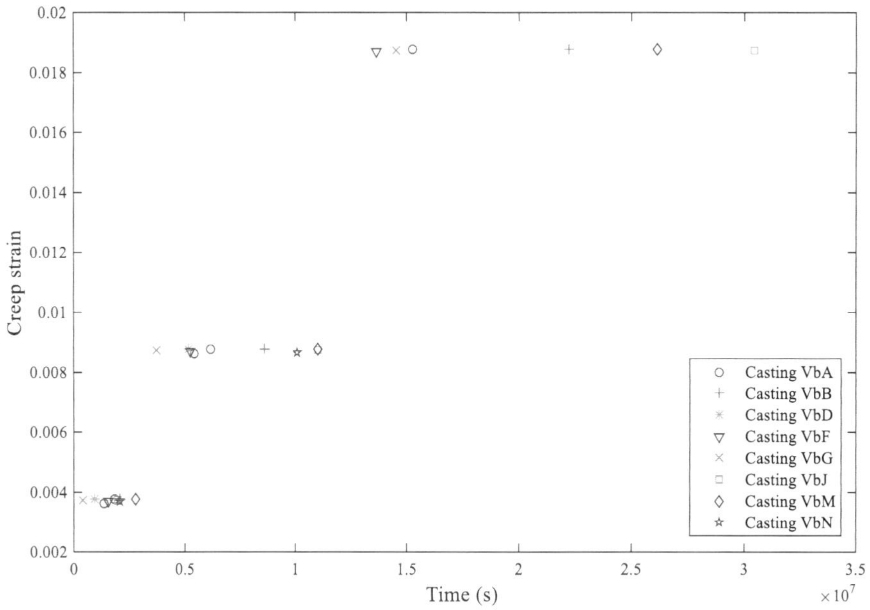

Figure 2: Scatter plot of NIMS experimental creep data for multiple castings at 550°C and 196MPa, adapted from [4]

A possible alternative to a single fit for the entire data set is to fit each casting separately. This will allow for optimal coefficients to be determined for a specific casting. A family of curves can then be generated that model the variety throughout the castings. This family of creep curves can then be used as input to Finite Element (FE) simulations, and the resulting variation analysed to identify worst cases.

The combined time hardening model in equation (1) models strain as a function of time, temperature and stress. The sum of least squares residual between the model and the experimental data can be modelled in a minimisation problem and is given by equation (2),

$$f = \sum_{i=1}^{n} \left(\frac{10^{x_1} \sigma_i^{x_2} t_i^{x_3+1} \exp\left(-\frac{x_4}{T_i}\right)}{x_3 + 1} + 10^{x_5} \sigma_i^{x_6} t_i \exp\left(-\frac{x_7}{T_i}\right) - \frac{\varepsilon_{data,i}}{100} \right)^2 . \tag{2}$$

For most castings, there are more data points than model coefficients. This means that multiple solutions are available to the same problem. It has already been mentioned that the experimental data in Figure 2 appears to follow the trend of a straight line with a positive offset on the y-axis. This is analogous to an extremely short lived primary creep response. To get an initial guess for model coefficients it is assumed that the amount of primary creep is not negligible, however the duration of primary creep is. This means that the combined time hardening model can be modified to represent a secondary creep model with an instantaneous primary creep. This can be modelled by equation (3),

$$\varepsilon = 10^{x_1}\sigma^{x_2} t \exp\left(-\frac{x_4}{T}\right) + 10^{x_5}\sigma^{x_6} t \exp\left(-\frac{x_7}{T}\right), \tag{3}$$

where the exponential time coefficient x_3 is omitted, with the primary term now solely representing the magnitude of primary creep.

The model coefficients in (3) can be determined by use of an unconstrained optimisation to minimise the sum of least squares residual between the experimental data and the model. Due to the nonlinear nature of the objective function, a downhill simplex method was used as the optimisation algorithm. The downhill simplex method, also called the Nelder-Mead method, requires only function evaluations and no derivatives [8]. The method works by constructing a simplex, which is a geometrical figure in N dimensions, consisting of $N + 1$ points (or vertices). In two dimensions this results in a simplex which is a triangle, and so on. The method works by evaluating the function at each of the simplex points and determining the maximum function value. The method converges to at least a local minimum through taking a series of steps, called reflections, by moving the point with the largest function value through the opposite face of the simplex. When the method reaches a local minimum it contracts itself in all directions until it converges to a result within the user defined convergence criterion. The downhill simplex method is shown schematically in Figure 3.

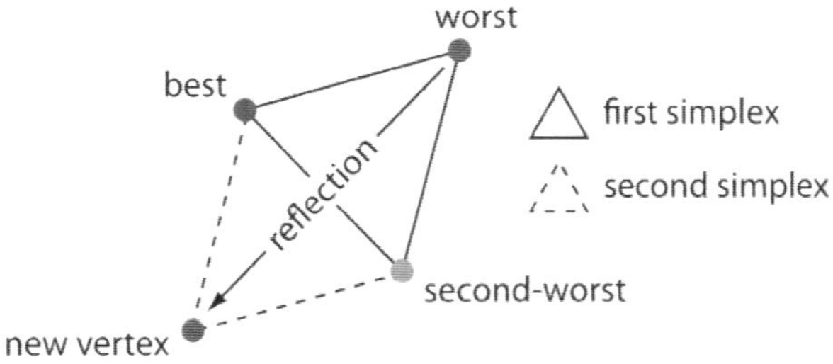

Figure 3: Schematic of downhill simplex method [9]

Due to the fact that the downhill simplex method easily converges to local minima, the problem is termed ill-posed. In order to try and locate a global minimum multiple random starting points are utilised and the best solution saved.

Optimisation of the modified combined time hardening law in (3) will lead to an initial guess for the coefficients. By biasing the coefficient results of the full combined time hardening model in (1) it is possible to obtain a solution capable of representing both the primary and secondary response for each casting. The use of Tikhonov regularisation is considered, where an extra term is added to the initial least squares residual objective function which penalises deviations of the coefficients from a chosen solution. The penalty term P is given by equation (4),

$$P = \boldsymbol{w}\left(\frac{\boldsymbol{x}}{\boldsymbol{x}^*} - 1\right)^2, \tag{4}$$

where P is written in terms of the weighting vector $\boldsymbol{w}$, the preferred solution $\boldsymbol{x}^*$ and the current solution $\boldsymbol{x}$.

Adding the penalty function from equation (4) to equation (2), a new objective function is formulated which will allow us to solve for the full model coefficients, but with a solution biased to that obtained from solving the modified model of equation (3). This new objective function f, given by

$$f = \sum_{i=1}^{n}\left(\frac{10^{x_1}\sigma_i^{x_2} t_i^{x_3+1} \exp\left(-\frac{x_4}{T_i}\right)}{x_3+1} + 10^{x_5}\sigma_i^{x_6} t_i \exp\left(-\frac{x_7}{T_i}\right) - \frac{\varepsilon_{data,i}}{100}\right)^2 + \boldsymbol{w}\left(\frac{\boldsymbol{x}}{\boldsymbol{x}^*} - 1\right)^2, \tag{5}$$

is minimised using the same downhill simplex optimisation algorithm. Again one must implement multiple random starting points to attempt convergence to a global minimum.

Although minimisation of the above objective function is unconstrained, one can choose to constrain the sign of specific coefficients with the use of absolute values. This is in order to force the coefficient solutions to be physically possible. For example, it is known that the stress sensitivity exponents are required to be positive. Additionally, in order for the creep model to predict an expected creep curve the coefficients $x_{1,3,5}$ are required to be negative, with all other coefficients positive. With this in mind, the objective function of equation (5) is modified to equation (6),

$$f = \sum_{i=1}^{n} \left(\frac{10^{-|x_1|} \sigma_i^{|x_2|} t_i^{-|x_3|+1} \exp\left(-\frac{|x_4|}{T_i}\right)}{-|x_3|+1} + 10^{-|x_5|} \sigma_i^{|x_6|} t_i \exp\left(-\frac{|x_7|}{T_i}\right) - \frac{\varepsilon_{data,i}}{100} \right)^2 + \boldsymbol{w}\left(\frac{\boldsymbol{x}}{\boldsymbol{x}^*} - 1\right)^2, \tag{6}$$

which constrains the signs of the coefficients.

RESULTS

The above formulations were optimised using the downhill simplex method as discussed above. The results of the best converged coefficients for the simplified combined time hardening model are tabulated in Table 5. The best solution was determined by that with the lowest sum of absolute residuals between the model and experimental data. Due to the fact that the model is primarily based on a secondary model, one would like to focus on the converged secondary constants, namely x_i (i = 5, 6, 7). It has already been seen that the variation among the castings is drastic, getting larger for higher strain rates. Thus it is expected to see a variation in stress sensitivity exponent x_6, which strongly controls the gradient of the creep curve in the secondary regime ($\dot{\varepsilon}_{min}$).

Table 5: Converged coefficient results for simplified combined time hardening model

Casting	$\sum\|Residual\|$ (%)	x_1	x_2	x_4	x_5	x_6	x_7
VbA	1.426	0.000	3.105	18670	-0.018	1.985	25603
VbB	1.321	-64.144	66.800	183600	-5.028	6.815	37259
VbD	2.172	-1.442	0.000	2160	-6.265	8.465	41769
VbF	2.071	0.000	2.949	17680	-0.002	0.812	20864
VbG	1.455	-0.179	0.034	4590	-7.017	10.755	50356
VbH	0.000	-0.454	0.168	4290	-10.936	1.229	6329
VbJ	0.938	-1.337	0.000	2690	-7.717	9.838	45645
VbM	0.879	-2.999	0.132	0	-10.417	14.360	60058
VbN	2.224	-1.143	0.011	2710	-11.817	12.810	50886

The results for specific castings are shown in Figure 4. The graphs depict the experimental data points, the model approximations, as well as the time to tertiary strain (TTS) as reported in Table 3. It can be seen from these results that non-sensible coefficients can result in a graph that appears to be correct. In the case of casting VbF, a good result appears to be obtained when looking at the graph. However, closer inspection of the coefficients reveals that the stress sensitivity for primary creep is larger than that for secondary creep, which is not a sensible solution.

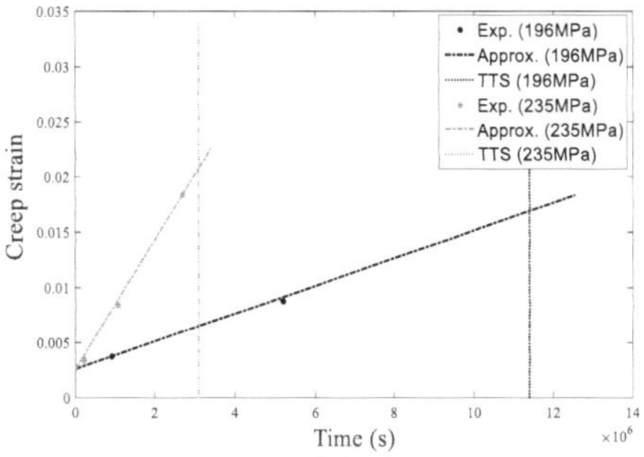

(a)

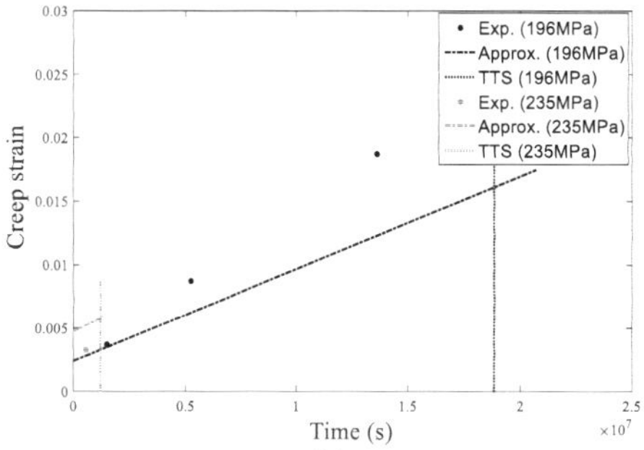

(b)

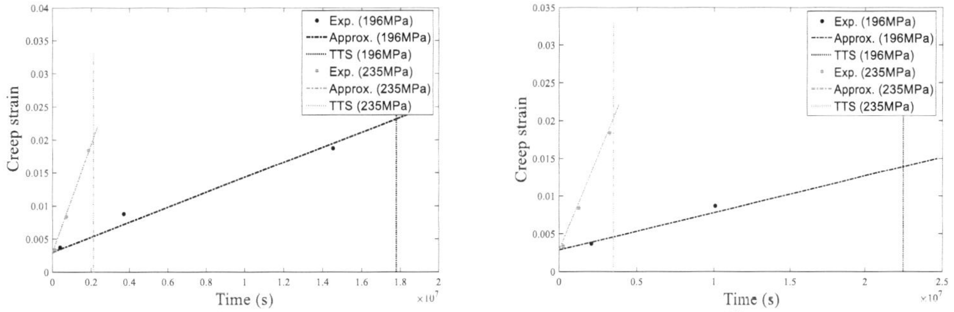

(c) (d)

Figure 4: Simplified combined time hardening model (3) approximation for (a) casting VbD at 550°C (b) casting VbF at 550°C (c) casting VbG at 550°C and (d) casting VbN at 550°C

From the results it is possible to choose a set of model coefficients $\boldsymbol{x}^*$ that represent a preferred solution. From the results above it is noticed that casting VbA, VbF and VbH all have non sensible coefficients. Considering the rest of the data it is possible to choose a set of coefficients that are an average representation of best converged results. There is no exact science to this process and the results are based mainly on trial and error and experimentation with various combinations of coefficients. With this in mind the preferred solution $\boldsymbol{x}^*$ and weighting vector $\boldsymbol{w}$ are tabulated in Table 6. Note that no weighting is applied to coefficient x_3 due to the fact that no information is available for it. Additionally, coefficients $x_{4,7}$ have weightings of zero to allow the optimisation algorithm slightly more flexibility when solving. Coefficients $x_{1,5}$ were set based on user preference as these are simply constant multiples in the equations.

Table 6: Preferred coefficients and associated weightings for application of Tikhonov regularisation

	x_1	x_2	x_3	x_4	x_5	x_6	x_7
$\boldsymbol{x}^*$	-6	1.5	-0.85	10000	-6	8	50000
$\boldsymbol{w}$	1	1	0	0	1	1	0

Optimising the full combined time hardening model results in the coefficients tabulated in Table 7. The results for four castings are shown thereafter in Figure 5. Notice that, as mentioned above, castings VbA, VbF and VbH appear to perform relatively poorly with completely different converged coefficient results.

Table 7: Converged coefficient results for full combined time hardening model

Casting	$\sum\lvert Residual\rvert$ (%)	x_1	x_2	x_3	x_4	x_5	x_6	x_7
VbA	1.956	-0.691	1.000	-0.924	10881	-0.040	1.311	22831
VbB	1.395	-2.378	1.755	-0.882	11345	-7.246	8.418	40234
VbD	2.184	-5.619	1.073	-0.721	3267	-3.106	8.729	48917
VbF	2.055	-5.022	1.291	-1.000	15674	-0.003	0.621	20063
VbG	1.734	-2.805	1.542	-0.932	9154	-6.338	9.974	48262
VbH	0.009	-5.368	1.479	-1.000	15547	-6.844	4.437	25895
VbJ	1.003	-5.342	2.146	-0.905	7950	7.566	9.922	46272
VbM	1.698	-0.050	1.318	-0.910	13415	9.946	13.212	56010
VbN	2.328	-3.433	1.669	-0.814	9177	11.915	12.986	51573

It can be seen from the graphs that the gradient for the secondary creep stage remains relatively constant for both the simplified and full model. This is due to the regularisation that was used to penalise deviations away from the preferred solution.

Another point worth noting is that the model approximations at 450°C were relatively poor in comparison to 500°C and 550°C. This can be seen for casting VbN in Figure 6 where the approximations at 450°C and 550°C are shown. This poor fit at 450°C can be attributed to the amount of data available. Less experimental data is available for this temperature, and thus these points contribute less to the total least squares residual function. Therefore although the fit looks mediocre at 450°C, it is the best possible solution with the data available.

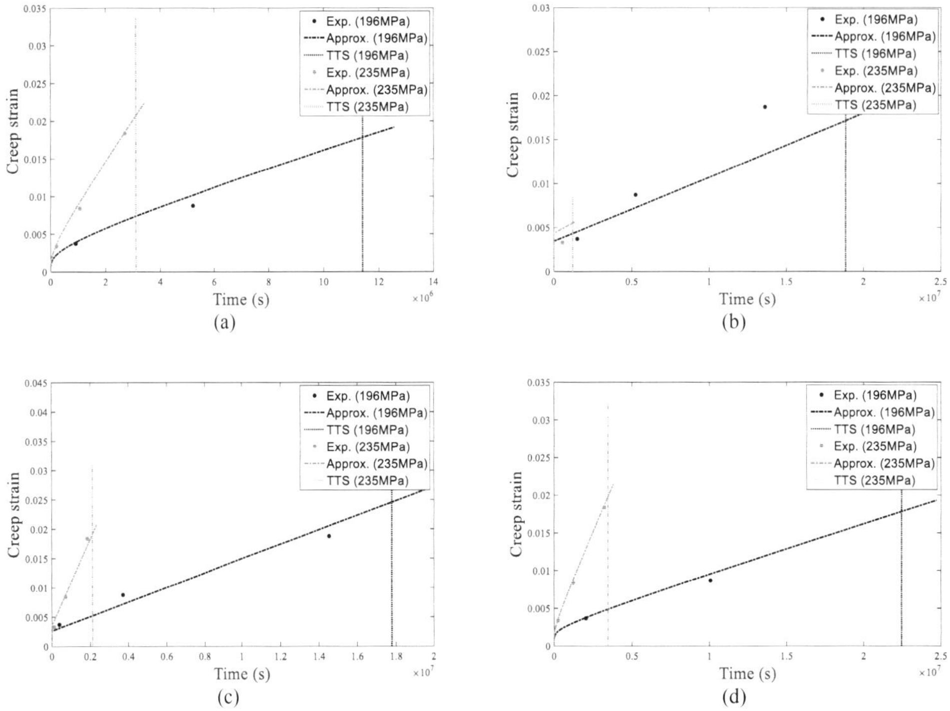

Figure 5: Full combined time hardening model (1) approximation for (a) casting VbD at 550°C (b) casting VbF at 550°C (c) casting VbG at 550°C and (d) casting VbN at 550°C

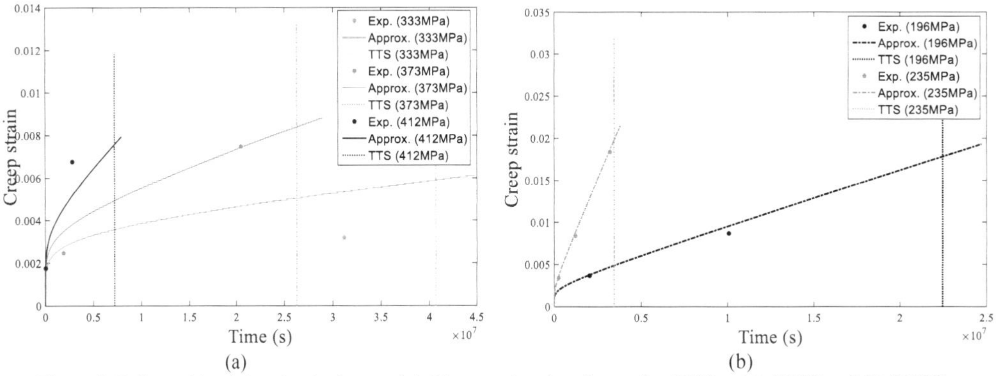

Figure 6: Full combined time hardening model (1) approximation for casting VbN at (a) 450°C and (b) 550°C

Lastly, consider the plot showing the creep curves for all castings in Figure 7. Notice that there is a clear band of generated creep curves, capturing the range of experimental creep data (shown by the solid black markers). These creep curves can be implemented into an FEA package and useful remaining life calculations done based on either a worst case or average scenario.

It is seen from Figure 7 that the variation of creep curves across the castings is quite severe, with some castings predicting up to three times more or less creep strain than others. It is a well-known fact that the variation of creep properties is massive, with no single set of properties being able to model a material. For this reason it is important to note the large confidence intervals when making creep life calculations, and use worst case or average properties conservatively. Again, the lack of data for 450°C is evident from the poor fit.

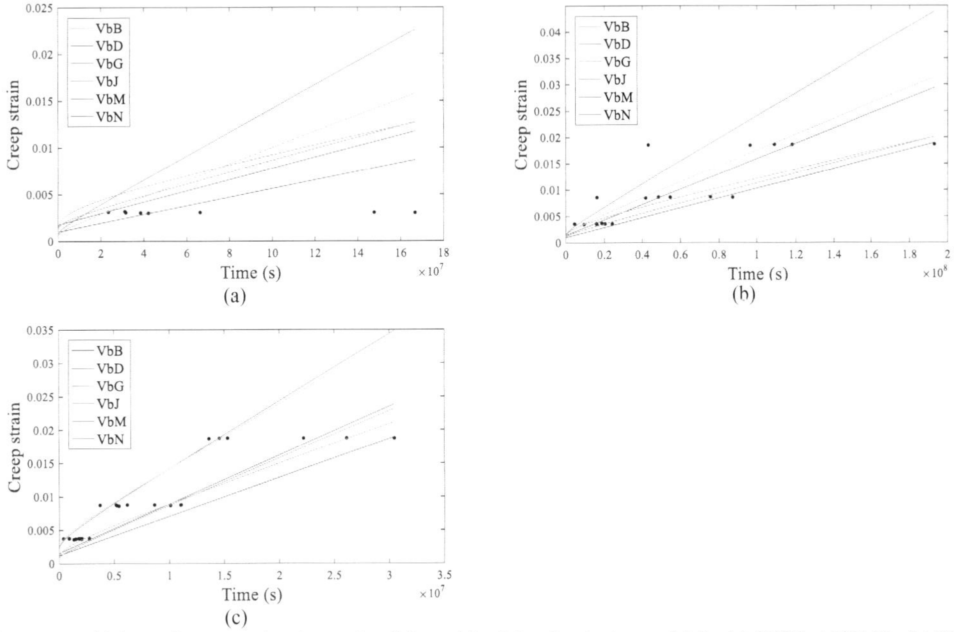

Figure 7: Multiple casting approximations using full combined time hardening model for (a) 450°C at 333MPa (b) 500°C at 235MPa (c) 550°C at 196MPa

CONCLUSION

An effective methodology to determine ANSYS creep model coefficients was proposed. The methodology made use of a simple unconstrained optimisation algorithm with Tikhonov regularisation. Due to the vast scatter of the data among castings, an approach was adopted that fits each casting separately, resulting in a family of creep curves. This family of creep curves can be used for remaining useful life estimations in commercial FEA software. It is worth noting that the method struggles for certain castings, where converged coefficients were non-sensible. Additionally, at 450°C the model struggled to accurately predict the experimental data. This is due to the limited data at lower temperatures. In addition to limited data, the vast scatter limited accurate results. It can thus be concluded that, given the available NIMS creep data, the fits are most accurate for temperatures in the range of 500°C and 550°C. The authors believe that the optimisation methodology proposed for solving model coefficients for the combined time hardening creep model yields results which are representative of the dataset.

REFERENCES

1 G. Merckling, T. U. Kem and K. Yagi, "Creep properties of Heat Resistant Steels and Superalloys," vol. 2B, 2004.

2 M. E. Kassner, Fundamentals of creep in metals and alloys, Elsevier, 2009.

3 W. Stumpf, *Mechanical Metallurgy,* 2015.

4 NIMS, "Data sheet 31B: data sheets on the elevated temperature properties of 1Cr-1Mo-0.25V steel castings for steam turbine casings," National Institute for Materials Science, Tokyo, Japan, 1994.

5 S. R. Holdsworth, Creep-Resistant Steels: Constitutive equations for creep curves and predicting service life, Woodhead Publishing, 2008.

6 A. A. Becker, T. H. Hyde, W. Sun and P. Andersson, "Benchmarks for finite element analysis of creep continuum damage mechanics," *Computational Materials Science,* vol. 25, 2002.

7 ANSYS, Mechanical APDL Materials Reference: Rate-Dependent Plasticity, ANSYS inc., 2014.

8 W. H. Press, Numerical Recipes: The Art Of Scientific Computing, Press Syndicate of the University of Cambridge, 1987.

9 S. Towers, "Polymatheia: Simplex Method," 14 July 2014. [Online]. Available: http://sherrytowers.com/2014/07/14/simplex-method/. [Accessed 10 February 2016].

Cyclic effects and recrystallisation in temperature and rate dependent state variable based plasticity

G.J. Jansen van Rensburg[1,2,*], S. Kok[2], D.N. Wilke[2]

[1]Advanced Mathematical Modelling, Modelling and Digital Sciences, CSIR, South Africa
[2]Department of Mechanical and Aeronautical Engineering, University of Pretoria, South Africa

*jjvrensburg@csir.co.za

ABSTRACT

A constitutive model that captures the dominant strengthening and softening physics or microstructural changes of metallic materials is needed within a numerical environment for the effective prediction and simulation of plasticity during hot working processes. Dislocation density or related stress values play the primary role of internal state dependent variables in the development of many unified constitutive material models using the Kocks-Mecking work hardening theory as foundation. This theory is based on an approach to model the viscoplastic behaviour of crystalline materials as a result of interaction between mobile and forest dislocations in the material microstructure. Dislocation density based state dependent models have the ability to not only provide a good description of the mechanical response of metallic materials in uniform loading, but also have good predictive capability. One popular model is the Mechanical Threshold Stress model while other dislocation kinetics based models have also been proposed in literature to include various physical phenomena. In this conference contribution, the development and implementation of temperature and rate dependent state variable based plasticity is presented. The model foundation as well as extensions to include cyclic effects and recrystallisation are coded into a user material subroutine for use in finite element packages.

KEYWORDS

Mechanical Threshold Stress, dislocation density based material modelling, combined hardening, recrystallisation

INTRODUCTION

Most Finite Element Analysis (FEA) software packages have a comprehensive list of material models already implemented to choose from. If a different material model is required, a user implemented material subroutine can be linked to Abaqus, CalculiX and Code_Aster for example. All three of these packages can make use of the Abaqus user material format coded in Fortran. Sophisticated state variable based models such as the Mechanical Threshold Stress (MTS) model [1] as well as dislocation density based combined isotropic and kinematic hardening [2] or models for static and dynamic recrystallisation [3] are available in the research space. Many of these models can be seen as implementations and extensions using the Kocks-Mecking hardening theory [4, 5, 6, 7] as foundation.

The development and implementation of state dependent user materials in computational plasticity is discussed in the Ph.D. thesis by Jansen van Rensburg [8]. In the thesis, an incrementally objective user material framework for purely isotropic and kinematic hardening J_2 plasticity is developed, implemented and verified against the native implementation of Abaqus. These user material frameworks can be extended by altering either the purely isotropic or combined hardening subroutines as well as the support functions called by the framework, resulting in an array of more sophisticated models.

Within the implemented framework for user material subroutines, the MTS model as well as dislocation density based extensions are presented in this paper. Coding of these scalar equations into purely isotropic or combined hardening subroutines result in incrementally objective implementations of the models for use in FEA [8].

THE MECHANICAL THRESHOLD STRESS MODEL

The MTS model was developed by Follansbee and Kocks [1] to describe the post yielding behaviour of metals. In the MTS model, the mechanical threshold ($\hat{\sigma}$) is the internal state variable. This threshold value represents a theoretical maximum flow stress at 0K. From there the material flow stress, σ_{Y}, is obtained by scaling the mechanical threshold to accommodate rate and temperature dependence. Following Nabarro et al. [9], the threshold flow stress is separated into an athermal $\hat{\sigma}_a$ and k different thermal components $\hat{\sigma}_T^k$. At different absolute temperature values T and plastic strain rates $\dot{\varepsilon}_{\mathrm{p}}$, the contribution to the flow stress σ_T^k is related to the threshold counterpart $\hat{\sigma}_T^k$ through the scaling function $S_k(\dot{\varepsilon}_{\mathrm{p}}, T)$, so that

$$\sigma_T^k = \hat{\sigma}_T^k \frac{\mu(T)}{\mu_r} S_k(\dot{\varepsilon}_{\mathrm{p}}, T). \tag{1}$$

Here, $\mu(T)$ and μ_r indicate the current and reference shear modulus values. The temperature dependent shear modulus is expressed by Varshni [10] as

$$\mu(T) = \mu_r - \frac{D_r}{\exp\left(\frac{T_r}{T}\right) - 1}, \tag{2}$$

where μ_r, D_r and T_r are constants. The reference temperature T_r has to be well below the temperatures of interest. The scaling functions take the form [1]

$$S_k(\dot{\varepsilon}_{\mathrm{p}}, T) = \left[1 - \left(\frac{k_{\mathrm{B}}T}{g_{0k}\mu b^3} \ln \frac{\dot{\varepsilon}_{0k}}{\dot{\varepsilon}_{\mathrm{p}}}\right)^{1/q_k}\right]^{1/p_K}, \tag{3}$$

where g_{0k} is the normalised activation energy for dislocations to overcome the obstacles, b is the magnitude of the Burgers' vector length and $k_{\mathrm{B}} = 1.38064852 \times 10^{-23}\mathrm{m^2kgs^{-2}K^{-1}}$ is the Boltzmann constant. $\dot{\varepsilon}_{0k}$ is a constant proportional to the density of mobile dislocations while p_k and q_k are statistical constants that characterise the shape of the obstacle profile. In the original MTS model, two thermal stresses are used, i.e. $\hat{\sigma}_T^1$ and $\hat{\sigma}_T^2$. The first is a constant thermal stress $\hat{\sigma}_i = \hat{\sigma}_T^1$. The second is an evolving threshold stress internal state variable $\hat{\sigma}_\varepsilon = \hat{\sigma}_T^2 = \frac{1}{2}M\mu b\sqrt{\rho}$ with M the average Taylor factor that accounts for the polycrystallinity of the material [5]. The equivalent flow stress σ_{Y} is finally expressed as

$$\sigma_{\mathrm{Y}} = \hat{\sigma}_a + \hat{\sigma}_i \frac{\mu(T)}{\mu_r} S_i(\dot{\varepsilon}_{\mathrm{p}}, T) + \hat{\sigma}_\varepsilon \frac{\mu(T)}{\mu_r} S_\varepsilon(\dot{\varepsilon}_{\mathrm{p}}, T). \tag{4}$$

The temperature and rate dependent evolution of the internal stress value in the presence of additional equivalent plastic strain according to the Kocks-Mecking work hardening theory can take the form [5]

$$\frac{d\hat{\sigma}_\varepsilon}{d\varepsilon_{\mathrm{p}}} = \frac{M\mu b}{4\sqrt{\rho}} \frac{d\rho}{d\varepsilon_{\mathrm{p}}} \quad \text{where} \quad \frac{d\rho}{d\varepsilon_{\mathrm{p}}} = Mk_1\sqrt{\rho} - Mk_2(\dot{\varepsilon}_{\mathrm{p}}, T)\,\rho. \tag{5}$$

The initial work hardening θ_0 as well as the temperature and rate dependent saturation stress $\hat{\sigma}_s(\dot{\varepsilon}_{\mathrm{p}}, T)$ can be introduced with

$$\theta_0 = \frac{M^2\mu b k_1}{4} \quad \text{and} \quad \hat{\sigma}_s(\dot{\varepsilon}_{\mathrm{p}}, T) = \frac{2\theta_0}{Mk_2(\dot{\varepsilon}_{\mathrm{p}}, T)}. \tag{6}$$

This means that the evolution of the threshold stress value $\hat{\sigma}_T$ could be expressed as a function of itself. This substitution results in the Voce-law [11]. Adding an exponent α results in the modified form of the Voce-law [11].

$$\frac{d\hat{\sigma}_\varepsilon}{d\varepsilon_{\mathrm{p}}} = \theta_0 \left(1 - \frac{\hat{\sigma}_\varepsilon}{\hat{\sigma}_s}\right)^\alpha. \tag{7}$$

Following the work of Chen and Gray [12], the threshold saturation stress $\hat{\sigma}_s\left(\dot{\varepsilon}_{\mathrm{p}}, T\right)$ is a function of both strain rate and temperature, through

$$\ln \frac{\dot{\varepsilon}_{\mathrm{p}}}{\dot{\varepsilon}_{0s}} = \frac{g_{0s}\mu b^3}{k_{\mathrm{B}}T} \ln \frac{\hat{\sigma}_s}{\hat{\sigma}_{0s}}, \tag{8}$$

where $\dot{\varepsilon}_{0s}$, g_{0s} and $\hat{\sigma}_{0s}$ are constants. The version of the MTS model contained in this section is implemented to work within the purely isotropic user material subroutine in [8]. The model is calibrated to hardening data at different strain rates and temperatures digitised from the work done by Tanner and McDowell [13] using a material point simulation. To compare the model response using a detailed Abaqus simulation to the data, compression of an axi-symmetric billet is modelled with material response provided by the user material. An axi-symmetric quarter of a billet is modelled using a 20×20 grid of quadratic axi-symmetric elements. The billet is compressed using a rigid analytical plane for the die contact surface. Hard normal contact between the billet and die is modelled with a Coulomb friction coefficient $\mu_{\text{frict}} = 0.2$.

Four different problems are simulated that correspond to the experimental test data for 1s^{-1} and 0.0004s^{-1} at 298K, 0.0004s^{-1} at 407K and 0.01s^{-1} at 542K. The billet section modelled has a radius of 2.5mm and a half length of 5mm, meaning a 100% reduction in length corresponds to a die displacement of $5\text{mm} \times (\exp(-1) - 1) = -3.1606\text{mm}$. This displacement is applied over 1s in the 1s^{-1} case, 100s in the 0.01s^{-1} case and 2500s in the 0.0004s^{-1}. The simulations are solved over 100 time increments.

The material property values used in the simulation are itemised below.

- The elastic property values of $\mu_r = 49.91\text{GPa}$, $D_r = 3.29\text{GPa}$ and $T_r = 200\text{K}$ in Equation (2) are used with a Poisson's ratio of $\nu = 1/3$.
- The stress values in Equation (4) are $\hat{\sigma}_a = 0\text{MPa}$, $\hat{\sigma}_i = 20.67\text{MPa}$ and $\hat{\sigma}_\varepsilon|_{\varepsilon_{\mathrm{p}}=0} = 0\text{MPa}$.
- The scaling function values in Equation (3) are $k_{\mathrm{B}}/g_{0i}b^3 = 1.402\text{MPa/K}$, $k_{\mathrm{B}}/g_{0\varepsilon}b^3 = 1.632\text{MPa/K}$, $p_\varepsilon = p_i = q_\varepsilon = q_i = 1$ and $\dot{\varepsilon}_{0\varepsilon} = \dot{\varepsilon}_{0i} = 10^7\text{s}^{-1}$.
- The power law hardening according to Equation (7) is used with $\theta_0 = 2611.94\text{MPa}$ and $\alpha = 2$.
- The saturation stress in Equation (8) have property values $\hat{\sigma}_{0s} = 689.12\text{MPa}$, $k_{\mathrm{B}}/g_{0s}b^3 = 0.5011\text{MPa/K}$ and $\dot{\varepsilon}_{0s} = 10^{10}$.

In Figure 1, the von Mises stress distributions after 100% reduction in length is displayed for the hardest and softest of the material responses modelled. In Figure 1(a) the 1s^{-1} at 298K results are given while (b) is the stress distribution in the 0.01s^{-1} at 542K case. From the Abaqus results, the true strain values at each increment is determined from the prescribed die displacement. The volume preserving area assumption is further used to calculate the true stress using the reaction force extracted at the mid plane nodes. In Figure 1(c) the effective stress versus strain curves for the four cases simulated are presented compared to the experimental data of Tanner and McDowell [13].

The Abaqus simulation data with simple post processing to estimate the true stress - true strain data compares well to the experimental test data. The non-smooth response starting at about $\varepsilon \approx 0.5$ is due to the contact between the die surface and rollover of elements at the corner. If greater accuracy between the finite element result and experimental test data is required however, further characterisation using numerical optimisation can be performed using finite element evaluations within the objective function evaluation.

DISLOCATION DENSITY BASED

An alternative choice on internal state variable could also be considered following the dislocation density based modelling approach of Estrin [2]. While the initial value of the true dislocation density values ($\rho_0 = \rho|_{\varepsilon_{\mathrm{p}}=0}$) make the choice of dislocation density itself a difficult choice for an internal state variable, a stress like constant $\sigma_0 = \frac{1}{2}M\mu b\sqrt{\rho_0}$ is be introduced so that the threshold stress value is now determined from

$$\hat{\sigma}_\varepsilon = \sigma_0\sqrt{Y}. \tag{9}$$

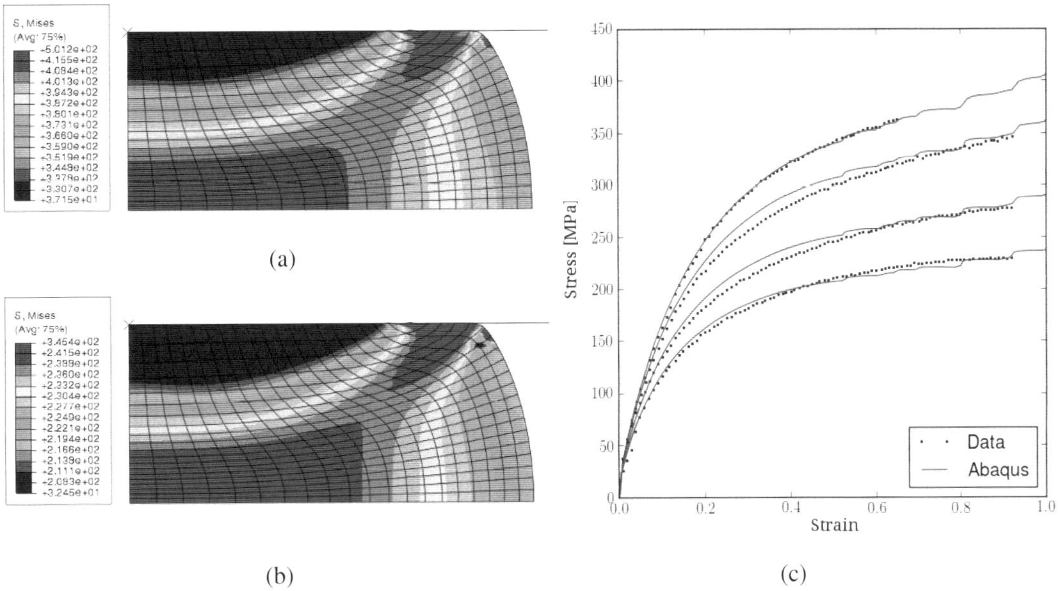

Figure 1: Von Mises stress distributions after 100% reduction in length for (a) 1s^{-1} at 298K and (b) 0.01s^{-1} at 542K. (c) Abaqus axi-symmetric simulation response using the material parameters estimated for OFHC Cu compared to the experimental test data in descending order for 1s^{-1} and 0.0004s^{-1} at 298K, 0.0004s^{-1} at 407K and 0.01s^{-1} at 542K.

This modification introduces the dislocation density ratio variable $Y = \frac{\rho}{\rho_0}$, a new internal state variable with typical initial value $Y|_{\varepsilon_\text{p}=0} = 1$. Equation (9) can be substituted into Equation (4) to formulate the MTS model in terms of the new internal state variable Y instead of the threshold stress value $\hat{\sigma}_\varepsilon$. Assuming no constant thermal stress component ($\hat{\sigma}_i = 0$MPa), the temperature and rate dependent flow stress in Equation (4) has the form

$$\sigma_\text{Y} = \hat{\sigma}_a + \sigma_0 \sqrt{Y} \frac{\mu(T)}{\mu_r} S_\varepsilon(\dot{\varepsilon}_\text{p}, T). \tag{10}$$

The constitutive formulation may now be completed by using the theory on dislocation density based modelling to evolve the dislocation density ratio in the process of plastic deformation instead of the threshold stress directly. By introducing $C_1 = Mk_1/\sqrt{\rho_0}$ and $C_2(\dot{\varepsilon}_\text{p}, T) = Mk_2(\dot{\varepsilon}_\text{p}, T)$, evolution of the dislocation density ratio internal state variable following Equation (5b) is now determined from [2]

$$\frac{dY}{d\varepsilon_\text{p}} = C_1\sqrt{Y} - C_2(\dot{\varepsilon}_\text{p}, T)\, Y. \tag{11}$$

The evolution of the dislocation density ratio in Equation (11) is determined by different values of the constant C_1 and form of the the dynamic recovery annihilation function $C_2(\dot{\varepsilon}_\text{p}, T)$. We use a function

$$C_2(\dot{\varepsilon}_\text{p}, T) = C_{20} \exp\left[-\frac{T}{a_{02}\mu} \ln\left(\frac{\dot{\varepsilon}_\text{p}}{\dot{\varepsilon}_{02}}\right)\right], \tag{12}$$

that is equivalent to the saturation stress based formulation in Equation (8) used by Chen and Gray [12]. In this function C_{20}, $\dot{\varepsilon}_{02}$ and a_{02} are constants. a_{02} is a constant to represent the convenient grouping of parameters $a_{02} = g_{0s}b^3/k_\text{B}$. Further inclusion of physical phenomena is possible by reformulating Equation (11).

Following the work done by Kok et al. [14] an evolving geometric substructure is included in the model by introducing the average slip plane lattice incompatibility $\bar{\lambda}$ as another internal state variable. Kok et al. [14] observed that the evolution of the average slip plane lattice incompatibility is inversely proportional to the grain size d_{x}. Using the proportionality constant C_λ, an evolution equation for this internal state variable is [14]

$$\frac{d\bar{\lambda}}{d\varepsilon_{\mathrm{p}}} = \frac{C_\lambda}{d_{\mathrm{x}}}. \tag{13}$$

Also using the formulation by Song and McDowell [15] as a further extension to the dislocation density ratio evolution equation, thermal recovery is added. The rate form of Equation (11) with large strain and thermal recovery effects is

$$\dot{Y} = \dot{\varepsilon}_{\mathrm{p}} \left(C_g \bar{\lambda}^{r_g} + C_1 \sqrt{Y} - C_2(\dot{\varepsilon}_{\mathrm{p}}, T)\, Y \right) - C_3(T) Y^{r_3}. \tag{14}$$

where C_g, r_g and r_3 are constants while C_3 is a temperature dependent function. If the constant C_{30} is introduced while $a_{03} = Q_{\mathrm{self}}/R$ is associated with the scaled activation energy for self diffusion. The static or thermal recovery term is modelled using [15]

$$C_3(T) = C_{30} \exp\left(\frac{-a_{03}}{T} \right). \tag{15}$$

The dislocation density based model up to here is a physically extended form of the reformulated isotropic MTS model with large strain effects and static recovery using two evolving internal state variables (Y and $\bar{\lambda}$). Cyclic effects are now added to the model formulation. Some of the dislocations are recoverable in the event of load reversal, making it necessary to distinguish between truly immobilised and recoverable stored dislocation density. A non-dimensional form of the partially recoverable part of the dislocation density ratio is given by the internal state variable Z following the work done by Estrin [2]. This internal state variable evolves following

$$\dot{Z} = \dot{\varepsilon}_{\mathrm{p}} \left(C_4 \sqrt{Y} - C_2(\dot{\varepsilon}_{\mathrm{p}}, T) Z \right) - C_5(T) Z. \tag{16}$$

Equation (16) contains a storage rate constant C_4 and dynamic recovery term $C_2(\dot{\varepsilon}_{\mathrm{p}}, T)Z$. $C_5(T)Z$ further represents the recoverable edge density loss rate due to thermally activated jog formation [2]. The last term is time rather than strain dependent and can take the form of the Arrhenius expression involving the activation energy for jog formation. Much like Equation (15), if $a_{05} = Q_{\mathrm{jog}}/R$ is used as a parameter representing the scaled activation energy associated with jog formation

$$C_5(T) = C_{50} \exp\left(\frac{-a_{05}}{T} \right). \tag{17}$$

The dislocation density variable Z is recovered upon each stress reversal. This corresponds to an internal state variable update

$$Y := Y - Z \qquad \text{and} \qquad Z := 0 \tag{18}$$

Estrin [2] further introduces a back stress σ_{B} into his chosen one dimensional kinetic equation. The evolution of the one dimensional back stress is considered to obey the equation

$$\frac{d\sigma_{\mathrm{B}}}{d\varepsilon_{\mathrm{p}}} = C_6 \sqrt{Y} - C_7 \sigma_{\mathrm{B}}, \tag{19}$$

where C_6 and C_7 are constants. The combined hardening model using Y, $\bar{\lambda}$, Z and back stress evolution equations is coded into a subroutine that works with the combined user material framework [8]. To illustrate the ability of the model, it is characterised to the deformation data for Alloy 800H and Inconel 738LC digitised from the work by Estrin [2].

Two cycles of the Alloy 800H material response is digitised and fit using the model. The deformation is modelled at 1123K and a strain rate of $2 \times 10^{-3}\mathrm{s}^{-1}$. The model response fit to the digitised data is presented in Figure 2(a). The

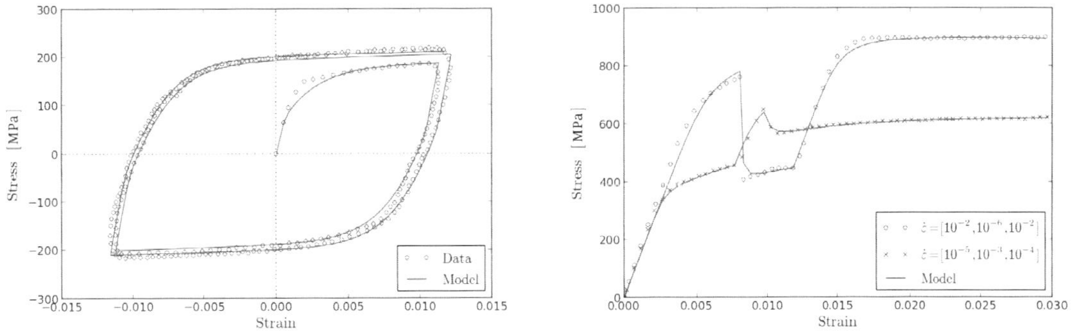

(a) (b)

Figure 2: (a) Cyclic data for Alloy 800H [2] versus the implemented model prediction. (b) Variable strain rate data for Inconel 738LC [2] versus model prediction.

elastic parameter values of $\mu_r = 60\text{GPa}$, $\nu = 0.3$ and $D_r = 0\text{GPa}$ so that $\mu = \mu_r$ are used. The athermal stress component is $\hat{\sigma}_a = 0\text{MPa}$ while the reference stress parameter is $\sigma_0 = 734.41\text{MPa}$. Scaling factor parameter values are $a_{0\varepsilon} = g_{0\varepsilon}b^3/k_\text{B} = 0.7424\text{K/MPa}$, $p_\varepsilon = 0.5$, $q_\varepsilon = 1.5$ and $\dot{\varepsilon}_{0\varepsilon} = 10^7\text{s}^{-1}$. The other material parameter values are $C_g = 0$, $C_{30} = 0\text{s}^{-1}$, $C_1 = 43.302$, $C_{20} = 32.469$, $a_{02} = 4.497\text{K/MPa}$, $\dot{\varepsilon}_{02} = 10^7\text{s}^{-1}$, $C_4 = 10.589$, $C_{50} = 0.7253\text{s}^{-1}$ and $a_{05} = 20000\text{K}$ with the parameters associated with the back stress evolution $C_6 = 64897.42\text{MPa}$ and $C_7 = 628.34$.

In Figure 2(b), the model prediction is presented along with the variable strain rate data for Inconel 738LC. In this case, the elastic parameter values of $\mu_r = 50\text{GPa}$, $\nu = 0.3$ and $D_r = 0\text{GPa}$ are used. The athermal stress component is $\hat{\sigma}_a = 277.184\text{MPa}$ with $\sigma_0 = 4410.47\text{MPa}$. Scaling factor parameter values are $a_{0\varepsilon} = 0.9392\text{K/MPa}$, $p_\varepsilon = 0.2551$, $q_\varepsilon = 0.755$ and $\dot{\varepsilon}_{0\varepsilon} = 10^7\text{s}^{-1}$. Furthermore $C_g = 0$, $C_{30} = 0\text{s}^{-1}$, $C_1 = 22.23$, $C_{20} = 75.67$, $a_{02} = 0.478\text{K/MPa}$, $\dot{\varepsilon}_{02} = 10^7\text{s}^{-1}$, $C_4 = 0.0039$, $C_{50} = 11.924\text{s}^{-1}$ and $a_{05} = 20000\text{K}$. The back stress is evolved using $C_6 = 30329.65\text{MPa}$ and $C_7 = 180.11$.

There is a good fit despite setting $C_g = 0$ and therefore ignoring the effect of $\bar{\lambda}$ on the dislocation density accumulation. This indicates that the carbide particles present in the alloy does not affect the cyclic behaviour of the material according to the data used. If more data is available at different rates and temperatures during the model characterisation, the parameter values should have other values since the Alloy 800H cyclic data is at isothermal conditions for a single rate for example.

RECRYSTALLISATION

A recrystallisation model similar to the one validated by Brown and Bammann [3] is implemented. Multiple waves of recrystallisation can be simulated with various recrystallised and unrecrystallised volume fractions existing simultaneously. The recrystallisation is based on the mobility of grain and subgrain boundaries. From the work by Cahn and Hagel [16], the recrystallised volume fraction growth rate is

$$\dot{f}_\text{x} = A_\text{x} v_\text{x}. \tag{20}$$

Here, A_x is the interfacial area between recrystallised and unrecrystallised regions and v_x is the average velocity of the interface. Average grain boundary velocity is calculated using [17]

$$v_\text{x} = MP, \tag{21}$$

where M is the grain boundary mobility and P the driving pressure. As the material deforms, the average misorientation across the geometrically necessary subgrain boundaries increase, which in turn increases the mobility of the boundaries. In the dislocation density ratio based model implementation, the average slip plane lattice misorientation $\bar{\lambda}$ internal state variable is assumed proportional to the average misorientation angle $\bar{\theta}$ [8]. The misorientation angle ratio used by Chen et al. [18] is therefore replaced and a reformulated equation using the average lattice slip plane incompatibility is introduced. The average subgrain boundary mobility is approximated as [8]

$$\bar{M} = M_0 \exp\left(-\frac{Q_M}{RT}\right)\left[1 - \exp\left(-C_{M\theta}\bar{\lambda}^{r_{M\theta}}\right)\right] \tag{22}$$

where Q_M is the activation energy for grain boundary mobility while M_0, $C_{M\theta}$ and $r_{M\theta}$ are constants. The pressure driving the subgrain boundary growth can in it's simplest form be expressed as $P = \frac{1}{2}\mu b^2 \rho$ [19]. Considering the dislocation density ratio used as evolving internal state variable, $Y = \rho/\rho_0$, the driving pressure is calculated by

$$P = \frac{1}{2}\mu b^2 \rho_0 Y. \tag{23}$$

Equations(22) and (23) means recrystallisation is modelled in terms of the internal state variables Y and $\bar{\lambda}$. An isotropic implementation using these two state variables is therefore considered.

The original unrecrystallised material has a volume fraction $f_{x_0} = 1$. f_{x_1} represents the material volume fraction that has undergone one recrystallisation cycle while the remaining unrecrystallised volume fraction is $1 - f_{x_1}$. Given multiple cycles of recrystallisation, $f_{x_i} - f_{x_{i+1}}$ represents the total volume fraction of material that has been recrystallised i times as in the model by Brown and Bammann [3]. Each volume fraction has it's own set of internal state variables. The variables Y_{x_i} and $\bar{\lambda}_{x_i}$ represent the internal state variables associated with the material volume fraction $f_{x_i} - f_{x_{i+1}}$. The interfacial boundary area A_x in Equation (20) is approximated between the f_{x_i} and $f_{x_{i+1}}$ volume fractions using the equation formulated by Brown and Bammann [3]

$$A(f_{x_i}, f_{x_{i+1}}) = f_{x_i}\left(\frac{f_{x_{i+1}}}{f_{x_i}}\right)^{r_{Rxa}}\left(1 - \frac{f_{x_{i+1}}}{f_{x_i}}\right)^{r_{Rxb}}\left(1 + C_{Rxc}\left(1 - f_{x_i}\right)\right), \tag{24}$$

where r_{Rxa}, r_{Rxb} and C_{Rxc} are constants. Using Equations (22), (23) and (24) a generalised recrystallised volume fraction growth rate for $f_{x_{i+1}}$ is formulated using the state variables for the volume fraction $f_{x_i} - f_{x_{i+1}}$. Equation (20) is formulated for each $f_{x_{i+1}}$ as

$$\dot{f}_{x_{i+1}} = Y_{x_i} C_{Rx0} C_{RxT}(T) C_{Rx\lambda}\left(\bar{\lambda}_{x_i}\right) A(f_{x_i}, f_{x_{i+1}}). \tag{25}$$

The function is written so that C_{Rx0} effectively contains all the pre-exponential constants. Similarly, $C_{RxT}(T)$ contains the temperature dependence in a single function

$$C_{RxT}(T) = \mu(T) \exp\left(-\frac{a_{0Rx}}{T}\right). \tag{26}$$

Here, a_{0Rx} is again the convenient grouping of the activation energy and universal gas constant as done in Equation (15). The function $C_{Rx\lambda}\left(\bar{\lambda}\right)$ contains the geometric effects in the rewritten function

$$C_{Rx\lambda}\left(\bar{\lambda}_{x_i}\right) = 1 - \exp\left(-C_{Rx\lambda 0}\bar{\lambda}_{x_i}^{r_{Rx\lambda}}\right), \tag{27}$$

where $C_{Rx\lambda 0}$ and $r_{Rx\lambda}$ replaces the constant $C_{M\theta}$ and exponent $r_{M\theta}$ in Equation (22) for subscript consistency.

Given a time increment δt, the first (f_{x_1}) and second (f_{x_2}) volume fractions can both progress, meaning region $f_{x_1} - f_{x_2}$ will increase by δf_{x_1} and decrease by δf_{x_2}. Assuming recrystallisation removes the dislocation structure, the dislocation density ratio within a newly recrystallised portion δf_{x_1} should be reset. This is taken into account by Brown and Bammann [3] using the rule of mixtures. Doing the same for the model in Jansen van Rensburg [8], the dislocation density ratio evolution in Equation (14) is replaced by

$$\dot{Y}_{x_i} = \dot{\varepsilon}_p\left(C_g\bar{\lambda}_{x_i}^{r_g} + C_1\sqrt{Y_{x_i}} - C_2(\dot{\varepsilon}_p, T)\,Y_{x_i}\right) - C_3(T)Y_{x_i}^{r_3} - \frac{\dot{f}_{x_i}}{f_{x_i} - f_{x_{i+1}}}Y_{x_i} \tag{28}$$

for the dislocation density ratio associated with volume fraction $f_{x_i} - f_{x_{i+1}}$. The average slip plane lattice incompatibility in Equation (13) is also replaced in the presence of recrystallisation by

$$\dot{\bar{\lambda}}_{x_i} = \dot{\varepsilon}_p C_{\lambda x} - \frac{\dot{f}_{x_i}}{f_{x_i} - f_{x_{i+1}}} \bar{\lambda}_{x_i} \quad (29)$$

for the $f_{x_i} - f_{x_{i+1}}$ material volume fraction. $C_{\lambda x}$ here is a constant related to C_λ / d_x in Equation (13) if the grain size itself is not specifically known. In the presence of recrystallisation, the equivalent threshold stress of Equation (9) is calculated using the average dislocation density ratio

$$\bar{Y} = \sum_{i=0}^{n_x - 1} Y_{x_i} \left(f_{x_i} - f_{x_{i+1}} \right), \quad (30)$$

where n_x is the total number of recrystallisation cycles. The dislocation density based user material with recrystallisation is implemented into a subroutine called by the isotropic user material framework [8]. The model is illustrated here on a larger set of the Copper data digitised from the experimental work done by Tanner and McDowell [13]. The material parameters resulting in the fit to the Copper data in Figure 3 are:

- $\mu_r = 43.8$GPa, $D_r = 4.7$GPa, $T_r = 252$K and $\nu = 1/3$ for the elastic properties using the shear model relationship in Equation (2).
- $a_{0\varepsilon} = g_{0\varepsilon} b^3 / k_B = 2.1037$K/MPa, $p_\varepsilon = 1$, $q_\varepsilon = 2$, and $\dot{\varepsilon}_{0\varepsilon} = 10^6 s^{-1}$ for the temperature and rate dependent scaling function in (3).
- The athermal yield stress component is $\hat{\sigma}_a = 12.519$MPa and reference stress is $\sigma_0 = 17.295$MPa using Equation (10).
- $C_{\lambda x} = 1$ is used for the evolution of $\bar{\lambda}$ according to Equation (29).
- $C_g = 6378.74$, $r_g = 0.769$, $C_1 = 278.87$, $C_{20} = 11.773$, $a_{02} = 0.904$K/MPa, $\dot{\varepsilon}_{02} = 4.0112 \times 10^{12}$, $C_{30} = 83.07 s^{-1}$, $a_{03} = 6370.675$K and $r_3 = 0.8079$ for the dislocation density ratio evolution in Equation (28).
- $C_{Rx0} = 9346.62 s^{-1}$ in Equation (25), $a_{0Rx} = 17633.796$K in Equation (26) while $C_{Rx\lambda 0} = 47.247$ and $r_{Rx\lambda} = 3.87$ in Equation (27). $r_{Rxa} = 0.1424$, $r_{Rxb} = 1.7677$ and $C_{Rxc} = 393.44$ in Equation (24).

The model gives valid responses in a large range of rates and temperatures as visible in Figure 3 with temperatures ranging from 25°C to 541°C and rates between $0.0004 s^{-1}$ and $6000 s^{-1}$. Thanks to the user material framework [8] it is also possible to use the recrystallisation model in a finite element simulation. In Figure 4, different internal state variables are displayed as a result of a simulation using the material parameters identified for Copper.

Compression of a Copper billet is modelled using an axi-symmetric simulation. A billet with diameter 5mm and height 7.5mm is compressed to 50% its original height in 2000 seconds. The compression is modelled at a constant temperature of 541°C. In Figure 4(a) the equivalent dislocation density using Equation (30) is illustrated with the volume fraction averaged slip plane lattice misorientation (b). Figure 4(c) shows the volume fraction recrystallised at least once while Figure 4(d) shows the material volume fraction that has undergone at least two recrystallisation cycles.

CONCLUSIONS

Thanks to the incrementally objective user material frameworks [8], sophisticated state dependent material models can be implemented into a finite element environment. Different model formulations are possible by altering the Fortran code called to represent the relevant one dimensional model formulation. This was illustrated using the MTS model as well as models for dislocation density based combined hardening and isotropic hardening with recrystallisation as an example [8].

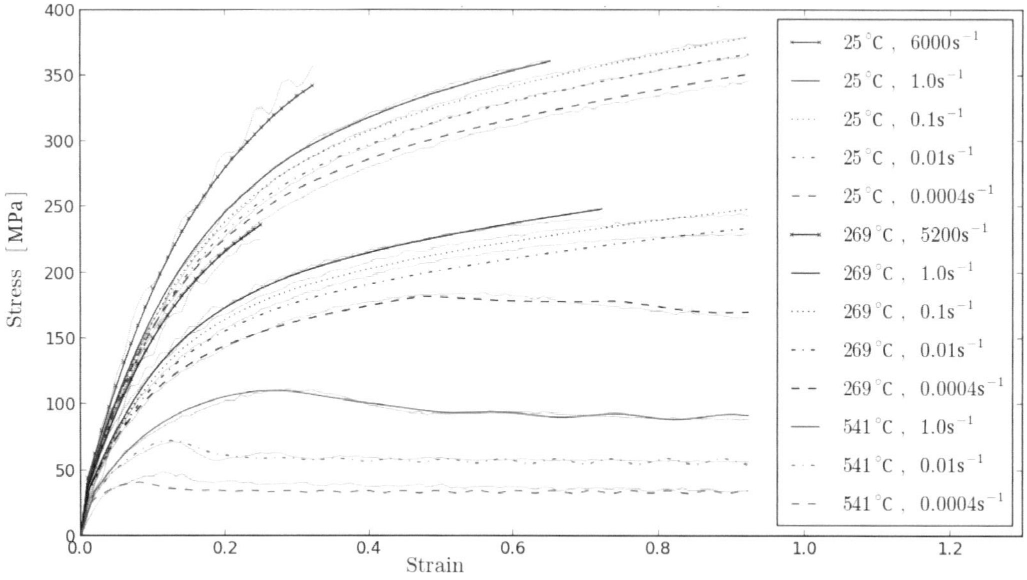

Figure 3: Numerical model (colored lines) calibrated to Tanner and McDowell's OFHC copper data [13] for different strain rates at 25°C, 269°C and 541°C.

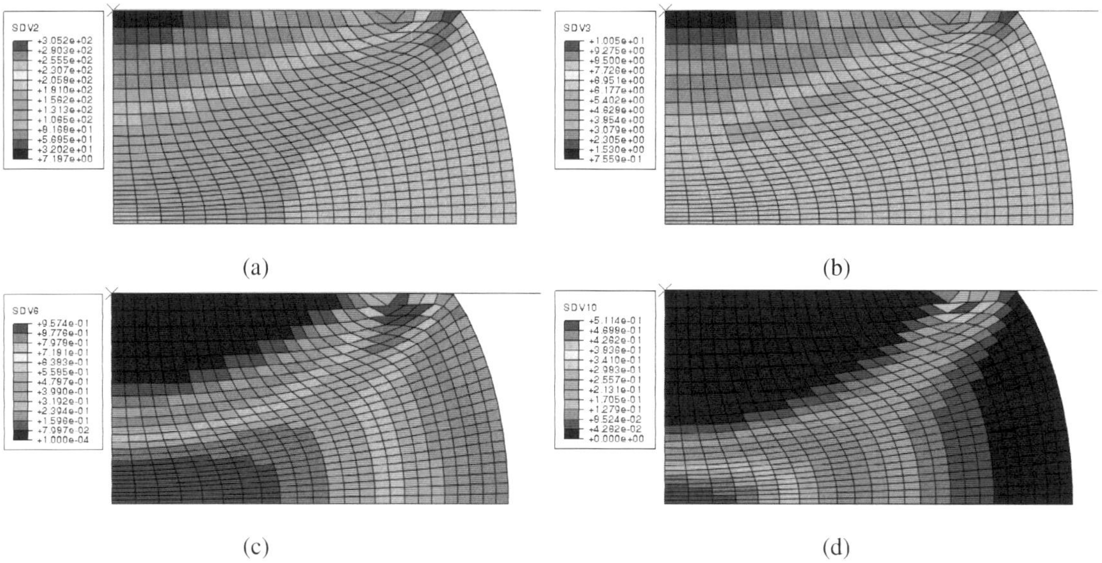

Figure 4: Internal state variables of a billet with recrystallisation compressed 50% in 2000 seconds at 541°C. (a) Equivalent dislocation density ratio and (b) Misorientation. Volume fractions associated with (c) the first and (d) second wave of recrystallisation.

REFERENCES

1 Follansbee, P. S. and Kocks, U. F., "A constitutive description of copper based on the use of the mechanical threshold stress as an internal state variable", Acta Materialia, Vol. 36, No. 1, pp. 81 – 93, 1998.

2 Estrin, Y., "Dislocation-density-related constitutive modeling", In: Krausz, A. and Krausz, K. (Eds.), Unified constitutive laws of plastic deformation, Ch. 2, pp. 69 – 106, Academic Press, San Diego, 1996.

3 Brown, A. A. and Bammann, D. J., "Validation of a model for static and dynamic recrystallization in metals", International Journal of Plasticity, Vol. 32-33, pp. 17 – 35, 2012.

4 Kocks, U. F., Argon, A. S., and Ashby, M. F., "Thermodynamics and Kinetics of Slip", Vol. 19, Pergamon, Oxford, 1975.

5 Kocks, U. F., "Laws for work-hardening and low temperature creep", Journal of Engineering Materials and Technology, Vol. 98, pp. 76 – 85, 1976.

6 Kocks, U. F. and Mecking, H., "A mechanism for static and dynamic recovery", In: Haasen, P., Gerold, V., and Korstorz, G. (Eds.), Strength of Metals and Alloys, pp. 345 – 350, Pergamon, Oxford, 1979.

7 Mecking, H. and Kocks, U. F., Acta Materialia, Vol. 29, pp. 1865 – 1877, 1981.

8 Jansen van Rensburg, G. J., "Development and Implementation of State Variable Based User Materials in Computational Plasticity", Unpublished Thesis, University of Pretoria, 2016.

9 Nabarro, F. R. N., Basinski, Z. S., and Holt, D. B., "The plasticity of pure single crystals", Advances in Physics, Vol. 13, No. 50, pp. 193 – 323, 1964.

10 Varshni, Y. P., "Temperature dependence of the elastic constants", Physical Review B, Vol. 2, No. 10, pp. 3952 – 3958, 1970.

11 Kocks, U. F., Tomé, C., and Wenk, H., "Texture and Anisotropy", Cambridge University Press, 1998.

12 Chen, S. R. and Gray, G. T., "Constitutive behavior of tantalum and tantalum-tungsten alloys", Metallurgical and Materials Transactions A, Vol. 27A, pp. 2994 – 3006, 1996.

13 Tanner, A. B. and McDowell, D. L., "Deformation, temperature and strain rate sequence experiments on OFHC Cu", International Journal of Plasticity, Vol. 15, No. 4, pp. 375 – 399, 1999.

14 Kok, S., Beaudoin, A. J., and Tortorelli, D. A., "On the development of stage IV hardening using a model based on the mechanical threshold", Acta Materialia, Vol. 50, No. 7, pp. 1653 – 1667, 2002.

15 Song, J. E. and McDowell, D. L., "Grain Scale Crystal Plasticity Model with Slip and Microtwinning for a Third Generation Ni-Base Disk Alloy", pp. 159 – 166, John Wiley & Sons, Hoboken, 2012.

16 Cahn, J. W. and Hagel, W. C., "Theory of the pearlite reaction", Decomposition of Austenite by Diffusional Processes, Vol. pp. 131 – 196, 1962.

17 Doherty, R. D., Hughes, D. A., Humphreys, F. J., Jonas, J. J., Juul Jensen, D., Kassner, M. E., King, W. E., McNelley, T. R., McQueen, H. J., and Rollett, A. D., "Current issues in recrystallization: A review", Materials Science and Engineering A, Vol. 238, No. 2, pp. 219 – 274, 1997.

18 Chen, S. P., Zwaag, S., and Todd, I., "Modeling the kinetics of grain-boundary-nucleated recrystallization processes after cold deformation", Metallurgical and Materials Transactions A, Vol. 33, No. 3, pp. 529 – 537, 2002.

19 Humphreys, F. J. and Hatherly, M., "Recrystallization and Related Annealing Phenomena", Elsevier, 2nd Edition, 1995.

Design and Evaluation of a Flutter-Suppression Control System for an Aircraft Wing

S. Jivan*, G. Venter†, S.J. van der Spuy‡
*16542665@sun.ac.za, †gventer@sun.ac.za, ‡sjvdspuy@sun.ac.za

ABSTRACT

This paper documents the design and evaluation of a flutter-suppression control system for a high-aspect-ratio wing of an aircraft. The goal of the control system is to suppress vibration, experienced by the wing, by using existing control surfaces. The project makes use of ANSYS software to simulate 2-way Fluid-Structure Interaction (FSI) where ANSYS Parametric Design Language (APDL) scripting was used to embed a Linear-Quadratic (LQ) controller within the FSI model. The wing's tip displacement was observed to progressively increase in the open-loop simulation whereas the close-loop simulation illustrated the LQ controller's capacity to subdue the wing's vibration. The control system was therefore; successful in suppressing the vibration thereby increasing the aerodynamic damping of the system.

KEYWORDS

Flutter, Fluid-Structure Interaction, LQ controller, ANSYS, APDL scripting

INTRODUCTION

Flutter is defined as an unstable and self-excited oscillation of a structure, at its natural frequency. Flutter arises from the fact that aeroelastic instability exits. This instability stems from the aerodynamic forces feeding energy into the structural system, by virtue of its position or configuration, as rapidly as it is dissipated by the structure's damping (Lakin and Haynes [1]).

The motion of a structural system can be considered as the superposition of the normal modes of vibration. The vibration of a structure would naturally be dissipated if not for the aerodynamic forces acting upon it. Depending on the speed of the air, the aerodynamic forces could decrease, maintain or increase the amplitude of the vibration (Scanlan [2]).

This phenomenon plays a significant role in the design, development and commissioning of vehicles in the aerospace industry. Flutter can be detrimental to the performance of an aircraft's structural components, with regards to its structural integrity and endurance. Flutter was first identified in 1918 on the Handley Page bomber when the flutter mechanism consisted of the fuselage torsion mode with an anti-symmetric elevator rotation mode (Garrick and Reed III [3]).

In modern era, aircraft design is focused on maximising performance by improving several facets of the aircraft. An example would involve structural weight reduction on a wing by using lighter and more flexible material (De Marqui Junior *et al.* [4]). Moreover; a means of developing a more efficient aircraft may involve the use of high-aspect-ratio wings. However; these wings tend to be less rigid (Wan *et al.* [5]). This, coupled with the fact that weight reduction may result in reduced stiffness, increases the susceptibility to aeroelastic instability (Marretta and Marino [6]).

A manner of resolving this aeroelastic instability would be to increase the damping of the wing. The wing's control surfaces provide a means of suppressing vibration, thereby increasing the aerodynamic damping of the system. Marretta and Marino [6] demonstrated the use of control surfaces to suppress flutter, using a mathematical model to perform the aeroelastic analysis.

Modelling a aeroelastic phenomenon requires a means of accounting for both the structure and fluid environments. This multi-physic phenomenon, known as a FSI, is to be modelled by means of numerical analysis using commercial software, ANSYS. The flutter-suppression control system was required to suppress vibration on a wing. Suppression of vibration would aid in the dissipation of energy, being fed into the system by the fluid, thereby increasing the aerodynamic damping of the system.

OBJECTIVES

- Establish a structure and fluid environment, in ANSYS, which constitutes a numerical FSI model
- Use a high-aspect-ratio wing in the FSI model
- Derive state-space models of the FSI, using Matlab's system identification toolbox, and implement the models in the control system
- Embed the control system within the FSI model using APDL scripting
- Use existing control surfaces in the control system
- Document the performance the control system

WING ASPECT RATIO

The wing's geometry has a significant influence on the aircraft's performance. High-aspect-ratio wings are long and narrow whereas low-aspect-ratio wings are short and stubby. Figure 1 illustrates the transition of an aircraft's wings as the aspect ratio varies from a low, to moderate and finally a high aspect ratio (Wan *et al.* [5]). Quantifying the aspect ratio of a wing is expressed by equation 1. Figure 2 depicts the top view of a wing planform, illustrating the wing's span and chord.

$$Aspect\ ratio = \frac{Span}{Chord} \tag{1}$$

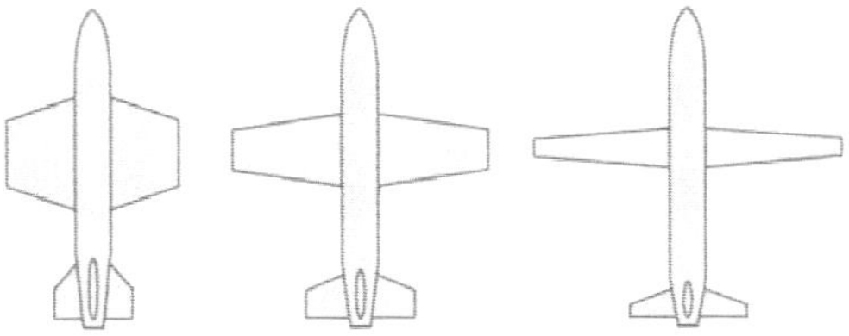

Figure 1: Wing aspect ratio
Science Learning [7]

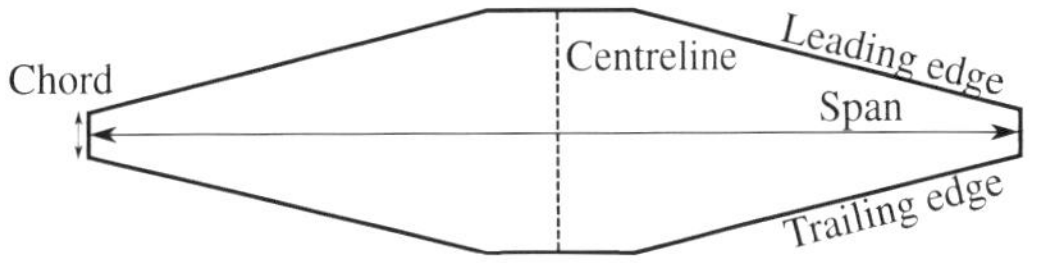

Figure 2: Wing planform

High-aspect-ratio wings are desirable since they benefit from a superior lift-drag ratio and are lightweight. An undesirable feature of these wings include a lack of manoeuvrability. As a result, these wings are not found on aerospace vehicles such as fighter aircraft but are used in civil and military applications such as high-altitude long-endurance (HALE) aircrafts and unmanned aerial vehicles (UAVs) (Wan *et al.* [5]).

AEROELASTIC INSTABILITY

Figure 3 depicts a structural system where a rotational spring and plunge spring constitutes a 2-degree-of-freedom (DOF) system. Despite the simplicity of the system, it represents an aerofoil and is fundamental to understand flutter behaviour. The plunge spring represents the wing's bending stiffness whereas the rotational spring represents the torsional stiffness. The shape of the aerofoil determines where the aerodynamic center is located.

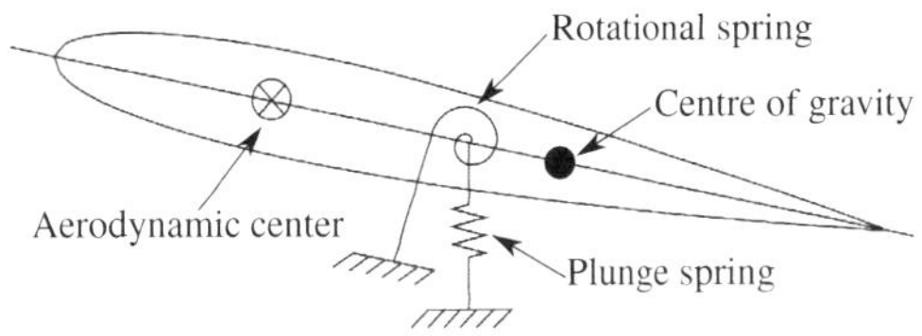

Figure 3: 2-DOF aerofoil system

Aerodynamic forces excite this spring/mass system. The critical speed of the wing is encountered when the energy input would equal that of the energy being dissipated by the structural system. Once the critical speed is exceeded, the 2-DOF system could excite one of its two natural modes or a coupled mode which is a coalescence of the two natural modes (De Marqui Junior *et al.*[4]).

The system above does not account for control surfaces that may exist on an actual aircraft. Aeroelastic instability may arise from the presence of control surfaces, as their DOF present their own natural modes. An example includes the previously mentioned Handley Page bomber where the rotational DOF of the elevator contributed to the flutter mechanism. This flutter mechanism illustrates how the coalescence of two natural modes are possible.

CONTROL SYSTEM

State-space control design is a technique in which the control engineer is able to design a dynamic compensation by working directly with the state-variable description of the system. A manner of representing a physical system is by using linear state-space models, as expressed by Franklin *et al.* [8] in equations 2 - 3. In most instances, the state variables of a system are not known or are not measurable by the control designer therefore; a state estimate ($\dot{\tilde{\mathbf{x}}}$) is computed by constructing a full-order model of the system dynamics, shown in equation 4 (Franklin *et al.* [8]).

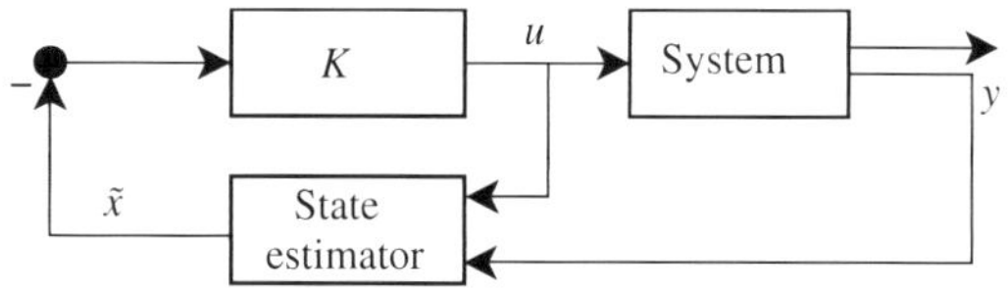

Figure 4: Closed-loop feed-back control system Kulakowski *et al.* [9]

Figure 4 illustrates how a state-estimator is utilised to compensate a system. The state-estimator is computed before the controller gain (**K**) compensates the input-signal (u) to the system.

$$\dot{\mathbf{x}} = \mathbf{F}\mathbf{x} + \mathbf{G}\mathbf{u} \tag{2}$$

$$\mathbf{y} = \mathbf{H}\mathbf{x} + \mathbf{E}\mathbf{u} \tag{3}$$

$$\dot{\tilde{\mathbf{x}}} = \mathbf{F}\tilde{\mathbf{x}} + \mathbf{G}\mathbf{u} \tag{4}$$

STRUCTURE GEOMETRY

The NACA 0012 aerofoil was chosen for its simplistic profile and the availability of additional test data over a wide range of test conditions which offers a means of validating the model (Abbott and Von Doenhoff [10]). In order to yield a high-aspect-ratio wing, the wing's chord and span were chosen as 0.07 m and 1.4 m, respectively, to generate an aspect ratio of 20. Figure 5 depicts the aerofoil generated from 102 data points.

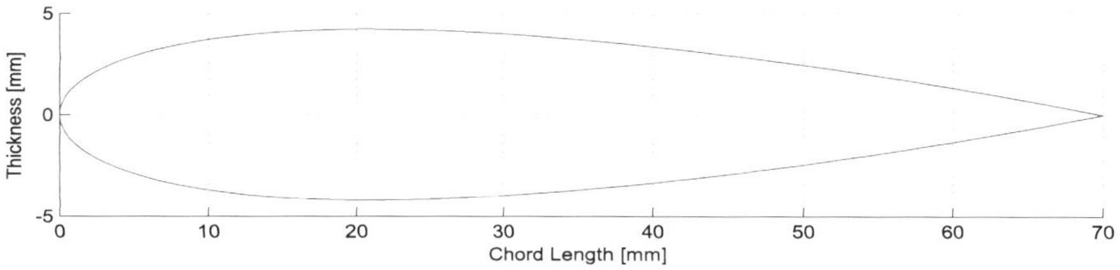

Figure 5: NACA 0012 wing profile

Although an aircraft may vary from one to another, the ailerons, flaps, slats and spoilers have been identified as possible options for the control surfaces to be used in the control system. Ailerons have previously been used in mathematical studies to suppress flutter (Huang *et al.*[11], Waszak[12]). As a result, the control system would make use of ailerons to suppress flutter.

The CAD models of the wing and ailerons were created using Autodesk Inventor. Figures 6 and 7 depicts the wing and aileron respectively. The areas devoid of material on the wing are use to accommodate the 6 ailerons. All the ailerons are identical in geometry and are placed along the span of the wing's trailing edge.

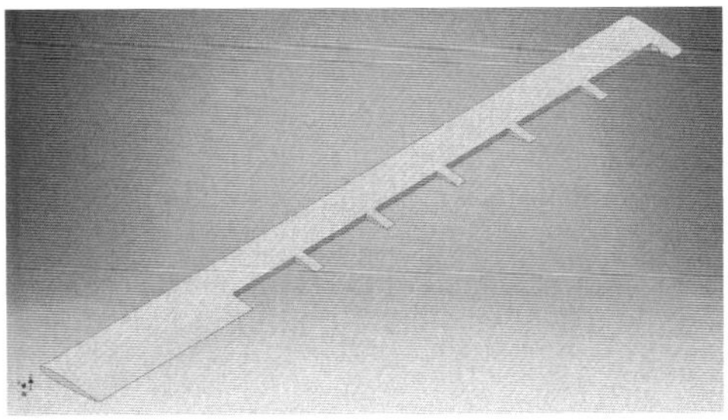

Figure 6: Wing without ailerons

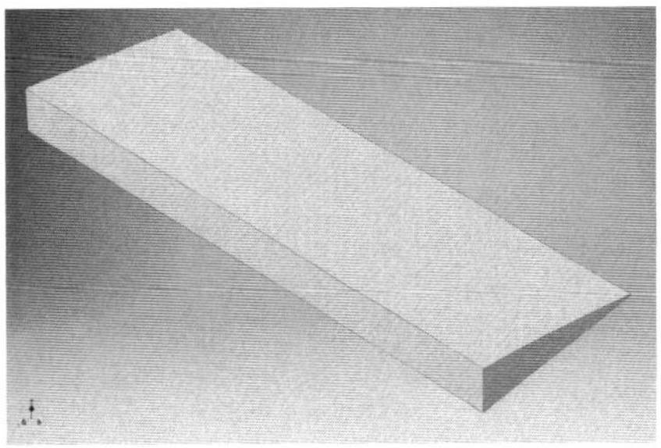

Figure 7: Aileron

The ailerons were remotely attached to the wing by means of the revolute connection. Figure 8 illustrates the revolute connection, provided in ANSYS, used to secure the ailerons to the wing. In this instance, two revolute connections, A and B, are applied to the sides of the aileron to locate them against the adjacent sides of the wing. A local coordinate system is defined against the edge of the aileron to allow the desired rotational DOF, about the z-axis.

The wing-aileron connection was not assigned a rotational stiffness to ensure that the ailerons did not contribute to the flutter mechanism.

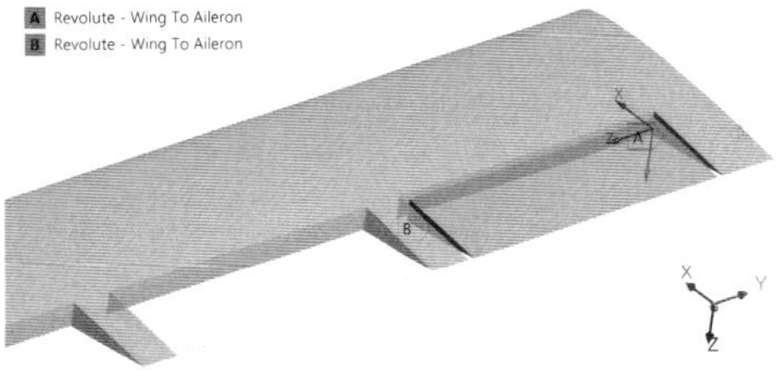

Figure 8: Revolute connection

FE MODEL

ANSYS provides solid shell elements (SOLSH190) which are applicable for structures with a thin to moderate thickness (ANSYSInc [13]). Therefore; discretization of the wing and ailerons made use of these elements. Meshing required two parallel faces which required the chord to be trimmed by 1% (0.5% at both the leading and trailing edge). Brandsen [14] documented the use of solid shell elements, in a FSI model using compressor blades, with promising results. Figure 9 depicts the meshed wing and ailerons, using the SOLSH190 elements. Table 1 lists the coarse, medium and fine meshes used in the wing's mesh refinement process.

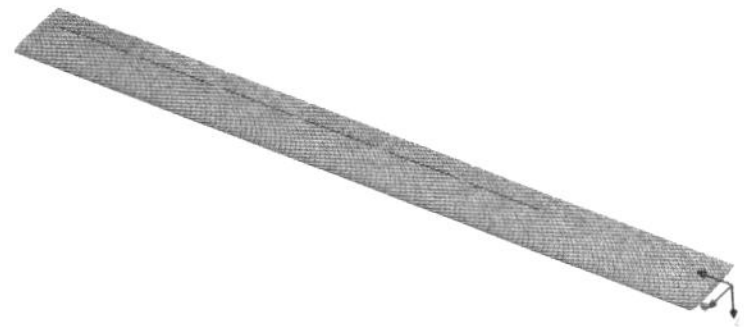

Figure 9: Meshed wing

Table 1: Mesh refinement

Mesh	Elements	Nodes
Coarse	2,212	1,583
Medium	6,212	4,252
Fine	12,228	10,546

Mesh refinement was performed until the tip displacement of the wing was observed not to vary. However; the discrepancies amongst the 3 meshes were small. Since the computational effort required for the FE model is negligible, compared to that of the CFD model, the fine mesh was selected for the simulations

MODAL ANALYSIS

A modal analysis was performed to determine the wing's natural modes and corresponding frequencies. As stated previously, the ailerons were constrained in a manner to ensure that their own natural modes did not contribute to the analysis. Therefore; the modal analysis computes the natural modes of the wing, without the influence of the ailerons' DOF. The first 5 natural modes were computed. Table 2 lists the natural frequencies of the wing and their corresponding dominant modes. Figure 10 depicts the first natural mode where bending is dominant.

Figure 10: First natural mode

Table 2: Natural frequencies of the wing

Mode	Frequency [Hz]	Dominant mode
First	1.4091	Bending
Second	8.7856	Bending
Third	11.273	In-plane
Fourth	24.483	Bending
Fifth	32.739	Torsional

Huang *et al.* [11] states that in-plane natural modes contribute much less to flutter. The in-plane mode at the third natural frequency was therefore neglected. The bending modes are most common amongst the 5 generated. This is due to the geometry of a high-aspect-ratio wing and is evident as Huang *et al.* [11] also documented the natural frequencies of a low-aspect ratio wing where the first and second frequencies are a bending and torsional mode, respectively. Bending would therefore be the dominate natural mode for a high-aspect-ratio wing. A base excitation was used to disturb the wing from its resting position and excite aeroelastic instability.

FLUID DOMAIN GEOMETRY

The geometry of the fluid domain uses dimensions 10×6 chord lengths after Mohammadi and Johari [15] observed that a domain of 7×4 was not adequate for modelling a wing. Figure 11 illustrates the fluid domain which makes use of the aforementioned dimensions where 6 chords lengths (denoted by c) are used from the leading edge and 10 chord lengths are used from the trailing edge of the wing. The fluid domain spans 1000 mm, allowing 300 mm to account for the free-end. The CAD model for the fluid domain was generated using AutoDesk Inventor.

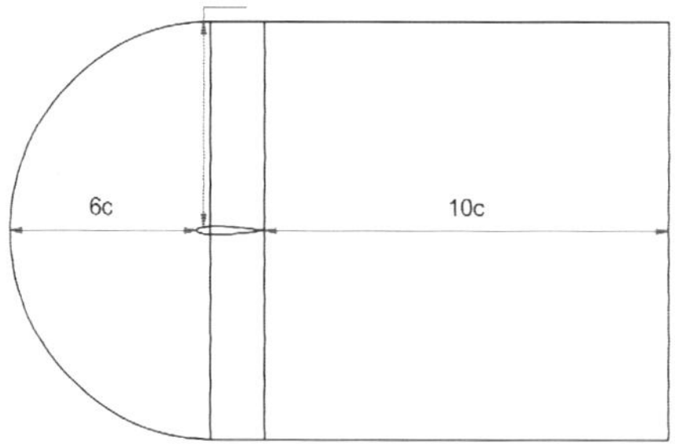

Figure 11: Fluid domain geometry

CFD MODEL

Figure 12 depicts the 3-dimensional model of the meshed fluid domain. The fluid domain was partitioned to allow better control with regards to the generation of the mesh. Partitioning of the fluid domain enabled particular sections to make use of a structured mesh, whereas the rest made use of an unstructured mesh.

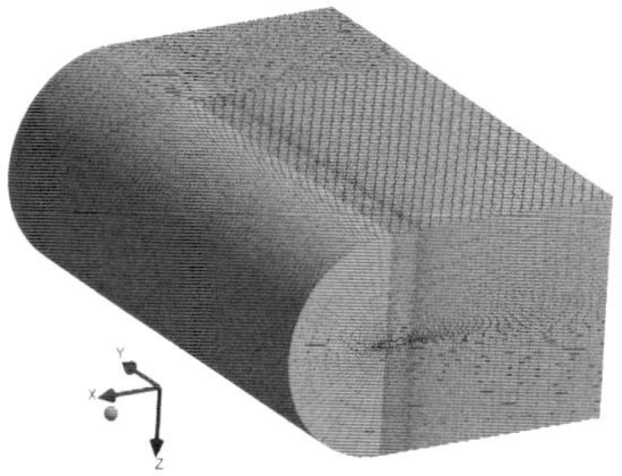

Figure 12: Meshed fluid domain

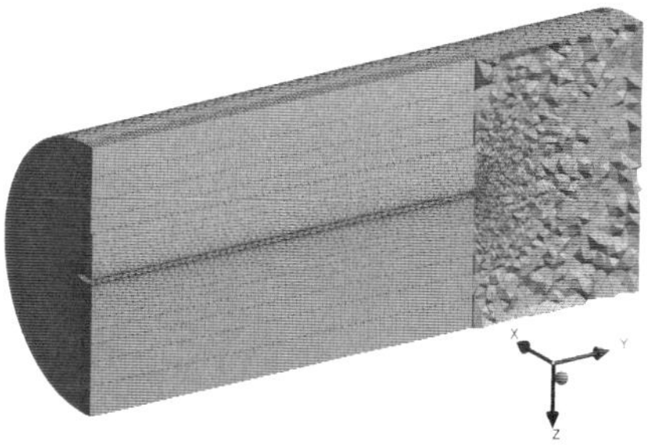

Figure 13: Sectioned fluid domain

Figure 13 depicts a sectioned view of the fluid domain, illustrating the sections which used a structured mesh and sections which used an unstructured mesh. The sections, in which the wing was located, consisted entirely of hexahedral elements. The adjacent section, downstream, also consisted of hexahedral elements. The sections which models the free-end made use of an unstructured mesh. Tables 3 and 4 lists the mesh refinement for the fluid domain and the properties of the air, respectively. The Reynold's number was determined to be $Re = 201,764.70$

Table 3: Mesh Refinement

Mesh	Elements	Faces	Nodes
Coarse	510,515	1,313,275	329,808
Medium	1,452,584	2,872,865	652,801
Fine	1,889,854	3,842,255	1,221,365

Table 4: Air properties

Named surface	Boundary type
Speed $[\frac{m}{s}]$	40
Density $[\frac{kg}{m^3}]$	1.225
Dynamic viscosity $[\frac{kg}{ms}]$	1.7×10^{-5}

TURBULENCE MODEL

The $SST\ k-\omega$ turbulence model was selected for the CFD model. The $SST\ k-\omega$ is a variation of the $k-\omega$ turbulence model which was developed by Menter [16] in an effort to blend the robustness and accuracy of $k-\omega$ model, at the near-wall region, and the $k-\varepsilon$ model that offers the free-stream independence in the far field. The $SST\ k-\omega$ turbulence model is thus more accurate and reliable for a wider range of fluid flow (Eleni *et al.* [17]). In addition to this, Eleni *et al.* [17] concluded that the $SST\ k-\omega$ model was an appropriate turbulence model when modelling a NACA 0012 aerofoil.

STATE-SPACE MODELS

A simplified state-space model of the FSI system is to be determined. This work utilizes Matlab's System Identification toolbox, which entails estimating the state-space model of a system by supplying the toolbox with the system's input-data and corresponding output-data (Takacs and Rohal-Ilkiv [18]). Assuming no particular knowledge of the internal structure of the system, a linear state-space model is estimated, which is well-conditioned. To compute the models of the FSI, an initial simulation was performed, forcing the ailerons to rotate with a known value (input-data). The subsequent wing-tip displacement (output-data) was then determined. The state-space models were then computed (shown below) using the relevant input- and output-data.

$$\mathbf{F} = \begin{bmatrix} 0.2630229307 & 0.0670387530 & 0 \\ 0.6156435799 & -0.2995311298 & 0 \\ 0 & 0 & 0 \end{bmatrix} \quad \mathbf{G} = \begin{bmatrix} -261.2986386568 \\ 3577.676056130 \\ 0 \end{bmatrix}$$

$$\mathbf{H} = \begin{bmatrix} 7.37730050e-05 & -4.72234999e-06 & 0 \end{bmatrix}$$

The controller gain ($\mathbf{K}$) and estimator gain ($\mathbf{L}$) were then computed using the *dlqr* and *kalman* commands, respectively, in Matlab.

$$\mathbf{K} = \begin{bmatrix} 0.0001665 & -0.0000845 & 0 \end{bmatrix} \quad \mathbf{L} = \begin{bmatrix} 9548.9 \\ 9386.8 \\ 0 \end{bmatrix}$$

APDL SCRIPTING

APDL is a scripting language that allows the user to parameterize models and automate tasks in ANSYS. Using commands, this scripting language allows the user to access the preprocessor and post-processor phase of the simulation. Logging into the preprocessor allows for the creation and/or altering of the FE model while logging into the post-processor allows for the retrieval of results from the FE model.

A LQ controller was chosen due to its successful implementation in smart structures (Nestorović [19], Nestorović and Trajkov [20]). This type of control requires the system to assume a state-space form where the state-estimate (at time-step N) is computed according to equation 5 whereas the input-signal, $\mathbf{u}^N$, is determined using equation 6 (adapted from Franklin *et al.* [8]).

APDL scripting was then used to embed the controller within the FSI model. Figure 14 illustrates how APDL scripting was used in computing a new state-estimate, controller gain and how the scripting updated the FE model. Once the simulation reaches the end of a time-step the controller computes a new state-estimate and controller gain, used in the following time-step. The simulation enters the following time-step and updates the FE model by compensating the input-signal, thereby altering the ailerons' rotational DOF. The process is repeated until the final time-step is reached and the simulation ends.

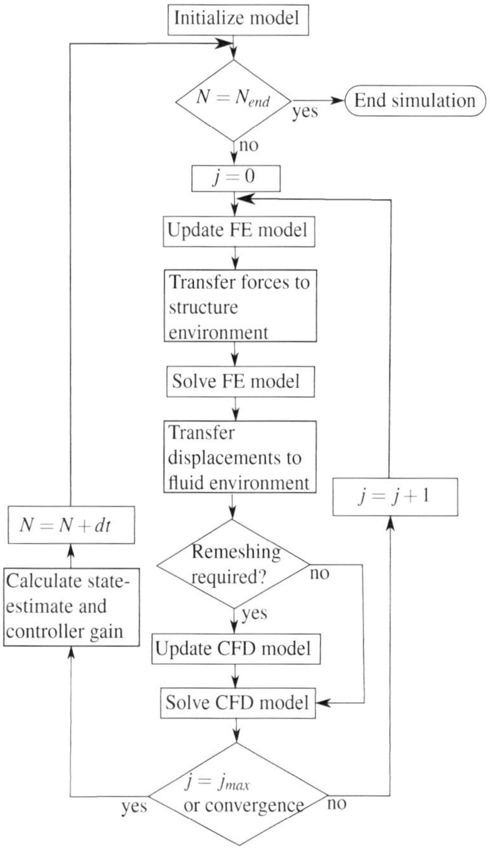

Figure 14: APDL logic

$$\tilde{\mathbf{x}}^N = \mathbf{F}\tilde{\mathbf{x}}^{N-1} + \mathbf{G}\mathbf{u}^N + \mathbf{L}(\mathbf{y} - \mathbf{H}\tilde{\mathbf{x}}^{N-1}) \tag{5}$$

$$\mathbf{u}^N = -\mathbf{K}\tilde{\mathbf{x}}^N \tag{6}$$

RESULTS

Figures 15, 16 and 17 represent the wing's tip displacement, velocity and acceleration, respectively. The open-loop simulation was performed before the close-loop simulation. The wing's tip displacement was observed to progressively increase in the open-loop simulation indicating that the structure's damping was not adequate to dissipate energy being fed into the system. The close-loop simulation illustrates the LQ controller's capacity to suppress the wing's vibration. The progressive trend witnessed in the open-loop simulation was subdued, thereby illustrating how use of the ailerons aided in increasing the aerodynamic damping of the system.

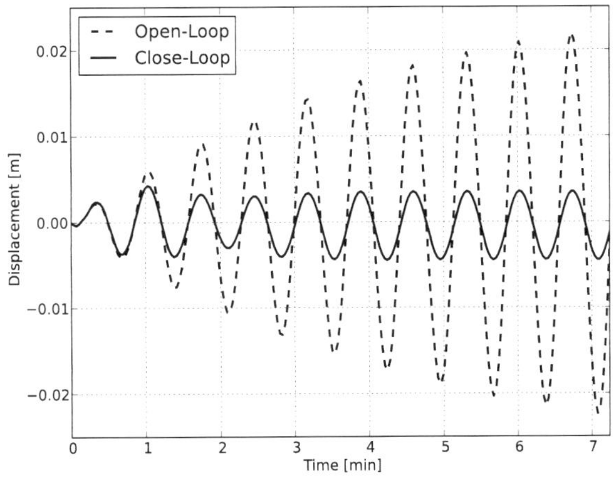

Figure 15: Wing-tip: displacement

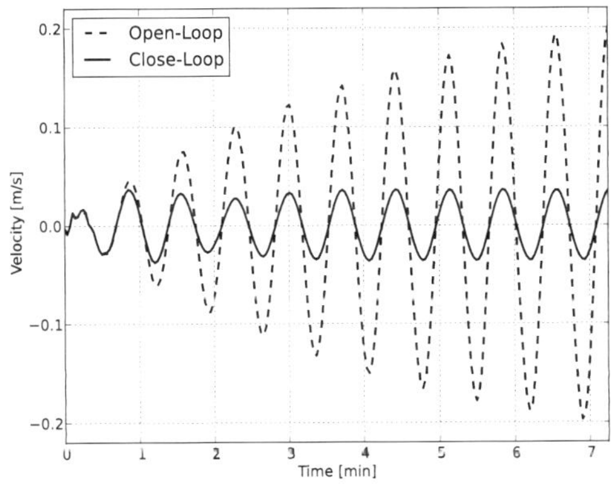

Figure 16: Wing-tip: velocity

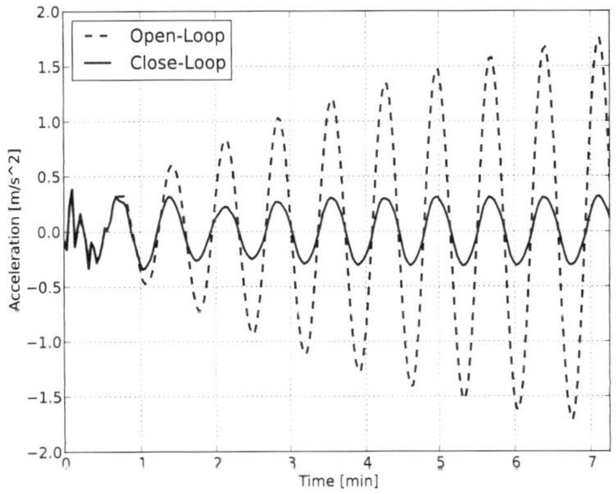

Figure 17: Wing-tip: accleration

A decrease in the wing's tip displacement would reduce the induced stress that the wing would experience. The von-Mises stress was determined for the open- and close-loop simulations. A maximum value of 0.13384 MPa was observed in the open-loop simulation, at the maximum thickness of the wing's root. The maximum value in the close-loop simulation was found to be 0.117 MPa, once again at the maximum thickness of the wing's root.

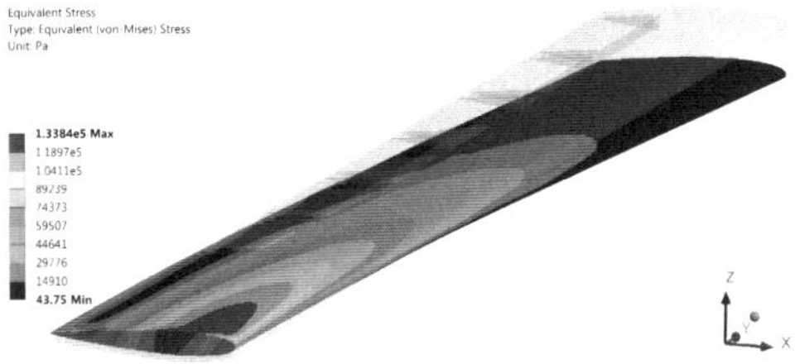

Figure 18: Von-Mises stress plot: open-loop

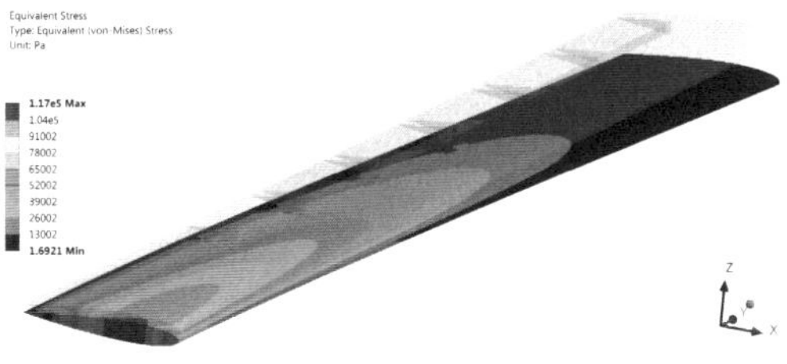

Figure 19: Von-Mises stress plot: close-loop

CONCLUSION

The design and evaluation of a flutter-suppression control system has been documented in this paper. A 2-way FSI model was established in ANSYS. A high-aspect-ratio wing was used in the FSI where ailerons were used in the control system to suppress the vibration that the wing experienced. A LQ controller was successfully embedded within the FSI model. This controller regulated the wing's tip displacement and suppressed the vibration experienced by the wing with the use of the ailerons. The closed-loop simulation illustrated how the controller was able to successfully suppress the vibration thereby increasing the aerodynamic damping of the system.

References

[1] Lakin, G. and Haynes, J.: *Flutter and Aeroelastic Stability*. 1997.

[2] Scanlan, R.: *Introduction to the study of Aircraft Vibration and Flutter*. The Macmillan Company., 1951.

[3] Garrick, I. and Reed III, W.H.: Historical development of aircraft flutter. *Journal of Aircraft*, vol. 18, no. 11, pp. 897–912, 1981.

[4] De Marqui Junior, C. *et al.*: Design of an experimental flutter mount system. *Journal of the Brazilian Society of Mechanical Sciences and Engineering*, vol. 29, no. 3, pp. 246–252, 2007.

[5] Wan, Z. *et al.*: Static aeroelastic analysis of a high-aspect-ratio wing based on wind- tunnel experimental aerodynamic forces. *Science China Technological Sciences*, vol. 54, no. 10, pp. 2716–2722, 2011.

[6] Marretta, R.M. and Marino, F.: Wing flutter suppression enhancement using a well-suited active control model. 2007.

[7] Learning, S.: Wing aspect ratio. 2011.
Available at: http://sciencelearn.org.nz/Contexts/Flight/Science-Ideas-and-Concepts/Wing-aspect-ratio

[8] Franklin, G. *et al.*: *Feedback control of dynamic systems*. Pearson, Upper Saddle River, N.J. Harlow, 2010. ISBN 978-0-13-500150-9.

[9] Kulakowski, B. *et al.*: *Dynamic moing and control of engineering systems*. Cambridge University Press, Cambridge New York, 2007. ISBN 9780521864350.

[10] Abbott, I.H. and Von Doenhoff, A.E.: *Theory of wing sections, including a summary of airfoil data*. Courier Corporation, 1959.

[11] Huang, R. *et al.*: Single-input/single-output adaptive flutter suppression of a three-dimensional aeroelastic system. *Journal of Guidance, Control, and Dynamics*, vol. 35, no. 2, pp. 659–665, 2012.

[12] Waszak, M.R.: Modeling the benchmark active control technology wind-tunnel model for application to flutter suppression. *AIAA paper*, vol. 3437, 1996.

[13] ANSYSInc: Ansys element reference. 2009.

[14] Brandsen, J.D.: Prediction of axial compressor blade vibration by modelling fluid-structure interaction. 2013.

[15] Mohammadi, M.A. and Johari, H.: Computation of flow over a high-performance parafoil canopy. *Journal of Aircraft*, vol. 47, no. 4, pp. 1338–1345, 2010.

[16] Menter, F.R.: Two-equation eddy-viscosity turbulence models for engineering applications. *AIAA journal*, vol. 32, no. 8, pp. 1598–1605, 1994.

[17] Eleni, D.C. *et al.*: Evaluation of the turbulence models for the simulation of the flow over a national advisory committee for aeronautics (naca) 0012 airfoil. *Journal of Mechanical Engineering Research*, vol. 4, no. 3, pp. 100–111, 2012.

[18] Takacs, G. and Rohal-Ilkiv, B.: Direct closed-loop active vibration control system prototype in ansys. In: *INTER-NOISE and NOISE-CON Congress and Conference Proceedings*, vol. 244, pp. 1–12. Institute of Noise Control Engineering, 2012.

[19] Nestorović, T.N.: Active control of mechanical structures in research and education. *Theoretical and Applied Mechanics*, vol. 40, no. 2, pp. 203–221, 2013.

[20] Nestorović, T. and Trajkov, M.: Active control of smart structures-an overall approach. *Facta universitatis-series: Architecture and Civil Engineering*, vol. 8, no. 1, pp. 35–44, 2010.

Formulation of a unit cell for a thermal-hydraulic systems model of a prismatic fuel block of a High Temperature Gas-cooled Reactor

*SN Khoza[1], CG du Toit[1] & PG Rousseau[2]

[1]North-West University, Potchefstroom

[2]University of Cape Town, Cape Town

*khoza.samun@gmail.com

ABSTRACT

Thermal-hydraulic analysis in high temperature prismatic block reactors involves determining the temperature distribution in the fuel blocks due to heat conduction, convection and radiation. It also includes the heat transfer between the graphite block, fuel rods, coolant channels and bypass gaps.

The formulation of a thermal-hydraulic analysis model of a section of a prismatic block using a 1-D network approach is given. The proposed model can serve as a basic building block for the analysis of the temperature distribution for the whole or any part of the core without too much computational expense or effort. In the proposed method, a basic unit cell containing part of the graphite, fuel rod and coolant channel was developed. A conduction shape factor was derived to account for the heat transfer from the fuel rods to the graphite and from the graphite to the coolant channels. The results are of the same magnitude as those reported in literature.

KEYWORDS

Thermal-hydraulic analysis, high temperature reactor, prismatic block, heat transfer, unit cell

INTRODUCTION

The active core of a prismatic high temperature reactor (HTR) is of annular design with an array of hexagonal fuel blocks. The fuel blocks must be maintained within a safe operating envelope which does not exceed about 1250 °C and below 1600 °C in the case of an accident [1]. The cross section of the fuel block has a complex geometry and high temperature gradients are expected between the fuel rods and the helium cooling channels. The complexity of the core thermal-hydraulic analysis in high temperature prismatic block reactors makes it difficult for fine mesh computational fluid mechanics (CFD) methods to be applied to the whole core. Implementation of CFD methods requires a very fine mesh for accurate temperature predictions. This has resulted in a number of studies only concentrating on part of a fuel block or a few fuel blocks or part of the core as investigated in a study by Tak et al. [2]. In recent years, there has been a drive to simplify the computational models for the HTRs so that they can be applied to larger areas of the core.

A simplification for the high temperature test reactor (HTTR) was introduced by Rousseau and Greyvenstein [3] to model the complex geometry of the HTTR using axi-symmetric 2D geometry. The HTTR is a prismatic reactor which has annular fuel rods inserted in vertical bore holes in the hexagonal blocks. The coolant flows through the annular gap between the fuel rods and the fuel rod bore holes. A thermal-fluid network model for the HTTR was developed using Flownet (a thermal-fluid network code) [3-5]. The graphite blocks that have the same flow path were grouped together. The fuel assembly blocks and the surface temperature of the flow passages in the fuel block containing fuel rods were represented by a single node. Thus the fuel blocks were represented by a single row of conduction elements and nodes. The graphite sleeve and fuel compact were discretised into a number of layers each represented by a single thermal mass node [4].

Another widely adopted method for simplification of prismatic block-type HTR models for heat transfer analyses is the unit cell heat transfer approach [2]. The unit cell method recognises a repeated pattern or arrangement of unit cells within the fuel block and uses these cells as a basic unit for models [1-2, 7-9].

Several types of basic unit cells have been used but without distinction between different regions and therefore no explicit fuel and graphite temperatures were computed. More recently the unit cell method has been improved by using a combination of pre-generated mesh cells to generate computational grids [7-8]. In a study done by Tak et al [8], the unit cell concept was implemented by using several types of basic unit cells for generating computational grids to model the heat transfer through the hexagonal blocks and a 1-D formulation to model the flow through the coolant channels. The computational grids were generated using a combination of pre-generated mesh cells and thus the fuel and coolant channels were explicitly modelled.

A previous study by Sambureni [9] done on a single prismatic block showed that Flownex (a system CFD code) can be used to build a network model of a prismatic block reactor. In the study, however, the heat conduction was investigated only in the radial and tangential directions and discretisation in the axial direction was not done. The heat generated in the fuel was uniformly distributed in the nodes within the control volume. A unit cell was used to calculate a conduction shape factor for heat transfer between the fuel rod and the coolant channel wall. The basic unit cell contains part of the graphite and the associated fuel rods and coolant channels.

This paper describes the conceptual formulation of a model for the conduction path between the fuel rods and the coolant channels for a central control volume of a sixth of a prismatic block using a unit cell as a foundation. In this approach, Flownex is used to predict the axial and radial temperature distribution within a single control volume of a hexagonal fuel block. The proposed model differs from the Flownet model [3-5] in that the prismatic fuel block is modelled in detail therefore employing multiple nodes to represent the fuel block, the fuel rods and the coated fuel particles. The discretisation of the coated fuel particle is important because the temperature of the kernel is a critical temperature in terms of fuel preservation.

DESCRIPTION OF A PRISMATIC FUEL BLOCK

A typical prismatic block type HTR hexagonal fuel block is shown in Figure 1. The geometric dimensions of the standard fuel block used were those of the PMR200 (200 MW_{th}) reactor core based on the design concept of a gas turbine-modular helium reactor (GT-MHR). A standard hexagonal fuel block has 204 fuel rods and 108 channels for helium coolant flow. The kernels, coated with layers of pyrolytic carbon and silicon carbide are mixed with a carbonaceous matrix and bonded into cylindrical fuel compacts. The nominal helium inlet and outlet temperatures are considered to be 490 °C and 950 °C [2].

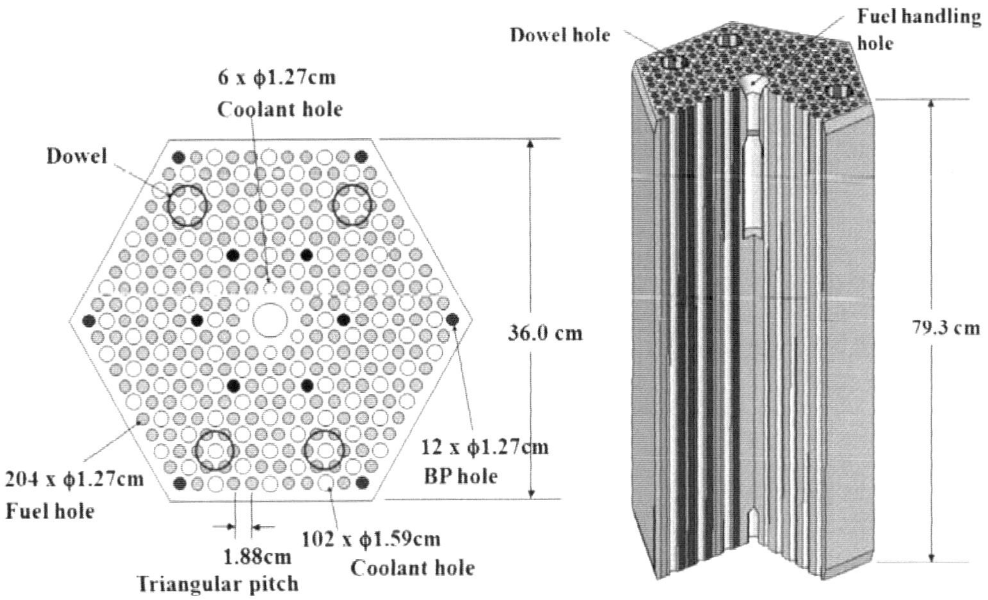

Figure 1. Standard fuel block for a typical prismatic reactor [2].

The boundary conditions for the PMR200 applied to the control volume are summarized in Table 1.

Table 1. Boundary conditions

Inlet temperature	490 °C
Inlet Pressure	7 MPa
Mass flow rate for the central control volume	0.0503 kg/s
Thermal power	21.65 kW

The moderator and coolant properties are listed in Table 2.

Table 2. Material properties

Graphite	Density (kg/m^3)	1740
	Thermal conductivity (W/m.K)	52.7
	Specific heat (kJ/kg.K)	1.5
Helium	Density (kg/m^3)	3.3931
	Thermal conductivity (W/m.K)	0.37
	Specific heat (J/kg.K)	5197.6
	Dynamic viscosity (Pa.s)	4.6x10^{-5}

SIMULATION METHODOLOGY

The purpose of the model described here is to lessen the computational effort required by fine mesh CFD methods to allow quick plant simulations. This was done by developing a network in Flownex for the combined conduction, radiation and convection heat transfer through a representative prismatic fuel block with internal heat generation and flow in coolant channels. This will be demonstrated using a simple model of a selected basic control volume within the fuel block.

In Flownex, the analysis of thermal-fluid networks is based on the numerical solution of the governing equations of fluid dynamics and heat transfer. Flownex solves the partial differential equations for mass, momentum and energy conservation to obtain the mass flow, pressure and temperature distributions throughout a network [10-11]. The model is based on the fundamental equations of mass, momentum and energy for compressible fluid flow.

Fig. 2a shows the arrangement of fuel rods and coolant channels in a part of the block shown in Fig. 1. The fuel rods are represented by the solid red circles and the coolant channels by the open blue circles. A typical prismatic block coolant channel (represented by C7 in Fig. 2a) is surrounded by six fuel rods (represented by F7, F8, F10, F11, F13 and F14). Fig. 2b shows the chosen unit cell b which depicts a representative conduction path in a typical prismatic fuel block. The unit cell is formed by the indicated symmetry planes, fuel rod surface and coolant channel surface [9].

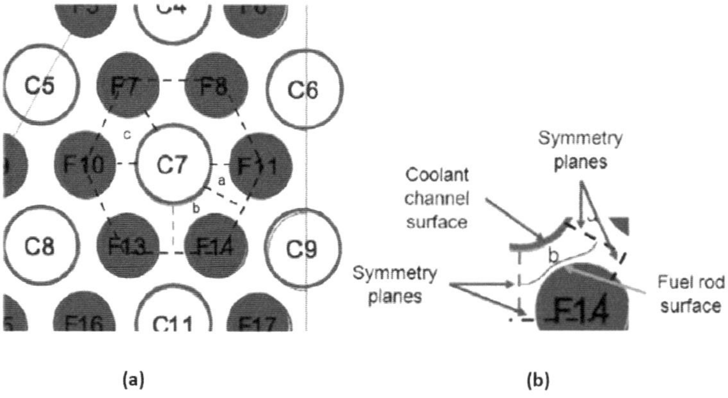

Figure 2. Fuel rods and coolant channels arrangement [9].

To determine the mass averaged graphite block temperature, a CFD (STAR-CCM+) simulation was conducted on unit cell b in a study by Sambureni [9]. This mass averaged graphite block temperature was used to determine the conduction shape factor for the heat transfer from the fuel rod surface to the graphite block node using Eqn. (1):

$$Q = S_{fs}k(T_f - T_s) \qquad \ldots (1)$$

Where:
S_{fs}= conduction shape factor for the heat transfer from the helium gap edge node to the graphite block node
T_f = temperature of the fuel rod surface, and
T_s= mass averaged temperature of the graphite block.

Similarly, the conduction shape factor for the heat transfer from the graphite block to the coolant channel wall was determined by employing Eqn. (2):

$$Q = S_{sw}k(T_s - T_w) \qquad \ldots (2)$$

Where:
S_{sw}= conduction shape factor for the heat transfer from the graphite block node to the coolant channel wall node, and
T_w = surface temperature of the coolant channel wall.

In Flownex the conduction shape factor for the heat transfer from the helium gap edge node to the graphite block node (S_{fs}) and the conduction shape factor for the heat transfer from the graphite block node to the coolant channel wall node (S_{sw}) can be expressed by implementing Eqns. (3) and (4):

$$S_{fs} = \frac{A_{fs}}{L_{fs}} \qquad \ldots (3)$$

and $$S_{sw} = \frac{A_{sw}}{L_{sw}} \qquad \ldots (4)$$

Where L_{fs} is the radial distance from the fuel rod surface to the graphite block node and L_{sw} is the radial distance from the graphite block node to the coolant channel wall. A_{fs} is the equivalent surface area of the fuel rods and A_{sw} is the equivalent surface area of the coolant channels. Eqns. (3) and (4) are used to obtain the relevant areas that must be specified in Flownex.

MODEL DESCRIPTION

The unit cell approach described above was implemented in a case study of a 1/6 of a fuel block discretized into a 3x3 grid as shown in Fig. 3. The control volume was also discretised into 3 axial nodes (in the vertical direction along Z). The nodes, shown by rectangles in Fig. 3, are equally spaced in the radial (along lines AE, AG and AI) and tangential directions (along lines CK and EI). The control volume associated with the central node, shown in red in the figure was isolated and modelled in Flownex. As a first approximation it was assumed that the fuel rods and coolant channels are uniformly distributed over the fuel block and all the coolant channels are the same diameter. (See Fig. 1 for the actual distribution and the sizes of the coolant channels). Thus, the total conduction area for the heat transfer from the fuel rod surface to the graphite associated with the actual number of fuel rods in the central control volume can be deduced. The total conduction area for the heat transfer from the graphite to the coolant channel wall associated with the actual number of coolant channels associated with the selected control volume can also be calculated. The mass associated with graphite block node was obtained from the conduction elements connecting the solid nodes. The new model differs from one developed by Sambureni [9] in that in the old model, the heat transfer was dominant in the radial direction which was modelled using conduction heat transfer (CHT) elements. The heat transfer in the tangential direction was modelled by employing cross-conduction elements. The old model did not include discretisation in the axial direction and also did not include discretisation in the fuel rod and the coated fuel particles. In the new model, the axial direction is the primary discretization direction for conduction which means the thermal mass can now be accurately predicted. Zero thermal mass CHT elements were used for conduction in the radial and tangential directions.

There are 8.5 fuel rods and 4.5 standard coolant channels associated with the selected central control volume. The edges of the central control volume were assumed to be adiabatic. Since this control volume is not positioned on the edge of the block, bypass flow was not included in the model. The flow network in the model was simplified by using a one dimensional pipe network approach with the appropriate area, wetted perimeter and mass flow rate to model the three dimensional fluid flow in the coolant channels. The fuel block used to build this model is assumed to be at the top of the active core, near the inlet plenum so that the inlet temperature is 490 °C.

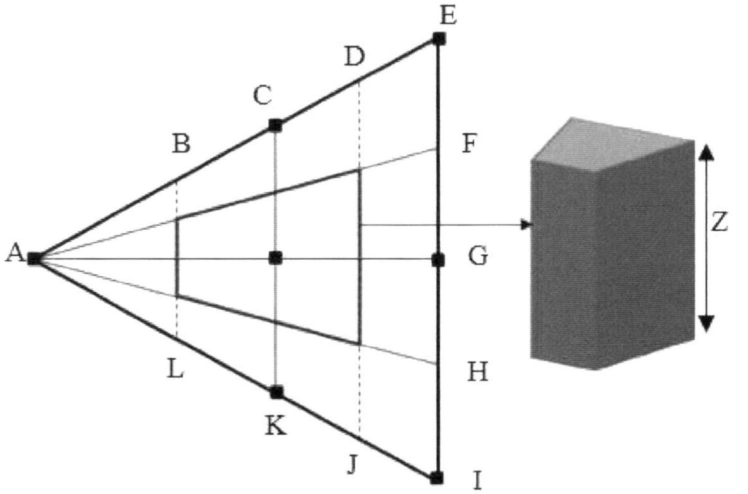

Figure 3. 1/6 of a fuel block discretized by a 3x3 grid.

In Flownex, the conduction shape factor can be implemented using the radial distance from the edge of the helium gap to the graphite block node and the radial distance from the graphite block node to the coolant channel wall and the associated areas. The Flownex model for the heat transfer from the edge of the helium gap to the coolant channel wall consists of a solid node and conduction elements. The conduction elements account for the heat transfer from the fuel rod surface to the graphite block node and from the graphite block node to the surface of the coolant channel wall respectively. The axial direction was treated as the primary conduction direction and therefore the heat transfer mass in the fuel rods and the graphite is distributed in the axial direction.

To simplify the model, the coolant channels and fuel rods were not modelled explicitly. Therefore, heat transfer within the fuel block and between the graphite and the coolant channels was modelled by defining representative coolant channels, fuel rods and coated fuel particle layers associated with each unit cell. The fuel rods were discretised into cylindrical rings for the fuel compact and helium gap layer, as shown in Fig. 4. A number of conduction elements account for the heat transfer from the centre of the fuel rod to the helium gap. The model for the helium gap includes radiation and conduction elements to account for the heat transfer through the stagnant helium in the gap.

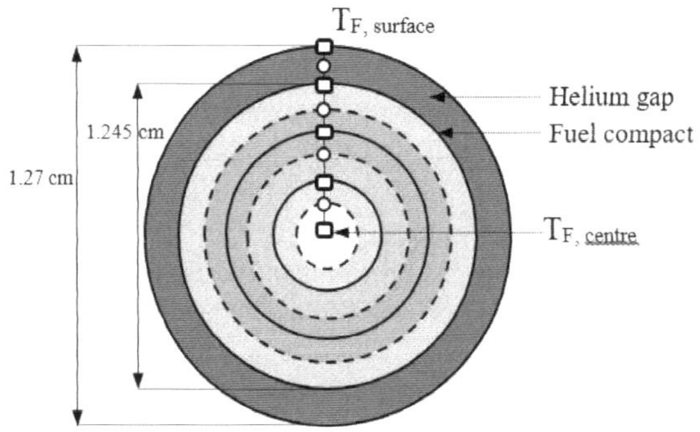

Figure 4. Fuel rod discretisation.

The model for the heat transfer from the centre of the fuel kernel to the surface of the coated fuel particle also consists of conduction elements. The conduction elements account for the conduction heat transfer from the fuel kernel centre through to each successive coating layer. As a first approximation, the surface temperature of the coated fuel particle was taken to be the same as the surface temperature of the surrounding graphite matrix within the fuel rod. Power was applied to the representative kernel volume associated with the radial layer of the central control volume.

Cylindrical coolant channels run along the height of the block. The helium flowing in the coolant channels has no direct contact with the fuel compacts. The heat generated in the fuel compacts is transferred through the graphite in the fuel block and eventually removed through convection by helium in the coolant channels. The helium flow network is then connected to the fuel rods via the convection elements which were in-turn connected to the coolant channel wall nodes. The interaction between the solid structure network and the flow network is via fluid-solid surface convection. The flow path is discretised in the axial direction and the pressure losses in the flow path were calculated using the Darcy-Weisbach equation.

The geometry of the block was maintained in the calculation of areas to ensure that the thermal mass is the same and that the flow paths have the same length. The thermal resistances in the graphite block matrix were adjusted to account for the porosity in the graphite block due to coolant channels.

The temperatures obtained for nodes represent the average mass temperature of a control volume. The thermal mass assigned to a node is computed from the specific heat and density of the material evaluated at the temperature of the node. The flow network was represented by a set of pipes and nodes.

The appropriate combination of the above makes it possible to perform thermal-hydraulic analysis in high temperature prismatic block reactors which involves determining the temperature distribution in the fuel blocks due to coupled heat conduction, convection and radiation. A Flownex model consisting of thermal and flow networks, schematically represented in Fig. 5, was set up. This was built through a set of nodes, thermal resistances, pipe elements and boundary conditions.

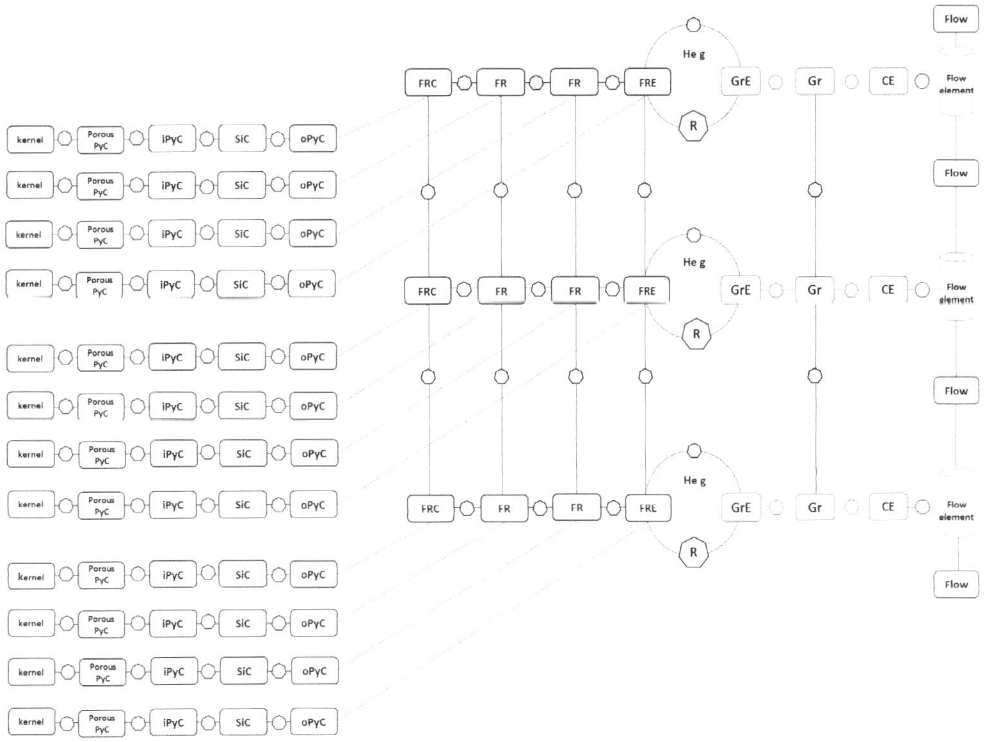

Figure 5. Schematic network representation of the heat transfer model in the central control volume.

In Fig. 5, the rectangles represent the nodes, the heptagons represent the conduction and radiation heat transfer elements, and the circles represent the convection heat transfer elements. Flow elements are represented by cylinders. The different layers of the coated fuel particle are represented in Fig. 5 as kernel, porous PyC (porous carbon buffer layer), iPyC (inner pyrolytic carbon layer), SiC (silicon carbide) and the oPyC (outer pyrolytic carbon layer). The coated fuel particle network is linked to the fuel rod via a node for the respective control volume in the fuel rod radial/cylindrical layer. The fuel rod nodes are indicated by FR with the centre/inner fuel rod node represented by FRC and the node at the edge of the fuel rod represented by FRE. The helium gap between the fuel rod and the graphite block (represented by He g in the figure) consists of radiation and conduction elements. The model for the heat transfer from the helium gap surface to the coolant channel wall consists of three solid nodes and two conduction elements. The first node (GrE in Fig. 5) represents the fuel rod surface temperature, the second (Gr in Fig. 5) represents the graphite block temperature and the third (CE in Fig. 5) represents the coolant channel wall temperature.

RESULTS

The results for temperatures in the central control volume are shown in Table 3.

Table 3. Axial and radial temperatures for a central control volume of 1/6 of a hexagonal block in °C

	Coated fuel particle					Fuel rod				Graphite block			He
H(m)	Kernel	Porous PyC	iPyC	SiC	oPyC	Inner radial layer	Mid radial layer	Outer radial layer	He gap	GrE	Gr	CE	Flow element
0.793	1191.3	1141.1	1098.3	1089.8	1087.6	1080.7	1079.9	1077.5	1073.5	1048.5	899.1	604.1	500.7
	1190.5	1140.2	1097.4	1089.0	1086.8	1079.9							
	1188.0	1137.8	1095.1	1086.6	1084.4	1077.5							
	1184.0	1133.9	1091.1	1082.6	1080.4	1073.5							
0.3965	1198.8	1148.5	1105.7	1097.3	1095.1	1088.2	1087.4	1085.1	1081.2	1057.1	913.5	632.8	532.2
	1198.1	1147.7	1104.9	1096.5	1094.3	1087.4							
	1195.7	1145.4	1102.6	1094.2	1092.0	1085.1							
	1191.8	1141.5	1098.8	1090.3	1088.1	1081.2							
0	1206.0	1155.6	1112.8	1104.3	1102.1	1095.3	1094.5	1092.3	1088.5	1065.2	926.9	659.0	562.9
	1205.3	1154.9	1112.1	1103.6	1101.4	1094.5							
	1203.0	1152.6	1109.8	1101.4	1099.2	1092.3							
	1199.2	1148.8	1106.1	1097.6	1095.4	1088.5							

The temperature of the fuel kernel is highest (1206 °C) at the centre of the fuel rod and the bottom of the control volume and 1095.3 °C for the surrounding graphite block matrix. The temperature of the coated fuel particle is the highest at the centre where the heat is generated and decreases with each coating layer of the TRISO particle and with each radial layer of the fuel rod. It can be seen that the temperature in the coated fuel particles and fuel rods increases with decreasing vertical position in the fuel rod/block. The temperature was observed to decrease gradually through the radial length of the fuel rod. There was, however, a drop in temperature of about 24.1 °C on average across the helium gap. The helium exit temperature for the central control volume was found to be 572.95 °C, showing an increase in the coolant temperature of 82.9 °C with the flow direction. If we ignore all other losses and assume that the increase will be 82.9 °C for each block and extrapolate over an assembly of 6 vertically stacked blocks, the outlet temperature of the coolant can be predicted at 987.7 °C which is higher than the outlet temperature of 950 °C given in Tak et al. [2]. This difference can be because this extrapolation does not take into account the heat losses to the graphite and metallic support structures that support the core and the coolant mixing that will occur at the outlet plenum.

A sensitivity study was done, to understand the effect of conduction and radiation across the helium gap. Results showed that if the effect of radiation across the gap is ignored, an increase of about 3 °C in the kernel and fuel rod centre nodes temperature was observed. When the conduction and radiation across the helium gap were included in the model, the fuel rod centre temperatures showed a decrease of about 24 °C as there is a decrease in thermal resistance in the system.

Sambureni [9] modelled 1/6 of a block using Flownex with a 98 kW heat input and a mass flow rate of 0.358 kg/s through the standard cooling channels was used. The helium exit temperature was found to be 542.94 °C showing an increase of 52.98 °C. It should be noted though that this study was done for a single horizontal layer of a fuel block since there was no discretisation in the axial direction.

Fig. 6a shows the predicted temperatures from Tak et al. [8] of a single unit cell which consists of a fuel compact, helium gap, and graphite region. The heat generation in the fuel compact was assumed to be uniform with a fixed surface temperature applied to the graphite. Fig. 6b shows the predicted temperatures for a central control volume modelled with Flownex. In Fig. 6b, the top blue line shows a plot of temperature distribution within the representative coated fuel particle associated with the central control volume and the red line shows the temperatures of the graphite matrix within the representative fuel rod, the helium gap and the surrounding graphite block matrix.

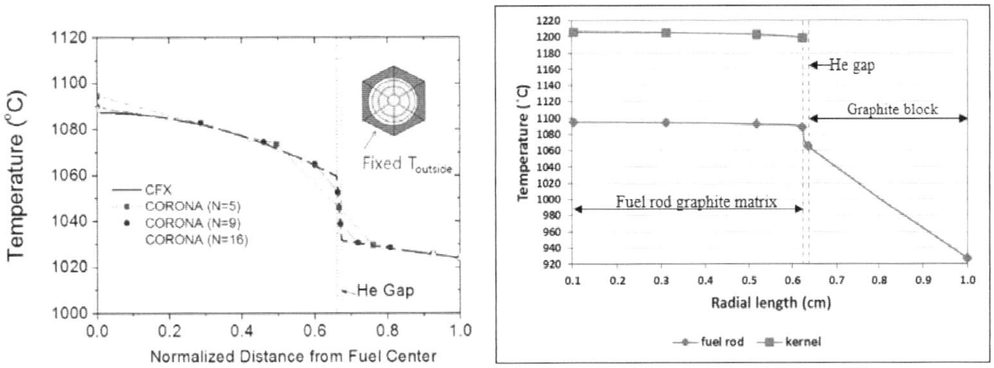

(a) **(b)**

Figure 6. Comparison of predicted radial temperature distribution in a section of a standard fuel block [8].

The results from Tak et al. [8] in Fig. 6a show that the maximum fuel temperature at the centre of the fuel compact was found to be around 1090 °C with a ~30 °C temperature drop across the Helium gap. The Flownex result for the maximum temperature of the kernel is 1206.1 °C and 1095.3 °C in the fuel rod with an average of 24.1 °C decrease across the helium gap. The fuel rod and coated fuel particle temperatures in Fig. 6b appear flatter with a maximum delta of ~6.7 °C for the coated fuel particle layers and the fuel rod graphite matrix. The model used in Fig. 6a does not include the coolant channel and therefore the temperature drop observed in the surrounding graphite matrix after the helium gap is less pronounced than in the results of the Flownex model in Fig. 6b. The heat removal in the Flownex model developed in this study for the central control volume is only through the representative coolant channels as the model does not include bypass and cross-flow between adjacent blocks which play a major role in the temperature distribution within the block. Although no direct comparison can be made, the fuel temperature is in the same order of magnitude and the temperature distribution follows a similar trend.

CONCLUSION

This paper presented the conceptual formulation of a model to simulate the thermal-fluid phenomena of the HTR block central control volume. It was shown that a collection of network elements could be used to construct a comprehensive multi-dimensional network model that can be used as a building block for the analysis of the temperature distribution for the whole or any part of the core. The network approach can be used to reasonably predict temperature distributions within a prismatic fuel block.

ACKNOWLEDGMENTS

This work is based on the research supported by the South African Research Chairs Initiative of the Department of Science and Technology and National Research Foundation of South Africa (Grant No 61059).

Any opinion, finding and conclusion or recommendation expressed in this material is that of the authors and the NRF does not accept any liability in this regard.

REFERENCES

1 MacDonald, P. E., "NGNP Preliminary Point Design – Results of the Initial Neutronics and Thermal-Hydraulic Assessments," Idaho National Engineering and Environmental Laboratory, INEEL/EXT-03-00870 Rev. 1, 2003.

2 Tak, N. I., Kim, M. H., Lee, W. J., "Numerical investigation of a heat transfer within the prismatic fuel assembly of a very high temperature reactor." Annals of Nuclear Energy, vol. 35, pp. 1892-1899, 2008.

3 Rousseau, P., Greyvenstein, G., "Modelling of HTTR in Flownet," Potchefstroom, South Africa: Department of Mechanical & Materials Engineering: Potchefstroom University for Christian Higher Education, Potchefstroom, South Africa, 2002.

4 Van der Merwe, J. J., "Simulation of heat and momentum transfer in the high temperature test reactor," North-West University, Potchefstroom, South Africa, 2003.

5 Emslie, F. N. "Flownex analysis of high temperature test reactor thermo-hydraulic benchmarks," North-West University, Potchefstroom, South Africa, 2005.

6 Tak, N. I., Kim, M. H., Lim, H. S. & Noh, J. M., "A practical method for whole-core thermal analysis of a prismatic gas-cooled reactor." Nuclear Technology, vol. 177., pp. 352-365, 2012.

7 Huning, A. J., "A Steady State Thermal Hydraulic Analysis Method for Prismatic Gas Reactors," Georgia Institute of Technology, Atlanta, 2014.

8 Tak, N. I., Lee, S.N., Kim, M.H., Lim H.S., Noh, J.M, "Development of a core thermo-fluid analysis code for prismatic gas cooled reactors," Nuclear Engineering and Technology, vol. 46(5), pp. 641-654, 2014.

9 Sambureni, P., "Thermal fluid model for a prismatic block in a gas-cooled reactor using FLOWNEX," North-West University, Potchefstroom, South Africa, 2015.

10 M-Tech Industrial, "Flownex theory manual," 2015.

11 M-Tech Industrial, Flownex SE Version, 8.5.0.2466, http://www.flownex.com, 2015.

Heat Transfer in Two and Three-dimensional Single Span Greenhouses

S Kruger[1], L Pretorius[2]

[1]University of Johannesburg
Faculty of Engineering and the Built Environment
Department of Mechanical Engineering Science
Johannesburg, South Africa
skruger@uj.ac.za

[2]University of Pretoria
Faculty of Engineering, the Built Environment and Information Technology, Graduate School of Technology Management
Pretoria, South Africa
Leon.Pretorius@up.ac.za

Email of Corresponding Author: skruger@uj.ac.za

ABSTRACT

The purpose of this paper is to investigate the heat transfer in two-dimensional and three-dimensional cavities representing a single span greenhouse. This investigation is conducted numerically using Computational Fluid Dynamics (CFD). The heat transfer and temperature fields driven by buoyancy forces are investigated. The research commences with the validation of a few fundamental geometries used as the building blocks for a large commercial greenhouse. The first fundamental geometry is a square filled with air. The CFD results for a square cavity are first evaluated against experimental results found in the literature for both two and three dimensional cavities. The heat transfer inside the cavities is then investigated and compared to those found in the literature. A reasonably good comparison between the numerical CFD results and the experimental results was found for both the two- and three-dimensional cavities. Based on the validated CFD models, two three-dimensional single span greenhouses containing a pitched roof were investigated to determine the effect of design alterations on the heat transfer within the cavity. The results were also compared to two-dimensional greenhouses with a 30 and 45 degree roof angle respectively. Results found that there are significant differences between the two and three-dimensional cases when the average Nusselt number is considered, especially for a greenhouse containing a roof angle of 45 degrees. Temperature distributions were also found to vary significantly throughout the three-dimensional greenhouses.

KEYWORDS

Greenhouse, Natural ventilation, Computational Fluid Dynamics

INTRODUCTION

Natural convection has received considerable attention due to its practical interest in a large number of engineering applications. A few examples are the natural or forced convection of buildings, and electronic device cooling. Baines and Turner [1] initially considered this problem by investigating the effect of a small buoyancy source on the environment in a closed space. During mid-season and sunny winters greenhouses are usually fully closed, with buoyancy effects due to warm soil and transpiring leaves the only driving forces present. This situation corresponds to the classic Rayleigh-Bénard convection where a cavity is heated from below and cooled from above. The thermal situation in a closed greenhouse containing a continuous crop canopy surface can be approximated by natural convection in a dry, floor heated cavity [2]. Numerous studies have been performed on three-dimensional cavities. In a study by Leong et al [3] the authors attempted a physically-realizable experiment in a laboratory to be used for CFD validation. The work was extended by Mamun et al [4] to include a new orientation. A cubical air-filled cavity with one pair of opposing faces at different temperatures, the remaining faces with a linear variation from the hot wall to the cold wall was considered. The cavity was tilted at various angles and Nusselt numbers were measured and found to be within an average deviation of 0.3% when compared to the simulation results with the CFD code used. A thorough study consisting of both experimental and numerical work was performed by Baïri [5]. A square cavity at various inclination angles and Rayleigh numbers was investigated. Results were found to agree reasonably well with results published in the literature. For all values of the Rayleigh number investigated, it was found that the inclination of the cavity played a major role in the convective exchanges. Various turbulence models in unsteady natural convection were assessed by Altaç and Uğurlubilek [6]. Their study revealed that 3D laminar and RANS models resulted in almost similar mean Nusselt number predictions up to $Ra = 10^{10}$. They also found these predictions to be compatible with results from 2D simulations, but is not the case for larger Rayleigh numbers. Overall their study indicated that 3D RANS models yield more accurate mean Nusselt numbers. In the present work, two and three-dimensional cavities representing a greenhouse with a zero degree roof angle are firstly evaluated against experimental data. The 2D and 3D models are then modified to investigate the effect of design alterations such as roof angle.

COMPUTATIONAL FLUID DYNAMICS

Numerical simulations using Computational Fluid Dynamics (CFD) have been employed quite frequently to investigate natural convection. But experimental work pertaining to natural convection is generally complicated and time consuming. With advances in computer technology, Computational Fluid Dynamics (CFD) has become a valuable research and design tool for the investigation of natural convection. Quantitative predictions of for example fluid flow are based on the laws of conservation (mass, momentum and energy).

In Computational Fluid Dynamics a numerical solution of partial differential equations, typically the Navier-Stokes Equations, is obtained [7]. The transport of mass, momentum and energy in moving fluids are governed by these equations. The transport of the mentioned quantities is governed by three laws: conservation of mass, Newton's second law of motion and the first law of thermodynamics [8].

If a general variable ϕ is introduced, the conservative form of all fluid flow equations can usually be written as shown in Equation (1) [9]:

$$\frac{d(\rho\phi)}{dt} + \operatorname{div}(\rho\phi\bar{u}) = \operatorname{div}(\Gamma \operatorname{grad} \phi) + S_\phi \tag{1}$$

Equation 1 is called the transport equation for property φ. This equation emphasizes the various transport processes: the first term on the left-hand side is the rate of change term, while the second term is the convective term. The first term on the right-hand side is the diffusive term (Γ= diffusion coefficient) and the last term is the source term.

In the approach followed in this study the finite volume discretization is the first step in solving these transport equations. This method subdivides the solution domain into a finite number of small control volumes, which corresponds to the cells of a computational grid. Discrete versions of the integral form of the continuum transport equations are applied to each volume. The objective of this method is to obtain a set of algebraic equations. An algebraic multi-grid solver such as the one described in [10] can then be used to solve the resulting equations. The integration of equation 1 over a three-dimensional control volume (CV) is a crucial step in the finite volume method:

$$\frac{d}{dt}\int_{CV} \rho\phi dV + \int_{A} \operatorname{div}(\rho\phi\bar{u})dV = \int_{CV} (\Gamma \operatorname{div}\phi)dV + \int_{CV} S_\phi dV \tag{2}$$

The first term on the left hand side indicates the rate of change of the total amount of fluid property φ in the control volume. The product in the second term expresses the flux component of the property φ due to fluid flow along the outward vector n, therefore the second term on the left hand side is the convective term, i.e. the net rate of decrease of fluid property φ of the fluid element due to convection. The first term on the right is the diffusive term and associated with a flux into the element. This term represents the net rate of increase of fluid property φ of the fluid element due to diffusion. The last term represents the rate of increase of property φ due to sources inside the fluid element [9]. A detailed description of a typical discretization procedure can be found in for example Patankar [11].

Two-dimensional Numerical Model – Square Greenhouse

The initial square cavity representing a single-span greenhouse with a zero-degree roof angle presented in this research paper was based on the cavity studied experimentally and numerically by Baïri [5]. The cavity has dimensions of 0.75m x 0.75m. The bottom and top walls were specified as isothermal (heated and cooled respectively), generating the buoyant flow in the cavity, while the vertical walls were adiabatic. The top wall was assigned a temperature of T_c=15°C, and the temperature on the bottom wall, T_H, was adjusted to achieve different values of the Rayleigh number (Ra). The flow field was assumed to be steady, and the fluid (air) incompressible. The commercial CFD code StarCCM+ [10] was used to perform the numerical analysis. The Rayleigh number was calculated using $\Delta T = T_h - T_c$ associated with the distance L between the two active walls and the following equation:

$$Ra = \frac{g\beta L^3 \Delta T}{\nu\alpha} \quad (3)$$

The properties of air used in the model were calculated using the average temperature between the hot and cold wall (T_{ave}) for each case.

Mesh

The square cavity was meshed for CFD purposes using the polyhedral meshing model included in the StarCCM+ software. As turbulent flow in a square cavity with natural convection is characterized by a thin boundary layer containing relatively large flow gradients, a large number of cells are required in this region. The advancing prism layer meshing model [10] was used to generate these cells in order to adequately capture boundary layer, turbulence effects and heat transfer near the wall boundaries. Four different base sizes were tested to determine mesh independence. The maximum simulated x- and y-velocity components in the CFD domain were monitored, and the average Nusselt number on the hot wall was calculated for each base size. Taking simulation time and available resources into account, it was decided to use a base of 0.016m for all the simulations in this section. The difference between the values is less than 1%, therefore it can be concluded that the results are independent of the cell size. The following models were used in the CFD software to model the physics for the fluid in the square: Ideal Gas, K-Epsilon Turbulence, Low y+ Wall Treatment, Reynolds-Averaged Navier-Stokes, Segregated Flow - Segregated Fluid Temperature, Standard K-Epsilon Low-Re, Steady, Two Dimensional, Gravity. The solution update for the segregated flow model is controlled according to the SIMPLE algorithm [9]. The segregated fluid temperature model was chosen as a companion to the segregated flow model. This model solves the total energy equation with temperature as the solved variable. The equation of state is then used to calculate enthalpy. The model is appropriate for simulations that do not involve combustion [10]. Since only the flow in the turbulent regime will be investigated, the k-epsilon turbulence model was activated. Accurately predicting turbulent buoyant flows are difficult to achieve, and have been successful in only a few cases as illustrated in [12]. This is due to the complexity of the numerical simulation of boundary layers adjacent to the walls, as well as the uncertainty of the general turbulence models for turbulent natural convection [12]. Two turbulence models were tested for the initial case considered here, namely the RANS k-epsilon realizable two-layer model, and the k-epsilon Standard low-Reynolds Number turbulence model. In a study by Henkes et al [13] it was mentioned that various low-Reynolds number models predicted wall heat transfer the closest to the experimental values. The Low-Reynolds number has additional damping functions which allow it to be applied to the viscous boundary layer adjacent to the walls, and is recommended for use in natural convection problems. The low Y+ wall treatment was used in conjunction with the low-Reynolds number turbulence model. It is assumed by this wall treatment that the viscous sublayer is resolved, and wall laws are unnecessary. This wall treatment can however be used only with a fine enough mesh. An initial simulation was done to determine whether the near wall spacing is appropriate for the simulation, i.e. whether the amount of prism layers created on the walls was sufficient. Ten prism layers were initially specified in the mesh setup, with a thickness of 0.01m. The dimensionless Wall Y+ was defined as follows [10]:

$$y^+ = \frac{yu^*}{\nu} \quad (4)$$

Where y is the normal distance from the wall to the wall cell-centroid, u^* is a reference velocity and ν the kinematic viscosity. The dimensionless Wall Y+ values were plotted as a scalar on the wall boundaries of the cavity. They were found to be overall less than 1 for the chosen mesh base size, which implies that the prism layer thickness and amount of prism layer cells are adequate for the chosen turbulence model.

RESULTS – SQUARE GREENHOUSE

Validation – Two-dimensional Square Greenhouse

The CFD simulation results have been published previously [14] and is repeated here for clarity. In order to assess the convective contribution of the heat exchange within the cavity, the surface average Nusselt number (ratio of convective to conductive heat transfer across a boundary) was calculated for the floor of the cavity. The Nusselt number is given by equation 5:

$$Nu = \frac{hL}{k} \tag{5}$$

The surface averaged Nusselt number on the hot floor across all the cells was calculated from the CFD results and compared to results found in the literature as shown in Figure 4.6. A custom user function was written in StarCCM+ which calculated the Nusselt Number for each cell on the hot wall using the boundary heat flux. The syntax can be seen in equation 6:

*(abs($BoundaryHeatFlux)/ΔT) * L/k* (6)

The equation essentially takes the absolute value of the boundary heat flux ($ implies it is a scalar function in StarCCM+), divides it by the temperature difference, multiplies the answer by the height of the cavity and lastly divides by the thermal conductivity of the air. The surface averaged Nusselt number was calculated as follows using the scalar-based report function in the software, where the surface averaged Nusselt number was calculated for the hot wall (cavity floor) as shown in equation 7:

$$\text{Surface Averaged Nusselt Number} = \frac{1}{a}\int Nu\, da = \frac{\sum_f Nu_f A_f}{\sum_f A_f} \tag{7}$$

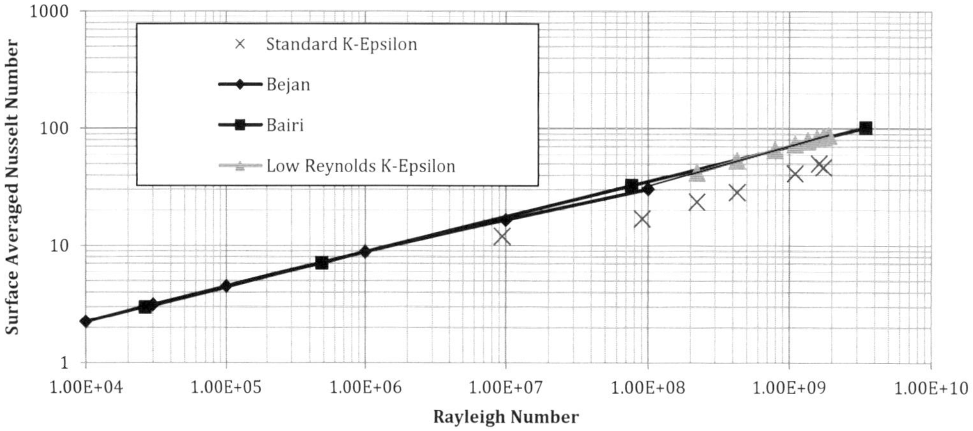

Figure 1: Average Nusselt number along the hot wall vs Rayleigh number

The graph in Figure 1 shows the numerical results for the Nusselt number when applying both the standard realizable k-epsilon turbulence model as well as the LRN k-epsilon turbulence model. Simulations applying the standard realizable k-epsilon turbulence model indicate a relatively large discrepancy when compared to experimental results, as well as the numerical results using the standard low-Reynolds number k-epsilon turbulence model. The results obtained for the low-Reynolds number k-epsilon are essentially in agreement with results published by Baïri [5]. The Nusselt-Rayleigh relationship that best fit the curve was found to be:

$$Nu = 0.0589 Ra^{0.3422} \tag{8}$$

Overall, a relatively good agreement was found between the numerical and experimental results when using the low-Reynolds number k-epsilon turbulence model.

RESULTS – SQUARE GREENHOUSE

Validation - Three-Dimensional Square Greenhouse

The previous numerical two-dimensional square greenhouse scale model was extended to a three-dimensional cubical cavity. The results were firstly compared to experimental results for a 200mm cube found in the literature by Küreckci and Özcan [15], who investigated the natural convection of a cubical cavity both experimentally and numerically. This is again done to increase confidence in the current CFD models developed for the greenhouse. The cubical enclosure numerically modelled has dimensions of 200mm x 200mm x 200mm to be able to compare to Kürekci et al [15]. One vertical wall of the cavity was specified a constant temperature of 69°C, while the opposite cold wall was kept at a constant temperature of 41°C. The remainder of the cube walls were adiabatic. The details of the experimental setup to which the current model can be compared can be found in [15]. The gravitational vector is in the negative y-direction in the CFD model. The CFD model was run in the laminar regime to be able to compare to previously published results. To monitor numerical CFD accuracy, a mesh sensitivity analysis was once again conducted. Variables such as temperature, velocity and surface average Nusselt number were monitored. As mentioned by Dol et al [16], three-dimensional high-Reynold number (turbulent) flow is rather complex, and requires a fine mesh in the regions adjacent to the solid walls, therefore 20 prism layers were once again selected, with a combined thickness of 0.02m. A base size of 0.003m was chosen as the temperature no longer varied with a decreasing mesh size. The numerical results obtained for the current 3D square scale model are compared to results obtained in the literature [17]. A difference of 5.04% was found between the numerical obtained Surface Averaged Nusselt number and the Nusselt number found by Kürekci [17]. Therefore additional confidence has again been established in the current three-dimensional CFD model of a cubical enclosure. A more detailed comparison can be found in [18]. To enable the comparison of the two-and three-dimensional cavities, a three-dimensional model with dimensions 0.75m x 0.75m x 0.75m was created in StarCCM+ for the current research. This was done in order to ascertain the influence of a third dimension on temperature and velocity distributions. The same approach as described previously was used to model the original 0.75m greenhouse with zero degree roof angle as a three-dimensional cube, and the results can then be compared to the two-dimensional case. The top and bottom walls were specified as 15.1°C and 55.1°C ($Ra = 1.35 \times 10^9$) respectively. The sidewalls were all specified as adiabatic. The base size, number of prism layer and prism layer thickness were all kept the same as in the two-dimensional case.

The simulated CFD temperature contour plots for the two cases are compared in Figure 2. For the three-dimensional case, the contour plot was taken in the $z/H = 0.5$ plane. The temperature contours for the two-dimensional case tends to be more circular, whereas the contours for the three-dimensional case tend to be slightly more of square shape. This is also visible in the velocity vector plots (Figure 3). The vector plot for the two-dimensional case also shows two secondary convective cells in the top left and bottom right corners. This is not present in the three-dimensional case. The temperature difference through the boundary layer to the uniform core region for the three-dimensional case is not as steep as for the two-dimensional case, which is visible in the temperature distribution plot (Figure 4). The centre of the cavity is at the same temperature for both cases. As far as velocity distribution in the centre of the cavity is concerned (Figure 5), quite a significant difference is noticed between the two-and three-dimensional case. Both cases exhibit the same trend of reaching a maximum adjacent to the walls, and decreasing towards the centre. The maximum velocity reached in the two-dimensional case is 0.33 m/s whereas a velocity of 0.22 m/s is reached in the three-dimensional case.

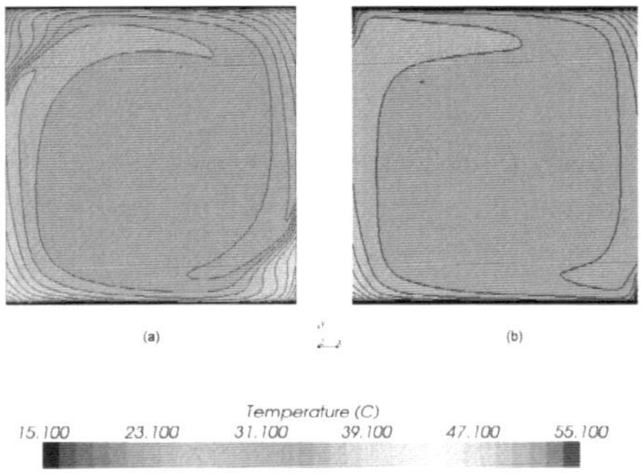

Figure 2: Temperature contour plot comparison for (a) Two-dimensional case (b) Three-dimensional case (z/H=0.5)

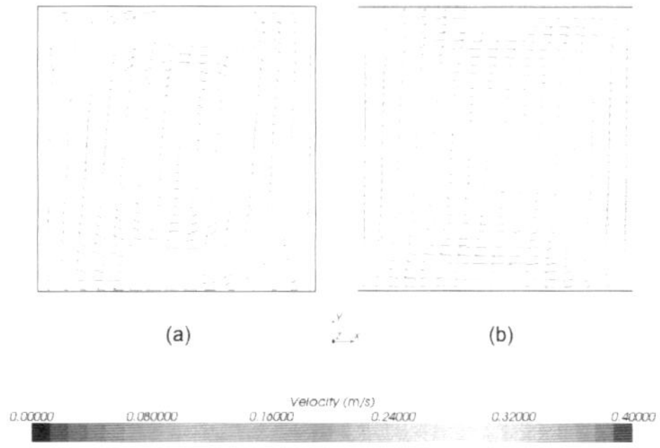

Figure 3: Vector plot comparison for (a) Two-dimensional case (b) Three-dimensional case (z/H=0.5)

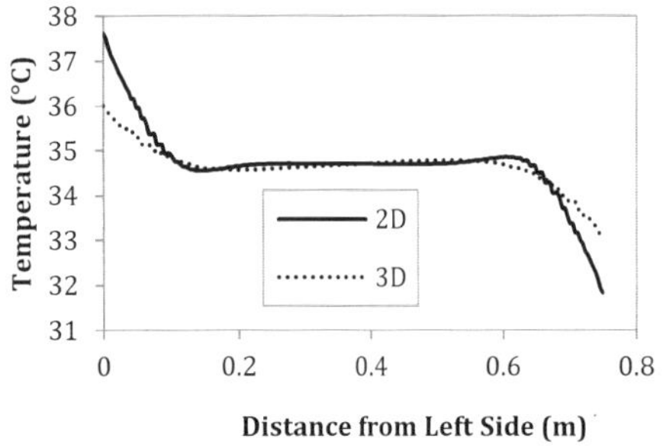

Figure 4: Comparison of temperature distribution in the centre of the cavity

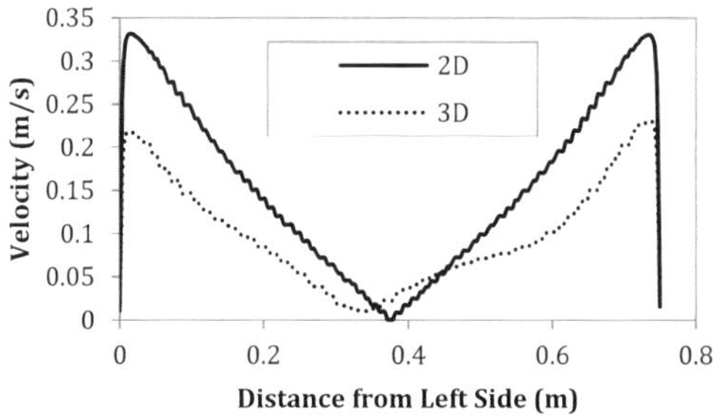

Figure 5: Comparison of velocity distribution in the centre of the cavity

The Nusselt number distribution calculated as indicated previously from CFD results is shown in the 3D surface plot in Figure 6. The maximum Nusselt number is visible towards the front right corner. The average Nusselt number was calculated to be 77.3. The Nusselt number for the two-dimensional case was 78.7, which amounts to a difference of 1.8%.

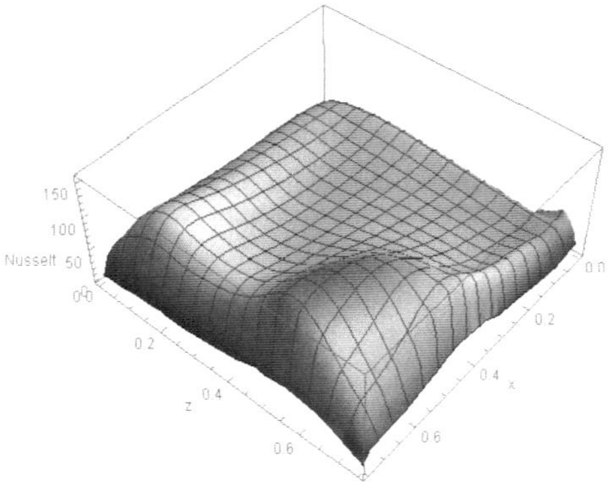

Figure 6: Nusselt Number distribution on the floor of the cube

Three-dimensional Modified Greenhouse

Two types of three-dimensional greenhouses have been numerically investigated. The first greenhouse had a roof angle of 30 degrees, whereas the second greenhouse had a roof angle of 45 degrees. The average Nusselt number on the floor of the cavity was calculated as shown in equation 6, and compared to the average Nusselt number for each of the two-dimensional cases. The results are plotted in Figure 7 and Figure 8 respectively. From Figure 7 it can be seen that there is not as significant difference in the calculated average

Nusselt Number on the floor of the cavity for the two and three-dimensional cases (for a greenhouse with a 30 degree roof angle). The difference increases with increasing Rayleigh number. For the highest Rayleigh number (1.9 x 109) the difference average Nusselt number is 2.35%. If the roof angle is increased to 45 degrees, the difference between the two-dimensional and three-dimensional average Nusselt numbers is quite noticeable. For the lowest Rayleigh number (2.22 x 10^8) the difference is 18.6% and for the largest Rayleigh number the difference increases to 51%.

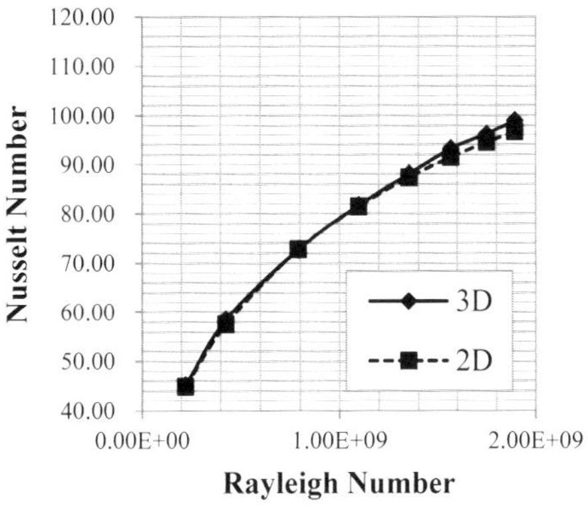

Figure 7: Average Nusselt Number on the Hot Wall for 2D and 3D for a 30 Degree Roof Greenhouse

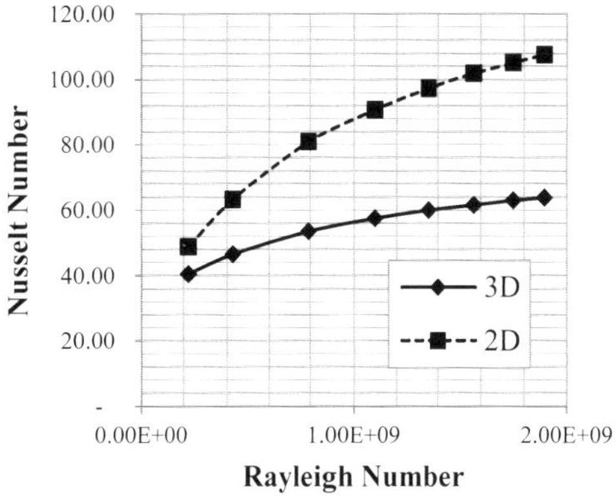

Figure 8: Average Nusselt Number on the Hot Wall for 2D and 3D for a 45 Degree Roof Greenhouse

The Nusselt-Rayleigh relationships that best fit the above simulated curves for the three-dimensional models have been deduced from the previous plots and are tabulated in Table 1.

Table 1: Nusselt-Rayleigh Relationships

Roof Angle	Nusselt-Rayleigh Relationship
30°	$Nu = 0.047Ra^{0.3584}$
45°	$Nu = 0.6646Ra^{0.2139}$

Three-dimensional Models with a 30 Degree Roof Angle

To continue the investigation of the heat transfer characteristics in a three-dimensional greenhouse, the case of a greenhouse with a 30 degree roof angle will be discussed in more detail here. It was decided to focus on only three Rayleigh numbers: 4.27 x 10^8, 1.35 x 10^9 and 1.9 x 10^9. These three Rayleigh numbers correspond to a temperature difference between the roof and the floor of the greenhouse of 10°C, 40°C and 70°C respectively. The isotherms in three different planes (front, middle and back) are shown in Figure 9 for the three chosen Rayleigh numbers. The temperatures for the smallest Rayleigh number (Figure 9a) are relatively low, with higher temperatures close to the floor. The temperatures are also homogeneously distributed throughout the height and width of the cavity. If the Rayleigh number is increased to 1.35 x 10^9 (Figure 9b) the temperatures are slightly higher compared to the lowest Rayleigh number, and also more non-uniform. The isotherms become distorted, with more densely spaced temperature contours at the bottom adjacent to the heated floor in the front of the cavity. At the back of the cavity the temperature is slightly higher compared to the rest of the cavity, indicating lower heat transfer. The temperature contours for the highest Rayleigh number, 1.9 x 10^9 indicates a high temperature at the bottom in front of the greenhouse, with a larger uniform temperature region in the centre of the greenhouse toward the back of the greenhouse. Figure 10 shows the temperature distribution at mid-height for the three locations in the greenhouse for the lowest Rayleigh number. This figure indicates that the temperature distribution varies quite significantly from the front to the back of the cavity. The center of the greenhouse has a homogenous temperature distribution, whereas the front of the greenhouse reaches it maximum temperature at approximately 0.2m from the left wall. The temperatures reach a minimum at the walls in the font of the greenhouse. At the back of the greenhouse temperatures are slightly higher, especially toward the left wall.

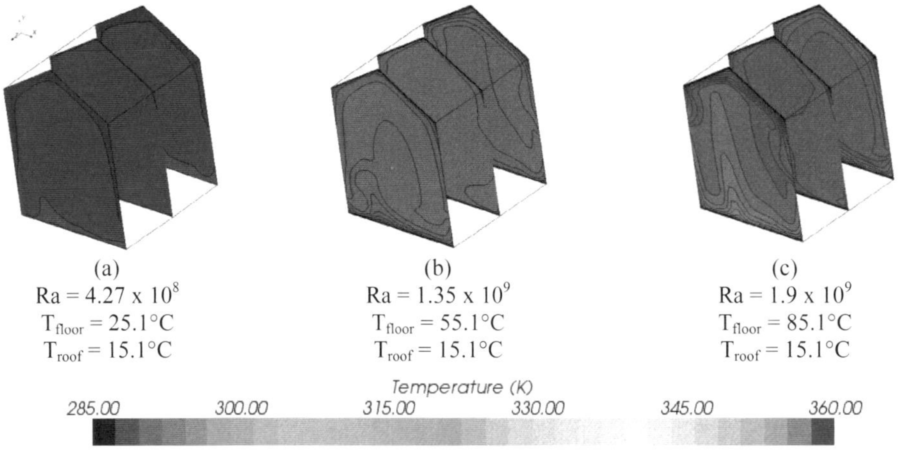

Figure 9: Temperature Distribution in three planes (30 Degree Roof Angle)

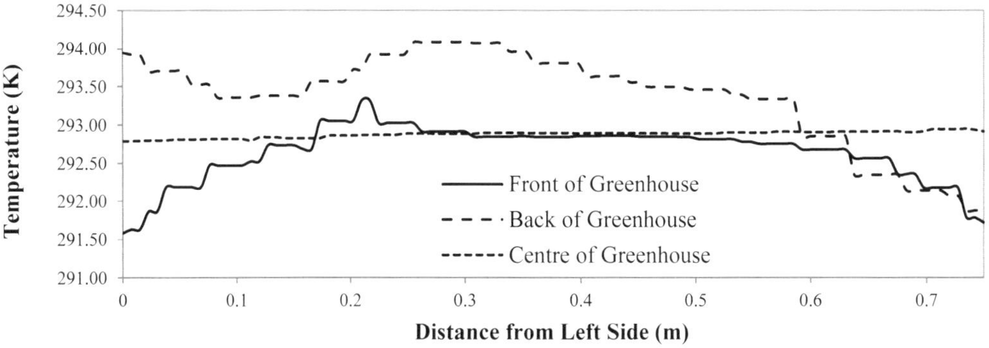

Figure 10: Temperature distribution in three planes (front, middle, back) for Ra = 4.27×10^8

The temperature distribution at mid-height for Ra = 1.35×10^9 are shown in Figure 11. The temperature in the front of the cavity at mid-height is characterized by a parabolic type distribution. The maximum temperature reached is similar to the temperature in the centre of the greenhouse (306K). The center of the greenhouse has a uniform temperature distribution at mid-height, whereas the temperature increases toward the back of the greenhouse with a maximum temperature of 311K. The temperature is lower at the back of the greenhouse adjacent to the right wall. This graph also shows the temperature plots are horizontal at the left and right wall, indicating that these walls are adiabatic.

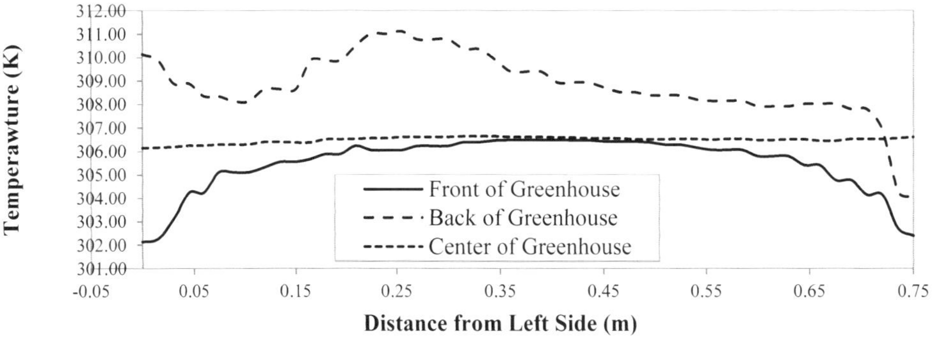

Figure 11: Temperature distribution in three planes (front, middle, back) for Ra = 1.35×10^9

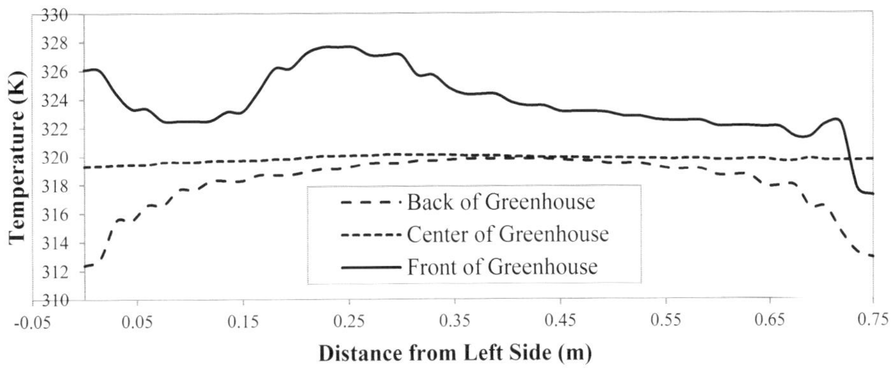

Figure 12: Temperature distribution in three planes (front, middle, back) for Ra = 1.9 x 10^9

The scenario changes for the highest Rayleigh number as shown in Figure 12. In this case the front of the greenhouse exhibits the highest temperature distribution, the center is once again homogeneous, and the back of the greenhouse shows a parabolic type temperature distribution. The maximum temperature reached in the back of the greenhouse is similar to the temperature reached in the centre of the greenhouse. The Nusselt number distribution for the three cases is shown in Figure 13. From this figure it can be seen that there are two areas of maximum heat transfer for each case. For the two lower Rayleigh numbers the two areas of maximum heat transfer – adjacent to the left wall and in the bottom right corner. For the maximum Rayleigh number the areas of maximum heat transfer moves to the two top corners, which is actually located at the back of the greenhouse. The location of the maximum heat transfer is due to convective recirculation (not shown here) impinging at these locations on the floor. These plots correlate with the temperature distributions given in the previous figures. The low heat transfer rate in the front of the greenhouse for the highest Rayleigh number is responsible for the higher temperature as seen in Figure 12.

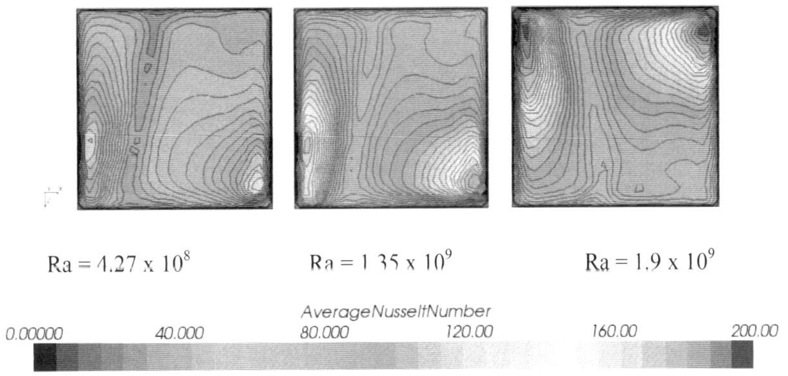

Figure 13: Nusselt Number Distribution on hot floor (30 Degree Roof Angle)

CONCLUSION

In this paper a numerical investigation was conducted to investigate the heat transfer in two and three-dimensional cavities corresponding to a model of a single span greenhouse. Initially the CFD models were verified using results found in the literature for a square and cubical cavity. The CFD results correlated well with the experimental results. The three-dimensional model was modified to include a roof angle of 30 degrees and 45 degrees respectively and compared to similar two-dimensional models. When the average Nusselt numbers on the floor (hot wall) were compared, it was found that the difference was minimal for the 30 degree roof angle, but was quite significant in the case of a 45 degree roof angle. Nusselt-Rayleigh relationships were also deduced for the 30 and 45 degree roof angle cavities. The heat transfer of the 30 degree roof angle cavity was further investigated in more detail, and it was shown that the maximum heat transfer was not always found in the same location on the floor of the three dimensional cavity. The temperature distributions for various plane sections through the cavity also indicated that temperatures varied throughout the greenhouse. These temperature variations can become important once crops are present, as non-uniform temperature distributions in a greenhouse can lead to non-uniform crop yield. The differences between the 3D and 2D heat transfer characteristics for the 30 degree and 45 degree greenhouse is a topic to be addressed in future research.

REFERENCES

[1] W. D. Baines and J. S. Turner, "Turbulent Buoyant convection from a source in a confined region," *Journal of Fluid Mechanics,* vol. 37, pp. 51-80, 1969.

[2] M. A. Lamrani, T. Boulard, J. C. Roy and A. Jaffrin, "Airflows and Temperature Patterns induced in a Confined Greenhouse," *Journal of Agricultural Engineering Research,* vol. 78, no. 1, pp. 75-88, 2001.

[3] W. H. Leong, K. G. T. Hollands and A. P. Brunger, "On a physically!realizable benchmark problem in internal natural convection," *International Journal of Heat and Mass Transfer,* vol. 41, pp. pp 3817 - 3828, 1998.

[4] M. A. Mamun, W. H. Leong, K. G. Hollands and D. A. Johnson, "Cubiical-cavity natural-convection benchmark experiments: an extension," *International Journal of Heat and Mass Transfer,* vol. 46, pp. 3655-3660, 2003.

[5] A. Bairi, "Nusselt–Rayleigh correlations for design of industrial elements: Experimental and numerical investigation of natural convection in tilted square air filled enclosures," *Energy Conversion and Management,* vol. 49, pp. 771 - 782, 2008.

[6] Z. Altec and N. Ugurlubilek, "Assessment of turbulence models in natural convection from two-and three-dimensional rectangular enclosures," *International Journal of Thermal Sciences,* vol. 107, pp. 237-246, 2016.

[7] I. H. Shames, Mechanics of Fluids, New York: McGraw-Hill, 2003.

[8] A. W. Date, Introduction to Computational Fluid Dynamics, New York: Cambridge University Press, 2005.

[9] H. K. Versteeg and W. Malalasekera, An introduction to Computational Fluid Dynamics, The Finite Volume Method,2nd Ed, England: Pearson Prentice Hall, 2007.

[10] CD-Adapco, "StarCCM+ User Guide," CD-Adapco, 2016.

[11] S. V. Patankar, Numerical Fluid Flow and Heat Transfer, New York: Hemisphere, 1980.

[12] M. Aounallah, Y. Addad and S. Benhamadouche, "Numerical investigation of turbulent natural convection in an inclined square cavity with a hot wavy wall," *International Journal of Heat and Mass Transfer,* vol. 50, pp. pp 1683-1693, 2007.

[13] R. A. Henkes, F. F. van der Vlugt and C. J. Hogendoorn, "Natural Convection flow in a square cavity calculated with low-Reynolds-number turbulence models," *International Journal of Heat and Mass Transfer,* vol. 2, pp. 377-388, 1991.

[14] S. Kruger and P. L, "Analysis of Natural Convection in a Mono-Span Greenhouse," in *South African Conference on Applied Mechanics*, Johannesburg, 2012.

[15] N. A. Kurekci and O. Ozcan, "An Experimental and Numerical Study of Laminar Natural Convection in a Differentially-heated Cubical Enxlosure," *Journal of Thermal Science and Technology,* vol. 32, no. 1, pp. 1-8, 2012.

[16] H. S. Dol and K. Hanjalic, "Computational study of turbulent natural convection in a side-heated near-cubic enclosure at a high Rayleigh number," *International Journal of Heat and Mass Transfer,* vol. 44, pp. 2323-2344, 2001.

[17] N. A. Kurecki and O. Ozcan, "An experimental and Numerical Study of Laminar Natural Convection in a differentially-heated Cubical Enclosure," *Journal of Therrmal Science and Technology,* vol. 32, no. 1, pp. 1-8, 2012.

[18] S. Kruger, A NUMERICAL EVALUATION OF MULTI-DIMENSIONAL HEAT TRANSFER EFFECTS IN GREENHOUSES, Johannesburg: University of Johannesburg, 2016.

[19] CD-Adapco, "StarCCM+ User Guide," CD-Adapco, 2012.

Aerodynamic analysis of a sailplane self-launch system: Propeller sub-model

J.P. le Roux*, Dr. J.H. Kruger, Dr. J.J. Bosman
School of Mechanical and Nuclear Engineering, North-West University
22143785@nwu.ac.za

ABSTRACT

Self-launching sailplanes are sailplanes equipped with a retractable propeller that can be used to launch the sailplane, as well as to extend flights. The engine-driven propeller is mounted on a pylon and is deployed out of the fuselage when needed. When the system is deployed, it breaks the streamlined shape of the sailplane and as a result, increases the drag of the sailplane considerably. The rotating propeller also disrupts the airflow towards the radiator, and the influence thereon needs to be investigated.

The ultimate goal of the project is to complete an aerodynamic analysis of the sailplane and self-launch system using the appropriate level of modelling techniques. The main computational fluid dynamics (CFD) model is obtained by combining various sub-models into an integrated CFD simulation. Obtaining confidence in these models, required validation of the models' predictions with experimental results. This paper focuses on the propeller sub-model, its results and model validation. The remaining sub-models are briefly discussed to explain how they will be integrated into the main CFD model. The details of other sub-models are not included in this paper, but will be completed at a later stage when it will be incorporated into the main CFD model.

Validation of the propeller sub-model was performed by mounting the self-launch system on a test bench and measuring the static thrust delivered by the propeller at different propeller rpm's (revolutions per minute). The static thrust of the propeller from the experimental tests and CFD results compared favorably, all within 7% of each other when experimental uncertainty is taken into account.

KEYWORDS

Computational fluid dynamics, propeller, sailplanes, self-launch system, radiator

NOMENCLATURE

L	Length
P_i	Inertial resistance
P_v	Viscous resistance
v	Velocity
ΔP	Delta pressure

1. INTRODUCTION

1.1 Background

A sailplane is a lightweight aircraft designed for primarily unpowered flight. It gains elevation by utilizing thermals and thereafter glides to the desired location. Sailplanes are most commonly used in the sport of gliding and for some military applications.

Sailplane manufacturers are in a continuous drive to deliver the versatile and high performance sailplanes. Several sailplane manufacturers offer an optional engine with a retractable propeller. This self-launch system enables the sailplane to take off without any assistance as opposed to the conventional way of using a winch or a tow-plane.

The self-launch systems are designed to be retractable to restore the aerodynamic shape of the sailplane. Therefore, after the sailplane has reached the desired height — using the assistance of the engine and propeller — the self-launch system is retracted.

The self-launch system also helps to avoid land-outs and makes cross-country flights more practical. This makes gliding more accessible to participants and is therefore becoming increasingly popular among the gliding community. A deployed self-launch system can be seen in figure 1a and the deployment route of the system can be seen in figure 1b.

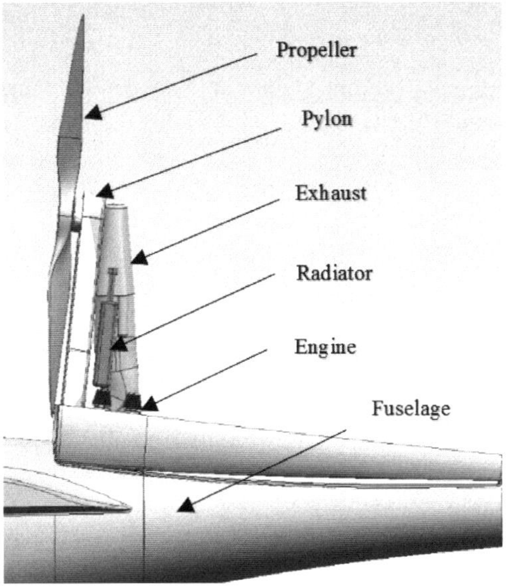

Figure 1a. Deployed self-launcher system

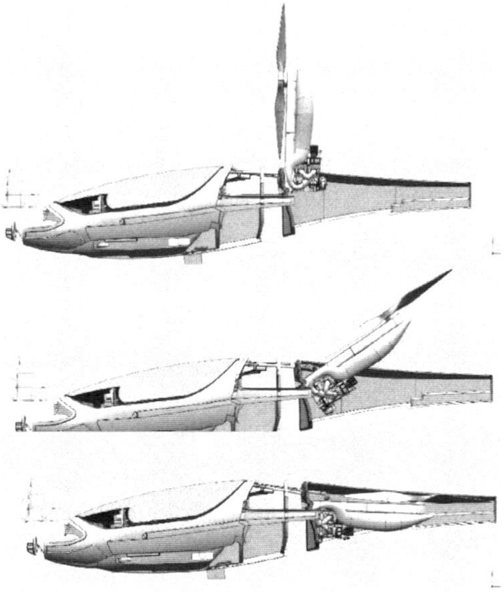

Figure 1b. Deployment locus of system

The propeller is mounted on the pylon and creates the required thrust needed for the sailplane to take off and also to extend flights. The propeller is driven by a two-stroke engine and is connected to the engine by means of pulleys and a belt. As a result, the propeller is rotating with a reduction ratio of 2.32 when compared to the engine's rpm.

1.2 PROBLEM DESCRIPTION

The convenience that a self-launch system offers, entails certain disadvantages. When the self-launch system is deployed, it breaks the streamlined shape of the sailplane and as a result, increases the drag of the sailplane considerably. The wake created by the propeller also has an effect on the handling characteristics of the sailplane. The increased drag created by the system leads to an increased sink rate of the sailplane. When the engine of the self-launch system fails to start in an emergency deployment, it can become dangerous for the pilot when at low altitude. To put it in perspective, a DG-1000T sailplane sink rate almost increases twofold —from 0.65 m/s to 1.1 m/s— at a speed of 90 km/h, when the self-launcher system is deployed with the propeller not running (Flugzeugbau [1]). A self-launch system must therefore not be regarded as a life insurance that will give you the needed thrust to gain lift, as engine failures or user errors do occur (Greenwell [2]). An aerodynamic analysis of a sailplane self-launch system is needed to determine the effects of the self-launch system on the sailplane's drag, sink rate etc.

The radiator is one of the main influential factors pertaining to the thrust delivered by the self-launch system. Internal combustion engines produce a large amount of heat and approximately 30% of the total energy entering an engine is transferred to the environment by means of heat rejection (Pulkrabek [3]). The radiator is required to ensure adequate cooling of the engine, thus keeping it within safe optimal operating temperatures for peak performance. A two-stroke engine is used in the sailplane for its lower weight, simpler design and its ability to work at acute angles when compared with a four stroke engine. Due to the constrained space available for the self-launch system when it is retracted, is it important to use a space-efficient radiator while ensuring adequate cooling is achieved. Therefore the size and position of the radiator are important design parameters.

The swirling airflow that reaches the radiator due to the rotation of the propeller, will influence the efficiency of the radiator. When the swirling airflow reaches the radiator at an angle, it leads to an increased pressure drop per unit length in the flow direction of the radiator (Du Toit [4]).

1.3 OBJECTIVE

The ultimate objective of the project as a whole is to complete an aerodynamic analysis of a deployed sailplane self-launch system. The increased drag created by the deployed system and the heat transfer capabilities of the radiator under these conditions are of main interest. The objective of this paper is to share the methods followed in developing and validating specifically the CFD sub-model of the propeller. A general overview of the project is however provided as context for this sub-model.

2. SIMULATION METHODOLOGY

2.1 System overview

CFD simulations have become a useful tool to analyse performance of sailplanes and obtain insight into flow structures (Hansen [5], Bosman [6]).

The main CFD model of this project integrates several validated sub-models. The sub-models were developed as efficiently as possible to keep the overall simulation's computational power required at acceptable levels while providing acceptable maintaining accurate results. These sub-models must be developed and validated before they can be integrated into the main CFD model. Figure 2 shows the different CFD models that will be used and integrated into the main CFD model.

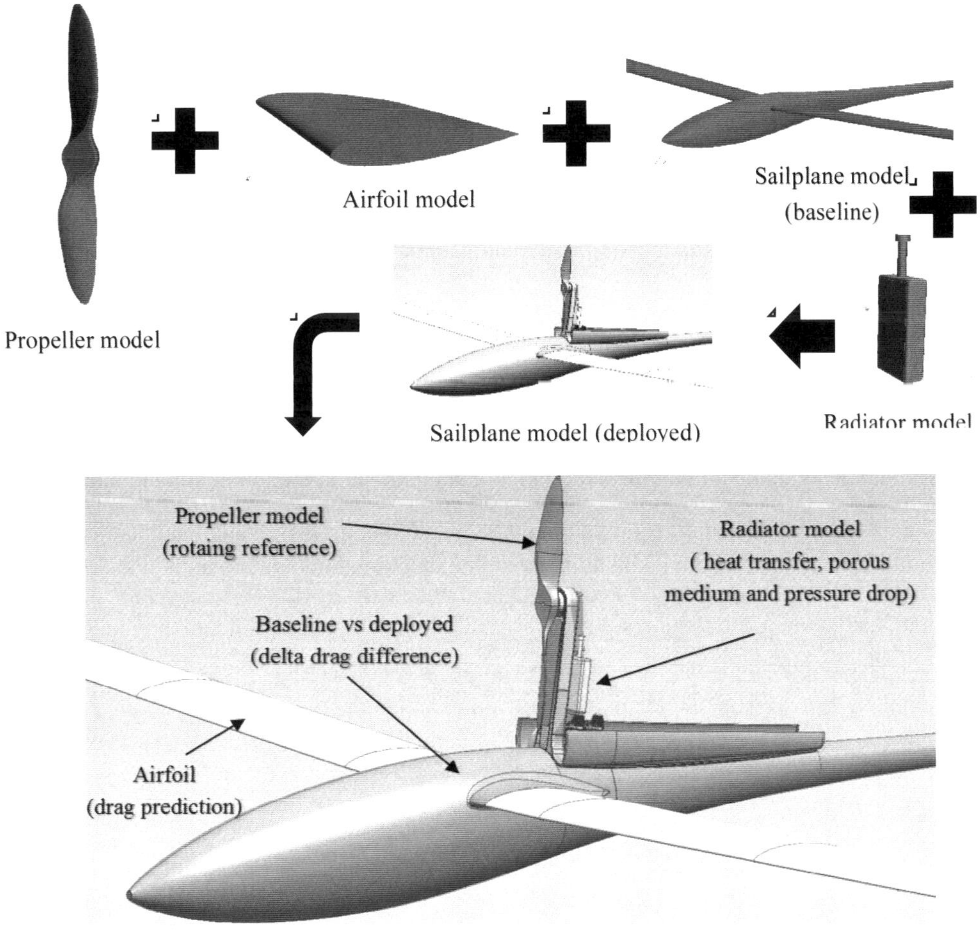

Figure 2. Simulation layout

2.2 CFD sub-models

The first sub-model consists of the propeller where the rotation of the propeller and its influence on the surroundings were simulated. The propeller was modelled using a rotating reference frame. This allowed the model to achieve a steady state solution, as this method incorporates the effects of motion without moving the vertices of the mesh (CD Adapco [7]). This model will be discussed in greater detail in the propeller simulation section, as it is the main focus of this paper.

The second sub-model consists of a wing section where the drag of the airfoil is predicted. The model is required to be able to predict transitional boundary layers and boundary separation to accurately determine the drag of the airfoil (Bosman [6]). Confidence in the model is obtained by validating the model's results with wind tunnel tests and XFoil calculations.

Assuming success validating the airfoil model, one can postulate that the drag of the sailplane with the retracted self-launch system (third sub-model) can be predicted. This result is then used as the baseline drag for the sailplane. Due to the symmetric nature of the geometry, the baseline can be simulated using only half of the sailplane and a symmetric plane. By simulating only half of the sailplane the number of cells in the mesh can be reduced to speedup calculations.

The fourth sub-model concerns the heat transfer between the airflow and the radiator. To simulate the heat transfer and pressure drop through the complex geometry of the radiator, the computational power required is very demanding. By simulating the radiator as a porous medium, the required computational resources can be reduced considerably, with accurate results still possible (Kadir et al. [8]). However, a detailed unit cell of the radiator needs to be simulated to obtain the porous coefficients needed by the porous model for the airflow through the radiator. The inertia and viscosity coefficients for the porous model used to represent the radiator, can be obtained by simulating a detailed section (unit cell) of the radiator as seen in figure 3.

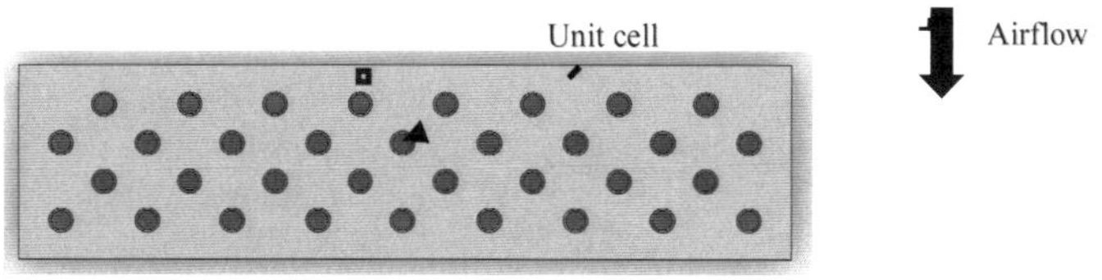

Figure 3. Radiator cut section with unit cell

By determining the pressure drop over the radiator at different velocities, the inertial and viscous resistances can be calculated using equation 1:

$$\frac{\Delta p}{L} = -(P_i|v| + P_v) \qquad \ldots (1)$$

After the pressure drop of the radiator is simulated successfully, a porous heat transfer model of the radiator must be developed and validated with experimental results. The inertia and viscosity coefficients of the radiator are needed to accurately determine the drag created by the deployed self-launch system.

In the final model the sailplane's drag are computed when the self-launch system is deployed. The results are then compared to the drag of the baseline sailplane. The delta drag between the models will then give an indication of the effect of the deployed self-launch system on the sailplane.

Lastly, the sub-models are then incorporated into one large main CFD model where the sub-models interact and influence each other. It will then be possible to determine the overall effect of the self-launching system on the sailplane and if the radiator delivers adequate cooling.

3. PROPELLER SUB-MODEL DEVELOPMENT

3.1 Geometry

The shape and main dimensions of the propeller can be seen in figure 4. The exact detail of the propeller's geometry is confidential, since it is a proprietary design of Jonker Sailplane. The propeller rotates clockwise when viewed from the cockpit.

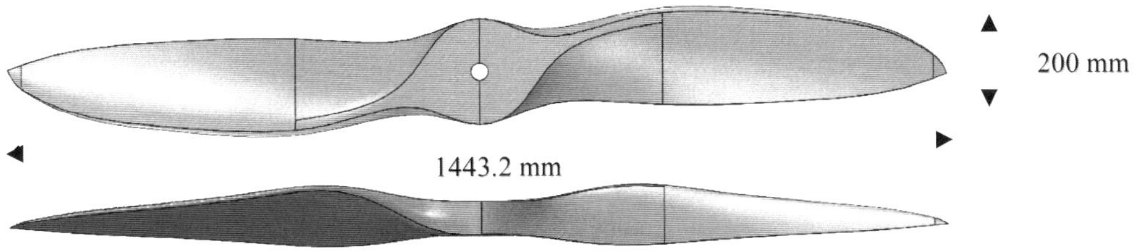

Figure 4. Propeller CAD model

3.2 Mesh

The polyhedral mesh type was chosen to be the best suited meshing model for the propeller/sailplane simulation. Although tetrahedral meshes has been use for decades in a large number of simulations, recently the use of polyhedral meshes have resulted in a number of simulations yielding more accurate results when compared to using tetrahedral meshes (Brezzi [9], Rodriguez [10])

Polyhedral meshes are more efficient when compared to tetrahedral meshes in term of cell count, as it uses up to five times fewer cells (Peric and Ferguson [12]). Convergence is also achieved faster with fewer cells when compared to a tetrahedral mesh (Balfaras [11]). Polyhedral cells have an average of 14 faces, which result in 14 neigbours. This gives it the ability to capture swirling and recirculating flow particularly good, because of a greater likelihood that local flow vectors can be orthogonal to connecting faces (Peric and Ferguson [12]).

In an attempt to capture the boundary layers of the flow, prism layers were used on the propeller's boundaries. The prism layers consisted of 16 layers (5mm thick) with a stretch factor of 1.3 as seen in figure 5b. A fine mesh was used to accurately capture the geometry of the propeller (figure 5b). A section cut of the simulation's mesh can be seen in figure 5b. Note that the outer boundaries of the mesh are coarse and as the mesh approaches the propeller, it becomes increasingly finer to capture the flow in detail near the propeller. This was achieved by using a surface growth rate of 1.3.

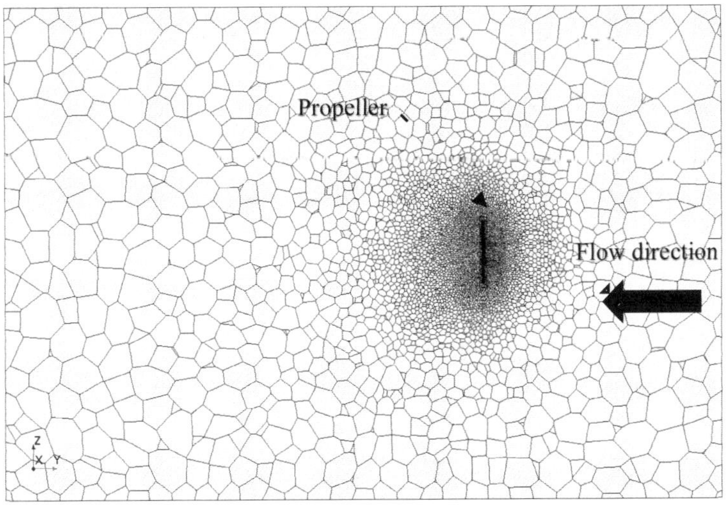

Figure 5a. Test domain mesh

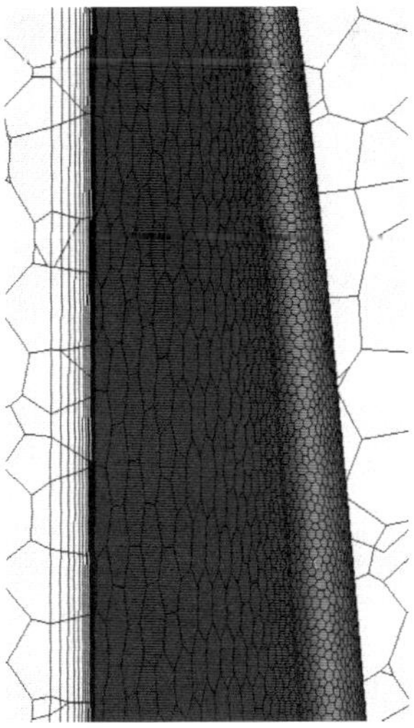

Figure 5b. Propeller's mesh with prism layers

When the mesh resolution is increased, the accuracy of the numerical solution is improved [13]. A mesh dependency investigation was performed to ensure that a good balance between simulation run time and accurate results was achieved. Details are given in the results section

In the main CFD model a mesh refinement of the wake behind the propeller will be implemented to predict the drag created by the deployed propeller. It is expected that the wake refinement may not be necessary when calculating the static thrust of the propeller. This statement has to be confirmed and a mesh refinement was applied as seen in figure 6a to determine the influence of mesh design on the static thrust of the propeller.

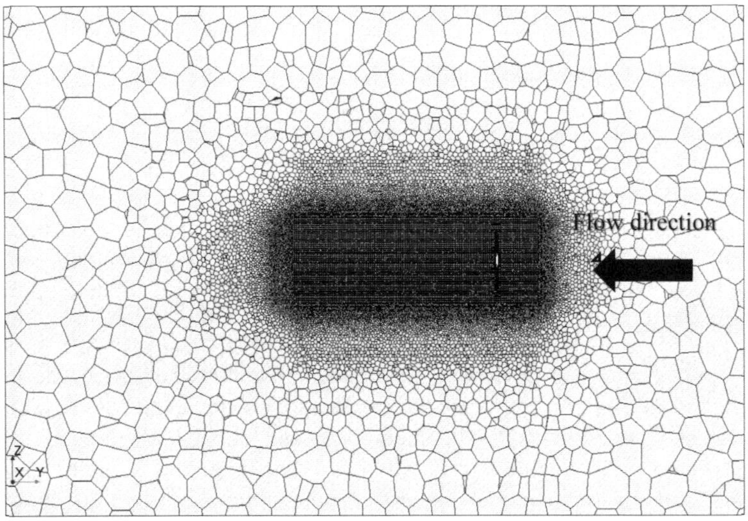

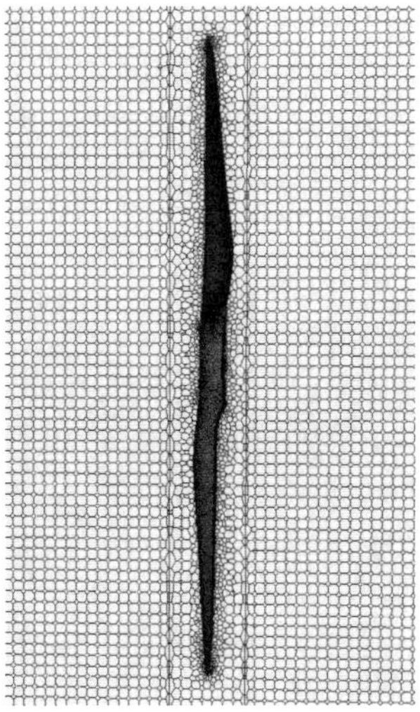

Figure 6a. Test domain mesh with wake refinement

Figure 6b. Details of the refined mesh

The influence of the number of cells on the static thrust of the simulations can be seen in figure 7. All mesh surface sizes were set to reference the base size, with the exception of the prism layers. As the base size was decreased, the cell count increased.

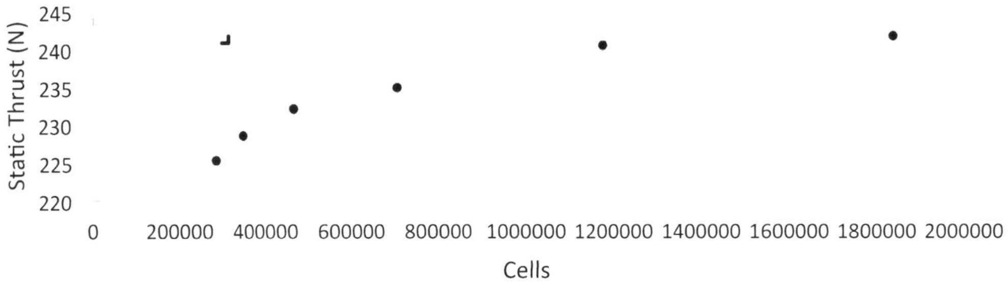

Figure 7. Mesh dependency investigation

A cell count of 1.2 million was determined to be adequate as a variance of only 0.5% in static thrust is observed between 1.2 million and 1.8 million cells.

The effect of the wake refinement behind the propeller on the mesh can be seen in table 1.

Table 1. Mesh wake refinement

	No of cells (Million)	Static thrust @ 3000 (rpm)	Static thrust @ 6000 (rpm)
Wake refinement	1.1	242	980
No wake refinement	3.8	241	976
% difference	110.2	0.675	0.375

The implemented wake refinement increased the number of cells considerably, with only an insignificant change in static thrust. The results in Table 1 confirm the statement made earlier that the wake refinement might not be necessary to predict static thrust. Therefore, the wake refinement was not included in the propeller CFD model's mesh. In the main CFD model a wake refinement behind the propeller will be implemented to account for fuselage effects when determining the drag of the system.

3.3 Domain size

The domain size needs to be large enough to simulate free stream conditions. The test domain consisted of a cylinder with a velocity inlet, a pressure outlet and slip sidewalls as seen in figure 8. The cylinder size was based on the diameter of the propeller (see table 2).

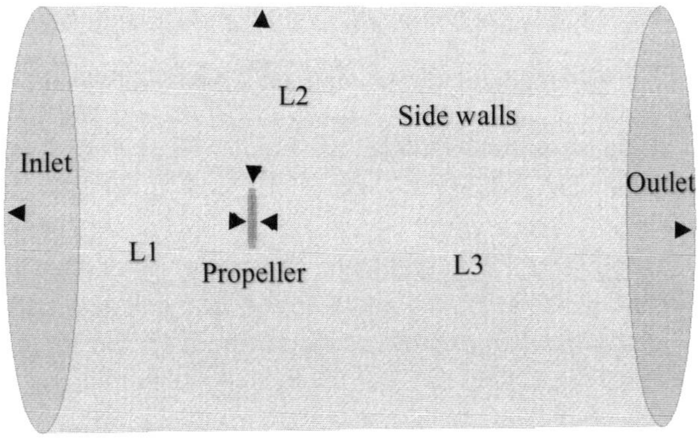

Figure 8. Domain size

Table 2. Domain size and parameters

#	L1 (m)	L2 (m)	L3 (m)	Static thrust (N)	Cell count
1	3	3	6	242.1	31218
2	4.5	4.5	9	241.1	76217
3	6	6	12	240.5	109418
4	7.5	7.5	15	240.8	193140

The domain size of case number three was determined to be of adequate size, as a variance in static thrust was negligible small.

3.4 Model physics

A steady state RANS approach was followed due to the fact that incompressible flow was assumed with Mach number under 0.3, and the effects of the unsteady wake were not of importance (CD Adapco [7]). The k-omega turbulence model, together with the Gamma-Re-Theta transition model, was used in an attempt to capture the transition of laminar to turbulent flow over the propeller. The Gamma Re-theta transitional model is a correlation-based transition model that is coupled with the SST (Shear Stress Transport) K-Omega turbulent model (Menter and Langtry [13]). It uses two transport equations, one for intermittency and the other for transition onset momentum-thickness Reynolds number.

The same physics models were used by Hansen [5] and Bosman [6] and delivered acceptable results in determining the performance of a sailplane. These models are widely used in the aerospace industry and have the ability to predict transitional boundary layers and boundary separation accurately (CD Adapco [7]).

4. EXPERIMENTAL SETUP

To validate the CFD model one can compare the simulated values with experimental measurements. The propeller's static thrust was chosen as the primary parameter with which to validate the CFD model, as it is an uncomplicated and cost efficient experiment. The tests were done on the self-launch system test bench as seen in figure 9a. The self-launch system was securely mounted on the test bench and the pylon was supported on hinges. A restraining cable connected to the pylon prevents it from falling forward as the propeller delivers thrust. The hinges and cable system is similar to the design used for the actual aircraft version.

The engine rpm was measured with a tachometer. A calibrated load cell was connected to the restraining cable to measure the thrust of the propeller. The load cell has a tolerance of 0.1kg. During the experiment the vibration on the test rig caused the readings of the load cell to oscillate, and together with a slow refresh rate of the load cell, the measurements had a tolerance of ±1 kg. The load cell measured the thrust of the propeller at an angle and the geometry of the system was thus used to construct a force diagram. The force diagram gave the relationship between the load cell thrust and the horizontal thrust as seen in figure 9b. The force diagram was confirmed by applying a known force with a second calibrated load cell behind the propeller and comparing the calculated value with the load cell reading connected to the restraining cable. The static thrust of the propeller was measured at propeller speeds varying from approximately 1100 to 2600 rpm. Equation 2 shows the relationship between the propeller's static thrust and the load cell measured force.

$$\frac{T_{propeller}}{F_{load\ cell}} = cos \propto \approx \frac{1454}{1759} \approx 0.83 \qquad \ldots(2)$$

As an example, when the load cell is measuring a force of 100N, the static thrust will be 83N.

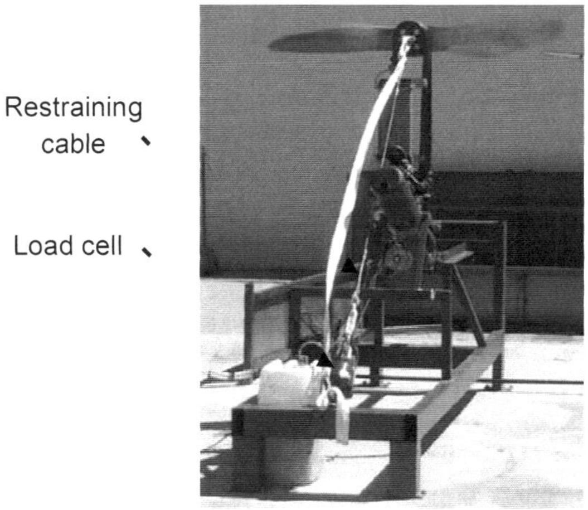

Figure 9a. Self-launch test bench (Grundling [14])

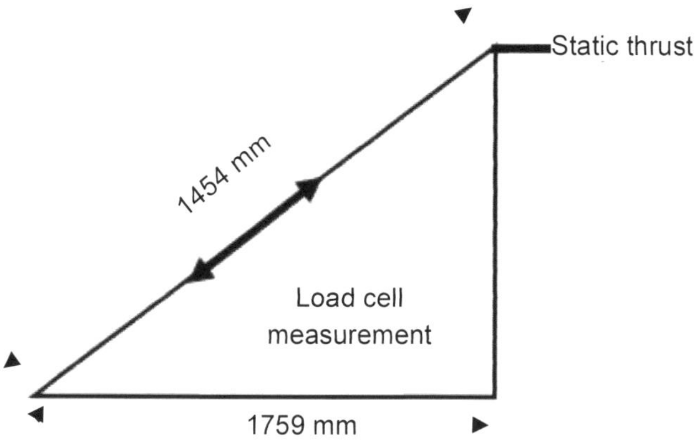

Figure 9b. Force diagram

5. RESULTS

5.1 Experimental and CFD results

The propeller's static thrust was simulated in STAR CCM+ at different propeller rpms. The static thrust values measured were then converted to kg and compared with the experimental measurements as seen in figure 10.

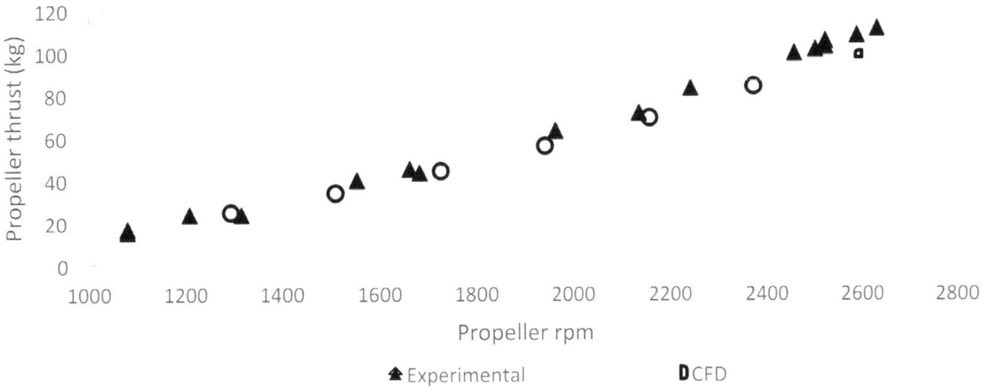

Figure 10: Static thrust comparison

The experimental and simulated static thrust compares favorably, especially at lower propeller rpm's. At higher rpm's the simulation under-predicts the propeller's static thrust, with the largest percentage difference at approximately 7%, when the propeller's speed is at 2586 rpm.

5.2 Airflow Characteristics

The airflow near the propeller was also investigated to obtain further confidence in the propeller's CFD model. The propeller consists of several airfoils and according to airfoil theory, is it known that the pressure at the top of a lifting airfoil is lower than at the bottom of the same airfoil (Anderson [15]). This can be seen in figure 11.

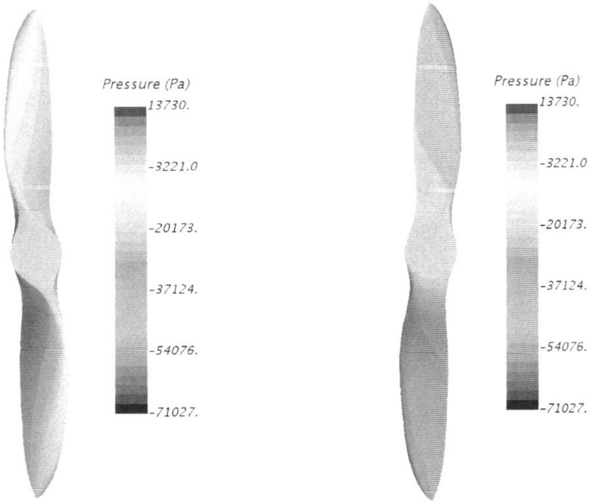

Figure 11. Left side: Pressure distribution on the front of the propeller
Right side: Pressure distribution on the back of the propeller

The negative pressure created by the rotating propeller, results in the air being sucked towards the propeller. The effect of the rotating propeller on the airflow near the propeller is represented by streamlines as seen in figure 12.

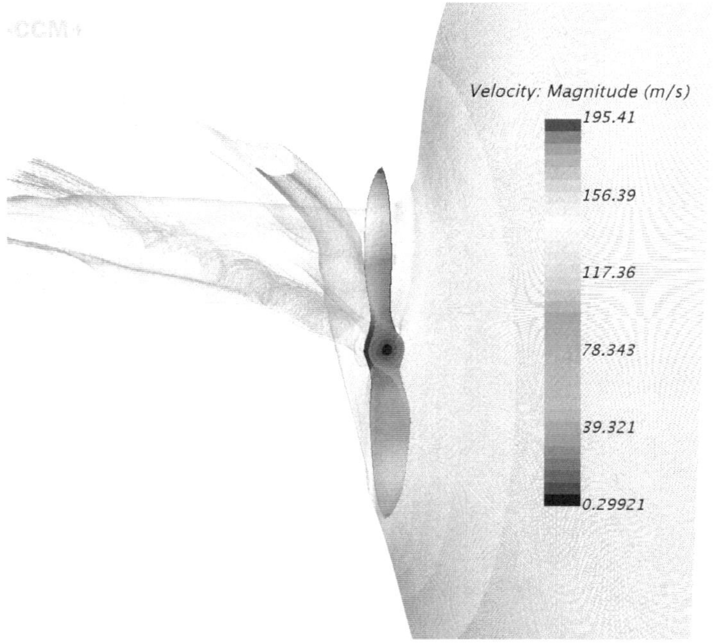

Figure 12. Airflow near the propeller

The velocity over the propeller increases as one moves further away from the axis of rotation. One can see how the airflow swirls behind the propeller. The streamlines also converge as they approach the propeller due to the suction (negative pressure) created by the rotating propeller.

Conclusion

Good progress was made in the aerodynamic analysis of the sailplane self-launch system. The rotating propeller was successfully simulated by the CFD sub-model and compared favorably with the experimental results (within 7% difference). The difference in static thrust can be due to the assumptions made in the CFD model. Assumptions such as constant air density, steady state conditions and modelling the propeller with a rotating reference frame may have contributed to the difference in static thrust. However these assumptions and simplifications make for considerably more efficient calculations that will ensure the final CFD model are kept within acceptable computational requirements. The vibration of the self-launch test rig during the experiments may also have contributed in the difference in static thrust. When the drag of the baseline sailplane is eventually compared with the deployed self-launch system sailplane, the delta drag will however be of more importance than the absolute values of the results.

Airflow characteristics near the propeller were also investigated to further increase the confidence in the model. The pressure distribution over the propeller was in accordance with airfoil theory where the pressure of at the top of a propeller's airfoil is lower than at the bottom of the same propeller's airfoil. The swirling effect of the airflow behind the propeller was captured and illustrates the ability of polyhedral cells to capture swirling airflow without using a highly refined mesh. This is important as the swirling airflow will affect the radiator's efficiency. The propeller CFD sub-model can be confirmed to be adequate and ready to be incorporated into the main CFD model. The process to create the main CFD model of the project has also been laid out, and by applying the basic methodology used by the propeller model, the rest of the sub-models together with the integrated CFD model can be completed.

Acknowledgements

I would like to thank the following people for their assistance during this project:

- Dr. Jan-Hendrik, for encouraging me to take part in SACAM and sharing his CFD knowledge.
- Dr. Johan Bosman, for his advice on how to simulate a propeller in STAR CCM+
- The NWU Unit for Energy and Technology Systems for providing funding to attend the conference.

References:

1 Flugzeugbau., Flight manual for the motorglider DG-1000T., 2005. [Online] Available at: http://www.dg-flugzeugbau.de/Data/Manuals/flugh-1000t-e.pdf [Accessed 16 May 2016].

2 Greenwell, E., A Guide to Self-Launching Sailplane Operation., 2004. [Online] Available at: http://www.sac.ca/website/index.php/en/documents/safety-and-training/instructor-resources/426-motorglider-ops-manual/file [Accessed 16 May 2016].

3 Pulkrabek, W. W., “Engineering Fundamentals of the internal Combustion Engine.”, Wisconsin: Pearson Prentice Hall, 2004.

4 Du Toit C.G., Kotze J.C.B., Du Plessis J.P., “The Modelling of the Air Flow through a Grill and Radiator Core”, Stellenbosch: University of Stellenbosch, 1990.

5 Hansen, T., “Modeling the Performance of the Standard Cirrus Glider using Navier-Stokes CFD.” *Technical Soaring 38.1*, pp.5-14, 2012.

6 Bosman, J. J., “Evaluation of new aerodynamic concepts using CFD for the improvement of a glider fuselage”, Potchefstroom: North West University, 2012.

7 CD Adapco, “STAR-CCM+ v.11.02 User Guide”, 2016.

8 Guler, Kadir G., Barbaros Cetin, and M. Haluk Aksel. "Simulation of thermal characteristics of radiators using a porous model." Ichmt digital library online. Begel House Inc., 2014.

9 Brezzi, Franco, et al. "A new discretization methodology for diffusion problems on generalized polyhedral meshes." Computer Methods in Applied Mechanics and Engineering 196.37 (2007): 3682-3692.

10 Rodriguez, R., “CFD Analysis on the Main-Rotor Blade of a Scale Helicopter Model using Overset Meshing”, Stockholm, 2012.

11 Balafas, Georgios. Polyhedral Mesh Generation for CFD-Analysis of Complex Structures. Diss. Master Thesis, Technische Universitat Munichen, 2014.

12 Peric M, Ferguson S. The advantage of polyhedral meshes. CD-adapco< http://www. cd-adapco. com/press_room/dynamics/24/testVspoly. html. 2005.

13 Oskam, B. Slooff, J.W Recent advances in computational aerodynamics at NLR. AIAA Paper, (98-0138), 1998.

14 Grundling, G., “Aerodynamic design of a propeller for a self-launch sailplane”, Potchefstroom: North West University”, 2016.

15 Anderson, J. D., “Fundamentals of Aerodynamics”, 3rd ed. New York: McGraw-Hill, 2001.

Experimental Observation of Computed Ultrasonic Guided Wave Dispersion Phenomena

Philip W. Loveday[1*] and Dineo A. Ramatlo[1,2]

[1]CSIR Materials Science and Manufacturing, [2]University of Pretoria

ploveday@csir.co.za

ABSTRACT

Guided wave ultrasound offers the ability to inspect large lengths of structural members from a single transducer location and such a system for monitoring rail track is being developed in South Africa. Guided wave inspection is complicated by the presence of numerous modes of propagation and dispersion and modelling is required to interpret experimental measurement results. The semi-analytical finite element method has been applied extensively for modelling elastic wave propagation in rail track. The method produces dispersion curves of wavenumber and group velocity versus frequency and mode shapes for propagating modes. Some interesting features have been observed in these predictions. The wavenumber – frequency curves of two modes sometimes appear to cross each other but closer inspection reveals that the curves approach each other and then repel each other without crossing. This feature was investigated and found to be analogous to that which occurs in plates. Measurements were performed on continuously welded rail track to investigate if these phenomena can be observed experimentally. An ultrasonic transducer was used to transmit energy along the rail which is reflected at welds in the rail track and then received at the transducer. The received time signals were analysed using the short time Fourier transform and the resulting spectrograms reveal the times of arrival of different frequencies. Numerically predicted arrival times were superimposed on the spectrograms and the accuracy of group velocity predictions was confirmed. It is believed that the repulsion phenomenon is visible in the experimental results.

KEYWORDS

Guided wave ultrasound, rail track, dispersion measurement, mode repulsion

INTRODUCTION

Guided wave ultrasound offers the ability to inspect large lengths of structural members from a single transducer location especially when the structure resembles a constant cross-section elastic waveguide. Continuously welded rail track is an example of such a structure and a system for monitoring rail track has been developed in South Africa over many years [1], [2]. The production version of the system comprises transmit and receive stations spaced alternately along the length of the rail line. This system is primarily a broken rail detector. The transmit stations transmit coded ultrasonic signals along the rail every 15 minutes and if these signals are not received at the neighbouring receive stations an alarm is triggered to halt train operation. Permanently installed ultrasonic transducers are used in the system and the transducers are spaced approximately 900m apart. The system is installed on 840 km of Transnet Freight Rail's iron export line. Further development has been conducted to increase the distance between stations to 2 km in future installations. Research is being conducted to improve the functionality of the system to detect cracks in the rail prior to complete rail breaks.

Guided wave inspection and monitoring is complicated by the presence of numerous modes of propagation and dispersion where different frequencies propagate at different group velocities and cause signals to become distorted especially if long ranges are used. These phenomena are not encountered in conventional ultrasound inspection and require modelling during system design and for the interpretation of experimental measurement results. This modelling includes modelling of the excitation of waves by transducers and transducer arrays, the propagation of waves and the interaction of the waves with defects. The propagation of guided waves in one-dimensional waveguides of constant cross-section may be modelled efficiently by using the semi-analytical finite element (SAFE) method. This method is used in the next section to model the modes of propagation in a free infinite rail. The predicted dispersion curves relating wavenumber and group velocity to frequency for the modes of propagation reveal some interesting features. The wavenumber – frequency curves of two modes sometimes appear to cross each other but closer inspection reveals that the curves approach each other and then repel each other without crossing. This sudden change in slope of these curves manifests as a sudden change in group velocity and the two modes appear to swop group velocity and also mode shape. Similar phenomena have been reported for the dispersion characteristics of Lamb waves in plates [3]. A perturbation analysis was used in [3] to show that coupling between longitudinal and shear waves causes repulsions between neighbouring symmetric or antisymmetric dispersion curves. It was also stated that the character of the Lamb waves switches across the repulsion region. The SAFE computed dispersion characteristics are analysed to see if similar conclusions can be drawn.

The question arises of whether these sudden changes are a feature of the assumptions made during mathematical modelling or if they actually occur in the physical world. Experimental measurements are performed to investigate whether these features can be observed in the physical world.

COMPUTATION OF GUIDED WAVE DISPERSION CHARACTERISTICS

Elastic waveguides with complex geometry such as a rail require the use of numerical modelling to predict the dispersion characteristics. Numerous research groups have implemented the SAFE method for this purpose [4]–[8]. These elements are formulated to include the wave propagation as a complex exponential in the element formulation and require only a two-dimensional mesh of the rail profile. The displacement fields (u, v, w) employed by Gavrić [4] were chosen to take the form:

$$\begin{aligned} u(x,y,z,t) &= u(x,y)\cdot e^{-j(\kappa z-\omega t)} \\ v(x,y,z,t) &= v(x,y)\cdot e^{-j(\kappa z-\omega t)} \\ w(x,y,z,t) &= w(x,y)\cdot e^{-j(\kappa z-\omega t-\pi/2)} = j\cdot w(x,y)\cdot e^{-j(\kappa z-\omega t)} \end{aligned} \quad , \qquad (1)$$

where, z is the coordinate in the direction along the waveguide, κ the wavenumber and ω the frequency. $u(x,y)$, $v(xy)$ and $w(x(y)$ are the interpolated displacements in the x, y and z directions respectively.

Application of the finite element method results in a system of equations of motion. For free vibration the equation of motion is:

$$M\ddot{U}+\left[\kappa^2\cdot K_2+\kappa\cdot K_1+K_0\right]U=0. \qquad (2)$$

It should be noted that in the formulation presented by Gavrić the mass (M) and stiffness (K_0, K_1, K_2) matrices are all real and symmetric. The equations of motion are similar to those obtained using conventional three-dimensional finite elements except for the presence of stiffness matrices that are proportional to the wavenumber and the square of the wavenumber. If we consider free harmonic vibration we obtain the eigenvalue problem:

$$\left[\kappa^2\cdot K_2+\kappa\cdot K_1+K_0\right]u=\omega^2 Mu\,. \qquad (3)$$

The propagating modes may be computed by setting the wavenumber, κ to a real value and solving the eigenvalue problem to obtain the frequencies, ω and mode shapes of the propagating modes. If we wish to compute the modes at a specified frequency this can be achieved by adding an identity to the eigenvalue problem to obtain the following equation which may be solved to compute the wavenumbers and mode shapes at a specified frequency:

$$\begin{bmatrix} K_0-\omega^2 M & 0 \\ 0 & -K_2 \end{bmatrix}\begin{Bmatrix} U \\ \kappa U \end{Bmatrix}+\kappa\begin{bmatrix} K_1 & K_2 \\ K_2 & 0 \end{bmatrix}\begin{Bmatrix} U \\ \kappa U \end{Bmatrix}=\begin{Bmatrix} 0 \\ 0 \end{Bmatrix}. \qquad (4)$$

This solution provides the non-propagating modes in addition to the propagating modes. These modes are required for some applications but are not considered in this paper. Equation 3 was solved to compute the dispersion characteristics for a UIC60 profile rail section as shown in figure 1.

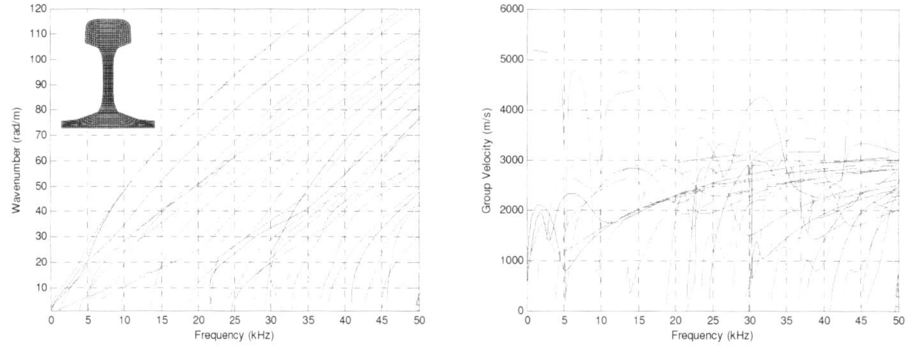

Figure 1. SAFE mesh and computed dispersion curves.

The rail profile is symmetric about the vertical axis. We can separate the modes into two sets; one that contains the symmetric modes and one that contains the antisymmetric modes and these sets are plotted in figure 2.

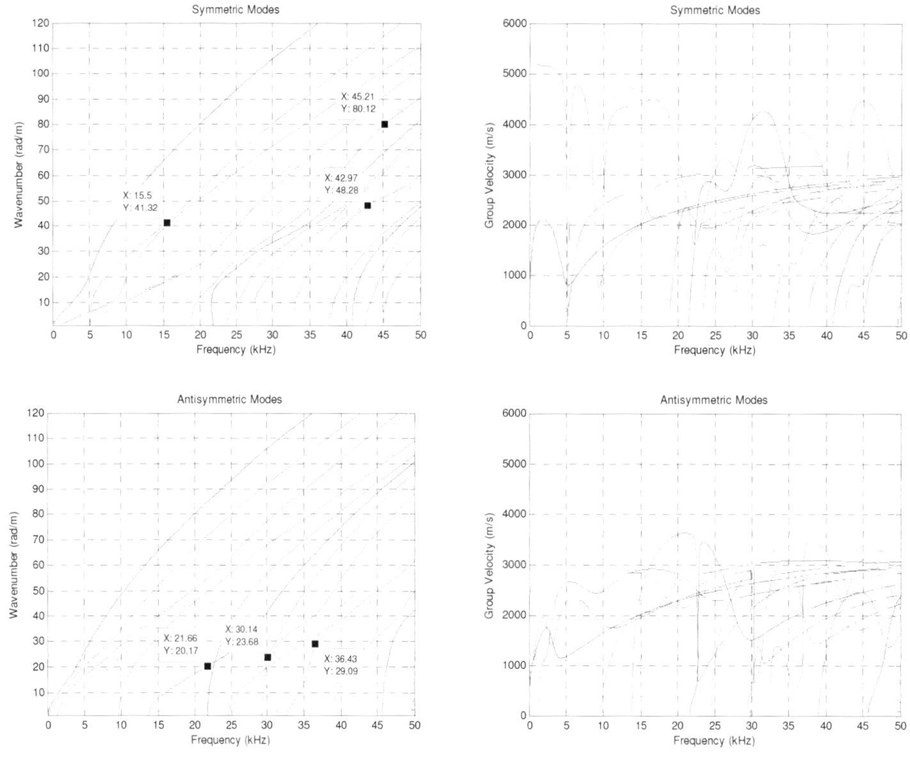

Figure 2. Computed dispersion curves for symmetric and antisymmetric modes.

Careful examination of the symmetric wavenumber versus frequency curves reveals that the curves for the modes do not cross. We see that the modes approach and then appear to repel each other at frequency – wavenumber points such as those marked in the figure 2. When this happens we see that the group velocity curves of the two modes cross each other and appear to then continue as though they have swopped. The same behaviour is seen in the curves for the antisymmetric modes. In this plot the data markers have been slightly offset from the point of interest so as not to obscure the feature.

Figure 3 shows one selected feature in better detail along with the mode shapes at 43 kHz and 47 kHz. These mode shapes were computed by setting the frequency and the solving the eigenproblem in equation 3. We see that while the wavenumber curves do not cross, the mode shapes swop. This repulsion region was selected for illustration of the phenomenon because it will be studied experimentally in the following section.

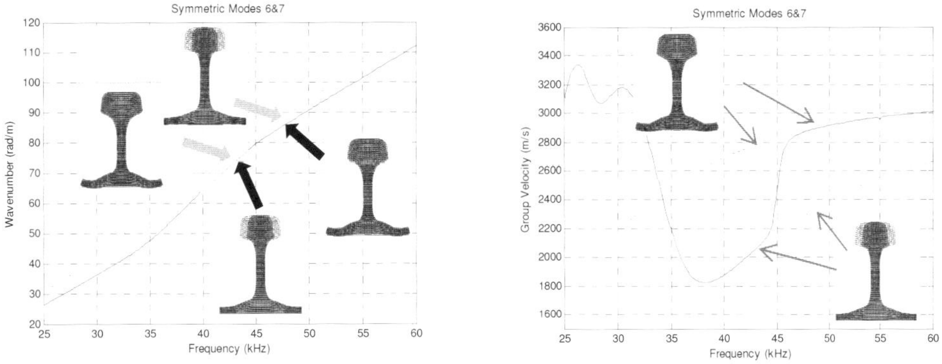

Figure 3. Selected dispersion curves (s_6 yellow, s_7 black) and mode shapes at 43 and 47 kHz.

The behaviour of the symmetric and antisymmetric dispersion curves and modes is analogous to that of the symmetric and antisymmetric Lamb waves in plates [3]. The fact that the wavenumber dispersion curves within the symmetric and antisymmetric mode families do not cross, means that the modes can be numbered in a simple manner in order of decreasing wavenumber as s_1, s_2, s_3 ... and a_1, a_2, a_3 ... for the two families.

EXPERIMENTAL MEASUREMENTS

Measurements were performed on Transnet Freight Rail's COALLINE at Overvaal Station. This is a heavy haul line for transporting coal from Ermelo to Richard's Bay for export. The rail type is UIC60, which indicates that it weighs 60kg per metre. The measurements presented were performed using transducers attached under the rail head as shown in figure 4. These transducers were custom designed to excite a symmetric mode with vertical motion of the rail head as shown in figure 4b. Note that this mode was computed using SAFE and the results from the two-dimensional mesh were extruded to provide the three-dimensional illustration.

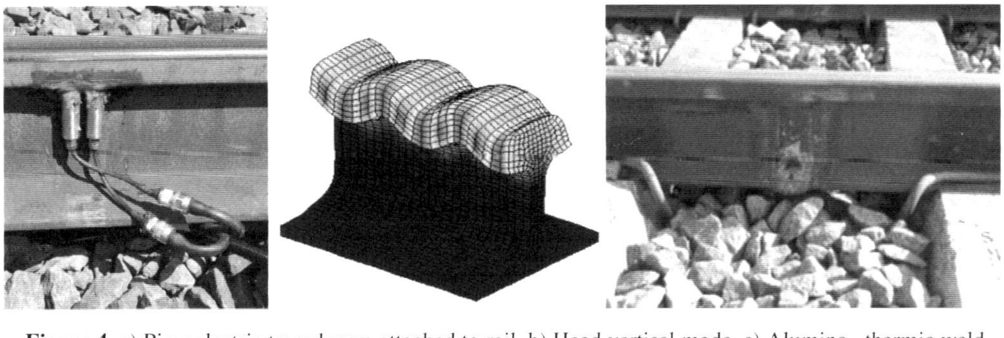

Figure 4. a) Piezoelectric transducers attached to rail, b) Head vertical mode, c) Alumino - thermic weld.

The transducers are driven in pulse – echo mode where a short-duration large voltage signal (800 Vpp) is applied and the response of the transducer is measured. Chirp signals have been used but for the results presented here tone bursts were used. These signals comprise 17.5 cycles of a constant frequency signal with a Hanning window applied. Different frequencies can be selected to examine different parts of the frequency spectrum. An example excitation signal and response signal are shown in figure 5. The response signal shows reflections of the guided waves from alumino – thermic welds that are used to join the rail during installation. One such weld is shown in figure 4c.

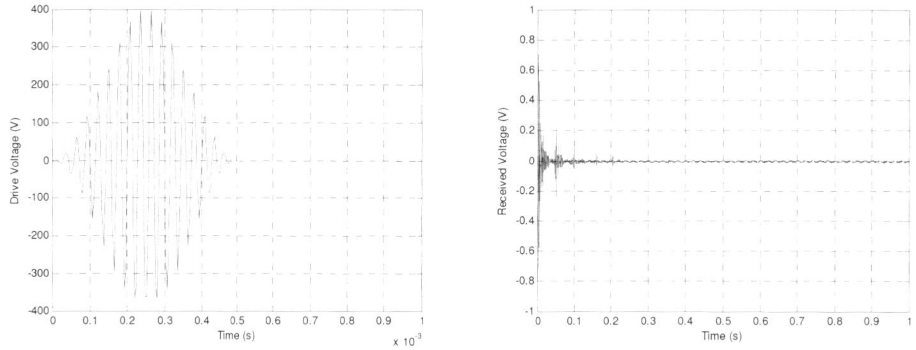

Figure 5. a) Tone burst excitation signal. b) Typical response signal.

When a transducer is used to excite guided waves a number of different guided wave modes will be excited. While the transducer used was optimised to excite a particular mode of propagation numerous other modes will also be excited. Some of these modes are near-field modes that decay exponentially over a short distance while others are the propagating modes whose dispersion characteristics were shown in the previous section. The majority of these propagating modes however do not propagate over large distances due to attenuation. The rail is supported on polymer pads on concrete sleepers on ballast rock and modes with energy in the foot of the rail are quickly damped.

When a mode of propagation reaches a discontinuity such as an alumino – thermic weld scattering occurs. The incident mode is reflected as the incident mode but also couples to reflect as other modes. Transmission past the discontinuity contains the incident mode and other coupled modes. First we consider a simplified problem where only two modes of propagation exit with mode A having greater group velocity than mode B. If we excite mode A and B which propagate to a weld and couple before returning to the transducer we expect to receive a signal corresponding to forward propagation of mode A and backward propagation of mode A (AA), forward propagation of mode A and backward propagation of mode B (AB), forward propagation of mode B and backward propagation of mode A (BA) and forward propagation of mode B and backward propagation of mode B (BB). The wave packet AA would arrive first, followed by AB and BA which would arrive together and finally BB. The problem becomes more complex when more than two modes are present, when dispersion occurs and when there are multiple reflectors.

One way to analyse a complex received signal is to use the short time Fourier Transform (STFT) as shown in [9]. This can be particularly effective when propagation over large distances is measured as is the case in this paper. The signal to be analysed was obtained by exciting the transducer with tone burst signals centred at 35, 40 and 50 kHz, capturing and averaging the received signals and summing these together. Figure 6 shows the spectrogram of this combined signal. The rail contained alumino – thermic welds at distances of approximately 82, 158, 254, 322 and 494 m from the transducer. Predicted times of arrival for energy propagating in selected mode combinations have been plotted on the spectrogram. These predictions are based on the SAFE computed group velocities and estimated distances to the welds. The material properties of the rail were adjusted to obtain good agreement with the measured data. A Young's modulus of 210 GPa, a Poisson's ratio of 0.3 and a density of 7800 kg/m^3 were used.

Mode s_3 (vertical head mode) was found to be propagated to the welds, reflected by the welds and propagated back to the transducer relatively strongly. The most vertical lines (dot dash lines) plotted on the spectrogram are for this mode propagating twice the distance to the weld. We can see this mode at a time of approximately 0.31s, which corresponds to a propagation distance of nearly 1000m to and from the weld at 494m. In this mode the higher frequencies propagate at slightly larger speed and the dispersion is relatively small.

The mode combination of s_3 (vertical head mode) and s_7 (horizontal head mode at lower frequencies) is plotted (dashed lines). This would include the s_3 mode propagating to the weld, mode conversion to the s_7 mode at the weld, and the s_7 mode propagating back to the transducer and also propagation of the s_7 mode to the weld, mode conversion to the s_3 mode at the weld, and propagation of the s_3 mode back to the transducer. This curve shows considerable dispersion with the energy at 38 kHz arriving last, which corresponds to the lowest group velocity of the s_7 mode. Even though the dispersion is large we can clearly see this mode combination reflecting from the weld at 322 m. This mode combination is visible up to a frequency of around 43 kHz.

The third set of time of arrival curves (dotted lines) are for the mode combination of the s_3 mode and the s_6 mode (which starts off as a mode containing foot motion and then becomes a head horizontal mode above 45 kHz). We can see this mode combination in the spectrogram at frequencies above 46 kHz. It makes sense that only the combinations including the mode shapes that have little foot motion are observed as these combinations would be expected to have low attenuation due to damping of the rail foot by the polymer pads and sleepers. Unfortunately these measurements were conducted for other purposes than observing the mode swopping and therefore an excitation signal centred at 45 kHz was not used in these measurements.

A second set of measurements were performed some 18 months after the first set. These measurements were performed on the same rail section as the first set but the location of the transducers differed by approximately 380m. In the second set of measurements two transducers were attached under the head of the rail but this time with one on the inside and one on the outside of the rail. In this way it was expected that adding the signals received from the two transducers would enhance the contribution of the symmetric modes and cancel the contribution of the antisymmetric modes in the response. The spectrogram shown in figure 7 was obtained by summing responses due to excitations with tone burst signals centred at 35, 37, 39, 41 and 45 kHz.

Once again we see the same mode combinations as before and the corresponding SAFE predicted arrival times are plotted for a new set of weld distances. In general the attenuation appeared to be larger in this set of measurements and it was noticed that the rail was sitting deeper in the ballast on this occasion. The spectrogram was obtained by summing the signals obtained from the two transducers to enhance the symmetric modes. While later reflections do not show evidence of the mode swopping phenomenon it does appear to be present in the second reflection at a time of 0.13s and a frequency of 45 kHz. Rail maintenance is performed periodically and one operation, known as tamping, involves lifting the rail and sleepers and repacking the ballast underneath. It is recommended that this measurement should be repeated after this maintenance has been performed. The transducers used were designed to operate at 35 kHz and therefore transmit less energy at higher frequencies and also are less sensitive at higher frequencies. Ideally, the transducer design should be modified to perform measurements at 45 kHz.

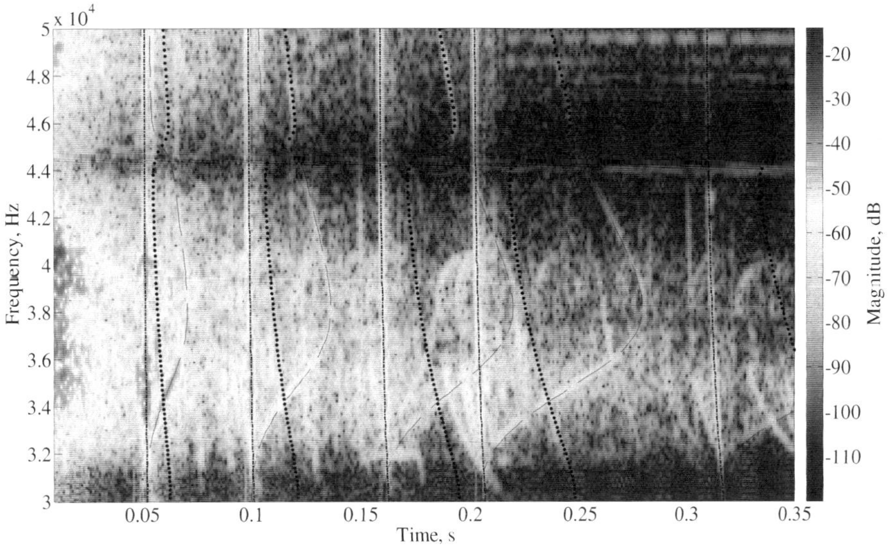

Figure 6. Spectrogram of first set of measurements performed using one transducer.

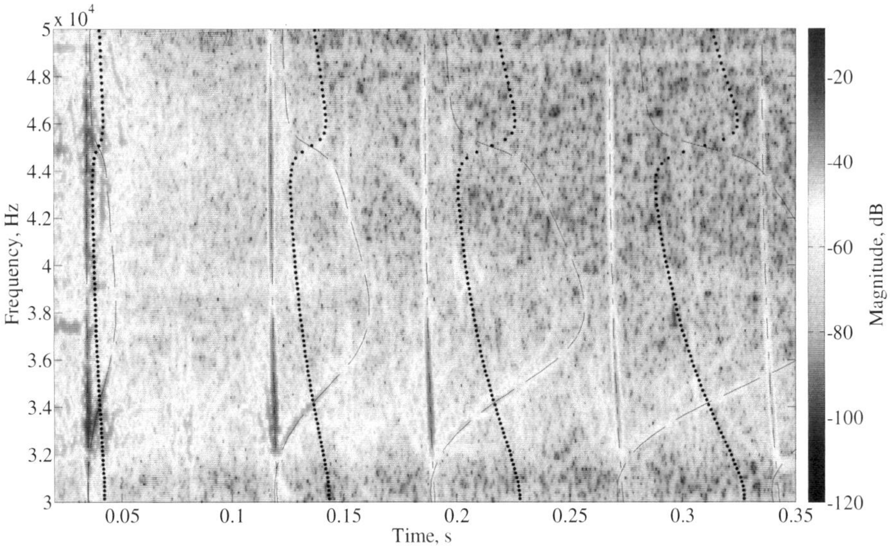

Figure 7. Spectrogram of second set of measurements performed using one transducer for excitation and two transducers for receiving.

CONCLUSIONS AND RECOMMENDATIONS

The phenomenon of mode repulsion was investigated numerically and experimentally. The SAFE computed dispersion curves were divided into a family of symmetric modes and a family of antisymmetric modes. It was shown that while wavenumber – frequency curves for symmetric modes can cross those for antisymmetric curves the curves for each family do not cross. This makes mode numbering and mode tracking simple. The curves within a family can approach each other and then they appear to repel each other. This behaviour is seen in the group velocity – frequency curves as the curves crossing suddenly. The sharpness of this change in group velocity can vary. Investigation of the mode shapes below and above the frequency of the repulsion region showed that the mode shapes effectively swop between the two curves. This is analogous to the behaviour for Lamb waves in plates.

Pulse – echo measurements of the reflection from welds were performed on an operational heavy haul line. Time frequency analysis of the received signals was used to produce spectrograms and the SAFE predicted times of arrival for selected mode combinations were plotted. These predictions showed remarkable agreement with the measurements and aid greatly in interpreting the spectrograms. It is believed that mode repulsion has been observed and that one mode combination was visible up to the repulsion region and the second mode combination was seen after the repulsion point. The experimental results could have been clearer and it is recommended that the measurements should be repeated after rail maintenance has been performed.

A procedure could be developed to automatically identify the rail material properties that produce the best agreement of the predicted arrival times with the measured arrival times. This inverse problem could be extended to include estimating the amount of rail grinding that has been performed by altering the geometry of the SAFE model. It may even be possible to estimate the axial load in the rail and to detect when rail tamping is required. Having an accurate model of each section of rail can be of use in the model-based signal processing that will be used for defect detection in future systems.

ACKNOWLEDGEMENTS

Access to the railway track for field measurements was provided by Transnet Freight Rail and is gratefully acknowledged. Funding for this project was provided by the CSIR, the Department of Science and Technology and the National Research Foundation of South Africa (Grant No's: 78858 & 85330).

REFERENCES

[1] P. W. Loveday, "Development of piezoelectric transducers for a railway integrity monitoring system," in *Proceedings of SPIE, Smart Systems for Bridges, Structures and Highways*, 2000, vol. 3988, pp. 330–338.

[2] P. Loveday, D. Ramatlo, and F. Burger, "Monitoring of Rail Track Using Guided Wave Ultrasound," in *19th World Conference on Non Destructive Testing*, 2016, pp. 1–8.

[3] H. Überall, "Repulsion of phase-velocity dispersion curves and the nature of plate vibrations," *J. Acoust. Soc. Am.*, vol. 96, no. 2, p. 908, 1994.

[4] L. Gavric, "Computation of propagative waves in free rail using a finite element technique," *J. Sound Vib.*, vol. 185, no. 3, pp. 531–543, Aug. 1995.

[5] T. Hayashi, W.-J. Song, and J. L. Rose, "Guided wave dispersion curves for a bar with an arbitrary cross-section, a rod and rail example," *Ultrasonics*, vol. 41, no. 3, pp. 175–183, May 2003.

[6] I. Bartoli, A. Marzani, F. Lanza di Scalea, and E. Viola, "Modeling wave propagation in damped waveguides of arbitrary cross-section," *J. Sound Vib.*, vol. 295, no. 3–5, pp. 685–707, Aug. 2006.

[7] V. Damljanović and R. L. Weaver, "Propagating and evanescent elastic waves in cylindrical waveguides of arbitrary cross section," *J. Acoust. Soc. Am.*, vol. 115, no. 4, p. 1572, 2004.

[8] P. W. Loveday, "Semi-analytical finite element analysis of elastic waveguides subjected to axial loads.," *Ultrasonics*, vol. 49, no. 3, pp. 298–300, Mar. 2009.

[9] M. Legg, M. K. Yücel, V. Kappatos, C. Selcuk, and T.-H. Gan, "Increased range of ultrasonic guided wave testing of overhead transmission line cables using dispersion compensation.," *Ultrasonics*, vol. 62, pp. 35–45, Sep. 2015.

MODELLING MOORING LINE BEHAVIOUR TO DETERMINE THE IMPACT FORCE ON SQUID EGG BEDS

Vutlhari A Maluleke*[1], Graeme J Oliver [2], Michael J. Roberts [3]

Organization(s): 1: Cape Peninsula University of Technology, South Africa; 2: Cape Peninsula University of Technology, South Africa; 3: Nelson Mandela Metropolitan University, South Africa

[1] vutlharimabasa@gmail.com
[2] oliverg@cput.ac.za
[3] mike.roberts@nmmu.ac.za

ABSTRACT

In South Africa, squid fishing vessels need to find and then anchor above benthic squid egg beds to effect viable catches. However, waves acting on the vessel produce a dynamic response on the anchor line. These oscillatory motions produce impact forces of the chain striking the seabed. It is hypothesised that this causes damage to the squid egg bed beneath the vessels. Different mooring systems may cause more or less damage and this is what is investigated in this research. The commercial package ANSYS AQWA is used in the study to model the incoming wave and moored vessel motion and forces. The dynamic motion of the mooring line and its impact on the seabed is investigated beyond what is output from ANSYS AQWA using ABAQUS finite element analysis. Initial results for the motion of mooring line shows that there is contact force between the seabed and the mooring line. The impact pressure of the anchor chain is obtained by solving a contact problem in ABAQUS. The dynamic structural solution in ABAQUS is obtained utilising the tension on the cable from the vessel motion for a particular mooring and wave excitation. The impact pressure is found in contact so that we can determine what kind of damage would be done to the seabed with various mooring options under different ocean conditions.

KEYWORDS

Mooring line, seabed, contact pressure, ocean waves, vessel.

1. INTRODUCTION

The South African chokka squid fishery is based in the Eastern Cape between Plettenberg Bay and Port Alfred, and is a major source of foreign revenue as the entire catch, on average some 8000 t, is exported to Europe. The fleet comprises 138 vessels ranging between 11 and 20 m in length, and cable of 3 weeks at sea (Figure 1a). Each vessel carries about 22 fishermen who land the squid using hand-held jigs on fishing line. This number excludes the number of crew who are not allowed to fish. Commercial squid fishing is only viable when the vessels are above spawning aggregations — formed in the water column above egg beds on the seafloor (Figure 1b). Egg beds comprise hundreds of thousands of translucent, slim and slimy egg capsules about 15 cm in length that are glued to the bottom substrate forming massive mats often spanning tens of meters. Hatching occurs about 3 weeks from spawning. Traditionally, the fishing vessels position themselves above an egg bed using a single mooring system with the anchor dropped upwind of the egg bed. A significant part of the chain lies on the seafloor over eggs.

Waves acting on the vessel set up dynamic behaviour in the mooring line which rapidly lifts the chain off the seabed, dropping it back with considerable force on the bottom (Sarkar & Taylor, 2001), and possibly damaging squid eggs. As sea and wind conditions change daily, vessels regularly pick up anchor and relay the anchor chain. In 2010, a new 'double anchor system' was introduced and used by about 10 vessels. This 'V' shaped anchor line configuration offers vessels greater position control over the egg beds but potentially doubles the impact of the chain on the eggs. In 2013, the chokka squid fishery crashed and has not recovered. Concern has been raised by both fisheries managers and boat owners that the chain impact — especially from the double anchor system — maybe causing excessive damage to the squid eggs reducing recruitment.

Figure 1. (a) A chokka squid fishing vessel anchored (b) above an active egg bed. These egg masses can extend over an area as large as 10 000 m^2

This study is the first part of an investigation into the impact of anchor chains on the seabed; it focuses on the mechanical chain system under various wave height scenarios. The second part of the investigation to be done after these results are available will then study the damage and consequences of the chain impact on the egg beds and hatching success. The numerical investigation of the behaviour of the mooring line and seabed interaction is performed here using *ANSYS AQWA* software to obtain structural motion and forces in the time domain, and by the *ABAQUS* finite element model to determine the impact force on the seabed.

2. MOORING SYSTEM DESCRIPTION

Figure 2 presents a schematic of a single mooring system which is made up of three components — the anchor, chain, and vessel (Vineesh et al., 2014).

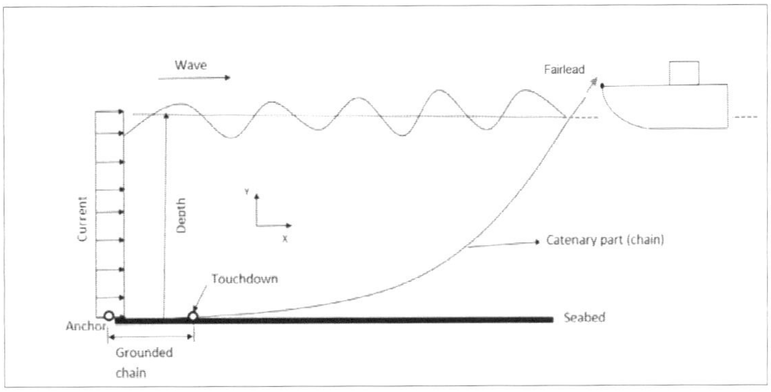

Figure 2. Mooring system

The Klusman 100–150 kg anchor is mostly used by the chokka fishery with about 100 m of 20 mm link diameter steel chain. Chain weight, therefore, ranges between 9 and 10.8 kg/m. The chain controlled by a winch on the vessel's foredeck that feeds chain through the fairlead on the vessel's bow. There are two types of chain links namely studless and studded. The studded chain link is designed to prevent knot formation but is more susceptible to fatigue failure than the studless link (ABC Moorings, n.d.). In mechanics, the chain component is characterised by strong catenary stiffness (effect), low elasticity, high breaking strength and is relatively heavy. As shown, the mooring system is subjected to varying wind, waves and current, all of which introduce dynamic behaviour into the mooring line.

The part of a mooring line that lies on the seabed is termed *grounded chain* while that suspended in the water column the *catenary*. The *touchdown point* is a position along the mooring where the chain begins lifting off the seabed. This point varies and is based on the sea conditions. Pellegrino and Ong, (2003) demonstrated that when the mooring chain is excited due to wave action loading, the chain dynamically interacts with the seabed; which creates a boundary condition that varies in time and in space. This is because of the significant change in the chain's longitudinal profile resulting in a corresponding lifting off or touchdown action in the chain's

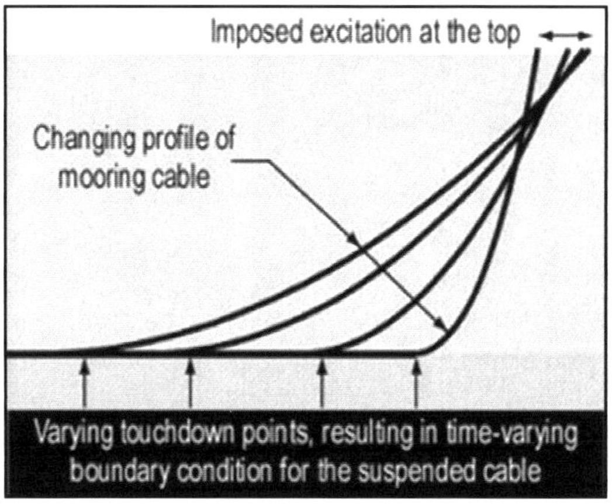

Figure 3. Mooring line touchdown points resulting in time-varying boundary conditions (http://www-civ.eng.cam.ac.uk/dsl/ppao2/)

touchdown position (Yu & Tan, 2006). This is illustrated in Figure 3 and can be modelled using a dynamic simulation which accounts for the application of loads on the system over time with consideration of wave inertia forces and structural damping.

3. MATHEMATICAL FORMULATION FOR ANALYSING THE MOORING SYSTEM

3.1. Structural dynamic equations

The coupled dynamic deterministic model involves the formulation of a non-linear stiffness matrix which reflects mooring line tension fluctuations due to variable buoyancy and other non-linearities. The model requires that the selection and solution of wave theory to reasonably represent the water particle kinematics to estimate drag and inertia for all the six degrees of freedom. The statically coupled problem is solved iteratively using the Newton-Raphson method. The time domain is numerically integrated to solve the equations of motion and to obtain the response time histories (Jameel et al., 2011).

The equation of motion describing the mooring system equilibrium between inertia, damping, restoring and exciting forces can be assembled as follows:

$$[\mathbf{M}]\{\ddot{\mathbf{x}}\} + [\mathbf{C}]\{\dot{\mathbf{x}}\} + [\mathbf{K}]\{\mathbf{x}\} = \{\mathbf{F}(\mathrm{t})\} \qquad \ldots(1)$$

Where,

$[\mathbf{M}]$= Total mass matrix (Structural mass and added mass)

$[\mathbf{C}]$= Damping matrix (simulated using Rayleigh Damping in Abaqus)

$[\mathbf{K}]$= Total Stiffness matrix (consist of Elastic and Geometric stiffness matrix)

$\{\mathbf{x}\}$= 6 DOF structural displacements

$\{\mathbf{F}(\mathrm{t})\}$= Excitation forces

The added mass of the structure occurs due to the water surrounding the entire structure. The overall damping to the system is being offered by structural and hydrodynamic damping. The major damping is induced due to the hydrodynamic effects. This structural damping is simulated by Rayleigh damping (Jameel et al., 2011). Hydrodynamic damping is dominant in case of oscillating slender surrounded by water.

3.2. Cable motion equations in ANSYS AQWA

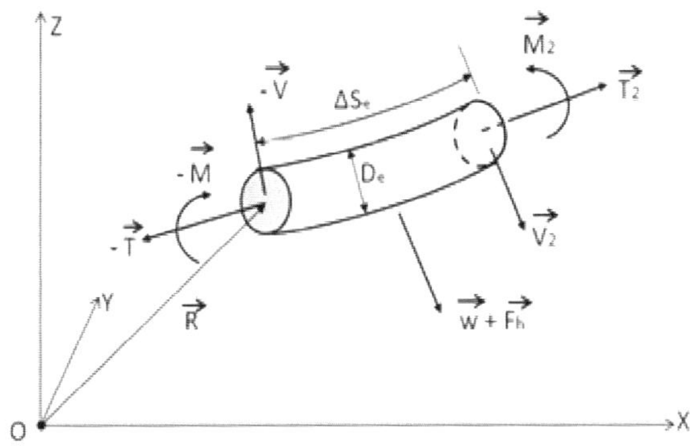

Figure 4. Forces on a Chain Element (Adapted from ANSYS Aqwa, 2015)

There are various forces that act on the mooring line; a single element of circular slender cable is shown in Figure 4, it shows the single element subjected to external hydrodynamic loadings and structural inertial loading (ANSYS Aqwa, 2015):

The motion equation of this cable element is

$$\frac{\partial \vec{T}}{\partial S_e} + \frac{\partial \vec{V}}{\partial S_e} + \vec{w} + \vec{F}_h = m\frac{\partial^2 \vec{R}}{\partial t^2} \qquad \ldots (2)$$

$$\frac{\partial \vec{M}}{\partial S_e} + \frac{\partial \vec{R}}{\partial S_e} \times \vec{V} = -\vec{q}$$

$\boldsymbol{m}$ – is the structural mass per unit length

$\vec{q}$ – is the distributed moment loading per unit length

$\vec{R}$ – is the position vector of the first node of the cable element

ΔS_e and ΔD_e – are the length and diameter of the element respectively

$\vec{w}$ and $\vec{F}_h$ – are the element weight and external hydrodynamic loading vectors per unit length respectively

$\vec{T}$ – is the tension force vector at the first node of the element

$\vec{M}$ – is the bending moment vector at the first node of the element

$\vec{V}$ – is the shear force vector at the first node of the element

The dynamic response of the cable with bending governed by Equation 2 is solved numerically by employing the discrete Lump-Mass model in ANSYS AQWA. The entire mooring line is treated as a slender structure where nonlinear partial differential equations are used to compute the cable dynamics (Yu & Tan, 2006).

4. MOORING LINE SIMULATION METHODOLOGY

In this study, the mooring line system is first modelled in *ANSYS AQWA* which models the incoming wave and moored vessel motion and forces in the time domain. The dynamic motion of the mooring line and its impact on the seabed is investigated beyond what is output from *ANSYS AQWA* using *ABAQUS* finite element analysis software. A two-dimensional finite element model is created in *ABAQUS* for the numerical analysis of the mooring line and seabed interaction. The model built in *ANSYS AQWA* accounts for the different directions of motion of the moored structure excited by different wave amplitudes. Figure 5 below shows the *ANSYS AQWA* model for the mooring system.

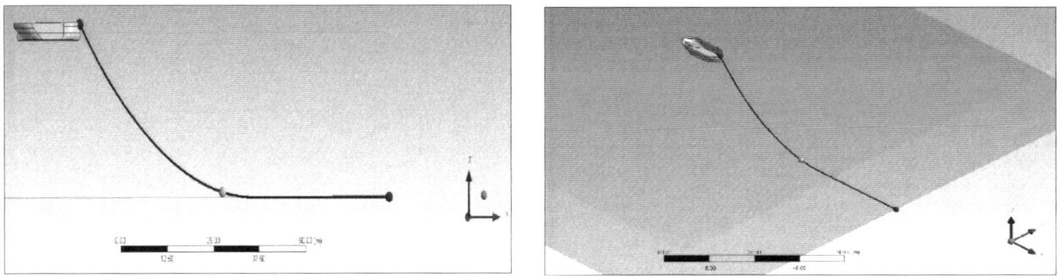

(a) (b)

Figure 5. *ANSYS AQWA* model of the mooring system in 2D (a) and 3D model (b)

Figure 5(a) shows the mooring line system attached to a vessel. The clump weight of an insignificant mass is shown by the round yellow object in the figure above for tracking the actual chain displacement to be used for *ABAQUS* simulation. The vessel and the mooring line are both under the influence of wave forces. The mooring line is shown to be in contact with the seabed and has a catenary profile. Table 1 shows the dimensions of the vessel, the point mass which is the mass inertia matrix that is defined via the Radius of Gyration of the vessel. The K_{xx} of the vessel is defined by K_{xx}=0.34*width of the vessel, K_{yy} = 0.25*Length, and K_{zz} = 0.26*Length. The mass of the vessel is 127 364 kg.

Table 1. Vessel data

Vessel Details		
Parameter	Value	Units
Vessel dimensions	17 (Length) x 6 (width) x 3 (depth)	m
K_{xx}	2.04	m
K_{yy}	4.25	m
K_{zz}	4.42	m
Mass	127 364	kg

Table 2 below shows the mooring line data. A studless chain link was used in this study. The test data obtained showed the mooring line weight per unit length to be 10 kg/m. The maximum allowable tension of the mooring is 36 kN. The mooring chain is solved as a catenary section with a circular diameter of 0.02 m and a cross-sectional area of 0.00031 m^2. The stiffness of the chain is defined by E which is the elastic modulus of the structure and the cross-sectional area. The Added Mass Coefficient and Longitudinal Drag Coefficient are the program's default values. The data in Table 1 and 2 is entered manually by either calculations or is obtained from literature who did similar work in similar conditions.

Table 2. Mooring line data

Mooring Line Details		
Parameter	**Value**	**Units**
Mooring Line Length	95	m
Mass/Unit Length	10	kg/m
Equivalent Cross-Sectional Area	0.006	m^2
Stiffness, EA	7 696 920.01	N
Maximum Tension	36 000	N
Added Mass Coefficient	1	-
Transverse Drag Coefficient	1	-
Equivalent Diameter	0.09	m
Longitudinal Drag Coefficient	0.025	-

The above data is required to fully define the *ANSYS AQWA* model and is solved using a "Time Response" approach which calculates the dynamic forces and motion of the structure in time.

5. NUMERICAL SIMULATION

The nature of the problem in this investigation suggests the use of coupled dynamic approach where the interaction of the mooring line and the vessel motion is modelled directly. *ANSYS AQWA* Time Response analysis offers this option.

5.1. Numerical simulation results in ANSYS AQWA

When a 1 m wave amplitude is applied for a period of 4.6 s for a duration of 600 s, the dynamic simulation results in Figure 6 are obtained. The wave condition described on the latter best describes the actual chain motion in water.

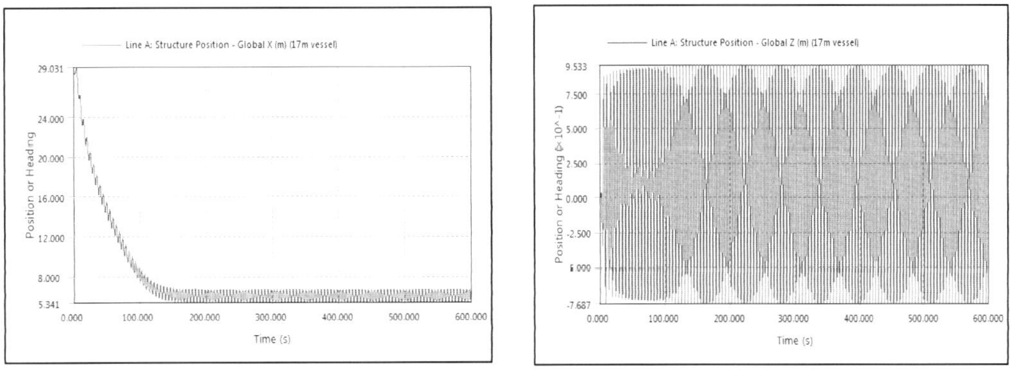

(a) (b)

Figure 6. Horizontal motion of the vessel (a) and vertical motion of the vessel (b)

As can be seen from Figure 6, the vessel's position is very unstable at the beginning of the simulation, the vessel starts to develop a more stable pattern after about 150 s. The vessel initial position is 29 m from its origin. The origin of the vessel is on it's on its stern while the mooring line is attached to the bow. The vessel then gradually moves closer to the origin where the mooring line takes the catenary profile which is typical for moored vessels. The catenary profile shows that the vessel is under tension because to the action of wave forces acting on it. A more stable pattern is developed after 150 seconds and the vessel oscillates between the range of 5 and 6 m.

Whereas the heave (vertical) motion of the vessel oscillates between a range of about 0.95 m and -0.77 m. The vertical oscillation of the vessel is directly proportional to 1 m wave excitation. Figure 7 below show the chain motion tracked on the clump weight. This point was tracked in the x and y directions to obtain results shown in figure 6 below and the data is used in the *ABAQUS* model to excite the mooring line.

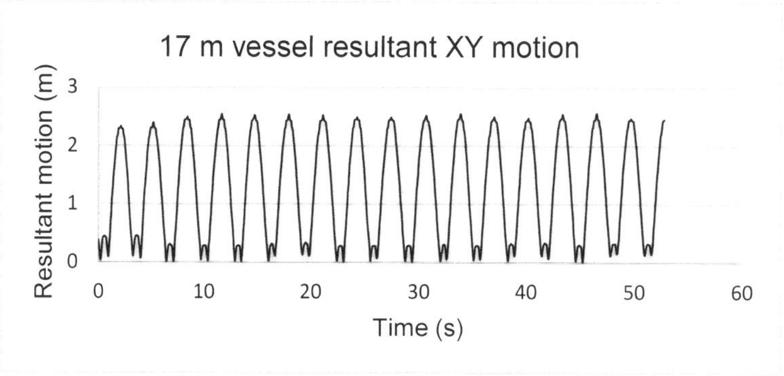

Figure 7. Chain motion in x and y directions

Table 3 below shows the ocean environment condition selected for the analysis. Linear wave (Airy wave) was selected for regular ocean waves. This is considered as the simplest ocean wave and is based on the assumption of homogeneous, incompressible, inviscid fluid and irrotational flow.

Table 3. Ocean environment data

Ocean Environment Properties		
Parameter	Value	Units
Water Density	1025	kg/m^3
Ocean Depth	30	m
Wave Type	Airy wave theory/ Regular wave	-

Figure 8 below shows how pre-tension varies along the length of the chain. As expected, the ground part of the chain is assumed to have zero tension. This graph shows pre-tension force decrease along the line as the depth of the sea increases. The touchdown point is at about 60 m away from the anchor point.

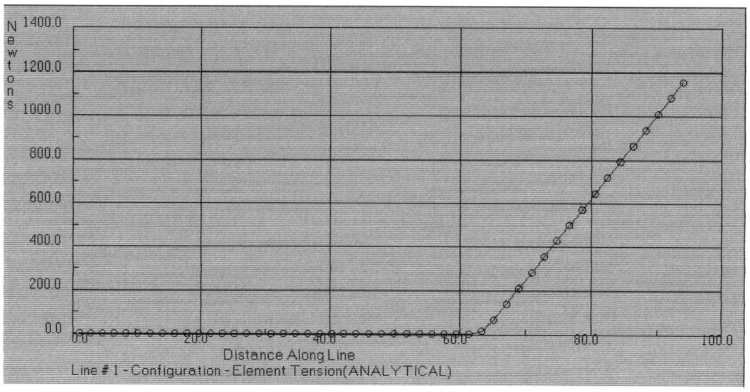

Figure 8. Pre-tension Force along the mooring line

The time response of the chain mooring line is shown in Figure 9 below which shows the time history of the tension force acting on the mooring line. The maximum tension force experienced by the mooring line is 22 kN and the minimum tension force is 650 N. The tension force is zero when the chain is slack.

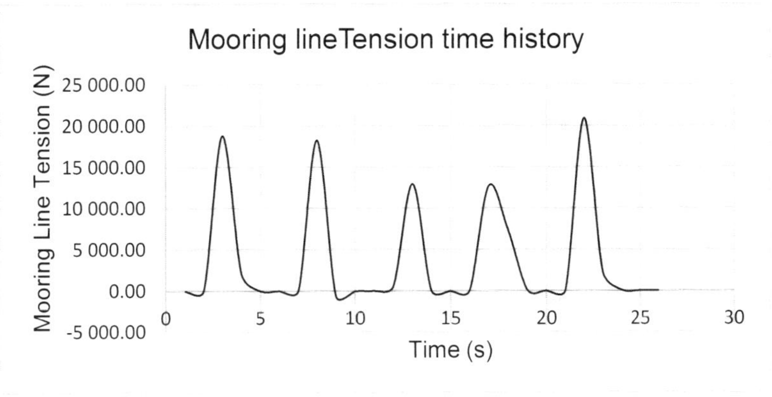

Figure 9: Tension Force along the mooring line

The mooring line force has been found to be of vital importance in this study. It has been found that the mooring line force increases as the wave amplitude increases. The horizontal forces on the vessel are greater than the vertical forces. This correlates with the purpose of mooring lines which is to limit the horizontal motion of a vessel. This observation was a result of applying a wave amplitude of 0.5 to 1.5 m. Wave height by definition is two times the wave amplitude. In general, the waves of 1 m to 3 m were evaluated in this study to be able to see the trend of the behaviour of the vessel at sea and the corresponding mooring forces. Figure 10 shows the mooring line forces when 1 m wave amplitude excites the vessel. The maximum mooring line force in the horizontal direction is 15 692.83 N while the minimum is -452.34 N; whereas the maximum vertical mooring line force is 389.75 N while the minimum force -17 222.05 N. Table 4 shows the summary of the results obtained from *ANSYS AQWA* when the wave amplitude was varied.

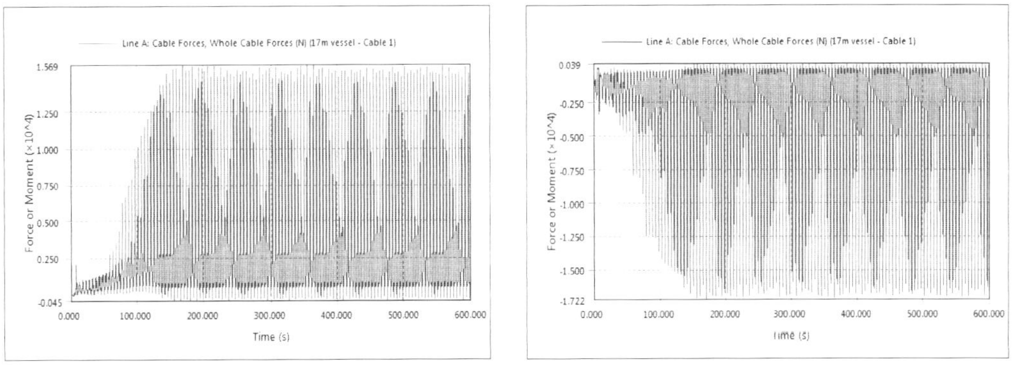

(a) (b)

Figure 10: Mooring line horizontal forces (a) and vertical forces (b) with 1 m wave excitation

Table 4. Mooring line forces with varying wave amplitude

	x	x	y	y
Wave amplitude (m)	**Max (N)**	**Min (N)**	**Max (N)**	**Min (N)**
0.5	902.64	-20.40	-220.78	-3 683.13
1	15 692.83	-452.34	389.75	-17 222.05
1.5	33 041.95	-1 278.40	807.90	-26 046.50

5.2. Numerical simulation model in ABAQUS

ABAQUS has a nonlinear option to solve dynamic problems. The model shows a section of the chain that was used for tracking using Tracker software; the model only analyses the length of the mooring up to the reference point which was used for tracking. The height of the reference point used for tracking is 15 . The finite element model shown in figure 11 has been built for analysing the mooring line and seabed interaction. The properties of the data used to build the model are shown in Table 2. The wave amplitude for the simulation is 1 m and wave period is 4.6 s. The computation lasts for 600 s.

Hybrid beam elements with quadratic formulation were used to model the chain mooring because of their ability to handle very slender non-linear geometries, where the axial stiffness of the beam is very large compared to the bending stiffness (Abaqus Theory manual, 2012). This formulation has better convergence and is a good approximation of the chain mooring line. Dongsheng Qiao (2014) also used hybrid beam elements to simulate the mooring line (Qiao et al., 2014).

Dashpots were used to account for mooring line damping with a coefficient of 0.3. The model is solved using ABAQUS/Standard dynamic implicit analysis taking geometric non-linearities into account. Displacement/Rotation boundary condition was used at the mooring line fairlead point and the displacement was applied as tabular amplitudes.

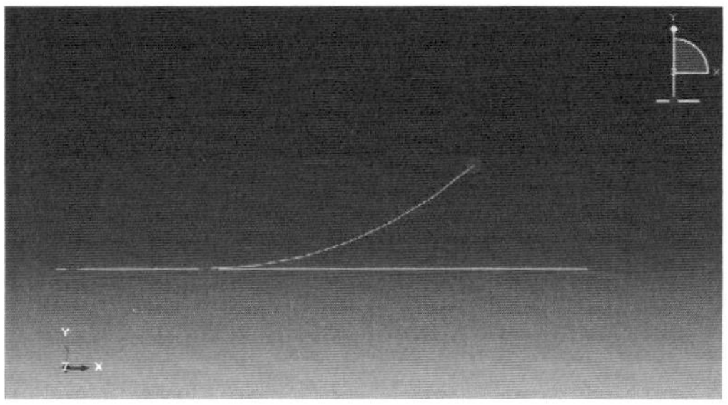

(a)

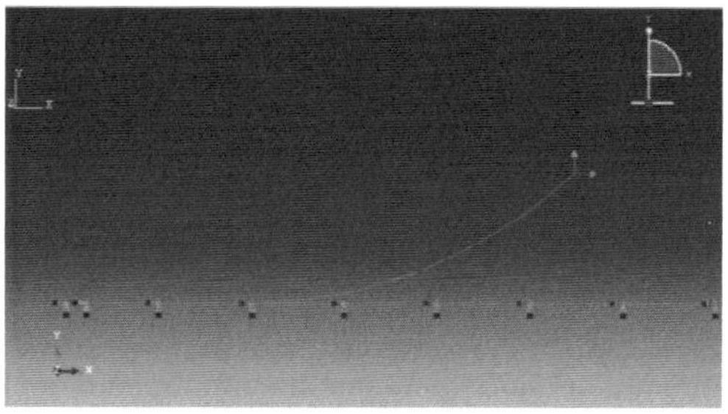

(b)

Figure 11. *ABAQUS* model with dashpots (a) and with loads and boundary conditions (b)

5.3. Contact problem

The contact between the mooring line and seabed is an important consideration in both theory and practical applications. When contact is analysed in the ABAQUS model in the contact zone, the final state of offsets and tractions depends on the entire loading history. The problem has to be solved by applying the load incrementally. The contact algorithms in ABAQUS/Standard are designed to handle a large class of contact problems for which small-sliding contact formulation and master/slave algorithm can be used. The seabed is modelled as a master surface and the mooring line as a slave surface. The seabed is modelled as a fixed boundary across its entire length. The formulation assumes that the surfaces may undergo arbitrarily large rotations but that a slave node will interact with the same local area of the master surface throughout the analysis. A tangential behaviour option with friction formulation was used to describe the kind of contact between the mooring line and seabed. The friction coefficient of 0.74 recommended by Taylor and Valent which describes the roughness of seabed sand was used (Taylor & Valent, 1984).

5.4. Mooring line Tracker data analysis in ABAQUS

This section presents results obtained by using a video motion analysis software Tracker for tracking the mooring line displacement under wave influence over time. The output from Tracker is shown in the Figure 12 below. These results are used as input for the ABAQUS model shown in Figure 13. The video analysis data showed the mooring line vertical displacement to be the most dominant compared to the horizontal motion. Furthermore, it was also noticed that part of the mooring line reached a maximum of about 2.5 m vertical displacement under wave excitation; the chain then dropped back down to the seabed and lies for a certain amount of time and then lift up again. This behaviour is what is modelled in ABAQUS in order to determine the contact force with the seabed.

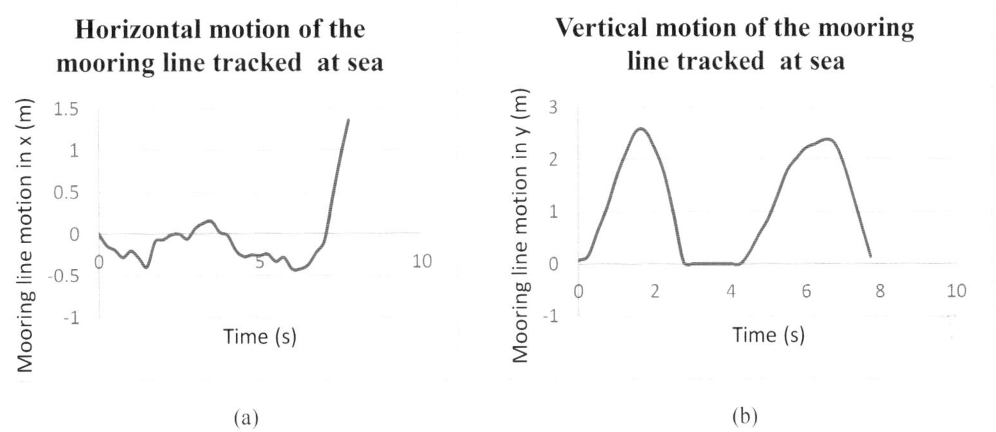

Figure 12. Horizontal motion of the mooring line (a) and vertical motion of the mooring line (b) tracked at sea

Figure 13 below shows the horizontal and vertical displacement of the mooring line when excited by the vertical motion obtained from Tracker. The vertical displacement reaches a maximum of 7.32 m with the corresponding horizontal displacement of 1.5 m.

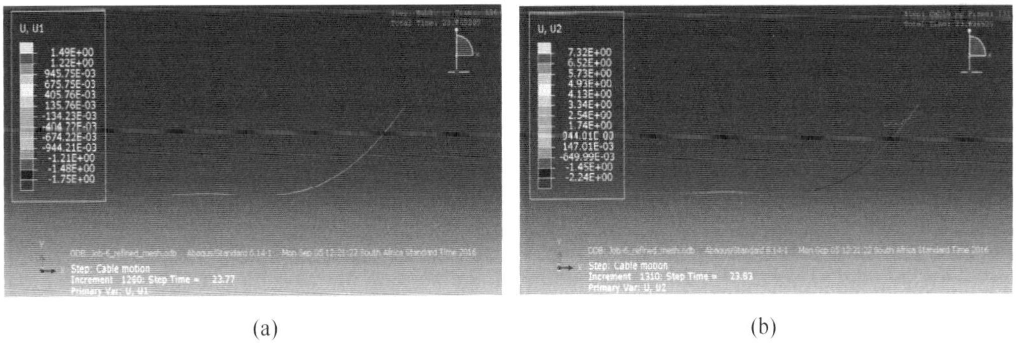

Figure 13. Horizontal (a) and vertical (b) mooring line displacement in *ABAQUS*

Figure 14 shows the maximum and minimum contact pressure on the seabed to be 68.30 kPa and 9.45 kPa respecticely. These result show that contact pressure on the seabed varies with time. The contact area of a single chain link varies depending on its orientation when in contact. When applying the pressure formula as follows:

$$Pressure = \frac{Force}{Area}$$

The project contact area of a single chain link is 0.001 m^2 and the calculated surface area is $0.006m^2$. This means that the contact force on the seabed will vary depending on the single chain link orientation in contact.

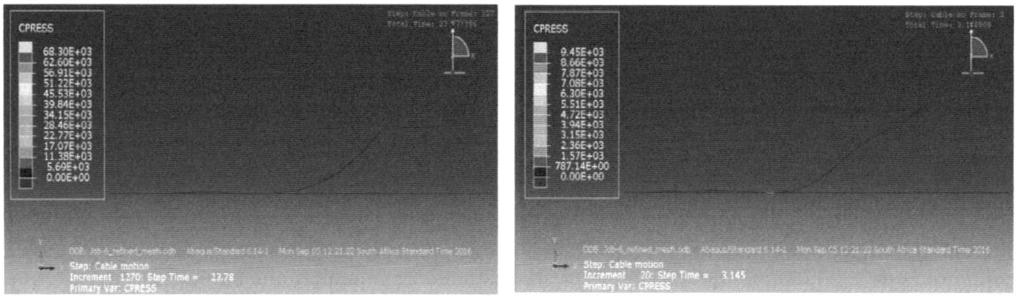

(a) (b)

Figure 14. Maximum (a) and minimum (b) seabed contact pressure

The *ABAQUS* model velocity showed to be in alignment with the measured velocity obtained Tracker. This shows that the *ABAQUS* model properly simulates the effect of the vertical motion of the mooring line contact with the seabed.

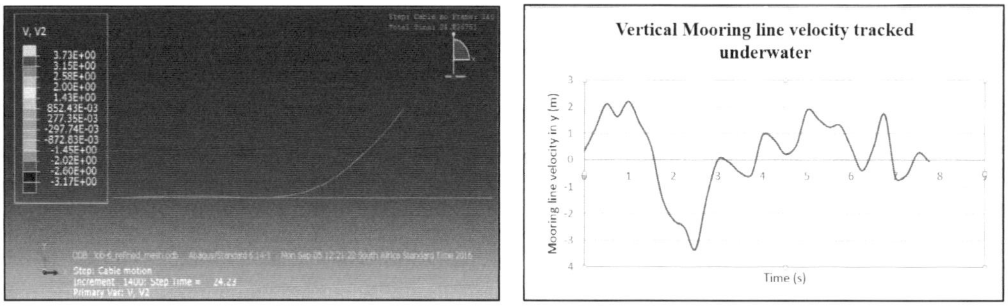

(a) (b)

Figure 15. Mooring line vertical velocity (a) and vertical velocity tracked at sea (b)

5.5. Mooring line ANSYS AQWA data analysis in ABAQUS

Figure 16 below shows the data obtained from *ANSYS AQWA* model. This data helps on further analysis of the effects of the hydrodynamic environment on the boat and the mooring line. The tracked data of the mooring line at sea was used to calibrate the *ANSYS AQWA* model in order to evaluate other possible wave environment conditions effect on the mooring line behaviour. The *ANSYS AQWA* model mooring line displacements are in agreement with the tracked motion; the difference is the vertical motion which peaks to 3.5 m in *ANSYS AQWA*. In Figure 16 shows the mooring line vertical displacement is dominant than the horizontal motion. This aligns well with the observation made from the data obtained by tracking the mooring line at sea as discussed in the previous section above. These results were obtained when a 1 m wave height with a period of 4.6 s for a duration of 600 s.

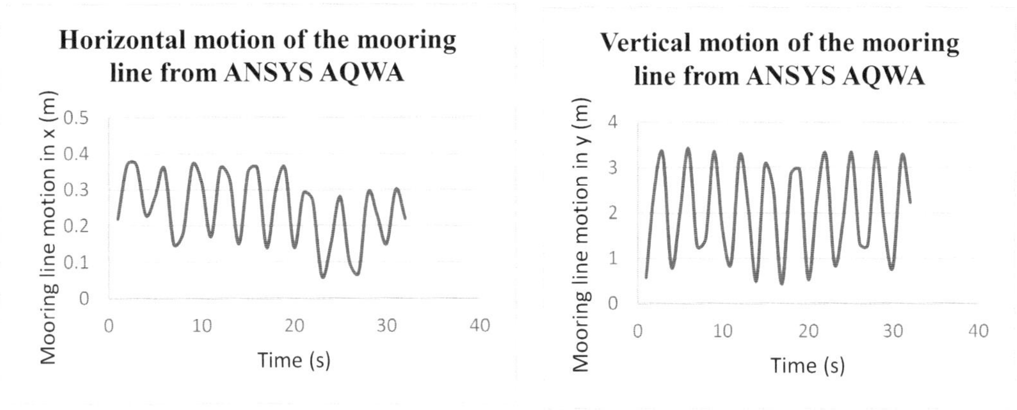

(a) (b)

Figure 16. Horizontal (a) and vertical motion (b) of the mooring line tracked using a clump weight marker

When the vertical excitation of the mooring line from *ANSYS AQWA* analysis is used to determine the contact pressure on the seabed. The contact pressure in Figure 17 is obtained showing the maximum and the minimum contact pressure of 138.46 kPa and 3.64 kPa.

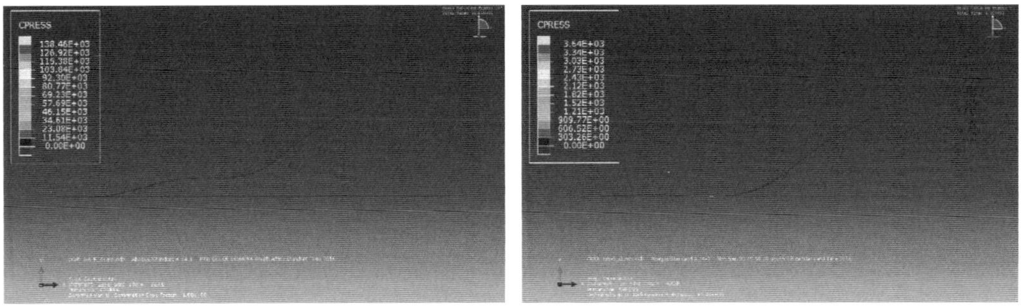

(a) (b)

Figure 17. Maximum (a) and minimum (b) seabed contact pressure obtained using *ANSYS AQWA* data

6. CONCLUSION

It is hypothesised that squid fishing vessels off the South African coast, destroy benthic squid egg beds as a result of mooring line movements. Given the large number of vessels in the fishery, this could influence recruitment and hence the squid stock biomass (and ultimately catches). The objective of this work has been to analyse the mooring line interaction with seabed under the influence of ocean waves. The motions and structural forces were first modelled in *ANYS AQWA* hydrodynamic software which accounts for the environmental loads. We have been able to develop a modelling system that allows us to study the mooring line contact with the seabed for different wave conditions as well as different chain types and single or two point moorings. The linked simulation procedures have enabled us to determine the effect of ocean conditions such as the wave amplitude on mooring line impact on the seabed. It is shown in this study that an increase in wave height leads to an increase in the forces acting on the mooring line in the horizontal direction. The mooring line motion showed to have greater vertical displacement than the horizontal displacement; whereas the horizontal forces of the mooring line were found to be significant and greater than the vertical forces. So the drag of the chain on the seabed is significant. The corresponding vertical mooring force also increases. The ABAQUS model has made it possible to quantify the amount of impact force. Frictional contact between the seabed and the mooring line has also been found to be a significant factor in this study. The tension along the mooring line decreases with the increase in depth. The mooring line forces significantly influences contact with seabed especially the horizontal component of the mooring force. Future work will focus on finding ways to develop seabed soil models for deformable analyses. This will include finding different seabed surfaces and compositions in order to determine the actual chain mooring line impact on the seabed. This could also relate to the position of the boat relative to the mooring line touchdown point to minimise tension as the boat rides a wave. This will also be looked at for a two-point mooring system.

REFERENCES

1 Abaqus Theory manual. 2012. Abaqus 6.1 2.

2 ABC Moorings. Mooring Lines. http://abc-moorings.weebly.com/mooring-lines.html 2 April 2015.

3 ANSYS Aqwa Theory Manual. 2015. Aqwa Theory Manual.

4 Chrolenko, M.O. 2013. Dynamic Analysis and Design of Mooring Lines.

5 Ebbesen, C. 2013. Analysis of Motions and Anchor Line Forces for Floating Production Units. Norwegian University of Science and Technology.

6 Jameel, M., Ahmad, S., Islam, A.S. & Jumaat, M.Z., 2013. Non-linear dynamic analysis of coupled spar platform. Journal of Civil Engineering and Management, 19(4), pp.476-491.

7 Pellegrino, S. & Ong, P.P.A. 2003. OMAE2003-37465. In Proceedings of OMAE'03 22nd International Conference on Offshore Mechanics and Arctic Engineering. 1–10.

8 Qiao, D., Yan, J. & Ou, J. 2014. Effects of mooring line with buoys system on the global responses of a semi-submersible platform. Brodogradnja, 65(1): 79–96.

9 Sarkar, A. & Taylor, R.E. 2001. Dynamics of Mooring Cables in Random Seas. Journal of Fluids and Structures, 16: 193–212. http://discovery.ucl.ac.uk/73002/.

10 Taylor R & Valent P, 1984. Design Guide for Drag Embedment Anchors, Naval Civil Engineering Laboratory (USA), TN No N-1688.

11 Vineesh, M. V, Sabu, N. V & Manju, P.M. 2014. Finite Element Analysis of Mooring Cable. International Journal of Engineering Research and Applications (IJERA) ISSN, (January): 13.

12 Yu, L. & Tan, J.H. 2006. Numerical investigation of seabed interaction in time domain analysis of mooring cables. Journal of Hydrodynamics, 18(4): 424–430.

Numerical modelling and validation of heat generation in haul truck tyres

J. Marais *, G. Venter ᶲ

Student, Supervisor

* 16105923@sun.ac.za, ᶲ gventer@sun.ac.za

ABSTRACT

The explosion of haul truck tyres due to internal heat generation during operation can lead to the loss of life and severe damage to the vehicle and its surroundings. In order to prevent this occurrence, a working knowledge of the heat generation in the tyre cross-section is required. This paper describes the development of a finite element model for the numerical modelling of a 23.5R25 earthmover tyre to evaluate the heat generation in the tyre under a combination of operating conditions using MSC.Marc/Mentat. These conditions include different inflation pressures, loads and rolling velocities. The purpose of these results is to determine the influence of the operating conditions on the heat generated in the tyre. The heat generated in the tyre is used to simulate the temperature distribution in a thermal analysis of the tyre. In order to verify the proposed model, the numerical results were compared to experimental data obtained from the tyre manufacturer. From the results obtained, it was concluded that the proposed model is viable and the results can be used to establish a relation between the tyre temperature and the stated operating conditions. In future work, this relation will be used to establish a safe operating envelope.

KEYWORDS

Finite element analysis, earthmover tyres, heat generation, temperature distribution

INTRODUCTION

Earthmover tyres are typically used for haul trucks in the mining industry. These trucks are designed to operate in severe environmental conditions while transporting extreme loads. In order to meet these requirements, earthmover tyres consist of large amounts of rubber. When rubber is subjected to cyclic deformation, such as the rolling of a tyre, the rubber tends to absorb more energy that it can dissipate. The lost energy is known as hysteresis loss and is stored in the rubber as heat. The heat stored in the rubber can cause an increase in the internal tyre temperature and consequently an increase in pressure.

At elevated temperatures rubber starts to degrade. This degradation process is known as either pyrolysis or thermo-oxidation. Pyrolysis can occur in the absence of oxygen, while thermo-oxidation can only occur in the presence of oxygen (Benoît et al. [1]). The gases produced during these reactions contribute to the pressure build-up inside the tyre. If a sufficient concentration of flammable gas, a temperature and pressure reading equal to or greater than the auto-ignition temperature and pressure of the flammable gas, and an oxygen concentration above 5.5% are present in the tyre, the tyre can explode violently (Benoît et al. [1]).

In the mining industry this occurrence is a major concern, since the shock waves and debris propagating from this explosion can lead to fatal injuries and severe damage to the surroundings and the vehicle itself. A working knowledge of the heat generation in the tyre cross-section is required to manage and eliminate this risk. The structural analysis of pneumatic tyres by finite element analysis is a cost-effective alternative to field tests that can be used to gain insight into the heat generation of a rolling tyre (Maritz [2]).

In the present work, the geometry and material properties of a 23.5R25 earthmover tyre is established from tests done on the actual tyre to develop a finite element model using MSC.Marc/Mentat. The numerical model is subjected to different inflation pressures, loads and rolling velocities in a structural analysis to calculate the heat generation rate in the tyre during rolling. The heat generation rate is implemented into a thermal model to simulate the temperature distribution in the tyre cross-section. The numerical results are compared to experimental data provided by the tyre manufacturer to confirm the accuracy of the proposed model.

TYRE GEOMETRY AND MATERIAL PROPERTIES

Tyres are composite products, consisting of a variety of rubber compounds and steel, textile or synthetic reinforcements. This section describes the results of the experimental tests done on a 23.5R25 radial earthmover tyre to numerically model its geometry and material properties. A segment of the earthmover tyre was donated to Stellenbosch University for the purpose of this study. A side view and cross-sectional view of the donated tyre segment are shown in Figures 1 (a) and (b), respectively.

The tyre geometry was established from measurements done on the donated tyre segment. This specific tyre segment had only one third of its tread remaining, of which the circumferential grooves in the tread have been worn out. Figure 1 (c) shows the remaining tread pattern. This study, however, considers the heat generation in a new tyre and therefore the measured geometry was adjusted to be consistent with that of a new tyre. These adjustments include the modelling of the circumferential grooves in the tread, based on the tread depth specified by the manufacturers. The tyre model and rim profile were developed in MSC.Marc/Mentat. The rim profile was based on the dimensions of the recommended rims specified by the manufacturers.

The structural components and rubber compounds were identified from the visual inspection of the donated tyre segment. The structural components included a radial steel body ply, five steel belt plies, and a bead bundle consisting of seventy-two individual wires. The rubber compounds included the inner liner, inner filler, soft apex, sidewall, shoulder, base and tread. The identified components can be seen in Figure 2. Only the inner liner, sidewall and tread rubber compounds were selected for material tests to simplify the numerical model. These compounds were selected based on either its volume representation, or by its unique composition.

The material behaviour of the selected rubber compounds were modelled using the two-term Mooney-Rivlin hyperelastic material model. This material model describes the stress-strain behaviour of hyperelastic materials in terms of a strain energy density function. The strain energy density is defined as the stored potential energy of a material during elastic deformation per unit volume (Smith et al. [3]). The two-term Mooney-Rivlin form of the strain energy density function is given as,

$$U = C_{10}(I_1 - 3) + C_{01}(I_2 - 3) \qquad \text{... (1)}$$

where U is the strain energy density (J/m^3) and, C_{10} and C_{01} are the Mooney-Rivlin material constants. I_1 and I_2 are the first and second order deviatoric strain invariants defined as,

$$I_1 = \lambda_1^2 + \lambda_2^2 + \lambda_3^2 \quad \text{and} \quad I_2 = \lambda_1^2\lambda_2^2 + \lambda_2^2\lambda_3^2 + \lambda_3^2\lambda_1^2 \qquad \text{... (2)}$$

where λ_i are the principal stretch ratios in the principal directions of the model. For incompressible materials, such as rubber, $\lambda_1 \lambda_2 \lambda_3 = 1$(Clark [4]). The stretch ratio is defined as the deformed length, l, divided by the undeformed length L:

$$\lambda = \frac{l}{L} \qquad \text{... (3)}$$

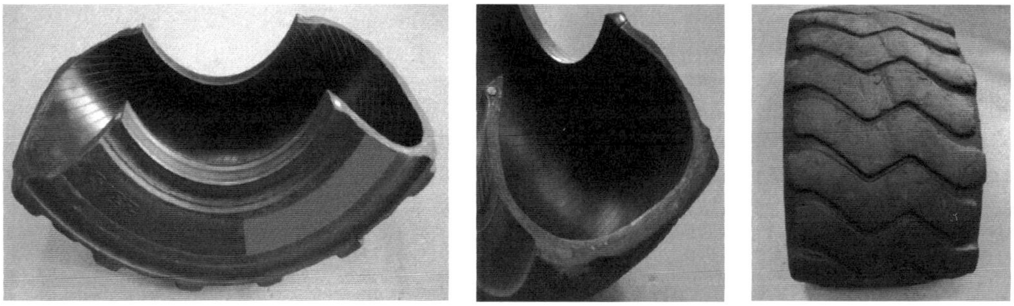

(a) (b) (c)

Figure 1. (a) Side view of the donated earthmover tyre segment; (b) Cross-sectional view and; (c) remaining tread pattern of the donated tyre segment.

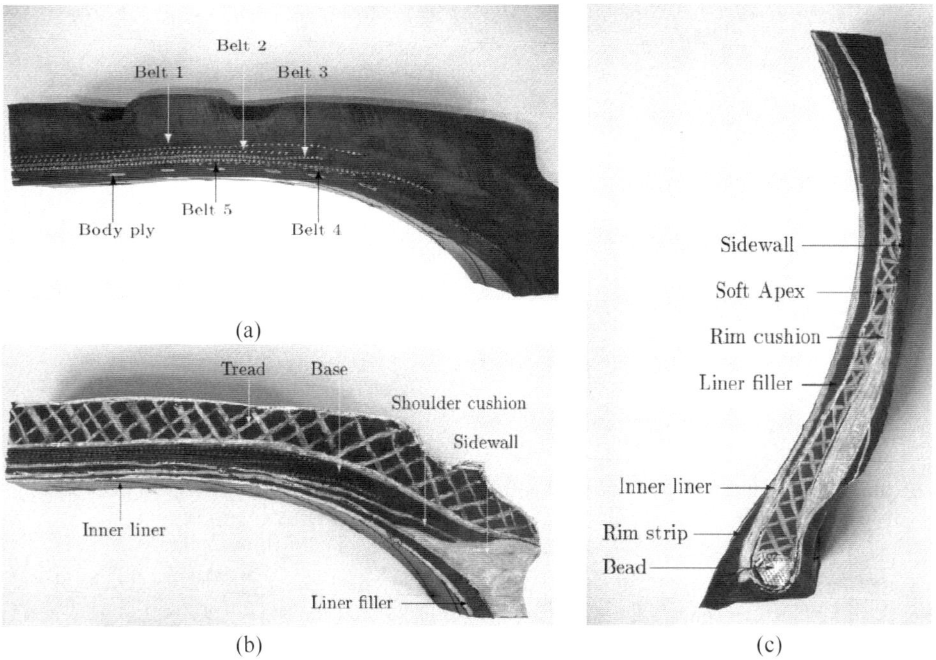

Figure 2. (a) Body ply and steel belt reinforcement and; (b) rubber compounds identified in a tread cut-out of the tyre; (c) Rubber compounds and bead bundle identified in a sidewall cut-out of the tyre.

The Mooney-Rivlin material fit is based on a polynomial fit of the material deformation during loading. The material behaviour of rubber can be described by tension and compression strain modes. Thus, three sets of uniaxial tensile and compression tests were performed on samples cut from the sidewall and tread rubber and three sets of uniaxial tensile tests were performed on samples cut from the inner liner rubber. Only tensile test were performed on the inner liner, since it was too thin for compression tests, but was modelled separately due to its unique composition (McGarry [5]). The average of the test data sets were used to obtain the material fits of the compounds. The resulting two-term Mooney-Rivlin fit of the inner liner, sidewall, rim strip and tread rubber is shown in Figure 3 (a) and (b), and Figure 4 (a) and (b), respectively. The material fit of the rim strip and tread are both based on the test data of the tread rubber, but at different strain ranges, since the rim strip typically experiences larger strains than the tread rubber.

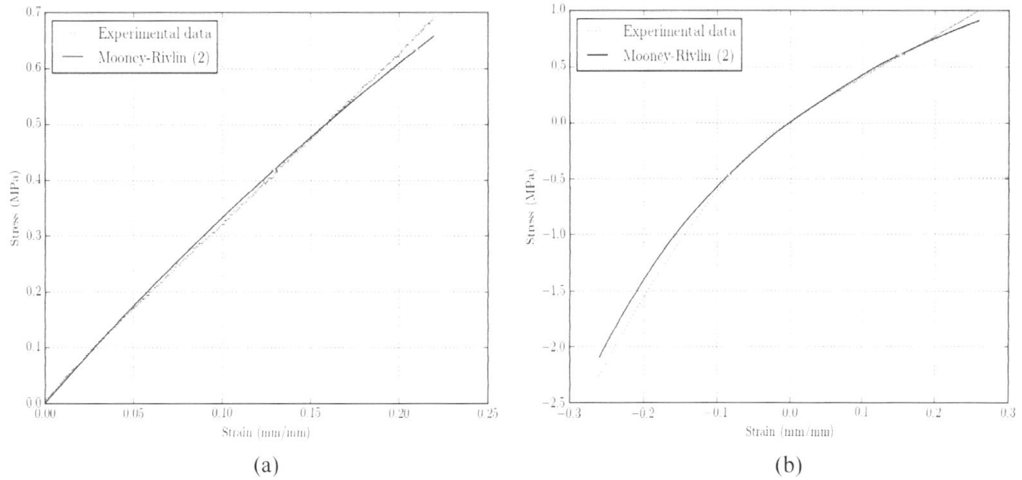

Figure 3. (a) Average of three tensile stress-strain test data sets and the two-term Mooney-Rivlin fit of the inner liner rubber test data; (b) Average of three tensile and compression stress-strain test data sets and the two-term Mooney-Rivlin fit of the sidewall rubber test data.

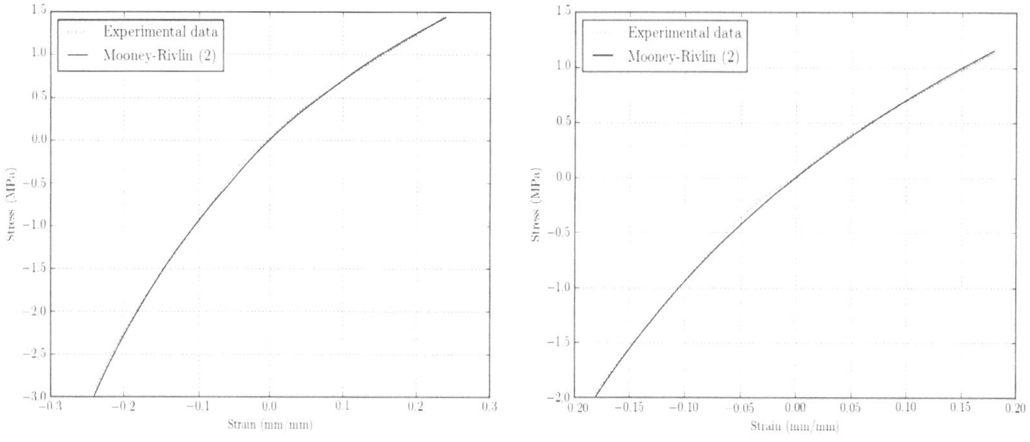

(a) (b)

Figure 4. Average of three tensile and compression stress-strain test data sets and the two-term Mooney-Rivlin fit of the (a) rim strip and; (b) tread rubber test data.

The material coefficients given in Equation 1, were automatically calculated in MSC.Marc/Mentat for each of the selected rubber compounds. These coefficients are given Table 1. The Poisson's ratio for rubber, which is assumed an incompressible material, is 0.5 (Gent [6]). The rubber density was determined from the volume and mass measurements of rubber samples cut from the donated tyre segment. The average of the result were used for the selected compounds. The resulting properties used for the numerical model are also given in Table 1. The numerical model of the tyre cross-section and its components are shown in Figure 5.

The belts and body consist of a number of steel cords. These cords were modelled using rebar elements that can be inserted into the rubber elements using the insert option in MSC.Marc/Mentat. These elements were defined according to the cord area, the number of cord layers per belt or ply, the number of cords per unit length and the cord angle with respect to the selected reference axis. The resulting properties used for the numerical model are given in Table 2. The rebar angle is defined with respect to the positive z-axis shown in Figure 5.

The cords used in tyres are typically manufactured from linear-elastic materials, such as stainless steel ranging from AISI types 302, 304, 303 and 316 (Bell [7]), or AISI 1005-1090 carbon steel, depending on its carbon content (Gent and Walter [8]). Although these materials have different tensile strengths, its Poison's ratio and density typically range from 0.27-0.3, 7 850-8 000 kg/m^3. Thus, a Poison's ratio of 0.3 and density of 7 850 kg/m^3 were selected for the structural components (Ghoreishy [9]). The Young's modulus of these components were established from the slope of its stress-strain curve in the linear region. Thus, three sets of uniaxial tensile tests were performed on each component to determine the slope of its stress-strain curve. The resulting Young's modulus, density and Poisson's ratio used for the structural components in the numerical model are given in Table 3.

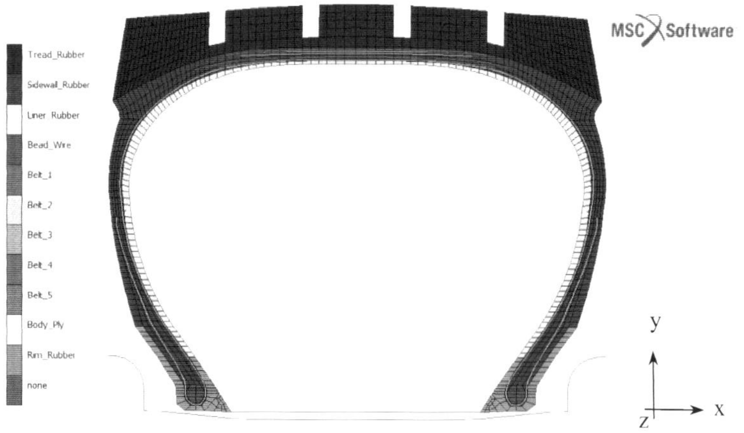

Figure 5. 2-D tyre geometry, rubber compounds, structural components and rim profile.

Table 1. Material properties of the inner liner, sidewall, rim strip, and tread rubber compounds.

	Mooney-Rivlin constant C_{10} (MPa)	**Mooney-Rivlin constant C_{01} (MPa)**	**Poisson's ratio**	**Density (kg/m^3)**
Inner liner	0. 5424	0.0706	0.5	1040
Sidewall	0. 3735	0. 4387	0.5	1040
Rim strip	0. 7103	0.6295	0.5	1040
Tread	0. 6925	0. 6516	0.5	1040

Table 2. Parameters used for the definition of the Rebar elements in the numerical model.

	Layers	**Rebar area (10^{-6} m^2)**	**Rebars per unit length (#/m)**	**Rebars angle**
Belt 1	1	1.7670	150	-45°
Belt 2	1	1.7670	150	30°
Belt 3	1	1.7670	222	-38°
Belt 4	1	1.7670	222	30°
Belt 5	1	0.7854	500	70°
Body ply	1	1.7670	200	90°

Table 3. Material properties of the bead wire and cords used for the belts and body ply.

	Young's modulus (GPa)	**Density (kg/m^3)**	**Poisson's ratio**
Bead wire	237.14	7870	0.3
Belt 1 & 2	32.34	7870	0.3
Belt 3 & 4	148.59	7870	0.3
Belt 5	148.51	7870	0.3
Body ply	148.59	7870	0.3

From the results obtained, it should be noted that the cords used in belts 1 and 2 have a very low Young's modulus. This low modulus is expected, due to the cord construction. The cords used in tyre manufacturing typically consist of multiple steel filaments with a specific geometrical arrangement, which influences its material behaviour (Gent and Walter [8]). The cords used in belts 1 and 2 had a multi-strand arrangement which requires a high level of twist. This arrangement allows it to have a large degree of flexibility during bends to absorb and distribute most of the road shock to protect the belts underneath and body ply (Gent and Walter [8]). The cords used in belts 3 and 4 and the body ply had a single strand arrangement. This arrangement requires a lower level of twist than a multi-strand arrangement, and consequently these cords have a higher modulus, as seen in Table 3. The high Young's modulus of the bead wire is also expected, since high tensile steel wires are used to properly seat and seal the tyre on the rim (Gent and Walter [8]).

THEORETICAL CALCULATION OF HEAT GENERATION IN A ROLLING TYRE

The derivation of the hyperelastic material models, including the Mooney-Rivlin model, are based on the assumption of perfect elastic material behaviour. This behaviour implies complete and immediate material recovery upon deformation. Rubber, however, is a hyper-viscoelastic material (Smith et al. [3]). Thus, when rubber is subjected to cyclic loading, such as the rolling of the tyre, its viscous component causes a lag in the material recovery and consequently the rubber tends to absorb more heat than it can dissipate. This lost energy is stored in the rubber as heat and known as hysteresis loss. Since the material model used to numerical model the elastic behaviour of rubber, it does not account for this viscous component and the lost energy must be obtained in another manner.

Dynamic mechanical analysis (DMA) can be performed on viscoelastic materials to determine its hysteresis coefficient, H. The hysteresis coefficient can be defined as the lost energy, U_{loss}, divided by the total energy, U_{total}, in a cyclic system (Lin et al. [10]):

$$H = \frac{U_{loss}}{U_{total}} \qquad \ldots (4)$$

Considering a cyclic system represented by a rolling tyre, the total energy in the system can be obtained from the sum of the strain energy density in one revolution using the Mooney-Rivlin material model. If it is assumed that the lost strain energy density completely contributes to internal heat generation in the tyre, the lost strain energy density can be calculated by rewriting Equation 4 as (Smith et al. [3]):

$$U_{loss} = HU_{total} \quad \text{... (5)}$$

For a numerical model consisting of a finite number of elements, the total strain energy density for the element node at each increment of one revolution can be calculated as,

$$U_{total\_node} = \sum_{i=1}^{n} \Delta U_i \quad \text{... (6)}$$

where U_{total_node} is the total strain energy density (J/m^3) of a node, n is the number of times the elastic strain energy density increases from increment to increment and ΔU_i is the increase in elastic strain energy density (J/m^3) from increment to increment. The total strain energy density of an element can be calculated by averaging the nodal values over the element (Martitz [5]). The heat generation rate in each element can be calculated from,

$$\dot{q}_g = U_{loss} f \quad \text{... (7)}$$

where $\dot{q}_g$ is the heat generation rate in the element (W/m^3) and f is the rolling frequency of the tyre (Hz). The rolling frequency of the tyre can be defined as the tyre velocity divided by the circumference of the rolling tyre,

$$f = \frac{v_c}{L_r} \quad \text{... (8)}$$

$$L_r = 2\pi R_r \quad \text{... (9)}$$

where v_c is the tyre velocity (m/s), L_r the circumference of the rolling tyre (m), and R_r is the radius of the rolling tyre (m). The frequency is proportional to the rolling velocity and therefore the heat generation will be proportional to the rolling velocity (Lin et al. [10]). This heat generation rate can be used in a thermal analysis to simulate the temperature distribution in the tyre cross-section during steady-state rolling.

FINITE ELEMENT MODELS AND ANALYSIS PROCEDURE

In order to simulate the temperature distribution in the tyre during steady-state rolling, a structural and thermal analysis of the finite element model were completed separately. Initially, a full 2D axisymmetric model of the tyre was developed to inflate the tyre using a cavity pressure load. The 2D model was expanded into a 3D model using the Axisymmetric Model To 3-D feature in MSC.Marc/Mentat. The 3D model was inflated using the results of the inflation model as an initial condition. The 3D model was loaded by applying a point load to the road surface. The vertical tyre deflection was obtained from the loaded tyre and was enforced as a displacement control condition in the rolling analysis to load the tyre.

During the rolling analysis the rim was fixed and the tyre could rotate freely on the rim as the road surface enforced the rotation of the tyre. The road velocity was specified at 15 km/h and a friction coefficient of 0.7, which is typically used for gravel and ground road contact (Chi and Caldas [11]), was specified for the tyre and road contact interaction. The rolling analysis boundary conditions are shown in Figure 6 (a).

The elastic strain energy density for one revolution of the segment specified in Figure 6 (a) was used to calculate the heat generation rate in the tyre according to the method described in the previous section. The hysteresis coefficient of the inner liner, sidewall and tread rubber were established from DMA completed by Makgatho [12] on samples cut from the donated tyre segment. The hysteresis coefficients used for the tread rubber was used for the rim strip as well. These coefficients, along with a summary of the material properties used for the thermal analysis are given in Table 4 (Gent [6]).

The heat generation rate of each element was implement as in volume flux in a thermal/structural analysis of a 2D axisymmetric finite element model of the tyre. The conduction and convention heat transfer boundary conditions were included in the analysis. The effects of heat transfer by radiation were ignored, since previous studies have shown its effects negligible (Nyaaba et al. [13]). A constant convective film coefficient as given by Equation 10, were defined for all exposed surfaces in the model (Nyaaba et al. [13]).

$$h = 5.9 + 3.7v_c \qquad \ldots (10)$$

Where h is the convection coefficient (W/m^2 K) on the outside of the tyre. A natural convection coefficient of 5.9 W/m^2 K was used for the convection on the inside of the tyre. A uniform initial temperature distribution of 38 °C were applied to all the nodes. The ambient temperature on the inside and outside of the tyre were specified as 38 °C, which is typically used for ambient temperatures in the mining industry, as specified by the manufacturers. The boundary conditions defined for the thermal model are shown in Figure 6 (b).

Table 4. Material properties and hysteresis coefficients used for the thermal analysis

	Thermal conductivity (W/m K)	**Specific heat (J/kg K)**	**Emissivity**	**Hysteresis coefficient**
Belts, body ply and bead wire	49.8	448	0	-
Inner liner, sidewall, rim strip and tread	0.25	1900	0.94	0.2132 0.1557 0.1864

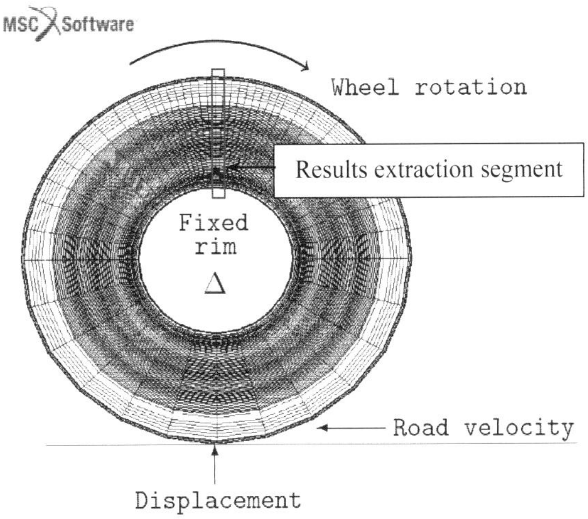

(a)

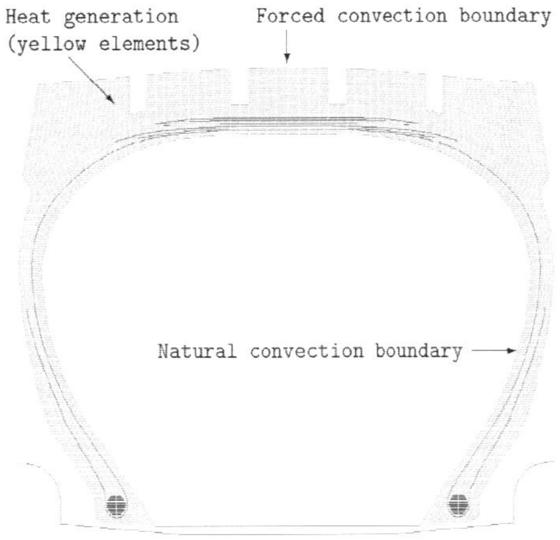

(b)

Figure 6. Boundary conditions for the (a) rolling analysis and; (b) thermal analysis.

OPERATING CONDITIONS

In order to describe the temperature distribution in the tyre as a function of its operating conditions, the numerical model was subjected to different inflation pressures and loads. These conditions were based on the typical conditions these tyres would be exposed to in industry and include under inflation, over inflation and at the correct inflation. The selected operating conditions are given in Table 5.

The tyre was inflated to each of the inflation pressures in Table 5. Each of the inflated models were loaded to each the loads given in the table. The heat generation rate for each of these models were calculated for a road velocity of 15, 30 and 45 km/h. The heat generation rate was calculated at different rolling velocities, since the elastic strain energy density obtained from the rolling analysis was a function of the tyre deformation, and not its velocity. The thermal analysis were simulated at steady-state heat transfer conditions at each combination of the stated conditions.

Table 5. Operation conditions defined for numerical analysis.

Tyre pressure (kPa)	Force (N)		
200	22 588	48 656	74 703
268	22 588	48 656	74 703
336	22 588	48 656	74 703
450	48 656	74 703	90 742

NUMERICAL AND EXPERIMENTAL RESULTS

Since the elastic strain energy density results used to calculate the heat generation rate, were based on the tyre deformation during the rolling analysis, the accuracy of the numerical model was evaluated by comparing the tyre deflection to experimental data provided by the tyre manufacturer. The experimental data included a prescribed set of specifications and therefore not all the conditions considered in this study could be experimentally verified. However, the data was considered sufficient for the evaluation of the proposed model. The numerical results are defined according to parameters shown in Figure 7. The tyre inflation results are discussed initially, followed by the tyre deflection, contact area and a brief description of the thermal results.

The numerical tyre radius at no-load for inflation pressures 200, 268, 336 and 450 kPa were 0.800, 0.803, 0.805 and 0.808 m, respectively. The specified tyre radius at an inflation pressure of 450 kPa and no-load is 0.806 m, and its overall width is 0.601 m. The overall tyre width at 450 kPa for the numerical model was 0.603 m. Thus, the numerical tyre inflation initial conditions were considered sufficient for the purpose of this study.

The experimental and numerical tyre deflection and contact area are given in Table 6 and 7. A percentage error of the values are given in the tables as well. It should be noted that the percentage error of the tyre deflection were based on difference the loaded tyre radius and not the difference of the experimental and numerical deflections stated in the table. Based on these results, it was concluded that the proposed numerical cloud be used for the calculation of the heat generation rate in the tyre cross-section.

Only some of the thermal models have been completed, and therefore the interpretation and validation of the results cannot be sufficiently verified. However, the temperature distribution in an axisymmetric cross-section of the tyre model inflated to 268 kPa and loaded with 74 703 N at 15, 30 and 45 km/h are shown in the Figure 8. This illustration can be used to verify the analysis procedure used to simulate the temperature distribution in the tyre cross-section. From this model it can also be seen that the maximum temperature occurs at the sidewall, which is as expected, since most of the tyre deformation occurs in this region. This result is also consistent with practice, since the tyre typically burst at the sidewall, since most of the tyre deformation also occurs in this region in practice. Furthermore, it can be seen that the temperature increases as the speed of the tyre increases, which is also as expected. The relevance of these results with regard to future work is described in the following section.

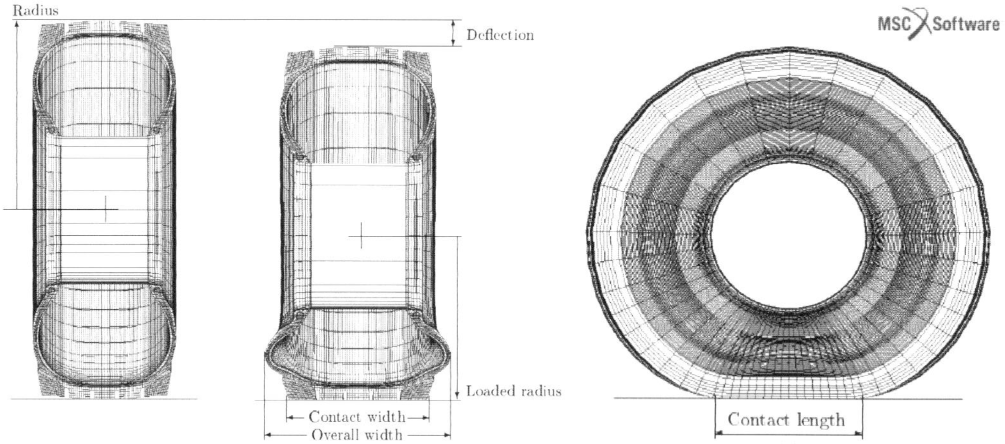

Figure 7. Numerical deflection and contact parameter definition.

Table 6. Experimental and numerical results of the tyre deflection and overall width.

Tyre pressure (kPa)	Force (N)	Road surface deflection (m)			Overall width (m)		
		Exp.	Num.	Error %	Exp.	Num.	Error %
200	22 588	0.042	0.039	0.382	0.640	0.645	0.788
	48 656	0.086	0.070	2.282	0.670	0.686	2.340
	74 703	0.127	0.100	3.939	0.699	0.717	2.466
268	22 588	-	0.053	-	0.636	0.636	0.076
	48 656	-	0.060	-	0.661	0.670	1.318
	74 703	-	0.085	-	0.685	0.699	2.033
336	22 588	-	0.032	-	0.633	0.629	0.686
	48 656	-	0.055	-	0.655	0.657	0.326
	74 703	-	0.075	-	0.676	0.685	1.143
450	48 656	0.050	0.048	0.503	0.649	0.645	0.647
	74 703	0.073	0.065	1.341	0.666	0.667	0.177
	90 742	0.087	0.075	1.893	0.676	0.678	0.349

Table 7. Experimental and numerical results of the tyre contact width and length.

Tyre pressure (kPa)	Force (N)	Contact width (m)			Contact length (m)		
		Exp.	Num.	Error %	Exp.	Num.	Error %
200	22 588	0.515	0.535	2.824	0.310	0.322	3.786
	48 656	0.527	0.535	0.485	0.433	0.402	7.244
	74 703	0.540	0.535	1.830	0.548	0.540	1.538
268	22 588	0.513	0.535	3.121	0.294	0.273	7.198
	48 656	0.524	0.534	1.154	0.397	0.397	0.138
	74 703	0.534	0.535	0.207	0.493	0.541	9.710
336	22 588	0.512	0.535	3.365	0.283	0.266	5.830
	48 656	0.521	0.535	2.512	0.374	0.391	4.530
	74 703	0.530	0.535	0.995	0.455	0.456	0.097
450	48 656	0.519	0.535	3.005	0.348	0.326	6.325
	74 703	0.526	0.535	1.780	0.416	0.392	5.714
	90 742	0.530	0.535	1.081	0.455	0.459	0.778

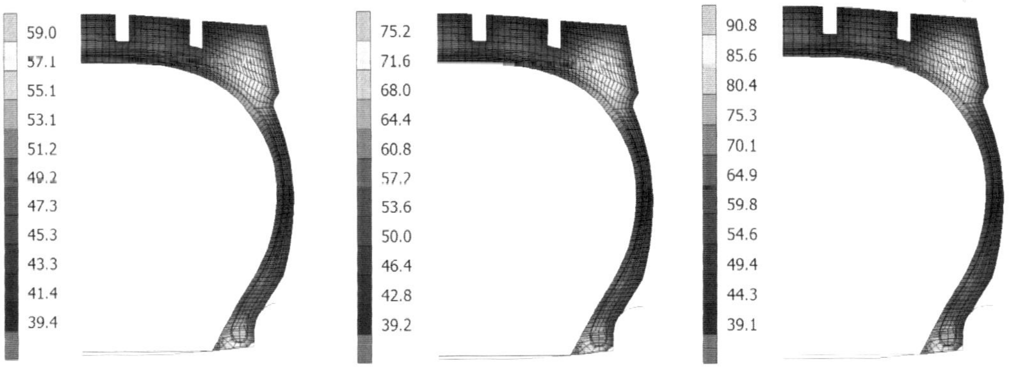

Figure 8. Temperature distribution in a 2D axisymmetric model of the tyre inflated to 268 kPa and loaded with 74 703 N at (a) 15 km/h; (b) 30 km/h and (c) 45 km/h.

FUTURE WORK

When the thermal models for all the operation conditions defined in Table 5 are completed, the models can be interpreted and verified. The results can then be used to describe the temperature distribution in the tyre as a function of its operating conditions. This function can be used to establish a safe operating envelope that would prevent the internal that generation in the tyre to reach critical conditions which can lead to the explosion of these tyres.

CONCLUSION

The finite element model of a 23.5R25 earthmover tyre is developed based on measurements and tests done on a segment of the actual tyre. The model was simulated at different inflation pressures and loads to evaluate its corresponding deflection and contact area. The numerical results were compared to experimental data provided by the tyre manufacturer to evaluate the accuracy of the numerical model. From the results obtained, it was concluded that the proposed numerical model can be used to compute the heat generation in the tyre. A thermal analysis of a 2D axisymmetric tyre model was completed to illustrate the temperature distribution in the tyre cross-section at steady-state heat transfer conditions. This illustration confirmed that proposed analysis procedure can be used to simulate the temperature distribution in the cross-section of a rolling the tyre. The thermal results must still be interpreted and verified to describe the heat generation in the tyre as a function of its operating conditions that can be used to establish a safe operating envelope.

REFERENCES

1. Benoît, R., et al., "Heavy vehicles tire blowout and explosion", Tech. Rep., Montréal: Association Sectorielle Transport Enterposage, 2009.
2. Maritz, J. C., "Numerical modelling and experimental measurement of the temperature distribution in a rolling tire", Thesis, Stellenbosch: Stellenbosch University, 2015.
3. Smith, R. E., et al., "Simulation of thermal signature of tires and tracks", Warren: Proceedings of the 2012 Ground Vehicle Systems Engineering and Technology Symposium (GVSETS), 2012.
4. Clark, S. K. (ed.), "Mechanics of pneumatic tires", Washington, D.C.: U. S. Dept. of Transportation, National Highway Traffic Safety Administration, 1981.
5. McGarry, P., "Haul truck tires and open pit mining applications", Thesis, Kingston: Queen's University, 2007.
6. Gent, A. N. (ed.), "Engineering with rubber: How to design rubber components", 3rd edn., Cincinnati: Hanser Publications, 2012.
7. Bell, S. L. (ed.), "Standard wire, rope and cable, or iron steel", Washington, D.C.: U. S. Dept. of Homeland Security, U. S. Customs and Border Protection, 2010.
8. Gent, A. N., and & Walter, J.D., "The pneumatic tire", Washington, D.C.: U. S. Dept. of Transportation, National Highway Traffic Safety Administration, 2006.
9. Ghoreishy, M. H. R., "Steady state rolling of a radial tyre: Comparison with experimental results", Proc. IMechE, Vol. 220. Part D: J. Automobile Engineering, pp. 713 – 721, 2006.
10. Lin, Y.-J., and Hwang, S.-J., "Temperature prediction of rolling tire by computer simulation", Mathematics and Computers in Simulation, Vol. 67, pp. 235-249, 2004.
11. Chi, S., and Caldas, C. H., "Development of an automated safety assessment framework for construction activities", Austin: Proceedings of 26th International Symposium on Automation and Robotics in Construction (ISARC), pp. 38-46, 2009.
12. Makgatho, G., Polymer Characterization Technician, Council for Scientific and Industrial Research (CSIR), Pretoria, South Africa, GMakgatho@csir.co.za, 2016.
13. Nyaaba, W., et al., "Thermo-mechanical stress loading and deformation of OTR dump truck tires in surface mining operations", International Journal of Advanced and Innovative Research, Vol. 4. No. 9, pp. 2278-7844, 2015.

A systematic comparison of two direct exchange area smoothing techniques for modelling radiation heat transfer in three-dimensional enclosures using the zonal method

W.O. Monnaemang[1,*], P.G. Rousseau[1], L. Jestin[1]

[1] Department of Mechanical Engineering, University of Cape Town, Private Bag X3, Rondebosch, Cape Town, South Africa

wm@mtechindustrial.com, pieter.rousseau@uct.ac.za, louis.jestin@uct.ac.za

ABSTRACT

The radiative transfer equation (RTE) has been the traditional way for solving radiation heat transfer in high temperature industrial systems for many years. This paper is based on the zonal method, which has been used extensively to solve radiation heat transfer problems. The application of the zonal method allows for the prediction of heat flux and temperature distributions on the walls, inside, and at the exit of enclosures. The basis for the zonal method is so-called direct exchange areas (DEAs). The evaluation of the DEAs is done by discrete numerical integration, which inherently leads to errors, and therefore requires smoothing. Two direct exchange area smoothing techniques are compared for two different enclosure geometries at different levels of discretisation. The generalised Lawson's smoothing technique was found to produce reliable DEAs and zonal method results only for cubic enclosures. The least-squares smoothing technique using Lagrange multipliers as proposed by Larsen and Howell was found to produce reliable DEAs and zonal method results for both cubic and rectangular enclosures. It was further found that without smoothing, the energy conservation is in fact never achieved for these test cases. It can therefore be concluded that smoothing is absolutely essential for all cases.

KEYWORDS

Radiation heat transfer, Zonal method, Direct exchange areas, Smoothing, Discretisation

INTRODUCTION

Thermal radiation is the dominant mode of heat transfer in high temperature systems. Its analysis is complicated by the presence of a participating medium inside the volume of the enclosure. The challenge in deriving an analytical solution for the integro-differential Radiative Transfer Equation (RTE) in the presence of a participating medium has led to the development of a variety of alternative numerical methods over the last few decades [1]. These include, but are not limited to the flux method, the zonal method and the Monte Carlo method. This paper is based on the application of the zonal method, which has been used extensively to solve radiation heat transfer problems.

The zonal method is known to offer a good compromise between accuracy and computational effort [2]. In this method, an enclosure is divided into a finite number of surface and volume zones that are assumed to be isothermal and have uniform radiative properties [3]. The different zones are then connected to one another via a network of radiation heat transfer elements, which are characterised by so-called direct exchange areas (DEAs). The DEAs are evaluated using discrete numerical integration. Since the numerical integration process inherently leads to errors, the calculated DEAs need to be "smoothed" to ensure that energy conservation will be achieved in the final results obtained from the network solution.

The focus of this paper is on the comparison of two techniques for smoothing the DEAs, namely the least-squares smoothing technique (LSS) proposed by Larsen and Howell [4] and the generalised Lawson's smoothing (GLS) technique suggested by Méchi et al. [5].

BACKGROUND

Zonal method

The zonal method dates back to 1958 when Hottel and Cohen [3] developed it for the purpose of analysing radiation heat transfer in enclosures containing emitting, absorbing and non-scattering gray media with a constant absorption coefficient. In 1967, Hottel and Sarofim [6] enhanced this method to cater for more complex three-dimensional geometries containing emitting, absorbing and scattering non-gray media with non-constant absorption and scattering coefficients. Since then, the zonal method has been widely used by researchers for modelling radiation heat transfer in high temperature systems.

In the zonal method, the net radiation heat transfer rate away from a surface zone or out of a volume zone is the difference between the rate of energy lost by emission of radiation and the rate of energy absorbed by way of radiation [7]. Equations (1) and (2) show the net radiation heat transfer rate for a surface zone ($Q_{s,i}$) and volume zone ($Q_{g,i}$) in terms of the DEAs, respectively [8]:

$$Q_{s,i} = \varepsilon_i \left(A_i E_{bs,i} - \sum_{j=1}^{N} \overline{s_i s_j} J_{s,j} - \sum_{k=1}^{K} \overline{s_i g_k} J_{g,k} \right) \quad (1)$$

$$Q_{g,i} = 4\beta_i V_i E_{bg,i} - \sum_{j=1}^{N} \overline{g_i s_j} J_{s,j} - \sum_{k=1}^{K} \overline{g_i g_k} J_{g,k} \quad (2)$$

where β, V, J, E_{bg}, A and ε are the extinction coefficient, volume of the zone, radiosity, gas emissive power, surface zone area and surface emissivity, respectively. Furthermore, $\overline{s_i s_j}$, $\overline{s_i g_k}$, $\overline{g_i s_j}$ and $\overline{g_i g_k}$ are the surface-to-surface, surface-to-volume, volume-to-surface and volume-to-volume DEAs, respectively. These DEAs are calculated by integrating over appropriate emitting and receiving surface and volume zones, while taking into account the absorbing and scattering characteristics of the medium separating the zones.

Direct exchange areas (DEAs)

The DEAs are the basis of the zonal method. They are a function of the enclosure geometry and the radiative properties of the walls and the participating medium that fills the enclosure volume. Three types of DEAs exist for an enclosure, namely surface-to-surface, surface-to-volume and volume-to-volume, as illustrated in Figure 1 [9]:

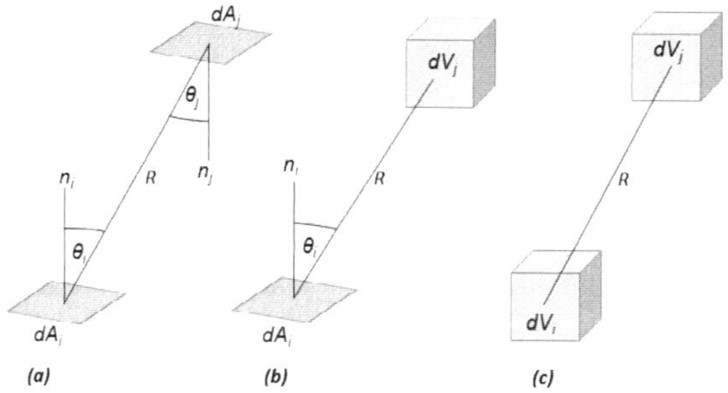

Figure 1: (a) Surface-to-surface, (b) Surface-to-volume & (c) Volume-to-volume DEAs [9].

In Figure 1, A is the area of the surface zone and V is the volume of the volume zone. R is the line connecting any two zones and θ is the angle between R and the surface normal vector denoted by n. The surface-to-surface, surface-to-volume and volume-to-volume DEAs are respectively mathematically expressed as [9]:

$$\overline{s_i s_j} = \int_{A_i} \int_{A_j} \exp(-\beta R) \frac{\cos\theta_i \cos\theta_j}{\pi R^2} dA_j dA_i \quad (3)$$

$$\overline{s_i g_j} = \int_{V_j} \int_{A_i} \exp(-\beta R) \frac{\beta \cos\theta_i}{\pi R^2} dA_i dV_j \quad (4)$$

$$\overline{g_i g_j} = \int_{V_i} \int_{V_j} \exp(-\beta R) \frac{\beta^2}{\pi R^2} dV_j dV_i \quad (5)$$

The extinction coefficient β represents the combined effects of the absorption and scattering coefficients of the medium. It is a measure of how the intensity of a beam of radiation passing through a medium changes with distance [10]. The term, "$\exp(-\beta R)$" forming part of Equations (3), (4) and (5) is the transmittance which gives the fraction of energy that passes between two points in a medium along a straight line of length R [6].
These DEAs must satisfy reciprocity relationships and therefore [6]:

$$\overline{s_i s_j} = \overline{s_j s_i} \qquad \overline{g_i s_j} = \overline{s_j g_i} \qquad \overline{g_i g_j} = \overline{g_j g_i} \tag{6}$$

Equations (7) and (8) show the relationships representing the energy conservation principles that must also be satisfied [6]:

$$\sum_{j=1}^{N_s} \overline{s_i s_j} + \sum_{k=1}^{N_g} \overline{s_i g_k} = A_i \tag{7}$$

$$\sum_{j=1}^{N_s} \overline{g_i s_j} + \sum_{k=1}^{N_g} \overline{g_i g_k} = 4\beta_i V_i \tag{8}$$

for $i = 1,\ldots,N_s$ in Equation (7) and $i = 1,\ldots,N_g$ in Equation (8) where N_s and N_g are the number of surface and volume zones, respectively.

Smoothing

It can be seen from Equations (3), (4) and (5) that the calculation of DEAs involves a large set of double area/area, area/volume and volume/volume integrals. These multi-dimensional integrals require numerical evaluation. Achieving high accuracy in performing the multi-dimensional integrals is challenging and the application of numerical techniques for evaluating the DEAs results in inherent numerical errors [5]. As a result, DEAs are expected to include random inaccuracies merely because the calculated DEAs are approximations to the correct values and may not satisfy conservation principles [11]. If the DEAs do not satisfy conservation principles, then the sum of the surface heat fluxes emanating from the ultimate network solution will be non-zero. This led to the need for techniques that could correct these errors, known as smoothing. Smoothing is therefore defined as the process of taking a set of approximated DEAs and adjusting it to satisfy the conservation and reciprocity relationships [12].
Larsen and Howell [4] and Lawson [12] proposed such smoothing techniques that would ensure that energy is indeed conserved in the final heat flux results.

Larsen and Howell [4] proposed the least-squares smoothing (LSS) technique based on Lagrange multipliers. This is essentially a constrained optimisation procedure that optimises the objective function with respect to the DEAs in the presence of energy conservation and reciprocity constraints. When using weighting factors of equal size, Larsen and Howell [4] found that all the DEAs were adjusted equally and that zero valued DEAs were disturbed. This results in negative values of DEAs. Smoothing using weighting factors that are equal to the square root of the DEAs were recognised to be effective and efficient since it resulted in DEAs that were adjusted in proportion to its original size [4], thereby leaving zero valued DEAs unchanged.

Lawson [12] also proposed a smoothing technique primarily for surface-to-surface DEAs in a transparent medium. Here, each DEA value is adjusted according to its size and it ensures that no adjusted DEA value is negative. Furthermore, the error in the conservation condition is firstly calculated for a surface by taking the difference between the area of the surface and the sum of all the DEAs from the same surface ($\delta_i = A_i - \sum \overline{s_i s_j}$). Following this, each DEA value is adjusted by adding the error difference relative to the sum of all the DEAs to the original DEA value, resulting in the following smoothing equation [12]:

$$\overline{s_i s_j}' = \overline{s_i s_j} + \frac{\overline{s_i s_j}}{\sum \overline{s_i s_j}} \cdot \delta_i = \overline{s_i s_j} \left(\frac{A_i}{\sum \overline{s_i s_j}} \right) \tag{9}$$

Méchi et al. [5] extended Lawson's smoothing technique to also cater for surface-to-volume and the volume-to-volume DEAs in a participating medium. They referred to the enhanced technique as the generalised Lawson's smoothing (GLS) technique. They further considered radiation heat transfer using the zonal method for an inhomogeneous non-gray participating medium in a two-dimensional black-walled rectangular enclosure. They numerically evaluated DEAs and compared the effect of two smoothing techniques on wall heat flux results. They concluded that the application of the generalised Lawson's smoothing and the least-squares smoothing techniques has no significant effect on the final results of the zonal method for homogeneous and inhomogeneous participating media [5].

Murty et al. [13] also studied the significance of DEA smoothing in zone modelling. The paper only addressed the least-squares smoothing using Lagrange multipliers technique proposed by Larsen and Howell [4]. They considered two different weighting factors, namely those that are equal to one and those that are equal to the square root of the DEAs in the smoothing technique. Furthermore, they only considered a cubic enclosure. They concluded that it did not matter what type of weighting factor was used in the smoothing technique since smoothing in itself added no appreciable contribution to the zonal method predictions [13].

The review provided above reveals an inconsistency that needs to be clarified. Although it is generally understood that smoothing is essential to ultimately ensure energy conservation, both Méchi et al. [5] and Murty et al. [13] concluded that the application of either of the smoothing techniques did not significantly affect the final zonal network results. However, Méchi et al. [5] only considered a two-dimensional black-walled rectangular enclosure and Murty et al. [13] considered only a cubic enclosure. This leaves one with the question whether their conclusions are also valid for three-dimensional rectangular enclosures, which are more representative of typical coal-fired furnace geometries. This paper will strive to attain more clarity regarding this question.

METHOD

Direct exchange areas approximation

The evaluation of DEAs involves a set of multi-dimensional integrals that are usually approximated through discrete numerical integration. The surface-to-surface DEA integral in Equation (3) is approximated as:

$$\overline{s_i s_j} = \sum_{i=1}^{N_s}\sum_{j=1}^{N_s} \exp(-\beta R)\frac{\cos\theta_i \cos\theta_j \Delta A_i \Delta A_j}{\pi R^2} \tag{10}$$

ΔA is the elemental area on each discretised surface and R is the beam connecting the midpoints of the discretised surfaces. The x, y and z-coordinates of the midpoints of the elemental surfaces are used to calculate the magnitude of R as follows:

$$|R| = \sqrt{(x_i - x_j)^2 + (y_i - y_j)^2 + (z_i - z_j)^2} \tag{11}$$

R forms polar angles, θ_i and θ_j with the surface normal vectors, n_i and n_j as seen in Figure 1a. The magnitudes of these angles are calculated using the dot product as:

$$\cos\theta_i = \frac{n_i \cdot R}{|n_i||R|} \quad \text{and} \quad \cos\theta_j = \frac{n_j \cdot R}{|n_j||R|} \tag{12}$$

Substituting Equation (12) into Equation (10) yields:

$$\overline{s_i s_j} = \sum_{i=1}^{N_s}\sum_{j=1}^{N_s} \exp(-\beta|R|)\frac{\left(\frac{n_i \cdot R}{|n_i||R|}\right)\left(\frac{n_j \cdot R}{|n_j||R|}\right)\Delta A_i \Delta A_j}{\pi |R|^2} \tag{13}$$

Equation 13 is used to approximate the surface-to-surface DEAs. The same procedure is followed for the surface-to-volume and volume-to-volume DEA approximations.

Least-squares smoothing (LSS) using Lagrange multipliers

In the LSS technique the initial set of DEAs are assembled to form one matrix as shown below [12]:

$$[X] = \begin{bmatrix} \left[\overline{ss}\right] & \left[\overline{sg}\right] \\ \left[\overline{sg}\right]^T & \left[\overline{gg}\right] \end{bmatrix} \tag{14}$$

The idea is to adjust $[X]$ to meet all the reciprocity relationships and energy conservation principles while minimising the disturbance to the originally calculated DEA values. Minimising the disturbance ensures that the geometric effects captured in the original values are essentially retained. An objective function (H) is defined as [12]:

$$H = \sum_{i=1}^{N_t} \sum_{j=1}^{N_t} \frac{1}{2w_{ij}} \cdot (x_{ij}^{'} - x_{ij})^2 \tag{15}$$

where N_t is the total number of surface and volume zones, w_{ij} are the arbitrary weighting factors, $x_{ij}^{'}$ are the adjusted DEA values and x_{ij} are the original DEA values. The energy conservation principles stipulated in Equations (7) and (8) serves as constraints and form part of the Lagrangian used to calculate the Lagrange multipliers. Using the Lagrange multipliers (λ), the adjusted DEA values are obtained from [12]:

$$x_{ij}^{'} = x_{ij} + w_{ij} \cdot \left(\lambda_i + \lambda_j \right) \tag{16}$$

In this paper, the weighting factors are set equal to the square root of the original DEA values.

Generalised Lawson's smoothing (GLS)

Using the GLS technique suggested by Méchi et al. [5], the smoothed surface-to-surface, surface-to-volume and volume-to-volume DEAs are respectively obtained from [5]:

$$\overline{s_i s_j}' = \overline{s_i s_j} \cdot \left(\frac{A_i}{\sum_{j=1}^{N_s} \overline{s_i s_j} + \sum_{k=1}^{N_g} \overline{s_i g_k}} \right), \quad \overline{s_i g_j}' = \overline{s_i g_j} \cdot \left(\frac{A_i}{\sum_{j=1}^{N_s} \overline{s_i s_j} + \sum_{k=1}^{N_g} \overline{s_i g_k}} \right), \quad \overline{g_i g_j}' = \overline{g_i g_j} \cdot \left(\frac{4\beta_i V_i}{\sum_{j=1}^{N_s} \overline{g_i s_j} + \sum_{k=1}^{N_g} \overline{g_i g_k}} \right) \tag{17}$$

where $\overline{s_i s_j}$, $\overline{s_i g_j}$ and $\overline{g_i g_j}$ are the DEAs before smoothing and $\overline{s_i s_j}'$, $\overline{s_i g_j}'$ and $\overline{g_i g_j}'$ are the DEAs after smoothing. The smoothed DEAs are ultimately used in the zonal method to predict heat fluxes and medium temperature distributions for cubic and rectangular enclosures.

RESULTS AND DISCUSSION

In this section, the effect of smoothing on the reciprocity relationships will be investigated, followed by its effect on the ultimate energy balances obtained after solving the zonal model network equations.

Reciprocity relationships

In addition to conservation principles, DEAs must satisfy reciprocity relationships as stipulated in Equation (6). In this section, the DEAs before and after smoothing using both the GLS and the LSS techniques are analysed to establish whether they actually ensure conformance to the reciprocity relationships. The analysis is conducted by evaluating a reciprocity error fraction which is evaluated for the DEAs between the front surface and the other surfaces making up a three dimensional enclosure. Figure 2 shows the positions of the six surfaces designated as front, back, bottom, top, left and right.

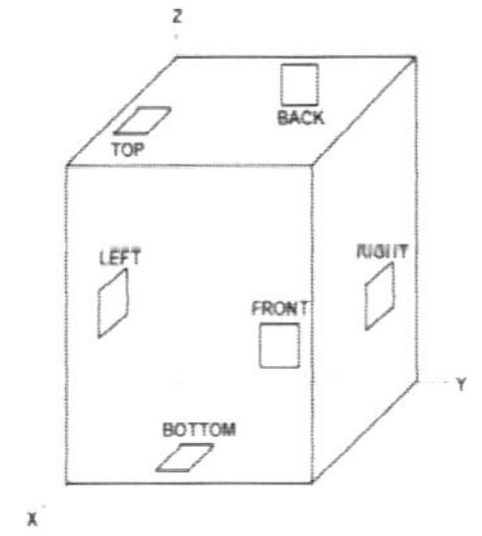

Figure 2: Enclosure surfaces.

The DEAs are also assembled into one matrix $[X]$ as shown in Equation (14). The reciprocity relationships require that each row in the $[X]$-matrix be equal to the transpose of each column in the $[X]$-matrix. The reciprocity error fractions are therefore obtained from:

$$R_{j,i} = \frac{\left| x_{i,j} - x_{j,i} \right|}{\max \; x_{i,j}, x_{j,i}} \quad \text{for} \begin{cases} i = 1, \ldots, N_t \\ j = 1, \ldots, F_n \end{cases} \tag{18}$$

N_t is the total number of nodes which is also equivalent to the total number of surface and volume zones combined for an enclosure and F_n is the number of nodes on the front surface. Figure 3 shows the results of the reciprocity error fractions before and after smoothing between the front and bottom surfaces as a function of variations in the enclosure dimensions for a constant grid size with four nodes per side.

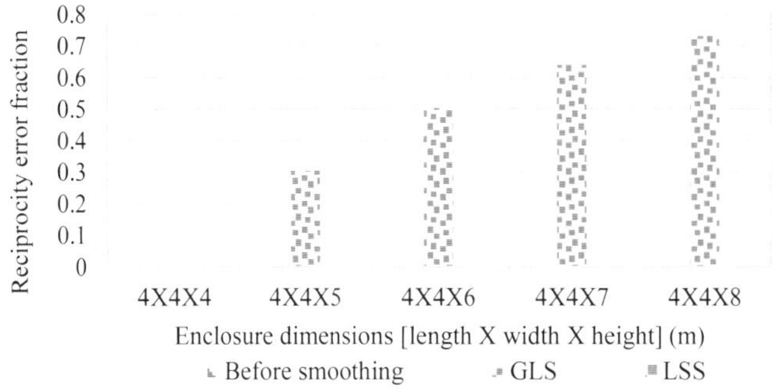

Figure 3: Reciprocity error fraction as a function of enclosure dimensions.

It is clear from Figure 3 that for the 4X4X4 m case the reciprocity error fractions are zero before smoothing as well as after application of both smoothing techniques. This indicates that all three sets of DEAs satisfy the reciprocity relationships for a cubic enclosure. As the shape of the enclosure becomes more rectangular and less cubic, the DEAs before smoothing and after LLS smoothing continue to satisfy the reciprocity relationships. However, the reciprocity errors increase for GLS smoothing as the enclosure dimensions become more rectangular. This reveals that the GLS technique produces DEAs that satisfy the reciprocity relationships for cubic enclosures only, whereas LSS satisfy reciprocity relationships for both cubic and rectangular enclosures, hence only the GLS error fractions are visible in Figure 3.

The last rectangular enclosure reported in Figure 3 with length, width and height of 4X4X8 m is now studied further to investigate the reciprocity compliance of the GLS technique for different levels of discretisation, while keeping the dimensions of the enclosure constant. Figure 4 shows results of the reciprocity error fractions between the front and back surfaces as well as the front and bottom surfaces as a function of grid size.

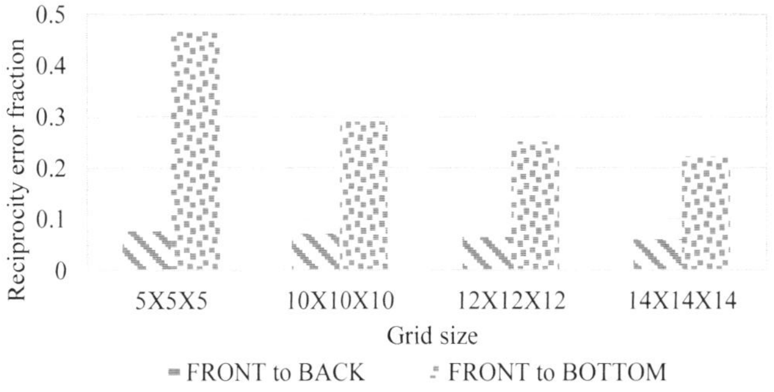

Figure 4: Reciprocity error fraction as a function of grid size.

In Figure 4 it can be seen that there is a decrease in reciprocity error fractions with an increase in grid size, i.e. finer discretisation. It can also be seen that the reciprocity errors are smaller for the DEAs of the surfaces that are directly facing one another (front to back) and larger for the surfaces that are perpendicular to one another (front to bottom).

While Figure 4 shows results where the enclosure is discretised into an equal number of nodes on each surface, Figure 5 shows the results where the number of nodes are in proportion to the physical dimensions of the enclosure. Figure 5 shows a trend that is similar to that of Figure 4, but with smaller errors.

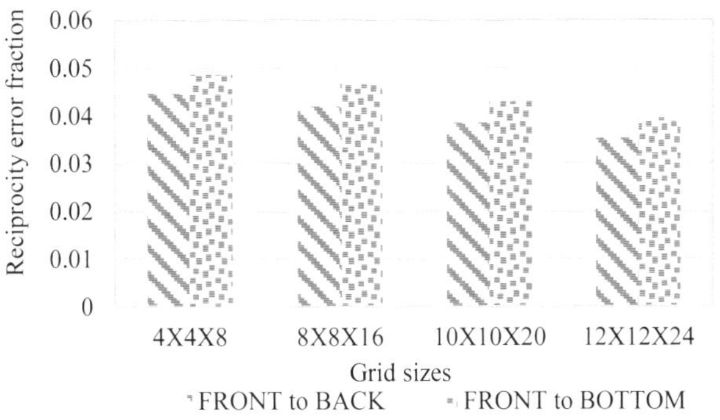

Figure 5: Reciprocity error fraction as a function of grid size.

The results provided above indicate that the application of the GLS technique disturbs the reciprocity relationships. The unwanted effect of the disturbance is however reduced by increasing the level of discretisation and/or by discretising in proportion to the physical dimensions of the enclosure. This implies that the resultant area zones become more square-shaped and the volume zones more cube-shaped.

Energy conservation

In the following sections the results for simple cubic and rectangular three-dimensional enclosures are studied further. Specific focus will be put on the resultant energy balances obtained from the zonal model, with and without smoothing. Furthermore, comparison of the temperature distribution results of the rectangular enclosure and an example available in literature will be shown. For both the cubic and rectangular geometries, the radiation heat transfer analysis is done using the zonal method with a homogeneous gray participating medium inside the three-dimensional enclosures.

Case 1 - Cubic enclosure: The first case is based on a cubic enclosure with side lengths of 4 m. The walls making up the cubic enclosure are all specified with a constant temperature of 700 ^{o}C and an emissivity of 0.85. The enclosure holds a medium with an absorption coefficient of 0.3 m^{-1} and a heat source of 5 kW/m^{3} is distributed uniformly inside the volume of the enclosure. For this case, DEAs were calculated and employed in the zonal method to calculate surface fluxes at different grid sizes in order to evaluate whether grid independence is obtained.

Figure 6 shows the sum of the DEAs on the front surface as viewed from all the other surfaces making up the enclosure, at different grid sizes. The results are shown for three approaches, namely (i) without smoothing, (ii) after smoothing using the GLS technique and (iii) after smoothing using the LSS technique. The enclosure was discretised up to a total grid size of 2197 nodes. At this grid size, the percentage difference between the DEA results of subsequent grid sizes is less than 0.5%.

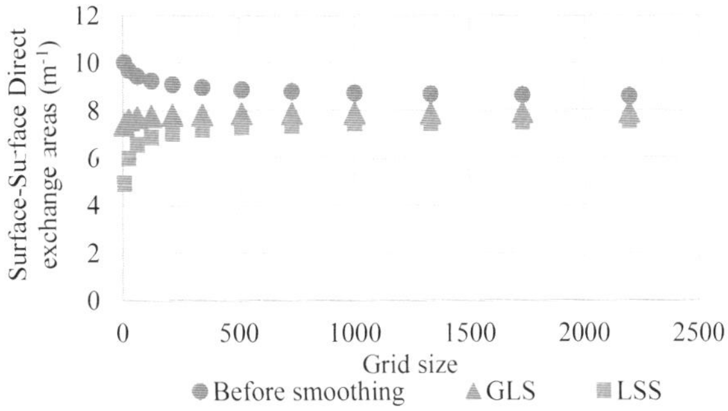

Figure 6: Front surface-to-surface DEAs as a function of grid size.

From the results it can be seen that the calculated DEAs are in fact influenced by the fineness of the grid and that it becomes more or less grid independent for grid sizes of more than 2000 nodes. Also note that the three approaches do not provide the same results, even after grid independence has been obtained.

For the same grid sizes, the corresponding front surface heat fluxes were calculated and are shown in Figure 7. The negative values of the heat fluxes indicate a net flow of heat towards the surface. The values calculated using the DEAs without smoothing show significant grid dependence but tend to become grid independent at around 2000 nodes. For both cases where smoothing is applied, the results are identical and grid independent. This is despite the fact that the DEAs are not grid independent in all cases, as illustrated in Figure 6.

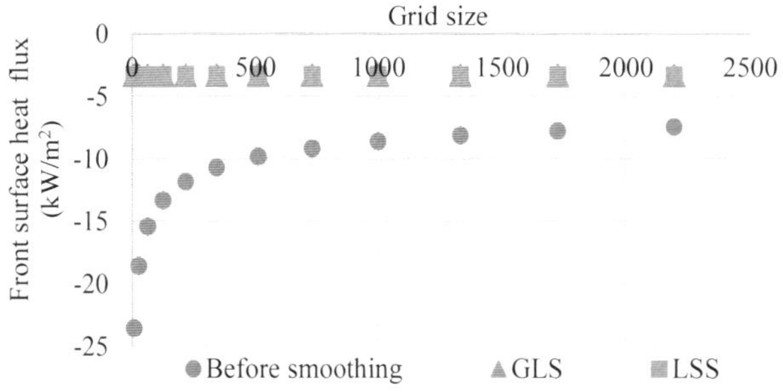

Figure 7: Front surface heat flux as a function of grid size.

The next question to be addressed is whether the total energy balance is grid independent or not and what the impact of smoothing on the energy balance is. Figure 8 shows the energy balance results obtained from using the DEAs with and without smoothing at different grid sizes. Note that the GLS results are not clearly visible since it lies directly underneath the LSS results, i.e. the results are identical.

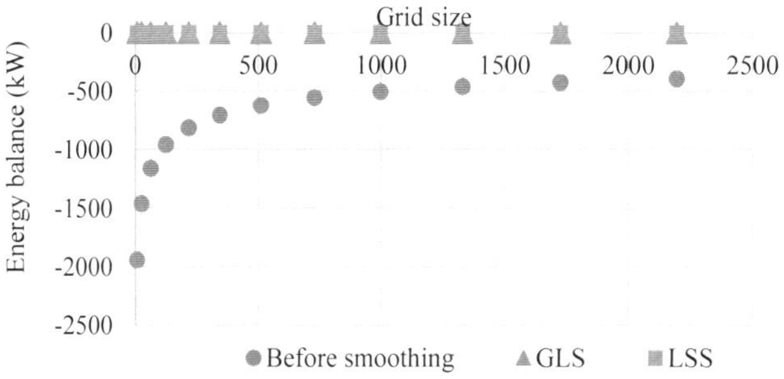

Figure 8: Energy balance on cubic enclosure.

From the results it is clear that without smoothing the energy balance is highly grid dependent and that energy conservation is in fact never achieved, even at very large grid sizes. However, as expected, smoothing ensures that an overall energy balance is obtained in all cases.

Hence, it can be concluded that smoothing is absolutely essential for this case, but that any of the two smoothing techniques can be used to ensure that energy balance is obtained. It was also shown that the heat flux results are essentially independent of grid size, even if the DEAs are not grid independent. However, since the cubic enclosure is essentially symmetric, the effects of asymmetry in the boundary values and geometry still need to be investigated.

Case 2 - Rectangular enclosure: The second case is based on a rectangular enclosure. For this case study, the enclosure is filled with a medium that has an absorption coefficient of 0.5 m^{-1} and a heat source of 5 kW/m^3 is distributed uniformly inside the volume of the enclosure. The boundary condition of the enclosure is taken as similar to those employed by Ebrahimi et al. [7] as tabulated in Table 1.

Table 1. Enclosure boundary conditions [7].

Firing end wall	Bottom surface (z = 0 m)	T = 1200 K	ε = 0.85
Exit end wall	Top surface (z = 4 m)	T = 400 K	ε = 0.70
Side walls	Other surfaces	T = 900 K	ε = 0.70

The first rectangular enclosure has length, width and height of 2X2X4 m. Figure 9 shows the sum of the DEAs on the bottom surface as viewed from all the other surfaces making up the enclosure, at different grid sizes.

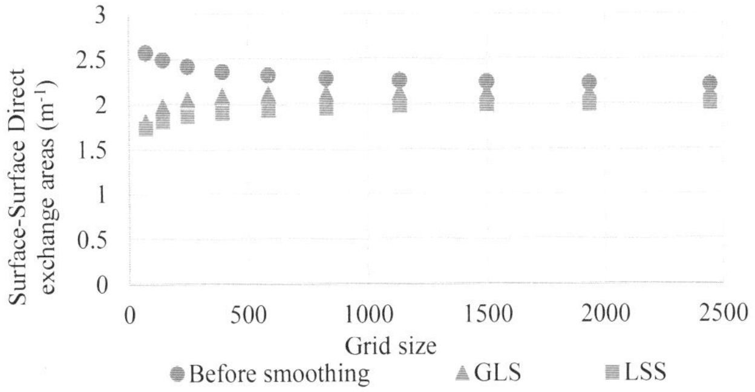

Figure 9: Bottom surface-to-surface DEAs as a function of grid size (2X2X4) m^3.

As in the cubic case, the results show that the calculated DEAs are initially grid dependent and become more or less grid independent with a grid size of over 2000 nodes. Again, the results without smoothing are different from those with smoothing, even after grid independence is obtained.

Figure 10 shows the bottom surface heat flux results obtained from the zonal method at different grid sizes after employing the DEAs reported in Figure 9. The positive values indicate a net flow of heat away from the surface.

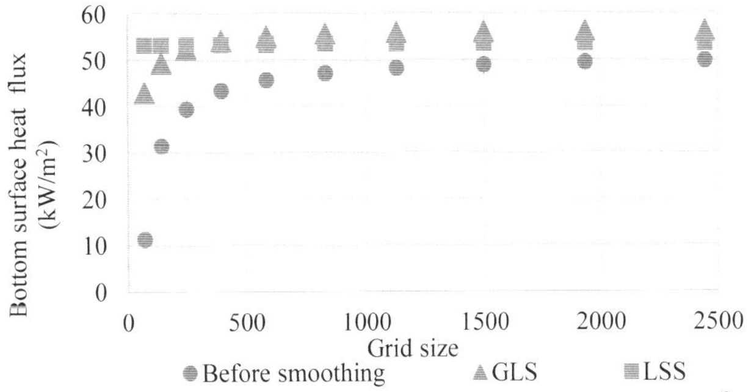

Figure 10: Bottom surface heat flux as a function of grid size (2X2X4) m^3.

As in the cubic case, the heat fluxes calculated using the DEAs without smoothing show significant grid dependence and only become grid independent at around 2000 nodes. Furthermore, the heat fluxes with GLS smoothing which were independent of grid size for the cubic case are now also grid dependent. Similar to the cubic case, the LSS smoothed values of heat flux prove to be independent of grid size despite the fact that the DEAs are not grid independent in all cases. This shows that the grid independence of the LSS results is not determined by the geometry or boundary values being fully symmetric, whereas the GLS results are impacted by the non-symmetry. The impact on the overall energy balance is demonstrated by the results shown in Figure 11.

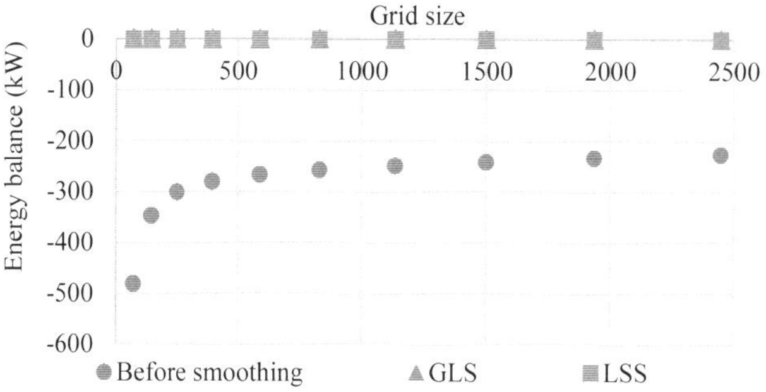

Figure 11: Energy balance on rectangular enclosure (2X2X4) m^3.

Figure 11 shows the energy balance results obtained from the application of the zonal method with and without applying smoothing. As in the cubic case the energy balance results are highly grid dependent for the case without smoothing and again energy balance is never achieved, even at very large grid sizes. As expected, the DEAs with smoothing again ensure that an overall energy balance is obtained for all cases.

The results provided above indicate that both the LSS and GLS smoothing techniques are highly successful in enforcing the energy balance requirements, for which they were initially derived. Furthermore, the resultant temperature distributions within the enclosure obtained from the zonal method after using the DEAs smoothed via the LSS technique is shown in Figure 12.

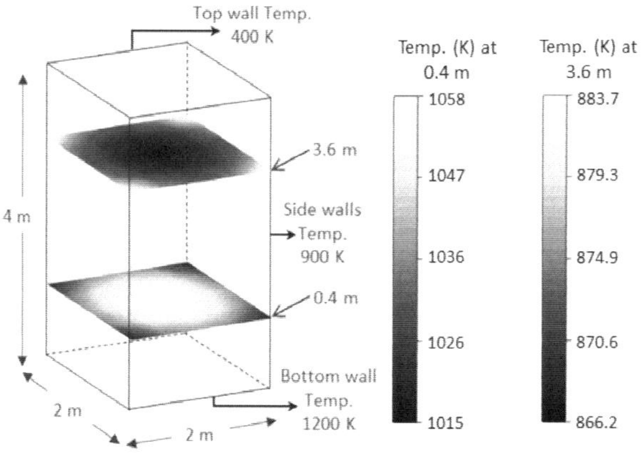

Figure 12: Medium temperature distribution at heights of 0.4 m and 3.6 m.

Figure 12 shows the medium temperature distributions at a height of 0.4 m and 3.6 m from the application of the LSS technique. Here, it can be seen that at a height of 0.4 m, the highest medium temperatures are found at the centre and gradually decrease towards the walls of the enclosure. This is due to the side wall boundaries set at fixed temperatures of 900 K with the bottom wall temperature set at 1200 K. It can also be seen that there is a steep decrease in temperature near the corners. This is due to the effect of the neighbouring low temperature walls surrounding the enclosure [14]. It is interesting to see that at a height of 3.6 m, the medium temperature profile is inverted. This is mainly due to the lower top wall temperature that is fixed at 400 K.

Case 3 – Extinction coefficient versus discretisation: For this case, only the DEAs smoothed via the LSS technique are used to solve the radiation heat transfer network. This investigation is done for the rectangular enclosure presented in Case 2, namely with dimensions 2X2X4 m. The solutions obtained for different values of extinction coefficient together with three different grid sizes are investigated namely 3X3X8 (72 nodes), 9X9X14 (1134 nodes) and 12X12X17 (2448 nodes). Here, the case with 2448 nodes serves as the benchmark fine grid while the case with 1134 nodes represents an average sized grid and the 72 nodes a coarse grid. Figure 13 shows the calculated total heat flux away from the front and bottom surfaces respectively for different values of extinction coefficient and for different grid sizes.

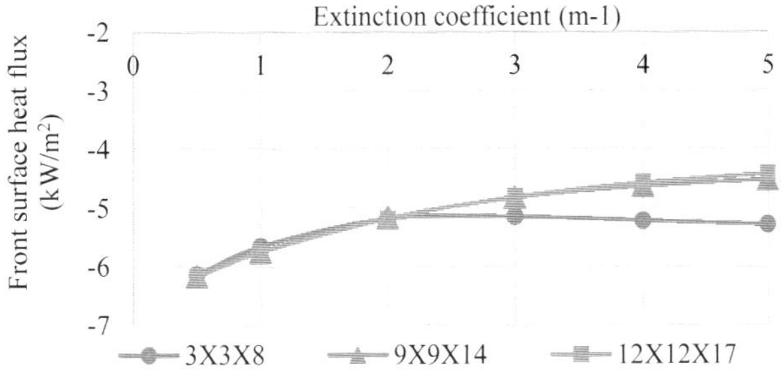

Figure 13: Surface heat flux as a function of extinction coefficient at different grid sizes.

Figure 13 shows that the net heat flux onto the front wall decreases continually as the value of the extinction coefficient is increased, and that the results are essentially the same for the fine and average grids. As the value of the extinction coefficient increases, less of the heat generation distributed within the medium actually reaches the wall.

Also evident in Figure 13 is that the coarse grid provides physically realistic results only for lower values of the extinction coefficient. For higher values the results become unrealistic with the heat fluxes increasing with an increase in the value of the extinction coefficient. The reason for this is that as the grid becomes coarser, the distances between the nodes representing the volume and surface zones become larger. Therefore, the values of the original unsmoothed DEAs are reduced proportional to $\exp(-\beta R)$. As the extinction coefficient increases the DEAs become even smaller and eventually approach zero.

These results show that although LSS smoothing provides DEAs that satisfy both the reciprocity and energy conservation constraints in all cases including very coarse grids, it does not necessarily guarantee physically realistic results if the level of discretisation is too coarse. Higher values of extinction coefficient require finer discretisation that properly accounts for the reduced effective conductivity between zones that are situated further apart.

CONCLUSION

The purpose of this paper was to systematically investigate the impact of two DEA smoothing techniques for modelling radiation heat transfer in three-dimensional enclosures using the zonal method.

For all the cases investigated here, the calculated DEAs are influenced by the fineness of the grid, but do eventually become grid independent for grid sizes in the order of 2000 nodes. The cases without smoothing and those where the GLS and LSS techniques were applied do not provide the same results, even after grid independence has been obtained. For the cubic geometry with symmetric boundary values, the heat fluxes derived from the two smoothing techniques are identical and independent of grid size. However, this is due to the symmetry of the geometry and boundary conditions. For all the cases investigated both the smoothing techniques ensure that an overall energy balance is achieved, as expected.

The LSS technique provides consistent results for all cases, including very coarse grids, and is therefore deemed to be a very reliable and convenient technique. However, it does not necessarily guarantee physically realistic results if the level of discretisation is too coarse relative to the value of the extinction coefficient. To ensure physically realistic results at larger values of extinction coefficient, the level of discretisation must account for the reduced effective conductivity between zones that are situated further apart.

For all the cases investigated here it is clear that without smoothing the overall energy balance is highly grid dependent and energy conservation is in fact never achieved, even for very large grid sizes, i.e. very fine discretisation. It can therefore be concluded that smoothing is absolutely essential for all cases.

ACKNOWLEDGEMENTS

The first author wishes to acknowledge the contribution of the relevant parties in the Eskom Power Plant Institute specialisation centre in Energy Efficiency at the University of Cape Town for the opportunity to participate in the programme as well as the THRIP programme and its industrial partner M-Tech Industrial (Pty) Ltd plus the NRF for providing the funding that enabled this opportunity. In addition, the authors wish to thank Dr Wim Fuls at the University of Cape Town for his useful contributions to this study.

REFERENCES

[1] K. Lari and S. A. Nassab, "Analysis of combined radiative and conductive heat transfer in three-dimensional complex geometries using blocked-off method," *Transactions of Mechanical Engineering*, vol. 35, pp. 107-119, 2011.

[2] A. Batu and N. Selcuk, "Modeling of Radiative Heat Transfer in the Freeboard of a Fluidized Bed Combustor Using the Zone Method of Analysis," *Turkish Journal of Engineering & Environmental Science*, vol. 26, pp. 49- 58, 2002.

[3] H. C. Hottel and E. S. Cohen, "Radiant heat ex-change in a gas-filled enclosure: allowance for non-uniformity of gas temperature," *AIChE Journal*, vol. 4, pp. 3-14, 1958.

[4] M. E. Larsen and J. R. Howell, "Least-squares smoothing of direct exchange areas in zonal analysis," *Journal of Heat Transfer*, vol. 108, pp. 239-242, 1986.

[5] R. Méchi, H. Farhat, K. Guedri, K. Halouani, and R.. Said, "Extension of the zonal method to inhomogeneous non-grey semi-transparent medium," *Energy*, vol. 36, pp. 1-15, 2010.

[6] H.C. Hottel and A.F. Sarofim, *Radiative transfer*. New York: McGraw-Hill, 1967.

[7] H. Ebrahimi, A. Zamaniyan, and J. S. M. Zadeh, "Three Dimensional Modeling of Radiative Heat Transfer in Furnaces," *9th Iranian Chemical Engineering Congress*, pp. 2609-2619, 2004.

[8] M.F. Modest, *Radiative heat transfer*, 3rd ed. New York: Elsevier, 2013.

[9] G. Martins, O.S.M Hernandez, and E.P.F. Bandarra, "Zonal method implementation to determine the thermal radiation heat transfer in bi-dimensional furnaces," *14th Brazilian Congress of Thermal Sciences and Engineering*, pp. 18-22, 2012.

[10] R Siegel and J. R. Howell, *Thermal radiation heat transfer*. New York: McGraw Hill, 1972.

[11] W. Tian and W. K. S. Chiu, "Calculation of Direct Exchange Areas for Nonuniform Zones Using a Reduced Integration Scheme ," *Journal of Heat Transfer*, vol. 125, pp. 839-844, 2003.

[12] D. A. Lawson, "An improved method for smoothing approximate exchange areas," *International Journal of Heat Transfer*, vol. 38, no. 16, pp. 3109-3110, 1995.

[13] C.V.S Murty and B.S.N Murty, "Significant of exchange area adjustment in zone modelling," *International Journal of Heat and Mass transfer*, vol. 34, no. 2, pp. 499-503, 1991.

[14] Y. Lei, Y. Zhongyuan, L. Ming, L.U. Pengfei, and L. Junjie, "Numerical simulation of radiative heat transfer in cylindrical enclosure," *Chinese scientific papers online*, pp. 1-9, 1989.

[15] R. J. Tucker, "Direct Exchange Areas for Calculating Radiation Transfer in Rectangular Furnaces," *Journal of Heat Transfer*, vol. 108, pp. 707-710, 1986.

Analysis of vibration inducing sources of a large-scale cooling system fan blade

Jacques Muiyser[1,*], Danie Els[1], Johan van der Spuy[1], Nico Basson[1]

Department of Mechanical and Mechatronic Engineering, Stellenbosch University

jmuiyser@sun.ac.za

ABSTRACT

In South Africa a large percentage of the country's power is generated using coal-fired power stations where condensation of the process steam is achieved by means of heat rejection to ambient air. This process occurs in what is called an air-cooled condenser where an array of fans forces the air through inclined fin-tube heat exchanger bundles. These fans are installed at heights of approximately 50 m and are subjected to a variety of inlet air flow conditions due to the presence of surrounding fans, buildings and the effect of the prevailing wind conditions. Strain gauge measurements have been recorded at such a fan during operation where large vibrations of the blades were found. It was expected that these vibrations were caused by the dynamic blade loading brought about by distorted inlet air flow conditions and the effect of the downstream fan bridge that acts as a flow obstruction. However, without the ability to measure for an extended period of time it is difficult to isolate and attribute certain dynamic measurements to their aerodynamic causes. It is for this reason that two sets of tests were conducted on scale representations of the fan system where the expected sources of vibration could be much better isolated and investigated. Tests were conducted on a single axial flow fan with a downstream flow obstruction as well as a fan installed in a multiple fan test facility where the platform height could be adjusted to alter the degree of inlet air flow distortion. The results showed that blade vibration is indeed caused, in part, by the identified aerodynamic sources and can be used as input parameters for further investigations or system verification tests.

KEYWORDS

Axial flow fans, air-cooled condenser, distorted inlet air flow, blade vibration

NOMENCLATURE

D_{fan}	Fan diameter [m]
E	Strain amplitude [µm/m]
E_{max}	Maximum strain amplitude [µm/m]
H_f	Fan platform height [m]
ε_{avg}	Measured strain [µm/m]
ε^*_{avg}	Maximum average measured strain for the reference case [µm/m]
ε_{dyn}	Dynamic strain [µm/m]
ε_{rms}	Rms of measured strain [µm/m]
ε^*_{rms}	Maximum strain rms per test case [µm/m]

INTRODUCTION

In coal-fired power stations water is boiled to produce steam which drives a turbine connected to a generator. Mechanical draft air-cooled condensers (ACCs) use air as a cooling medium to extract heat from, and thereby condense, the process steam once it exits the low pressure side of the turbine. In arid regions the use of air as a cooling medium is preferable to water as it has significant environmental and financial benefits [1]. In large-scale forced draft ACCs there are generally in excess of 200 fans installed at a height of 50 m above ground. These fans are approximately 9 m in diameter and force air through inclined finned tube heat exchanger bundles.

Each fan draws in air from below the ACC and is suspended from a bridge where the motor and gearbox is mounted. As shown in Figure 1, a typical fan bridge structure is constructed from steel I-beams, has a square mesh walkway and is located approximately 800 mm above the fan tip's plane of rotation. Due to this configuration the fan blows air upwards through the bridge which then acts as a downstream flow obstruction.

Additionally, when operating within an ACC, a fan is subjected to distorted inlet air flow conditions as well as the effect of various flow obstructions. These conditions have been shown to reduce the performance of the fans and cause the blades to vibrate as a result of the variation in aerodynamic loading. Research has been conducted to quantify the effect of the presence of a fan bridge on the performance of a cooling system fan [2]. The effects of various components in a cooling system on the fan's performance were analysed and it was found that the presence of a fan bridge results in a drop in fan static pressure rise over the fan for a specific flow rate. Furthermore, full-scale measurements showed that there is a decrease in local air flow downstream of the fan bridge due to it acting as a flow obstruction [3]. The magnitude of the decrease was found to be dependent on the wind speed and direction and is difficult to quantify.

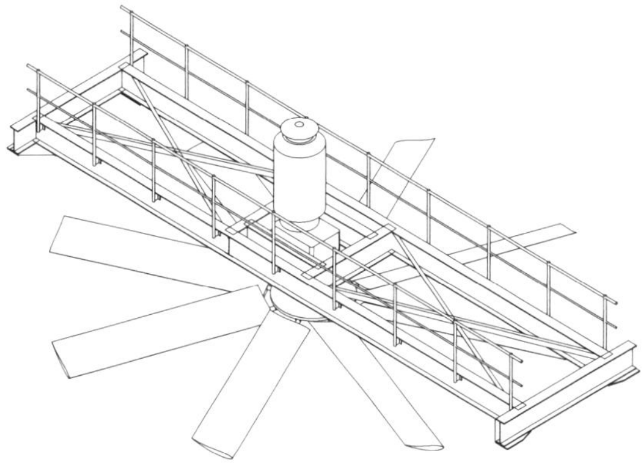

Figure 1. Typical cooling system fan installation

Numerical work was performed by Hotchkiss et al. [4] where a simplified fan model, known as the actuator disc, was used to determine the blade torque and thrust variation on a fan that is subjected to off-axis inflow. It was found that there was an increase in blade loading variation as the off-axis angle increased up to 45°. Bredell et al. [5] used a similar simplified fan model to numerically investigate the bending moments exerted on large-scale fan blades under distorted inlet air flow conditions. The distorted inlet air flow conditions were induced by simulating multiple ACC fans and reducing the platform height, thereby increasing the cross-flow magnitude. The bending loads were presented as a function of the blade's rotational position and it was found that, in general, fan blades experience higher bending moments on the windward side of the casing. However, these researchers did not include the effect of the fan bridge on blade loading.

In addition to published work, the presentations at the annual ACC user's group conferences indicate that the problem of reduced performance and increased blade vibration during windy conditions is of great importance to plant operators worldwide. Maulbetch and Filipino [6] reports on the findings of a study conducted at the Caithness power station. The goal was initially to improve the performance of the power station by installing wind shields, which are made of a porous material that reduces the amount of cross flow at the inlets of fans located on the edges of an ACC. However, the researchers have shown that, in addition to improving the performance of the plant, the wind screens also reduce the loading on the fan blades under windy conditions, thereby showing that the cross flow velocity undoubtedly has an effect on fan system vibration.

Investigating various mitigation strategies for blade vibration, Romano [7] used CFD simulations to determine blade loading under conditions of varying cross wind velocity. These loads were then used to determine the dynamic response of the fan blade and investigate several vibration mitigation techniques. The research showed that wind screens as well as a different fan shaft design reduce blade vibration.

As a result of industry demands a measurement project was conducted on a full-scale fan that showed that, in addition to the effects of distorted inlet air flow conditions, the presence of a fan bridge is an additional source of aerodynamic loading on the cooling system fan blades [8]. These measurements were obtained by attaching strain gauges to a full-scale ACC fan blade and recording measurements during operation. Through analysis of the measured data in the frequency domain it was discovered that a source of aerodynamic loading existed at twice the rotational speed of the fan and was assumed to be as a result of the blade passing the fan bridge. As the gauges were attached directly to the fan blades and no acceleration signals were recorded from stationary components such as bearings no system level vibrations such as that at once and twice the blade passing frequency were observed in the results.

Due to the findings of Muiyser et al. [8] it was decided to further investigate certain hypothesized flow induced vibrations in such a way that their effects may be included in further computation analyses of fan system dynamics. In this paper the effect of the bridge and distorted inlet air flow conditions on the vibration of a cooling system fan blade is experimentally investigated. The effect of the bridge is investigated by performing experiments on a 1.542 m diameter fan where a downstream flow obstruction has been installed. In this way the effect of the distance between the rotor and obstruction is investigated as well as that of the obstruction's porosity. In addition to this set of experiments, the effect of distorted inlet air flow conditions is investigated using a 630 mm fan installed in a multiple fan test facility. This test facility aims to emulate the installed conditions by having multiple fans draw in air past an edge fan, thereby creating distorted inlet air flow conditions. As neither of these fans is the same size or possess the same stiffness as the full-scale counterpart only changes in blade loading could be investigated and not the effect on blade response.

THE EFFECT OF A FAN BRIDGE ON BLADE VIBRATION

To investigate a number of different bridge configurations it was decided to perform the investigation on a 1.542 m diameter experimental fan. The test facility, shown in Figure 2, was designed by Venter [2] and is referred to in the BS848 standard as a type A configuration [9]. Air flow enters the wind tunnel through a fibre glass bell mouth inlet which has a diameter of 1 m. The bell mouth possesses four pressure taps, evenly spaced around its circumference, that allow the measurement of the average static pressure experienced at the inlet. A throttling device is used to regulate the volumetric flow rate of the air through the wind tunnel. The throttling device controls the position of an array of louvres, which restrict air flow and can be adjusted to a total of 21 positions by means of a mechanical arm situated on the outside of the wind tunnel. An auxiliary fan is located downstream of the throttling device and is used to overcome the pressure losses caused by the various components of the wind tunnel. The settling chamber contains a set of three rectangular mesh screens which ensure a uniform inlet air flow velocity profile. The settling chamber pressure is measured by four static pressure taps that are located on the walls of the chamber.

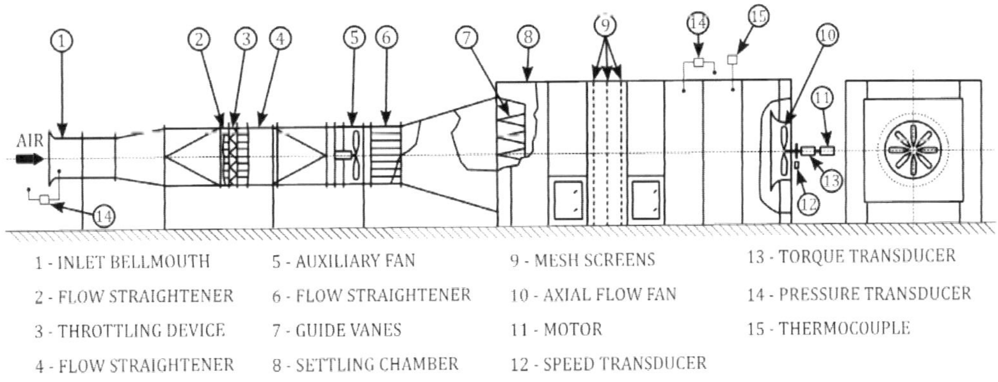

Figure 2. Standard axial flow fan test facility

The 1.542 m diameter test fan is housed in a shroud and is located at the outlet of the settling chamber. The test fan is driven by an electric motor connected through a torque transducer located outside the settling chamber, downstream of the fan, and is controlled by means of a variable speed drive. For the purpose of this investigation a flat plate fan was used with aluminium blades. The blade profiles were machined in accordance with the dimensions given by Riegels [10]. As expected, the flat plate fan does not have the same cross-sectional profile as a typical full-scale cooling fan and as such does not have the same lift and drag characteristics either. However, for the purpose of this investigation it is not necessary to replicate the performance of the full-scale fan, but rather to investigate the effects of the fan bridge on a fan's performance and vibration.

The axial flow fan used to conduct the investigation is 6 times smaller than a typical full scale cooling system fan and as such the bridge was designed by scaling all full scale dimensions by a factor of six as well. The scale model of the bridge was constructed with 20 mm steel square tubing representing the structural I-beams of the full-scale bridge. Non-structural components, such as the hand rails and the walkway supports were omitted due to the assumption that their effect on air flow would be negligible. Furthermore, the scale model of the bridge was designed to be installed at an adjustable distance from the fan rotor, which enabled the investigation of the effect that variations in the distance between the fan bridge and rotor has on fan performance. Two distances between the bridge and fan were investigated. These were selected based on the 1:6 scale of the fan being tested. The selected distances were 150 mm ($0.1D_{fan}$) and 250 mm ($0.16D_{fan}$), representing the cases where the bridge is placed close to the fan and far from the fan, respectively. Finally, two different inserts were used that correlated to high and low porosity. These can be seen in Figure 3 where a bridge containing a mesh and plate insert is shown.

Figure 3. Fan bridge with mesh inserts (left) and plate inserts (right)

To measure blade loading, a set of strain gauges was attached near the root of a single blade of the fan being investigated. As shown in Figure 4, by using a blade profile with flat pressure and suction surfaces, the strain gauges were easily attached and connected in a full bridge configuration to measure the bending strain near the root of the blade. By using a full Wheatstone bridge configuration, maximum measurement sensitivity is achieved and the axial forces exerted on the blade due to centrifugal loading are compensated for. To record the strain measurements from the rotating fan, a MicroStrain SG-Link wireless bridge amplifier was used. This is the same amplifier that was used for a full-scale investigation and is capable of wirelessly transmitting data up to a rate of 736 Hz. An HBM Spider8 was used for the data acquisition from the pressure and torque transducers as well as the rotary encoder used to measure the fan's rotational speed.

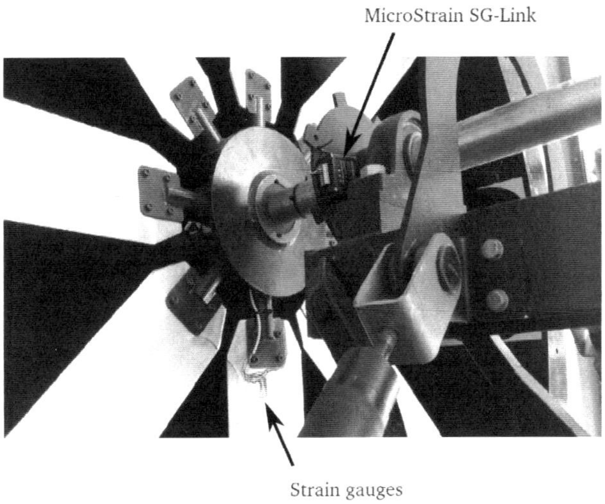

Figure 4. Strain gauges and Microstrain SG-Link installed on fan

It can be assumed that the average blade loading is proportional to the fan static pressure and as such it is expected that the average blade load would also increase with decreasing flow rate. Figure 5 shows the average normalised blade loading for each bridge configuration at a variety of flow rates. The blade loading for this figure has been normalised with respect to the maximum average load for the reference case where there is no bridge present. It can be seen that the average blade load does indeed increase with decreasing flow rate. There is effectively no significant difference between the different loading curves, which indicates that the average blade loading is not sensitive to the bridge configuration.

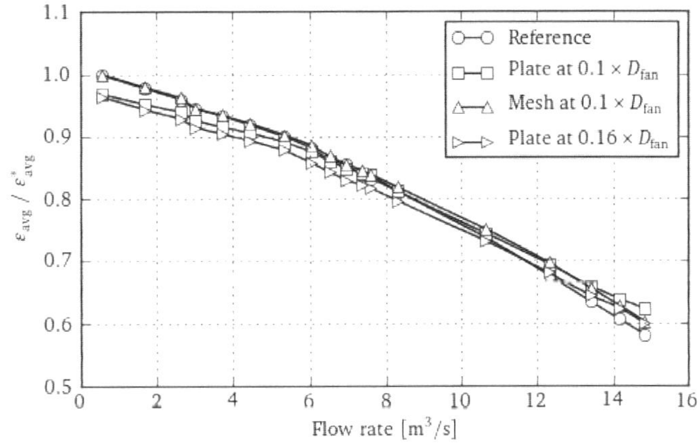

Figure 5. Normalised average blade loading versus flow rate

Upon further investigation of the measured data, it was found that the amplitude of the measured blade loading differs greatly for different bridge configurations and flow rates. Figure 6 shows two cases where strain measurements were recorded while the bridge with plate inserts was installed. The graph on the left shows the measurements recorded at a high flow rate while the graph on the right shows measurements recorded for the same installation at a lower flow rate. Both of these measurements have been normalised with respect to the average loading in each case to illustrate the difference in amplitude only.

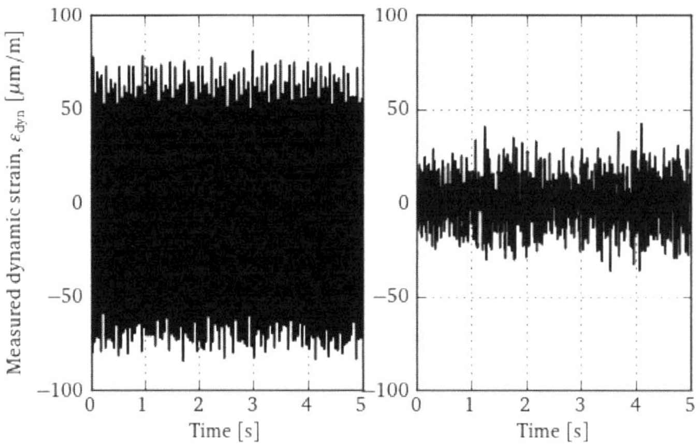

Figure 6. Normalised strain measurements with plate inserts at a high flow rate (left) and a low flow rate (right)

To obtain the amplitude of blade loading for each combination of flow rate and bridge configuration the root mean square (rms) value was calculated for the dynamic component of each set of measurements. This dynamic component was obtained by simply subtracting the average load from the measurements to disregard the effect of increasing or decreasing average blade load as a function of flow rate. The results of this analysis is shown in Figure 7 and shows that there is indeed an increase in the blade loading amplitude for every bridge configuration at higher flow rates. As can be expected, the largest amplitudes are experienced when the bridge is placed close to the fan, with plate inserts and the amplitudes decrease with the decreasing levels of air flow obstruction of the other bridge configurations.

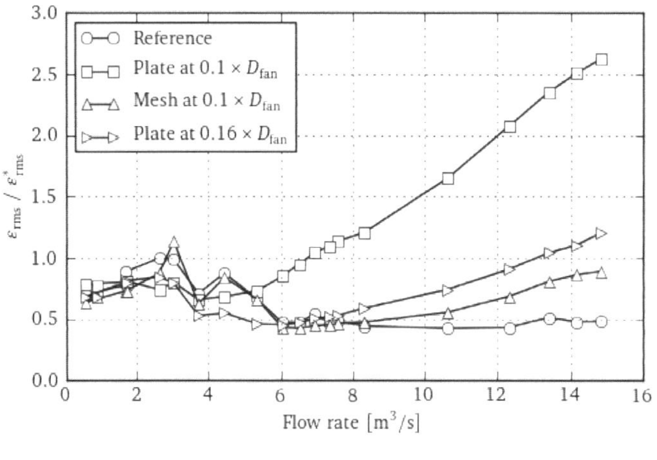

Figure 7. Normalised amplitude of vibration versus flow rate

As can be seen in Figure 7, at low flow rates there is an unexpected increase in blade loading amplitudes for all of the bridge configurations as well as the reference curve. To analyse this occurrence, a FFT was performed on two data sets where the bridge was placed close to the fan with plate inserts and for the case where there was no bridge present. The two data sets corresponded to measurements recorded at a high and low flow rate, respectively. Figure 8 shows the FFTs for the case where the bridge was placed close to the fan and contained plate inserts. The figure exhibits three distinct peaks at 12 Hz, 18 Hz and 24 Hz where the peak at 12 Hz corresponds to the rotational speed of the fan causing a variation in blade loading due to the effect of gravity. The peak at 24 Hz corresponds to the bridge passing frequency, which is twice the rotational speed of the fan. Note that the peak at the bridge passing frequency is much larger at higher flow rates, which is to be expected considering the results shown in Figure 7. In contrast, the peak at the rotational speed of the fan remains the same irrespective of the flow rate. Alternatively, the third peak at approximately 18 Hz only appears at low flow rates, which also coincides with the observation in Figure 7. In addition to these peaks there are also other peaks coinciding with the harmonics of the fan's rotational speed. However, these peaks are very small when compared to the effect of the fan bridge at twice the rotational speed and other components such as the blade passing frequency are not present.

Note that the working fluid for these test cases is ambient air and that no steam is present in the system. As such, there is a negligible temperature differential over the fan. Furthermore, it is assumed that the reason for the increase in vibration at low flow rates is due to the onset of stall instead of some mechanical interaction. Research conducted on axial flow fans operating at low flow rates have shown the transient nature of flow separation during stall [11] which may cause the more broadband excitation observed at 18 Hz. However, the detailed investigation of air flow phenomena falls outside the scope of the current project and was not analysed further.

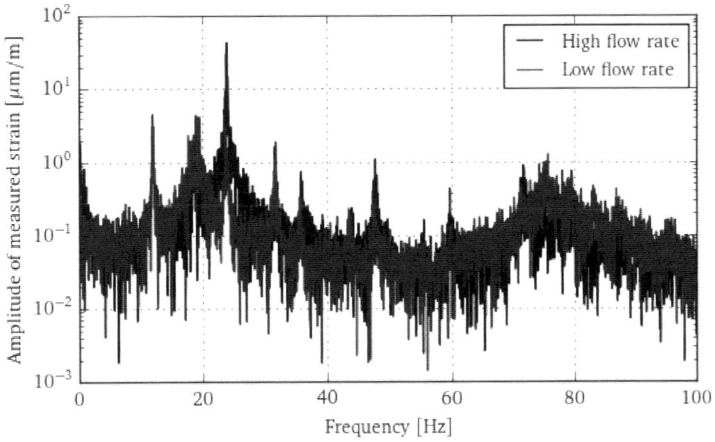

Figure 8. FFTs of measured strain

THE EFFECT OF DISTORTED INLET AIR FLOW CONDITIONS ON FAN BLADE VIBRATION

To investigate the effect of varying levels of cross flow Visser [12] constructed a multiple fan test facility where up to 6 fans of 630 mm diameter could be tested. This test facility is shown in Figure 9. In order to alter the severity of fan inlet crossflow, the platform height could be adjusted. Visser [12] found that the edge fan experienced the greatest reduction in volumetric flow rate at low platform heights. However, this was improved with the addition of a 150 mm wide walkway to the windward side of the edge fan.

Salta and Kröger [13] made use of the same multiple fan test facility to show that fans located on the periphery of the ACC are more susceptible to distorted inlet air flow conditions than the interior fans. They published graphs showing the decrease in each of the 6 fans' volumetric effectiveness where it was clear that the edge fan experiences the greatest decrease at decreasing platform heights.

Conradie [14] reduced the number of fans in the test facility from six to three after which Van der Spuy [15] made use of the same multiple fan testing facility to evaluate the effectiveness of various simplified fan models. The actuator disk and pressure jump methods were evaluated at low platform heights by measuring the volumetric effectiveness of each 630 mm fan as well as the inlet air flow distribution by means of particle image velocimetry (PIV). These measurements showed clear distorted inlet air flow at the fan inlets when the platform height is decreased, with the edge fans affected most severely. Van der Spuy [15] also developed the extended actuator disk method in order to account for the inception of radial flow at low platform heights.

Figure 9. Multiple fan test facility

Finally, Fourie [16] used the multiple fan testing facility to investigate the effect of wind on the performance of axial flow fans in air-cooled steam condenser systems. The test facility was used to validate numerical models which were then used to predict the performance of a complete ACC operating under windy conditions.

The fan to be tested was the same N-fan used by Van der Spuy [15] and Fourie [16]. The reason for this is that the expected natural frequency of the slender blades would be low enough to conduct a test that is comparable to full-scale measurements. The geometry of the fan is such that it is a 630 mm replica of a 9.216 m fan, currently being used in the power generation industry. The exact details of the fan geometry are described by Conradie [14].

Strain gauges were attached to a single Polyvinyl chloride (PVC) fan blade using cyanoacrylate. The 350 Ω HBM gauges were configured to measure full-bridge bending strain and can be seen in Figure 10. The gauges were then connected to a MicroStrain SG-Link, which is a wireless bridge amplifier used to measure the bending strain during operation. Two SG-Link units were attached to the hub of the fan to minimise any unbalance and can be seen in Figure 10. The brackets on which the SG-Link units were attached do not extend more than a few mm past the edges of the hub and as such is expected not to influence the air flow into the fan. Measurements were recorded at a rate of 512 Hz, which was deemed acceptable when compared to the 15.5 Hz (930 rpm) speed of the fan. This speed, which is lower than the 1000 rpm used by Van der Spuy [15] and Fourie [16], was selected because it is approximately half of the blade's natural frequency, which was measured to be equal to 29 Hz.

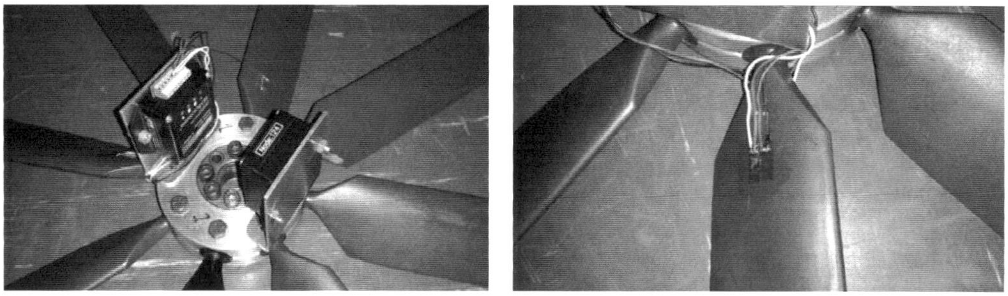

Figure 10. SG-Link amplifiers installed on fan hub (left) and strain gauges attached to fan blade (right)

To simulate the effects of increasing cross flow at the instrumented edge fan, the floor height was adjusted to reduce the cross-sectional inlet area of the multiple fan installation. Tests were conducted at dimensionless platform heights, H_f, of $4.5D_{fan}$, $3.5D_{fan}$, $2.5D_{fan}$ and $1.5D_{fan}$. At each height, as well as a reference point with no floor installed, the blade loading was measured for approximately 30 s.

To investigate the effect of cross flow on blade vibration the root mean square (rms) value of the blade loading was calculated for the various platform heights. It was found that calculating the rms of the dynamic portion of the signal, i.e. when the mean has been subtracted, provides a much better indication of the increase in vibration as the platform height is reduced. Figure 11 clearly shows the increase in the measured vibration as the platform height is reduced from $4.5D_{fan}$ to $2.5D_{fan}$. At a platform height of $1.5D_{fan}$ a slight decrease in bending load is observed.

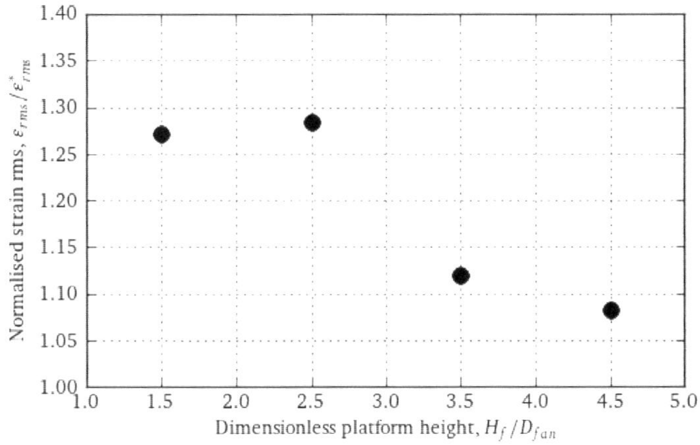

Figure 11. Dimensionless blade loading versus platform height

The decrease in blade loading observed at a platform height of $1.5D_{fan}$ is counter-intuitive due to the fact that the cross flow velocity, and distorted inlet air flow, increases with decreasing platform height. Figure 12 shows that the measured response of the blade differs dramatically for each of the case where $H_f = 2.5D_{fan}$ (left) and $H_f = 1.5D_{fan}$ (right). Such a change in response would indicate that there is a change in the excitation mechanism.

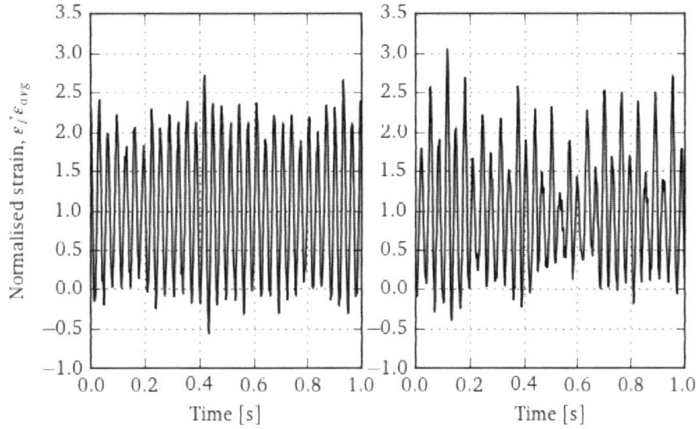

Figure 12. Normalised bending strain for $\boldsymbol{H_f = 2.5D_{fan}}$ (left) and $\boldsymbol{H_f = 1.5D_{fan}}$ (right)

Figure 13 shows FFTs of the measured strain for the case where the platform height is at $2.5D_{fan}$ and $1.5D_{fan}$. To allow for comparison between these two cases the results have been normalised with respect to the amplitude of the response at twice the rotational speed of the fan measured for each case. This normalisation also allows for comparison of the relative magnitudes of the other harmonics. The results show that there is a definite change in excitation mechanism as the platform height is reduced beyond a height of $H_f = 2.5D_{fan}$ due to the fact that the ratio between the harmonics has changed. At a platform height of $1.5D_{fan}$ the response component at a frequency of twice the rotational speed is not as dominant as it is at a height of $2.5D_{fan}$. The dotted line plotted on Figure 13 shows the normalised magnitude of the response component measured when there is no floor installed where one can see that the normalised magnitude of this component is comparable to that of the case where the platform is at the low height of $1.5D_{fan}$. This correlation indicates that the excitation mechanism at a low platform height and an infinite platform height may be similar.

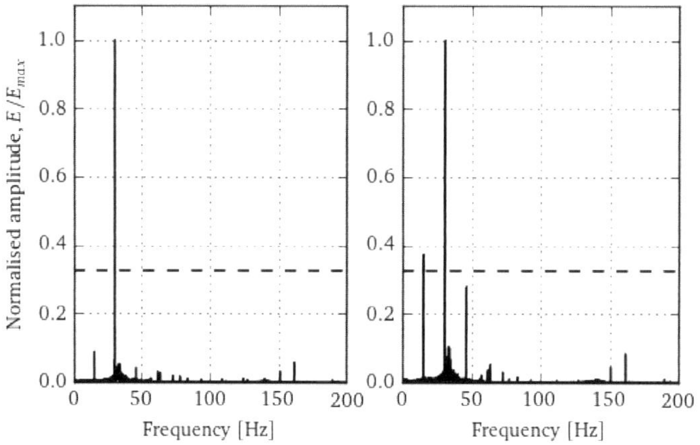

Figure 13. FFT of blade loading when $\boldsymbol{H_f = 2.5D_{fan}}$ (left) and $\boldsymbol{H_f = 1.5D_{fan}}$ (right)

To visualise the change in excitation mechanism the results of two numerical simulations performed by Van der Spuy [15] using the actuator disk model for a platform height of $H_f = 2.5D_{fan}$ and $H_f = 1D_{fan}$ are shown in Figure 14. The air flow distributions indicate that at a platform height of $1D_{fan}$ the cross flow velocity becomes so great that the area of flow separation bypasses the edge fan entirely. In contrast, at a height of $2.5D_{fan}$ the inlet air flow at the edge fan is clearly distorted and one would expect the blade's to experience a much greater variation in loading. Note that these observations only apply to the edge fan and that higher cross flow velocities may cause the area of flow separation at the edge of the ACC to extend to interior fans, resulting in increased vibration levels for these fans as well.

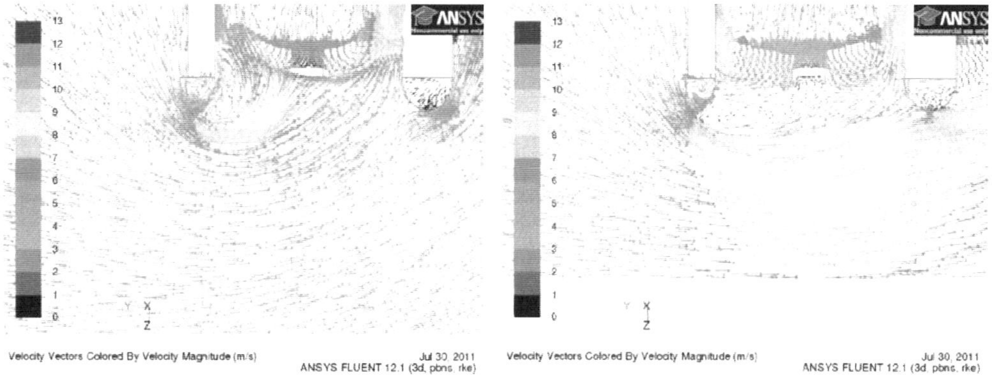

Figure 14. Simulation results obtained using the actuator disc when $\boldsymbol{H_f = 2.5D_{fan}}$ (left) and $\boldsymbol{H_f = D_{fan}}$ (right). Source: Van der Spuy [15]

CONCLUSION

The investigation has shown that even if a fan is expected to operate under steady air flow conditions, it cannot be assumed that the blade loading will also be steady if there is a downstream flow obstruction such as a fan bridge present. As such, fan bridges need to be designed in such a way to minimise its effect on blade vibration. As a result of these observations, in order to minimise blade vibration, it is recommended that if the fan bridge is constructed from a porous material, the bridge distance should be at least 0.16 times the fan diameter. This distance corresponds to the furthest setting that was tested and as such is the only value that may be proposed based on this investigation's findings.

Furthermore, the results obtained verify that increasing wind speeds increase the levels of vibration experienced by fans installed in an ACC. This correlates well with the findings of Bredell et al. [5] who made use of the actuator disc to model a fan operating under distorted inlet air flow conditions. It is therefore recommended that the vibration mitigating effects of the wind shields investigated by Maulbetch and Filipino [6] be quantified through similar experimentation. Additionally, the investigated platform heights should be translated to effective cross flow speeds in order to determine whether or not prevailing winds will cause excessive fan blade vibrations.

Finally, even though the results presented in this investigation have not been compared to any numerical simulations, it is expected that the findings may be used to improve future models. These results may also be used to simplify the excitation mechanisms included in detailed numerical models, similar to the combined potential flow and finite element approximation of a fan operating under distorted inlet air flow conditions of Muiyser et al. [17].

REFERENCES

1 Kröger, D.G., "Air-cooled Heat Exchangers and Cooling Towers", Penwell Corp. Tulsa, 2004.

2 Venter, S. "The Effectiveness of Axial Flow Fans in A-Frame Plenums", Ph.D. thesis, Department of Mechanical and Mechatronic Engineering, University of Stellenbosch, South Africa, 1990.

3 Van Aarde, D., "Vloeiverliese deur 'n A-raam vinbuisbundel in 'n lugverkoelde kondensator", Master's thesis, Department of Mechanical and Mechatronic Engineering, University of Stellenbosch, South Africa, 1990.

4 Hotchkiss, P et al., "Numerical investigation into the effect of cross-flow on the performance of axial flow fans in forced draught air-cooled heat exchangers", Applied Thermal Engineering, Vol. 26, pp 200-208, 2006.

5 Bredell, J. et al., "Numerical investigation into aerodynamic blade loading in large axial flow fans operating under distorted inflow conditions", R&D Journal of the South African Institute of Mechanical Engineering, Vol. 22, pp 11-17, 2006.

6 Maulbetsch, J.S. and DiFilippo, M.N., "Caithness windscreen study", 7th Annual ACC users group conference, Gettysburg, PA, 2015.

7 Romano, N. "Wind loads on fan blades & blade dynamics", 7th Annual ACC users group conference, Gettysburg, PA, 2015.

8 Muiyser, J. et al., "Measurement of air flow and blade loading at a large-scale cooling system fan", R&D Journal of the South African Institute of Mechanical Engineering, Vol. 30, pp 30-38, 2014.

9 BS848 Part 1: Performance testing using standardized airways, Fans for general purposes, 1997.

10 Riegels, F. "Results from Wind Tunnel Investigations", London Butterworths, 1961.

11 Louw, F.G. et al. "Lift and drag characteristics of an air-cooled heat exchanger axial flow fan", Journal of Fluids Engineering, Vol. 137, pp. 1-9, 2015.

12 Visser, J. "Die invloed van Versteurde Inlaatvloeipatrone op Aksiaalwaaiers", Master's thesis, Department of Mechanical and Mechatronic Engineering, University of Stellenbosch, South Africa, 1990.

13 Salta, C. and Kröger, D.G., "Effect of inlet flow distortions on fan performance in forced draught air-cooled heat exchangers", Heat Recovery Systems & CHP, Vol. 15, pp 555-561, 1995.

14 Conradie, P. "Edge fan performance in air-cooled condensers", Master's thesis, Department of Mechanical and Mechatronic Engineering, University of Stellenbosch, South Africa, 2015.
15 Van der Spuy, S.J. "Perimeter fan performance in forced draught air-cooled steam condensers", Ph.D. thesis, Department of Mechanical and Mechatronic Engineering, University of Stellenbosch, South Africa, 2011.
16 Fourie, N. et al., "Simulating the effect of wind on the performance of axial flow fans in air-cooled steam condenser systems", Journal of Thermal Science and Engineering Applications, Vol. 7, pp. 1-12, 2015.
17 Muiyser, J. et al. "Investigation of large-scale cooling system fan blade vibration", ASME Turbo Expo, Düsseldorf, Germany, 2014.

Multi-scale simulation of droplet-droplet interactions and coalescence

[1,2]**Ndivhuwo M. Musehane***, [1]**Oliver F. Oxtoby and** [2]**B. Daya Reddy**

1. Aeronautic Systems, Council for Scientific and Industrial Research, P.O Box 395, Pretoria, 0001, South Africa.
2. University of Cape Town, Department of Mathematics and Applied Mathematics and Centre for Research in Computational and Applied Mechanics, Rondebosch, 7701, South Africa.

* nmusehane@csir.co.za

ABSTRACT

In this work, droplet-droplet interaction is modelled using a multiphase flow computational fluid dynamics approach coupled with a reduced order surface thin-film model. An existing multiphase flow model based on a multiple marker volume of fluid method is extended to include the thin-film model that is derived to simulate film drainage on the sub-grid scale. This allows for the prediction of coalescence or rebound of colliding droplets from first principles. The model is implemented into the open source tool set OpenFOAM® and tested against experimental results of colliding hydrocarbon droplets in the literature. It is found to produce accurate interface deformation results for the duration of the collision, and to consistently predict the outcome of the collision.

KEYWORDS
Coalescence, Droplets, VOF, Film drainage

INTRODUCTION

This work consists of the development, numerical implementation and validation of a reduced order surface thin film model coupled to the multiple marker volume of fluid (VOF) method [1, 2], to model droplet-droplet interactions and the resultant coalescence or rebound. Droplet-droplet interactions are found in various industrial processes and their coalescence influences the hydrodynamic and fluid transport phenomena that govern the overall flow behaviour. It is found to affect the droplet size distribution and shape which influence the overall optimal operating conditions. However, there are difficulties in modelling the collision process computationally.

The various multiphase flow modelling techniques available differ in their description and treatment of the droplet-fluid interface. Euler-Euler and Euler-Lagrange type methods lack in that they rely heavily on prior knowledge of the flow structure in the form of empirical correlations when modelling coalescence [3]. On the other hand, interface tracking methods [4] can be computationally intensive due to the mesh deformations required and, due to mesh entanglement, struggle to capture large topology changes that result when droplets interact. This work endeavours to eliminate the need to use empirical correlations based on phenomenological models by developing a multi-scale model that predicts the outcome of a collision between droplets from first principles but at a reduced computational cost.

The equations that govern incompressible immiscible droplets dispersed in a continuous fluid phase undergoing isothermal and laminar Newtonian three-dimensional flow are the volume averaged conservation of mass, momentum, and liquid-gas interface advection equations,

$$\nabla \cdot (\rho \mathbf{u}) = 0, \tag{1}$$

$$\frac{\partial(\rho \mathbf{u})}{\partial t} + \nabla \cdot (\rho \mathbf{u}\mathbf{u}) = -\nabla p + \nabla \cdot (\mu \nabla \mathbf{u}) + \rho \mathbf{g} + \sum_{\theta=1}^{N} \sigma \kappa_\theta^s \nabla \alpha_\theta \text{ and} \tag{2}$$

$$\frac{\partial \alpha_\theta}{\partial t} + \nabla \cdot (\alpha_\theta \mathbf{u}) = 0 \tag{3}$$

where

$$\kappa_\theta^s = \frac{\nabla \alpha_\theta^s}{|\nabla \alpha_\theta^s|}, \tag{4}$$

$$\alpha_\theta^s - \alpha_\theta = \nabla^2 (\eta \alpha_\theta^s). \tag{5}$$

Here ρ is the density, $\mathbf{u}$ is the velocity, p is the pressure, μ is the dynamic viscosity, $\mathbf{g}$ is the gravitational acceleration, N gives the number of droplets, σ is the surface tension coefficient, and α_θ is the volume fraction field assigned different values for multiple droplets containing the same fluid.

The last term on the RHS of equation (2) is the surface tension force obtained from the Continuum Surface Force (CSF) formulation [5] where κ_θ^s is the signed curvature calculated from smoothed volume fraction field values α_θ^s to improve the numerical calculation of second order gradients [6].

The formulation above is known as the multiple marker VOF method, which uses separate indicator functions α_θ for each droplet and is essential to prevent premature numerical coalescence. Although simple and straightforward to implement computationally, as it stands it lacks in that it prevents coalescence in all circumstances [2]. Thus, to model coalescence with this model this work extends the method to include a reduced order surface thin film model that will correctly predict the outcome of droplet-droplet collisions. Once coalescence is detected, the separate indicator functions are merged.

COALESCENCE MODEL

The collision process between droplets can be described in three stages based on the film drainage theory of coalescence [7]. Consider two droplets; Stage 1 is when they approach each other, and as they do so a thin film of the surrounding fluid is trapped between them; Stage 2 is when this fluid film drains out; and lastly in Stage 3 the droplets will either coalesce if the thin film attains a critical thickness h_{Crit}, becomes unstable and ruptures; or bounce apart if the fluid film does not attain h_{Crit}.

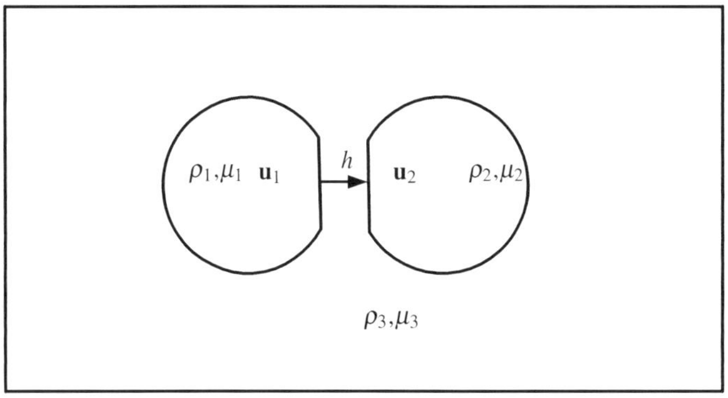

Figure 1: Two droplets with radius R approaching each other with velocities $\mathbf{u}_1$ and $\mathbf{u}_2$ forming a locally planar thin film

The thin film thus is an important aspect in modelling coalescence. The method developed here is aimed at modelling the dynamics of the thin liquid film during Stage 2 and Stage 3 of the collision process. It calculates the thickness of the thin film and compares it to a critical film thickness to decide whether coalescence should occur. Due to the extremely small scale of the thin film (typically h_{Crit} is of the order of micrometres) a multiscale model is essential, where calculation of the film thickness occurs on the sub-grid scale.

The sub-grid model is based on the Reynolds equation for two moving interfaces from lubrication theory [8]. It is derived based on the following assumptions: the thin film thickness h is assumed to be much smaller than the radius R of the droplets ($h \ll R$). This implies that the region between the interfaces of the two droplets can be assumed to be locally planar (1) with the thin film locally resembling either a box in three dimensions or a tube in two dimensions. In this region, the effects of gravity and surface tension are negligible. The pressure gradient across the thickness of the thin film is assumed to be constant. With these assumptions it can be shown that the momentum equations can be reduced to

$$\frac{\partial(\rho h)}{\partial t} + \mathbf{Q} \cdot [\nabla \rho h - (\nabla \rho h \cdot \mathbf{n})\mathbf{n}] = \nabla \cdot \left[\frac{-\rho h^3}{12\mu} [\nabla p - (\nabla p \cdot \mathbf{n})\mathbf{n}] \right], \tag{6}$$

where $\mathbf{n}$ is defined as a unit vector normal to the interface and

$$\mathbf{Q} = \frac{\mathbf{u}_1 + \mathbf{u}_2}{2}, \tag{7}$$

is the interface velocity vector.

NUMERICAL IMPLEMENTATION

The model equations are made up of equations (1)–(7). Equations (1)–(4) govern the flow between colliding droplets while equations (6)–(7) determine the outcome of the collision by evaluating the thickness of the thin liquid film.

A Finite Volume Method (FVM) of discretization is followed, where the integral form of the governing equations on a finite number of non-overlapping control volumes is used to obtain a system of coupled nonlinear equations.

The solution to equation (6) gives the thickness h of the thin film between two interfaces. In this study, the pressure obtained from the solution of the Navier-Stokes equation is used together with an initial film thickness estimate h_{Est} to obtain a new value of the film thickness. Equation (6) is only valid when the droplets are close to each other.

The pressure used in equation (6) is obtained from the flow simulation since it is primarily determined by the dynamics on the grid scale, i.e. by the impulse acting to decelerate the colliding droplets. An alternative approach is taken by Mason *et al.* [9], where the rate of change of film thickness is obtained from the velocity field of the flow solution, and the pressure is solved for in the film equation (in keeping with traditional applications of the Reynolds equation). This approach has been avoided since the film-thinning velocity is extremely small and difficult to detect compared to the larger flow velocities superimposed on it. Indeed, the rate of film-thinning is not reliably calculated by the flow equations as it stems from the film-drainage mass balance in the sub-grid region.

The solution procedure thus involves firstly devising a way to measure the initial separation distance h_{Est} between the droplets that is valid for separation of a few mesh cells or greater. A graphical approach based on measuring the distance from each volume fraction is used to obtain the separation h_{Est} between the particles.

Consider two interfaces that enclose fluids described by the volume fraction fields α_1 and α_2 which are close to each other in a computational domain. Consider lines which start from the interfaces pointing in the direction of the outward unit normal to the interface $\mathbf{n}_1^s$ and $\mathbf{n}_2^s$ given by

$$\mathbf{n}_1^s = \frac{\nabla(\alpha_1^s)}{|\nabla(\alpha_1^s)|}, \qquad \mathbf{n}_2^s = \frac{\nabla(\alpha_2^s)}{|\nabla(\alpha_2^s)|}. \tag{8}$$

Define the length from a point on the interface $\mathbf{x}_i$ to a point $\mathbf{x}$ on a ray by the path integral along the ray in the unit normal directions,

$$h_{Est_1} = \int_{\mathbf{x}_i}^{\mathbf{x}} (1 - \alpha_1)\mathbf{n}_1^s \cdot d\mathbf{l}, \tag{9}$$

and

$$h_{Est_2} = \int_{\mathbf{x}_i}^{\mathbf{x}} (1 - \alpha_2)\mathbf{n}_2^s \cdot d\mathbf{l}. \tag{10}$$

The path integrals (9) and (10) start from the interface so a switch is used to set the initial value at the interface to zero. To achieve this, the above two equations need to be nullified and replaced with $h_{Est_1} = 0$ and $h_{Est_2} = 0$ inside the droplet, so we multiply both sides of equations (9) and (10) by boolean switches $[1 - \text{pos}(\alpha_1 - 0.95)]$ and $[1 - \text{pos}(\alpha_2 - 0.95)]$ respectively, and add $\text{pos}(\alpha_1 - 0.95)h_{Est_1}$ and $\text{pos}(\alpha_2 - 0.95)h_{Est_2}$ to the respective equations so that when $\alpha_1 > 0.95$ and $\alpha_2 > 0.95$ we end up with $h_{Est_1} = 0$ and $h_{Est_2} = 0$. Here the function $\text{pos}(s)$ is a positive boolean given by

$$\text{pos}(s) = \begin{cases} 1 & s \geq 0, \\ 0 & s < 0. \end{cases} \tag{11}$$

Rewritting equations (9) and (10) in differential form, also considering the above arguments, the equations for h_{Est_1} and h_{Est_2} become

$$[1 - \text{pos}(\alpha_1 - 0.95)]\,(\nabla h_{Est_1} \cdot \mathbf{n}_1^s) = [1 - \text{pos}(\alpha_1 - 0.95)]\,(1 - \alpha_1) + \text{pos}(\alpha_1 - 0.95)h_{Est_1} \tag{12}$$

and

$$[1 - \text{pos}(\alpha_2 - 0.95)]\,(\nabla h_{Est_2} \cdot \mathbf{n}_2^s) = [1 - \text{pos}(\alpha_2 - 0.95)]\,(1 - \alpha_2) + \text{pos}(\alpha_2 - 0.95)h_{Est_2}. \tag{13}$$

In the region between the two interfaces, the sum of h_{Est_1} and h_{Est_2} gives the separation distance between the interfaces. Thus the estimate h_{Est} is obtained from

$$h_{Est} = h_{Est_1} + h_{Est_2} \tag{14}$$

since this gives the sum of the distance to interface 1 along the ray in direction $\mathbf{n}_1^s$ and the distance to interface 2 along the ray in direction $\mathbf{n}_2^s$ from any given point. This approach cannot however be used to obtain distances that are smaller than a mesh cell spacing. The value of h_{Est} is used as an initial condition for the Reynolds equation.

From this initial separation distance, a region in space can be calculated that determines where the Reynolds equation is valid. For this purpose we define a switch β that turns the computation of the Reynolds equation on when the separation estimate h_{Est} spans a few mesh cells and turns it off for larger separations. A multiple of the mesh spacing estimate is used to restrict and the mesh cells in which equation (6) is actually calculated. This restricts the calculation of the equation to areas that are very close to the thin film region. The boolean switch which identifies the region is defined by

$$\beta = \text{pos}(\xi\,\delta h - h_{Est})\,\text{pos}(\alpha_1^s - 0.001)\text{pos}(\alpha_2^s - 0.001) \tag{15}$$

where ξ is a parameter that determines the width of the region where the thin film equation is solved as a multiple of the mesh spacing δh.

There is a numerical discontinuity between the values of h inside the thin film region β and those outside that have been pegged to h_{Est}. Evaluation of the gradient of h would be incorrect for computational cells that surround the thin film region. To avoid the discontinuity of the gradient of h at the boundary of the thin film region the gradient of h is evaluated in a reduced thin film region β^* which excludes the outermost layer of cells in the region defined by β.

The discretized form of equation (6) evaluated using an implicit Euler scheme is given by

$$\begin{aligned} &\beta \frac{\rho_P h_P^{m+1} - \rho_P h_P^m}{\Delta t} V_P + \beta \mathbf{Q} \cdot \left[\nabla \rho_P h_P^{m+1} - \left(\nabla \rho_P h_P^{m+1} \cdot \mathbf{n}\right) \mathbf{n}\right] V_P + (1-\beta) h_P^{m+1} V_P = \\ &\beta \frac{-\rho_f (h_f^{m+1})^3}{12\mu_f} \sum_f \mathbf{A}_f \cdot (\overline{\nabla p_t}^s)_f^{m+1} + \beta^* \frac{-3\rho_f (\overline{\nabla p_t}^s)_f^{m+1} (h_f^{m+1})^2}{12\mu_f} \mathbf{A}_f h_f^{m+1} + (1-\beta) h_{EstP}^{m+1} V_P, \end{aligned} \tag{16}$$

where subscript $(\cdot)_P$ and $(\cdot)_f$ are quantities evaluated at the cell centres and cell faces of the control volume while $(\cdot)^{m+1}$ and $(\cdot)^m$ are values evaluated at the new and old time step. V_P is the volume of the computational control volume, and $\mathbf{A}_f$ is the surface area vector that points in the unit normal direction to the interface.

For simplicity the tangential pressure term has been assigned a new symbol $\nabla p_t = \nabla p - (\nabla p \cdot \mathbf{n})\mathbf{n}$. In order to calculate the pressure gradient in the sub-grid region between the droplets, the surface tension pressure jump is removed by subtracting the surface tension force from the tangential pressure as follows:

$$\overline{\nabla p_t}^s = \nabla p_t - \sum_{\theta=1}^{N} \sigma \kappa_\theta^s \nabla \alpha_\theta. \tag{17}$$

The solution procedure of the model is as follows:

- Set-up initial and boundary conditions
- Calculate the available fluxes and the multiple VOF fields
- Assemble the momentum equation
- Assemble and solve the pressure equation using a pressure correction algorithm PISO
- Calculate initial h_{Est}
- Where h_{Est} is within a certain number of mesh cells
  - Solve the reduced order surface thin film equations to obtain h
  - If $\min(h) < h_{Crit}$
    * Merge the volume fraction fields
  - Where h_{Est} is too large
    * Set $h = h_{Est}$
    * Do not solve the reduced order surface thin film equations
- Move on to the next time-step and repeat until final time is reached

NUMERICAL EVALUATION

In this section the implementation of the reduced order surface thin film model coupled to the multiple marker VOF method is tested against the experiment of [10]. The experiment provides detailed time resolved interface deformation images of the head on collision of two identical Tetradecane ($C_{14}H_{30}$) droplets in 1 atm Nitrogen gas. The material properties are shown in Table 1.

	Liquid (Tetradecane)	Gas (Nitrogen)
Density ($\mathrm{kgm^{-3}}$)	762	1.225
Viscosity (μm)	107.2	170.6
Surface Tension ($\mathrm{kgs^{-2}}$)	0.0265	

Table 1: Material Properties for Tetradecane and Nitrogen at 20°C [10]

Quantity	Case I	Case II	Case III	Case IV
U_0 ($\mathrm{ms^{-1}}$)	0.302	0.24	0.496	0.596
R (μm)	107.2	170.6	167.6	169.7
We	2.25	2.26	9.33	13.63
T (ms)	0.415	0.831	0.811	0.826

Table 2: Physical Quantities [10]

Four test cases were studied: the first two result in droplet coalescence and rebound with minimal droplet deformation while the last two produce the same results with large deformations in the droplet shapes.

The experimental parameters, viz. individual droplet velocity U_0, droplet radius R and the Weber number $We = \frac{2\rho U_0^2 D}{\sigma}$ for each case are shown in Table 2. The experimental results were normalized by the droplet oscillation period $T = 2\pi\left(\frac{\rho_l R^3}{8\sigma}\right)^{1/2}$ [10] where ρ_l is the density of Tetradecane.

Since the radii of the droplets is of the order $O(10^{-6})$, gravity is negligible compared to surface tension forces and the term $\rho\mathbf{g}$ in the momentum equation is neglected. The simulations are performed on an axisymmetric structured, Cartesian and uniform mesh. The domain dimensions are $4D \times 2D$ where D is the corresponding droplet diameter. The droplets are initialised a distance of $1.5D$ apart. The boundary conditions for the pressure are set to zero gradient at the walls with an internal air pressure set to 1 atm. A no slip velocity boundary condition is set at the walls with an initial internal gas velocity of zero and velocity within each droplet set to U_0. The computational domain is shown in figure 2.

Figure 2: Axisymmetric case set-up

The solution procedure as mentioned before involves firstly obtaining the initial thin film thickness estimate to define the thin film region β and secondly determining the critical film thickness h_{Crit}.

In the determination of the thin film region β, the numerical parameter ξ is required to determine the mesh cell spacing at which to switch from the geometric determination of the initial separation distance between droplet interfaces h_{Est} to the solution through the surface thin film model. The restrictions on ξ are that it should be big enough that the geometric determination spans several mesh cells to make it valid, but it should not be too big (resulting in β spanning too many mesh cells) because the surface thin film model is valid when $h \ll R$.

It is thus essential that the results are not sensitive to the value of ξ used in the simulations. The switch-over point from the geometric determination to the solution of the surface thin film equations should take place at a separation distance where both methods are valid. A necessary condition for this is that the results should not be sensitive to the switch-over point in this range.

The selection of ξ is determined by considering the sensitivity of the minimum predicted thin film thickness to the selection of ξ. Therefore, to determine ξ, the value of h_{Crit} is set to zero, and the numerical solution to Cases I-IV is obtained with $\xi = 3, 4, 5$. Table 3 gives the numerical values of the minimum film thickness for different values of ξ. An increase in the value of ξ improves the estimate to the thin film thickness. A very small change is observed between $\xi = 4$ and $\xi = 5$. It is therefore inferred that the results become independent of ξ in the region of $4 \leq \xi \leq 5$, and a value of $\xi = 4$ is chosen.

ξ	Case I	Case II	Case III	Case IV
3	3.363×10^{-7}	5.619×10^{-7}	4.716×10^{-7}	2.078×10^{-7}
4	1.618×10^{-7}	3.872×10^{-7}	2.573×10^{-7}	1.382×10^{-7}
5	1.649×10^{-7}	3.296×10^{-7}	2.146×10^{-7}	1.094×10^{-7}

Table 3: Minimum film thickness h_{Min} (m) for Case I, Case II, Case III and Case IV

The one physical parameter present in the model is the critical film thickness h_{Crit} at which film rupture and, therefore, coalescence, occur. This is a universal parameter for a given pair of fluids, but is not known beforehand. We will therefore set this value to be consistent with the results of the four experiments. Effectively, the parameter is set to match the result of one pair of experiments, and the other pair is used as validation.

Similar to the determination of ξ, h_{Crit} is initially set to zero and the numerical results at $\xi = 4$ are obtained for the four test cases (see 3). A global h_{Crit} that is applicable to all four test cases is sought: a value such that 1.618×10^{-7}m $< h_{Crit} < 2.573 \times 10^{-7}$m must be selected. The critical film thickness selected to obtain the numerical results is $h_{Crit} = 1.650 \times 10^{-7}$m. Thus according to the solution procedure of the surface thin film model outlined in the previous section, when $\min(h) < h_{Crit} = 1.650 \times 10^{-7}$m the volume fraction fields are merged.

The time series interface deformation plots of the droplets for selected times (for brevity) are given in figures 3–6. The results are presented at nondimensional time normalised by the droplet oscillation period T.

The interface deformation agrees qualitatively in the initial stages of the collision. At this point the surface thin film model is not applicable since the distance between the interfaces needs to be much smaller than the radius of the droplets for the model to be applicable. As the droplets continue to move closer the thin film equation becomes valid and the surface thin film model is used to calculate the film evolution.

The numerical coalescence time or rupture time has been somewhat underestimated by the model, leading to an earlier coalescence. This results in the evident delay in the dynamics after coalescence has taken place, but good qualitative agreement with the experimental results. The numerical results are obtained at a reasonable computational cost without the need to excessively refine the computational mesh.

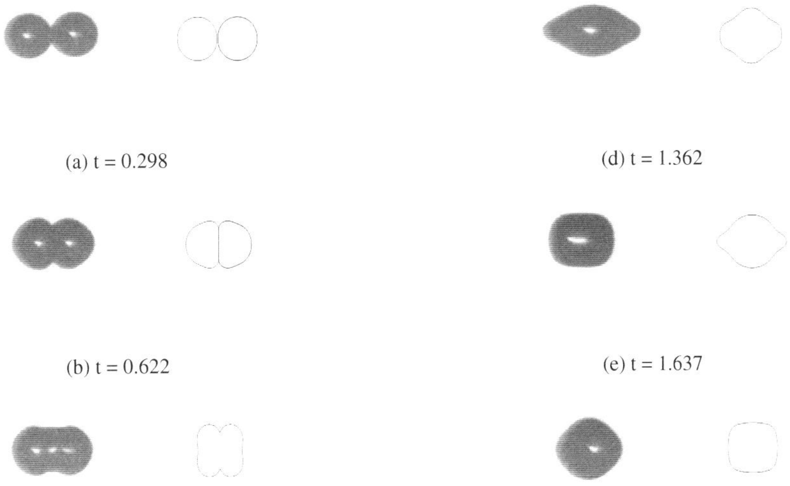

(a) t = 0.298 (d) t = 1.362

(b) t = 0.622 (e) t = 1.637

(c) t = 1.013 (f) t = 1.944

Figure 3: Time series interface shape deformations for Case I

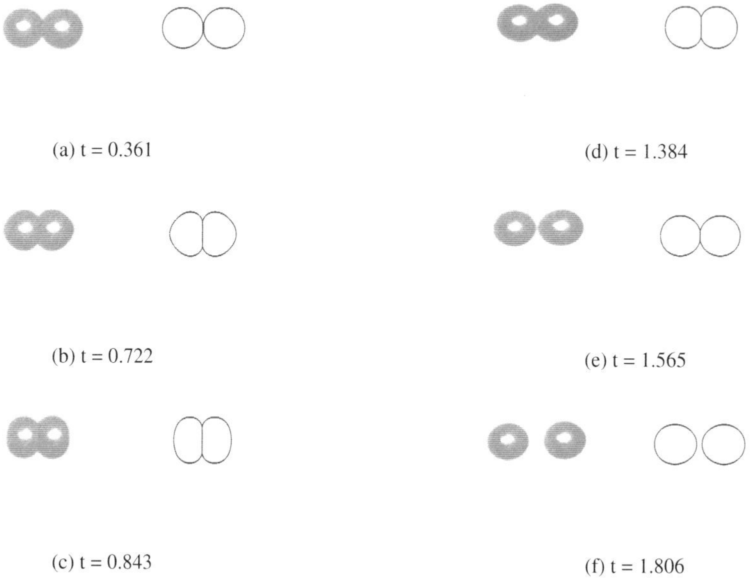

(a) t = 0.361

(b) t = 0.722

(c) t = 0.843

(d) t = 1.384

(e) t = 1.565

(f) t = 1.806

Figure 4: Time series interface shape deformations for Case II

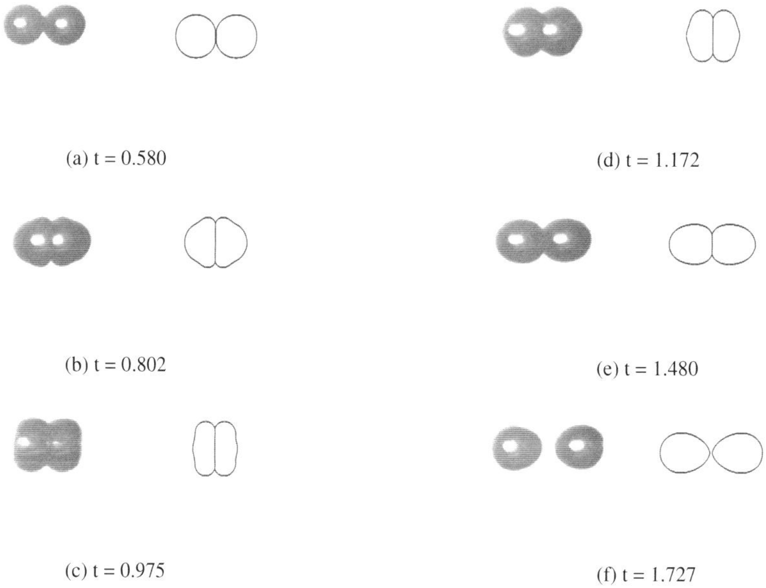

(a) t = 0.580

(b) t = 0.802

(c) t = 0.975

(d) t = 1.172

(e) t = 1.480

(f) t = 1.727

Figure 5: Time series interface shape deformations for Case III

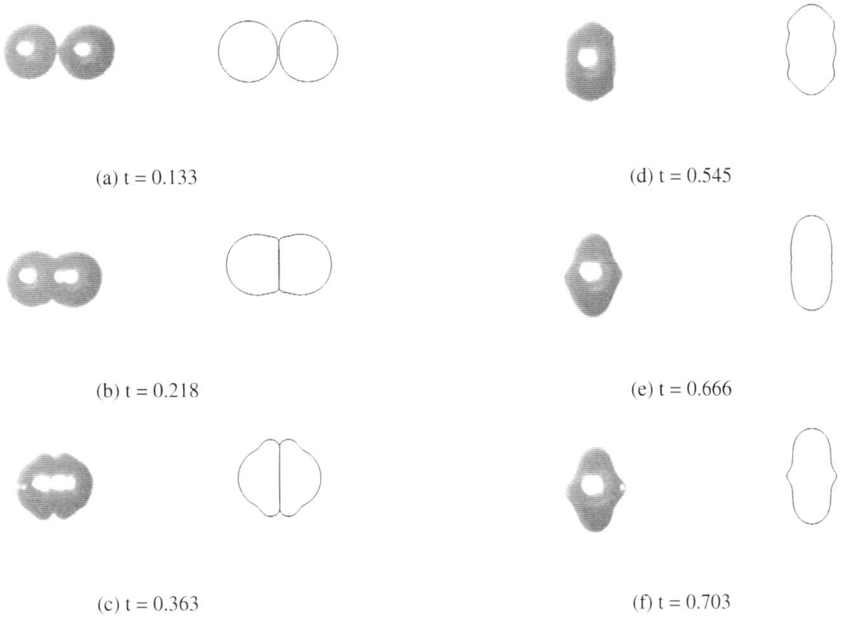

Figure 6: Time series interface shape deformations for Case IV

CONCLUSION

To model droplet-droplet coalescence a reduced order surface thin film model coupled to a multiple marker VOF method is presented. The model is derived based on the thin film drainage theory which assumes the presence of a thin liquid film between interacting droplets. The governing set of equations are discretized using a FVM and numerical solution obtained using an implicit scheme with pressure correction.

The method is tested by simulating the binary collision of identical Tetradecane droplets in 1 atm Nitrogen and it is found to correctly predict the behaviour of the droplets before and after the duration of the collision. The numerical model slightly under predicts the instant of coalescence which results in the interface deformations after coalescence having a slight delay compared to the experimental time series.

The results of the current study demonstrated the predictive power of the surface thin film model. Although the critical film thickness was set based on the experimental data, the results agreed with all four sets of experiments of [10] under different collision conditions.

ACKNOWLEDGEMENTS

Funding for this work was provided by the CSIR under the CSIR masters studentship.

References

[1] C.W. Hirt and B.D. Nichols. Volume of fluid (VOF) method for the dynamics of free boundaries. *Journal of Computational Physics*, 39(1):201–225, 1981.

[2] E. Coyajee and B.J. Boersma. Numerical simulation of drop impact on a liquid-liquid interface with a multiple marker front-capturing method. *Journal of Computational Physics*, 228(12):4444–4467, 2009.

[3] H.A. Jakobsen, H. Lindborg, and C.A. Dorao. Modeling of bubble column reactors: Progress and limitations. *Industrial & Engineering Chemistry Research*, 44(14):5107–5151, 2005.

[4] M.R. Nobari, Y.J. Jan, and G. Tryggvason. Head-on collision of drops–a numerical investigation. *Physics of Fluids*, 8(1):29, 1996.

[5] J.U. Brackbill, D.B. Kothe, and C. Zemach. A continuum method for modeling surface tension. *Journal of Computational Physics*, 100(2):335–354, 1992.

[6] J.A. Heyns and O.F. Oxtoby. Modelling surface tension dominated multiphase flows using the VOF approach. In *6th European Conference on Computational Fluid Dynamics*, 2014.

[7] J.-D. Chen, P.S. Hahn, and J.C. Slattery. Coalescence time for a small drop or bubble at a fluid-fluid interface. *American Institute of Chemical Engineers Journal*, 30(4):622–630, 1984.

[8] J.-D. Chen. A model of coalescence between two equal-sized spherical drops or bubbles. *Journal of Colloid and Interface Science*, 107(1):209–220, 1985.

[9] L.R. Mason, G.W. Stevens, and D.J.E. Harvie. Multi-scale volume of fluid modelling of droplet coalescence. In *Ninth International Conference on CFD in the Minerals and Process Industries*, 2012.

[10] K.-L. Pan, C.K. Law, and B. Zhou. Experimental and mechanistic description of merging and bouncing in head-on binary droplet collision. *Journal of Applied Physics*, 103(6), 2008.

A Comparative Study of the Thermo-Hydraulics of the SNUF Test Facility Core using FLOWNEX AND RELAP5

V.V.Naicker*, C.S.Zikhali, JH Kruger
School of Mechanical & Nuclear Engineering
North-West University, South Africa

*20242840@nwu.ac.za

ABSTRACT

Given the current possibility of a nuclear power new build endeavour in South Africa, attention also needs to be focussed on the extent to which localization of the nuclear build will be implemented. This localization can extend into many different areas of expertise. One such area is the numerical simulation of the reactor core. This in turn can also be further subdivided into fields of expertise, with the thermo-hydraulic simulation of the reactor core being one such field. Computer codes are normally used in these thermo-hydraulic simulations, and FLOWNEX®, a locally developed code, is one such code. As such, further development of this code as applied to nuclear systems can assist with the new build localization effort.

In this paper, the capability of FLOWNEX® is tested by comparing its results with those from RELAP5, a benchmark thermo-hydraulic simulation code in the nuclear industry. Computer models of the heater core of the Seoul National University Facility (SNUF) were built. The aspects of the model that were studied in detail were the heat transfer coefficient of the fluid and the power distribution applied to the heater elements of the core. In the case of FLOWNEX®, a user-defined C# code was implemented to model the two-phase Chen correlation so that both codes produced comparable results. For the power distribution, both surface sources and volume distributed sources were applied to the heat structures used in FLOWNEX®. The limitations of these models were investigated and evaluated.

KEYWORDS

FLOWNEX®, heat transfer coefficient, Chen correlation, distributed heat source

1. INTRODUCTION

Electricity production in South Africa using nuclear power has been in existence since 1982, when Eskom deployed two nuclear power units of 900 MWe each at Koeberg, Capetown. In 2010 the South African government through the issue of the IRP2010 (DOE [1]) planned for a further 9.6 GW of power to be added to the electricity grid by 2025. This would most probably be of the pressurized light water reactor type (PWR), considering the successful operating history at Koeberg. Given the status that South Africa is a developing nation, a partnership between foreign and local role players will most probably be employed. This will include the manufacture of plants, construction and operation and the parallel tasks of licensing and regulation. From a local context, this will involve the power plant vendor and operator, academic institutions, the South African Nuclear Energy Cooperation (NECSA) and the National Nuclear Regulator (NNR).
One of the localization tasks that is being actively pursued is that of reactor analysis. This includes criticality, burnup, shielding and accident analysis of the reactor.
Reactor analysis is most commonly carried out by using computer codes. An important component of reactor analysis is the characterization of the thermal-hydraulic behaviour of the nuclear reactor core. A locally developed thermal-hydraulic/thermal fluid code FLOWNEX® (M-Tech Industrial [2]) can be considered for use in this characterization. It is therefore important to test whether FLOWNEX®, in its present release, can model a PWR reactor core appropriately or whether further development on the code is required, in terms of the phenomena addressed in this paper. Since this a challenging task, a simpler system was investigated in this paper as a possible starting point for this task. Both FLOWNEX® and RELAP5 (Nuclear Safety Analysis

Division [3]) models were built of the heated core of the simpler system. The results from both sets of calculations were compared, and an assessment on the use of FLOWNEX® to model the more complicated PWR core was made. It is noted that this assessment applies only to the heat transfer coefficient of the fluid and the thermal conductivity of the heat structures of the heated core. It is further noted that a comparison with the experimental results from the SNUF Test Facility was not made. To this effect, RELAP5 is considered to be a verified and validated code, and the results of RELAP5 were used as the standard with which the FLOWNEX® results were compared.

2. THE SNUF TEST FACILITY

The heated core of the Seoul National University Facility (SNUF) (Kim et.al [5]) was used as the case study for the comparison calculations. SNUF is an integral test facility of the Seoul National University in South Korea. It is a scaled down facility of the APR1400, the Advanced Power Reactor 1400 MW_E (Kim et.al. [5]). The results and analysis of the tests performed on the facility was part of a research program to develop a coupled calculation methodology from the initiation of a design-basis accident (DBA) to a severe accident (SA) involving core melting and relocation. The facility consists of two loops, with each loop consisting of one hot leg and two cold legs, as shown in Figure 1. One loop is designed to be the intact loop and the other to be the broken loop, in the accident scenario of a large break loss of cooling accident (LBLOCA). The test vessel as shown in Figure 1 consists of the four inlets for the cold legs and two outlets for the hot legs. The fluid, on entering the vessel through the cold legs, is directed to the bottom of the vessel via the annulus (also known as the downcomer). The fluid then moves upwards through the heater section, and exits the vessel through the hot legs. The heater section consists of 24 heaters made up of nichrome wire surrounded by a stainless steel casing. In this study, a suitable state of the system was required, and the initial conditions for a prototype analysis as given in Bae et.al. [6] were used. For this state, the inlet pressure was 0.8 MPa, and the inlet and outlet temperatures were 422.74 K and 434.20 K respectively. The core power was 0.244 MW_{TH}.

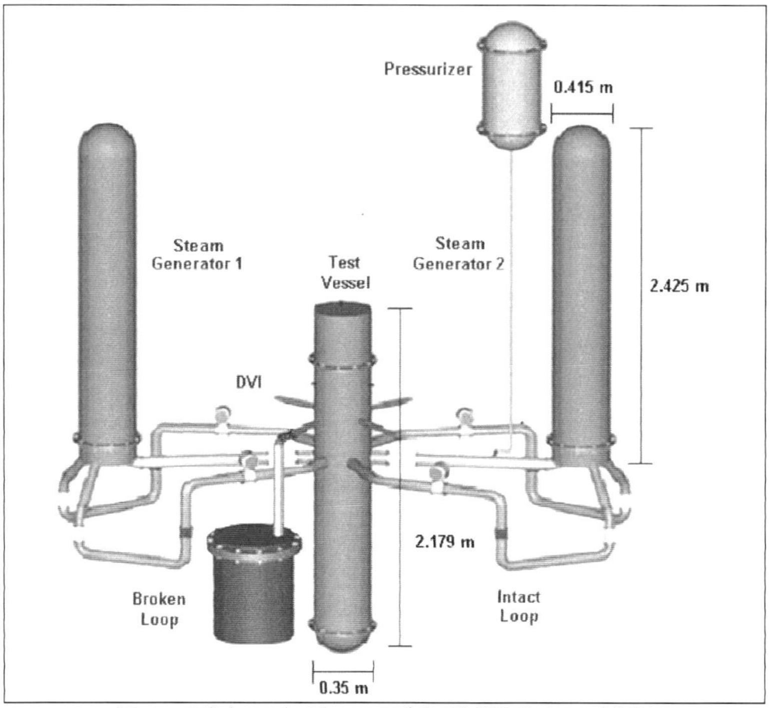

Figure 1 Schematic Diagram of the SNUF (Lee et. al [6])

3. COMPUTER CODES

3.1. FLOWNEX®

The code FLOWNEX® is a system level simulation tool for thermodynamic modelling of thermal-fluid systems (M-Tech Industrial [2]). FLOWNEX® can be used to calculate flow rates, temperatures, pressures and heat transfer rates using both steady state and transient models. The flow is treated as one dimensional. When treating two phase flow, the homogenous equilibrium model is used. This model is a mixture model in which the conservation equations of mass, momentum and energy are solved under the assumption that the vapour and liquid velocities are equal and the vapour and liquid temperatures are the same over the cross-sectional area of the flow.

Heat transfer through the solid structures can be modelled using many different components, all based on a generic solid structure component that accounts for heat conduction through various different materials layered on top of each other. The heat conduction model accounts for the different values of thermal conductivity and heat capacity of each material layer. The model also incorporates the convection heat transfer between exposed surfaces and adjacent fluids on both sides of the solid. Two conduction components were used in this study. The first component is made up of two elements, labelled as "conduction" and "convection" which are connected in series to the fluid structure, shown in Figure 2a. For the second component, labelled as "distributed heat source", as shown in Figure 2b, distributed heat sources within the volume can be modelled.

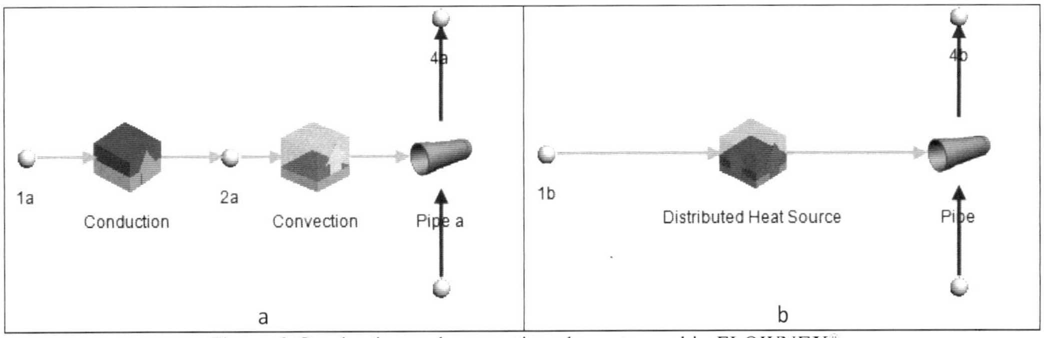

Figure 2 Conduction and convection elements used in FLOWNEX®

The default built-in correlation for the convection heat transfer coefficient at the solid surfaces in internal flows is the Dittus-Boelter correlation. This default correlation can be modified by employing different values for the coefficients and exponents in the generic correlation. Other correlations may also be employed via a user written C# script, as was carried out in this work.

Fluid flow in pipes, ducts and channels can be modelled via a generic pipe element for which the cross sectional area, hydraulic diameter, length and surface roughness may be specified. In this pipe model the flow path may be discretized into any number of discreet elements positioned in series along the length of the flow path.

Provision for nuclear heat generation including the point kinetic approach is made by embedding C# scripts into the FLOWNEX® platform.

3.2. RELAP

The Reactor Excursion and Leak Analysis Program (RELAP5) is a generic transient analysis code for thermal hydraulic systems using a fluid that may be a mixture of steam, water, non-condensable media, and a non-volatile solute (Nuclear Safety Analysis Division [4]). The hydrodynamic model is a one dimensional, transient, two-fluid model for flow of a two-phase steam-water mixture that can contain non-condensable components in the steam phase and/or a soluble component in the water phase (Nuclear Safety Analysis Division [3]).

The LWR applications for which the code is intended include accidents initiated from small break loss-of-coolant accidents, operational transients such as anticipated transients without SCRAM, loss of feed, loss-of-offsite power, and loss of flow transients (Nuclear Safety Analysis Division [3]).

FLOWNEX® and RELAP5 can thus be viewed as codes used to calculate fluid parameters such as temperatures and pressure differences and the heating elements parameters such as temperature differences.

4. MODEL DEFINITION

The modelling approach taken was to model only the heated core of the SNUF facility, with the view that subsequent projects would consider additional components until the entire facility is modelled using FLOWNEX$^{(R)}$. A MARS input model for the facility was obtained from (Cho [7]). The schematic nodalization is shown in Figure 3. It was assumed that this model represented the SNUF facility adequately within the constraints and approximations that are normally used when building a system level numerical model. For example, it was assumed that the hydraulic diameter and secondary losses were adequately represented. This input file was modified so that only the core and the downcomer surrounding the core were modelled. This is the region within the blue border as shown in Figure 3. The downcomer was kept in the model, so that pressure losses could also be accounted for. In the MARS input the downcomer was modelled as five axial layers. Each layer was in turn partitioned into six segments so that cross flow in the tangential directions in the plane of the layers could be modelled. In the models used for this paper, these six segments in each layer were modelled as a single volume, with the result that the downcomer was modelled as five vertical single volumes connected in series, without cross-flow.
The boundary conditions required were the inlet mass flow rate, pressure and temperature. However, the inlet pressure and initial cold and hot leg temperatures were available, as listed above. These values were 0.8 MPa, 422.74 K and 434.20 K respectively. Using the inlet pressure and cold leg temperature, RELAP5 calculations were performed, and the mass flow rate was adjusted to yield the required hot leg temperature. The mass flow rate calculated in this manner was found to be 4.92 kg/s. The power that was supplied was 0.244 MW.

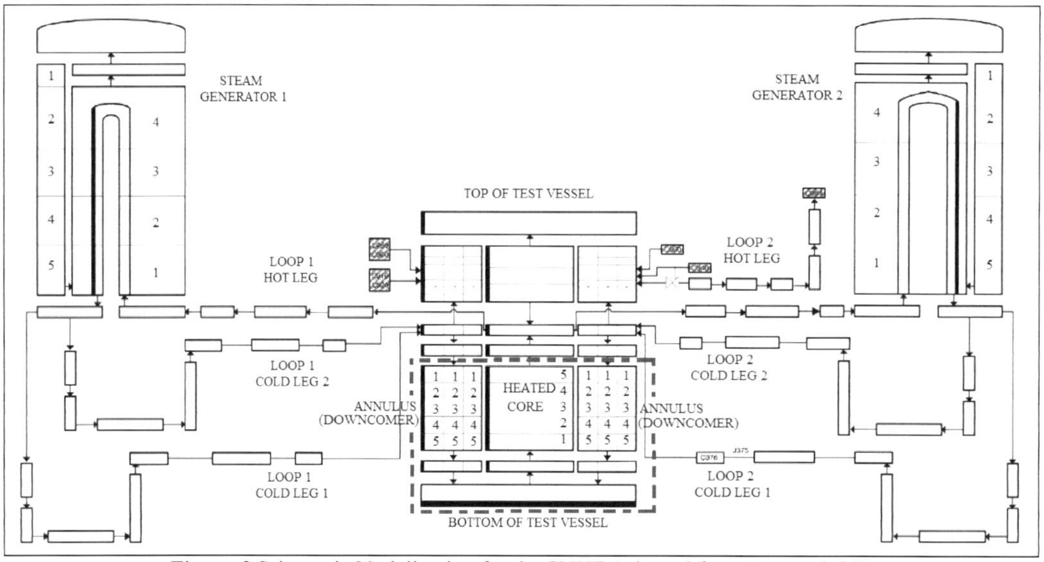

Figure 3 Schematic Nodalization for the SNUF (adapted from Lee et.al. [6])

5. CALCULATION BASE

5.1 Heat Transfer Correlation Coefficient

Fluid flowing up through a fuel assembly in a PWR core during normal operation typically experiences nucleate boiling at some axial point of the assembly (Todreas and Kazimi [8]). Correlations have been developed to calculate the heat transfer coefficient at such conditions. The Chen correlation is one such correlation, and this correlation is used in RELAP5 (Nuclear Safety Analysis Division [9]). The Chen correlation is mainly used for saturated boiling. However, application to nucleate boiling is possible with modifications. The Chen correlation used for nuclear boiling is characterized by the formula (Todreas and Kazimi [8])

$$q'' = h_c(T_w - T_f) + h_{NB}(T_w - T_{sat}) \quad \dots (1)$$

where T_w, T_f and T_{sat} are the wall, bulk fluid and fluid saturation temperatures, respectively.
h_c and h_{NB} are given by

$$h_C = 0.023\left(\frac{G(1-x)D_e}{\mu_f}\right)^{0.8}\left(Pr_f\right)^{0.4}\frac{k_f}{D_e}F \text{ and}$$

$$h_{NB} = S(0.00122)\left[\frac{\left(k^{0.79}c_p^{0.45}\rho^{0.49}\right)_f}{\sigma^{0.5}\mu_f^{0.29}h_{fg}^{0.49}\rho_g^{0.24}}\right]\Delta T_{sat}^{0.24}\Delta p^{0.75}$$

with

$$S = \frac{1}{1+2.53\times10^{-6}Re^{1.17}}$$

$$Re = Re_l F^{1.25}$$

$$\Delta T_{sat} = T_W - T_{sat}$$

$$\Delta p = p(T_W) - p(T_{sat})$$

$$F = \begin{cases} 1 & \text{for } \frac{1}{X_{tt}} < 0.1 \\ 2.35\left(0.213 + \frac{1}{X_{tt}}\right)^{0.736} & \text{for } \frac{1}{X_{tt}} > 0.1 \end{cases}$$

$$X_{tt} = \left(\frac{1-x}{x}\right)^{0.9}\left(\frac{\rho_g}{\rho_f}\right)^{0.5}\left(\frac{\mu_f}{\mu_g}\right)^{0.1}$$

In the above equations, the following definitions apply:

G	is the mass flux (kg/m²s)	x	is the flow quality
D_e	is the equivalent diameter (m)	μ_f	is the fluid dynamic viscosity (kg/ms)
Pr_f	is the fluid Prandtl number	Re_f	is the fluid Reynolds number
k_f	is the fluid thermal conductivity (W/mK)	$c_{p,f}$	is the fluid specific heat capacity (J/kgK)
ρ_f	is the fluid density in the liquid phase (kg/m³)	ρ_g	is the fluid density in the vapour phase (kg/m³)
T_W	is the wall temperature (K)	T_{sat}	is the fluid saturation temperature (K)
T_f	is the fluid bulk temperature (K)		

The range of water conditions for the Chen correlation is (Todreas and Kazimi [8]):
Pressure: 0.17 – 6.9 MPa
Liquid inlet velocity: 0.06 – 4.5 m/s
Heat flux up to 2.4 MW/m²
Quality: 0 – 0.7

As shown in equation 1 the Chen correlation is based on the three temperatures T_W, T_{sat} and T_f. However, in FLOWNEX®, the heat transfer coefficient is related to the temperatures T_W and T_f using the equation (M-Tech Industrial [2])

$$q'' = h_{FN}\left(T_w - T_f\right) \quad \dots (2)$$

Here, h_{FN} is the heat transfer coefficient and q'' is the heat flow through the heat transfer area. (Note that in (M-Tech Industrial [2]) different symbols are used in defining equation 2).
h_{FN} can be determined as follows:

(i) Internally calculated by FLOWNEX® using the Dittus Boelter correlation,
(ii) Given as a constant by the user,
(iii) Calculated using a user written script.

In this study, option (iii) was used. In order to include T_{sat} in the calculations, h_{FN} was calculated using the formula

$$h_{FN} = \frac{h_C(T_W-T_f)+h_{NB}(T_W-T_{sat})}{T_W-T_f}$$

which was obtained by equating 1 and 2.
However, since T_w is in turn calculated using h_{FN} given q'' and T_f, an iterative solution must be employed to determine T_w and h_{FN} consecutively until convergence is achieved.

5.2 Heat Conduction in the Heater Elements

Assuming that the heat source in a cylindrical heater element is evenly distributed throughout the volume, the temperature at the centre of the heater element T_{max} is given by the following equation in cylindrical co-ordinates (Bergman et.al [10]):

$$T_{max} = T_{fo} + \frac{q''' r_{fo}^2}{4k} \qquad \ldots (3)$$

T_{fo} is the temperature at the surface of the element, r_{fo} is the radius of the element, q''' is the heat supplied per unit volume, and k is the thermal conductivity. k is taken to be constant throughout the volume.
The equivalent expression for slab geometry is

$$T_{max} = T_{fo} + \frac{q''' a^2}{2k} \qquad \ldots (4)$$

where a is the thickness of the slab.
In FLOWNEX®, a distributed heat source is implemented in the "distributed heat source" element (Leeb [11]). However, when using the "conduction" element, the heat source must be applied at the conduction element surface.
For heat conduction in a material without an internal heat source, such as the stainless steel casing of the heater element, two types of heat transfer modelling is possible in FLOWNEX®. One is used for linear one-dimensional heat transfer, and the other is used for cross-conduction heat transfer. Only the first type was implemented in this study. For one-dimensional heat transfer, the modelling geometry of the element is shown in Figure 4, and the heat transfer across the element is calculated using the equation

$$\dot{Q} = kA_m \left(\frac{T_{N1} - T_{N2}}{L}\right) \qquad \ldots (5)$$

$\dot{Q}$ is the total rate heat supplied to surface A_1, k is the thermal conductivity, A_m is the average area between the surface areas A_1 and A_2 and L is the length of the heat conduction path.

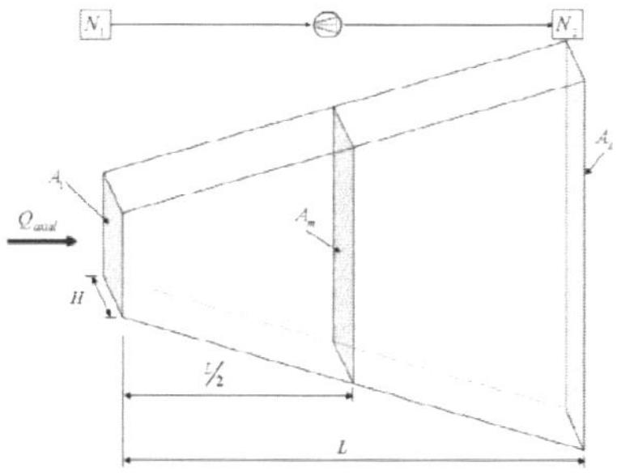

Figure 4. FLOWNEX® Conduction Heat Transfer Element Layout (M-Tech Industrial [2])

If alternate calculations are required to check the FLOWNEX and RELAP calculations, using for example an EXCEL calculation, then equation 5 is also applicable for slab geometry. However, one should use the following equation for cylindrical geometry (Bergman, et.al. [10])

$$\dot{Q} = \frac{2\pi hk(T_{N1} - T_{N2})}{ln(r_2/r_1)} \qquad \ldots (6)$$

where h is the height of the annulus, r_1 and r_2 are the inner and outer radii of the annulus respectively, with T_{N1} and T_{N2} being the temperatures on the corresponding surfaces.

6. RESULTS AND DISCUSSION

6.1. Fluid Results

Since the system studied was in the steady state, the process of heat conduction through the heater element could be analysed separately from the convection heat transfer from the wall to the fluid. This was because the heat that passed through the wall of the coolant was then constant.

In this section, the calculation of the heat transfer coefficient, and hence the wall surface temperature is described. For the RELAP5 case, with the Chen correlation being an inbuilt feature, the increase in the bulk fluid temperature and the wall temperature is shown in Table 1 for each volume element of the heated channel. It can be seen that the increase in temperature of the bulk temperature of the fluid is as expected, since the fluid is in the sub-cooled state for the entire length of the channel. Furthermore, the wall temperature remained constant for the length of the channel, since the fluid adjacent to the wall was expected to be at the fluid saturation temperature, and remain fairly constant, provided that the pressure drop is small.

In terms of the FLOWNEX® cases, two sets of results are presented in Table 1. In the first case, the heat transfer coefficient *h* using the Dittus Boelter correlation is used, and in the second case, the Chen correlation is used. In the case where *h* was calculated using the Dittus Boelter correlation, the wall temperature increased over the length of the heated channel, as was expected, since nucleate boiling was not considered. However, the values were extremely high, which meant that *h* was being under estimated.

However, the case in which the Chen correlation was used showed good agreement with the RELAP5 results, with the differences generally being less than 1 K. It is however noted that the pressure at which the calculation was done was outside the range of the Chen correlation. These values should therefore be compared with experimental values to test the validity of the Chen correlation in this range. These results were unfortunately not available. It is also noted that thermo-hydraulic models of the PWR cores using RELAP5 have been built. See for example (Aghaie [12]).

To model the Chen correlation, an EES application was embedded within the FLOWNEX® platform. The fluid properties were obtained from either the EES or FLOWNEX® data banks as appropriate. Since both the heat transfer coefficient and the wall temperature are required in the Chen correlation, an iterative procedure was employed. To do this, two C# scripts were used. The first transferred the wall temperature of each node from the FLOWNEX® calculation to the EES calculation, and the second transferred the calculated heat transfer coefficient from the EES calculation back to the FLOWNEX® calculation. In effect, in the FLOWNEX® calculations, *h* was treated as a constant for each step of the iteration. However this process led to divergence of the fluid temperature and therefore could not be automated. To overcome this, a limit based approach on the wall temperature was used. For a given mass flow rate, bulk fluid temperature, pressure and power, the wall temperature and the heat transfer coefficient would be correspond to a specific pair of values, which can be called the limiting values, T_0 and h_0. By starting with two initial guesses of the wall temperature T_L and T_U on either side of the expected value, *h* for these two values can be calculated, and the wall temperatures T'_L and T'_U subsequently calculated. These two values can then be used as the next set of starting values T_L and T_U to calculate the wall temperature T'_L and T'_U. In this way the iterative scheme can be continued, until the difference between the T'_L and T'_U reaches the required precision. In Figure 5, a number of iteration pairs for (T_L, T'_L) and (T_U, T'_U) are shown. It is noted that this is a user intervention method, and will not be suitable for transient calculations. It has been communicated however (Leeb [11]), that the Chen correlation was at one stage of the FLOWNEX® development included as an intrinsic heat correlation for calculating *h*. It is therefore recommended that this is re-introduced should FLOWNEX® be used in PWR applications.

Table 1. Bulk and Wall Temperatures

	Bulk Temperatures			Wall Temperatures		
	RELAP (Chen)	FLOWNEX (Dittus Boelter)	FLOWNEX (Chen)	RELAP (Chen)	FLOWNEX (Dittus Boelter)	FLOWNEX (Chen)
1	423.85	423.89	423.89	464.17	2066.37	465.24
2	426.08	426.19	426.19	464.17	2096.95	465.15
3	428.34	428.49	428.49	464.16	2129.38	465.08
4	430.59	430.78	430.78	464.16	2163.85	464.05
5	432.96	433.07	433.07	464.17	2200.61	465.03

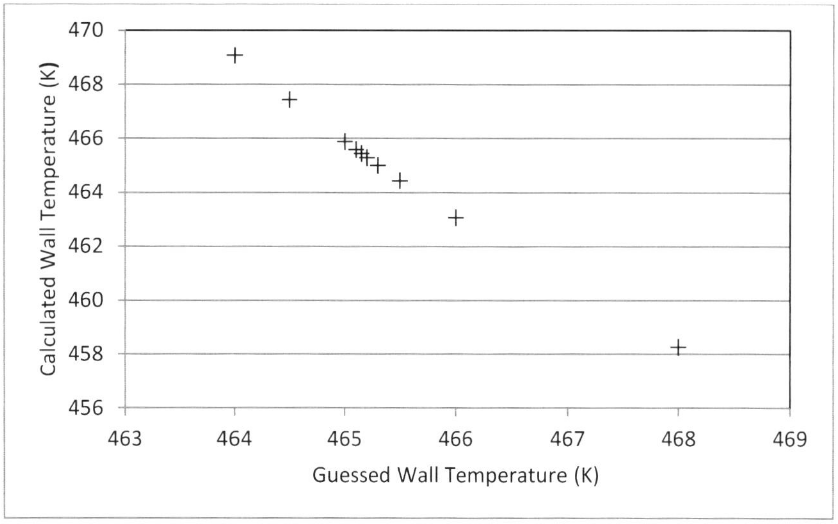

Figure 5. Iteration pairs to calculate the wall temperature

6.2. Heat Structure Results

6.2.1. Conduction without internal heat source

With the heat source applied to the nichrome wire, the stainless steel enclosing this wire acts as cylindrical conductor between the heat source and the coolant. The stainless steel is therefore a heat conductor without an internal heat source.

Although this is a cylindrical system, slab models for both RELAP5 and FLOWNEX® were first developed. The area of both the slab surfaces was taken to be the heat transfer area between the stainless steel surface and the coolant. The difference between the inner and outer surfaces of the stainless steel slab was calculated to be 137.42 K for all axial increments for both codes. Furthermore, a calculation using equation 5, which is applicable for a steady state calculation, also yielded the same value. In these calculations, the conductivity of the stainless steel was taken to be a constant.

In the cylindrical models, the stainless steel was modelled as a single annulus, with inner and outer radii to be 0.0025 m and 0.0050 m respectively. The difference between inner and outer surface temperatures of the annulus was found to be 183.23 K. Direct calculations, using the average area as shown in Figure 4, and equation 5, also yielded this value. However the calculation using equation 6 yielded a value of 190 K.

Since in this first set of calculations, only a single layer for the stainless steel was modelled, the effect of radial discretization was therefore further investigated. With two and four radial increments implemented, the difference between the inner and outer surface temperatures were 188.45 K and 189.96 K as shown in Figure 6. It could therefore be concluded that in using the solution method shown in Figure 4, at least four radial increments were required in the radial direction to yield the theoretical value calculated using equation 6. It is noted that the FLOWNEX® component "conduction" used to model the heat structure does not allow multiple structures to be modelled in series implicitly. Each heat structure was modelled explicitly, and it is therefore recommended to the code developer that this feature be introduced to the "conduction" component.

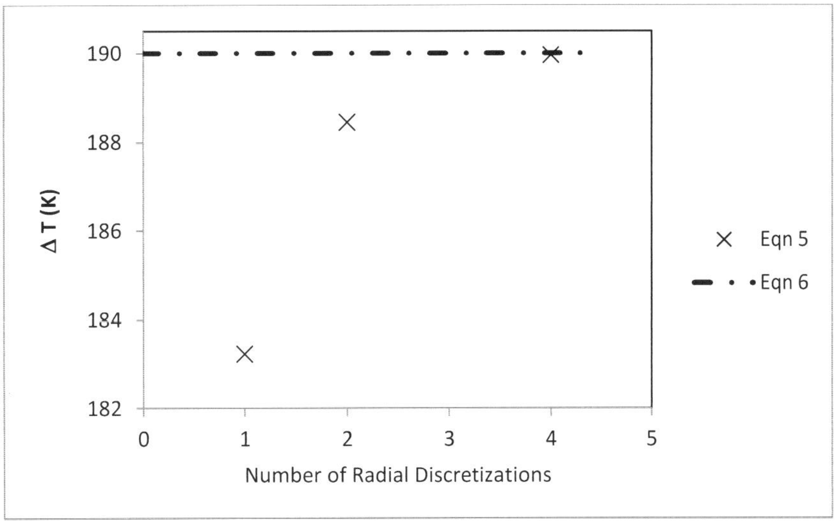

Figure 6 ΔT as a function of radial discretizations for the stainless steel casing

6.2.2. Conduction with Internal Sources: Slab geometry

The next conduction analysis was performed on the nichrome wire, to which in the context of the SNUF experiments, an electric current was applied, thus producing an internal heat source. It was assumed that the power distribution was both radially and axially uniform.

The slab geometry was modelled for both the RELAP5 and FLOWNEX® cases. In the RELAP5 case, the power is modelled as a distributed heat source, and direction comparison with equation 4 can be made. It was found that the temperature difference between the outer surface and the centreline was 104.9 K, in both the RELAP results and the calculation using equation 4. For the FLOWNEX® model, first the "conduction" heat structure was used. In this case, the heat source had to be applied to the surface of the heat structure, i.e. not as a distributed heat source within the heat structure. It was therefore found that a number of heat structures in series had to modelled, with equal surface areas. An equal fraction of the heat source was then applied to each "left" surface of the heat structures, with the total heat applied being conserved. In effect equation 5 was being used many times. It was found that as the number of slab discretizations was increased; that the temperature difference across the nichrome "slab" calculated using FLOWNEX® approached that of the RELAP calculation as shown in Figure 7. However, it was found that even with 10 slab discretizations modelled, the difference using Flownex was 115.4 K, which was still quite different from the RELAP and calculated value of 104.9 K. An analytical solution of equation 5 using the spread sheet package EXCEL was therefore done as a function of radial discretization and the result as a function of the number of discretizations modelled is shown in Figure 7 as the solid line. It was found that modelling 100 discretizations each with a thickness of 0.00025 m, yielded a temperature difference of 105.0 K, which was about 1 K difference from the RELAP and calculated value using equation 4.

It is however noted that modelling 100 heat structures in FLOWNEX® to achieve this accuracy is not optimal. The recommendation is to rather include a distributed heat source in the definition of the heat structure, or to have the number of sub-units to be modelled as a built in function.

The distributed heat source which is another heat structure element of FLOWNEX® as mentioned earlier was also tested. When only the nichrome wire was modelled, without the stainless steel casing attached, it was found that the FLOWNEX® distributed heat sources yielded as difference of 104.5 K, which was in excellent agreement with the RELAP5 values of 104.9 K.

However, when the stainless steel casing was included in the definition of the heat structure, the temperature difference across the stainless steel block was found to be 149.5 K, which was larger than the value calculated in RELAP5 by 12.7 K. This effect was found to persist, even with increasing the number of nodes in the radial direction of the stainless steel casing. This was found puzzling to understand, with the possibility that the heat structure was modelled incorrectly by the user, i.e. the interpretation by the user of the input manual was incorrect. Further correspondence with the code supplier will be made to resolve this. It is noted that the "distributed heat structure" element will conceptually be a better option to use rather than the "conduction" element.

For the case of the cylindrical system, only the RELAP5 model was investigated since the FLOWNEX® heat structures for the slab analysis did not yield satisfactory results. A difference of 207.5 K between the outer surface of the nichrome wire and the centreline was found. The value calculated using equation 3 was found to be 209.K, with a difference of 2.4 K. It is possible that this difference is due to the modelling of the heat structure using Figure 4. This requires further investigation.

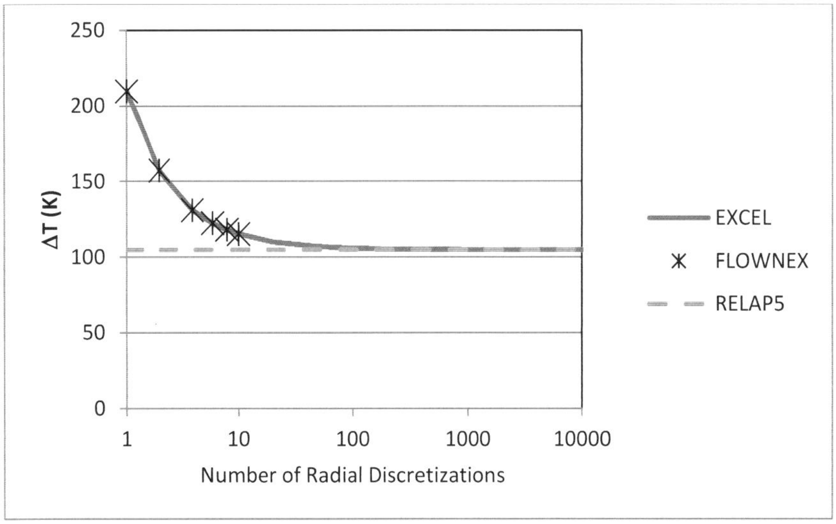

Figure 7 ΔT as a function of radial discretizations for the nichrome wire

6.3. Model Discretization

In addition to the radial discretization, as was discussed above, the axial discretization of the fluid channel was also studied. Discretization of 10 and 20 subunits, in addition to the 5 units of the given model were carried out for the RELAP5 model. The temperature as a function of axial height is plotted in Figure 8. With the difference in temperature at comparable axial positions being less than 0.1 K, it was considered that the core channel was adequately discretized using 5 units.

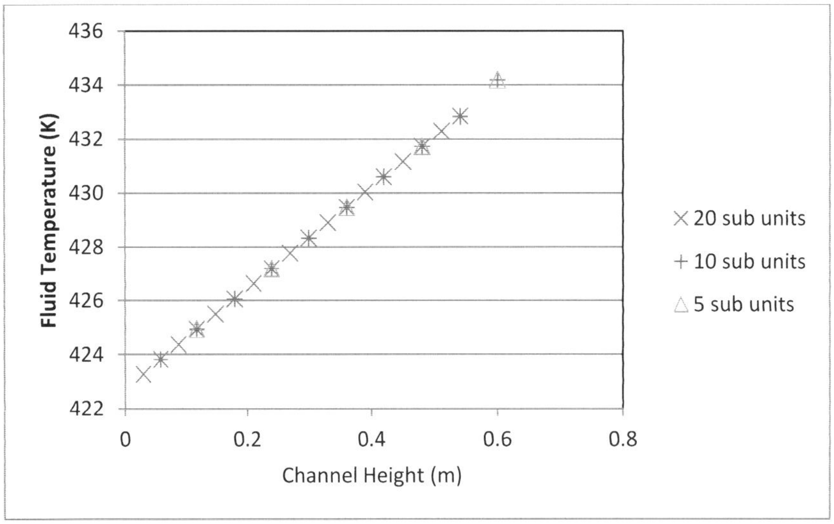

Figure 8 Axial discretization of the heated channel

7. CONCLUSIONS

The heat transfer and thermo-hydraulic comparison between the FLOWNEX® and RELAP5 models of the core element of the SNUF facility was successfully performed, with a number of conclusions being drawn together with suggestions made for adding more features to FLOWNEX®. It was found that using the Dittus Boelter correlation in FLOWNEX® produced unrealistic wall temperatures for the set of conditions used. Using a user written C# script, the Chen correlations produced good agreement with the RELAP5 results. However, it is suggested that the Chen correlation be a built in correlation of FLOWNEX®, rather than a user coded script, since automation of the calculation would be required in transient analysis.
With regards to the heat transfer components, it was found that discretization of the component in the radial direction for both RELAP and FLOWNEX® must be properly applied to get good results. The "conduction" element which required surface sources did not give satisfactory results unless a large number of discretizations were modelled. When the "distributed heat source" element of FLOWNEX® was used to model the nichrome wire only, without the stainless steel casing, the FLOWEX® results and the RELAP results had excellent agreement. However, when the stainless steel casing was included, the results differed, and this warrants further investigation.

ACKNOWLEDGMENT AND DISCLAIMER

This work is based on the research supported by the South African Research Chairs Initiative of the Department of Science and Technology and the National Research Foundation of South Africa (Grant No 61059).
Any opinion, finding and conclusion or recommendation expressed in this material is that of the authors and the NRF does not accept any liability in this regard.

REFERENCES

1 DOE 2013, Integrated Resource Plan for 2010-2030, http://www.doe-irp.co.za/content/IRP2010_updatea.pdf, Date of Access: 12 April 2016
2 M-TECH INDUSTRIAL, *FLOWNEX® Library Manual*, Potchefstroom, South Africa, June 2015
3 Nuclear Safety Analysis Division, Information Systems Laboratories, *RELAP5 Mod 3.3, Code Manual, Volume I, Code Structure, System Models, and Solutions Methods*, Rockville, Maryland, USA, 2001
4 Nuclear Safety Analysis Division, Information Systems Laboratories, *RELAP5 Mod 3.3, Code Manual, Volume II, User's Guide and Input Requirements*, Rockville, Maryland, USA, 2001
5 Kim Y.S., et.al., RELAP5/MOD3.3 analysis of coolant depletion tests after safety injection failure during a large-break loss-of-coolant accident, *Nuclear Engineering & Design* **235**, pp 2375-2390, 2005
6 Bae B.U., et.al., Scaling methodology for a reduced-height reduced-pressure integral test facility to investigate direct vessel injection line break SBLOCA, *Nuclear Engineering and Design*, 238, pp 2197-2205, 2008
7 Cho, H.K. *Personal communication*, Seoul National University, South Korea, 2014
8 N. E. Todreas & M.S. Kazimi, Nuclear Systems, Vol I, CRC Press, Boca Raton, FL, 2012
9 Nuclear Safety Analysis Division, Information Systems Laboratories, *RELAP5 Mod 3.3, Code Manual, Volume IV, Models and Correlations*, Rockville, Maryland, USA, 2001
10 Bergman T.L, et. al, *Fundamentals of Heat and Mass Transfer*, 7th Ed, John Wiley and Sons, USA, 2011
11 Leeb N., Private Communication, *M-Tech Industrial*, Potchefstroom, 2016
12 Aghaie M., et.al., Coupled neutronic thermal-hydraulic transient analysis of accidents in PWRs, *Annals of Nuclear Energy*, **50**, pp 158 – 166, 2012

CFD Simulation of a Single Burner Pulverised Coal Combustion Test Facility

Reshendren Naidoo*[1], Priven Rajoo[2], Walter Schmitz[3]

[1] Eskom Specialisation Centre for Combustion Engineering, School of Mechanical and Industrial and Aeronautical Engineering, University of the Witwatersrand, Johannesburg

[2] Eskom Research, Test and Development

[1] Eskom Specialisation Centre for Combustion Engineering, School of Mechanical and Industrial and Aeronautical Engineering, University of the Witwatersrand, Johannesburg

E-mail: Reshendren Naidoo@wits.ac.za

ABSTRACT

The combustion of fossil fuels and specifically pulverised coal is the predominant method of generating electricity in South Africa. By virtue of the sheer mass of fuel consumed, even small improvements in the combustion process can translate into significant cumulative gains over the operation period of a coal fired power station. Coal as a fuel, by nature, has significant variability in its combustion properties depending on where and how the coal was sourced. This variability requires the frequent testing of coals on a pilot scale, before it is burnt on a large scale at an actual power plant. Tests were conducted on a 1 MW combustion test facility and several temperature and gas species measurements were taken. Two distinct coals were tested. A detailed three dimensional CFD model was subsequently developed in a commercial code. The full range of combustion physics and chemistry was considered. This included detailed particle tracking of the coal particles including the heat and mass transfer from the dispersed phase to the continuous phase. Turbulent combustion was accounted for by means of a hybrid eddy break up scheme. Radiative heat transfer was accounted for using the discrete ordinates radiation model. Specific coal physics such devolatilisation and char burnout were also modelled using the best available techniques. The CFD simulation provided significant insight into the complex flow field inside the test furnace, with specific emphasis on the interaction of reacting particles in a swirling flow field. The results show that peak temperatures are predicted within a margin of 2% for coal A and coal B. The results from the code follow the experimental trend and also show good quantitative agreement. The O_2 and CO_2 concentrations on the outlet corresponded well with the test data. The results show the accurate modelling of char combustion is a challenge and the CFD code tends to under predict char burnout for the current study.

KEYWORDS

Pulverised Coal Combustion, CFD, 1 MW Coal Combustion Test Facility

INTRODUCTION

The combustion of fossil fuels and particular coal has been the predominant energy source in many modern economies since the beginning of the industrial age. Despite significant uptake of nuclear and more recently renewable energy, electricity production from the combustion of fossil fuels is still prevalent. This is particularly relevant in the South African context. Eskom, the South African national power utility, had a total installed capacity of 41 194 MW in 2011, of which 34 952 MW was generated by coal fired power stations [1]. Hence the understanding and efficient execution of coal combustion processes in South Africa is of significant importance. A consequence of such heavy reliance on coal combustion is that South Africa is a major emitter of carbon dioxide amongst industrialised nations. Along with this are other harmful emissions in the form of nitrogen oxides (NOx) and sulphur oxides (SOx) as well as particulate matter. Hence it is of paramount importance to combust coal efficiently and as cleanly as possible from an environmental perspective. Efficient coal combustion also contributes to increasing the profitability of a coal fired power station as less coal is consumed for the same amount of electricity produced. Furthermore, the ability to tailor the combustion process to avoid adverse conditions such as fireside corrosion, greatly improves the availability of power producing assets. Finally, from a safe operation point of view, it is essential to ensure that complete combustion is attained to prevent a situation where explosive quantities of volatile gases reaches the back end of boiler system.

COAL COMBUSTION

Coal combustion is a complex process. It is a multi-faceted problem that covers many fundamental physical and chemical processes. There are several sub-processes that are understood to occur simultaneously during combustion of coal. These processes were summarised by Smoot et al [2] and are shown graphically in Figure 1.

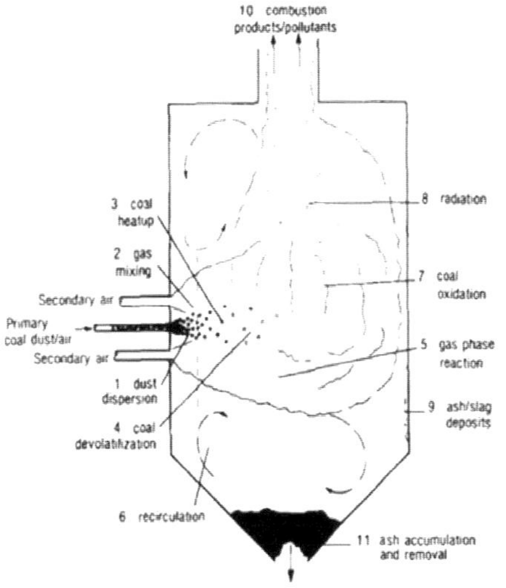

Figure 1: Physical Processes that Occur During the Combustion of Pulverised Coal [2]

The processes of interest are:

Pulverised Fuel (PF) Dispersion. This relates to pneumatic transport of the PF particles through burner ducting and it's delivery into the furnace region.

Gas Mixing. The flow at the mouth of a pulverised coal burner is typically of a high Reynolds number and hence the region may be considered one of turbulent mixing and this will have a bearing on the gas phase combustion as well as the heat transfer rates.

Coal Heat-up. This process refers to the modes of heat transfer between the environment and coal particles. This is important as the chemical reactions are temperature dependent.

Coal Devolatilisation. This refers to the phase of the coal combustion process where the particles are heated to a certain temperature and they release volatile gaseous materials into gas phase.

Gas Phase Reaction. This is where the combustible materials in the gas phase react with each other. This is typically hydrocarbon combustion and the rates and intensities of these reactions must be known.

Recirculation. The flow field with a boiler is complex and the various circulatory flows are essentially to bring back the combusting constituents back to the burner mouth to increase the heat rate to the fresh incoming PF and so enhance the devolatilisation rate.

Coal Oxidation. This is the portion of the combustion process where a large part of heat is released after devolatilisation ended. It is essentially the oxidation of the Char that is present in the Raw Coal and is governed by oxygen diffusion within the char particle through the channels that have been created during devolatilisation. Unlike the gaseous phase combustion, this is a surface/volume related process and hence it requires its own set of rate parameters to be modelled.

Radiation. This is most dominant heat transfer mode when considering how heat is transferred from the fireball to un-combusted particles that are being introduced into the furnace region. It is quite complex as the gaseous combustion region is not optically clear due to the presence of the particulates and hence it is considered a participating media in the radiation process.

Ash and Slag Deposits. This is an issue of operational importance as the slag that attached to the boiler tubes in the furnace region reduces the heat transfer capacity of the said tubes. The production of the slag is dependant of the furnace temperature, the flow field and also the properties of the coal being burnt.

Combustion Products and Pollutants. Apart from the products of the major heat generating reactions, there are also secondary processes that form pollutants such as Nitrous Oxides (NOx) and Sulphurous Oxides (SOx). It is important to understand how these are produced because they cannot be released into the atmosphere through the back-end of the boiler system. They had adverse environmental effects and must be controlled.

Ash Accumulation and Removal. A necessary by-product of coal combustion is the production of significant quantities of ash. The ash is not combustible and it must be removed at a rate close to what it is produced to ensure that boiler is mechanically able to sustain the combustion process. The issues are: understanding the properties of ash as it differs from raw coal and being able to predict its transport and also the contribution it makes to radiation within the furnace.

Each of the sub processes involved in the combustion process must be adequately addressed to predict combustion behaviour to a reasonable degree of accuracy. Some of the processes occur simultaneously such as particle devolatilisation and char combustion as well as the gas phase reactions. Since a continuous supply of fuel is coming into the furnace, at any given time the combustion zone will contain particles at varying stages of combustion i.e. that the time history of a coal particle is crucial to tracking the combustion process. For instance, a fresh coal particle that enters the furnace will have just devolatilised, whilst one that has spent a few seconds in the furnace would likely be completely burnt out and have been reduced to mostly ash.

Description of Test Facility

Eskom operates a 1 MW coal fired test furnace at its Research, Test and Development division. The facility is primarily used to test different coal types to determine its combustion characteristics. The facility comprises of a 1 MW Low NOx burner that is attached to a refractory lined furnace. The flue gas leaving the back end of the furnace is cooled by water fed heat exchanger tube bundles before finally being ducted into a bag filter plant and finally expelled into the atmosphere via a chimney. A dedicated milling plant is used to pulverise raw coal to the required size specification. The pulverised coal (PF) is then pneumatically conveyed to the burner and is introduced to the furnace along with the primary air flow. The primary, secondary and tertiary air flows are preheated by electrically operated heat exchangers. Lesser air flows are introduced via a core air pipe (after the oil lance has been removed) and two small ports that house the flame detectors. A schematic of the plant is shown in Figure 2.

The plant is instrumented with thermocouples to measure wall temperatures and has several ports that provide access to the furnace during operation. These ports are typically used to insert suction pyrometers and thermocouples to measure the in-flame temperatures and take samples of the flue gases and particulates, particularly for the estimation of combustion efficiency.

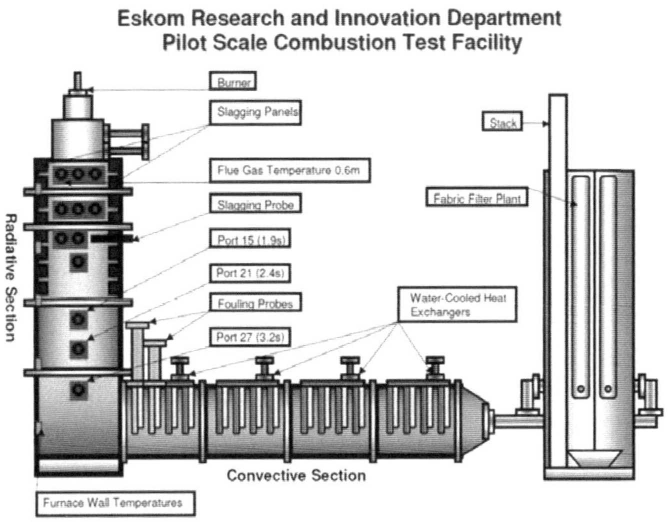

Figure 2: Eskom Pilot Scale Combustion Test Facility Layout

Research Method

The combustion behaviour of two distinct South African coals was studied by simulating the coal combustion process using the commercial CFD code Star CCM+ (V 9.04). The process involved the creation of a representative computational model. This is included detailed modelling of the geometry in particular the swirlers of the burner and it primary, secondary and tertiary air flow paths. Furthermore, the characteristics of the fuel were interpreted from proximate and ultimate test data, into a form consistent with the code implementation. Coal particle devolatilisation and char combustion was modelled using previously obtained test data. The results obtained from the CFD code were compared to experimental measurements and contextualised given the comprehensive level of insight gained from the model i.e. detailed understanding of the near burner region aerodynamics as well as the transport of the coal particles taking into account their respective time histories.

Computational Framework

Conservation

The coal combustion problem was simulated in the finite volume based CFD code; Star-CCM+ using a two phase Euler – Lagrangian methodology. The simulations were conducted in steady state. Specifically the reacting gas species was treated as a continuous phase incorporating a multispecies gas model. The coal particles where treated as a discrete phase and their trajectories were modelled using a Lagrangian particle tracking model that allowed for mass and energy transfer with the continuous phase. The non-premixed combustion scenario is adopted as the fuel (pulverised coal) and oxidiser (air) are not well mixed before sufficient enthalpy is available to initiate combustion.

The following equations govern the fundamental conservation quantities of the flow field [3] (differential equations are stated in compact suffix notation):

Continuity:

$$\frac{\partial \rho}{\partial t} + \frac{\partial}{\partial x_i}\left(\rho u_i\right) = 0 \qquad (1.1)$$

Momentum:

$$\frac{\partial}{\partial t}(\rho u_i)+\frac{\partial}{\partial x_i}(\rho u_i u_j)=-\frac{\partial p}{\partial x_i}+\frac{\partial \tau_{ij}}{\partial x_i}+F_i \qquad (1.2)$$

Where:

$$\tau_{ij}=\mu\left(\frac{\partial u_i}{\partial x_j}+\frac{\partial u_j}{\partial x_i}-\frac{2}{3}\delta_{ij}\frac{\partial u_k}{\partial x_k}\right) \qquad (1.3)$$

τ_{ij} is the viscous stress tensor whilst F_i is body force which accounts for effects of gravity amongst other forces. At this stage is should be noted that a combustion flow field inherently contains large temperature gradients and as such one cannot consider the mixture density to be constant. In fact careful consideration must be given as to how the mixture properties that are a function of pressure, temperature and species concentration is considered.

Energy:

$$\frac{\partial}{\partial t}(\rho h)+\frac{\partial}{\partial x_i}(\rho u_i h)=\frac{\partial}{\partial x_i}\left[\frac{\mu}{\sigma_h}\frac{\partial h}{\partial x_i}+\mu\left(\frac{1}{Sc_k}-\frac{1}{\sigma_h}\right)\sum_{k=1}^{N}h_k\frac{\partial Y_k}{\partial x_i}\right]+\frac{\partial p}{\partial t}+S_{rad} \qquad (1.4)$$

Where, σ_h is the Prandtl Number, and Sc is the Schmidt number.

Species Transport:

$$\frac{\partial}{\partial t}(\rho Y_k)+\frac{\partial}{\partial x_i}(\rho u_i Y_k)=\frac{\partial}{\partial x_i}\left(\rho D_k\frac{\partial Y_k}{\partial x_i}\right)+\dot{\omega}_k \qquad (1.5)$$

The combustion process inherently consumes oxygen and forms carbon based oxides in the course of the process. Hence it is important to ensure species are transported in a conservative manner and does not violate mass conservation principles. The last term in equation 1.5 describes the volumetric source or sink that determine the production or destruction of a given species respectively.

Turbulent Combustion and Global Chemistry

The simulations where conducted in steady state conditions using a RANS approach. The realizable k-ε turbulence model was used for this study. It provided a good balance between resolving key flow characteristics whilst still being computationally economical. It is acknowledged that better models are available, however it has been demonstrated by Peters and Weber [4], that even the simplest form of the k-ε turbulence provides adequate prediction of the bulk flame structure.

The hybrid eddy break up model [5] was used to model the turbulent combustion features. This model considers both turbulent mixing as well as finite rate kinetics to determine the respective combustion rate. The purpose of combustion models is to calculate the reaction state space, i.e. the concentrations of the various species present in a chemical reaction, and the quantities they influence, viz., density, viscosity, and temperature. A large chemical reaction set, such as found in hydrocarbon combustion, can span a wide range of time scales. In addition to these time scales, the turbulent flow field imposes its own limits on length and time scales, ranging from the Kolmogorov and Batchelor time scales at the low end to the large, energy-containing eddy time scales at the high end.

Resolving all the length and time scales affecting the grid-mean properties in a reacting flow system demands computational resources beyond those currently available. We therefore need combustion models to account for the processes that occur at length and time scales below what we can resolve on a numerical simulation grid.

The reaction space considered was:

$$\begin{aligned}
&Volatiles \rightarrow C + CO + H_2 + N_2 + SO_2 \\
&C + \frac{1}{2}O_2 \rightarrow CO \\
&CO + \frac{1}{2}O_2 \rightarrow CO_2 \\
&H_2 + \frac{1}{2}O_2 \rightarrow H_2O \\
&Char + O_2 \rightarrow 2CO + CO_2
\end{aligned} \tag{1.6}$$

Where the *volatile* species is a numerical construct used by the code to facilitate the mass transfer of the volatile matter to the continuous medium from the discrete particle medium. Thereafter the volatile species decomposes into its assumed constituents by means of the turbulent combustion model.

Coal Combustion Specific Models

Specific aspects related to the evolution of a coal particle during the combustion must be modelled and incorporated into the general mechanisms previously described. The code assumes a coal particle to be comprised of raw coal, char, ash and inherent moisture. This is a simplification of the true composition of coal that is more complex. The raw coal is considered to be the sum of the fixed carbon and volatile material fractions obtained from proximate analysis. Hence the char component is only explicitly needed in cases when the particle has already undergone a gasification process. This particle composition is shown in Figure 3.

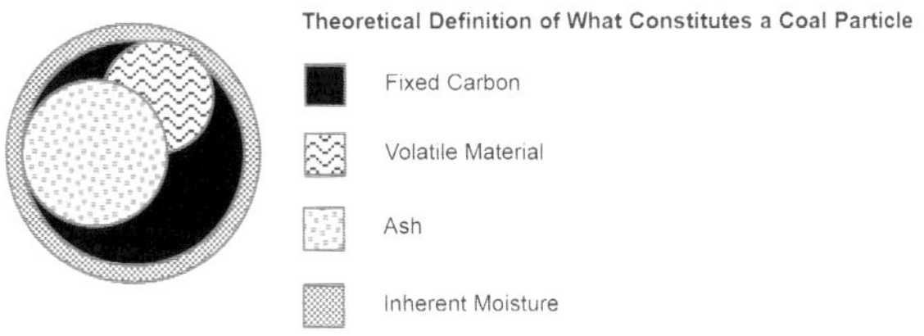

Figure 3: Coal Particle Definition

During the combustion process the particle is assumed to undergo three fundamental processes i.e. (i) dry out or moisture evaporation, (ii) devolatilisation and finally (iii) char combustion (see Figure 4). Typically a particle in an industrial furnace will experience heating rates in the order of 15000 K/s. Hence the particle is subjected to high levels of heat flux. Hence the code assumes that the three processes above occur sequentially for a given particle i.e. that dry must be complete before devolatilisation occurs (both in order of ms), and the second process must complete before char combustion (typically seconds) occurs. This assumption may not be consistent with true combustion behaviour.

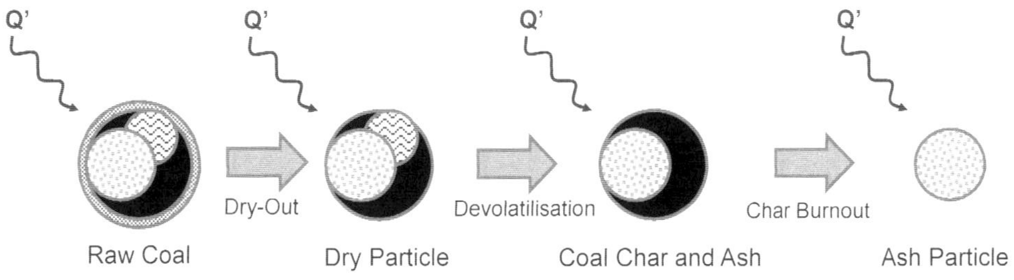

Figure 4: Coal Particle Combustion Processes

Coal Particle Moisture Evaporation

The inherent moisture initially specified with the particle is transferred to the continuous phase via a quasi-static single component evaporation model. Convective heat transfer is modelled using the Ranz-Marshall formulation with is valid for spherical particles with low slip velocities [5]

Coal Particle Devolatilisation

Devolatilisation is an important process in the context of the complete combustion process. Volatile materials in the form of hydrocarbon gases and tars are released during a gasification process as the particle is physically cracked due to the temperature gradients it is subjected to. Typically the quantity of the volatile materials released is a characteristic of a given coal and is determined by the proximate analysis. However, it was noted that under higher heating rates typically experienced in real furnaces, the volatile material release is higher than that stated by the proximate analysis. This addition data is derived by high temperature (up to 1800 K) devolatilisation tests conducted in a drop tube furnace in an inert environment. The concept was implemented in the current model by incorporating the work of Kobayashi [6] who first formulated the concept of the *high temperature volatile yield.* The premise of the model is that additional devolatilisation can occur at higher temperatures and this is accounted for by using two competing rates. Each rate is in turn presented in an Arrhenius form with rate coefficients derived from drop tube test data.

$$RawCoal\begin{Bmatrix} \overset{k_1}{\rightarrow} YY_1\left(volatile\right)+\left(1-YY_1\right) \\ \overset{k_2}{\rightarrow} YY_2\left(volatile\right)+\left(1-YY_2\right) \end{Bmatrix} \tag{1.7}$$

Where the reaction rates are defined as:

$$k_i = A_i e^{-Ea_i/RT_P} \tag{1.8}$$

YY_1 and YY_2 are the respective mass stoichiometric coefficients [5].

Char Combustion

The coal particle undergoes a surface char reaction with the oxygen in the gas phase. For the purposes of this study the half order char oxidation model was used. Char is assumed to react heterogeneously with gaseous oxidizers that diffuse to the particle surface from the bulk gas phase. Two rate-limiting steps are considered for this process [5]:

- Gas phase diffusion
- Half order heterogeneous reactions

The rate of surface char combustion is given by:

$$k_{pl} = A_{pl} T_P^n e^{-E_{pl}/RT_P} \tag{1.9}$$

And the overall char combustion rate calculated from the following implicit formulation:

$$r_{hpl} = \phi_l k_{cpl} A_P M_{hp}\left[C_{\text{olg}} - \left(\frac{r_{hpl}}{\phi_l k_{pl} A_P M_{hp} \zeta_{pl}}\right)^2\right] \tag{1.10}$$

Typically data obtained from coal char runs in a drop tube furnace is used to calibrate the coefficients in the above formulation. Further details are provided in the code user manual [5].

Particle Heat Transfer and Radiation Modelling

The heating rate of particle is crucial to the modelling of the combustion processes. The heat balance on a single coal particle (spherical) may be stated as:

$$\frac{4}{3}\pi r_P^3 \rho_P c_p \frac{dT_p}{dt} = 4\pi r_P^2 h(T_\infty - T_P) + 4\pi r_P^2 \sigma(\varepsilon_f T_f^4 - \varepsilon_P T_P^4) \quad (1.11)$$

The convective portion of the particle heat transfer is calculated using the Ranz-Marshall correlation. The radiative heat transfer is more complex and requires the solution of the radiation transfer equation. This is done is study be using the discrete ordinates radiation model. Since the flue gas contains significant quantities of CO_2 and H_2O, it participates in radiation as well. This effect is incorporated by adjusting the absorption coefficient of the flue gas mixture by implementing the weighted sum of grey gases model. The code has an additional model that accounts for the scattering and absorption of thermal radiation due to the presence of particles in the participating medium. The particle is treated as grey and hence the absorption coefficients are constant.

CFD Model

Geometry and Input Data

A detailed CFD model of the Pilot Scale Combustion Test Facility was generated in the code using an unstructured polyhedral mesh with prismatic boundary layers. The data presented in this study uses a model with a cell count of 1 959 509. The model geometry including the boundary definitions as well as the detailed swirler configurations are shown in Figure 5.

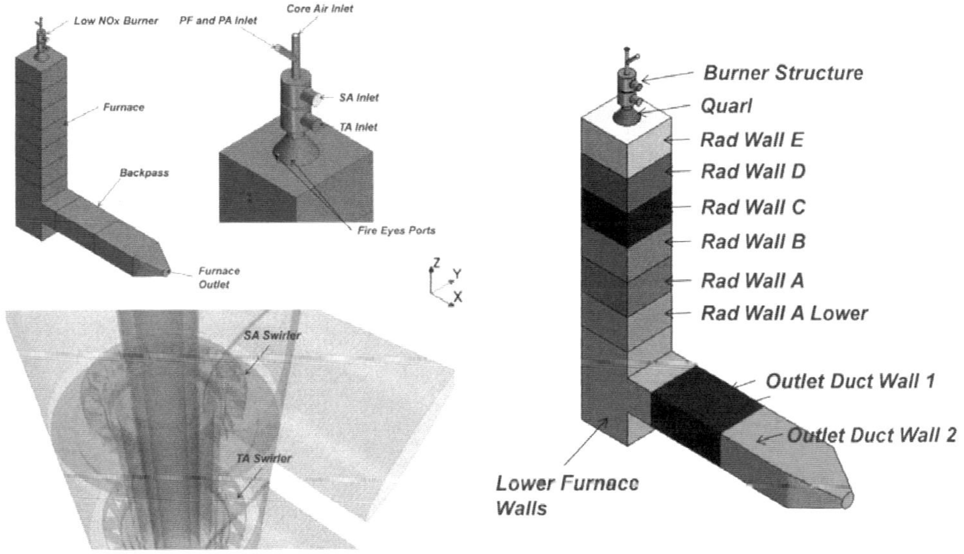

Figure 5: PSCTF Model with Boundary Definitions

The proximate and ultimate analysis of Coal A and Coal B are given in Table 1 on an as received basis. The particle size distribution obtained from a sieving analysis is also shown [7], [8]. The flow boundary conditions are consistent with the readings from the test rig as shown in Table 2. The furnace wall temperatures (see Figure 1, Rad Wall A, etc.) were set to fixed temperatures obtained from thermocouple measurements on the test rig [7], [8].

Table 1: Coal Specification and Particle Size Distribution

Proximate Analysis	Weight %		Ultimate Analysis	Weight %	
	Coal A	Coal B		Coal A	Coal B
Inherent Moisture	2.4	2.4	Carbon	55.04	56
Ash	28.8	26	Hydrogen	2.35	2.78
Volatile Matter	22.3	27.4	Nitrogen	1.2	1.22
Fixed Carbon	46.5	44.2	Total Sulphur	0.74	1.13
			Carbonate (as CO2)	2.81	2.87
			Oxygen (by difference)	6.67	7.57
GCV [MJ/kg]	21	23.31			
High Temperature Volatile Yield	1.11	1.24			

Particle Size [μm]	Fraction []
300	0.00
150	0.02
106	0.07
75	0.16
38	0.75

Table 2: Flow Boundary Conditions

	Coal A		Coal B	
Coal Flow [kg/s]	0,044		0,043	
Primary Air	0,098	[kg/s]	0,098	[kg/s]
	359,15	[K]	357,15	[K]
Secondary Air	0,232	[kg/s]	0,238	[kg/s]
	580,150	[K]	576,150	[K]
Tertiary Air	0,00028	[kg/s]	0,00250	[kg/s]
	486,15	[K]	444,15	[K]
Core Air	0,002	[kg/s]	0,002	[kg/s]
	359,15	[K]	359,15	[K]
Fire Eye Ports (x2)	0,011	[kg/s]	0,011	[kg/s]
	298,15	[K]	298,15	[K]
$(M_a/M_f)_s$	7,164		8,377	
Excess Air	13,242	%	12,605	%

The devolatilisation parameters for the two step competing rate model was extracted from a previous study [9], in which coals with similar properties were studied. The same devolatilisation characteristics were applied to both coals in this study. The char kinetics for the coals was extracted from experimental data generated specifically for the PSCTF. This data was generated using a drop tube furnace in an air environment. The pulverised coal particles were introduced to the furnace at the outlet of the primary air pipe. The size distribution of the particles was obtained [7], [8] from a seizing characterisation. Discrete size fraction particle injectors were used.

Table 3: Devolatilisation and Char Kinetics and Particle Size Fractions

Kobayashi Competing Rate Devolatilisation Model Parameters

		Coal A	Coal B
First Step	Activation Energy [J/kmol]	4.20E+07	4.20E+07
	Pre Exponent [s^{-1}]	4.60E+04	4.60E+04
	Mass Stoichiometric Coefficient[]	0.3	0.3
Second Step	Activation Energy [J/kmol]	8.40E+07	8.40E+07
	Pre Exponent [s^{-1}]	4.00E+06	4.00E+06
	Mass Stoichiometric Coefficient []	1	1

Char Combustion Kinetics

	Coal A	Coal B
Activation Energy [J/kmol]	5,12E+07	7,10E+07
Pre Exponent [s^{-1}]	62,43	74,55

Results

The CFD simulations successfully produced realistic pulverised coal flames. Temperatures at various locations in the model were calculated and compared to measurements obtained during the actual combustion tests runs. Points 1 to 6 are thermocouple measurements near the wall of the furnace. The outlet and O_2 and CO_2 values were also compared to experiment. These values were used as a convergence criterion, since once they stabilised it indicated that the combustion chemistry has reached a steady state.

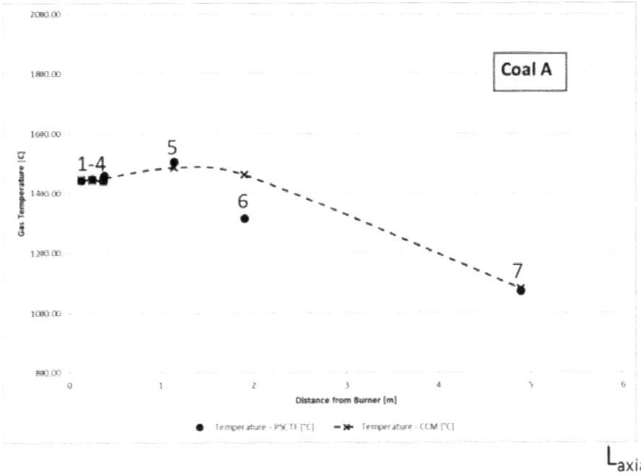

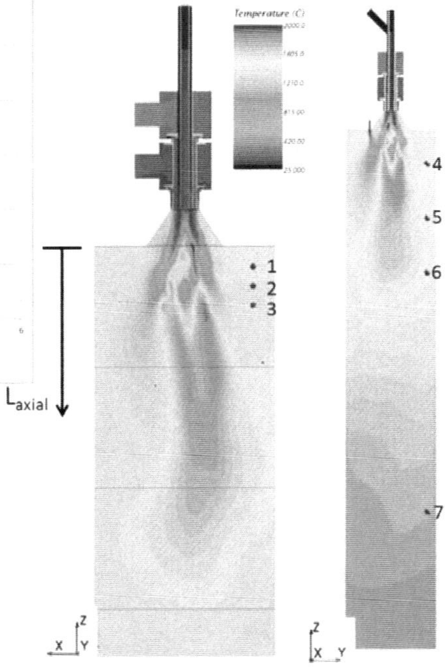

	L_{axial} - Distance from Burner Exit [m]	Index	Temperature - PSCTF [°C]	Temperature - CCM [°C]
	0.13	1	1443.69	1445.95
	0.25	2	1448.99	1444.76
	0.37	3	1444.02	1442.76
Port 4	0.381	4	1459.47	1450.89
Port 10	1.143	5	1505.82	1486.59
Port D	1.899	6	1314.45	1462.68
Port 28	4.899	7	1070.50	1079.68
Outlet O_2 [% Vol]			3.58	3.15
Outlet CO_2 [% Vol]			14.13	14.12

Figure 6: Coal A Combustion Results

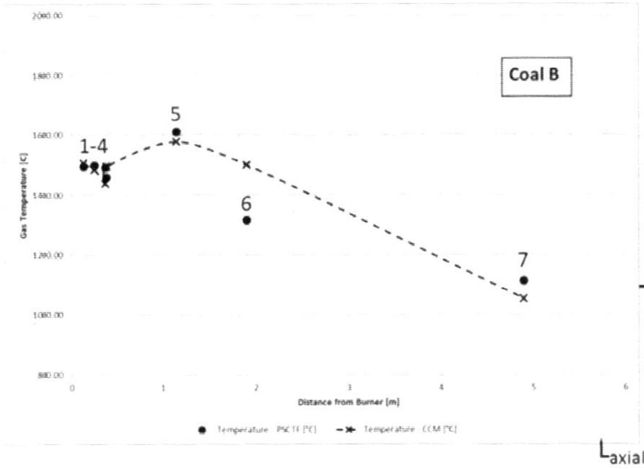

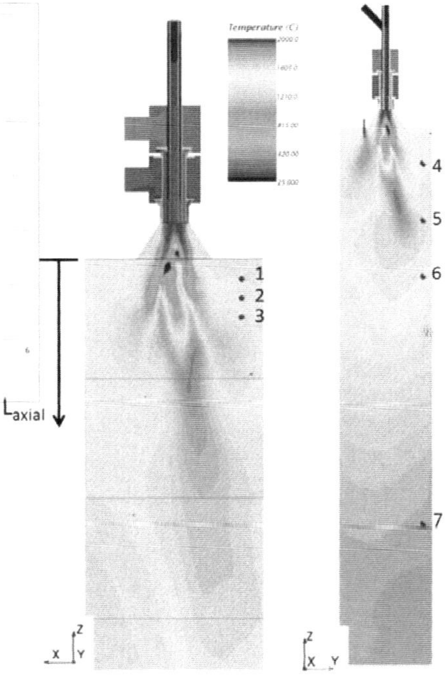

	L_{axial} - Distance from Burner Exit [m]	Index	Temperature - PSCTF [°C]	Temperature - CCM [°C]
	0.13	1	1495.84	1507.24
	0.25	2	1499.44	1484.85
	0.37	3	1492.86	1439.18
Port 4	0.381	4	1459.47	1496.60
Port 10	1.143	4	1610.81	1577.06
Port D	1.899	5	1316.95	1501.71
Port 28	4.899	6	1112.38	1052.30
Outlet O_2 [% Vol]			3.58	2.24
Outlet CO_2 [% Vol]			14.13	15.27

Figure 7: Coal B Combustion Results

Iso-surfaces of temperature (1650 °C) are shown in Figure 8. These shows the extent of the most reactive region of the combustion domain and also emphasises the complex flow field that exists by virtue of the three dimensional shape of flame.

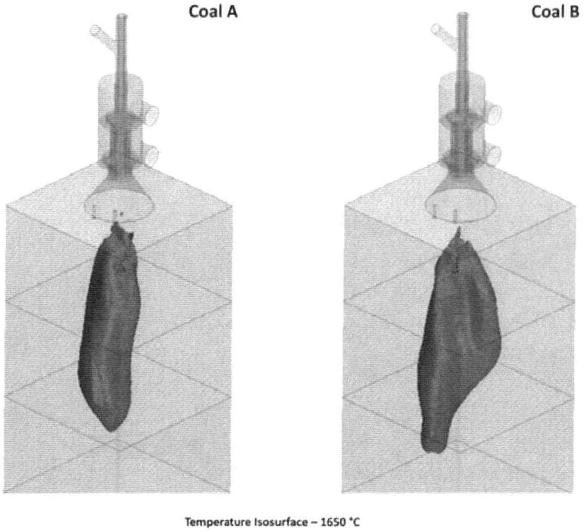

Figure 8: Flame Iso-surfaces

Combustion efficiency measurements were also conducted on the PSCTF by inserting a suction pyrometer at three different port heights, into the centre of the furnace and sampling particulates. It is essential that the sampling be done iso-kinetically. If this is not the case the accuracy of the device declines significantly. The combustion efficiency is defined as the proportion of residual carbon measured as compared to the amount introduced in the raw coal. This is shown in Figure 9 and Table 4.

Table 4: Combustion Efficiency

Distance from Burner Exit	Coal A		Coal B	
	Measurement [%]	CFD [%]	Measurement [%]	CFD [%]
0.661m	92.65	87.57	93.00	86.89
1.639m	96.05	96.41	97.22	91.13
2.174m	96.91	99.72	97.74	98.82

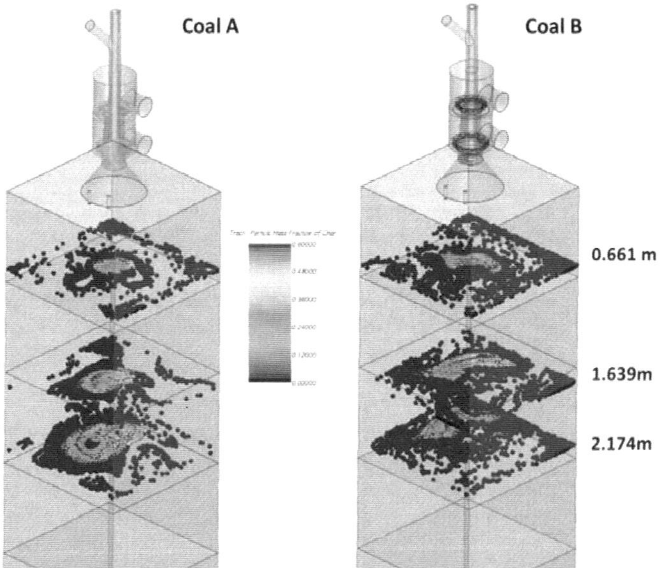

Figure 9: Mass Fraction of Char in Particles

Discussion of Results

A complex CFD model of a 1 MW single burner combustion test facility was successfully simulated using a commercial CFD code. Suitable adjustments we made to the coal specific sub-models to tailor the reactive behaviour to South African coals. The in furnace temperature measurements shown in Figure 6 and Figure 7, show that the simulation predicts the correct temperature trend, and there is good quantitative agreement with all thermocouple measurements except for point 6 in both cases.. The code does however show that coal B produces a higher peak temperature than coal A, which is a trend seen in the experimental data. This is promising as it demonstrates the implementation in the code can distinguish between coals based on their basic analysis. It should be noted that a typical pulverised coal flame is not steady and in fact fluctuates to varying degrees. This is due to the instabilities in the coal supply (i.e. the transportation of particles to the combustion zone) and noting that the shear layer of the flame is inherently unstable as there are large temperature gradients present in this region due to reactive flow. Hence the steady values stated should be treated as an average value. The outlet O_2 and CO_2 values correspond well with the measured data. This provides confidence that the overall stoichiometry of the test rig has been modelled consistently. The rate of char combustion is shown in Figure 9. The CFD code does produce realistic values for the amount of char consumed but it appears to under estimate the gross value as compared to the experimental data. This may be due to several factors such as the accurate modelling of the particle heating rates, slower than expected devolatilisation rates and also lower char combustion rates that are modelled. It is also important to note that suction probe method used to extract particles could also be prone to significant error and representative cross section of particles used to calculate the char burnout is not fully consistent between experiment and CFD. Further work is needed in this regard.

In conclusion, the CFD model shows great promise as a virtual test bed to test different scenarios such as different coals, stoichiometric ratios and even burner configurations with an emphasis on low power operations. It could be used to develop better devolatilisation and char kinetic models as well as improved NOx and SOx predictions.

Nomenclature

Abbreviations

CFD	Computational Fluid Dynamics
NOx	Nitric oxides
PF	Pulverised Fuel
PSCTF	Pilot Scale Combustion Test Facility
SOx	Sulphur oxides

Symbols

Φ_1	Char combustion stoichiometric ratio
Ai	Devolatilisation pre-exponent [s^{-1}]
A_{pl}	Char combustion pre-exponent [s^{-1}]
C_{olg}	Concentration of oxidiser [$kmol/m^3$]
E_{ai}	Devolatilisation activation energy [J/kmol]
E_{pl}	Char combustion activation energy [J/kmol]

k_i	Devolatilisation rate [ ]
k_{pl}	Mass transfer coefficient [ ]
r_{hpl}	Overall combined char reaction rate [ ]
Y_k	Mass fraction of species k [ ]
YY_i	Devolatilisation mass stoichiometric ratio [ ]

REFERENCES

[1] Eskom, "EPPEI Program 2012/2013," 2013.

[2] L. D. Smoot, "Modeling of Coal-Combustion Processes," *Prog. Energy Combust. Sci.*, vol. 10, no. 4, 1984.

[3] H. K. Versteeg and W. Malalasekera, *An Introduction to Computational Fluid Dynamics*, vol. M. 2007.

[4] A. a. F. Peters and R. Weber, "Mathematical Modeling of a 2.4 MW Swirling Pulverized Coal Flame," *Combust. Sci. Technol.*, vol. 122, no. 1–6, pp. 131–182, Apr. 1997.

[5] CD-Adapco, "Star CCM+ V9 User Manual." 2014.

[6] H. Kobayashi, "Devolatilization of Pulverized Coal at High Temperatures," *PhD Thesis*, 1976.

[7] P. Rajoo, "Eskom PSCTF Test Report: Coal A," 2013.

[8] P. Rajoo, "Eskom PSCTF Test Report: Coal B," 2013.

[9] S. Peta, "Ignition Characteristics of Low Volatile Quality High Ash Coal," University of the Witwatersrand, 2014.

Preload and amplitude dependency on dynamic properties of a rubber mount

C.B. Nel, Senior Lecturer, School of Mechanical and Nuclear Engineering
North-West University, Private Bag X6001, Potchefstroom, 2520, South Africa,
carlbnel@netactive.co.za

J. Bloem*, School of Mechanical and Nuclear Engineering
North-West University, Private Bag X6001, Potchefstroom, 2520, South Africa,
bloem.johann@gmail.com

ABSTRACT

An equivalent rubber mount comprising two industrial mounts connected in series was characterised with the use of an electrodynamic Shaker. This elastomeric mount was bolted between a variable rigid body mass and the base of the test assembly. The base of the test assembly was firmly bolted to a Shaker table. Linear guide bearings along four shafts guided the moving mass in the vertical plane with minimal friction. The rigid body mass was varied by adding steel plates of 5 kg each. The preload was varied from 30 kg to 70 kg in 5 kg increments. The amplitudes for each preload condition were varied from 0.1 to 1.4 mm. The preload and amplitudes were varied to determine the rubber mount characteristics to be used for computer modelling of a vibration screen application with different loading conditions and also different excitation amplitudes. Vibration responses with phase differences were measured at the base plate and the rigid body mass in the vertical direction. An FFT Analyzer was used to measure the time and frequency domain signals at a forced frequency of 45 Hz. The data was used to compute the dynamic properties of the mount by solving appropriate mathematical equations of motion. Amplitude and preload dependency on dynamic stiffness and damping ratio were investigated.

KEYWORDS

Stiffness, damping, rubber, preload dependency, Shaker

INTRODUCTION

Previous work that has been done showed that in general rubber dynamic properties depend on static preload, vibration amplitude, excitation frequency and temperature [1]. Strong amplitude dependence on the dynamic characteristics during cyclic loading of elastomeric mounts was reported by several researchers [2, 3, 4, 5, 6].

An optimisation approach to mounting characterisation was used to determine engine rubber mount dynamic properties *in situ* at a front wheel drive vehicle based on measured natural frequencies, and measured responses when the engine was running [7].

Amplitude dependency of the dynamic characteristics of engine rubber mounts was investigated. This was done because of different excitation levels caused by multiple engine operational conditions for vehicles [5, 8, 9]. These dynamic stiffness coefficients and mount loss factors were obtained by measuring the load-deflection properties of an individual mount by using a dynamic testing machine, comprising a hydraulic actuator, a load cell and digital read-out instrumentation. Other researchers also used servo controlled hydraulic actuators, and some used electrodynamic Shakers for excitation to do experimental characterisation of dynamic properties of rubber mounts [2, 3, 4, 5, 6, 10, 11, 12].

Preload and amplitude dependency on the properties of screen mounts were considered necessary for the design analysis and optimisation of a vibration screen [13]. The screen could be empty or fully loaded, and screen mounts are typically exposed to large amplitudes during start-up and shut-down conditions. In addition, different excitation amplitudes could be adjusted for screening of different materials.

MATHEMATICAL MODEL

A variable rigid body mass m, which is supported by a spring element k and viscous damping element c which represents the equivalent rubber mount for a vibration screen comprising two mounts connected in series, and firmly bolted on the base of an electrodynamic Shaker as shown in Figure 1, is considered [14]:

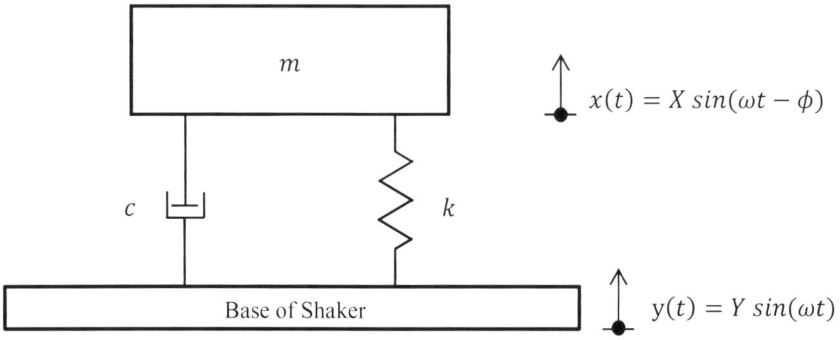

Figure 1. Shaker test representation.

When $y(t)$ denotes the displacement of the base of the Shaker table, and $x(t)$ the rigid body mass displacement from equilibrium position at time t, then the relative displacement of the spring element will be $(x-y)$ and the relative velocity of the two ends at the damper $(\dot{x}-\dot{y})$.

The response of an assumed viscous damped system under harmonic motion of the base can be described by the following equation of motion:

$$m\ddot{x}+c(\dot{x}-\dot{y})+k(x-y)=0 \tag{1}$$

with Y the displacement amplitude of the Shaker base and ω the forced frequency

$$y(t)=Y\sin(\omega t) \tag{2}$$

Equation (1) becomes

$$m\ddot{x}+c\dot{x}+kx=ky+c\dot{y}=kY\sin(\omega t)+c\omega Y\cos(\omega t) \tag{3}$$

Also

$$x(t)=X\sin(\omega t-\phi) \tag{4}$$

where the amplitude ratio is given by

$$\frac{X}{Y}=\left[\frac{k^2+(c\omega)^2}{(k-m\omega^2)^2+(c\omega)^2}\right]^{\frac{1}{2}} \tag{5}$$

and the phase angle

$$\phi=\tan^{-1}\left[\frac{mc\omega^3}{k(k-m\omega^2)+(c\omega)^2}\right] \tag{6}$$

The response X represents the amplitude of excitation and ϕ represents the phase angle between these X and Y displacement vectors.

The displacement amplitudes X and Y can also be written as

$$X = \frac{\ddot{X}}{\omega^2} \quad (7)$$

and

$$Y = \frac{\ddot{Y}}{\omega^2} \quad (8)$$

where $\ddot{X}$ and $\ddot{Y}$ are the corresponding acceleration amplitudes.

The critical damping coefficient

$$c_c = 2\sqrt{km} \quad (9)$$

The damping ratio ζ is defined as the ratio between the actual damping coefficient c and the critical damping coefficient c_c as

$$\zeta = \frac{c}{c_c} \quad (10)$$

COMPUTER IMPLEMENTATION

The mathematical model for the forced excitation Shaker approach was implemented in a computer program in a MATLAB environment and the measured data as recorded with an FFT Analyzer was then used as input data.

With the measured amplitudes of two acceleration signals as well as the phase difference between these two signals at a certain forced frequency, the dynamic stiffness and damping properties were determined, where equations (5) and (6) were simultaneously solved.

EXPERIMENTAL CHARACTERISATION

The equivalent rubber mount for a screen application comprising two industrial mounts was characterised with the use of a Ling Dynamic Systems V724 electrodynamic Shaker connected to a DPA4 Amplifier and SPC4 Signal Controller, which allowed different excitation amplitudes. The test assembly at the Shaker is shown in Figure 2. Linear polished shafts guided a moving mass with minimal friction. This mount was bolted between the mass and the base of the test assembly. The test assembly was then firmly bolted to the base of the electrodynamic Shaker. The test assembly and electrodynamic Shaker are shown in Figure 2. The dynamic properties of the rubber mount was characterised for different excitation amplitudes as well as different preload conditions. This allowed both preload and amplitude dependency to be investigated. The excitation amplitudes were varied between 0.1 and 1.4 mm, and the preload varied between 30 and 70 kg for each excitation amplitude. The preload was increased with the addition of different steel plates of approximately 5 kg each at the test assembly. The excitation amplitudes were varied with adjustments made at the SPC4 Signal Controller. All the tests were performed at a fixed frequency of 45 Hz.

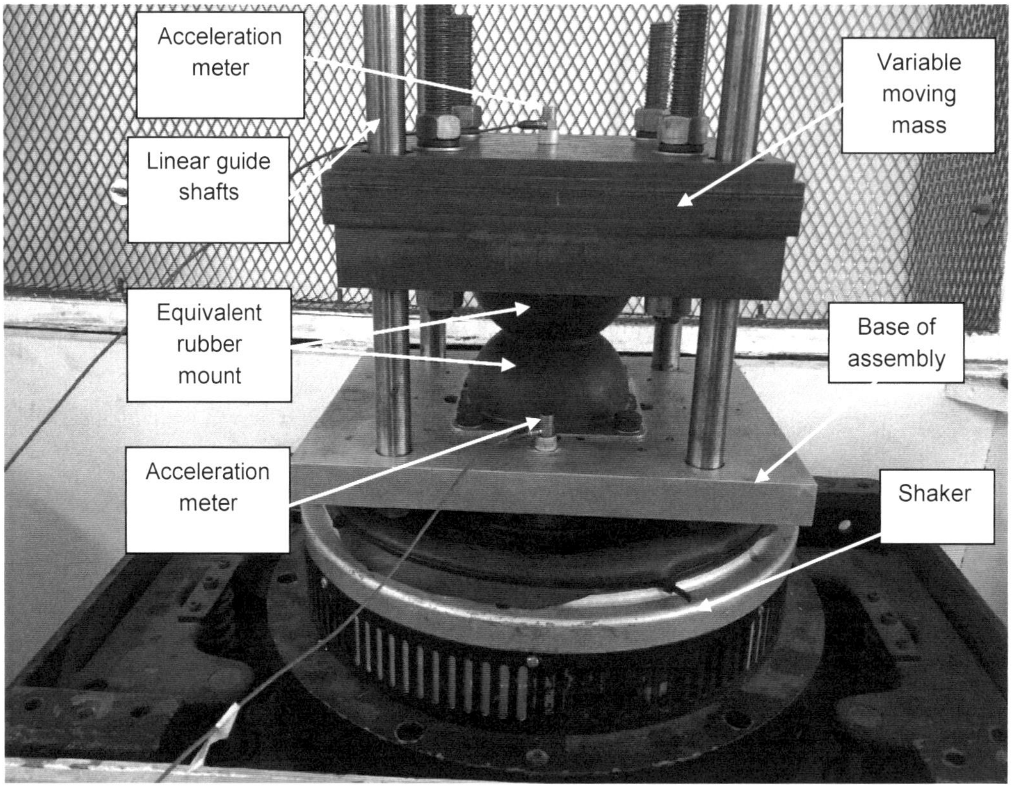

Figure 2. Ling dynamic systems V724 Shaker test setup with rubber mount assembly.

Two 100 mv/g PCB acceleration meters coupled to a Diagnostics Instruments 2200 FFT Analyser were used to measure the acceleration signals simultaneously at the base and moving mass of the test assembly. This measured data was then downloaded to a laptop for further analysis. The measured data was then used by a computer program implemented in a MATLAB environment for computations, in order to determine the equivalent mount stiffness and damping properties. This program was also used to filter out the noise of the measured signals generated at 50 Hz by the electrodynamic Shaker. The computer program converted the measured time domain signals into corresponding frequency domain signals, with phase angles. The dynamic stiffness, damping coefficient and damping ratio were then computed with this data. As an example, Figure 3 shows two simultaneously measured and filtered time domain signals respectively with a 30 kg preload and also 0.1 mm relative excitation amplitude. The Fourier coefficients (amplitudes at corresponding frequencies and phase angles) of the measured signals are shown graphically in Figure 4.

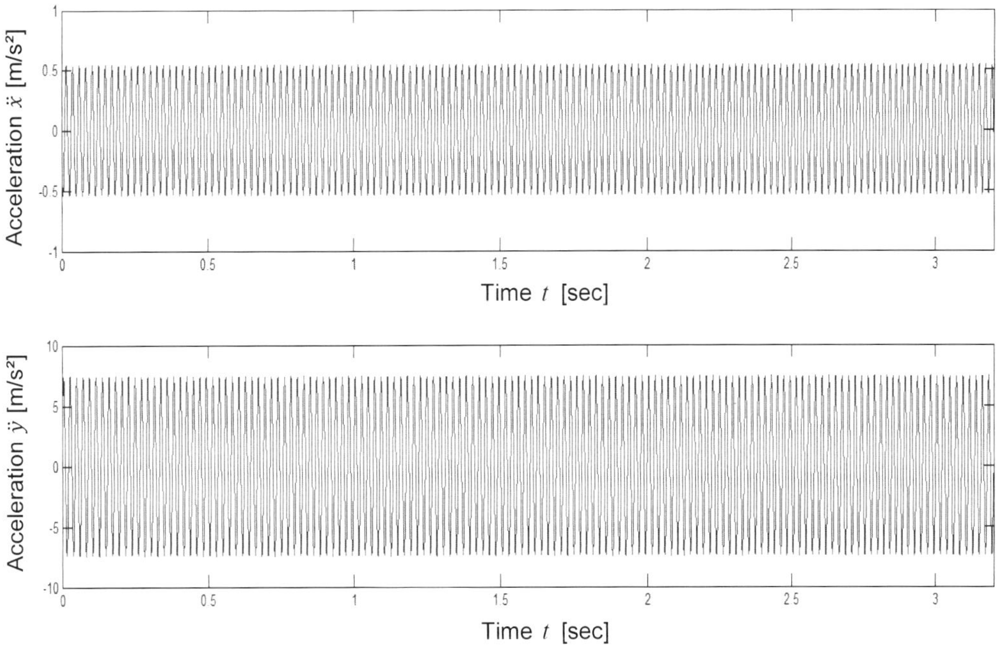

Figure 3. Filtered time domain acceleration signals, $\ddot{x}$ and $\ddot{y}$.

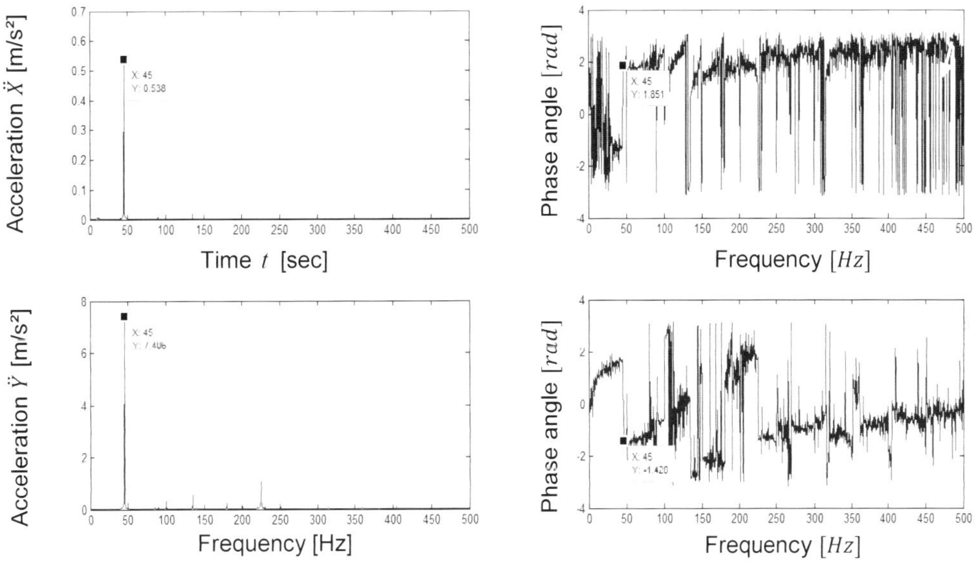

Figure 4. Graphic representation of Fourier coefficients (acceleration amplitudes at corresponding frequencies and phase angles)

The measured time domain signals were also used to compute all the dynamic forces acting on the equivalent rubber mount. As an example, the magnitudes of the dynamic forces and the phase angle ϕ were computed with the output from the computer program. The force magnitudes and phase angle were then used to construct a vector diagram as shown in Figure 5. This vector diagram comprises the four different dynamic force vectors and the phase angle ϕ for a 30 kg preload with 0.1 mm excitation amplitude. These dynamic forces which are in equilibrium proved the validity of the mathematical model.

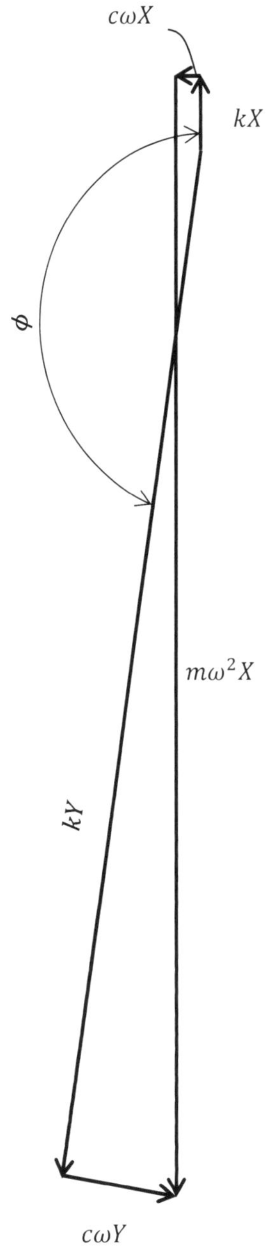

Figure 5. Vector diagram indicating the dynamic forces and phase angles.

RESULTS

Figure 6 shows a graphical representation of the equivalent rubber mount stiffness obtained from the different tests performed. The dynamic stiffness is a maximum for the lowest excitation amplitude and corresponding maximum preload. The graph shows that the dynamic stiffness is a maximum for the lowest excitation amplitude of 0.1 mm and also with largest preload of 70 kg. For all the different preload conditions, the dynamic stiffness magnitudes decreased with an increase in excitation amplitude. Strong amplitude dependency was obtained for this 0.1 to 1 mm amplitude range, with a smaller amplitude dependency between 1 to 1.4 mm. The dynamic stiffness also increased slightly with an increase in preload.

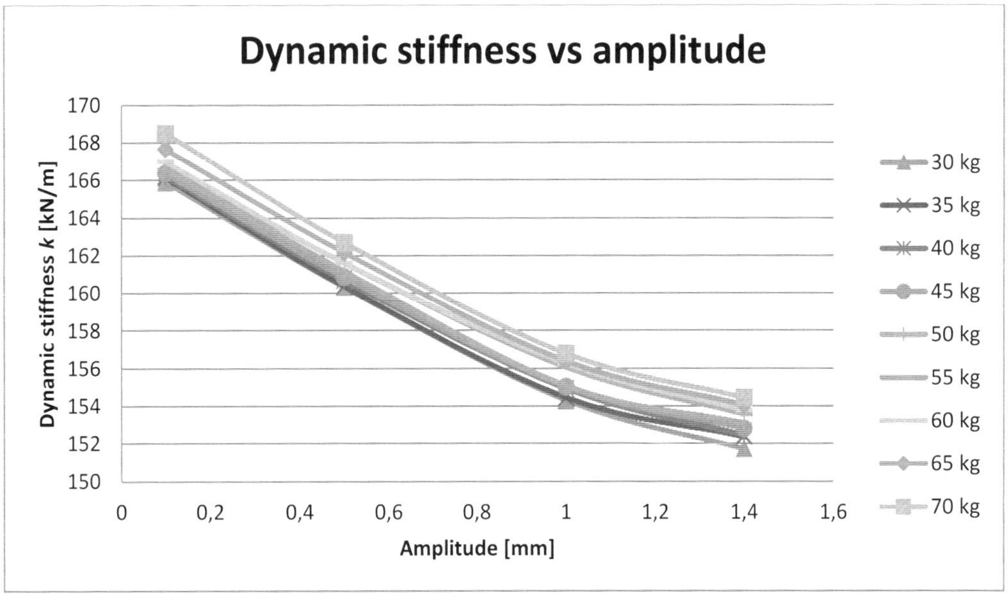

Figure 6. Dynamic stiffness properties of the rubber mount for different amplitude excitations and preload.

A graphical representation of the damping ratio ζ is shown in Figure 7. Low magnitudes for damping ratio were obtained. The damping ratio is a maximum for the smallest excitation amplitude of 0.1 mm and also with smallest preload of 30 kg. The damping ratio decreased with an increase in preload and also an increase in excitation amplitude.

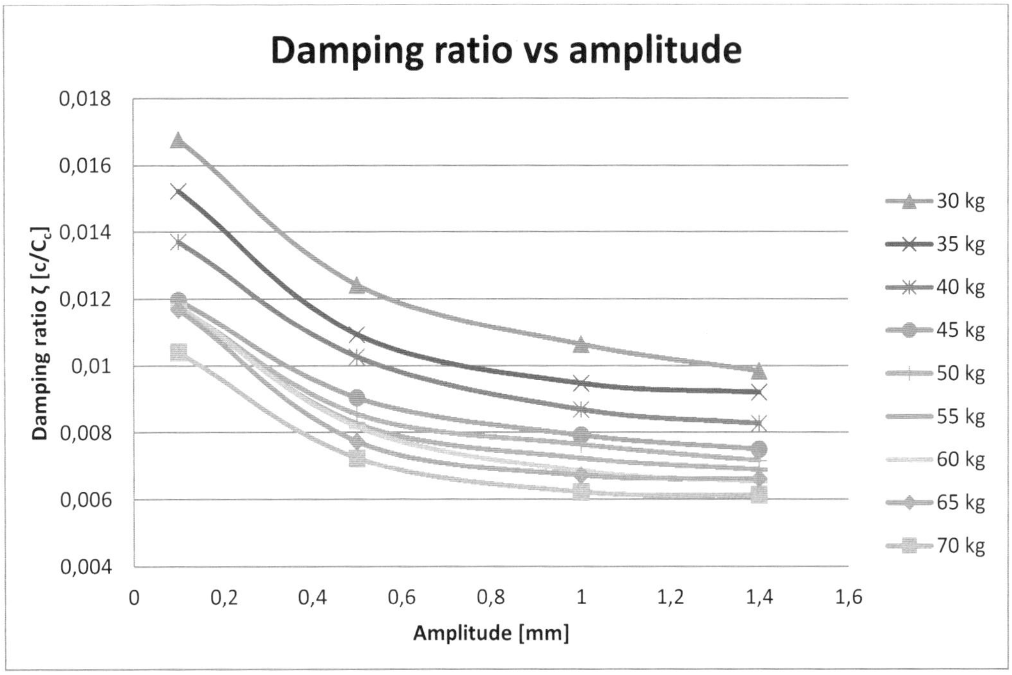

Figure 7. Damping ratio properties of the rubber mount for different amplitude excitations and preload.

CONCLUSIONS

Strong excitation amplitude dependency on the dynamic stiffness and also damping properties of a rubber mount system for a vibration screen application was obtained. Preload showed a significant effect on the damping properties of the mount. The increase in preload showed a slight increase in dynamic stiffness. The different properties as results obtained are very meaningful, and will be used in future work for computer modelling of the vibration screen with different loading conditions and also different excitation amplitudes. This modeling will entail the numerical integration of the equations of motion for the vibration screen assumed as a rigid body and supported by elastic rubber mounts. These accurate rubber mount properties as obtained will be used for simulations of response and dynamic forces transmitted.

REFERENCES

1 Lin, T.R., Farag, N.H. & Pan, J., Evaluation of frequency dependent rubber mount stiffness and damping by impact test, Applied Acoustics, 66:829-844, 2005.

2 Austrell, P. & Olsson, A.K., Considering amplitude dependence during cyclic loading elastomers using an equivalent viscoelastic approach, Polymer Testing, 31:909-915, 2012.

3 De-Wei, S., Zhi-Gang, C., Guang-Yu, Z. & Eberhard, P., Modeling and parameter identification of amplitude- and frequency-dependent rubber isolator, J. Cent. South Univ. Technol, 18:672-278, 2011.

4 Colgate, J.E., Chang, C.T., Chiou, Y.C., Liu, W.K. & Keer, L.M., Modelling of a hydraulic engine mount focusing on response to sinusoidal and composite excitations, Journal of sound and vibration, 3(184):503-528, 1995.

5 Nel, C.B. & Steyn, A.J., Stiffness and damping characterization for a Hydraulic Engine Mount. International Conference on Modal Analysis, Proceedings in IMAC-XXX, Florida, Jacksonville, USA, (Conference Proceedings of the Society for Experimental Mechanics. p. 129-135), Topics in Modal Analysis II, Volume 6: Proceedings of the 30th IMAC, A Conference on Structural Dynamics, New York, NY: Springer New York. p. 129-135), 2012.

6 Nel, C.B. & van Wyngaardt, J., Amplitude Dependency on Dynamic Properties of a Rubber Mount. International Conference on Modal Analysis, Proceedings in IMAC-XXXII, Florida, Orlando, USA, Dynamics of Coupled Structures, Volume 1, Chapter 42, Pages 447-454, ISBN: 978-3-319-04500-9, (Print) 978-3-319-04501-6 (Online), Springer, 2014.

7 Nel, C.B. & Heyns, P.S., An optimisation approach to mounting characterisation. Proceedings of Noise and Vibration, South Africa: University of Pretoria, November, 1995.

8 Nel, C.B., Optimisation of engine mount systems at front-wheel-drive vehicles for multiple operational conditions, ISMA25, Belgium: Katholieke Universiteit Leuven, 2000.

9 Nel, C.B. & Heyns, P.S., Experimental verification of an optimisation program for a front wheel drive engine mountsystem, ISMA21, Belgium: Katholieke Universiteit Leuven, 1996.

10 Kim, G. & Singh, R., Resonance, isolation and shock control characteristics of automotive nonlinear hydraulic engine mounts, ASME, 44:165-180, 1992.

11 Singh, R., Kim, G. & Ravindra, P.V., Linear analysis of automotive hydro-mechanical mount with emphasis ondecoupler characteristics, Journal of sound and vibration, 158(2):219-243, 1992.

12 Richards, C.M. & Singh, R., Characterization of rubber isolator nonlinearities in the context of single- and multi-degree-of-freedom experimental systems, Journal of sound and vibration, 247(5):807-834, 2011.

13 Bloem, J., Design optimisation and experimental evaluation of a grain vibration screen. Magister Thesis, North-West University, 2016.

14 Rao, S.S., Mechanical Vibrations, 4th ed., Prentice Hall, Pearson, 2004.

Finite Element Modal Analysis for a cooler fan support structure

C.B. Nel *, Senior Lecturer, School of Mechanical and Nuclear Engineering
North-West University, Private Bag X6001, Potchefstroom, 2520, South Africa,
carlbnel@netactive.co.za

ABSTRACT

Newly installed air cooler fans have experienced excessive vibration since commissioning in a chemical plant, which resulted in costly production losses. Measured vibration signals obtained from bump impact tests done at a fan support structure revealed the presence of near resonance conditions. A Finite Element Analysis (FEA) model was constructed and used for modal analysis. The first seven natural frequencies at corresponding mode shapes were obtained as output results of the FEA. The numerically predicted natural frequency magnitudes were compared to the corresponding measured magnitudes. The FEA model for the existing design of the support structure was then also changed, until a feasible modified design was obtained with the resonance problems then removed. The FEA model, as constructed for the modified design, entailed the use of additional structural steel stiffening members that were later practically also implemented in the plant. This modified design was successful to reduce the undesired vibration levels.

KEYWORDS

Finite Element Analysis, cooler fan, support structure, resonance, mode shapes, natural frequencies

INTRODUCTION

Previous work done indicate that FEA could be used to simulate natural frequencies at corresponding mode shapes to avoid resonance conditions for different steel structures in production plants, for example inadequate low stiffness at vibration screens [1, 2, 3]. Excessive vibration as a result of these resonance problems could cause structural damage with possible fatigue failures [4]. FEA was also successfully used for several other steel structures, for example at vehicle inner doors [5]. Newly installed air cooler fans have experienced excessive vibration since commissioning, which resulted in costly production losses. Possible sources of vibration such as dynamic balancing, misalignment, mechanical looseness, rotor blade angles and clearances were checked and found to be within specifications. Measured vibration signals taken at a fan support steel structure indicated the presence of near resonance conditions. FEA models were constructed to numerically predict the mode shapes and corresponding natural frequencies. FEA was used for models constructed to simulate mode shapes at corresponding natural frequencies. This was done for the existing design as well as for a modified design. Additional structural steel members were added to the existing design to shift the natural frequencies, in order to avoid most of the undesired resonance conditions.

VIBRATION MEASUREMENTS

Measured vibration signals were obtained from bump tests done at a fan support structure. A Diagnostic Instruments 2200 FFT Analyzer with a 100mV/g CTC accelerometer was used as instrumentation for these bump tests. The accelerometer was sequentially placed at selected positions on the elastic steel support structure and also the electrical motor. These positions of relatively more vibration during maximum speed were practically identified at a similar adjacent running fan. Impact loads were applied sequentially at several of these points on the support structure, with a combination of a rubber mallet and a steel hammer to excite a few selected mode shapes at corresponding frequencies. This was done in a limited time approach to investigate the possibility of resonance, before any FEA was done. The measured time and frequency domain signals were then recorded with the FFT Analyzer. The magnitudes of the measured natural frequencies for the existing design are indicted in Table 1. Other vibration measurements taken confirmed that the maximum running speed of the electrical motor was 16.5 Hz, with the first three harmonics at 33.0 Hz, 49.5 Hz and 66

Hz. The fan rotor speed is 3.78 Hz, with the first two harmonics at 7.56 and 11.34 Hz. The blade pass frequency was 26.46 Hz, with the first two harmonics at 52.92 Hz and 79.38 Hz. The fundamental belt pass frequency was 2.64 Hz, with the first two harmonics at 5.28 Hz and 7.92 Hz. The frequency domain measurements revealed the presence of near resonance conditions for some of these forced frequencies as described in a later paragraph. The simple bump test approach was a quick and cost effective method to confirm the suspected resonance problems, without a comprehensive experimental modal analysis.

EXPERIMENTAL CHARACTERIZATION FOR FINITE ELEMENT ANALYSIS MODELS

A Finite Element Analysis (FEA) model regarding the existing design was constructed and used for modal analysis. Drawings of the structural steel sections used at the fan support structure were used to create three dimensional drawings. The detailed structure was designed in a Solidworks version 2012 environment and the three dimensional CAD model was used to perform the Finite Element Analysis [6]. To simplify the analysis, the detail regarding bolts and small holes were neglected. The model was thus simulated as if the fan steel components of the assembly were firmly bonded together (see Figure 1). A so called solid mesh was selected for the FEA dynamic model simulations. The fan rotor blades with relatively small mass were neglected in this model. Dimensions and the mass property for the electrical motor were used to include a simplified equivalent steel cylinder to represent this motor as mounted to the elastic support structure in the FEA model. The FEA model for the existing design of the support structure was then also changed for a modified design, until a feasible design was obtained with the resonance problems then removed. The modified design comprised several attempts with some additional structural steel sections used to provide sufficient stiffness, in order to shift the natural frequencies of the existing design. This was a challenge, because mass was then also added to the system.

MODAL ANALYSIS AND VALIDATION OF FEA MODEL FOR EXISTING DESIGN

The first seven numerically predicted natural frequencies from the FEA for the existing design are indicted in Table 1. Four measured natural frequency magnitudes of the fan assembly are also indicted in Table 1. These four measured natural frequency magnitudes at the fan assembly corresponded well to the magnitudes obtained from FEA. The positions of some of the anti-node points for the numerically predicted mode shapes were also similar to the corresponding points, with relatively more vibration as practically identified for the existing design. This FEA model is thus regarded as accurate and reliable. It was found that the fan's existing support structure design was too soft, and that the bending stiffness of some of the elastic structural components used, is too small. Figure 1 shows a three dimensional view of the FEA model used for the existing fan assembly design. Figures 2 to 8 show the FEA results regarding the first 7 natural frequencies at corresponding mode shapes for the existing design. Figures 2, 5 and 7 clearly show some of the rotational bending modes of the electrical motor at 12.9 Hz, 35.7 Hz and 54.8 Hz respectively. When any one of these motor bending modes is exited, it will cause large horizontal vibration responses of the C-channel section at the position where the permanent vibration monitoring sensor is installed. The measured frequency domain signals revealed the presence of near resonance conditions regarding the third order maximum speed of the electrical motor at 49.5 Hz, and also the second order blade pass frequency at 52.92 Hz, thus both close to the measured natural frequencies at 50.75 Hz and 56.75 Hz (see Table 1). Relatively large or dominant vibration amplitudes were observed in measured vibration signals at these frequencies as taken during maximum speed at the similar adjacent fan. The FEA bending mode of the electrical motor at 54.8 Hz is close to resonance with the third order maximum speed of the electrical motor at 49.5 Hz, and also the second order blade pass frequency at 52.92 Hz. Figures 3 and 4 show clearly some of the other FEA rotational bending modes of the electrical motor at 21.5 Hz and 24.8 Hz respectively, but these electrical motor vibration bending mode vibration movements are 90 degrees with respect with the measured direction at the C-channel section where the permanent vibration monitoring sensor is installed.

Figure 6 shows another FEA combined rotational, horizontal and vertical mode of the electrical motor at 51.2 Hz. Although this electrical motor vibration rotational mode movement is mainly angular about the motor's axis of rotation, both vertical and horizontal movements are also observed. This combined rotational, horizontal and vertical mode is thus even closer to the third order maximum speed of the electrical motor at 49.5 Hz, and also the second order blade pass frequency at 52.92 Hz, thus near resonance. Figure 8 shows a FEA vertical mode of the electrical motor at 64.4 Hz, with movements mainly up and down, and thus with undesired local buckling of the C-channel section beam where the permanent vibration sensor is installed. The vibration measurements also revealed the presence of another near resonance condition regarding the fourth order maximum speed of the electrical motor at 66 Hz, but a relatively small vibration amplitude at this frequency was measured during maximum speed at the adjacent fan (see Table 1).

The presence of few other undesired vibration modes at corresponding lower natural frequencies were also exposed to resonance during start-up and also during sudden speed changes for the existing support structure design. These resonance conditions occasionally also caused large vibration levels, and under these conditions the vibration monitoring system tripped the fan to avoid damage, with production losses then as a result. Rotational and bending modes at the electrical motor were typically also excited during sudden changes in the speed of the fan as controlled by a Variable Speed Drive, which then lead to sudden and large electrical motor torque fluctuations transmitted to the flexible elastic steel support structure near the fan's electrical motor [7, 8]. These typical rotational excitations of the motor during sudden changes in speed were also observed in the plant.

Table 1. Natural frequencies [Hz]

Mode	Existing design - measured at fan assembly [Hz]	Existing design - predicted by FEA [Hz]	Modified design - predicted by FEA [Hz]
1	-	12.9	33.1
2	-	21.5	35.2
3	-	24.8	39.7
4	33.5	35.7	61.5
5	50.75	51.2	72.5
6	56.75	54.8	72.8
7	64.25	64.4	78.2

The FEA model for the existing fan assembly design was also modified for certain structural stiffening components then added (firmly bolted in practice), but that would still allow sufficient air flow through the air cooler fan on a trial and error basis. Several FEA computer run iterations were carried out for different possible designs, in order to finally obtain a feasible modified design. Figure 9 shows a three dimensional view of the FEA model used for this modified fan assembly design, with the structural stiffening components added to the existing design in blue as indicated. Figures 10 to 16 show the FEA results regarding the first 7 natural frequencies at corresponding mode shapes for the modified design. These natural frequency magnitudes for the modified design are also indicted in Table 1. All these natural frequencies are larger, compared to those of the existing design (magnitudes all shifted for the stiffer design). The feasible modified design was regarded as stiff enough, in order to avoid the undesired resonance conditions. Figure 10 shows the first bending mode of the electrical motor at 33.1 Hz which corresponds to the second order maximum speed of the electrical motor at 33 Hz. For this resonance, the main vertical vibration movement at this mode is 90 degrees with respect to the horizontal measured direction of the permanent vibration monitoring sensor. The fan rotor is driven by the motor with pulleys and belts. For the modified system, a vertical amplitude at second order maximum speed of the electrical motor could possibly be excited by small mechanical looseness such as bearing clearances in the vertical direction. This is however regarded as not so severe compared to the horizontal response that could be excited by small looseness for the existing design (see Figures 5 and 10). The motor's first bending mode at 33.1 Hz of the modified structure design is regarded as better, compared to the corresponding relatively softer mode at 12.5 Hz of the existing structure design, for excitation caused by motor changing torque fluctuations transmitted during start-up and sudden speed changes (see Figures 2 and 10).

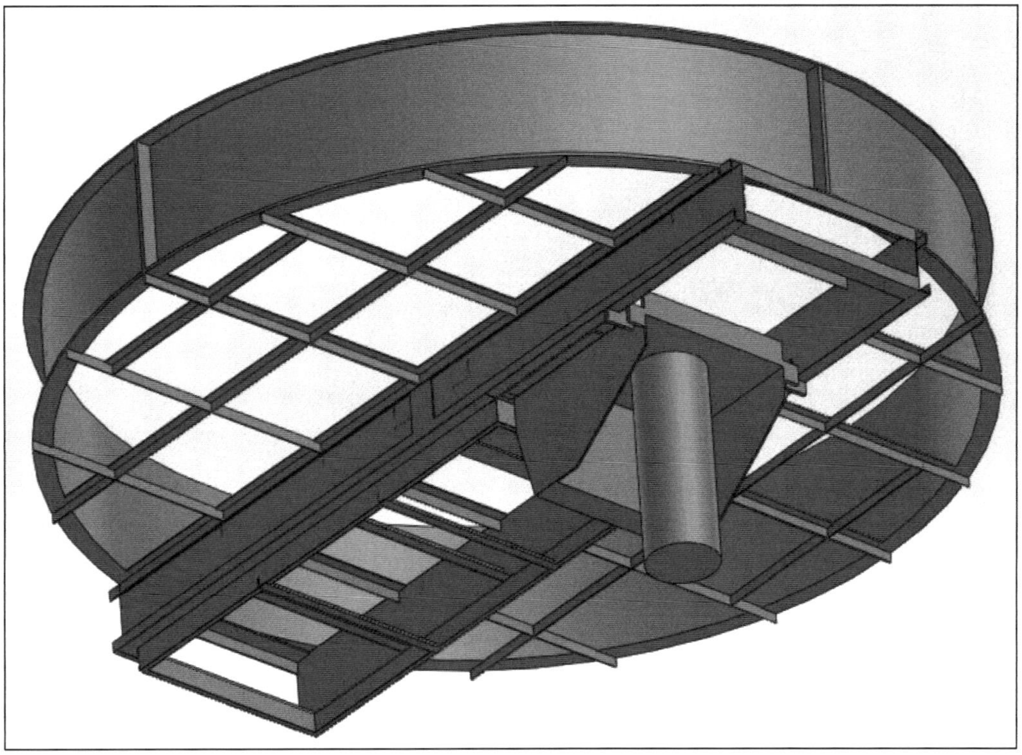

Figure 1. Three dimensional view of the FEA model for existing design.

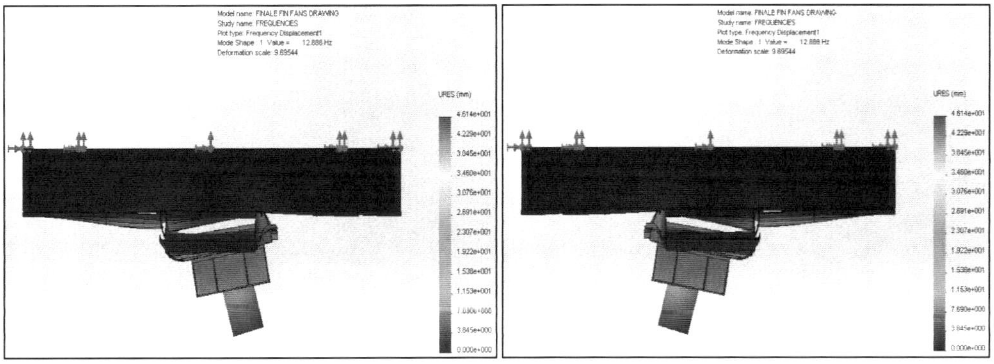

Figure 2. Mode 1 at 12.9 Hz for existing structure design.

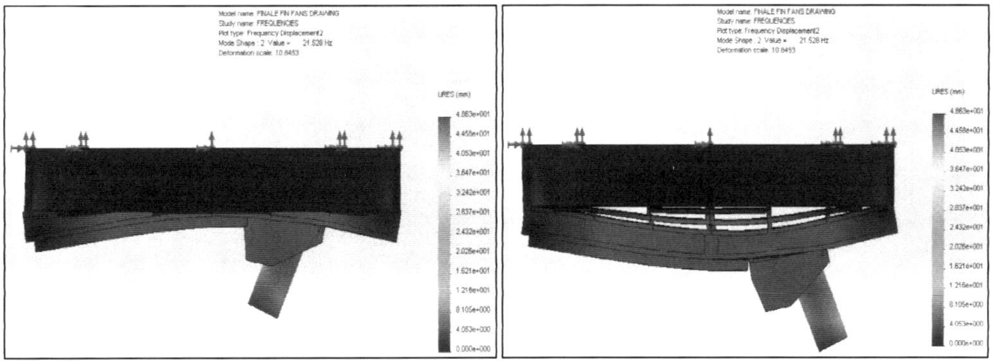

Figure 3. Mode 2 at 21.5 Hz for existing structure design.

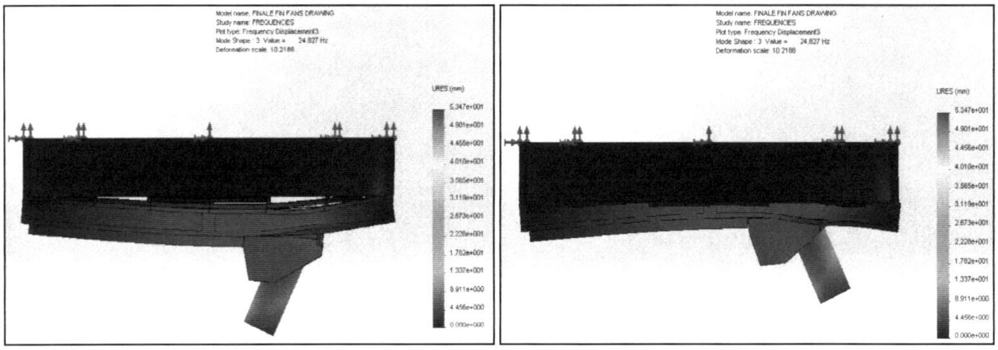

Figure 4. Mode 3 at 24.8 Hz for existing structure design.

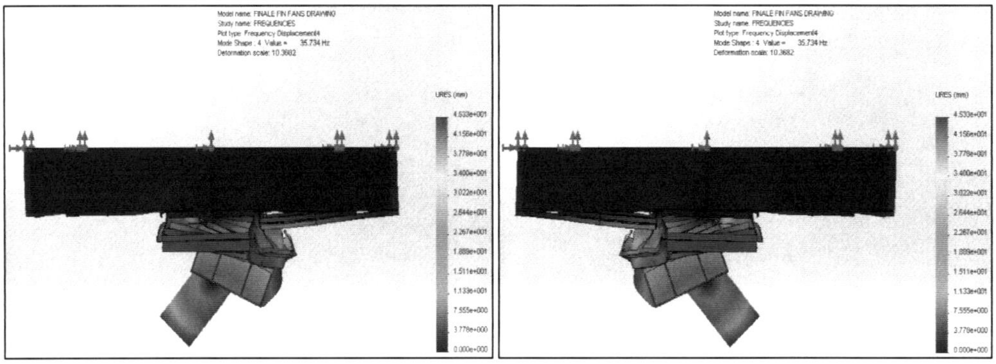

Figure 5. Mode 4 at 35.7 Hz for existing structure design.

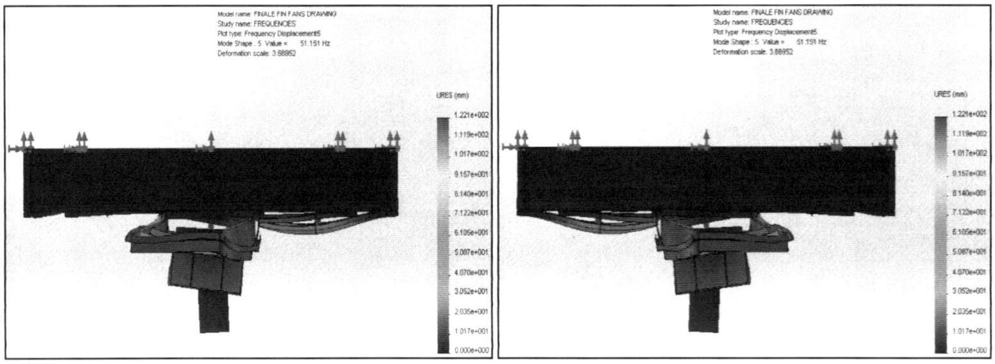

Figure 6. Mode 5 at 51.2 Hz for existing structure design.

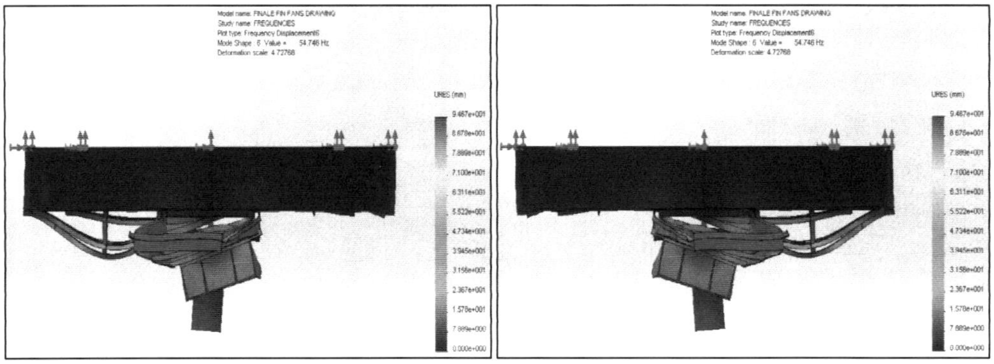

Figure 7. Mode 6 at 54.8 Hz for existing structure design.

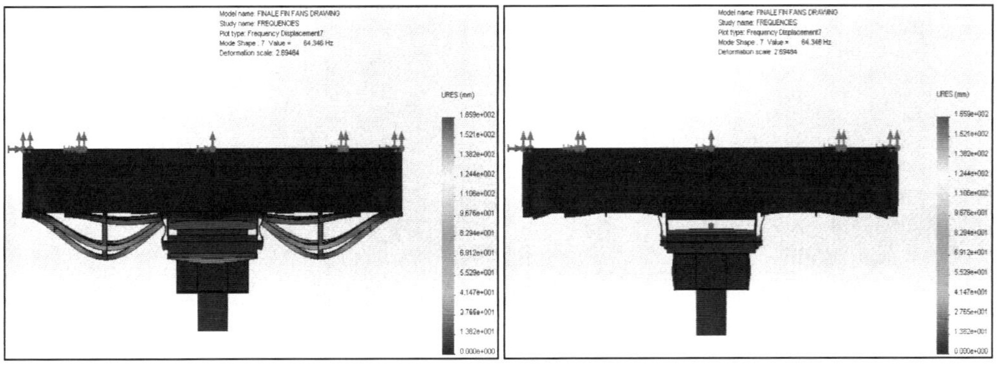

Figure 8. Mode 7 at 64.4 Hz for existing structure design.

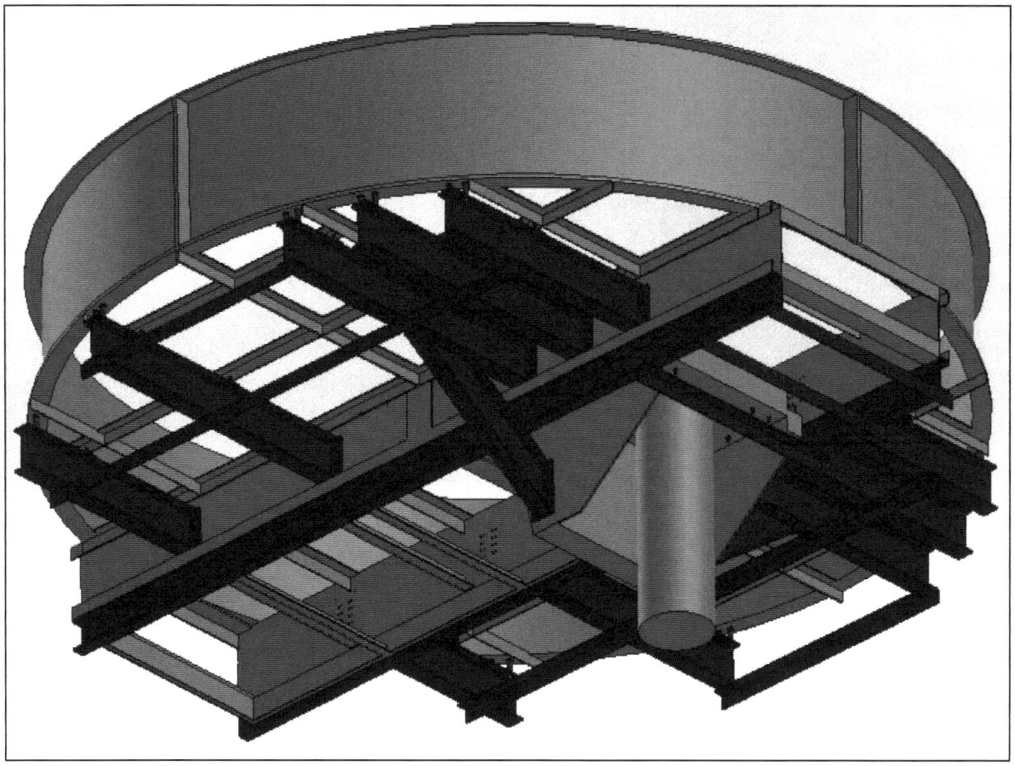

Figure 9. Three dimensional view of the FEA model for modified design.

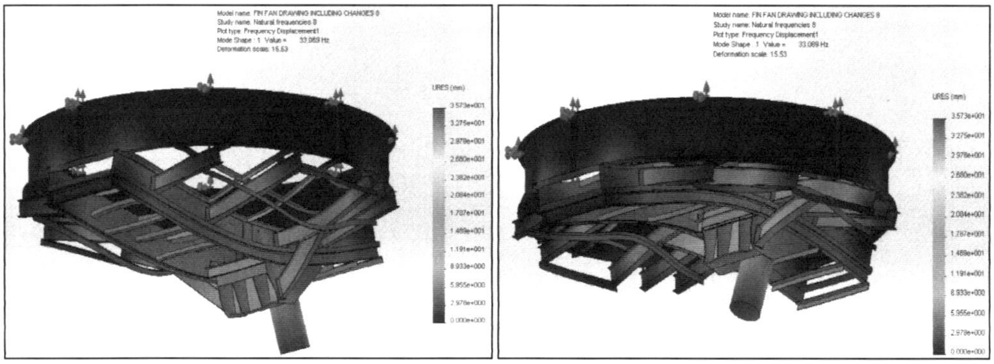

Figure 10. Mode 1 at 33.1 Hz for modified structure design.

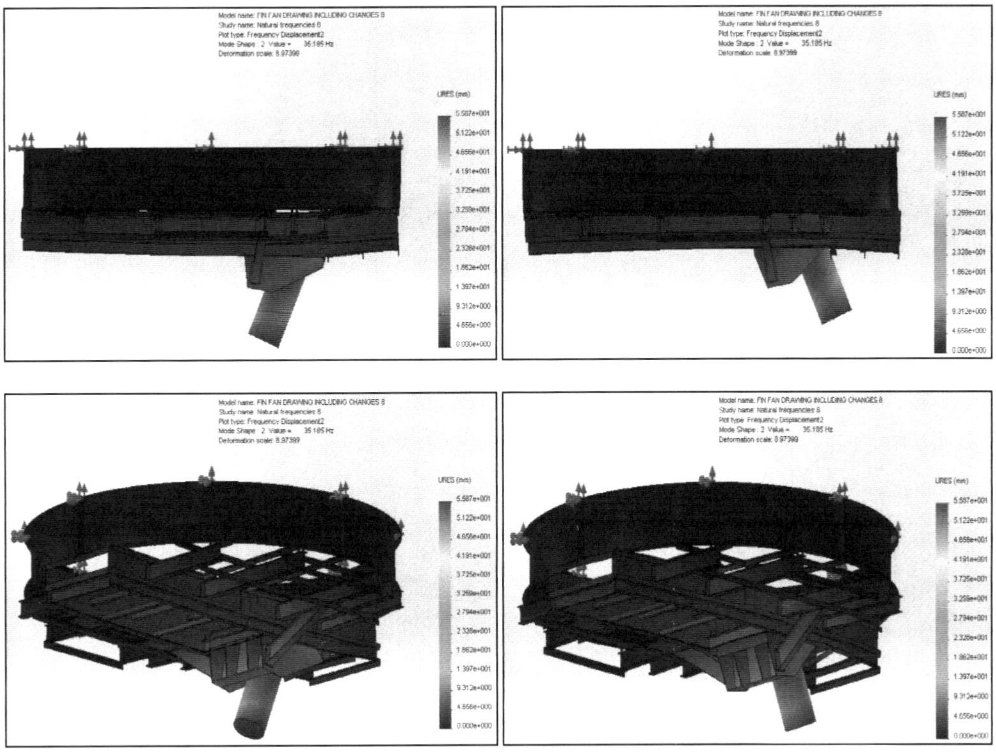

Figure 11. Mode 2 at 35.2 Hz for modified structure design.

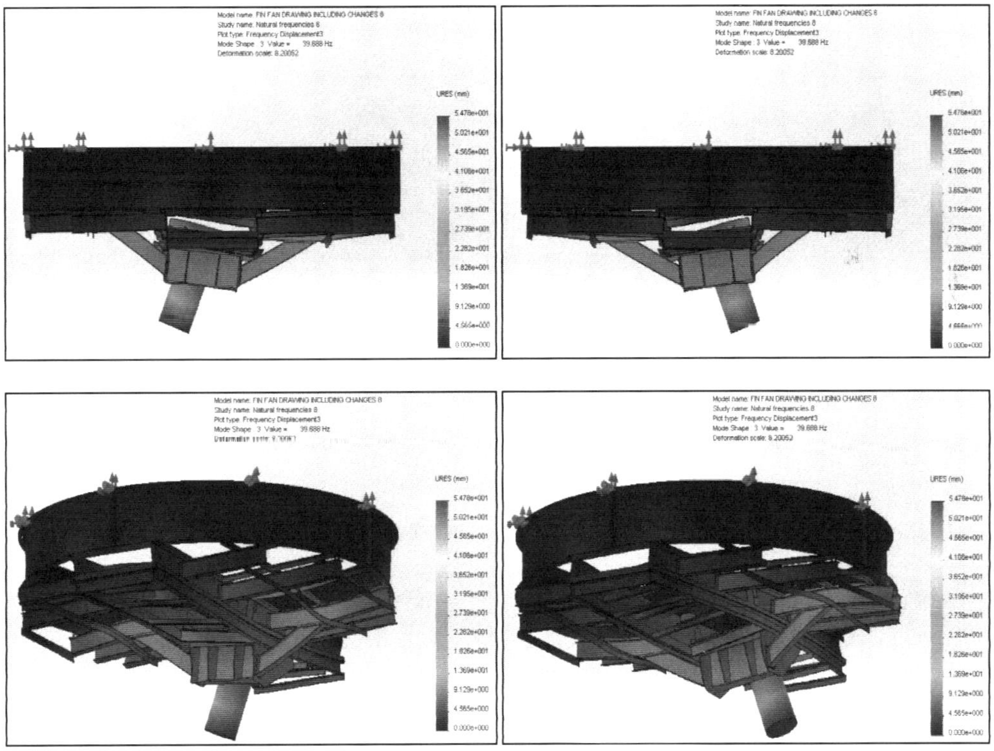

Figure 12. Mode 3 at 39.7 Hz for modified structure design.

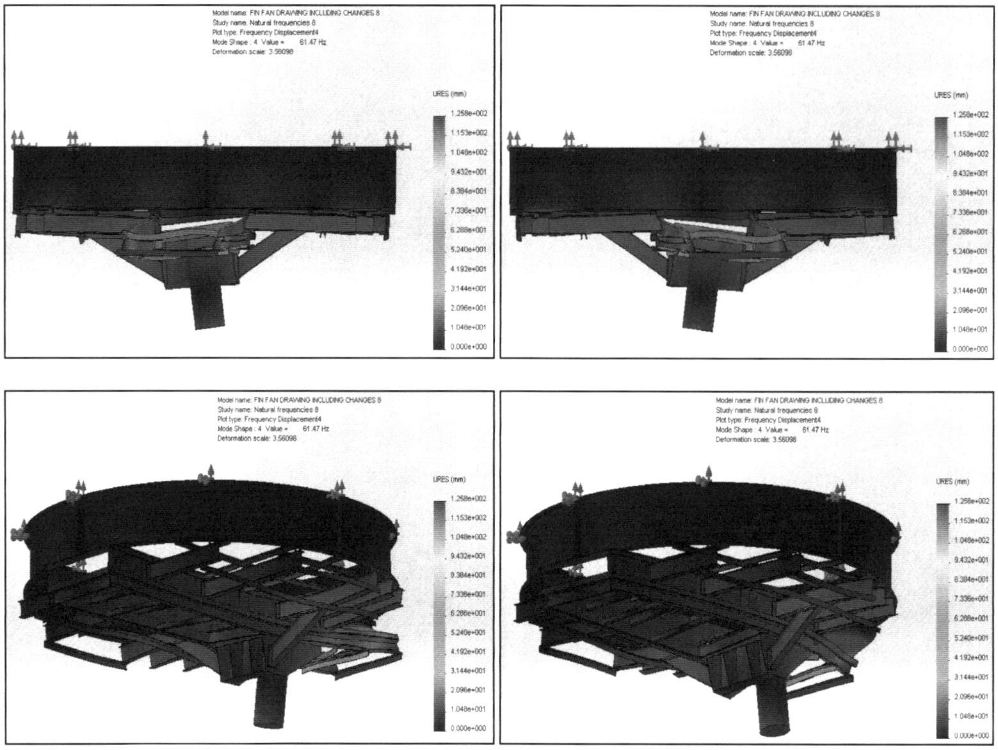

Figure 13. Mode 4 at 61.5 Hz for modified structure design.

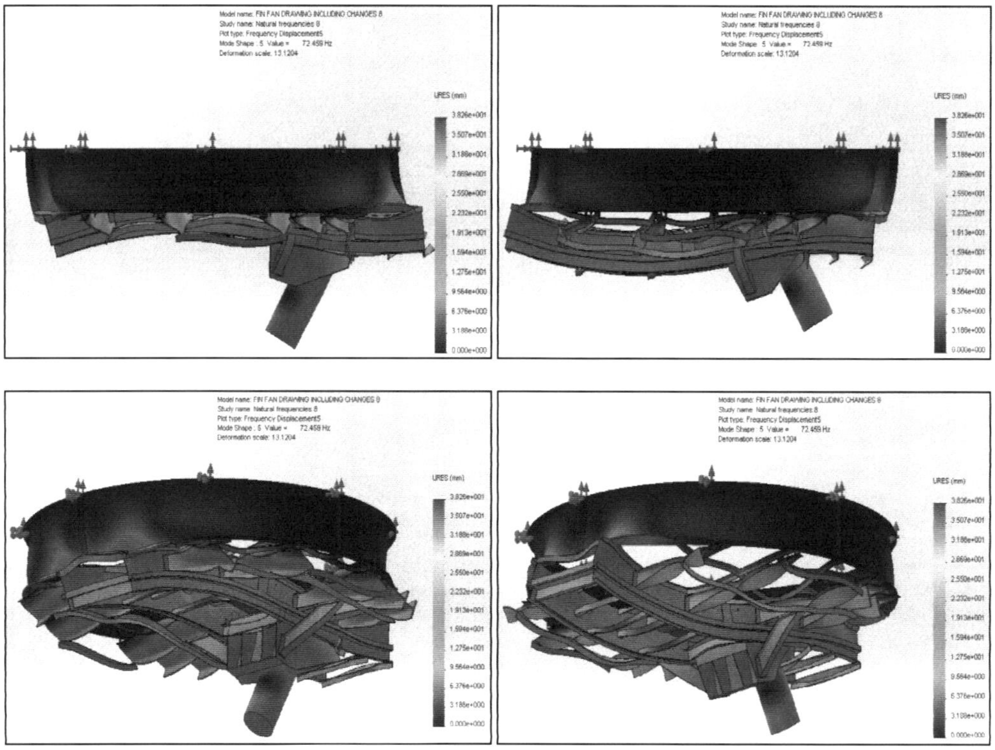

Figure 14. Mode 5 at 72.5 Hz for modified structure design.

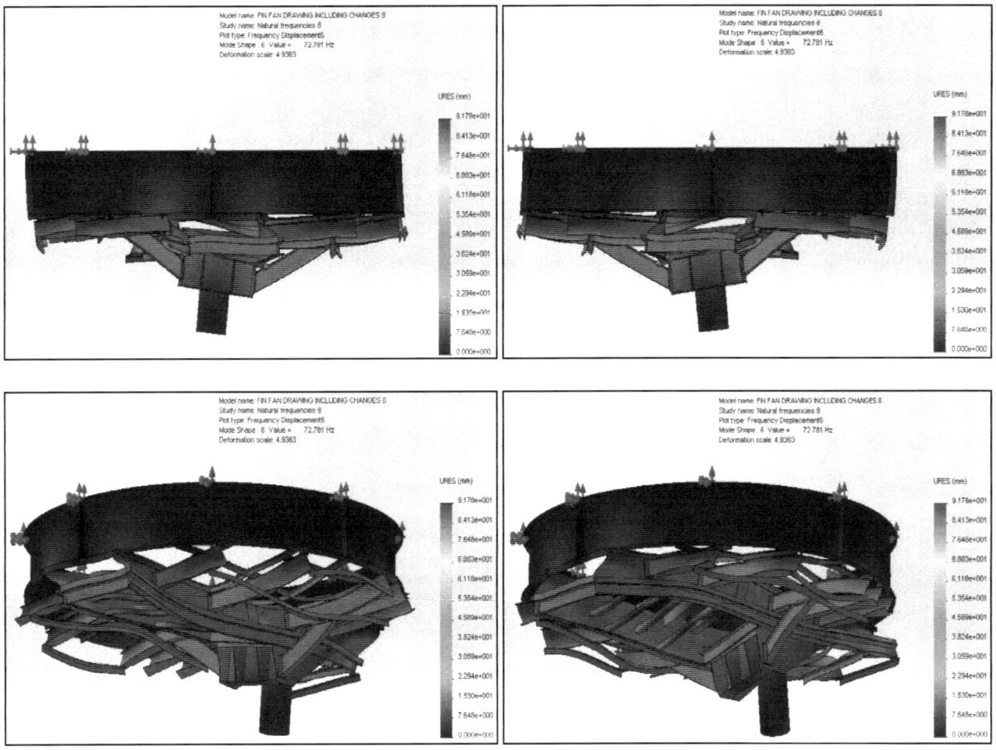

Figure 15. Mode 6 at 72.8 Hz for modified structure design.

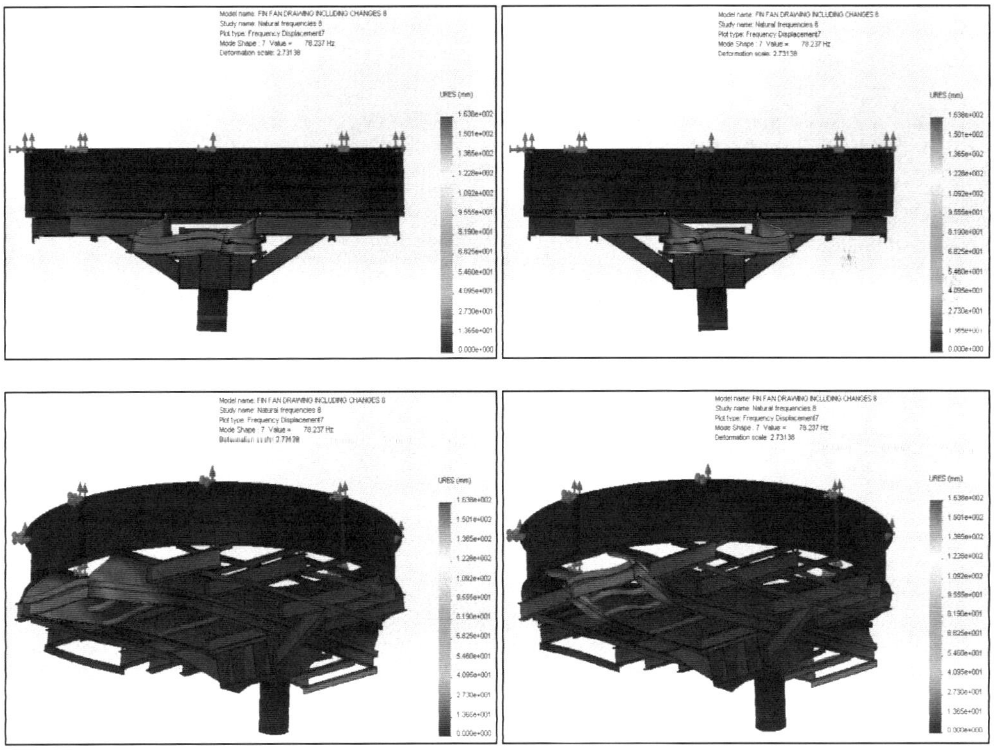

Figure 16. Mode 7 at 78.2 Hz for modified structure design.

CONCLUSIONS

Measured vibration signals obtained from the simple bump test approach was a quick and cost effective method to confirm the suspected resonance problems at the fan support structure, without a comprehensive experimental modal analysis. Finite Element Analysis (FEA) models were constructed and used for simulation to predict the natural frequencies at corresponding mode shapes for the existing fan assembly design. Some positions at the structure of relatively more vibration at maximum speed were practically identified and used during experimental bump tests, and these points corresponded to similar positions of anti-node points as obtained from the FEA. This was thus regarded as accurate and reliable. The FEA model for the existing fan assembly design was also modified for certain structural stiffening components then added in order to obtain a feasible modified design. The feasible modified design was regarded as stiff enough, in order to avoid most of the undesired resonance conditions. The feasible modified design was later practically implemented in the plant, and was then observed as successful to significantly reduce the undesired vibration levels during start-up, sudden speed changes and for maximum speed.

REFERENCES

1 Du, X., Modal analysis and design on the base plate of large-scale linear vibration screen. *Applied Mechanics and Materials,* 155-156, 514-518, 2012.

2 Yue-min, Z., Chu-sheng, L., Xiao-mei, H., Cheng-yong, Z., Yi-bin, W. & Zi-ting, R., Dynamic design theory and application of large vibrating screen. (*In.* Procedia Earth and Planetary Science organised by. p. 776-784), 2009.

3 Zhou, N., Dynamic characteristics analysis and optimization for lateral plates of the vibration screen. *Journal of Vibroengineering*, 17(4):1593-1604, 2015.

4 Rao, S. S., Mechanical Vibrations, 4th ed., Prentice Hall, Pearson, 2004

5 Ming Yu, M.C., Xuewu, H., Zhipeng, G. & Sizhuo, Z., Study on Vibration Characteristics for Vehicle Inner Doors. *Key Engineering Materials*, 693:227-232, 2016.

6 Dassault Systèmes Solidworks Corp., Solidworks Premium ®, 2012 x64 Edition.

7 Heyns, P. S., Nel, C.B., Experimental evaluation of an optimization program for a front wheel drive engine mount system, Proceedings. ISMA21. Noise and vibration harshness. Katholieke Universiteit Leuven, Belgium, September 1996.

8 Nel, C.B., Optimisation of engine mount systems at front-wheel-drive vehicles for multiple operational conditions, Proceedings. ISMA25, Katholieke Universiteit Leuven, Belgium, September 2000.

Fatigue Analysis of Class 5M Railway Bogie

*Mr Brendon Mark Nickerson[1], Prof Gerhard Venter[1]

1. Department of Mechanical and Mechatronic Engineering, Stellenbosch University

16603087@sun.ac.za

ABSTRACT

This study focuses on the creation of a numerical simulation process that can be used to determine the fatigue life of the Commonwealth railway bogie used on the class 5M train in South Africa. The research is supported and funded by the Passenger Rail Agency of South Africa (PRASA) Engineering Research Group at Stellenbosch University. A finite element (FE) model of the bogie has been created using an accurate three dimensional computer aided drawing (CAD) model, supplied by the PRASA Engineering Research Group. Dynamic load cases will be determined from standards such as EN 13749, and from simulation of simple low degree of freedom models of the bogie. The load cases will then be applied to the FE model to obtain stress histories. A script has been coded in Python to determine fatigue life from stress histories, and been successfully applied to a case study. The entire process consists of three black boxes: determining dynamic load cases, determining stress histories from these cases, and determining fatigue from the stress histories. The process will be designed in such a way that each black box can be altered and refined, allowing future research opportunities and refinements such as expanding the process to be applied to other bogies or allowing actual measured track data to be used as input for determining dynamic loads.

KEYWORDS

Railway bogie, fatigue, finite element model

NOMENCLATURE

C – Damping coefficient

E – Modulus of elasticity

K – Spring constant

G – Shear modulus

g – Gravitational acceleration constant

J – Polar moment of inertia

I – Area moment of inertia

L – Length

M – Mass

N_s – Number of sleepers

N_m – Number of mode shapes

q – Mode time coefficient

S_e – Endurance limit

S_{UT} – Ultimate tensile stress

S_y – Yield stress

Y – Mode shape function

w – Displacement in vertical direction

σ_a – Cyclic stress amplitude

σ_{ac} – Mean stress corrected cyclic stress amplitude

σ_m – Cyclic mean stress

ψ – Angular displacement

INTRODUCTION

The class 5M train has been in operation in South Africa since 1958 (Metrorail [1]). Most of these trains are still in use today, although they have been refurbished and updated (Metrorail [1]). Some of the challenges currently faced by the rail industry in South Africa are the poor levels of reliability and predictability and the high costs of maintenance due to decades of underinvestment (PRASA [2]).

One of the most critical structural components of a train is the bogie. A bogie is a chassis structure used for carrying the wheelsets, attached to the train car body. It consists of many components and while most of these are often replaced during its life cycle, the cast frames remain intact. Due to challenging operating conditions, the design life of a bogie is often placed under question.

The importance of bogie fatigue in railway rolling stock has been recognised as a contributing factor in structural failure (Oyan [3]). Luo et al. [4] state that the structural component most affected by dynamic loads is the bogie frame. Dietz et al. [5] state that for dynamically loaded structures, fatigue strength is an important design criterion. This is because the dynamic loads resulting in alternating stress cycles which can lead to fatigue failure. Due to its importance, this study will focus on the prediction of the fatigue life of the class 5M bogie and its current progress is presented in this paper.

The fatigue experienced by the bogie will be as a result of varying stresses over time. Theses stresses are caused by the various dynamic loads experienced by the bogie during operation. The dynamic loads are mostly as a result of track irregularities. These irregularities exist in both the vertical and lateral planes affecting track geometry parameters such as unevenness, cross levels, alignment, gauge and twist (Mundrey [6]). Figure 1 shows an example of vertical sinusoidal track irregularity.

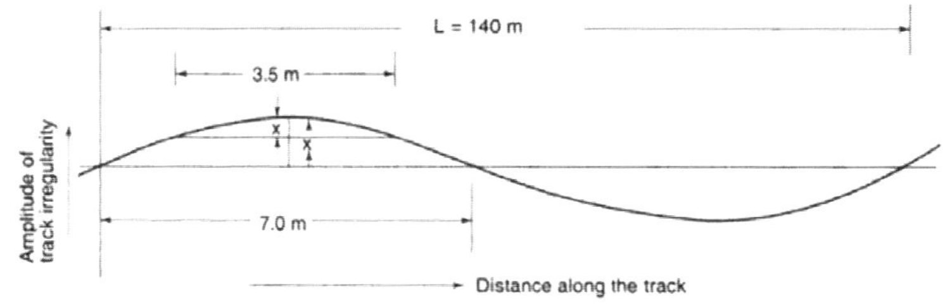

Figure 1. Example of a vertical track irregularity (Mundrey [6])

OBJECTIVES AND MOTIVATION

The aim of this study is to design and implement a numerical process that can be used to predict the fatigue life of a bogie. The objectives outlined to achieve this aim are:

- To provide a method to determine dynamic loads representing the operating conditions of the bogie;

- To determine critical points for fatigue failure on the bogie and extract stress histories, as a result of the dynamic loads, at these points;
- To provide a prediction of the fatigue life of the bogie based on the obtained stress histories.

This research is supported by the PRASA Engineering Research Chair at the Department of Industrial Engineering, Stellenbosch University. The organisation is interested in the results as it could be used in future to predict remaining life of bogies currently in service, and at which locations they are likely to fail. This in turn can lead to better maintenance planning and evaluation of current market assets.

NUMERICAL SIMULATION PROCESS

The numerical simulation process under construction for this research is shown in Figure 2. The process consists of three boxes, each of which is aimed at achieving one the objectives of the study. The boxes are designed to be independent of one another and, although intended to work together for this study, each process can take its inputs from or give its outputs to another source. For example, the fatigue life prediction box could also take as input a stress history converted from measured strain data from experimental testing of the bogie.

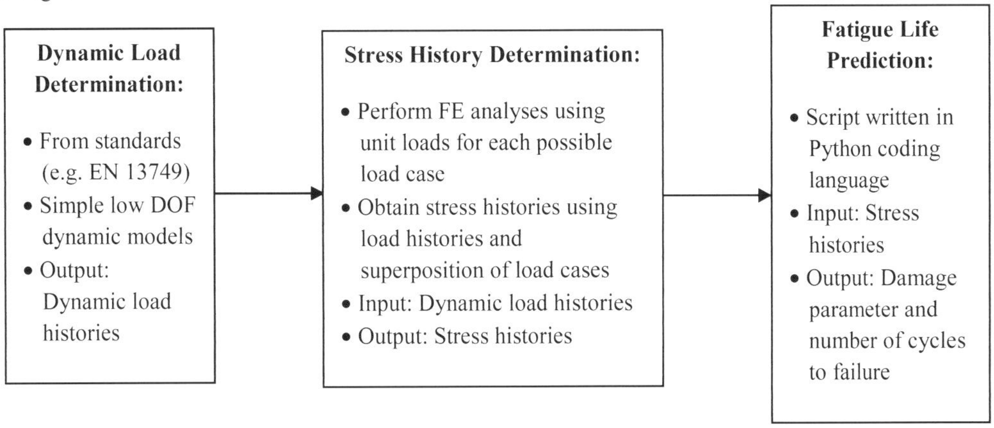

Figure 2. Numerical Simulation Process

FINITE ELEMENT MODELLING

A full three dimensional model of the 5M bogie assembly has been provided by the PRASA Engineering Research Chair, which is shown in Figure 3.

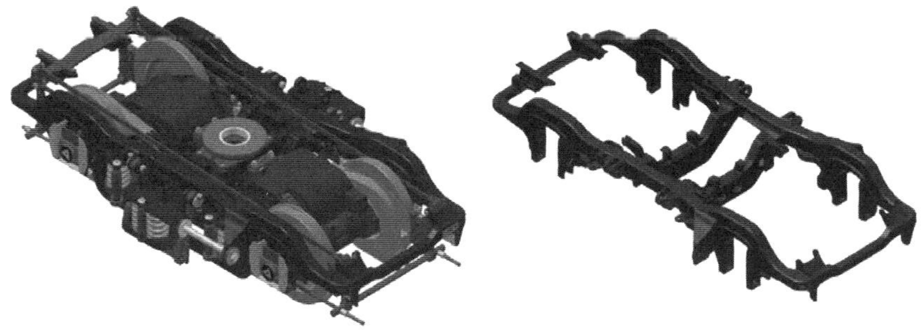

(a) *(b)*

Figure 3. *(a)* Three dimensional model of 5M bogie assembly and *(b)* bogie frame which forms the focus of this research

The model of the bogie frame has been used to create an FE model, using the bogie frame midsurfaces which have been meshed with shell elements. The latest software from MSC Software Corporation, MSC Apex, was used to create the model due to the ease with which geometry can be manipulated and used to extract the midsurfaces. The choice of using shell elements over solid elements was due to the large number of solid elements that was necessary for an acceptable mesh of the complex geometry, leading to longer solution times. With shell elements, the geometry could be greatly simplified and represented using comparatively few elements. Figure 4 shows a half symmetric model of the bogie frame that has been meshed with Q4 elements 30 mm in size.

Figure 4. Symmetric finite element model of 5M bogie using size 30 mm Q4 elements.

FATIGUE PYTHON SCRIPT

A script has been written, using the Python coding language, to predict fatigue life from a given stress history. The stress-based approach was selected for the analysis due to its simplicity. The stress-based approach is also appropriate because of the expectation of high cycle fatigue. Strain-based fatigue is typically preferred for low cycle fatigue. At long lives, where elastic strain dominates, the strain-based approach is equivalent to the stress-based approach (Dowling, [7]). However, it should be noted that the script can be easily expanded to include strain-based and even fracture mechanics approaches.

The stress-based approach to fatigue makes use of a stress-life (S-N) curve, which predicts the number of cycles to failure at a specified stress level. For the script an estimated S-N curve is created, based on the ultimate tensile stress of the selected material, using processes detailed in Budynas and Nisbett [8].

As the stress history that is imported will most likely consist of cycles at various amplitudes, it is necessary to implement a cycle counting method to determine how many cycles are experienced at each stress level. A number of different cycle counting methods exist, and the one chosen for this script was the rainflow cycle counting method. The reasons for this are:

- Fuchs and Stephens [9] state that all good counting methods count every part of every overall range once and only once, and also count smaller ranges down to some predetermined threshold once and only once. They go on to say that the three well documented methods that achieve this are: range-pair, rainflow, and racetrack counting.
- Dowling [10] shows that methods other than range pair and rainflow can result in inconsistencies and large differences between predicted and actual fatigue lives.

- Both Dowling [7] and Suresh [11] state that the rainflow method is the most widely used.

The rainflow cycle counting method was coded into the script using the algorithm provided by ASTM Standard E1049-85 [12].

The stresses that are imported are equivalent stresses (such as Von Mises stress) calculated from the multiaxial principal stresses, as recommended by Dowling [7] and Suresh [11]. These equivalent stresses can then be compared directly with an S-N curve. The effect of mean stresses has also been included as tensile mean stress will reduce fatigue life (Frost et al. [13]). This means the stress amplitude is 'corrected' based on the value of the mean stress. The mean stress correction can be applied in this script using either the Goodman relation, given in Equation 1, which is recommended for brittle metals or the Gerber relation, given in Equation 2, which is recommended for ductile metals (Suresh [11])

$$\sigma_{ac} = \frac{\sigma_a}{1 - \frac{\sigma_m}{S_{UT}}} \tag{1}$$

$$\sigma_{ac} = \frac{\sigma_a}{1 - \left(\frac{\sigma_m}{S_{UT}}\right)^2} \tag{2}$$

The various stress cycles are then combined into a cumulative damage calculation. The cumulative damage is calculated using Palmgrin-Miner's linear damage summation rule, Equation 3, which sums the ratios of cycles experienced at a certain corrected stress amplitude to cycles needed to fail at that stress level. When the ratio D is equal to 1 then failure is predicted to occur.

$$\sum_{i=1}^{m} \frac{n_i}{N_i} = D \tag{3}$$

The Palmgrin-Miner Rule was used as the initial failure rule due to its simplicity, however the code can be expanded to include non-linear damage summation rules.

The script is built up out of a number of smaller functions each with a specific task, namely:

1. Define material properties (S_{UT}, S_y, S_e)
2. Import stress history data from a .csv file
3. Extract the maxima and minima from the stress history for use in cycle counting.
4. Perform cycle counting using the rainflow method.
5. Perform cumulative damage analysis on counted cycles to determine damage parameter and cycles to failure.

The script has been tested using a study presented by Chen and Wang [14] in which a cantilever beam made of roller quenched and tempered RQC-100 steel is subjected to a suddenly applied tip load. The beam is allowed to oscillate to equilibrium and a fatigue analysis is performed. The study was recreated by creating an FE model of the beam using MSC Patran/Nastran. The stress history at the fixed end of beam was then exported to the fatigue script. The fatigue life was calculated using a 99 % reliability on the S-N curve. A comparison of the results for a 100 N tip load at a 0.025 damping ratio are given in Table 1.

Table 1. Comparison of Chen and Wang [14] and fatigue script

	Chen and Wang [14]	Fatigue script	Analytical
Finite Element Model			
First natural frequency	40.74 Hz	40.74 Hz	40.75 Hz
Second natural frequency	255.29 Hz	255.17 Hz	255.35 Hz
Third natural frequency	714.68 Hz	713.82 Hz	714.98 Hz
Max. static stress	291.80 MPa	292.00 MPa	291.82 MPa
Max. displacement	19.32 mm	19.32 mm	19.32 mm
Fatigue			
Damage parameter	5.05e-6	5.66e-6	-
Cycles to failure	198 170	176 575	-

The cycles to failure predicted by the fatigue script is 11 % less than those of the study conducted by Chen and Wang [14]. This is acceptable as predicting fatigue to the correct amount of cycles is near impossible and it is generally more important to predict the cycles to the correct order. It is planned to apply the fatigue script to more cases studies to determine its accuracy.

DYNAMIC MODELLING

A numerical dynamic model has been created for the purpose of determining the dynamic load cases on the bogie. The model is based on models found in literature such as those by Zhai et al. [15] and Sun and Dhanesekar [16]. The model is presented in Figure 5.

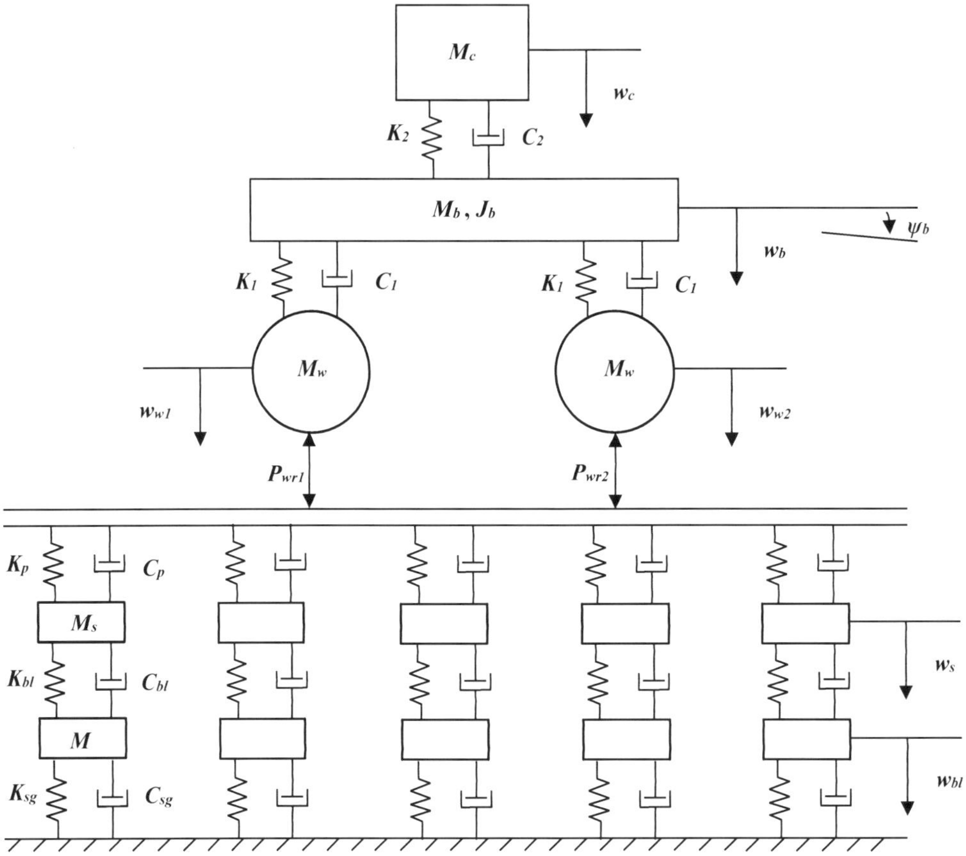

Figure 5. Numerical dynamic model of wagon and track systems

For the model, it is assumed that the responses are symmetric in the longitudinal plane and thus the vehicle system is modelled as a quarter car. The vehicle system consists of a quarter of the mass of the wagon body (M_c) supported on the bogie, while half of the mass of the bogie (M_b) is supported on two wheelsets. The wagon body is connected to the bogie through the secondary suspension and damping (K_2 and C_2 respectively) and the bogie to the wheelsets through the primary suspension (K_1 and C_1 respectively).

The track model is a three layer model consisting of a beam representing the rail which is discretely supported. The length of the beam modelled is such that the responses at the ends are negligible, thus simulating an infinitely long beam. At each discrete support, the mass of the sleeper and ballast is modelled. The sleeper is connected to the rail through the rail-pad stiffness and damping (K_p and C_p respectively), and to the ballast through the ballast stiffness and damping (K_{bl} and C_{bl} respectively). The ballast is connected to ground through the subgrade stiffness and damping (K_{sg} and C_{sg} respectively).

The wheel-rail contact is modelled using the non-linear Hertz contact theory, as shown in Equation 4, where C_H is the Hertz contact coefficient, w_w is the displacement of the wheel at the contact point, w_r is the displacement of the rail at the contact point, and w_i is due to the irregularity.

$$P_{wr} = \begin{cases} C_H(w_w - w_r - w_i)^{3/2}, & \text{if } (w_w - w_r - w_i) > 0 \\ 0, & \text{if } (w_w - w_r - w_i) < 0 \end{cases} \tag{4}$$

The rail is modelled using the Euler-Bernoulli beam formulation. The equation for the beam is given as Equation 5, where F_{si} is the force between the ith sleeper and the rail and δ is the Dirac-delta function.

$$EI\frac{\partial^4 w_r(x,t)}{\partial x^4} + M_r\frac{\partial^2 w_r(x,t)}{\partial t^2} = -\sum_{i=1}^{N_s} F_{si}(t)\delta(x - x_i) + \sum_{j=1}^{2} P_{wrj}(t)\delta(x - x_j) \tag{5}$$

The deflection of the rail can be derived from modal superposition, as shown in Equation 6.

$$w_r(x,t) = \sum_{i=1}^{N_m} Y_i(x)q_i(t) \tag{6}$$

Equation 6 is then substituted into Equation 5 which yields a set of ordinary differential equations in q, which can be solved in conjunction with the wagon and track equations of motion.

The vehicle model contains five degrees of freedom, while the track model contains two degrees of freedom per discrete support plus one degree of freedom for each mode shape used to model the rail. Each degree of freedom has an associated equation of motion derived from its mass and the stiffness and damping properties applied to it. The equations of motion are combined into a matrix form, Equations 7 and 8 for the wagon and track respectively, and then solved at each time step using the Newmark-Beta integration scheme.

$$M_W\ddot{x}_W + C_W\dot{x}_W + K_W x_W = F_W \tag{7}$$

$$M_T\ddot{x}_T + C_T\dot{x}_T + K_T x_T = F_T \tag{8}$$

Where,

$x_W = \{w_c, w_b, \psi_b, w_{w1}, w_{w2}\}$ and $x_T = \{q_1, q_2, \dots, q_{Nm}, w_{s1}, w_{s2}, \dots, w_{sNs}, w_{bl1}, w_{bl2}, \dots, w_{blNs}\}$

The model is capable of capturing the contact forces in the vertical direction, while the wagon system moves at a constant velocity, as well as the dissipation of these forces through the various suspension elements. This can therefore be used to determine the forces in the primary and secondary suspensions that the bogie experiences due to wheel or track irregularities. The model is capable of simulating various vertical irregularities such as wheel flats, rail corrugations and raised or dipped rail joints. The model is not able to simulate lateral irregularities or capture lateral responses.

The model has been tested against data provided by Zhai et al. [15]. Zhai et al. [15] developed a full car model for use with their research and provide both the experimentally measured and simulated results for a freight train with a wheel flat on the leading wheelset moving at 27 km/h. The depth and length of the wheel flat are provided as 1 mm and 52.8 mm respectively. Most of the wagon and track parameters are provided by Zhai et al. [15] except for the primary suspension parameters and the Hertz contact coefficient. Figure 6 shows the results provided by Zhai et al. [15] and Figure 7 gives the results from the model developed above under the same conditions. It can be seen that the model developed here provides a slightly different response, and this is due to the unknown parameters that were not provided by Zhai et al. [15]. It was seen that the model can be tuned to produce results closer to that found in literature, and it is important to obtain the correct parameters in order to properly replicate others' results.

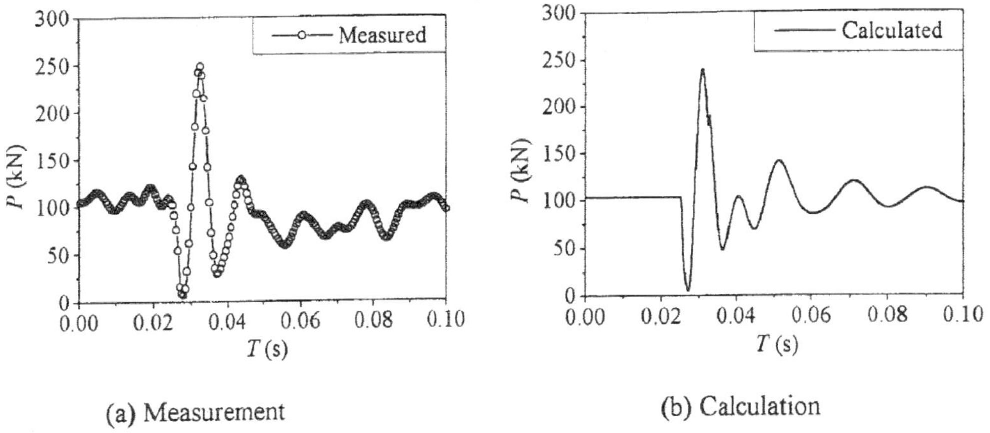

Figure 6. Results from Zhai et al. [15] showing the contact force at the leading wheelset with a wheel flat

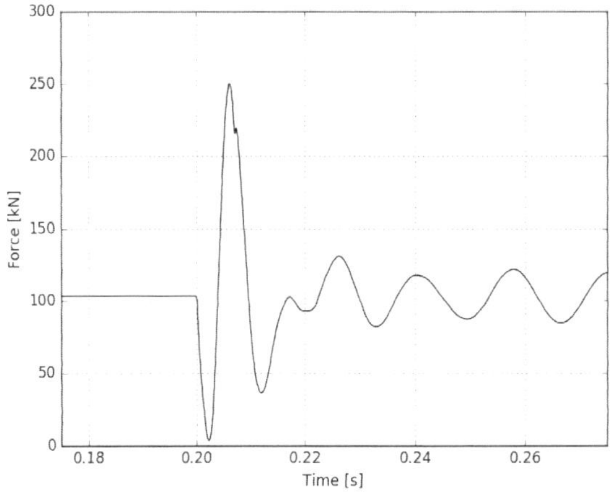

Figure 7. Contact force at leading wheelset with wheel flat using model developed for this research

PLANNED WORK

The work that remains is to integrate the three black boxes. Each black box has been constructed and must now be linked to perform the full simulation. The plan is as follows:

- Final validation of bogie FE model, fatigue script, and dynamic model.
- Apply single unit loads at each point on the bogie FE model to obtain the stress field for each force. Assuming that the structure remains in the linear range, these stress fields can then be multiplied and superimposed.
- Adjust the stress fields for each load based on the load histories determined by the dynamic model simulation. This will result in a stress history for each point in the stress field.
- These stress histories will be passed to the fatigue script, which will return the damage parameters and cycles to failure for each point.
- The damage parameters will then be read back into the FE software to display the areas that have a high risk of fatigue failure.

The integrated simulation will be performed through the use of a Python script.

CONCLUSIONS

An FE model of the class 5M railway bogie, a Python script used for fatigue prediction, and a dynamic model of the wagon and track have all been completed, although further testing of these components are still planned. The integration of the entire simulation process must also still be completed.

Once developed, this research will allow further research opportunities such as:

- Further expansions to the fatigue script,
- Applying the fatigue script to additional FE models,
- Expanding the wagon and track models (e.g. full car vehicle system, lateral irregularities, more track sublayers, etc.)
- Use of multibody dynamic simulations to determine dynamic load cases or stress histories,
- Using experimentally measured data to compare to the numerical results.

REFERENCES

1. Metrorail. (2007). *Rolling Stock.* Online: http://www.metrorail.co.za/RollingStock_2.html. Accessed: 16 June 2015
2. PRASA. (2014). *Prasa: Corporate plan 2014/15 – presentation to portfolio committee on transport.* Online: http://pmg-assets.s3-website-eu-west-1.amazonaws.com/140902prasa.pdf. Accessed: 16 June 2015
3. Oyan, C. (1998). Structural strength analysis of the bogie frame in Taipei rapid transit systems. *Proceedings of the Institution of Mechanical Engineers, Part F: Journal of Rail and Rapid Transit,* vol. 212, no. 3, pp. 253-262. ISSN 0954-4097
4. Luo, R.K., Gabbitas, B.L., Brickle, B.V. and Wu, W.X. (1998). Fatigue damage evaluation for a railway vehicle bogie using appropriate sampling frequencies. *Vehicle System Dynamics: International Journal of Vehicle Mechanics and Mobility,* vol. 29, pp. 404-415. ISSN 0042-3114
5. Dietz, S., Netter, H. and Sachau, D. (1998). Fatigue life prediction of railway bogie under dynamic loads through simulation. *Vehicle System Dynamics: International Journal of vehicle Mechanics and Mobility,* vol. 29, no. 6, pp. 385-402. ISSN 0042-3114
6. Mundrey, J.S. (2000). *Railway Track Engineering.* 3rd ed. New Delhi: Tata McGraw Hill.
7. Dowling, N.E. (2007). *Mechanical Behaviour of Materials: Engineering Methods for Deformation, Fracture and Fatigue.* 3rd ed. Upper Saddle River: Pearson Prentice Hall.

8. Budynas, R.G. and Nisbett, J.K. (2011). *Shigley's Mechanical Engineering Design.* 9th ed. New York: McGraw Hill.
9. Fuchs, H.O. and Stephens, R.I. (1980). *Metal Fatigue in Engineering.* New York: John Wiley & Sons, Inc.
10. Dowling, N.E. (1971). *Fatigue Failure Predictions for Complicated Stress-Strain Histories.* Urbana: Department of Theoretical and Applied Mechanics, University of Illinois.
11. Suresh, S. (1991). *Fatigue of materials.* Cambridge: Cambridge University Press.
12. ASTM Standard E1049-85. (2011). *Standard Practices for Cycle Counting in Fatigue Analysis.* West Conshohocken: ASTM International.
13. Frost, N.E., Marsh K.J. and Pook, L.P. (1974). *Metal Fatigue.* London: Oxford University Press.
14. Chen, C. and Wang, I. (1997). Fatigue Analysis of a Beam subjected to Dynamic Loading. *15th International Modal Analysis Conference.*
15. Zhai, W.M., Cai, C.B., Wang, Q.C., Lu, Z.W. and Wu, X.S. (2001). Dynamic effects of vehicles on track in the case of raising train speeds. *Proceedings of the Institution of Mechanical Engineers, Part F: Journal of Rail and Rapid Transit*, vol. 215, pp 125-135. ISSN 0954-4097
16. Sun, Y.Q. and Dhanesekar, M. (2002). A dynamic model for the vertical interaction of the rail track and wagon system. *International Journal of Solids and Structures*, vol 39, pp 1337-1359. ISSN 0020-7683
17. BS Standard EN 13749. (2011). *Railway applications – Wheelsets and bogies – Method of specifying the structural requirements of bogie frames.* Brussels: European Committee for Standardization.

Simulation of natural convective flow in an experimental reactor cavity cooling system facility

P. Niemand*[1], C.G. du Toit[1], N.I. Tak[2], M.H. Kim[2].
[1]North-West University, Private Bag X6001, Potchefstroom, South Africa
[2]Korea Atomic Energy Research Institute, Daedeokdaero 989-111, Yuseong, Daejong, Republic of Korea
Corresponding author: pfniemand@gmail.com

ABSTRACT

The very high temperature reactor (VHTR) has many safety features. One of these features is the reactor cavity cooling system (RCCS). This system is intended to remove decay heat from the reactor cavity during upset conditions. The Korea Atomic Energy Research Institute (KAERI) constructed a facility that represents a ¼ scale model of the RCCS of a VHTR. The preliminary testing on the facility has been completed and a simulation model has been set up for the facility, using the program GAMMA+. This study aims to simulate the facility using the 1D CFD program Flownex SE for comparison with the results obtained with GAMMA+. Since the fundamental equations used in both programs are the same, the results show a high degree of correlation. The Flownex simulation was set up as close as possible to the GAMMA+ model by using the same geometry and initial- and boundary conditions. The surface temperatures and the mass flow rate in the riser duct system were used as a measure to determine the level of agreement between the two programs. The results obtained by using the two programs show very good agreement.

KEYWORDS

Reactor cavity cooling system, experimental facility, thermal radiation, natural convection, mixed convection.

INTRODUCTION

The very high temperature reactor (VHTR) has many safety features. One of these features is the reactor cavity cooling system (RCCS). Korea has chosen the 200 MWth PMR (PMR200) General Atomic prismatic core VHTR reactor as the candidate reactor for the Nuclear Hydrogen Development and Demonstration (NHDD) project. The inlet temperature to the core is considered to be 490°C and the outlet temperature from the core to be 950°C [1]. Apart from removal of the residual and decay heat, it also requires the proper cooling of the reactor cavity under normal operating conditions. The RCCS is the only safety-related system for the residual and decay heat removal and its performance impacts directly on the maximum temperatures of the fuel, reactor pressure vessel (RPV) and concrete during off-normal conditions. The General Atomic air-cooled RCCS design concept, which is a completely passive system, has not yet been deployed in a VHTR. The Korea Atomic Energy Research Institute (KAERI) constructed a facility that represents a ¼ scale model of the RCCS of a VHTR. The natural cooling experimental facility (NACEF) has been simulated by using the code GAMMA+ [2]. This study aims to compare the results obtained by using GAMMA+ with the results obtained by using Flownex.

DESCRIPTION OF THE NACEF

Figure 1 is a representation of the NACEF. Air enters at the bottom through two inlet pipes and then enters the inlet plenum, from where it is directed into six riser tubes. The risers run vertically through the cavity, where a ceramic heater is installed. After exiting the riser tubes, the air enters the outlet plenum which directs it into the two chimneys.

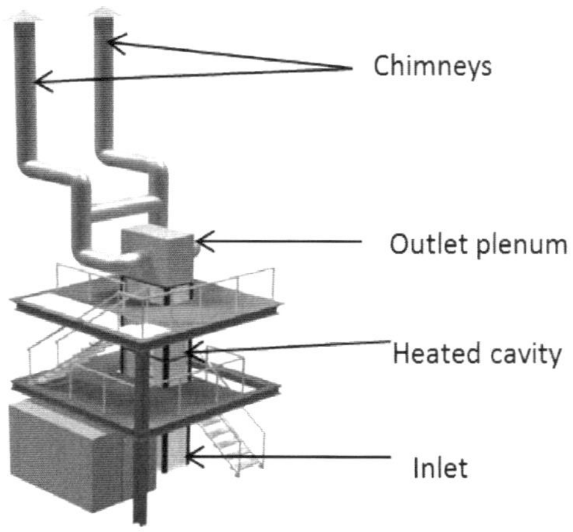

Figure 1. The natural cooling experimental facility (NACEF). [3]. Note the highlighted heated section.

Figure 2 shows a cross sectional view of the heated cavity. The ceramic heater provides a maximum power of 52 kW. A stainless steel plate is mounted at the front and rear of the cavity. The side walls are made of Calcium-Silicate. Various layers of insulation minimize heat loss to the environment.

The cavity contains six riser tubes as shown in Figure 2.

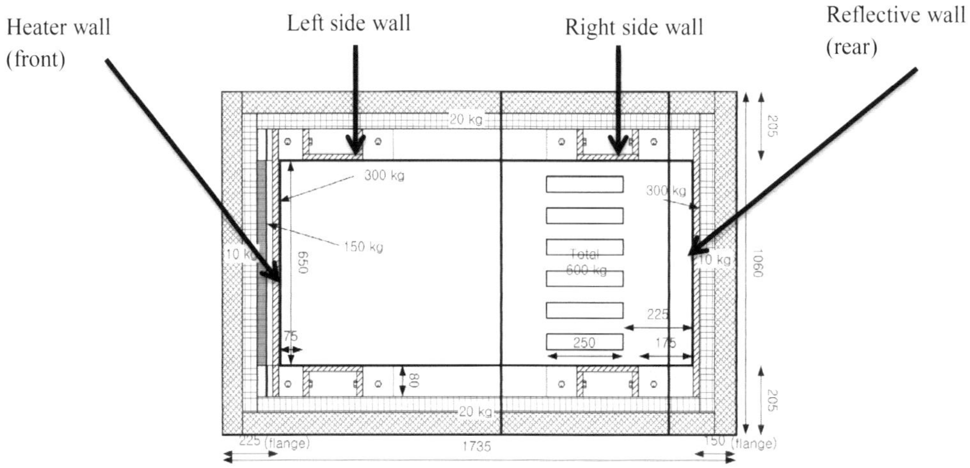

Figure 2. Cross sectional view of the heated section in the NACEF. Adapted from [2].

DESCRIPTION OF MODEL

The six riser tubes (Figure 2) were grouped together as follows: The two tubes closest to the side walls were modelled as a single tube (referred to as the outer tubes) and the four remainder tubes were modelled as a single tube (referred to as the inner tubes). The inner and outer tubes are ultimately simplified to a single representative tube.

The side walls were segmented so as to form two surfaces, as shown in Figure 2. The opposing walls are grouped together into a single surface. The result is two model surfaces, i.e. the left side wall and the right side wall. The cavity walls were assigned a thickness of 30 mm for the purposes of the simulation. The outside surfaces of the left and right side wall and the reflective wall are considered to be adiabatic. The solid parts of the model were specified as stainless steel.

The flow regime in the riser tubes was found to be mixed convection. The Nusselt number for upward flow was therefore calculated by interpolating between the free convection and forced convection Nusselt numbers [4]:

$$Nu_{mixed} = \left(Nu_{forced}{}^{3} - Nu_{free}{}^{3}\right)^{\frac{1}{3}} \quad \textbf{(1)}$$

Where Nu_{mixed}, Nu_{forced} and Nu_{free} are the mixed, forced and free convection Nusselt numbers respectively.

The convection coefficient was then calculated by [5]:

$$h_{mixed} = \frac{Nu_{mixed} * k}{D_h} \quad \textbf{(2)}$$

Where h_{mixed} is the convection heat transfer coefficient in W m^{-2} K^{-1}, k is the thermal conductivity in W m^{-1} K^{-1} and D_h the hydraulic diameter in m. All the fluid properties required to calculate the Nusselt numbers and the heat transfer coefficient are based on the film temperature, taken as the average of the wall and bulk fluid temperatures

GAMMA+ MODEL

GAMMA+ (General Analyzer for Multi-component and Multi-dimensional Transient Application) is a simulation code intended for use in the simulation of phenomena in high temperature gas cooled reactors. It is capable of predicting chemical reactions and thermal hydraulic phenomena [6]. Some benchmark tests have been done with the aim of validation of the code [7]. It has also been compared with Flownex in a code-to-code comparison, in which the results showed good agreement [8].

The GAMMA+ model of the NACEF is shown in Figure 3. The GAMMA+ model consists of fluid regions (F1 in Figure 3) and solid regions (W2, W3 in Figure 3). The fluid regions are made up of control volumes referred to as fluid blocks, which are connected by the junction blocks. The flow in one of the chimneys is choked, such that the mass flow rate is zero in the said chimney. The thermal radiation heat transfer in the cavity and within the riser tubes is accounted for in the model. The boundary conditions are applied as shown (BC in Figure 3). The heater, the cavity walls and the risers are all divided into 20 vertical increments of equal length.

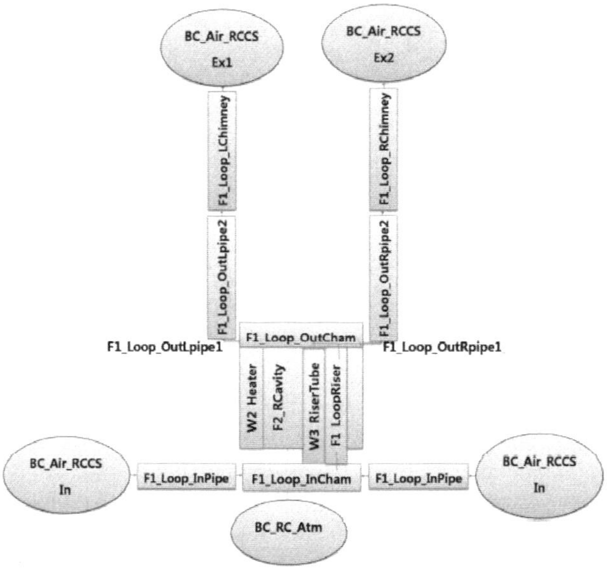

Figure 3. GAMMA+ model of the NACEF [2].

FLOWNEX MODEL

Flownex is a systems CFD code that is capable of solving thermal-fluid systems consisting of heat transfer, combustion, mechanical components etc. [9].

The inlet of the system as modelled in Flownex is shown in Figure 4. The temperature and pressure boundary conditions as stated in Table 4 are applied as shown in Figure 4.

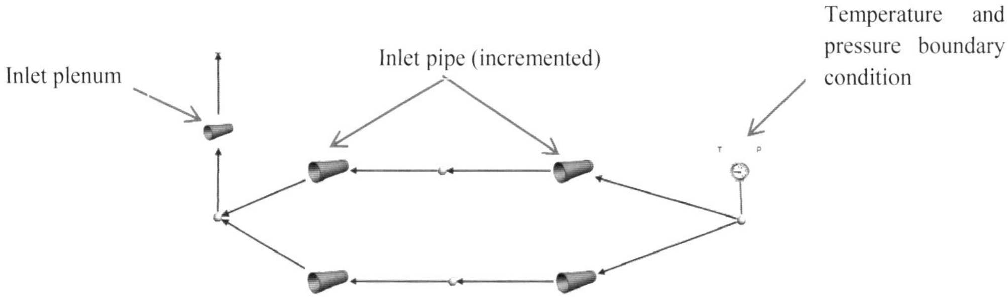

Figure 4. Inlet of the NACEF as modelled in Flownex.

The pipe network in Figure 4 connects to the riser tube, of which a single increment is shown in Figure 6. As in the GAMMA+ model the riser tube consists of 20 increments of equal length.

The flow element (Figure 6) models the momentum transfer of the air as it rises through the tube. The convection elements account for the convection heat transfer on the cavity walls, as well as, the inner and outer surfaces of the riser tubes. The conduction elements account for the conduction through the stainless steel walls. The compound components are defined by the user. In this case these components model radiative heat transfer. The compound components are illustrated in Figure 8 and Figure 9.

The scripting components calculate the convection heat transfer coefficient for the air in the riser tubes. The mixed convection theory described in equations (1) and (2) was programmed into the scripts. The GAMMA+ model contains similar theory to solve mixed convection flow. The convection heat transfer coefficients for the cavity walls and the riser tube outer walls could not be calculated in Flownex. The coefficients calculated in GAMMA+ were therefore used.

The axial conduction in all the walls, not shown in the figure, (as function of elevation of the cavity) has also been accounted for. The tangential conduction heat transfer between the riser tubes` walls was also accounted for.

The descriptions for the Flownex components used in Figure 6 are shown in Figure 5.

Figure 5. Description of the components used in Figure 6.

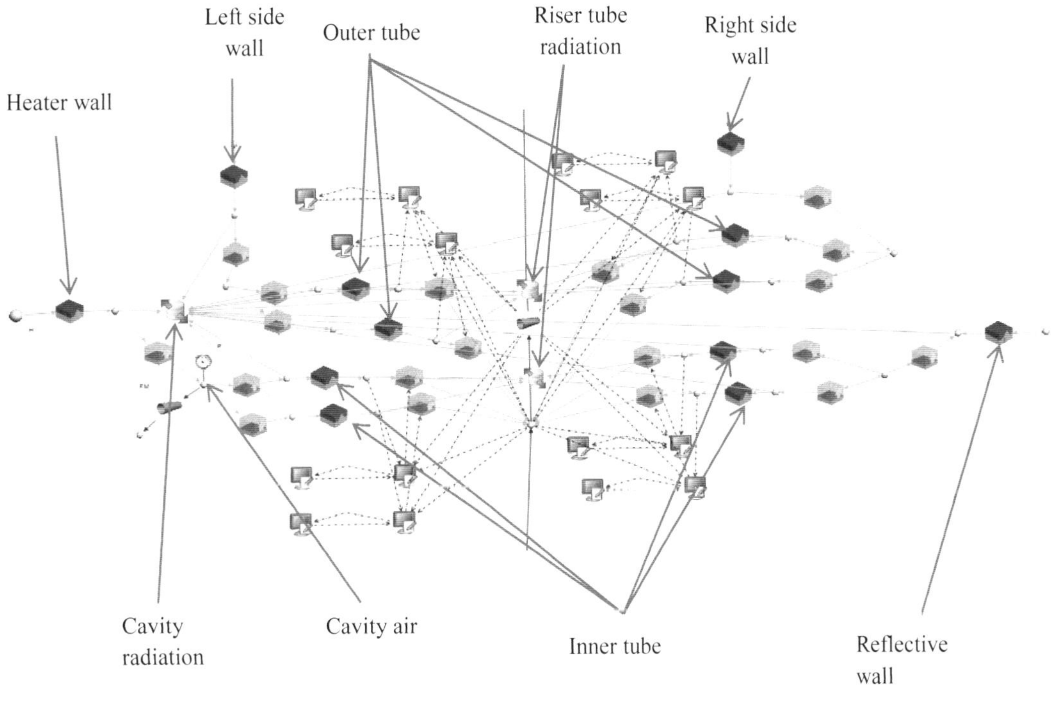

Figure 6. A single increment of the riser tube, as modelled in Flownex.

The chimneys, as modelled in Flownex, are shown in Figure 7. Flow in the right chimney was choked in accordance with the GAMMA+ model. The pipe increments were given the same lengths as to correspond exactly with the GAMMA+ model.

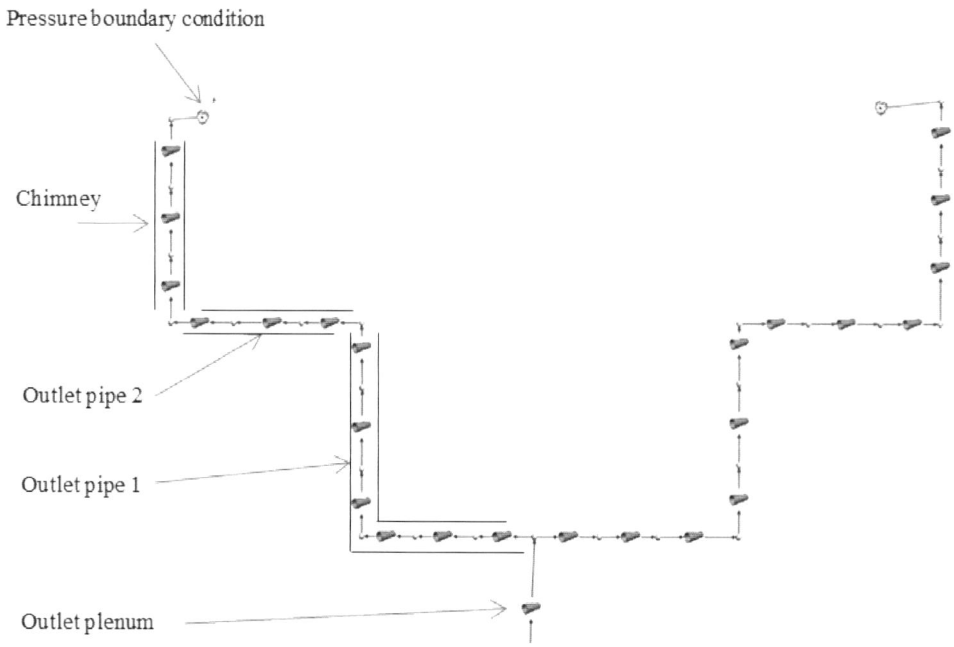

Figure 7. Outlet chimneys of the NACEF as modelled in Flownex.

The thermal radiation heat transfer in Flownex is accounted for by using surface and spatial radiation heat transfer elements. The network accounting for the thermal radiation heat transfer between the cavity surfaces is shown in Figure 8. The radiation view factor matrix for the cavity surfaces that was used in both Flownex and GAMMA+ is shown in Table 2:

Table 2. Radiation view factor matrix [10].

	Heater wall	Left side wall	Right side wall	Reflec-tive wall	Outer tube front surface	Outer tube side surface	Outer tube rear surface	Outer tube side surface	Inner tube front surface	Inner tube rear surface	Inner tube side surface	Inner tube side surface
Heater wall	0.00E+00	7.31E-01	3.24E-02	3.34E-02	4.00E-02	3.18E-04	0.00E+00	2.63E-02	8.90E-02	0.00E+00	1.27E-02	3.53E-02
Left side wall	2.58E-01	5.15E-01	1.04E-02	2.35E-04	3.65E-02	2.56E-02	0.00E+00	1.16E-02	6.83E-02	0.00E+00	4.22E-02	3.20E-02
Right side wall	2.22E-02	2.03E-02	1.25E-01	2.31E-01	0.00E+00	4.70E-01	3.49E-02	3.66E-03	0.00E+00	3.59E-02	4.11E-02	1.49E-02
Reflective wall	3.34E-02	6.65E-04	3.38E-01	6.15E-02	0.00E+00	3.91E-03	8.86E-02	6.07E-02	0.00E+00	2.26E-01	7.49E-02	1.13E-01
Outer tube front surface	2.60E-01	6.72E-01	0.00E+00	0.00E+00	6.81E-02	0.00E+00	0.00E+00	0.00E+00	0.0.	0.00E+00	0.00E+00	0.00E+00
Outer tube side surface	4.13E-04	9.42E-02	8.92E-01	5.09E-03	0.00E+00	8.49E-03	0.00E+00	0.00E+00	0.00E+00	0.00E+00	0.00E+00	0.00E+00
Outer tube rear surface	0.00E+00	0.00E+00	3.31E-01	5.76E-01	0.00E+00	0.00E+00	9.35E-02	0.00E+00	0.00E+00	0.00E+00	0.00E+00	0.00E+00
Outer tube side surface	3.42E-02	4.27E-02	6.94E-03	7.89E-02	0.00E+00	0.00E+00	0.00E+00	2.11E-02	0.00E+00	0.00E+00	8.16E-01	0.00E+00
Inner	2.89E-	6.29E-	0.00E	0.00E+	0.00E+00	0.00E+	0.00E+	0.00E+	8.20E-	0.00E+	0.00E+	0.00E+

tube front surface	01	01	+00	00		00	00	00	02	00	00	00
Inner tube rear surface	0.00E+00	0.00E+00	1.70E-01	7.34E-01	0.00E+00	0.00E+00	0.00E+00	0.00E+00	0.00E+00	9.56E-02	0.00E+00	0.00E+00
Inner tube side surface	8.22E-03	7.77E-02	3.89E-02	4.87E-02	0.00E+00	0.00E+00	0.00E+00	4.08E-01	0.00E+00	0.00E+00	1.08E-02	4.08E-01
Inner tube side surface	2.29E-02	5.90E-02	1.41E-02	7.34E-02	0.00E+00	0.00E+00	0.00E+00	0.00E+00	0.00E+00	0.00E+00	4.08E-01	4.23E-01

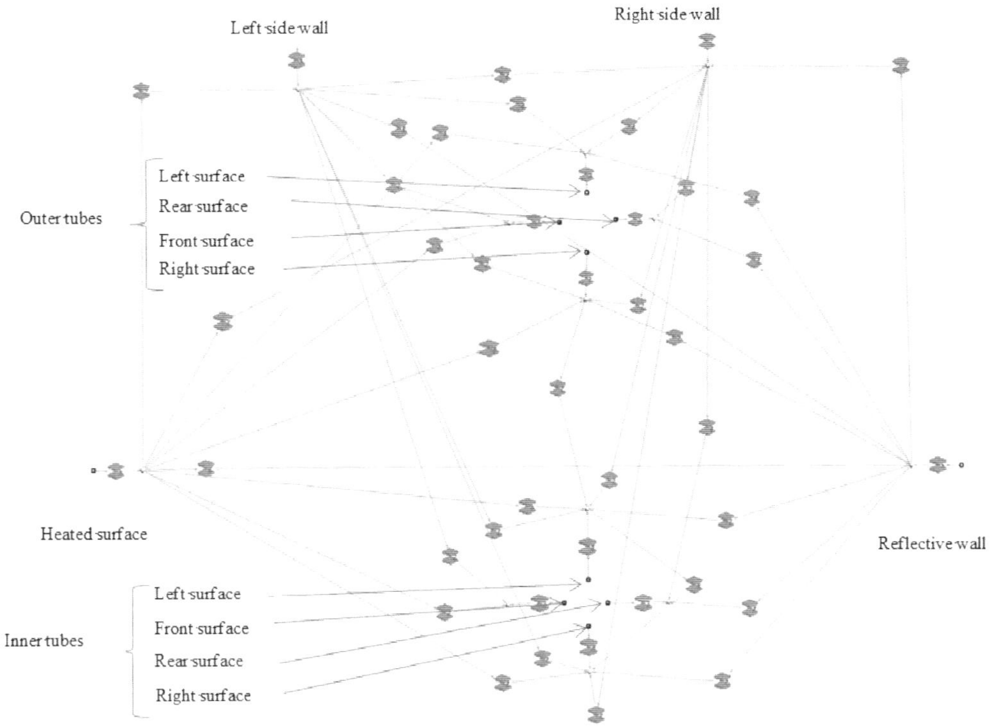

Figure 8. Flownex compound component describing the radiative heat transfer in the cavity.

The radiation heat transfer between the riser tube inner surfaces was modelled using the network as shown in Figure 9. The radiation view factor matrix for the riser tube inner surfaces used in both Flownex and GAMMA+ is shown in Table 3:

Table 3. Thermal radiation view factors for the riser tube inner surfaces [8].

Surface	Front	Side	Side	Rear
Front	0	0.45049	0.45049	0.09902
Side	0.090098	0	0.819804	0.090098
Side	0.090098	0.819804	0	0.090098
Rear	0.09902	0.45049	0.45049	0

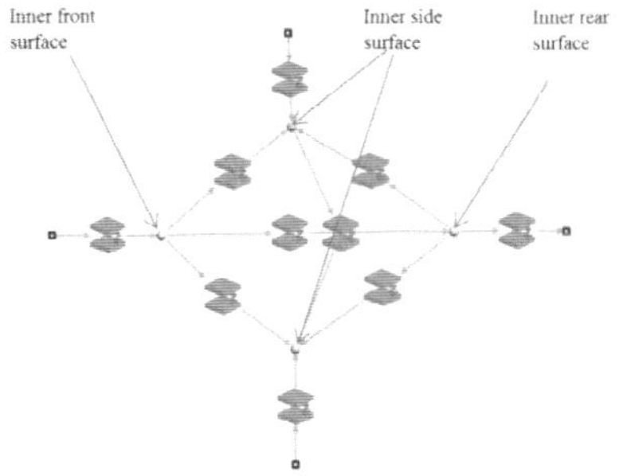

Figure 9. Radiative heat transfer network for the riser tube inner surfaces.

The boundary conditions for the model are given in Table 4. The inlet pressure was chosen to reflect purely hydrostatic flow.

Table 4. The boundary conditions for the model.

Inlet temperature	24° C
Inlet pressure	101.46 kPa
Outlet pressure	101.3 kPa
Heater flux	4.34 kW/m^2
Heater wall emissivity	0.75
Riser tube surface emissivity	0.8
Side walls emissivity	0.1

RESULTS

The air temperature is plotted as function of the elevation in the riser tube in Figure 10. The results obtained by Flownex and GAMMA+ are compared in the figure. The average difference between the results is 0.24%. The air temperature shows a gradual increase with an increase in elevation.

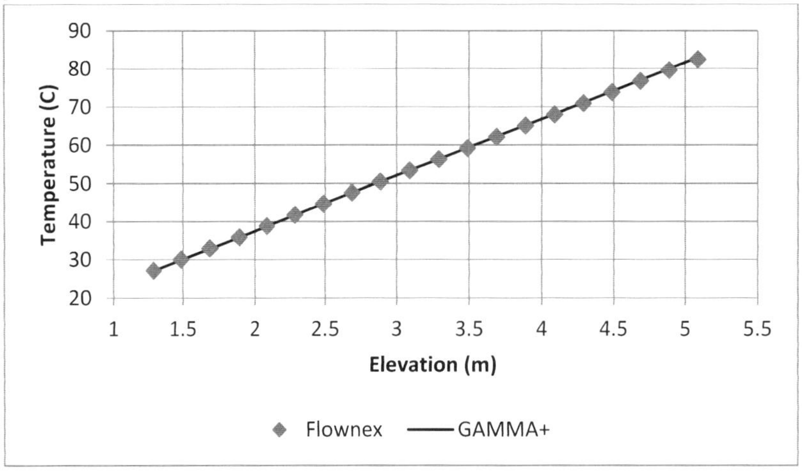

Figure 10. Riser tube air temperature as a function of elevation.

The heater wall and reflective wall temperatures calculated by GAMMA+ and Flownex are compared in Figure 11. The difference between the results is 0.23% and 1.39% respectively. The wall temperatures show an increase with an increase in elevation.

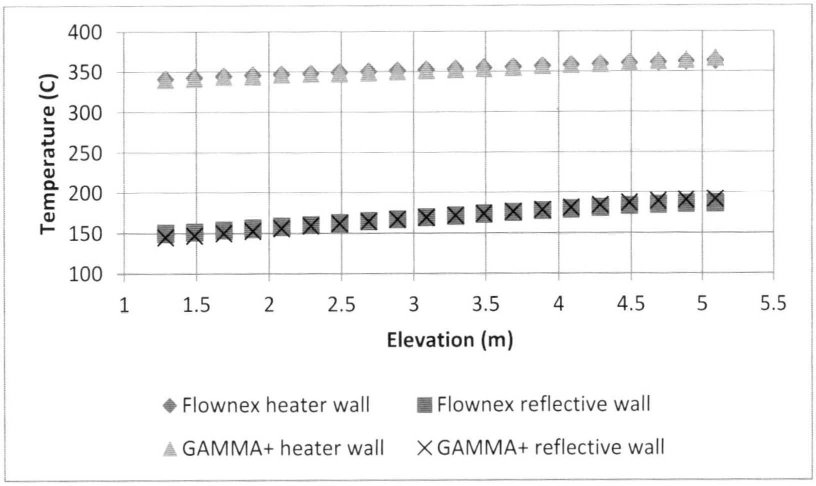

Figure 11. Heater- and reflective wall temperature as a function of elevation.

The left-side wall and right-side wall temperatures are compared in Figure 12. The average difference between the results is 0.69% and 0.9% respectively. The results show the largest deviation at the highest elevation.

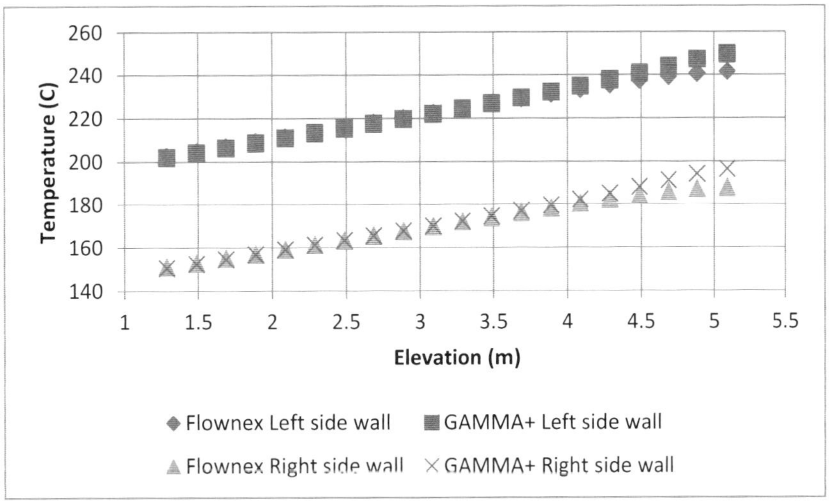

Figure 12. Left-side- and right side wall temperature as a function of elevation.

The convection heat transfer coefficient, as calculated in GAMMA+ and Flownex is compared in Figure 13. The average difference is 0.49%. The convection heat transfer coefficient at the lowest elevation is higher than the second; where after a gradual increase is experienced with an increase in elevation. The reason for this occurrence is as follows: in the first increment, the effect of the free convection is comparatively small. This means that equation 1, results in a higher Nusselt number and consequently a higher convection heat transfer coefficient is calculated in equation 2. There are two unheated increments of the riser tube below the lowest elevation in Figure 13. Some of the heat from the first increment escapes downward into the unheated section of the riser tube, by mechanism of conduction. This results in a lower wall temperature, which in turn influences the Nusselt number and subsequently the convection heat transfer coefficient. From the second elevation, the effect of the free convection is larger, resulting in a smaller coefficient to be calculated. For the rest of the tube, both the forced convection and free convection have a gradually increasing influence on the Nusselt number.

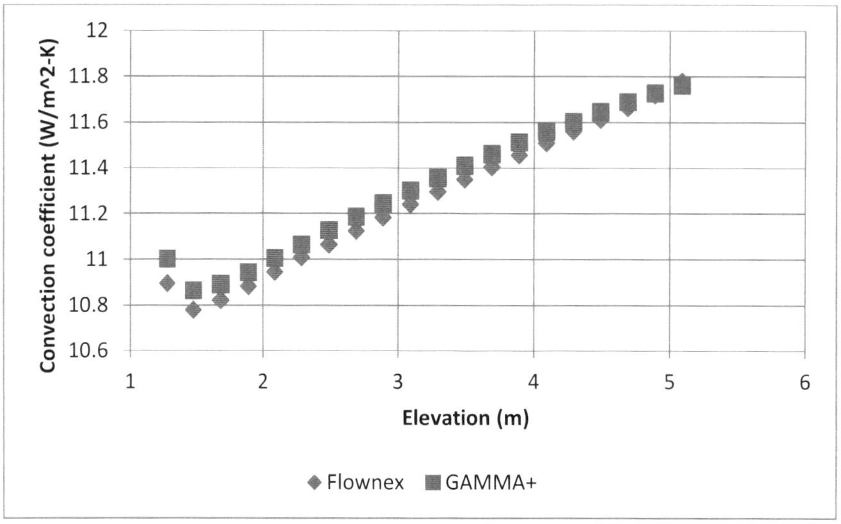

Figure 13. Convection heat transfer coefficient as a function of elevation.

The Flownex model calculated a mass flow rate of 0.192 kg/s in the riser tube. The GAMMA+ model calculated a mass flow rate of 0.189 kg/s. The mass flow rates show good agreement, with a difference of 1.77% between the codes.

The differences between the GAMMA+ and Flownex results can be explained as follows: GAMMA+ represents the cavity air by four sub fluid-blocks (connected control volumes) per increment, while Flownex uses a single node (control volume) per increment to represent the cavity air. The Flownex model calculates an average value for the air temperature in the cavity, while GAMMA+ can account for local air temperatures, as well as air circulation. The fluid properties calculated by GAMMA+ and Flownex are also slightly different due to the fact that they employ different correlations to obtain the values of the properties.

CONCLUSION

The reactor cavity cooling system is a safety feature built into the high temperature gas-cooled reactor. The system is intended to remove decay heat under upset conditions. The natural cooling experimental facility (NACEF) is a scale model of a RCCS. The NACEF was simulated by using Flownex and the results were compared with the results obtained by GAMMA+. There is very good agreement between the results. The reasons for the differences have been discussed.

ACKNOWLEDGEMENT

This work is based on the research supported by the South African Research Chairs Initiative of the Department of Science and Technology and National Research Foundation of South Africa (Grant No 61059), as well as supported by the Nuclear R&D Program of the NRF of Korea grant funded by the Korean government (Grant code : NRF-2012M2A8A2025679).

Any opinion, finding and conclusion or recommendation expressed in this material is that of the authors and the NRF does not accept any liability in this regard.

REFERENCES

1. Kim, M.H., Tak, N.I, Lim, H.S., "Thermal-fluid assessment of the design options for reactor vessel cooling in a prismatic core VHTR" Annals of Nuc. Eng., 37, pp. 1774-1782, (2010).
2. Kim, J., Bae, Y., Hong, S., Kim,C. & Kim, M. 2014. Results of the preliminary test in the ¼ scale RCCS of the PMR200 VHTR. Transactions of the Korean Nuclear Society Autumn meeting. Pyeongchang, Korea, October 30-31, 2014.
3. Kim, J. 2014. Experiments of Reactor Cavity Cooling System. 5th KAERI-NWU joint research collaboration meeting. 2014-11-04.
4. Celata, G.P., D'Annibale, F., Chiaradia, A. & Cumo,M. 1998. Upflow turbulent mixed convection heat transfer in vertical pipes. *International journal of heat and mass transfer,* 41(1998): 4037-4054.
5. Incropera, F.P., Dewitt, D.P., Bergman, T.L. & Lavine, A.S. 2013. Principles of heat and mass transfer. 7 th ed. Wiley.
6. Korea Atomic Energy Research Institute. 2014. Transient gas multicomponent mixture analysis code for analysing the thermo-fluid transients in an VHTR; Vol. 2 GAMMA+ theory manual.
7. Khoza, S., Tak, N., Lim, H., Lee, S., Cho, B & Kim, J. 2015. Preliminary Analysis of the KAERI RCCS experiment using GAMMA+. Transactions of the Korean Nuclear Society spring meeting. Jeju, Korea, May 7-8, 2015.
8. Rousseau, P.G., du Toit, C.G., Jun, J.S. & Noh, J.M. 2015. Code-to-code comparison for analysing the steady-state heat transfer and natural circulation in an air-cooled RCCS using GAMMA+ and Flownex. *Nuclear engineering and design*, 291:71-89.
9. M-Tech Industrial. 2015. Flownex simulation environment. General user manual.
10. Khoza, S. 2015. Preliminary analysis of KAERI RCCS experiment using GAMMA+. Nuclear hydrogen project calculation note. Doc. No.: NHDD-RD-CA-14-009, KAERI, 2015.

10th South African Conference on Computational and Applied Mechanics
Potchefstroom
3 – 5 October 2016

CFD Simulation of a Fluidized Bed Reactor for Petroleum Wastewater Photodegradation

Nhlanhla Nyembe*, Aoyi Ochieng[1], Letsabisa Lerotholi[2]

Nhlanhla Nyembe*, Aoyi Ochieng[1] and Letsabisa Lerotholi[2] are with the Centre for Renewable Energy and Water, Vaal University of Technology, Vanderbijlpark, 1900, RSA (e-mail: : nhlanhlan@vut.ac.za, ochienga@vut.ac.za, letsabisal@vut.ac.za).

ABSTRACT

A comprehensive study was conducted on the hydrodynamics in a fluidized bed reactor (FBR) used for the photodegradation of petroleum wastewater. The aim of this study was to apply computational fluid dynamics (CFD) and experimental techniques to optimize the hydrodynamic operating conditions in a FBR. The focus of this study was on modelling the hydrodynamic parameters and their substantial impact on the design and performance of the FBR. The CFD simulations were carried out by varying the grid size, discretization scheme, time step size and the sum of square error (SSE) being the response variable. The simulation results were validated with experimental data obtained from literature. From the optimization study, it was found that the 2nd order discretization scheme along with a fine mesh (500 000 cells) gave the lowest error. The maximum photodegradation efficiency (62.5%) was obtained at a superficial gas velocity of 17.3 mm/s. It was evident that the developed model predicted the experimental results fairly, displaying its effectiveness and applicability to study other phenomena such as mass transfer.

KEYWORDS

Fluidized bed reactor, Hydrodynamics, Photodegradation, Superficial gas velocity

NOMENCLATURE

AOP	Advanced oxidation process
BIT	Bubble induced turbulence
FBR	Fluidized bed reactor
SSE	Sum of square error
d_g	Bubble diameter, mm
$\vec{F}_k^D$	Drag force, N
C_D	Drag force coefficient
$\vec{g}$	Gravitational acceleration, m/s^2
$\vec{F}_k^L$	Lift force, N
C_L	Lift force coefficient
G	Production of turbulent kinetic energy
∇P	Pressure gradient, Pa
$\vec{F}_k^T$	Turbulent dispersion force, N
C_{TD}	Turbulent dispersion force coefficient
k	Turbulent kinetic energy, J/kg
$\vec{n}_w$	Unit normal pointing away from the wall
$\vec{u}_g$	Velocity vector of gas phase, m/s
$\vec{u}_l$	Velocity vector of liquid phase, m/s
$\vec{u}_k$	Velocity vector of phase k, m/s
$\vec{F}_k^{WL}$	Wall lubrication force, N
C_w	Wall lubrication force coefficient

Greek letters

$\mu_{BIT,l}$	Bubble induced viscosity, Pa.s
ρ_k	Density of phase k, kg/m^3

$\mu_{eff,l}$	Effective viscosity, Pa.s
$\in_k$	Volume of fraction of phase k
μ_l	Molecular viscosity, Pa.s
$\sigma_k, \sigma_\varepsilon$	Prandtl numbers
σ_{lg}	Schmidt number
τ_k	Shear stress term
ε	Turbulent kinetic energy dissipation rate, m^3/s^3
$\mu_{T,l}$	Turbulent viscosity, Pa.s

Subscripts

g	Gas
l	Liquid
k, j	Phase index

1. INTRODUCTION

Chemical contamination has posed a serious threat to most water sources in many countries including South Africa. Literature has shown that the main cause of water pollution are industrial operations such as: pharmaceutical, mining and petroleum refineries (Abhang et al. [1]; Shet and Vidya [2]; Khune et al. [3]). Wastewater from petroleum refineries often contains high concentration of organic pollutants, which are often released into the environment. These organic pollutants include aliphatic, aromatic and polyaromatic compounds and they are highly toxic to human, aquatic, micro-organism life and the environment itself (Zulfakar et al. [4]). It has been shown that the application of advanced oxidation process (AOP) for the degradation of organic pollutants is efficient due to the fact that it converts most pollutants into less harmful products (Qi et al. [5]). This process is attractive due to its ability to be operated at ambient temperatures without any addition of oxidant chemicals (Wang et al. [6]). In recent years, fluidized bed reactors (FBR) have been extensively used in AOPs either as immobilized reactors (Shet and Vidya [2]) or circulating reactors (Lim et al. [7]). This is due to the fact that FBRs have high mass transfer coefficients, good catalyst-to-pollutant contact and guarantee better catalyst particle light exposure (Boyjoo et al. [8]). Since most reactors employed in AOP (for the treatment of wastewater) are suspended in nature, the design and operation of photoreactors is based on the fundamentals of FBR. Effective application of the FBR for photocatalysis requires accurate understanding of reactor performance amongst other factors such as light distribution, reactor geometry and hydrodynamics. Furthermore, factors such as superficial gas velocity, fluid properties, material of construction and sparger design (Kulkarni et al. [9]) are also important for assessing the effectiveness of the reactor. Fluidized bed reactors employed as photoreactors are not widely used at a commercial scale due to design, scale-up and construction difficulties, size limitations and fouling on lamp walls (Mohajerani et al. [10]; Boyjoo et al. [8]).

Therefore, accurate modelling of the hydrodynamic behaviour of the FBR is essential in simplifying the complex hydrodynamics of multiphase systems, as to assist in the design, scale-up and optimization of the reactor and even evaluating the performance. In the past decades, computational fluid dynamics (CFD) has been a useful tool in enhancing our knowledge of the complex hydrodynamic phenomena that occur in multiphase systems (Ochieng et al. [11]). In addition, advancements in computing power and numerical techniques has been found very useful in simulating multiphase flow (Pourtousi et al. [12]). Wang et al. [6] conducted a CFD study to model a photocatalytic odor abatement process and found that the results were useful for the design, scale-up and optimization of the process. Qi et al. [5] modelled the hydrodynamics and degradation kinetics inside a slurry photocatalytic reactor for wastewater treatment. They reported that the simulation results based on solid particle concentration, gas hold-up, pollutant concentration and light intensity revealed that the AOP process is controlled by light distribution and local hydrodynamics. Evidently, literature has shown that computational, mathematical and numerical methods can produce suitable design, scale-up and optimization techniques for FBRs employed in wastewater treatment. Several research has been conducted on the advanced oxidation process in terms of the design of the photoreactor, synthesis of novel catalysts and optimization of operating conditions (Qi et al. [5]). However, to the best our knowledge, limited information is available in literature regarding the understanding of the hydrodynamics, including the bubble induced turbulence, in FBRs employed as photoreactors. Therefore, the aim of this study was to apply CFD and experimental techniques to optimize the hydrodynamic operating conditions in a fluidized bed reactor. The main focus of this study was on modelling the flow characteristics due to their substantial impact on the design, scale-up and performance on the FBR. The simulation results were compared to experimental data obtained from Kulkarni et al. [9].

2. MATHEMATICAL MODELLING

2.1. Governing equations

The Eulerian-Eulerian approach was used to describe the flow characteristics of each phase. In this approach, liquid is considered to be the continuous phase, whereas gas is dispersed suggesting that the phases are treated as interpenetrating continua (ANSYS [13]). Furthermore, this approach permits the modelling of multiple separate, yet interacting phases for which equations such as mass and momentum are solved for each phase (ANSYS [13]; Boyjoo et al. [8]). The governing differential equations are defined for each phase in the fluidized bed reactor, neglecting the interphase mass transfer:

The continuity equation is given as:

$$\frac{\partial}{\partial t}\left(\in_k \rho_k\right)+\nabla\left(\in_k \rho_k \vec{u}_k\right)=0 \qquad \ldots (1)$$

where $\in_k$, ρ_k and $\vec{u}_k$ are the volume fraction, density and velocity vector of phase k, respectively. The subscript *k* represents liquid, gas or solid phase.

Momentum equation is given as:

$$\frac{\partial}{\partial t}\left(\in_k \rho_k \vec{u}_k\right)+\nabla\left(\in_k \rho_k \vec{u}_k \vec{u}_k\right)=-\in_k \nabla p-\nabla\left(\in_k \tau_k\right)+\in_k \rho_k \vec{g}+\vec{F}_k \qquad \ldots (2)$$

where $\vec{g}$ and p are the gravitational acceleration and pressure gradient, respectively. $\vec{F}_k$ represents the interfacial forces exerted on phase *k* due to the presence of phase *j*. τ_k is the stress term of gas bubbles and is given by:

$$\tau_k=-\mu_{eff,k}\left[\left(\nabla \vec{u}_k\right)+\left(\nabla \vec{u}_k\right)^T-\frac{2}{3} I\left(\nabla \vec{u}_k\right)\right] \qquad \ldots (3)$$

where $\mu_{eff,k}$ is the effective viscosity of the liquid which in turn consists of three terms namely: molecular viscosity (μ_l), turbulent viscosity ($\mu_{T,l}$) and bubble induced turbulence ($\mu_{BIT,l}$).

$$\mu_{eff,l}=\mu_l+\mu_{T,l}+\mu_{BIT,l} \qquad \ldots (4)$$

2.2. Interfacial forces

An interaction exists between the gas and liquid phases that influences the interfacial forces. Several models for interfacial forces have been reported in literature (Pourtousi et al. [12]; Masood et al. [14]; Pourtousi et al. [15]) and correct modelling of the fluidized bed reactor is dependent in the selection of the correct interfacial forces. It has been reported that the use of a pair and/or all the interfacial forces improves the accuracy of the flow pattern prediction (Zhang et al. [16]). Therefore, in the current study, a combination of all the interfacial forces was used and are described by the following equations:

$$\vec{F}_k=\vec{F}_k^D+\vec{F}_k^L+\vec{F}_k^{WL}+\vec{F}_k^T \qquad \ldots (7)$$

where $\vec{F}_k^D$, $\vec{F}_k^L$, $\vec{F}_k^{WL}$ and $\vec{F}_k^T$ are the interfacial forces due to drag, lift, wall lubrication and turbulent dispersion, respectively. The source of the drag force is caused by the resistance experienced by the bubbles as they move through the liquid. The pressure distribution that surrounds the bubble creates form drag and the viscous stress creates skin drag (Tabib et al. [17]). The interphase momentum transfer due to the drag force is given as:

$$\vec{F}_{lg}^{D} = \frac{3}{4}\frac{C_{D,lg}}{d_g}\rho_l \in_g \left|\vec{u}_g - \vec{u}_l\right|\left(\vec{u}_g - \vec{u}_l\right) \quad \dots (8)$$

where C_D is the drag coefficient and d_g is the bubble diameter. The Grace relation was chosen to compute the drag coefficient.

The lift force acts on the particles of the secondary phase (bubbles) and is mainly due to the velocity gradients in the primary phase. Therefore, the lift force becomes more significant for larger bubble sizes. This force is created by the net effect of pressure and stress exerted on the surface of the secondary phase. The lift force acting on the primary (*l*) and secondary (*g*) phases can be calculated as:

$$\vec{F}_{lg} = -C_L \rho_l \in_g \left(\vec{u}_g - \vec{u}_l\right) \mathrm{x} \left(\nabla \mathrm{x}\, \vec{u}_l\right) \quad \dots (10)$$

The sign convention of this force depends on the orientation of the slip velocity with respect to the gravity vector. C_L is the lift coefficient and was computed using the Saffman-Mei lift model.

The wall lubrication force drives the secondary phase away from the wall resulting in the secondary phase concentrating in the area near but not immediately to the wall (ANSYS [13]). The wall lubrication force exerted on the secondary phase is given by:

$$\vec{F}_{w,l} = C_w . \rho_l \in_g \left|\left(\vec{u}_l - \vec{u}_g\right)\right|^2 \vec{n}_w \quad \dots (12)$$

where $\vec{n}_w$ is the unit normal pointing away from the wall; C_w is the wall lubrication coefficient and is computed using the Antal model.

The effect of turbulent fluctuations on the liquid velocity, caused by the turbulent eddies in the continuous phase, interact with the bubbles by transporting them from a region of high concentration to a region of low concentration (ANSYS [13]). This phenomena was described, in this study, by using the Burns et al. model and can be derived from the Favre averaging of the drag term:

$$\vec{F}_{TD,l} = -\vec{F}_{TD,g} = C_{TD} K_{lg} \frac{D_l}{\sigma_{lg}}\left(\frac{\nabla \in_g}{\in_g} - \frac{\nabla \in_l}{\in_l}\right) \quad \dots (14)$$

where C_{TD} is the turbulent dispersion coefficient and was assigned a value of 0.2 for the current study. σ_{lg} is the Schmidt number and a value of 0.9 is typically used.

2.3. Turbulence modelling

Alongside the governing equations and interfacial forces, turbulence models are a key factor in capturing the hydrodynamic behaviour (Pourtousi et al. [15]). *k-ε* models fall under a class of models referred to as two-equation models which allow the determination of both the turbulent length scale and time scale. This is done by solving two separate transport equations (ANSYS [13]). In the last two decades, *k-ε* models have been extensively used to describe the flow pattern in fluidized bed reactors. In addition, incorporating turbulence model with the correct interfacial force models leads to an accurate solution of the hydrodynamic parameters (Pourtousi et al. [12]). The great preference of this model by researchers is due to its simplicity, low computational requirement and relatively inexpensive and is preferred to predict the liquid velocity and gas hold-up profiles under low superficial gas velocity (Pourtousi et al. [12]). When the standard *k*-ε model is applied, the turbulent eddy viscosity ($\mu_{T,l}$) is define as:

$$\mu_{T,l} = \rho_l C_\mu \frac{k^2}{\varepsilon} \quad \ldots (15)$$

where *k* and ε are the turbulent kinetic energy and turbulent kinetic energy dissipation rate, respectively, and are calculated from their governing equations:

$$\frac{\partial}{\partial t}\left(\rho_l \in_l \mathrm{k}\right) + \nabla\left(\rho_l \in_l u_l \mathrm{k}\right) = -\nabla\left(\in_l \frac{\mu_{eff,l}}{\sigma_k}\nabla \mathrm{k}\right) + \in_l \left(G - \rho_l \varepsilon\right) \quad \ldots (16)$$

$$\frac{\partial}{\partial t}\left(\rho_l \in_l \varepsilon\right) + \nabla\left(\rho_l \in_l u_l \varepsilon\right) = -\nabla\left(\in_l \frac{\mu_{eff,l}}{\sigma_\varepsilon}\nabla \varepsilon\right) + \in_l \frac{\varepsilon}{k}\left(C_{\varepsilon 1} G - C_{\varepsilon 2}\rho_l \varepsilon\right) \quad \ldots (17)$$

where $\sigma_k = 1$ and $\sigma_\varepsilon = 1$ are the Prandtl numbers. The standard values of all the model constants are $C_\mu = 0.09$, $C_{\varepsilon 1} = 1.44$ and $C_{\varepsilon 2} = 1.92$. The term G in the above equation is the production of turbulent kinetic energy.

3. NUMERICAL SIMULATION

Computational fluid dynamics has a substantial contribution in the modelling of chemical industrial equipment due to its ability to model complex hydrodynamic phenomena efficiently. The shortfalls regarding CFD are the endless parameters that need to be adjusted in order to achieve accuracy and convergence at minimal computational expense. Previous simulations have shown that the most influential parameters that affect accuracy and convergence in transient simulations are: grid size, discretization scheme and time step size. Furthermore, 12 simulations were carried out by varying the grid size, discretization scheme and time step size and comparing them to experimental results from Kulkarni et al. [9] as a reference case.

3.1. Geometry and grid generation

The geometry consisted of a cylindrical shaped body comprising of 4 side ports and a conical shaped top section suited for degassing (Fig. 1). The cylindrical shaped section has an inside diameter of 40.8 mm with a height of 600 mm and the conical section has an inside diameter of 80 mm with a height of 100 mm. Three types of grids (fine, medium and course) were created, as shown in Fig. 1, for a grid independent study The finer grid consisted of 500 000 cells, medium grid consisted of 166 000 and the coarser grid consisted of 80 000 cells. All grids were consisted of hexahedral cells structurally placed throughout the computational domain. From the grid independent study, it was found that there was no significant difference, in simulation results, between the medium and coarse grids. Therefore, only the fine and course grids were considered in the optimization study.

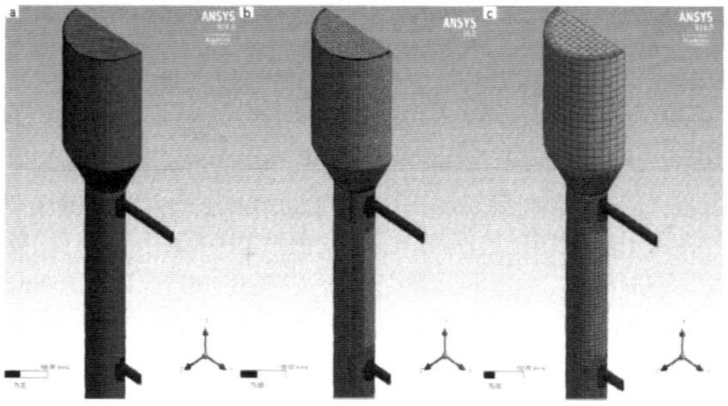

Figure 1. Simulation grid for a fluidized bed reactor: (a) fine cells and (b) coarse cells and (c) coarse cells.

3.2. Initial and boundary conditions

The walls were defined as a stationary wall with a no slip condition. The air inlet was defined as a fully open inlet where the gas flow rate was computed based on the superficial gas velocity of 20 mm/s with an air volume fraction of 1. For solution initiation, $\in_g$, k, ε, $\vec{u}_g$ and $\vec{u}_l$ were specified as 0 (at $t = 0$) in the whole computational domain. The pressure outlet boundary condition was applied at the liquid outlet and the degassing for the gas outlet. The bubble diameter at the inlet of the gas distributor was assumed to be 1.96 mm based on experimental data (Kulkarni et al. [9]). The backflow volume fraction was specified as 1 to reduce the entrainment of liquid (Boyjoo et al. [8]).

3.3. Numerical procedure

The simulations were carried out using ANSYS Fluent 16.0 commercial software and the equations of motion and continuity were solved, coupled with k-ε equations, to obtain complete flow pattern by scrutinizing the gas holdup and liquid velocity profiles. The phase coupled SIMPLE algorithm, which is an extension of the SIMPLE algorithm for multiphase flow, was used for the pressure-velocity coupling and correction. The time step size was varied and all the simulations were computed for 30 s to obtain steady state. The Under-Relaxation parameters were kept constant along with the convergence criteria of 1 x10^{-3} (Kulkarni et al. [9]).

4. PHOTODEGRADATION EXPERIMENTS

4.1. Reactor operation

The photocatalytic degradation of petroleum wastewater was performed using a solar photocatalytic fluidized bed reactor. All the photodegradation experiments were conducted outdoors (rooftop from 09:00 to 15:00) to utilize the sunlight as an ultraviolet (UV) source and the reactor was operated in batch mode. Figure 2 shows a schematic diagram of the photoreactor which consisted of four strategically placed side-ports, where, the top (wastewater outlet) and bottom (wastewater inlet) side-ports were used for wastewater recirculation by a peristaltic pump. A compressor was used to feed air into the reactor, through the bottom, where it was evenly distributed by a sintered plate type distributor with a pore size of 10 – 15 μm.

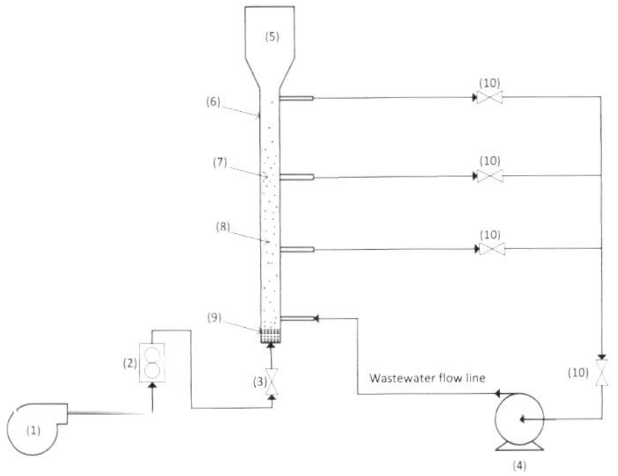

Figure 2. Experimental setup of a solar photocatalytic fluidized bed reactor. (1) Compressor; (2) Flowmeter; (3) Non return valve; (4) Peristaltic pump; (5) Degassing zone; (6) Fluidized bed photoreactor; (7) Catalyst particles; (8) Air bubbles (9) Gas distributor and (10) Ball valves.

5. RESULTS AND DISCUSSION

5.1. Numerical optimization

The CFD simulations were carried out by varying the number of cells, discretization scheme and time step size to determine the best combination to obtain accurate simulation data with minimal simulation time. The comparison between the simulation and experimental data was done by evaluating the axial liquid velocity and gas hold-up. In addition, the sum of square error (SSE) was used to assess the accuracy of all the simulations. From the SSE data, it was seen that the QUICK scheme (500 000 cells) and 2nd Order scheme (80 000 cells) had the least error (data not shown). Li and Zhong [18] conducted a CFD simulation study on the hydrodynamics in a three-phase bubble column. They varied the different discretization schemes, times step size and grid-size. Although their grid-sizes were relatively coarse (32 436, 56 865 & 95 788), they reported that the 2nd order discretization scheme and time step size of 0.001 had the least error in both the radial and axial positions. The experimental data used for axial liquid velocity and local gas hold-up profiles was obtained from Kulkarni et al. [9] as a result of similar reactor geometries. Therefore, the simulation results of axial liquid velocity and gas hold-up compared to experimental data obtained from literature (Kulkarni et al. [9]) are shown in Figures 3 and 4.

5.2. Axial liquid velocity profiles

Figure 3 shows the comparison of CFD and experimental axial liquid velocity profiles at different locations (*H/D* = 0.2, *H/D* =2.6 and *H/D* = 5.0) of the reactor. The plot for axial liquid velocity profile displayed a radial variation for all three graphs, whereby, the liquid had an upward movement at the centre and a downward near the wall. This was caused by the recirculation of the liquid as the gas flows up the column. The 2nd Order scheme accurately captured the axial liquid velocity profiles in all three locations and gave the best prediction. However, the CFD prediction of the axial liquid velocity was poor towards the wall and this is due to the presence of small eddies which limit the ability of the standard k-ε model to capture near the wall turbulence. It has been reported that the 2nd order scheme uses a multidimensional linear reconstruction method thus resulting in second-order accuracy (ANSYS [13]) and good convergence. This method uses the vertex value of a cell the compute the gradients of scalar variables at the cell center. The QUICK scheme uses the weighted average of the central interpolations and the scalar quantity of the second-order upwind, thereby, resulting in a more accurate solution that does not converge readily and requires more processing power.

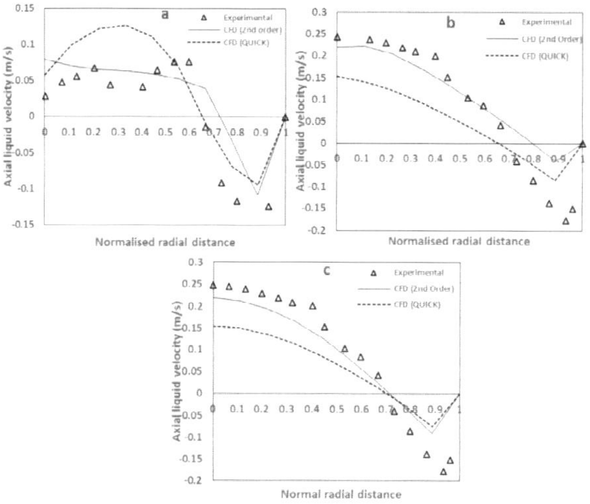

Figure 3. Comparison between the simulated and experimental axial liquid profiles at different axial positions. (a) *H/D* = 0.2; (b) *H/D* = 2.6 and (c) *H/D* = 5.0.

5.3. Local gas hold-up profiles

An efficient and simple way to understand mixing within the reactor is to study the variation of gas hold-up. Figure 4 shows the plot of local gas hold-up at different locations with a superficial gas velocity of 20 mm/s. The simulations were carried out in the homogeneous regime, implying that the bubbles move with virtually similar velocities and sizes and with minimal coalescence and break-up. This creates an added advantage of specifying a single bubble diameter (1.96 mm). The simulation results (Figure 4) gave a high gas hold-up towards the centre and a radial decrease when approaching the wall with an average radial gas hold-up of ~ 0.116. This is comparable to that of Kulkarni et al. [9] and Tabib et al. [17] who obtained an average radial gas hold-up of ~ 0.085 and ~ 0.10, respectively. The gas hold-up profile for the 2nd Order scheme compared well with the experimental data in Figs. 4b and c.

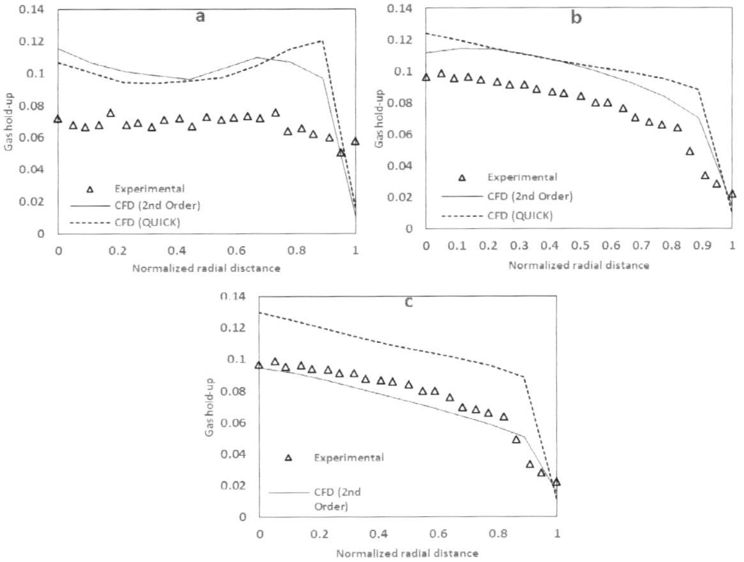

Figure 4. Comparison between the simulated and experimental gas hold-up profiles at different axial positions. (a) H/D = 0.2; (b) H/D = 2.6 and H/D = 5.0.

5.4. Effect of superficial gas velocity on the global gas hold-up

The optimization study was used to determine the optimum CFD setting methods. To verify the consistency of the CFD settings, they were further evaluated by investigating the global gas hold-up inside the photoreactor. The CFD data was compared to experimental global gas hold-up results determined by the author in reference. The global gas hold-up measurements were carried out by observing the displaced height of the clear liquid during sparging. The volume of the displaced liquid at different superficial gas velocities signifies the overall gas hold-up inside the photoreactor. Although this method approximates the global gas hold-up, it has been effectively used in literature (Li & Zhong [18]; Ellenberger & Krisha [19]; Ochieng et al. [20]) to characterize the hydrodynamic behaviour of the reactor. The optimal setting methods which were grid size of 500 000 cells, 2nd Order discretization scheme and the time step size of 0.001 were used for all the simulation cases from here onwards. The global gas hold-up measurements were carried out to validate the performance of the developed model and to characterize the behaviour of the reactor. Figure 5 shows the effect of superficial gas velocity on the global gas hold-up in the reactor. The simulation data was compared with the experimental results. The simulation data showed a fair prediction of the experimental results from a superficial gas velocity of 2.89-17.32 mm/s. It was found that an increase in superficial gas velocity results in an increase in the global gas hold-up. This similar trend was also reported by Pourtousi et al. [12] where they determined the effect of superficial gas velocity on the overall gas hold-up. At a superficial gas velocity of 15 mm/s, they obtained an overall gas hold-up of ~ 0.048, whereas in the current study, a superficial gas velocity of 14.44 mm/s gave a global gas hold-up of 0.052 and 0.050 for experimental and simulation, respectively. This agreement of the simulation data with the experimental results shows that the developed model is effective and can be applied to study other phenomena such as mass transfer.

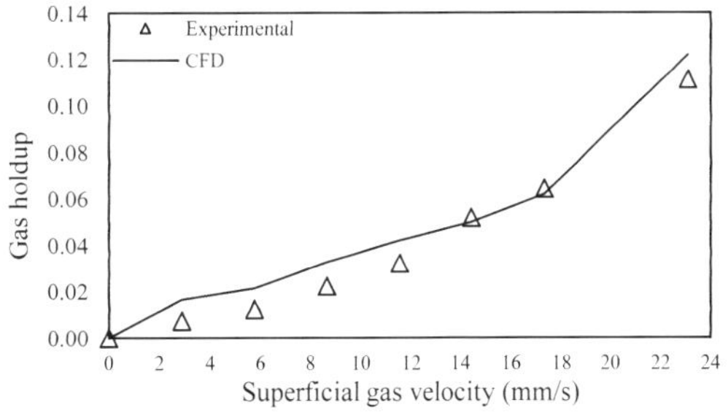

Figure 5. Effect of superficial gas velocity on the global gas hold-up: comparison between CFD simulation and experimental results.

5.5. Effect of Superficial gas velocity on the photodegradation of phenol

The central idea of a fluidized bed reactor is fluidization and air was used in the current study for fluidization and agitation. The results in Fig. 6 show the photodegradation experiments as affected by the (superficial gas velocity). The results were analyzed and processed using a statistical software called Design Expert®. Since the reactor hydrodynamics were dealt with in *section 5.3*, this section shows the effect of the hydrodynamics (superficial gas velocity) on the experimental photodegradation efficiency. The results show that an increase in the superficial gas velocity results in an increase in the photodegradation efficiency. The maximum superficial gas velocity was found to be 17.30 mm/s which gave a maximum photodegradation efficiency of 62.5%. The superficial gas velocity was varied within the homogeneous regime (8.66-17.30 mm/s) due to the fact that the catalyst requires a low superficial gas velocity for it to be fluidized. According to the hydrodynamic study, the bubble size distribution is uniform and the global gas hold-up is below 0.13 (Fig. 5). The trend in Fig. 6 indicates that an increase in the superficial gas velocity beyond 17.30 mm/s will result in an increase in the photodegradation efficiency. However, an increase beyond 17.30 mm/s will result in a heterogeneous bubble regime (high bubble size distribution) thus creating a more complex flow field. It is worth noting that the catalyst used has a small particle size (μm) thus requiring low fluidization.

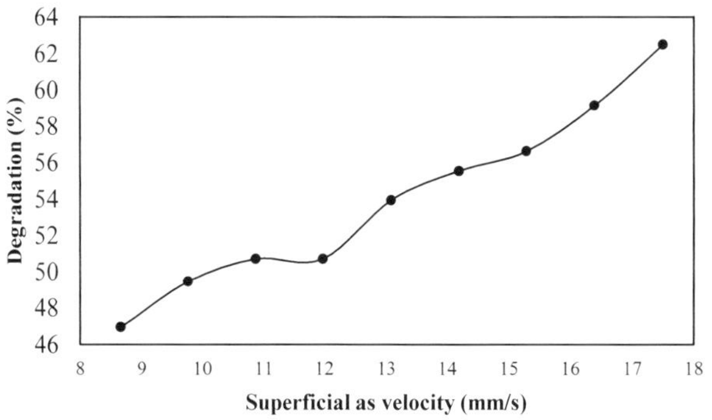

Figure 6. Effect of superficial gas velocity and catalyst loading on the photodegradation of petroleum wastewater.

6. CONCLUSION

The purpose of this study was to use CFD techniques to investigate the flow dynamics inside the fluidized bed reactor. Furthermore, it was shown that the hydrodynamics play an important role in the photodegradation of petroleum wastewater. The optimum CFD settings were obtained by investigating the hydrodynamic conditions such as axial liquid velocity and radial local gas hold-up profiles. The optimization study showed that a time step size of 0.001, grid resolution of 500 000 cells and 2nd order discretization scheme results in reduced numerical error. The CFD simulation showed a good agreement with the experimental data obtained from Kulkarni et al. [9] and an analysis of the hydrodynamic conditions gave an insight into the flow pattern inside the reactor. The developed model was validated by studying the effect of superficial gas velocity on the global gas hold-up and the CFD data was compared to the global gas hold-up experimental results obtained by the author in reference. From the photodegradation experiments, it was shown that a superficial gas velocity of 17.30 mm/s resulted in a maximum photodegradation efficiency. It was evident that the developed model predicted the experimental results fairly, displaying its effectiveness and applicability to study other phenomena such as mass transfer.

ACKNOWLEDGEMENT

The authors would like to acknowledgment the Water Research Commission (WRC) of South Africa for funding this project.

REFERENCES

1 Abhang, R.M., et al., "Design of Photocatalytic Reactor for Degradation of Phenol in Wastewater", International Journal of Chemical Engineering and Applications, Vol. 2. No. 5, pp. 3144 – 3148, 2011.

2 Shet, A., Vidya, S.K., "Solar Light Mediated Photocatalytic Degradation of Phenol using Ag Core – TiO2 Shell (Ag@TiO2) Nanoparticles in Batch and Fluidized Bed Reactor", Solar Energy, Vol. 127, pp. 67 – 78, 2016.

3 Khune, M., et al., "Solar Photodegradation of Phenol using a Composite Catalyst of Silica and TiO_2", International Conference on Chemical Engineering & Advanced Computational TEchnologies, Vol. 24-25, pp. 2 – 5, 2014.

4 Zulfakar, M., et al., "Photocatalytic Degradation of Phenol in a Fluidized Bed Reactor Utilizing Immobilized TiO2 Photocatalyst: Characterization and Process Studies", Journal of Applied Sciences, Vol. 11. No. 13, pp. 2320 – 2326, 2011.

5 Qi, N., et al., "CFD Modelling of Hydrodynamics and Degradation Kinetics in an Annular Slurry Photocatalytic Reactor for Wastewater Treatment", Chemical Engineering Journal, Vol. 172. No. 1, pp. 84 – 95, 2011.

6 Wang, Z., et al., "CFD Modeling of a UV-LED Photocatalytic Odor Abatement Process in a Continuous Reactor", Journal of Hazardous Materials, Vol. 215-216, pp. 25 – 31, 2012.

7 Lim, T.H., Kim, S.D., "Photocatalytic Degradation of Trichloroethylene (TCE) over TiO_2/Silica Gel in a Circulating Fluidized Bed (CFB) Photoreactor", Chemical Engineering and Processing: Process Intensification, Vol. 44. No. 2, pp. 327 – 334, 2005.

8 Boyjoo, Y., et al., "Some Aspects of Photocatalytic Reactor Modeling Using Computational Fluid Dynamics", Chemical Engineering Science, Vol. 101, pp. 764 – 784, 2013.

9 Kulkarni, A.A., et al., "On The Development of Flow Pattern in a Bubble Column Reactor: Experiments and CFD", Chemical Engineering Science, Vol. 62. No. 4, pp. 1049 – 1072, 2007.

10 Mohajerani, M., et al., "Computational Fluid Dynamics (CFD) Modeling of Photochemical Reactors", Applied Computational Fluid Dynamics, pp. 155 – 176, 2012.

11 Ochieng, A., Onyango, M., Kiriamiti, K., "Experimental Measurement and Computational Fluid Dynamics Simulation of Mixing in a Stirred Tank: A Review", South African Journal of Science, Vol. 105, pp. 421 – 426, 2009.

12 Pourtousi, M., et al., "Effect of Bubble Diameter Size on Prediction of Flow Pattern in Euler-Euler Simulation of Homogeneous Bubble Column Regime", Measurement: Journal of the International Measurement Confederation, Vol. 76, pp. 255 – 270, 2015.

13 ANSYS, "ANSYS Fluent Theory Guide", p.80, 2010.

14 Masood, R.M.A., et al., "CFD Simulation of Bubble Column Flows: An Explicit Algebraic Reynolds Stress Model Approach", International Journal of Multiphase Flow, Vol. 66, pp. 11 – 25, 2014.

15 Pourtousi, M., et al., "Effect of Interfacial Forces and Turbulence Models on Predicting Flow Pattern Inside the Bubble Column", Chemical Engineering and Processing: Process Intensification, Vol. 75, pp. 38 – 47, 2014.

16 Zhang, D., et al., "Numerical Simulation of the Dynamic Flow Behavior in a Bubble Column: A Study of Closures for Turbulence and Interface Forces", Chemical Engineering Science, Vol. 61. No. 23, pp. 7593 – 7608, 2006.

17 Tabib, M. V., et al., "CFD Simulation of Bubble Column-An Analysis of Interphase Forces and Turbulence Models", Chemical Engineering Journal, Vol. 139. No. 3, pp. 589 – 614, 2008.

18 Li, W., Zhong, W., "CFD Simulation of Hydrodynamics of Gas-Liquid-Solid Three-Phase Bubble Column", Powder Technology, Vol. 286, pp. 766 – 788, 2015.

19 Ellenberger, J., Krishna, R., "Shaken, Not Stirred, Bubble Column Reactors: Enhancement of Mass Transfer by Vibration Excitement", Chemical Engineering Science, Vol. 58. No. 3-6, pp. 705 – 710, 2003.

20 Ochieng, A., Odiyo, J.O., Mutsago, M., "Biological Treatment of Mixed Industrial Wastewaters in a Fluidised Bed Reactor", Journal of Hazardous Materials, Vol. B96, pp. 79 – 90, 2003.

Finite Element Formulation and Analysis of the Composite Structure of Overhead Transmission lines Conductors

Evans. E. Ojo*
Electrical Power Engineering, Faculty of Engineering and the Built Environment, Durban University of Technology,
P O Box 1334, Durban 4000, South Africa
E-mail: st_evansojo@yahoo.co.uk

Abstract

Mechanical oscillations of overhead transmission conductors occur when conductors receives energy from dynamic forces of nature such as the winding loading. Wind-induced vibration can be in the form of Aeolian vibration, Conductor galloping and Wake-induced vibration. The understanding of the dynamic behaviour of power lines has remained a challenge and many researchers in this field have carried out investigations in order to adequately model this phenomenon. This has proven to be very difficult. The inability to obtain a model the can adequately represent this problem is due to factors like its complex geometry, as well as variables associated with the geometry such as the inter-strand contacts and strands slippage around the contact points resulting in variable flexural rigidity.

This paper presents a developed finite element analysis model for the conductor as a composite structure. The use of the finite element analysis (FEA or FEM), the conductor is modeled using a new modeling approach, in which the layers of its discrete structure of helical strands were model as a bundled structure i.e. assemblies of helical strands, arranged in layers. This model was used to formulate the FEM equations for the systems and also done was the characterization of its self-damping. A compute program was used simulate the conductors response and then the dynamic properties evaluated.

Keywords: Wind-induced vibration, finite-element model, composite Structure, helical strand, self-damping

1. Introduction

Conductors are used in overhead transmission lines for the transfer power. To suit this application, conductor are normally constructed as a bundle of strands wrapped along a helical path around the central core. This means, they are usually constructed from assemblies of strands, arranged in a helical geometry and are in contact with each other either as strands in the same layer or in adjacent layer. In the 1920s, the phenomenon of conductor vibration was observed on overhead power lines which were mainly caused by wind loading. The excitation mechanism is the formation of vortices, at the top and bottom of leeward side of the conductor due to pressure differences caused as the wind flow passed the conductor as shown in figure (1). This thus induces an aerodynamic instability on the conductor and the conductor then oscillates tangentially to the direction of the prevailing wind. There are many forms of motion that result from this aerodynamic effect [1, 2], but the main concern of this study has to do with the vortex-induced vibration also known as Aeolian vibration. When the conductor is subjected to the wind excitation, as it undergo the Aeolian vibrations, with transverse of small amplitude and the displacement corresponding to the conductor bending. In such a motion, individual strands relative slip is restricted, up to a certain point, by friction forces, beyond a critical curvature, the conductor strands slip in relative to strands in the adjacent layer.

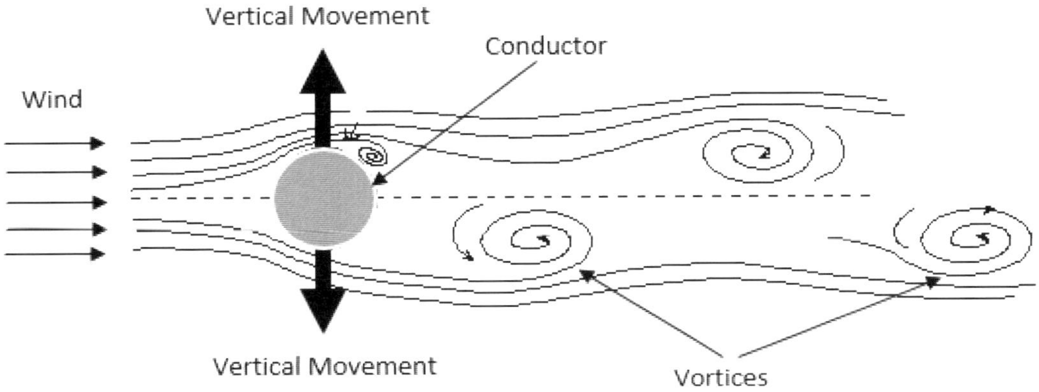

Figure 1: Vortex formation behind a conductor

For Aeolian vibration, the amplitudes of the vibrations are relatively small with maximum amplitude in the order of 1 diameter of the conductor. The frequency of vibration ranges from 5 to 150 Hz. Due to the complex structure of conductor and their nonlinearity in term of geometry, modelling this phenomenon has only been partially achieved analytically. This approach has often resulted in complicated formulae which additionally rely on over simplifying assumptions in order to obtain a closed form solution.

The use of finite element analysis for the predictions of conductor's dynamic behaviour has offered an alternative in modelling this problem. The advantage of using this approach is due to developments in finite element software as well as growth in computer technology, thus, in this respect have been an encouraging means of implementing the FEM.

The paper present an approach in modelling conductor structure as a bundle using the finite element analysis with the goal to simulate its transverse vibration. This can be used to obtain variable such as natural frequency, mode shapes and damping. The FEM predictions are compared with proven analytical method and test results.

2. The Concept of Finite Element Analysis

Finite element analysis (FEA) is a numerical tool that is implemented based on the principle of discretization and numerical approximation in order to provide an approximate solution to scientific and engineering problems. FEA is an important tool for scientific and engineering research. It is one of the most important numerical techniques for obtaining approximation solution to complicated geometric problems which is difficult or impossible to obtain a closed form solution. The application of this method ranges from deformation, flux and stress analysis for both static and dynamic systems. To implement this tool, the structure under consideration is discretized into smaller geometric domain and the material properties are analysed over these subdomain elements known as finite elements. The assemblies of these finite element can be used to find solution to the entire system. An assembly process duly considering the loading and constraints results in a set of equations. The global mass matrix and stiffness matrix of the conductor system as composite are obtained using the finite element method (FEM).

The purpose of this project is to use the FEA approach analyzing the conductor vibration that occurs in power lines. A finite assemblage of discrete elements leads to the formulation of the associated eigenvalue equation in the deformed configuration in matrices forms. The equation of motion of the whole vibrating system is then defined. Solution of the equation gives the approximate behaviour of the conductors. This equation was simulated to provide the natural frequencies and the mode shapes of the system.

3. Description of the Conductor Geometry

Conductor manufacturers produce various forms of conductors that come with different numbers of layers, strands sizes and materials that are applicable to various power transfer conditions. The conductors chosen for this FEA analysis are those with helical strand having a circular cross-section.

In such a conductor, the centreline of generic strands can be described by a circular helix path, with radius R, and lay angle, α.Typical cross-section of the conductor is made of strands, arranged as represented in figure (2). Cross-sections of the strands are usually round shaped. Consider figure (2), which is a cross-section of a conductor, made up of strands of circular cross-section. The stranded bare conductor cross-section with total a cross-section area A_T, consisting of i-layers ($i = 0, 1, 2, \ldots, N$), of strands of radius r_i ($d_i = 2r_i$) with lay angle α_i, wrapped over the centrally located core strand, of radius r_0 ($d_0 = 2r_0$).

Each particular layer consists of n number of strands. Using the compact arrangement, the number of strands in each layer will be defined as $n_i = 0, 6, 12, 24, \ldots$, The position of each strand in a given layer is defined as $P(i, j)$, where ($j = 0, 1, 2, \ldots, n_i$). $P(0, 0)$ defines the position of the core strand and others positions of strands in each layer are define on yz-plane in figure (2) in the anti-clockwise direction along the x-axis. These positions strand as located at its cross-section can be determined by:

$$P(i, j) = R_i \cos \varphi_i \qquad \text{......................} (1)$$

Where $\phi_i = 0 : \dfrac{360^o}{n_i} : 360^o$

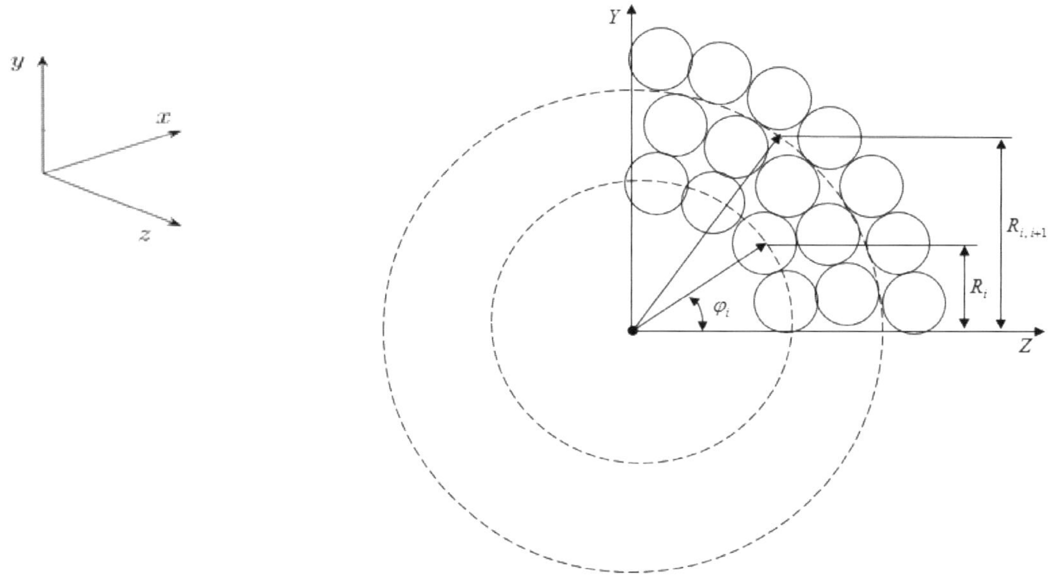

Figure. 2: The conductor cross-section.

The outer radius and the conductor cross-section can be calculated respectively by

$$r_N = \frac{d_N}{2} + \sum_{i=1}^{N-1} d_i + \frac{d_0}{2} \qquad \text{......................} (2)$$

$$A_T = \pi (r_N)^2 \qquad \text{......................} (3)$$

The radius between layers and the radius of the circular path that passes through the centre of each strands located at i-layer (excluding single layer conductor) from the origin, are calculated by

$$R_{i,i+1} = r_0 + \sum_{i=1}^{N-1} d_i \text{ and } R_i = r_0 + \sum_{i=1}^{N-1} d_i + r_{i+1} \qquad \text{......................} (4)$$

In the case of a single layer conductor, these values are obtained as:

$$R_{o,1} = r_0 \text{ and } R_1 = r_0 + r_1 \qquad \text{......................} (5)$$

For the strands arrangement in various layers, parallel to the x-axis, where the distance for the strand centre are defined from the centre of the core. The position of the centre of each strand in a given layer and the distance along these centres in the curvilinear axis of helix from the conductor neutral axis which as located parallel to the x-axis are given as:

$$R_{i,i+1} = r_{i,i+1}\sin(\varphi_i) \text{ and } R_i(x) = r_i \sin\left(\frac{2\pi x}{P_L} + \varphi_i\right) \quad \text{.........................} (6)$$

Where φ_i incremental helix angle (wrap angle) is the angular position from the z-axis in the anti-clockwise direction and P_L is the pitch length.

4. General FEM Modelling Approach

Conductors are commonly modelled as a distributed system in which the conductor is assumed to be homogeneous, and isotropic in nature with continuous parameters [3-6]. Most FEM formulation for the conductor found in literature were mostly done with the straight beam model. This has restricted to the use of the classical beam theory formulation for the conductors as an initially straight beams of constant cross-section. This model has been extensively been used and has been implemented by several researches investigating conductor dynamics [6, 7]. This simplified model can be accepted in modelling the conductor to some degree of accuracy, but in real problems, where the curvature becomes significant, this model becomes inadequate. In this case for the conductor, the linear variation of strain over the cross section is no longer valid. This clearly makes it unfeasible to model the conductor strands using the curved beam element. The individual strands can therefore be modelled as beam sections integrated with nonlinear material behaviour.

Many researchers [7, 8, 9, 10] have chosen to generate meshes for conductors by using the beam elements due to the relative ease of meshing and its implementation. The advantage of these models is that it can easily be implemented in a one dimensional topology. This model simplify the conductor model as continuous solid structure, this is in contrast to the true nature of the conductor.

Also, some researchers [4] have done some FEM using commercial software in which the helical (rod or beam) library is used to build the conductor structure and then implemented. In the implementation phase, a circular beam section is chosen from the library for the beam section shapes and the section's dimensions can then be defined, in this case, a circle with a radius. A material definition is used to define the material properties of the section. A linear elastic material of stress-strain relation is assumed. The problem with these forms of solutions is that because these softwares uses a generic algorithms, this poses a problem with the uniqueness and convergence. This is because these softwares are designed like a black box, thus implementing a user developed algorithm is virtually impossible.

In order to develop a more adequate representation of the conductor as a bundle, the use of curved beam model becomes appropriate. Furthermore, curved beams are known to be more efficient in transfer of loads than straight beams, because the load transfer is affected by bending, shear, and axial action. Thus, the conductor strands as the form of the sub-structure were modelled by the curve truss beam or the curved beam element. Therefore, in this paper the use of the curved beam model for the conductor strands is presented which its assemblies was used to modelled the conductor as a composite structure.

5. FEM Composite Structure Modelling of Conductors

The conductor can be modelled as a collection of helical strands. This form of conductor structure, as a bundle is referred to as "the composite structure" of a conductor. Contrast to the approach discussed in the last section, the conductor actually is composite structure formed by the assemblies of helical strands in successive layers. Based on this fact, formulating the conductor in one dimension as one continuum does not reflect the actual geometry of the conductor. The form of modelling of conductor as a composite structure enables the analysis of the concepts associated with the internal geometry of conductors such as contact and frictional effect around the contact regions. Hence, a two or three dimensional (2D or 3D) topology for the formulation of the problem becomes the best option. The major problem of adopting higher dimension topology is the difficulty of generating the finite element meshes that can accurately approximates the domain and also the efficient integration numerically with the boundary conditions to obtain the FEA model.

5.1. FEM Modelling Concept

To find the solution in a closed form for the complicated bundle structure of conductors is very difficult and almost impossible. This is due to the fact that the formulation and integration of these equations as a bundle is general more complicated than the solution of a single ordinary partial differential equation, as an example is the governing equation for an equivalent single continuous beam model for the conductor. The use of the discretised continuous beam FEM model is a simplified model which may be used to represent the conductor. Due to the mathematical simplicity, the dynamic analysis of conductor as a solid continuous system has limited use in practice. Contrary to the use of equivalent beam model, the composite model of the assemblies of strands was developed. This method of analysis of conductor structure is very complex because each conductor strand model is model as a beam and subjected to the coupling effect of axial, torsional, and bending. This composite structure concept for the conductor was achieved by the iso-parametric interpolation of the referenced curve thin beam for each individual strands.

5.2. The Curved Beams Model

In this study, the Euler-Bernoulli curved beam theory was used, thus there was still the assumption of plane cross sections remaining plane after deformation is still valid. The *xy*- longitudinal plane has been assumed to be the plane of symmetry with the bending load being applied. The conductor strands deformations are symmetric; this is why the helical strand can be regarded as an example of a curved beam with restrained sections at the ends. From the conditions of symmetry, the distribution of stress in one quadrant of the helical strands can be balanced by the same continuous strand in the opposite quadrant.

Consider figure 3, let any point on the cross-section of the beam be displaced from its unstrained position through small components of displacements v and u representing the tangential and radial displacements of points on the cross section, and y represent the transverse displacement in a direction normal to the plane of the beam. The transverse displacement at any point (x, y) may be expressed directly in term of θ(x) which is the rotation of the normal to the centroidal axis due to flexural effect, and u(x) which is the axial displacement due to the axial effect and the curvature. The axial displacement for the curved beam is given as:

$$u(x,y) = u_0(x) + \frac{v}{R} - y\left(\theta_x + \frac{u(x)}{R}\right) \quad \text{.......................... (7)}$$

Where $\boldsymbol{y}$ is the distance from the centroidal axis of the cross section to the point (x, z). The normal rotation θ(x) is equal to the slope of the centroidal axis minus a rotation and can be obtained as:

$$\theta_x = \frac{\partial v(x,y)}{\partial x} \quad \text{.......................... (8)}$$

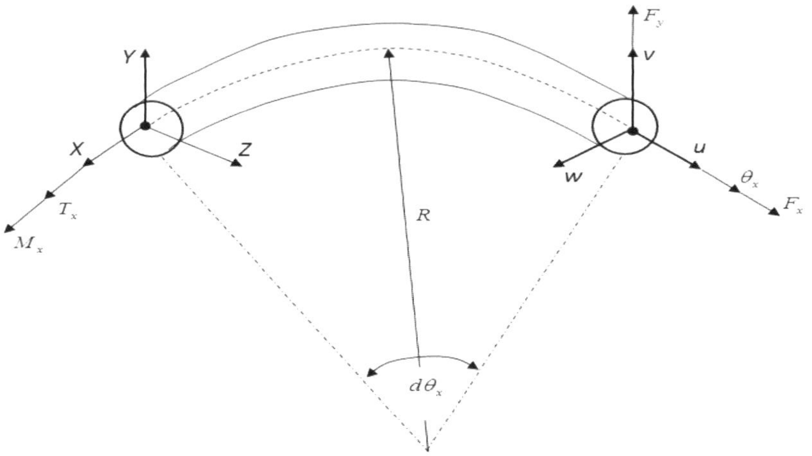

Figure 3: The curved beam

The axial strain equation is given as:

$$\varepsilon = \frac{\partial u}{\partial x} \qquad \text{......................(9)}$$

If the curved beam obeys Hook law, the stress-strain relation can be obtain by multiplying the strain by Young's Modulus to yields the stress equation as:

$$\sigma = E\frac{\partial u}{\partial x} \qquad \text{......................(10)}$$

The axial force and moment can be formulated long the curved length, s, using the following:
The axial tension becomes:

$$T_x = \int_A \sigma dA = E\frac{\partial u}{\partial s} \qquad \text{......................(11)}$$

The moment becomes:

$$M_x = -\int_A z\sigma dA = EI\frac{\partial^2 u}{\partial s^2} \qquad \text{......................(12)}$$

Using the Euler-Bernoulli curved beam and the shear strain is neglected. Therefore, since we are assuming the beam is elastic, the relevant strains in this case are the axial strain and the bending strain, where the bending strain now depends only on the radial displacement, u. The axial and bending strain are given as:

$$\varepsilon_A = \frac{du}{ds} - \frac{v}{R} \qquad \text{......................(13)}$$

$$\varepsilon_B = \left(-\frac{d^2 v}{ds^2} - \frac{du}{Rds}\right) + \left(-\frac{dv}{ds} - \frac{u}{Rds}\right) \qquad \text{......................(14)}$$

The mathematical expression for Hamilton's principle is given as:

$$\int_{t_1}^{t_2} \left(\delta(T-U) + \delta W_{nc}\right)dt = 0 \qquad \text{......................(15)}$$

The axial strain energy stored in the element is given as:

$$U_A = \frac{1}{2}\int_{-1}^{1} \sigma_x \varepsilon_x Adx \qquad \text{......................(16)}$$

Using the relation defined in equation (13)
The axial strain energy stored in the element is given as

$$U_A = \frac{1}{2}\int_{-1}^{1} EA\left(\frac{du}{ds} - \frac{v}{R}\right)Adx \qquad \text{......................(17)}$$

The bending strain energy stored in the element is given as:

$$U_B = \frac{1}{2}\int_{-1}^{1} z\sigma\left(\frac{du}{ds} - \frac{v}{R}\right)Adx$$

Using the relation defined in equation (14):
The bending strain energy stored in the element is given as:

$$U_B = \frac{1}{2}\int_{-1}^{1} EI\left(-\frac{d^2 v}{ds^2} - \frac{du}{Rds}\right) + T\left(-\frac{dv}{ds} - \frac{u}{R}\right)Adx \qquad \text{......................(18)}$$

The kinetic energy of the beam element is given as:

$$T = \frac{1}{2}\int_{-1}^{1} \rho A\left(\dot{u}^2 + \dot{v}\right)dx \qquad \text{......................(19)}$$

The virtual work for the element is express as:

$$\delta W = \frac{1}{2}\int_{-1}^{1} F_x \delta u dx \qquad (20)$$

The total energy and work for the curved beam element equations was therefore obtained by substituting equations (16-19) into equation (15). The total energy and work is obtained as:

$$\int_{t_1}^{t_2}\left[\int_{-a}^{a} \rho A \frac{dv}{dt}\left(\delta \frac{dv}{dt}\right)dx - \int_{-a}^{a} E d\varepsilon(\delta\varepsilon)dx + \int_{-a}^{a} P_y \delta v dx\right]dt = 0 \qquad (21)$$

8.1 Discretization using Shape and Trial Functions

The discretization of the conductor structure entails converting its continuous body containing infinite number of nodal points i.e. infinite degree of freedom into a discrete model with a defined number of nodes at the ends of the sun-domain. The discretization of structure are usually modelled using the shape functions. The approximating function also known as the basis function are defined in terms of the values of the field variables at the nodes, for the beam, thus a polynomials was used.

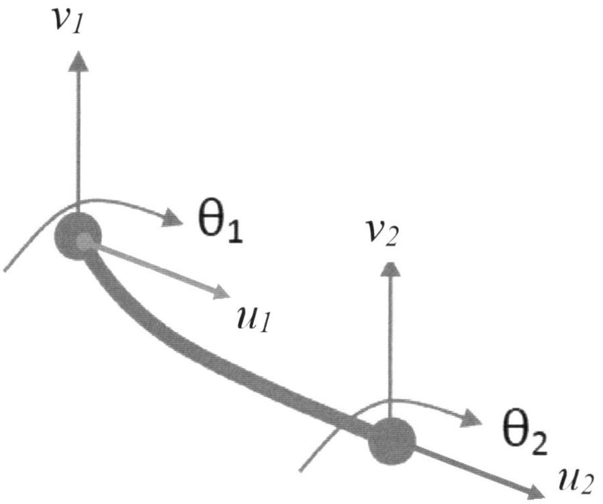

Figure 4: The curved beam finite element

For the curved beam shown in figure 4, there are three variables *per node* that are used to define the element behaviour. At a typical node is defined by displacements and a rotation $[u_i\ v_i\ \theta_i]$ where $i = 1, 2$ defines the axial, rotation and bending components at the node. The shape functions that were defined for the axial, transverse and rotation at the nodes, are expressed as follows:

$$u(s) = b_0 + b_1 s$$

$$v(s) = c_0 + c_1 s + c_2 s^2 + c_3 s^2 \qquad (22)$$

$$\theta = \frac{dv(s)}{ds} = c_1 + 2c_2 s + 3c_3 s^2$$

In a matrix form, the equation (22) can be represented for u and v:

$$\begin{bmatrix} u \\ v \end{bmatrix} = \begin{bmatrix} 1 & s & 0 & 0 & 0 & 0 \\ 0 & 0 & 1 & s & s^2 & s^3 \end{bmatrix} \begin{bmatrix} b_0 \\ b_1 \\ c_0 \\ c_1 \\ c_2 \\ c_3 \end{bmatrix} \qquad \text{.........................(23)}$$

Solving for the constants and substituting into equation (23), the shape functions equation was obtained:

$$\begin{bmatrix} v \\ u \\ \theta \end{bmatrix} = \begin{bmatrix} N_1 & 0 & N_2 & 0 & 0 & 0 \\ 0 & N_3 & 0 & N_4 & N_5 & N_6 \\ 0 & 0 & N_4 & & N_5 & N_6 \end{bmatrix} \begin{bmatrix} v_1 \\ u_1 \\ \theta_1 \\ v_2 \\ u_2 \\ \theta_2 \end{bmatrix} \qquad \text{.........................(24)}$$

The values for the shape function as given in equation (24) are defined as follows:

$N_1 = \frac{1}{2}(1-\xi)$ $N_2 = \frac{1}{2}(1+\xi)$ $N_3 = \frac{1}{4}(2-3\xi+\xi^3)$ $N_4 = \frac{1}{4}(1-\xi-\xi^2-\xi^3)$ $N_5 = \frac{1}{4}(2+3\xi-\xi^3)$

$N_6 = \frac{1}{4}(-1-\xi+\xi^2+\xi^3)$

Thus, the axial and transvers displacement can then be evaluated as:

$$u(s) = N_1 u_1 + N_2 u_2 \qquad \text{.........................(25a)}$$

$$v(s) = N_3 v_1 + N_4 \theta_1 + N_5 v_2 + N_6 \theta_2 \qquad \text{.........................(25b)}$$

Since we have three degrees of freedom per node, that is v u and θ, then the axial, bending, strains can be define using equations (13) and (14) as

$$\varepsilon_A = \hat{N}_A d \text{ and } \varepsilon_B = \hat{N}_{B,EI}\, d + \hat{N}_{B,T} \qquad \text{.........................(26)}$$

Therefore, the element stiffness matrix was form as a contribution from the axial and bending stiffness, such that the expression for the strand stiffness was given as:

$$[K] = \left[\frac{d}{2}\int_{-1}^{+1} \hat{N}_A^T (EA) \hat{N}_A \frac{d}{2} d\xi\right] + \left[\frac{d}{2}\int_{-1}^{+1} \hat{N}_{B,EI}^T (EI) \hat{N}_{B,EI} \frac{d}{2} d\xi\right] + \left[\frac{d}{2}\int_{-1}^{+1} \hat{N}_{B,T}^T (T) \hat{N}_{B,T} \frac{d}{2} d\xi\right] \qquad \text{.......(27)}$$

Substituting the displacement functions of equation (25) into the energy function of equation (19), the mass matrix expression was obtained as:

$$[M] = \int_{-1}^{+1} \frac{1}{2}\rho A(\dot{u}^2 + \dot{v}^2)dx = \frac{1}{2}\int_{-1}^{+1} \dot{u}^T \rho A \dot{u} + \frac{1}{2}\int_{-1}^{+1} \dot{v}^T \rho A \dot{v} \qquad \text{.........................(28)}$$

The load vector expression was given as:

$$\delta W = \frac{1}{2}\int_{-1}^{1} F_x \delta u dx \qquad \text{.........................(29)}$$

To implement iso-parametric mapping, the integration scheme that was employed for the numerical integration and the Gauss-Legendre numerical was used. Thus, natural coordinates range from -1 to +1, for a function f(x), was used to obtain the integral:

$$I = \int_{-1}^{+1} f(x)\,dx \qquad \text{.......................... (30)}$$

8.2. Iso-parametric Mapping

After the successful formulation of the finite element equations, then followed was the formation of the global model in terms of global Cartesian coordinates. Based on the derivation of the curved beam model, the finite element modelling of the conductor as a composite structure is new approach. This formulation entails the mapping of the finite element into the discretized conductor helical layered structure and the finite analysis model was developed for the conductor structural system on of the basis of the mapping of the normalized single curved beam. To implement this, all mapping is done with respect to the conductor neutral axis which was assumed to coincide with the centre line of the core. To carry out the iso-parametric mapping for the strands in each layer was developed with cylindrical coordinate systems along the curvilinear helical path defined for each strand path.

9. The System Equation

The geometric mapping and the coordinate transformation was used to assemble the element equations for mass, stiffness and load vector for the system. The assembled equation is the global equation system for the whole solution region. In other words, the system equation was obtained by the assembly of the local element equations for all elements used for the discretization. The elements connectivities were used for the assembly process. The above transformation resulted to the damped equation for the transverse vibration of conductor. The system matrix equation was obtained as:

$$\begin{bmatrix} M_{11} & M_{12} \\ M_{21} & M_{22} \end{bmatrix} \begin{pmatrix} \ddot{u} \\ \ddot{d} \end{pmatrix} + \begin{bmatrix} C_{11} & C_{12} \\ C_{21} & C_{22} \end{bmatrix} \begin{pmatrix} \dot{u} \\ \dot{d} \end{pmatrix} + \begin{bmatrix} K_{11} & K_{12} \\ K_{21} & K_{22} \end{bmatrix} \begin{pmatrix} u \\ d \end{pmatrix} = \begin{pmatrix} F_1 \\ F_2 \end{pmatrix} \qquad \text{.......................... (31)}$$

At this stage of investigation, the damping matrix in equation (31) was obtained by assuming the proportional damping i.e. the expression for the damping matrix as a linear combination of the mass and stiffness matrices, obtained as,

$$C = \alpha[M] + \beta[K] \qquad \text{...................... (32)}$$

Before solution, boundary conditions were imposed. However, the beam does not experience any bending moments, and was free to rotate; therefore the bending moments were zero at both ends, hence the assumption it was simply supported at both ends was used. To carry-out the vibration analysis for the conductor, the direct integration method was be used. This method employs a step-by step numerical integration method for the evaluation of the dynamics response and its algorithms usually assume continuity of displacements, velocities and acceleration. This form of analysis was done base the Newmark integration scheme [11 12]. Representing equation (31) in a more compact form as employed for the dynamic analysis, with this equation for the system expressed as:

$$[M]\{\ddot{y}\} + [C]\{\dot{y}\} + [K]\{y\} = [F] \qquad \text{.......................... (33)}$$

10. FEM Implementation and Results

The development of FEM modelling for the conductor as a composite structure is still ongoing process. The first results are promising and it looks like the conductor global properties can be predicted reasonably precise. However a detailed study about how parameters like the material definition, the mesh density, the finite element size and the numerical step time influences the results is still an ongoing investigation.

The experiments to validate the results for this study were conducted in an indoor laboratory. This indoor laboratory is the Vibration Research and Testing Centre (VRTC) which is situated at the University of KwaZulu-Natal, Westville campus. This experimental tests were relatively complicated and cumbersome and therefore not very often undertaken. The conductor used for this study is the ACSR conductor with the code name Pelican. The Pelican has two layers made of aluminium as shown in figure (5), with the centre core made of steel. Table 1 reports the main physical characteristics of the conductors. The bending stiffness values are given as: Minimum Bending Stiffness (EI_{min}): 8.66 N. m^2; Maximum Bending Stiffness (EI_{max}): 433 N. m^2.

Conductors	Pelican
Total area(mm^2)	255.77
Conductor diameter (mm)	20.70
Number of wires on each layer Steel	1
Number of wires on each layer Aluminium	6-12
Diameter of aluminium wires (mm)	4.14
Diameter of steel wires (mm)	4.14
Linear mass (kg/m)	0.771
Rated tensile strength (kN)	51.15

Table 1: Physical Properties of Pelican Conductor

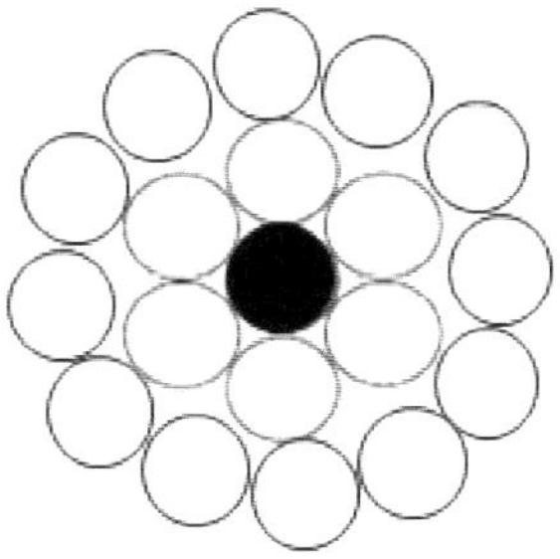

Figure 5 Pelican conductor

A sweep test was conducted. The sweep test was made to resonant the conductor at the natural frequencies and then identify the dominant resonance frequencies as shown in figure (6) for the Pelican conductor at 20 % UTS.

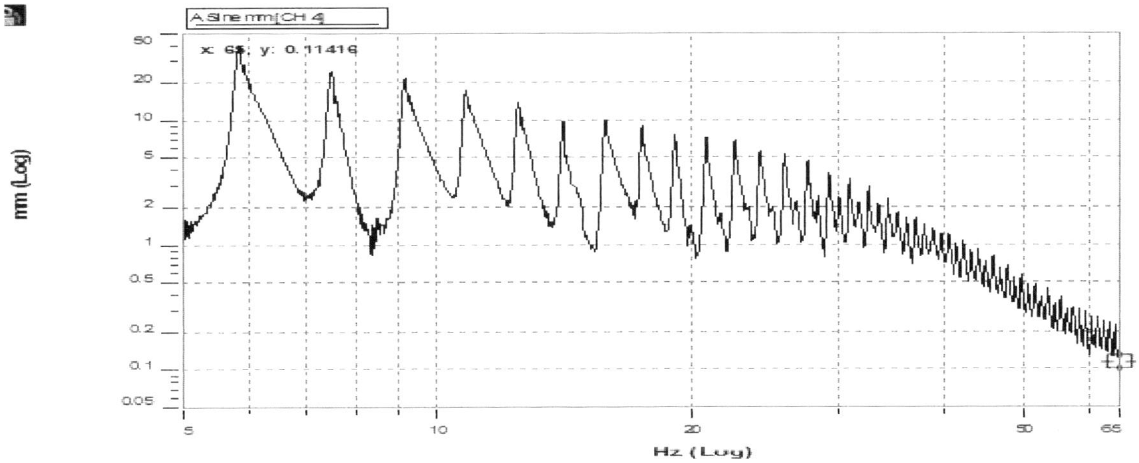

Figure 6: Sweet test graph done Pelican conductor at 20% UTS

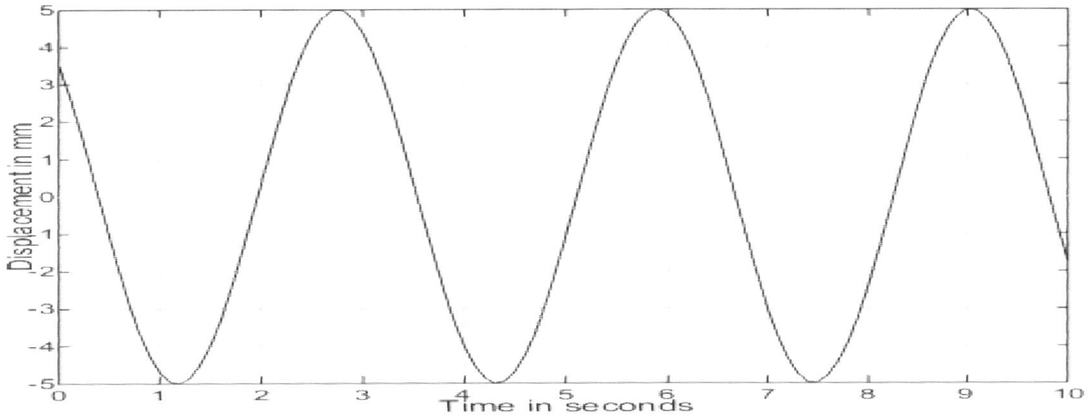

Figure 7: The conductor steady state response

Based on the Matlab code developed above, the FEM model was simulated in the Matlab environments. The dynamic response of the conductors simulated with the forced harmonic excitation with the frequencies using the Newmark numerical scheme. For reliability of results the analytical frequencies was used as a comparison and these FEM frequencies were based on similar system used for the laboratory tests. The FEA response of particular modes as a function of time can be seen in figure (7).

The various eigen frequencies were computed by the FEM analysis for the composite model. Tables 2. Showed the comparison between the natural frequencies obtained from the analytical, FEA models in comparison with experimental results.

Natural Frequency (Hz) 20 % UTS			Natural Frequency (Hz) 25 % UTS			Natural Frequency (Hz) 30 % UTS			Natural Frequency (Hz) 35 % UTS		
Analyt. Model	FEM	Exp. value	Analyt. Model	FEM	Exp. value	Analyt. Model	FEM	Exp. value	Analyt. Model	FEM	Exp. value
4.311	4.375	5.058	4.819	4.892	5.058	5.277	5.3591	5.692	4.930	5.680	5.948
8.66	8.752	9.053	9.676	9.785	8.924	10.590	10.718	10.014	9.861	11.362	11.898
13.106	13.130	13.320	14.610	14.679	14.067	15.973	16.079	14.734	14.793	17.044	17.848
17.674	17.510	17.717	19.659	19.575	19.373	21.461	21.442	19.934	19.727	22.728	23.801
22.408	21.893	22.183	24.858	24.474	25.622	27.087	26.80	25.110	24.663	28.415	29.756
27.346	26.279	26.908	30.241	29.376	31.349	32.882	32.175	30.283	29.601	34.106	35.714
29.890	30.671	31.123	32.009	34.282	35.167	37.002	37.548	34.000	34.544	39.800	41.677
34.678	35.068	35.965	36.234	39.194	39.976	42.123	42.924	38.987	39.490	45.499	47.646
38.789	39.472	39.893	40.123	44.111	44.789	47.689	48.307	43.895	44.442	51.205	53.620
43.001	43.883	44.103	45.234	49.036	49.789	53.211	53.696	48.789	49.401	56.918	59.603

Table 2: The comparison of the natural frequencies from the analytical, FEM and experimental studies

As tabulated, the steady state frequency responses were obtained for the analytical, FEM and the experimental studies. Comparison, the results showed a good agreement between the analytical and the FEM frequencies for the four different axial tensions as percentages of ultimate tensile strength (UTS). Still under investigation is the dynamic analysis to ascertain if the damping energy is small enough to neglect its influence on the conductor dynamic behaviour.

11. Conclusion

The FEM formulation, implementation was done using a more realistic model for the conductor as composite structure using the iso-parametric interpolation. The paper shows the modelling of conductor's dynamic analysis using curved beam element for the representation for the individual strands. This method allows the construction of the complex geometry and efficiently modelling the entire conductor as composite. As an example a conductor with code name Pelican was chosen and the comparison with calculations analytically, FEM and the experimental behaviour shows good correlation. This work forms part of the goal to develop a finite element toolbox which will specifically designed to model and analyse the static and dynamic behaviour of power line conductors

Acknowledgment

The authors gratefully acknowledge the support of this work by the Research office of Durban University of Technology, Durban, South Africa.

References

[1] E.S Duocy et al, "Transmission line reference book: wind-induced conductor motion," Gilbert/Commonwealth, Jackson, MI (USA)1979.

[2] EPRI, *Wind-Induced Conductor Motion:* : Transmission Line Reference Book, 2006.

[3] R. Claren and G. Diana, "Mathematical analysis of transmission line vibration," *IEEE Transactions on power apparatus and systems,* vol. 12, pp. 1741-1771, 1969.

[4] E. E. Ojo, "Dynamic Characteristics of Bare Conductors," University of KwaZulu-Natal, Durban, 2011.

[5] C. B. Rawlins, "Flexural self-damping in overhead electrical transmission conductors," *Journal of Sound and Vibration,* vol. 323, pp. 232-256, 2009.

[6] T. Slethei and J. Huse, "Conductor vibration—theoretical and experimental investigations on a laboratory test span," in *Proceedings of the Institution of Electrical Engineers*, 1965, pp. 1173-1179.

[7] N. Barbieri, O. H. de Souza Júnior, and R. Barbieri, "Dynamical analysis of transmission line cables. Part 1—linear theory," *Mechanical Systems and Signal Processing,* vol. 18, pp. 659-669, 2004.

[8] A. Nawrocki and M. Labrosse, "A finite element model for simple straight wire rope strands," *Computers & Structures,* vol. 77, pp. 345-359, 2000.

[9] J.-B. Dastous, "Nonlinear finite-element analysis of stranded conductors with variable bending stiffness using the tangent stiffness method," *Power Delivery, IEEE Transactions on,* vol. 20, pp. 328-338, 2005.

[10] E. E. Ojo and S. Shindin, "Finite Element Analysis of the Dynamic Behaviour of Transmission Line Conductors Using MATLAB," *Journal of Mechanics Engineering and Automation,* vol. 4, 2014.

[11] N. M. Newmark, "A method of computation for structural dynamics," *Journal of the engineering mechanics division,* vol. 85, pp. 67-94, 1959.

[12] K.-J. Bathe and E. L. Wilson, "Numerical methods in finite element analysis," 1976.

THE EVALUATION OF 3D CFD APPROACHES TO MODEL THE FLOW AND HEAT TRANSFER INSIDE AN UNSTRUCTURED BED OF UNIFORM SPHERES

MC Potgieter[1*], CG du Toit[1], JH Kruger[1]

[1]North-West University, Potchefstroom, South Africa

[*]20268890@nwu.ac.za

ABSTRACT

The physical phenomena associated with the modes of heat transfer within a packed pebble bed can be modelled by explicitly solving the associated partial differential equations using a finite volume method approach. Convection from the spheres to coolant, conduction between spheres as well as radiation between various components are the main modes of heat transfer with several lesser, albeit contributing phenomena also present. A representative bed was generated with DEM and the flow domain modelled using the CFD software package Star-CCM+ taking into account all primary modes of heat transfer as well as a specified approach to take secondary effects into account. By using such a fully integrated explicit model, a deeper understanding may be gained into the role of the underlying phenomena and their respective contributions to the overall heat transfer process.

Effects in the bulk, near-wall and wall contact regions of an unstructured packed bed vary and must be taken into account. A smaller representative cylindrical packed bed was simulated using three approaches to fully account for all flow phenomena and heat transfer occurrences. The best approach to model a larger section of annular packed bed as found in the PBMR-400 core is then evaluated. The pressure drop, velocity distribution as well as temperature distributions in the spheres and fluid were calculated and compared with experimental results from published literature.

KEYWORDS

Computational Fluid Dynamics, Discrete Element Method, Unstructured packed bed, Pebble bed, Heat transfer, Effective thermal conductivity.

INTRODUCTION

Modern nuclear power system designs must adhere to the highest standards for safety to be considered for demonstration and implementation. The locally designed and developed PBMR-400 is a Generation IV graphite-moderated gas-cooled 400 MWth reactor, capable of reaching very high core outlet temperatures. Fuel elements in the annular core consist of 330 000 fuel pebbles containing TRISO particles and an additional 110 000 graphite spheres [1]. The hydrodynamic and heat transfer phenomena present in an unstructured bed of spherical fuel elements have previously been studied with the German AVR and THTR-300. Observations from the operation of these reactors formed a benchmark for the requirements as set by the Nuclear Safety Standards Commission (KTA) through several standards for high temperature reactors (HTR) with pebble bed cores. Whilst adequate at the time and for the required purpose, these standards modelled or described the core using a purely implicit approach.

Advances in computational power and computational fluid dynamics (CFD) software capabilities have made it possible to model problems with complex geometry and integrated physics with greater ease and higher accuracy. By modelling a section of pebble bed core by using explicit methods, greater understanding and insights might be gained of the underlying phenomena contributing to the overall pressure loss and heat transfer in a pebble bed core. This in turn allows the adjustment of less computationally-intensive implicit methods when applied to an entire core, the results of which ultimately result in better compliance with required safety standards.

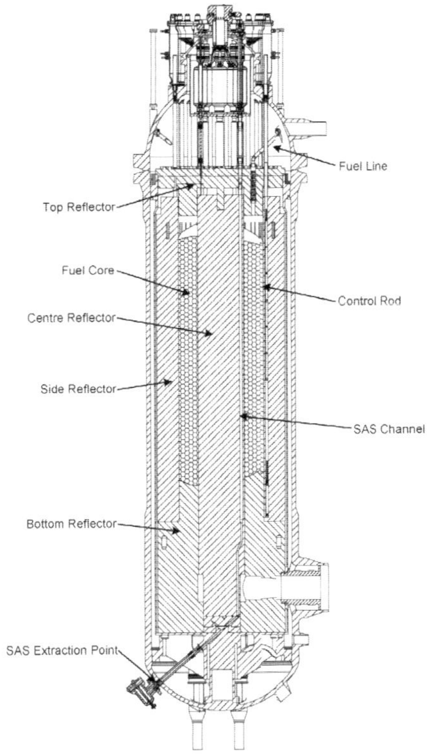

Figure 1: Vertical schematic section of the PBMR-400 reactor unit [2].

The study outlined in this article is concerned with modelling a representative section of unstructured packed bed with heat generated within the spheres and forced convective heat transfer due to coolant flow over the bed. Three possible methods of modelling using the CFD software package Star-CCM+ are explored and compared. The primary modes of heat transfer are modelled explicitly with the relevant physics models available in the software. For each method, the pressure drop and temperature distribution in the bed is determined and compared to that found in larger beds from relevant literature.

PACKED BED GEOMETRY AND PARAMETERS

The unstructured annular packing as found in the PBMR-400 pebble bed core can be seen in Fig. 1 which illustrates the core within the reactor pressure vessel. To investigate different approaches to simulate such an unstructured bed within a reasonable timeframe, a small scale cylindrical packed bed representative of a larger section of core, was generated using a DEM packing in the CFD software package Star-CCM+. A three dimensional view of this packing is shown in Fig. 2 (left). The computational grid characteristics for each of the approaches to be discussed remains the same, except in the large cell DEM. A mesh independence study was undertaken for each approach for validation, but is not reported on in this article. As future work the best approach identified here may consequently be applied to a larger annular bed section in order to better represent the actual reactor core.

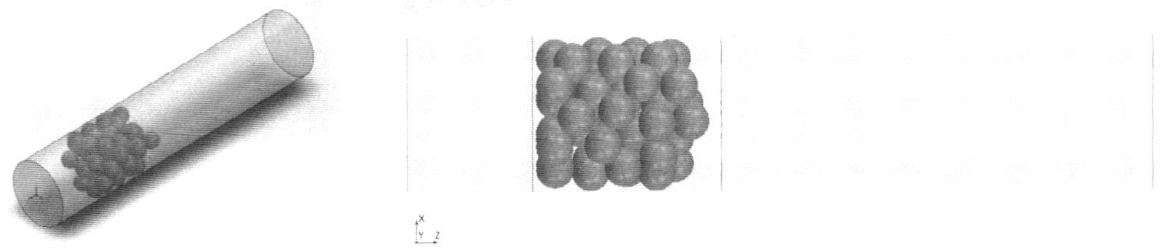

Figure 2: Three-dimensional (left) and side-view of cylindrical geometry (right).

The boundary geometry used with each approach was kept the same. It consisted of a cylinder with diameter 240 mm and a sufficient length to allow flow to fully develop before and after the bed section with spheres. The packed section had a length of 300 mm and was filled to contain 50 uniform spheres, each with a diameter of 60 mm. After packing with the discrete element method, considering Lagrangian particles with appropriate mechanical interactions using material properties of nuclear graphite, the centroids of each sphere was recorded. This in turn was used to generate geometry in a CAD package to be used in the discrete approach. The resulting bulk porosity after packing was found to be 0.581 and was required as input value for the porous medium approach.

Table 1: Boundary values of packed bed section for all cases.

Number of spheres (not applicable to 3rd approach)	50
Porosity ε (from discrete approach)	0.581
Bed section aspect ratio (d to D)	4*
Cylinder inlet temperature	250 °C
Cylinder inlet mass flow	1.182 kg/s
Outer cylinder boundary	Smooth, adiabatic surfaces

An energy source for the spheres was chosen similar to that found in the PBMR-400 core. For 330 000 spherical fuel elements containing 15 000 TRISO particles with 7.5 % enrichment, the heat generated per pebble on average during the fuel cycle is roughly 1.2 kW. For this study the heat generated per 50 spheres was taken as 58 480 W, gradually ramped up to this value after a certain amount of time steps. This equals an energy density of 10.34 MW/m^3.

*The KTA Safety Standard 3102.3 states that the minimum acceptable aspect ratio or bed diameter ratio D/d is limited by the modified Reynolds number [3]. In the case of the models considered where the modified Reynolds number $\mathrm{Re}/(1-\varepsilon) > 10^4$ then the minimum ratio is 4.

MODELING APPROACHES

Three approaches were investigated through which the section of bed could be modelled in STAR-CCM+ (version 11.02.009 with mixed precision was used in the study). For each, the basic container geometry remains the same and the way in which the spheres are represented mathematically, varies. The thermal contact resistance between spheres is not treated in detail for the sake of brevity. It is, however, of importance for larger discrete geometries and different methodologies to incorporate these resistances are discussed in literature [4, 5]. The three approaches are:

- DEM modeller with two-way coupling approach
    - Large Cell
    - Large Particle
- Discrete geometry approach
- Porous medium approach

Each approach is discussed in this order, providing background information to the method and its applicability. The goal is to model a packed bed with approaches with varying degrees of accuracy and evaluate the best method for achieving this in the least amount of time.

i) DEM with two-way coupling

In the first approach, the DEM modeller is firstly used to generate a packed bed of 50 spheres in the cylindrical container. Their positions within the fluid domain and the volume of cells being occupied are then used to calculate the porosity for the particular region on a per-cell basis. The DEM modeller "imprints" the position of spheres onto the computational grid in terms of porosity. For instance, if a cell in the continuous phase is occupied by a sphere element (from the discrete element solver); it is assigned porosity > 0. This value may be limited to a set threshold in order to avoid numerical instability. Any hydraulic calculations are then performed by the flow solver using this porosity value.

Secondly, the velocities of the spheres elements are frozen whilst all other models remain enabled. A two-way coupling solver is enabled which allows interaction amongst particles themselves and between the particles and fluid domain. This allows energy transfer to take place according to equations for conduction and convection. The energy source term per sphere element is distributed to cells which contain the particular sphere, after inter-particle heat transfer has been considered. This iterative process results in heat transfer being calculated explicitly and momentum transfer implicitly, relative to the bed. With this method it is possible to generate a large bed and solve both the pressure drop as well as primary modes of heat transfer in a short amount of time.

The DEM solver in Star-CCM+ requires that the size of cells in the continuous phase be larger than particles for stability, especially where two-way coupling is done. For the instances where there are only a few particles considered or the size of particles relative to the bed is such that this results in a very coarse grid, there is an option to imprint an element over several cells. This option to "smooth large particles into cells" is available at the expense of requiring much greater computational resources.

There are therefore two possible sub-methods for application of this approach. Both these cases, one where the smoothing option was enabled and used with a fine computational grid (referred to as 'Large Particle' in this paper) as well as disabled and used with a rough grid ('Large Sphere), were investigated. For each case and especially the former, the original particle's centre point and that of the cell or cells representing it within the flow field are misaligned which further contributes to the computational error made.

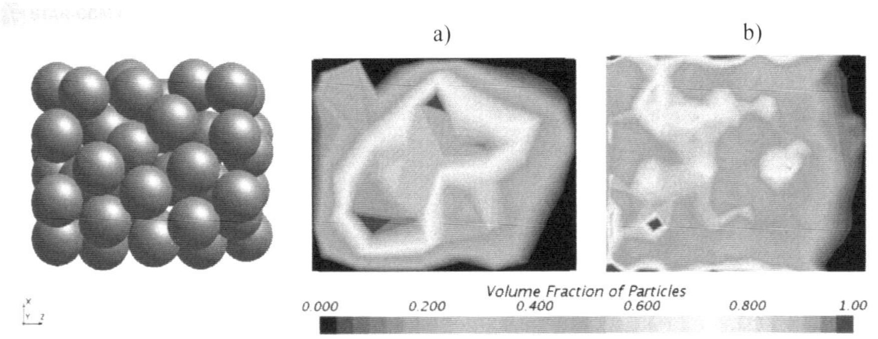

Figure 3: Volume fraction of particles per cell in the continuous phase for a) Large Cell DEM and b) Large Particle DEM approaches.

When using the first method with a smaller scale pebble bed, the resulting size of the computational grid might not be suitable in terms of the continuous phase to successfully treat pressure, temperature and turbulence gradients numerically. The effect of volume fraction for both the first a) Large Cell and second b) Large Particle sub-methods is shown in Fig. 3.

ii) Discrete geometry

The second approach involves a more direct method for working with the bed geometry and fluid domain. For this study a bed was generated with the DEM modeller in STAR-CCM+ and the coordinates of each sphere recorded. These coordinates were then used in a CAD package to construct the physical geometry of the packed bed. From this the negative volume is then taken from a Boolean operation to represent the fluid domain of the bed. Several treatments for the contacts between spheres may be followed from a heat transfer perspective [4] and is also of importance for the fluid domain. One such treatment is placing a fillet on the resulting edge of spheres in contact and is also beneficial for mesh generation in small gaps. The specific configuration concerning the contact treatments in this study are not discussed here in further detail.

This approach is fully explicit in its solving of both flow and all modes of heat transfer, at the expense of simulation set-up time and computational resources.

iii) Porous medium

For the third approach, a purely implicit model of the packed bed section is set-up with a porous medium region to represent the packing. It is assigned a variable porosity (calculated from the geometry used in the discrete approach) with averaged thermal characteristics in order to solve conductive and convective heat transfer across its boundaries. This method is the least computationally expensive of the three although no insights may be gained on particle level of the phenomena contributing to heat transfer. An entire reactor core could be modelled roughly with this approach and is useful for determining several constraints or boundary values which may then be used in more detailed models. The method is presented here to serve as benchmark for the former two.

The turbulence model used with all approaches is the Reynolds-Averaged Navier Stokes (RANS) method. Where small-scale phenomena are of importance the discrete geometry approach is required with better resolving turbulence models e.g. a Large Eddy Simulation (LES) method [6, 7].

RESULTS

The results obtained with each approach is summarised and compared in this section. The merit of each with regards to accuracy and computation time is discussed qualitatively and several quantitative values are also summarised in Table 2 for comparison.

i) DEM with two-way coupling

The DEM-driven approach offers a fair balance between computation time and representation of the physical phenomena of a moderately sized bed. Whilst the spheres are fully represented in terms of physical packing structure and interaction between them, their imprint unto the continuous phase computational grid is done in an approximated manner. A volume fraction is assigned per cell where occupied by a sphere and therefore has an effect on the momentum terms of the continuity equation. The energy source specified for the spheres is divided amongst cells accordingly and the conservation of energy is also affected by this convention of void fraction. Where a sufficiently large bed is considered these local effects have a negligible influence on the total effects. This approach can thus be described as an explicit sphere implicit heat transfer model of a packed bed.

ii) Discrete geometry

The discrete approach relies on the fluid domain to be represented explicitly in the computational mesh. As explained, the geometry is obtained using centre points as generated during the DEM approach. The fine cell structure near and around spheres result in a significantly higher cell count and this approach is therefore the most computationally expensive of all. If not limited by computational resources, a moderately sized bed discretized to several million cells can be simulated with convergence obtained within several days. The small bed used for this section contained 60 000 cells and required 12 CPU-hours on high-end workstation type hardware.

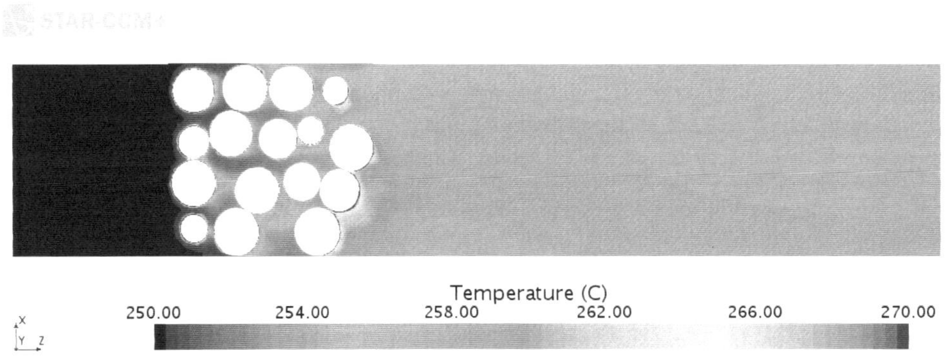

Figure 4: Temperature field variation through spheres along central plane for the discrete approach.

The porosity of a packed bed is observed to follow an oscillatory radial distribution in a cylindrical geometry [8]. This results in the defining of a wall, near-wall and bulk region in a packed bed. Due to the small aspect ratio of the models in this study, this phenomenon was not validated. The average porosity of the section was however calculated as 0.581 with the DEM approach which falls within the acceptable range.

With this approach the main advantage is small-scale flow phenomena such as eddies within cavities as well as boundary effects on velocity variations can be solved explicitly. These in turn contribute to the down-stream flow as well as affecting local heat transfer. The effect on the velocity field may be clearly seen in the graph from Figure 6.

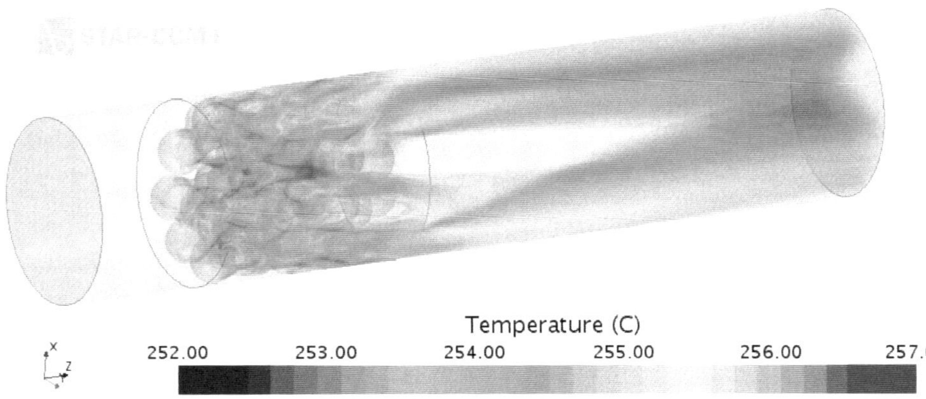

Figure 5: Volumetric rendering of discrete method temperature field for downstream temperatures exceeding 252°C.

The temperature variation across a planar surface through the centre of the cylinder is illustrated in Fig. 4 above. Local hot-spots occur within the temperature boundary layer of each sphere but are most profound around downstream spheres as concentrated and semi-circular wheres they are seen to be smaller and circular for those upstream. It also becomes apparent that more closely adjacent spheres contribute more to the increase in coolant temperature than isolated spheres.

An insightful graphical representation using the volume rendering technique in Star-CCM+ is presented in Fig. 5, showing a volumetric variation of temperatures above the initial inlet temperature of 250 °C. This corroborates the observation of ΔT temperature rise in the coolant from Fig. 4. It should be noted that the void section without colour downstream of the bed exceeds temperatures of 257 °C and it was limited in order to create the rendering with a definite distinction in colours. The average temperature found 300 mm downstream of the bed is 256.8 °C as summarised in Table 2.

iii) Porous Medium

The third approach which was considered involves the packed bed mid-section of the geometry being replaced by a porous medium region assigned the same porosity distribution as calculated from the previous discrete geometry approach. The specific heat and thermal conductivity of the solid phase is identical to that of the nuclear graphite used thus far.

Whilst this approach is by far the least computationally intensive between the three, very little detail of the flow and temperature distributions become apparent. This is most apparent in Fig. 6 where the axial velocity profile follows a typical distribution for turbulent flow through a pipe or duct.

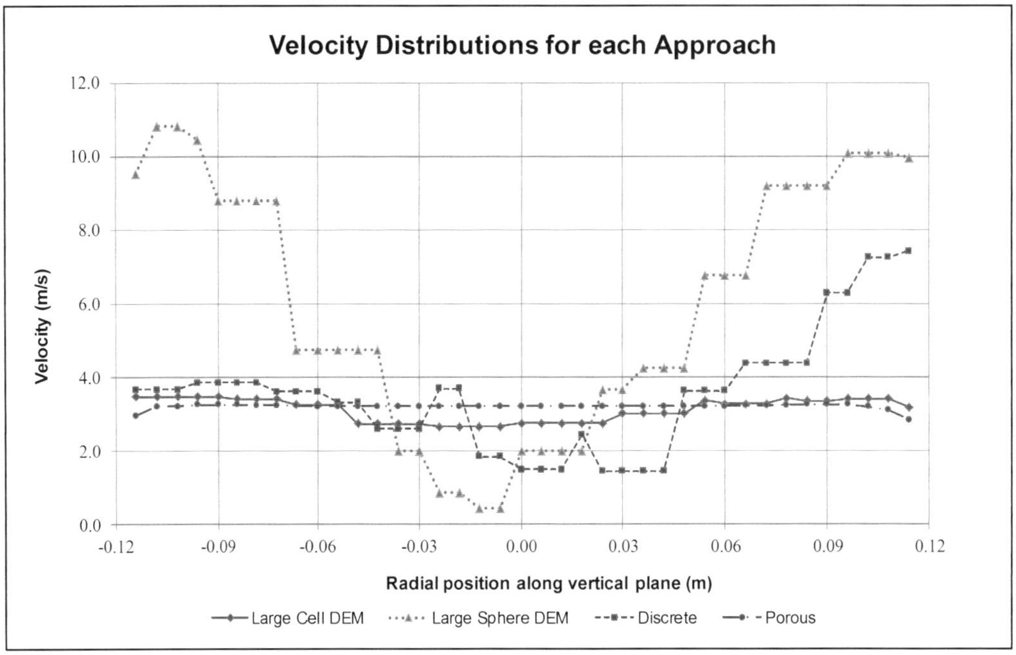

Figure 6: Velocity distribution in the radial direction on the centre vertical plane for each approach.

Comparison

For all cases an equal mass flow inlet was specified. The outlet mass flow is then compared in order to evaluate the conservation of mass and serves as indication of convergence. A comparison of certain solved parameters for each approach is summarised in Table 2 to further evaluate accuracy. Values are reported as surface-averaged values over a plane one bed diameter behind the packed bed in each case. Good agreement is found for the velocity and Reynolds number between all approaches. There is a distinct difference in temperature, pressure drop and dissipation of turbulence. This is due to the inherent suitability of the different approaches to simulate each of these phenomena.

The local turbulence effects with approaches where local phenomena are resolved "more explicitly" can be seen in Fig. 7, which gives a graphical representation of the data plotted in Fig. 6 above.

Table 2: Implicit-unsteady results at a converged time step for the different approaches in Star-CCM+.

Parameter (surface averaged, at one bed-diameter downstream of spheres)	DEM Approach		Discrete Spheres	Porous Medium
	Large Cells	Large Spheres		
Velocity (m/s)	3.198	3.318	3.122	3.162
Reynolds number (Re)	324×10^3	335×10^3	315×10^3	318×10^3
Temperature (°C)	249.3	251.0	260.0	252.7
Turbulent kinetic energy (J/kg)	0.012	0.008	0.935	0.007
Pressure drop from inlet (kPa)	-0.216	-0.109	-1.289	-0.083

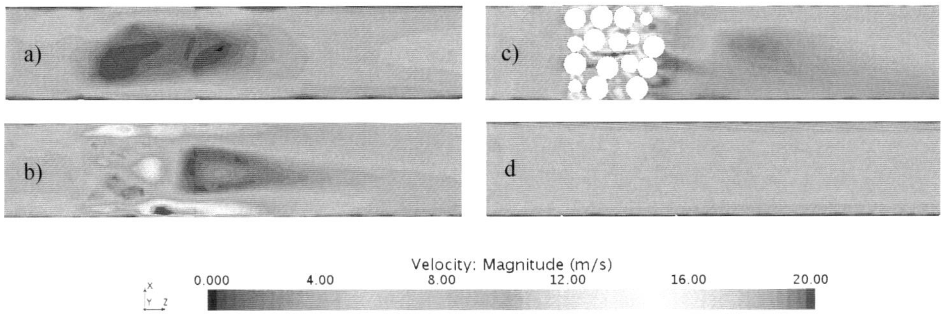

Figure 7: Velocity field found along the centre of the cylinder for the a) Large Cell, b) Large Sphere, c) Discrete Geometry and d) Porous Medium approaches.

CONCLUSION

The best approach to model a small to medium-scale unstructured packed bed of uniform spheres was found to be the discrete geometry approach with moderate mesh fineness. A choice of suitable meshing parameters is, however, subjective and should be evaluated against the specific problem and which physical phenomena are of importance.

The porous medium approach is completely lacking in detail and is only to be used for conceptual-phase calculations requiring a fast turn-around or to roughly estimate the expected results for more detailed simulations. It is applicable to large beds or entire cores and would be advantageous to model the core in a primary cooling loop, for example. Also, careful adjustment is needed of the parameters describing momentum change and energy if this is to be attempted.

Any approach using DEM with two-way coupling is best suited for packed beds where the local effects and contribution of braiding between single spheres and smaller eddies to the bulk flow are not of great importance. The scale of physics present in a packed bed makes the representation of 60 mm spheres by discrete particles unsuitable, given the purpose for which DEM models were developed. The size cells needed, even with large particle rounding, is therefore better suited to large scale beds. What should not be overlooked is the convenience of having inter-particle conductance modelled explicitly with the option of user-specified correlations which is available in the software package Star-CCM+.

To summarize, it is necessary to evaluate the relative size of packed bed, physical phenomena to be investigated as well as the scale of the latter, when selecting an approach.

ACKNOWLEDGMENTS

This work is based on the research supported by the South African Research Chairs Initiative of the Department of Science and Technology and National Research Foundation of South Africa (Grant No 61059).

Any opinion, finding and conclusion or recommendation expressed in this material is that of the authors and the NRF does not accept any liability in this regard.

REFERENCES

1 Nuclear Energy Agency, "PBMR Coupled Neutronics/Thermal-hydraulics Transient Benchmark The PBMR-400 Core Design," NEA/NSC/DOC(2013)10, pp.10, (2013).

2 Venter, P.J., Mitchell, M.N. "Integrated design approach of the pebble BeD modular reactor using models," Nucl. Eng. Des, 237, pp. 1341-1353, (2007).

3 German Nuclear Safety Standards Commission (KTA), "Reactor Core Design of High-Temperature Gas-Cooled Reactors. Part 3: Loss of Pressure through Friction in Pebble Bed Cores," KTA 3102.3, p 2, (1981).

4 Bu, S.S., Yang, J., Zhou, M., Li, S.Y., Wang, Q.W., Guo, Z.X., "On contact point modifications for forced convective heat transfer analysis in a structured packed bed of spheres," Nucl. Eng. Des, 270, pp. 21-33, (2014).

5 Ferng, Y.M., Lin, K.Y., "CFD investigation of thermal-hydraulic characteristics in a PBR core using different contact treatments between pebbles," Ann. Nucl. Energ, 71, pp. 156-165, (2014).

6 Shams, A., Roelofs, F., Komen, E.M.J., Baglietto, E., "Large eddy simulation of a randomly stacked nuclear pebble bed," Comput. Fluids, 96, pp. 302-321, (2014).

7 Shams, A., Roelofs, F. Komen, E.M.J., Baglietto, E., "Numerical simulation of nuclear pebble bed configurations," Nucl. Eng. Des, 290, pp. 51-64, (2015).

8 Van Antwerpen, W., Du Toit, C.G., Rousseau, P.G., "A review of correlations to model the packing structure and effective thermal conductivity in packed beds of mono-sized spherical particles," Nucl. Eng. Des, 240, pp. 1808, (2010).

A ROW-BY-ROW AXIAL TURBINE PROCESS MODEL BASED ON A ONE-DIMENSIONAL THERMOFLUID NETWORK APPROACH

Roelof Pottas[1], Pieter Rousseau[2]

[1,2] University of Cape Town; Mechanical Engineering

pottasrj@eskom.co.za

ABSTRACT

Turbine plant engineers are often required to analyse operational anomalies for which conventional turbine performance curves or generalised lumped correlations do not provide sufficient information. CFD simulations provide significant detail but require specialist knowledge and are highly resource intensive. There is therefore a need for a sufficiently accurate turbine process model that strikes a balance between the computational resources required and the accuracy and flexibility with which it can be applied in an industrial setting. For this purpose a turbine process model was developed based on a row-by-row discretisation using a one-dimensional (1D) network of flow elements. These elements represent various stationary and rotating "channels" for which the fundamental mass, energy and momentum conservation equations for compressible flow are solved in an integrated manner. The closure relations are obtained from loss coefficients for nozzles, rotors and leakage flows which are based on correlations available in the literature. The model is implemented in a custom-developed code written in Scilab. It utilises a pressure correction scheme to solve for combined mass and momentum conservation and superimposes energy conservation to account for work and heat transfer. Results reported in literature from experiments on a four stage axial flow gas turbine were used to validate the model. Three models with different discretisation schemes were tested. The turbine is discretised on a stage-by-stage basis with each stator and rotor flow path modelled by one or more radial flow elements. Leakage paths are also modelled via discrete flow elements with appropriate loss correlations. The results of the model agree well with the test data for the nominal case. The maximum error in total pressure is <2% and the total enthalpy values are predicted with a maximum error of less than 1%.

KEYWORDS

(Turbine process model, Thermofluid network)

INTRODUCTION

The turbine is a major component of thermal power plants and the overall performance of the plant is heavily influenced by the performance of the turbine. Various tools exist to assess the performance of turbines and to identify causes of deviations from optimal performance parameters. These range from simple hand calculations to complex fully 3D (three-dimensional) CFD (computational fluid dynamic) calculations. Maintenance of the plant performance is the responsibility of the plant engineer. However, these tools are not applied as often in the industrial setting as in the design- and research settings. This could probably be attributed to the nature of the industrial setting and the limitations presented by many available modelling tools. Simplified models typically map input variables to outputs while omitting many intermediate variables. Characteristic maps are often not very accurate or not available to the plant engineer and they are not well suited to the exploratory aspect of simulation [1]. Full 3D CFD analysis is a resource intensive activity that does not lend itself well to be applied as a general plant engineering tool in a practical manner. Proper CFD analysis requires an expert in the field to perform the work and it is therefore reserved for detailed studies where the time and effort can be justified by the expected outcomes. An innovative coupled 2D Blade Element Momentum/Reynolds Averaged Navier-Stokes (BEM-RANS) approach was proposed by Dominques, et al. [2] for application to a water turbine analysis. Although significant resource cost reduction was reported over typical high-fidelity unsteady CFD techniques, it has not yet been proven for multiple row analysis or compressible flows. Generally the time invested in 3D CFD analysis is several orders of magnitude larger than 1D and 2D analysis [2] thus rendering it impractical as a general engineering tool. Simplified calculations again do not provide the necessary accuracy required for real time simulation and control [3]. Often in industrial applications, anomalies go unnoticed or are difficult to assess due to the hurdles described above. Leonid, et al [2] illustrated that the use of 1D and 2D models based on validated empirical models provide sufficient accuracy for use in many design problems in practice. Typical anomalies that could be investigated in an industrial setting include:

- Degradation of turbine seals and glands.
- Deposits on blades.
- Erosion of blades.
- Gas path blockages.
- Deviations from nominal operating conditions.

The one-dimensional thermofluid network approach lends itself well to applications in this setting and commercial software packages using this approach are widely used in the energy industry. The goal of the present work is towards broadening the application of commercial thermofluid network software to the row-by-row modelling of turbines in an industrial setting.

OVERVIEW OF APPROACH

Turbines consist of stages of stationary and rotating blades. The stationary nozzles accelerate the steam and direct them at the rotating blades where the kinetic energy of the steam is converted to power delivered at the output shaft. Together with the main gas path mentioned above, there are also several smaller flow passages which include gas leakage flows past nozzle diaphragms and rotor blade tips, shaft seals, balance holes and leak-offs. In Figure 1 some of the minor flow paths are illustrated together with the main gas path through the blade passages.

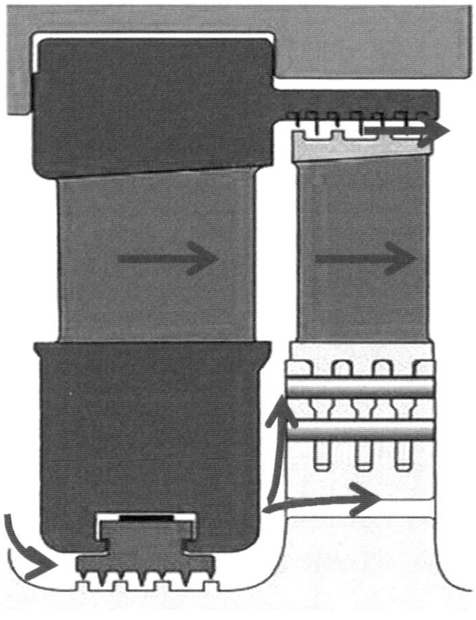

Figure 1. Individual steam paths in a typical steam turbine stage.

Each one of these flows in the complete turbine can be identified and characterised as a flow resistance with an associated pressure drop and mass flow in a thermofluid network. Deviations from design conditions can affect the resistance to the flow offered by an element and the way that it performs work on the fluid. By modelling these flows as part of a network of 1D elements many design and off-design conditions can be explored and analysed. Single elements can be modified for exploratory work to assess the impact of such a deviation on the complete network.

DESCRIPTION OF ONE-DIMENSIONAL FLOW ELEMENTS

The model is developed as a network of discretised nodes and 1D elements with the elements forming the connection between nodes. The complete turbine is discretised on a row by row basis with each stator, rotor and leakage path modelled by a single flow element and interconnected by nodes. An example domain of a single element is shown in Figure 2. The flow through this passage is characterised by the losses caused by the blade geometry at the mean radius and the work that the flow performs on the blades. Heat transfer has is not included in the present work. As the blade geometry is not uniform from root to tip, it is also possible to discretise the blade passage along the length using multiple elements, Figure 2. In the multiple element approach the blade geometry at the mean radius of the element is used.

Figure 2. A representation of the flow domain modelled by a single 1D element (left). A representation of the flow domain modelled by multiple 1D elements (right).

The fluid enters and exits the element, is redirected, accelerated and may perform useful work. The flow element is defined as a control volume and the process is governed by the laws of conservation because, mass, momentum and energy must be conserved. In terms of the governing equations the 1D flow element is fully described by the inlet and exhaust areas and mean angles as well as the length and volume of the element, as illustrated in Figure 3.

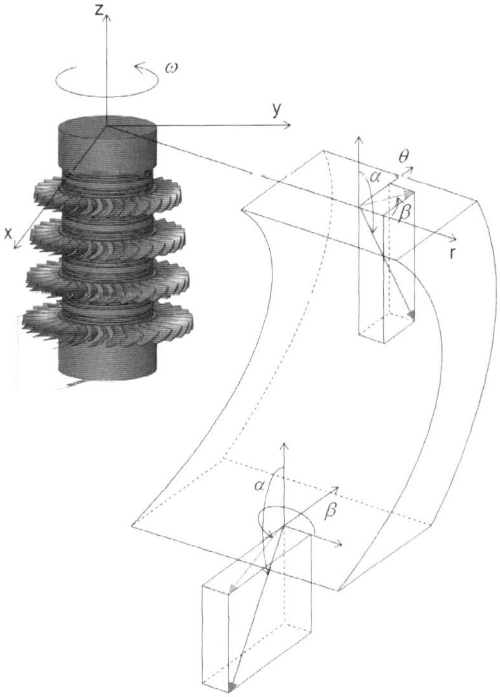

Figure 3. A representation of the control volume and the flow vectors at the inlet and outlet of a 1D element.

The governing equations below are presented for a steady state solution of a rotating element within a 1D non-inertial frame of reference. Nozzles and other stationary flow paths are modelled using the equations for a stationary element which are of a similar, but simpler form since the rotational terms vanish. The derivation is presented by [4].

Conservation of mass

$$\sum_i(\dot{m}_i)_{out} - \sum_i(\dot{m}_i)_{in} = 0 \quad (1)$$

with the subscripts 'in' and 'out' referring to the inlet and exhaust mass flow rates respectively.

Conservation of momentum

$$\frac{p_{avg}}{p_{0avg}}\left(p_{0,out} - p_{0,in}\right) + \frac{1}{2}\rho_{avg}c_{avg}^2\frac{1}{T_{0avg}}\left(T_{0,out} - T_{0,in}\right) + \Delta p_{0,loss} + \rho_{avg}g\left(z_{out} - z_{in}\right) = \rho_{avg}\omega X \quad (2)$$

with p static pressure, p_0 total pressure, T_0 total temperature, ρ density, C absolute velocity, z elevation and ω rotational speed. The subscripts 'in' and 'out' represent conditions at the inlet and exhaust respectively and the subscript 'avg' represents the average value over the control volume [4].

$$X = \omega\left(r_{out}^2 - r_{in}^2\right) + \left(\sin\alpha_{out}\sin\beta_{out}v_{out}r_{out} - \sin\alpha_{in}\sin\beta_{in}v_{in}r_{in}\right) \quad (3)$$

Equation (3) essentially represents the net change in specific angular momentum between the inlet and the outlet of the flow passage.

Conservation of energy:

$$M_z\omega = \dot{m}\left(h_{0,out} - h_{0,in}\right) \quad (4)$$

With M_z the moment about the z-axis and h_0 the total specific total enthalpy. The subscripts 'in' and 'out' represent conditions at the inlet and exhaust respectively and the subscript 'avg' represents the average value over the control volume. The moment , M_z is defined as:

$$M_z = \dot{m}X \quad (5)$$

Modelling pressure losses

The term $\Delta p_{0,loss}$ in equation (2) refers to the loss of total pressure attributed to inefficiencies inherent to the design, due to degradation or due to deviations from nominal operating conditions. This pressure loss coefficient within an element is defined as:

$$Y_{loss} = \frac{p_{0,in} - p_{0,\text{out}}}{p_{0,out} - p_{out}} \quad (6)$$

Thus the total pressure loss can be defined as:

$$\Delta p_{0,loss} = Y_{loss}\left(p_{0,\text{out}} - p_{\text{out}}\right) \quad (7)$$

The pressure loss can then be rewritten as a function of the Mach number in compressible flow by using the ratio of static to total pressure:

$$\Delta p_{0,loss} = Y_{loss}p_{out}((1 + \frac{\gamma-1}{2}M_{out}^2)^{\frac{\gamma}{\gamma-1}} - 1) \quad (8)$$

The total pressure loss is thus a function of the velocity and if the density is known, of the mass flow rate. Therefore we can say:

$$\Delta p_{0,loss} = f(\dot{m}) \tag{9}$$

Empirical correlations for the pressure loss coefficient, Y_{loss}, were originally developed by authors Soderberg [5] and Ainley *et al.* [6] and more recently by authors including Craig *et al.* [7], Kacker *et al.* [8], Denton [9], Benner *et al.* [10]. The total pressure loss representing the losses in the stator and rotor passages was obtained by applying the profile loss correlation of Kacker *et al.* [8] and the secondary losses described by Benner *et al.* [10]. The stator and rotor leakage flows were modelled using one flow element to represent each seal. The Saint Venant Wantzel model for a labyrinth seal is applied to model the leakage through inter-stage the seals. The losses associated with a diverging flow path and with the disturbance caused by the labyrinth seals is applied to the gaps between stages as described by Craig *et al.* [7].

NETWORK SOLUTION

If we substitute (9) into the momentum equation, (2), we obtain a relationship between element inlet and outlet pressures and element mass flows for the network,

$$\dot{m} = f\left(p_{0,in}, p_{0,\mathrm{out}}\right) \tag{10}$$

We can also write conservation of mass, equation (1), for each node and substitute in equation (10) to obtain an equation for each node where the connected element pressures are the unknowns. The element inlet and outlet total pressures correspond to nodal pressures in the network, so the equations can be used to simultaneously solve nodal pressures in the network. A pressure correction scheme based on the Newton-Raphson method is utilised in the present work to solve the non-linear set of equations.

The properties of the flow that are required to complete the equations are calculated using the initial guess values and are updated during each iteration. Therefore, the fluid properties are treated as constants during each iteration while solving for the total pressures. The mass flow rate for each flow element is then calculated from the pressure loss equation for each element. Once the pressure and mass flow rate distributions have converged, an energy balance is performed at each node to account for work done on the fluid and the associated change in enthalpy is accounted for. The pressure correction scheme is then repeated.

The network for the most complex case modelled was made up of 55 elements and solves in less than 1 minute on a personal computer. A typical mesh of one 3D CFD model developed by Gardzilewicz, et al. [11] of the last stage of a turbine required around 2 million finite volumes.

CASE STUDY FOR VALIDATION

Detailed information about the turbine geometry and experimentally measured pressure, temperature and velocity distributions are required to properly validate the proposed method. A data set was obtained from measurements on a four stage turbine in an open loop configuration with exhaust to atmosphere published by Fottner [12]. The turbine is coupled to a dynamometer and the rotational speed can be controlled by voltage control. The turbine blades are designed with varying degrees of reaction. The last stage blades are designed with approximately 50% reaction at the mean radius.

Table 1. Nominal parameters for the turbine [12].

Coupling power	703kW
Rotational speed	7500rpm
Mass flow rate	7.8kg/s
Inlet pressure	2.6bar
Inlet temperature	413K
Outlet pressure	1.022bar
Outlet temperature	319K
Efficiency	0.913
Exhaust hub ratio	0.525

VALIDATION RESULTS

In Figure 4 the modelled efficiencies show a good agreement with the test data for a broad range of relative mass flow rates. At nominal conditions, the efficiency has an error of <1%. The efficiency curve at off-design conditions is represented well, but appears to be over predicted for the lower mass flow rates at nominal rotating speed.. The results of three models were compared to the AGARD test data namely; model 1, model 3 and model 5. Each model differs in the discretisation scheme used to model the stator and rotor passages. A discretisation approach similar to that illustrated in Figure 2 was applied. The numeral refers to the number of radial discretisations of the stator and rotor passages in the respective models. The results are shown in Figure 4. The original data from Fottner [12] is labelled 'AGARD' and the separate sets of results from this work, each with a different discretisation along the nozzle / blade lengths are labelled; model 1, model 3 and model 5.

Results of the models discretised into one, three and five radial elements are shown and the results for this exercise show that there is very little difference in the results obtained by the three models. The models with more elements do appear to predict the losses at severe off-design conditions slightly better, predicting a lower efficiency for those mass flow rates.

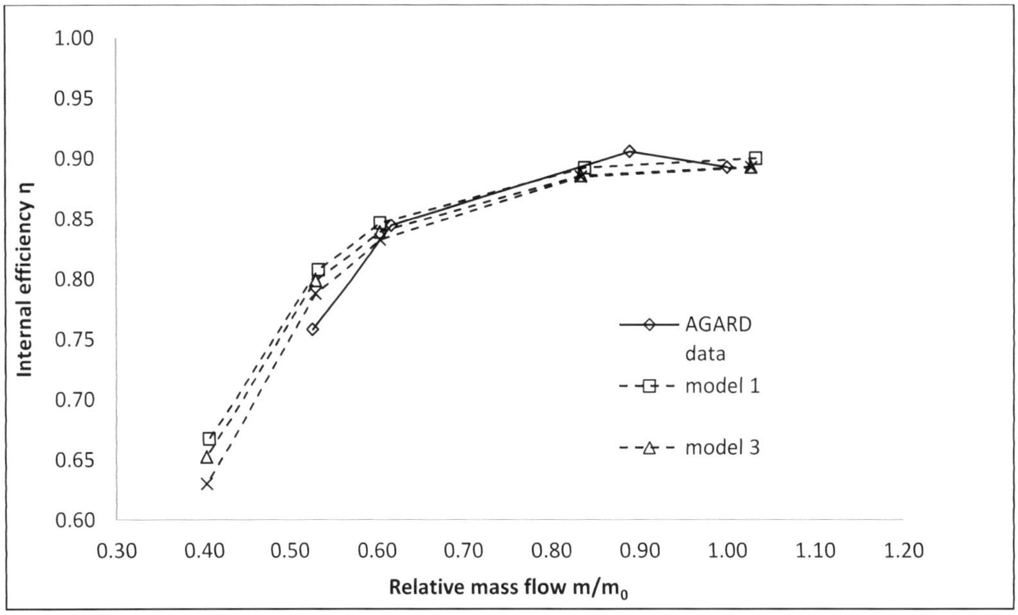

Figure 4. Plot of turbine efficiency data and modelled results for a relative speed of $n/n_0=1$.

Figure 5 shows that all three models slightly over predict the mass flow rate at the nominal condition. This improves for smaller pressure ratios. The accuracy of the calculated mass flow rate is within <3% of the test data for all three models. There is practically no difference in the results obtained by the three models.

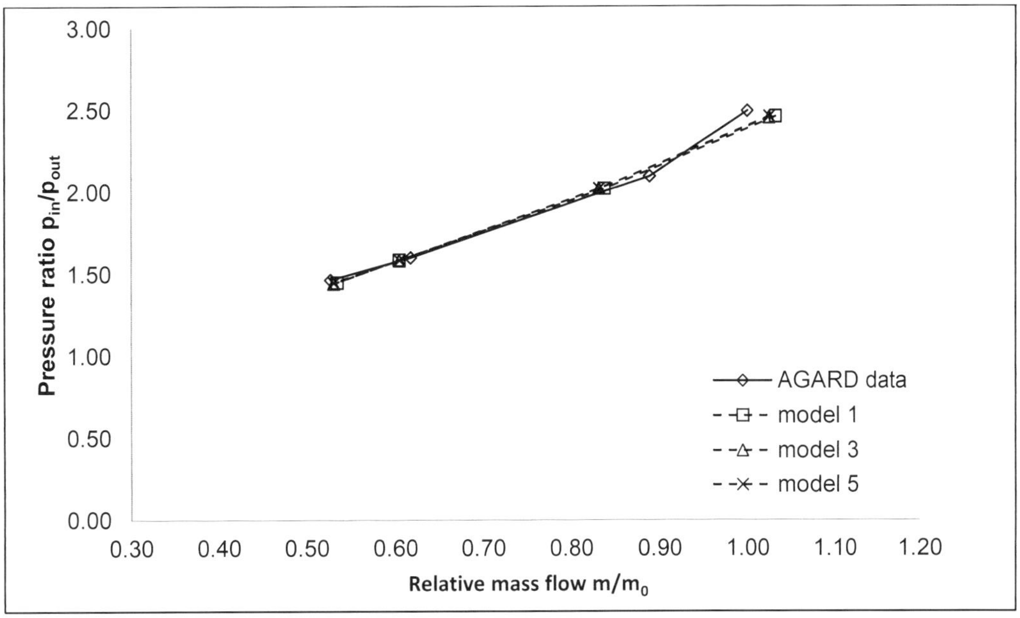

Figure 5. Plot of turbine pressure ratio data and modelled results at a relative speed of n/n0=1.

A comparison of internal flow parameters is also presented here for the three models at the design condition. The percentage difference is calculated as the difference between the modelled results and the test data at 5 stations in the turbine. Station 0 and 4 represent the turbine inlet and outlet respectively and stations 1-3 represent points between each stage.

Table 2. Description of model names.

Model name	Condition	Number of radial elements
TF401	nominal	1
TF402	nominal	3
TF403	nominal	5

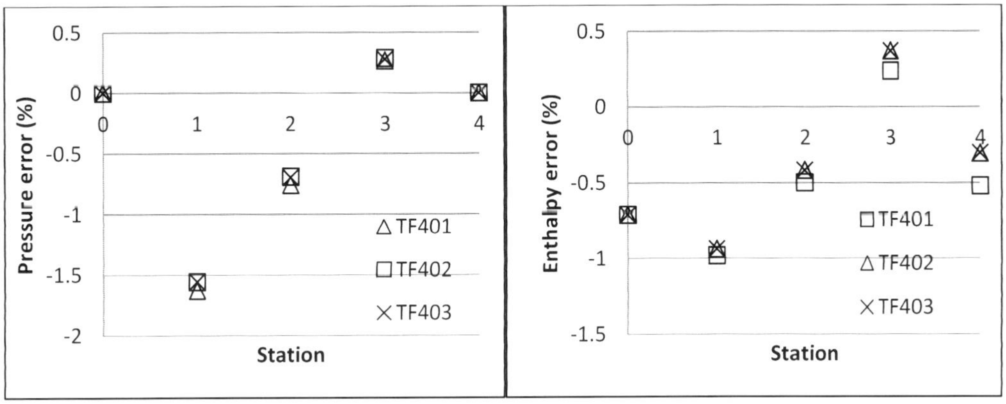

Figure 6. Percentage difference between measured data and modelled results at five measuring stations of total pressure(left) and total enthalpy(right).

In Figure 6 the measured values of absolute total pressure and total enthalpy are compared to modelled results. It appears that there is a good agreement between the modelled values and the measured data. The maximum deviation in total pressure of 1.6% occurs at station 1 and the total enthalpy values are predicted with an error of <1%. The results obtained by all three models are similar.

The static pressures, Figure 7, have an accuracy with a maximum error of <1.7% occurring at station 4.

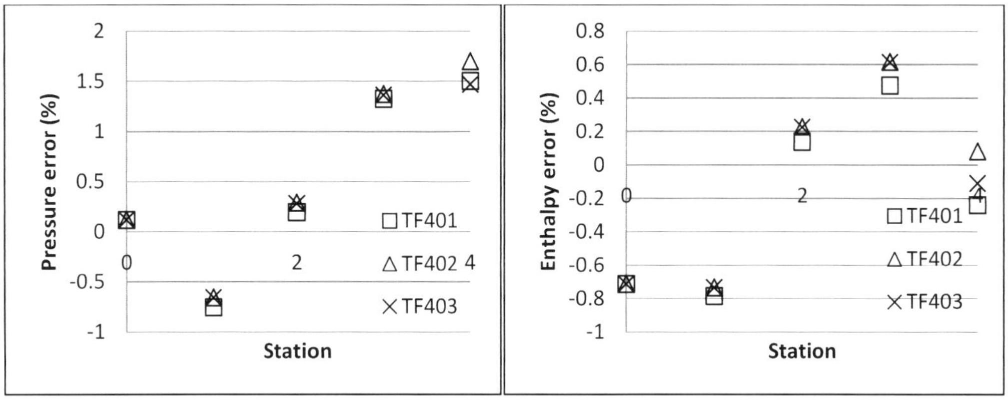

Figure 7. Percentage difference between measured data and modelled results at five measuring stations of static pressure(left) and enthalpy(right).

The model results of the enthalpy are considered to be a good representation of the model's ability to represent the real turbine's characteristics. The static enthalpy deviates less than 3.5kJ/kg which is an error of <1%.

CONCLUSION

A methodology was developed to solve the fundamental mass, momentum and energy equations as applied to a turbine consisting of a network of stationary and rotating flow elements. The methodology was implemented in Scilab and the correct implementation of the applicable mathematics and physics was verified against a broadly validated commercial software package.

A test case obtained in the literature was used to validate the model. The test case consisted of measurement data and geometric data of a real turbine setup in a laboratory. Four cases were modelled, consisting of a nominal case and off-design cases with varying degrees of pressure ratio. The model results showed good agreement with the test data.

The goal of the present work is not to provide a replacement for CFD, but an alternative approach that can be applied more readily in an industrial setting. With further development of the user interface this provides a feasible solution to many plant maintenance engineering applications.

NOMENCLATURE

α	angle measured from the z-axis	rad
β	angle in the r-θ plane	rad
γ	ratio of specific heats	
ρ	density	kg/m^3
ω	shaft rotation speed	rps
c	absolute velocity	m/s
g	gravitational acceleration	m/s^2
h	specific enthalpy	J/kg
$\dot{m}$	mass flow rate	kg/s
M	Mach number	
p	static pressure	Pa
p_0	total/stagnation pressure	Pa
r	radial distance	m
Re	Reynolds number	
T	temperature	K
T_0	total/stagnation temperature	K
v	relative velocity	m/s

X	change in specific angular momentum	
Y_{loss}	pressure loss coefficient	m^2 / s

REFERENCES

[1] O. Leonard and O. Adam, "A quasi-one-dimensional CFD model for multistage turbomachines," *Journal of thermal science,* vol. 17, no. 1, pp. 7-20, 2008.

[2] F. Domingues, J. Achard, J. Zanette and C. Core, "Fast power output prediction for a single row of ducted cross-flow water turbines using a BEM-RABS approach," *Renewable Energy,* vol. 89, pp. 658-670, 2016.

[3] M. Leonid, Y. Govoruschenko and P. Pagur, "Axial turbine stages design: 1D/2D/3D simulation, experiment, optimization," Reno-Tahoe, 2005.

[4] A. Chaibakhsh and A. Ghaffari, "Steam turbine model," *Simulation and modelling practice and theory,* vol. 16, no. 9, pp. 1145-1162, 2008.

[5] P. Rousseau, Thermal-fluid systems modelling 1 course notes, Potchefstroom: North-West University, 2013.

[6] C. R. Soderberg, Unpublished notes, Gas turbine laboratory, Massachusetts Institute of Technology (qouted in reference [Dixon, 2005]), 1949.

[7] D. Ainley and G. Mathieson, "A Method of performance estimation of axial flow turbines," 1951.

[8] H. Craig and H. Cox, "Performance estimation of axial flow turbines," 1970.

[9] S. Kacker and U. Okapuu, "A mean line prediction method for axial flow turbine efficiency," *Journal for engineering for power,* vol. 104, pp. 111-119, 1982.

[10] J. D. Denton, "Loss mechanisms in turbomachines," *Journal of turbomachinery,* vol. 115, 1993.

[11] M. W. Benner, S. A. Sjolander and S. H. Moustapha, "An empirical prediction method for secondary losses in turbines-Part II: A new secondary loss correlation," *Journal of turbomachinery,* vol. 128, pp. 281-291, 2006.

[12] A. Gardzilewicz, J. Swirydczuk, J. Badur, M. Karcz, R. Werner and C. Szyrejko, "Methodology of CFD computations applied for analysing flows through turbine exhaust hood," *Transactions of the institute of fluid-flow machinery,* vol. 113, pp. 157-168, 2003.

[13] L. Fottner, Test cases for computation of internal flows in aero engine components, Essex: AGARD, 1990.

Optimal Design of a Piezoelectric Transducer to excite Ultrasonic Guided Waves

Dineo A. Ramatlo[1*], Daniel N. Wilke[2] and Philip W. Loveday[3]

[1,2] University of Pretoria; [1,3] CSIR Materials Science and Manufacturing

*DRamatlo@csir.co.za

ABSTRACT

An Ultrasonic Broken Rail Detection (UBRD) system has been developed to detect complete breaks in a rail. There is currently a need to upgrade the functionality of the system to detect cracks in the web section of the rail. A transducer capable of exciting a guided wave mode with energy concentrated in the web thus needs to be developed. It is required that the transducer must strongly excite the web mode at the operational frequency of the UBRD system. This study aims to optimize the transducer to excite the web mode. The interaction of the transducer with the rail structure was modelled using the Semi-Analytical Finite Element – 3-Dimensional (SAFE-3D) method. An objective function was selected to maximize the energy transmitted by the transducer to the web mode. The identified design variables were the dimensions of the transducer components. We used a response surface-based optimization approach with a Latin Hypercube sampled design of experiments (DoE) that required SAFE-3D analyses at the sampled points. A Nelder-Mead optimization algorithm was then used to find an optimal transducer design on the response surface. The performance of the optimized transducer predicted by the response surface was found to be in good agreement with that computed from SAFE-3D.

KEYWORDS

Guided wave, Rail track, Transducer, Optimal design, SAFE-3D

INTRODUCTION

Regular monitoring of engineering structures like rail tracks has become of paramount importance in the field of non-destructive evaluation. The employment of inspection and monitoring systems can help to detect unsafe rails, and ultimately prevent catastrophic train derailments. An existing example of such a system is the Ultrasonic Broken Rail Detection (UBRD) system [1] developed by the Council for Scientific and Industrial Research (CSIR) and the Institute for Maritime Technology (IMT) to detect complete breaks in rail tracks. The system, currently installed on the Sishen-Saldanha Ore line, uses ultrasonic guided waves transmitted between transmit and receive transducers. If the transmitted ultrasound signal is not detected by the receive station, an alarm is triggered indicating a broken rail.

The current state of the UBRD system operates by employing a propagating guided wave mode with energy concentrated in the head of the rail (denoted the head mode). Such a selected mode is depicted in figure 1. The head mode was identified by Long and Loveday [2] in their study, where it was demonstrated that it is suitable for long range propagation. Their study further demonstrated that the head mode can be used to distinguish between thermite welds and various cracks in the head of the rail; while a second identified mode with energy concentrated in the web of the rail (denoted the web mode) is suitable for detecting welds and damage in the web of the rail. The web mode is highlighted in Figure 1.

To extend the functionality of the UBRD system to detect welds and damage in the web section before the occurrence of complete breaks, a piezoelectric transducer capable of strongly exciting a mode with energy concentrated in the web of the rail is required.

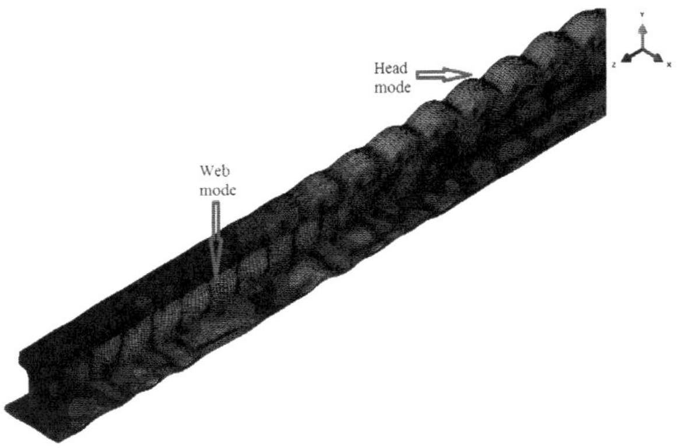

Figure 1. The head and web modes propagating through a rail track.

To date, various numerical modelling techniques for guided waves have been presented in the literature. Examples of such techniques include the finite element method, the finite difference method, the boundary element method and the Semi-Analytical Finite Element (SAFE) method. Many of the numerical modelling techniques are inefficient due to their computational demand. As an example, to model wave propagation using the traditional finite element method, a complete 3D geometry of the waveguide structure need to be modelled. The required cost of computation is thus very high, and increases proportionally with the size or length of the waveguide. More over, the complexity of the problem increases as care has to be taken for the free boundary reflections of the wave guide. Researchers in [3] and [4] have highlighted the reliability and efficiency of the SAFE method; hence the method has recently become widely adopted for studying wave propagation. In the SAFE method, only a 2D cross-sectional mesh of a waveguide is employed in the computations, and the waveguide length terms are treated analytically. The method is therefore attractive due to its improved computational efficiency.

To develop a transducer capable of optimally exciting the web mode, a resonant transducer is required to ensure that sufficient energy can be transmitted to the rail web. The transducer thus needs to be modelled in 3D. A rail-transducer model comprising of a 2D cross-sectional mesh of the rail modelled using the SAFE method and a 3D mesh of the transducer modelled using finite element method is created and the combination of the two methods is referred to as SAFE-3D in this paper. The SAFE-3D method has been demonstrated to allow for effective modelling of the interaction between the rail and transducer models when the transducer is used to excite guided-waves in the rail [5& 6].

The main objective of this paper is to employ the SAFE 3D method to design a transducer that excites the web mode optimally. In design optimization, an objective function quantifying the performance of a design as a function of input variables is required. For the design problem in this paper, the objective function is not an analytical function but can be evaluated at design points by solving the SAFE-3D model. The use of gradient-based optimization methods is thereby restricted since a well-defined continuous function does not exist. By employing the response surface based method, a meta-model of the objective function can be constructed through a Latin Hypercube Design of Experiments (DoE), which requires SAFE-3D analyses at the sampled points. The capability of using SAFE-3D in DoE with many sampling points makes the SAFE-3D method highly attractive due to its reliability and efficiency.

In this paper, an optimal transducer is designed by employing a Nelder-Mead optimization algorithm which searches for design variables leading to optimal conditions on the constructed response surface meta-model.

GUIDED WAVE MODELLING USING SAFE METHOD

Although the SAFE method only requires a 2D cross-sectional model of the rail, the propagation of guided waves is modelled in 3D. It is assumed that the displacements of a point on the cross-section are functions of position and time. The displacement, strain and stress fields are therefore interpolated over the cross-section, while the length terms are treated analytically. However, the method is applicable only if a wave guide cross-section is constant throughout its length.

With the xy coordinate system parallel to the cross-section of the waveguide and z being the direction of wave propagation, the displacement field of an element can be expressed as:

$$\begin{aligned} u(x,y,z,t) &= u(x,y)e^{-j(\kappa z-\omega t)} \\ v(x,y,z,t) &= v(x,y)e^{-j(\kappa z-\omega t)} \\ w(x,y,z,t) &= jw(x,y)e^{-j(\kappa z-\omega t)} \end{aligned} \qquad (1)$$

where κ is the wavenumber, ω is the wave natural frequency and the variable t denotes the time. The parameter j is the complex number equivalent to $\sqrt{-1}$.

The finite element dynamic equations can be derived from Hamilton's principle by employing the strain energy and virtual work relations [4]. For a waveguide subjected to external forces, the response at a distance z from the excitation point is given by [7]:

$$U(z) = \sum_{r=1}^{3N} \alpha_r \psi_r \qquad (2)$$

The eigen-vectors ψ_r and corresponding eigen-values κ_r are resolved from the system equation of motion, and the modal amplitude due to the external force is evaluated as:

$$\alpha_r(z) = -j\frac{\psi_r^T F}{\psi_r^T B \psi_r} e^{-j\kappa_r z} \qquad (3)$$

WEB MODE EXCITATION USING SAFE-3D METHOD

The SAFE-3D method is explained in detail in reference [5]. The SAFE-3D method can be briefly explained as constituting of two main objectives. The first objective is the modelling of the interaction between the waveguide and the transducer, while the second objective is the computation of system responses from dynamic equations.

A schematic representation of the first part of the method is illustrated in Figure 2. From the 3D infinite rail track, a 2D cross-section of the rail is extracted and modelled using 4 noded elements in SAFE. To excite ultrasonic guided waves in the web of the rail, the 3D transducer model is attached to the edge of the web cross-section, at the centre where it will excite propagation of the web mode. For effective modelling of the problem, model data will need to be accurately mapped between the 2D SAFE model and the 3D transducer model. A significant aspect of the SAFE-3D method is thus the computation of a nodal interaction matrix that allow for the interpolation of model data. A process flow chart outlining the details of the SAFE-3D method is shown in figure 2. The nodal interaction matrix is constructed by first identifying the unique z-planes of the 3D interface nodes (in contact with the SAFE mesh). Treating the length terms in the SAFE model analytically, the corresponding SAFE interface nodes interacting with each of the 3D interface nodes are identified. A nodal interaction matrix of dimension $[m \times n]$ (where m and n are respectively the number of 3D and SAFE interface nodes), consisting of components ranging between zero and one is then calculated. The number one signifies a complete interaction between a 3D node and the SAFE node, indicating that the nodes are coincident.

Now to excite guided waves with the transducer and compute the associated response, the identified SAFE interface nodes are first excited with unit forces. The corresponding modal amplitudes and the response of the waveguide are evaluated using equations 3 and 2, respectively. A compliance (receptance) matrix relating the applied normalised forces to the response is constructed through the relation:

$$U_{in} = R_w F_{in} \tag{4}$$

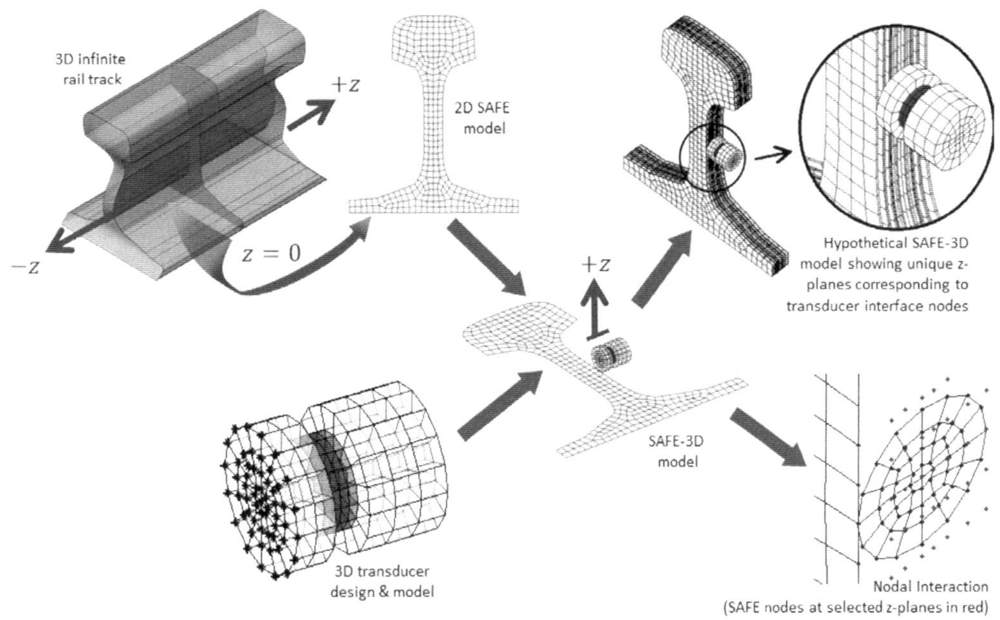

Figure 2. Schematic representation of the first part of the SAFE-3D method.

where U_{in} and F_{in} are respectively the displacement and applied forces at the SAFE interface nodes, and R_w is the receptance of the rail.

The dynamic stiffness of the rail is then computed by inverting the receptance:

$$D_w U_{in} = F_{in} \tag{5}$$

To impose the boundary condition due to the rail on the transducer interface, the dynamic stiffness of the rail D_w is added to appropriate terms in the dynamic stiffness of the transducer denoted as D_p. The response of the transducer and actual interface forces are then solved using the equation:

$$\begin{bmatrix} D_{p,nn} & D_{p,nin} \\ D_{p,inn} & D_{p,inin} + \widetilde{D}_w \end{bmatrix} \begin{Bmatrix} u_n \\ u_{in} \end{Bmatrix} = \begin{Bmatrix} f_n \\ f_{in} \end{Bmatrix} \tag{6}$$

where the subscripts in and n refer to the transducer interface and non-interface nodes respectively. The rail dynamic stiffness is now denoted as $\widetilde{D}_w$ indicating that it is interpolated to the 3D transducer model.

Finally, the computed actual forces at the interface are interpolated back to the 2D model. The forces are used to correctly scale the normalised SAFE modal amplitude and the actual response of the waveguide can be computed.

OPTIMIZATION PROBLEM FORMULATION

The formal design optimization approach as explained in the textbook of Arora [8] is explained. The optimization problem formulation for the design of the web transducer is stated next.

Problem description

The goal is to obtain an optimal transducer design for exciting the web mode. The design is of a sandwich type with a structural configuration shown in Figure 3. The piezoelectric transducer will be driven by a $35kHz$ centre frequency Hanning-windowed tone burst signal. The amount of energy that the transducer can transmit to the rail web at this operational frequency must be maximized.

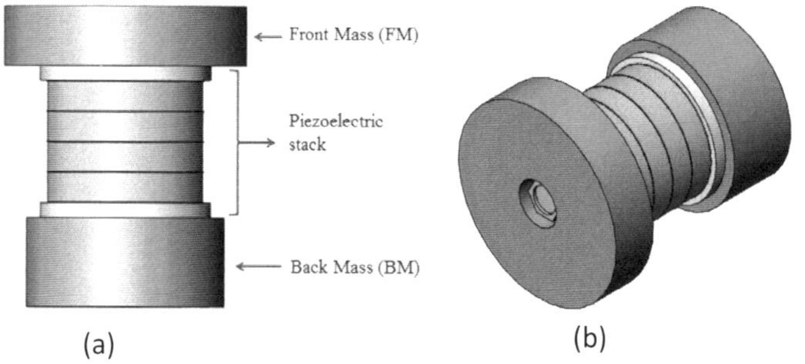

Figure 3. The (a)side and (b)isometric views of the transducer showing its structural composition.

Data and information collection

The front and back masses of the transducer are made of steel material while the piezoelectric stack is PZT ceramic. The aim of the design problem is to evaluate the effect of the front mass and back mass sizes. The geometry of the piezoelectric stack is therefore kept constant.

Definition of design variables

To allow for the use of available transducer housing, the diameter of the front mass is kept constant at a size compatible with the housing. The design problem therefore contain only three design variables; the thickness of the front mass (FM), the thickness of the back mass (BM) and the diameter of the back mass. Although fixing the diameter of the front mass poses a limitation on the design problem, it provides notable advantages which include: reduced problem complexity and improved computational cost and time.

Optimization criterion

The optimization criterion is the area under the frequency response function of the web mode which is representative of the amount of energy transmitted to the web mode at the required operational frequency of the transducer.

Formulation of constraints

The bound constraints defining the design space are as follows:

$$\begin{aligned} 4mm &\le T_{FM} \le 10mm \\ 4mm &\le T_{BM} \le 10mm \\ 12.05mm &\le D_{BM} \le 15.5mm \end{aligned} \tag{7}$$

where T is the thickness and D is the diameter.

RESPONSE SURFACE METHOD

The response surface method is explained in detail in the textbook of Arora [8]. In this paper, a radial basis function approach using a Gaussian basis is followed to approximate the design objective function. A design of experiments was constructed using Latin hypercube sampled points in the design space, and SAFE-3D method was employed to evaluate the objective function at the sampled design points. The response surface was then approximated by fitting a linear combination of Gaussian basis functions over the sampled designs.

The accuracy and numerical stability of the response surface was controlled by selecting a good value of the RBF parameter using the leave-one-out cross validation method.

In RBF interpolation, the interpolant is usually associated with numerous errors at and closer to the edges of the design space. To improve the accuracy of the response surface at the bounds of the design space, it was decided to perform the interpolation by also including data points that lie outside the actual design space. A second domain, referred to as the surrogate space, was thus determined. The surrogate space is an over estimated design space that is used to improve the accuracy of the objective function at the bounds of the actual space. The surrogate space was sampled with 125, 250 and 500 data points respectively, to evaluate the effect of sample size. The SAFE-3D model was solved at each of those design points. An illustration of the actual design space and the surrogate space is shown in Figure 4.

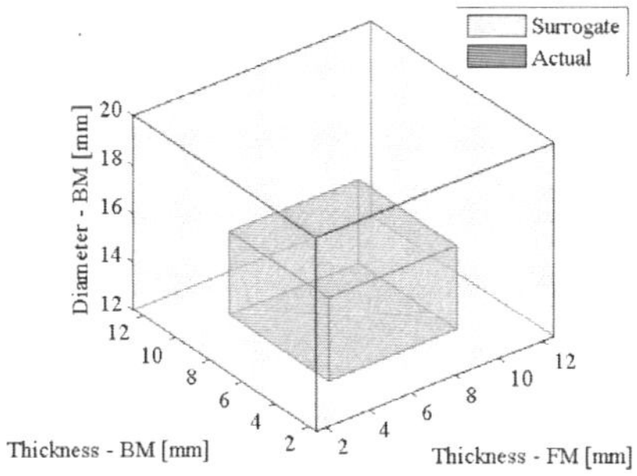

Figure 4. The surrogate space and actual design space.

RESULTS

The three sets of sampling data used in DoE and associated objective function values computed using the SAFE-3D method are plotted in Figure 5. One attractive feature of DoE is that it provides the designer with informative output data. It is clear from Figure 5 that the data gives a good indication of where to expect the optimum design.

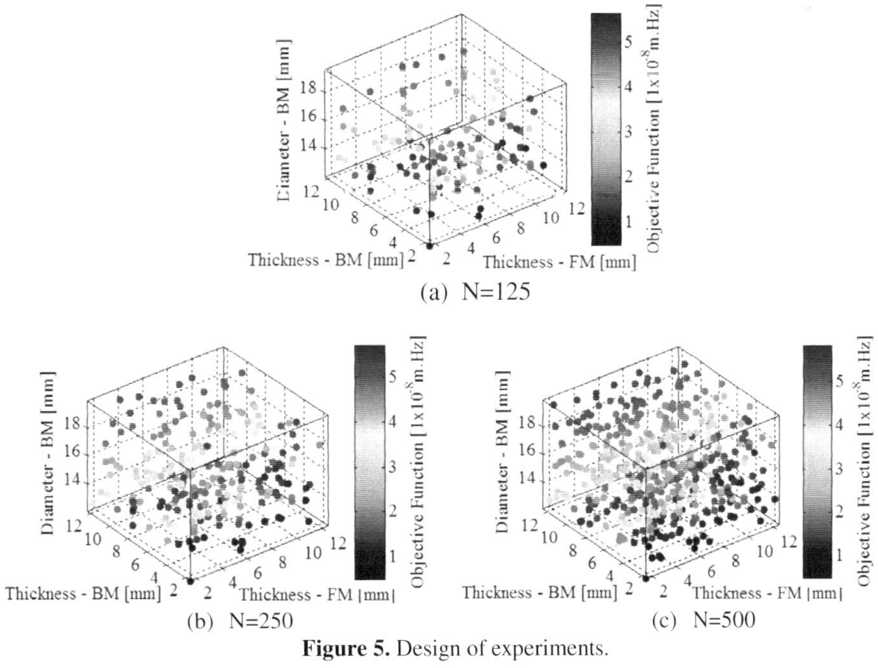

Figure 5. Design of experiments.

The response surface constructed by RBF interpolation using 500 data points is plotted in Figure 6. Within the design space, it was found that there are three categories of transducer designs; those with the modal amplitude peak of the web mode at a frequency of less than $35kHz$, greater than $35kHz$ and at or approximately at $35kHz$. As expected, designs with the web modal amplitude peak closer to 35kHz are associated with high function values, thereby indicating a good performance at $35kHz$.

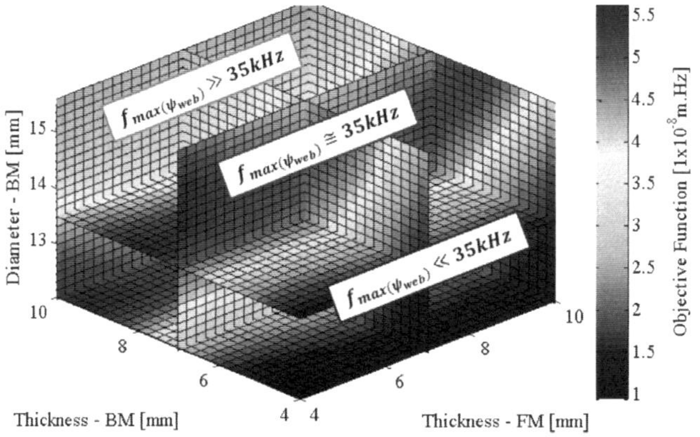

Figure 6. RBF predicted objective function.

To evaluate the accuracy of the response surface values predicted by RBF interpolation, 27 design points arranged in the form of a grid over the surrogate space were selected for error analysis. The RBF predicted function values were compared to the function values computed from the SAFE3D method. The error percentages associated with the selected points are plotted in Figure 7 for the three data sets.

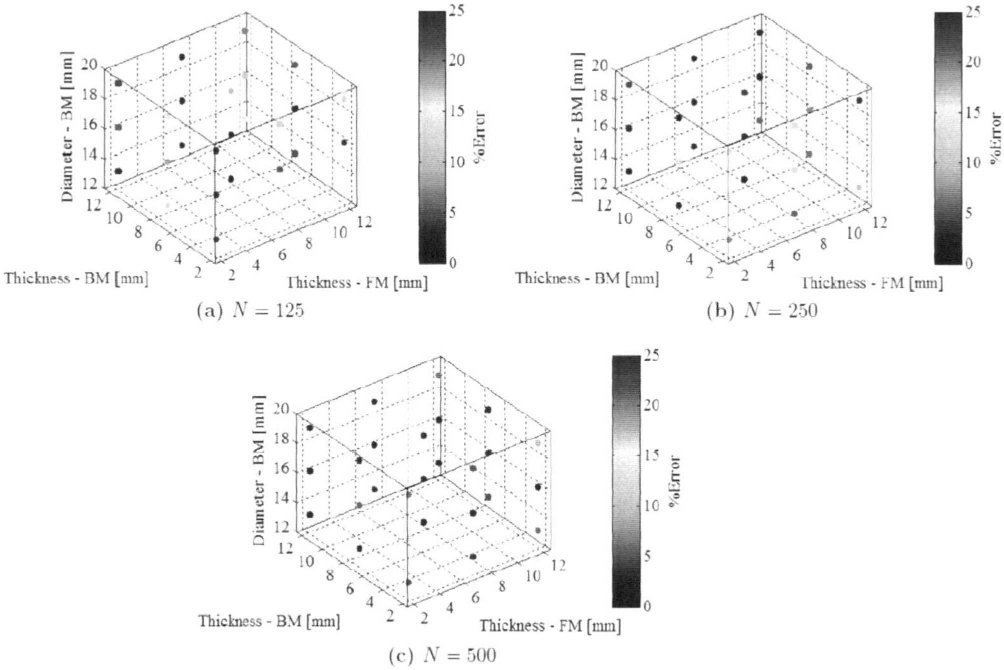

Figure 7. Errors within the surrogate space associated with the constructed objective function.

It is clear from Figure 7 that the errors are high at the boundary of the design space as compared to the interior. The amount of the errors for the three cases decreases as more data points are used in the approximation, thus improving the RBF predicted function values.

Now to demonstrate the significance of using a surrogate space to approximate the response surface, we evaluate the accuracy of a number of points that are within the actual design space. Figure 8 shows the errors associated with the actual design space. The errors within the design space are significantly smaller when compared to those within the surrogate space.

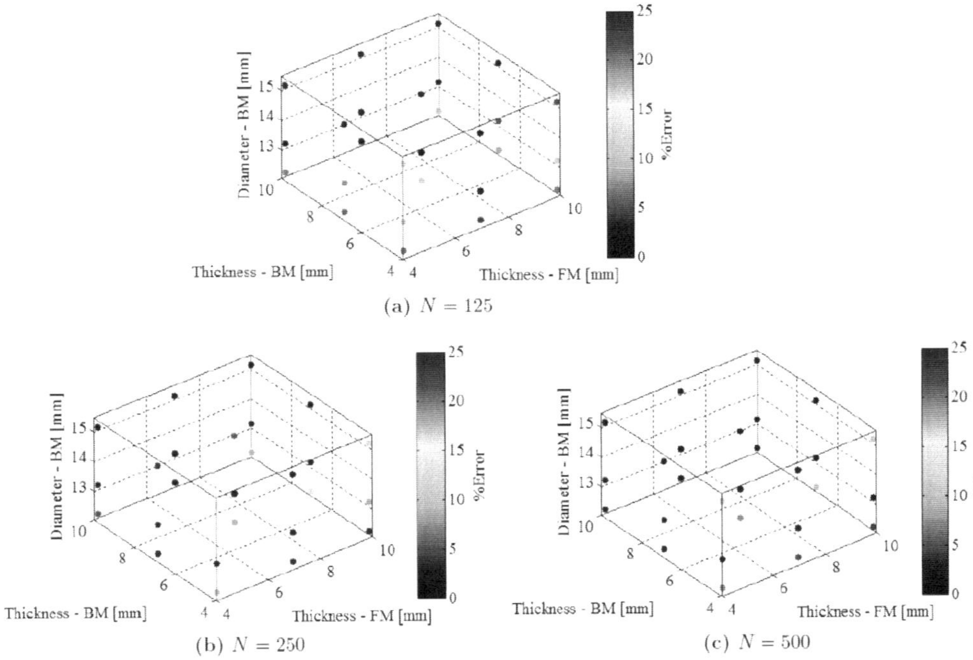

Figure 8. Errors within the actual design space associated with the constructed objective function.

The response surface indicating the performance of different transducer designs has been successfully constructed. The errors associated within the design space are found to be acceptable.

In optimum design, it is not always the case that the design will converge to the global maximum when more than one local maxima exists. Convergence is usually driven towards one of the local maxima, depending on where the starting point of the algorithm is situated. To capture all the local maxima in the objective function, the optimization procedure was carried out 100 times with the starting points randomly selected within the design space. Plotted in Figure 9 is a histogram of the local optimum designs predicted from the 100 runs. Each bin in the histogram represents the cost function value of a local optimum point, with the height of the bin indicating the number of times that the optimization convergence was to a local optimum. The three sets of data points in RBF interpolation are all considered for comparison.

A single bin in the histogram may contain different types of designs, for example, bins K1, K8 and B2 respectively contain designs K1a and K1b, K8a and K8b and B2a and B2b. A similar characteristic between different designs of the same bin is the value of the objective function, implying that they have the same energy. The histogram indicates that there exist a global maximum with a function value of 5.6e-8mHz, that was predicted in bins B9 and K8. It can be seen in Figure 10 that the designs B9 and K8b are the same, however due to computational errors, the two predictions are slightly different.

RBF interpolation predicted that the energy associated with the two global optimum designs (K8a and K8b plotted in Figure 11) is approximately 5.6e-8mHz. To justify the accuracy of this prediction, the SAFE3D method was used to compute the objective function of these two designs. The results from RBF interpolation and the SAFE3D method are compared in Table 1.

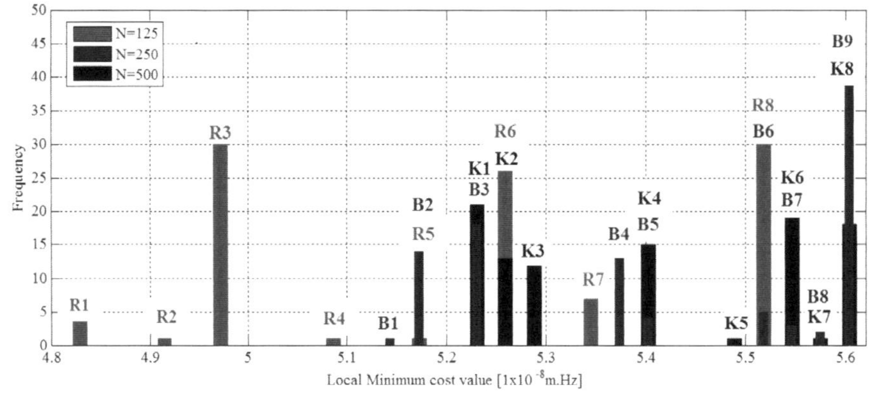

Figure 9. A histogram of the predicted local optimum designs.

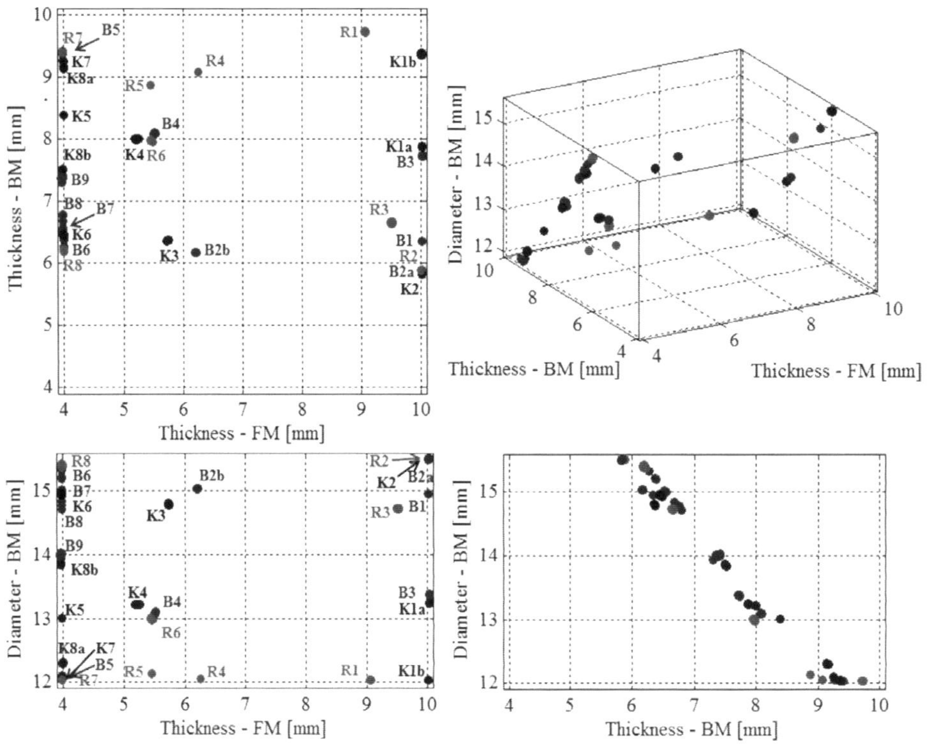

Figure 10. The location of the predicted local optima in the design space.

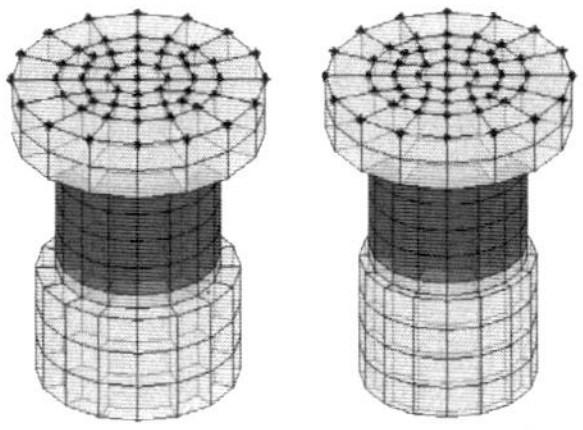

(a) K8a (b) K8b

Figure 11. The global optimum designs.

Table 1. Validation of the global optimum designs.

	Design K8a	**Design K8b**
FM thickness	4	4
BM thickness	7.6	9.1
BM diameter	13.8	12.3
RBD predicted energy	5.6114	5.6013
SAFE-3D predicted energy	5.5968	5.5712
$\|\%Error\|$	0.261	0.5403

It is evident from the table that the agreement between the RBF method and the SAFE3D method is exceptionally good.

CONCLUSION

This paper has demonstrated that the SAFE-3D modelling method combined with optimization by DoE and response surfaces can be used to efficiently design transducers. The effectiveness of the response surface method when solving practical design optimization problems was also demonstrated. The method was found to be efficient and reliable. Another notable aspect of optimizing using the response surface method is that it does not only output an optimum design but also produces informative plots, which provide insight into the system being explored prior to optimization.

REFERENCES

1 Burger, F. A., "A Practical Continuous Operating Rail Break Detection System Using Guided Waves", in *18th World Conference on Nondestructive Testing*, 2012.

2 Long, C. S. and Loveday, P. W., "Prediction of Guided Wave Scattering by Defects in Rails Using Numerical Modelling", *AIP Conference Proceedings*, Vol. 1581. pp. 240 – 247, 2014.

3 Hayashi, T., Song, W. J. and Rose, J. L., "Guided Wave Dispersion Curves for a Bar With Arbitrary Cross-section, a Rod and Rail Example", *Ultrasonics*, vol. 41. pp. 175 - 183, 2003.

4 Gavric, L., "Computation of Propagative Waves in Free Rail Using a Finite Element Technique", *Journal of Sound and Vibrations*, vol. 185. No. 3, pp. 531-543, 1995.

5 Ramatlo, D. A., Long, C. S., Loveday P. W. and Wilke, D. N., "SAFE-3D analysis of a piezoelectric transducer to excite guided waves in a rail web", in *Review of Progress in QNDE*, 2015.

6 Loveday, P. W., "Simulation of Piezoelectric Excitation of Guided Waves Using Waveguide Finite Elements," *IEEE Transactions on Ultrasoncs, Ferroelectrics, and Frequency Control*, vol. 55. No. 9, pp. 2038-2045, 2008.

7 Damljanovic, V. and Weaver, R. L., "Forced Response of a Cylindrical Waveguide with Simulation of the Wavenumber Extraction Problem," *Journal of Acoustical Society of America*, vol. 115. no. 4, pp. 1582-1591, 2004.

8 Arora, J. S., “Introduction to Optimum Design”, *Academic Press*, Boston, Ed.3, 2012.

Computational modelling of slag foaming in pyrometallurgical furnaces

Quinn G. Reynolds*, Markus W. Erwee

Pyrometallurgy Division, Mintek, Private Bag X3015, Randburg 2125

quinnr@mintek.co.za

ABSTRACT

In pyrometallurgical furnaces, raw materials are processed into products of value at high operating temperatures (1500°C or more). The materials are typically in a molten state, and will separate into two or more immiscible liquid phases due to differences in chemical composition. One such phase is slag, which contains predominantly metal oxides and is generally the least dense of the liquid phases present. Expansion and foaming of the slag layer is occasionally observed during furnace operation, and occurs as a result of the injection or generation of gases in the molten bath. Uncontrolled slag foaming may cause damage to the furnace interior or, in extreme cases, catastrophic failure. Controlled slag foaming can however be used to great effect in limiting energy losses from furnaces and improving their thermal efficiency.

A computational modelling study of gas passing through a slag layer using the OpenFOAM® platform is presented. Existing multiphase fluid flow solvers based on the volume of fluids (VOF) method were extended and used to model the influence of operating conditions and material properties on the extent and nature of slag foaming. The resulting model was successfully validated against empirical data and correlations from literature. A dimensional analysis was conducted on the slag foaming problem, and the effect of various key dimensionless groups on the computational model was studied.

KEYWORDS

Pyrometallurgy, Multiphase, Foaming

INTRODUCTION

The discipline of pyrometallurgy deals with the treatment of raw materials at high temperatures in order to create products of increased value. Pyrometallurgical furnaces are used extensively for the production and recycling of many commodities including iron and steel, ferro-alloys, titanium dioxide, platinum group metals, and others (Jones and Curr [1]). Because of the very high operating temperatures, the process materials inside furnaces are generally in the molten state, and typically separate into two immiscible phases, slag and metal. The slag phase contains predominantly metal oxides and generally floats above the denser metal phase.

Furnace processes may be classified in many ways, including but not limited to whether any chemical reactions are utilised to convert feed materials from one form to another (melting vs smelting), and whether or not the molten bath of process material inside the vessel is exposed to the gas atmosphere (open-bath) or covered by a layer of solid feed material (closed-bath). Some examples in the case of direct-current electric furnaces are shown in Figure 1. In the present study, only open-bath smelting furnaces are considered.

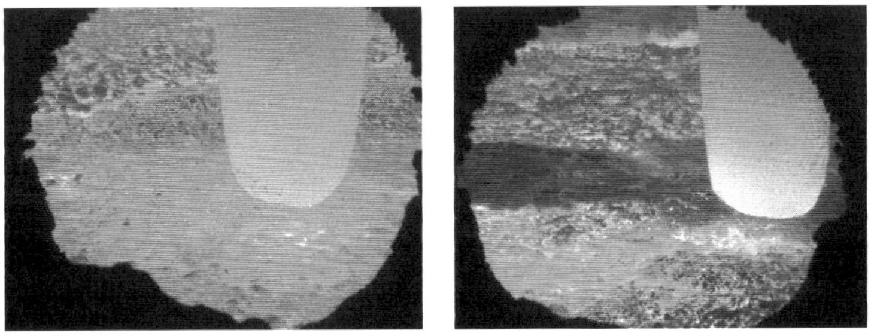

Figure 1. DC pilot-scale furnace interior, open bath (left) and partially closed bath (right) conditions

During the chemical reactions occurring in smelting furnaces, which typically involve the reaction of metal oxides with a carbonaceous reductant to produce metal products, gases such as carbon monoxide are evolved. Depending on the mechanism of the reduction reactions, the gas may be generated at the contact surfaces between solid reductant particles and molten slag (which in general are dispersed throughout the slag layer), or at the interface between the metal and slag liquid phases. The generated gas forms bubbles and rises through the slag in order to leave the furnace vessel causing the molten slag layer to be partially displaced and increasing its effective thickness. If the displacement is appreciable, this process is referred to as slag foaming.

Slag foaming is very much a double-edged sword for furnace operations – if it can be managed and controlled, the foamy layer of slag can help to shield the hot bath surface and coat the walls and roof of the furnace, reducing energy losses and protecting the refractory materials lining the vessel. If however uncontrolled foaming events occur, they can lead to increased wear and damage to the refractories and ultimately to premature furnace failure. Understanding slag foaming as it relates to a given process is therefore of considerable value to furnace operators.

Slag foaming behaviour is generally grouped into two distinct phenomena – expanded slags and true foaming. In slag expansion, the slag layer increases in thickness primarily as a result of the time taken for the gas bubbles to rise through it. In true foaming, the slag layer thickness increases primarily as a result of the formation of a layer of bubbles with high void fraction which accumulates near the slag surface. The difference is illustrated in Figure 2.

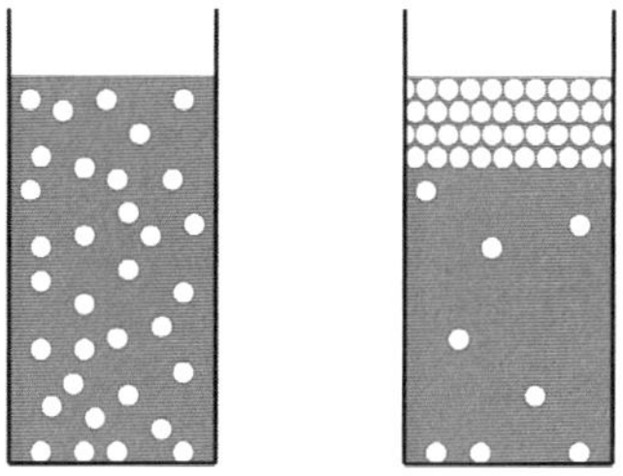

Figure 2. Comparison of slag expansion (left) and true foaming (right)

In real furnace systems the distiction between true foaming and expanded slags is often somewhat academic. Expansion is in general associated with higher and true foaming with lower gas flowrates (Seetharaman et al [2]), but it is quite possible for both phenomena to be occurring to some degree at the same time (Ozturk and Fruehan [3]). In the present paper however, the focus is on true slag foaming phenomena and the development of possible methods to model them.

One indicator of how likely a particular metallurgical slag is to generate a foam layer is the slag foaming index Σ, which has units of time and is defined as the rate of change of the foam layer thickness H_f with superficial gas velocity (Seetharaman et al [2]). Several authors have raised concerns over the use of the slag foaming index as a variable (Lotun and Pilon [4]), however, it remains valuable as a guidance tool for many industrial furnace operations where foaming occurs. A number of experimental studies of the effect of changing various parameters of interest on the slag foaming index have been performed; much of this body of earlier work also presents the results of dimensional analyses of the slag foaming problem, and empirical relationships between the dimensionless groups identified. Some examples are shown below (see Table 1 for definitions of the symbols used; the K_i are constants fitted from experimental data).

$$\Sigma = K_1 \frac{\mu}{\sqrt{\rho\sigma}} \qquad \text{(Jiang and Fruehan [5])} \qquad (1)$$

$$\Sigma = K_2 \frac{\mu^{1.2}}{\rho\sigma^{0.2} d_b^{0.9}} \qquad \text{(Zhang and Fruehan [6])} \qquad (2)$$

$$\Sigma = K_3 \frac{\mu^{0.8}}{(\rho\sigma)^{0.8} d_b^{2.6} u_g^{0.2}} \qquad \text{(Pilon et al [7], assuming zero minimum foaming velocity)} \qquad (3)$$

Although there is considerable variation in the exact values of the powers and empirical constants, the degree to which a slag is prone to foaming is generally proportional to the slag's viscosity, and inversely proportional to its density, surface tension, and the diameter of the gas bubbles.

Following the general approach of Pilon et al [7] but retaining the slag foaming index, the parameters shown in Table 1 are identified in relation to the slag foaming problem. It is assumed that the gas phase density and viscosity are low relative to the slag, and have comparatively little effect on bubble and foam behaviour.

Table 1. Variables of interest in slag foaming

Slag density	ρ, kg/m^3	Bubble diameter	d_b, m
Slag viscosity	μ, Pa.s	Interfacial surface tension	σ, N/m
Superficial gas velocity	u_g, m/s	Gravitational acceleration	g, m/s^2
Slag foaming index	$\frac{dH_f}{du_g} = \Sigma$, s		

Dimensional analysis using these variables yields the following functional relationship between four dimensionless groups:

$$N_\Sigma = \frac{g\mu\Sigma}{\sigma} = f(N_{Mo}, N_{Ca}, N_{Ar}) \quad (4a)$$

$$N_{Mo} = \frac{g\mu^4}{\rho\sigma^3} \quad (4b)$$

$$N_{Ca} = \frac{\mu u_g}{\sigma} \quad (4c)$$

$$N_{Ar} = \frac{g\rho^2 d_b^3}{\mu^2} \quad (4d)$$

Here, N_{Mo} is the Morton number, N_{Ca} the Capillary number, and N_{Ar} the Archimedes number. The effect of these groups in determining the functional dependence in (4a) will be studied further during the course of the paper.

To date, it would appear that little work has been performed in the field of computational modelling of slag foaming. This can be attributed to the success of theoretical and empirical correlations such as (1) – (3) and others in indicating foam formation within narrow parameter ranges for particular processes. Another contributing factor is that for design purposes, there is generally more interest in modelling the effects of foaming at the scale of industrial furnaces (tens of meters) rather than at the scale of the gas bubbles in the slag (a few millimeters). The foam layer is therefore often completely ignored in computational models, or treated in an approximated manner – for example as a continuum phase requiring considerable empirical input into the calculation (Sattar et al [8]), or using Euler-Lagrange-type formulations which represent bubbles or groups of bubbles as individual particles of a set shape, usually spherical (Lain et al [9]). Although they have great engineering utility, such approaches are not entirely satisfactory for a more fundamental exploration of slag foam formation and evolution at smaller scales.

MODEL DESCRIPTION

In order to study foaming phenomena in slags, a computational solver was implemented using the OpenFOAM® open source framework (OpenFOAM [10]). An existing volume-of-fluids (VOF) based multiphase flow solver was extended and enhanced for this purpose.

Several interesting challenges arise when attempting to model bubble phenomena using VOF methods. Perhaps the most important is the large difference in scale between the bubbles, which are typically a few millimeters in diameter in the slag foaming case, and the interfacial films between them, which may be as little as a few nanometers in thickness. As the VOF method can by definition only model an interface between phase fields which is of the order of the mesh resolution and no sharper, direct simulation of the film would require excessively fine and computationally-expensive meshes. If coarser meshes are used, particularly with materials such as slags having high surface tension, film structures in models using this method rupture prematurely – this makes VOF unsuitable in its standard form for modelling phenomena such as foams.

The aim of the present solver was to address some aspects of this problem in order to permit direct modelling of slag foam dynamics. To this end, the following modifications to the multiphaseInterFoam solver in OpenFOAM® were made:

- Each bubble in the model is treated as a separate phase, with each of these phases having physical properties identical to the main gas phase. Interfacial surface tension between the bubble phases is set at double the surface tension between the slag and gas phases to approximate the presence of a double-walled film of slag.
- A bubble bursting algorithm based on a film drainage sub-model was implemented. This involves testing of the interfaces between all pairs of bubble phases in the model at each time step. If the two phases are in sufficiently direct contact with each other, the film thickness between each of the involved phases is calculated using a film drainage sub-model. Once the film thickness for a particular bubble phase drops below a specified value (the minimum stable film thickness), and if the phase concerned is in contact with the main gas atmosphere, the bubble is burst by sweeping the model region, adding the bubble phase fraction to the main gas phase fraction, and setting the bubble phase fraction to zero. This has the effect of instantaneously removing the film separating this particular bubble from the main gas atmosphere, and merging its contents with the surrounding gas phase.
- By way of example a relatively simple film drainage sub-model based on flow driven by gravitational and capillary forces was implemented, and relates the film drainage time to the density, viscosity and surface tension of the slag as well as a representative scale measure calculated from the film perimeter length at every time step. Considerably more sophisticated approaches based on theoretical and experimental treatments of thin films (Bazlekhov et al [11], Chan et al [12]) are possible, and may be integrated into the present framework in a similar way.

As part of the standard multiphaseInterFoam solver algorithm, the VOF method is used to take account of the phase separation, requiring solution of a separate convective transport equation for the volume fraction field of each phase. This permits the use of a single velocity and pressure field to describe all phases in the multiphase system. The Pressure Implicit with Splitting of Operators (PISO) predictor-corrector algorithm is used for calculation of the flow fields at each time step. Gradient-limited discretisations were used for all divergence terms with the exception of those related to the phase fraction fields, for which the van Leer compression scheme together with the Multidimensional Universal Limiter for Explicit Solution (MULES) solution algorithm was specified. Adaptive time stepping was applied to keep the mesh Courant number below 0.75 in order to ensure acceptable numerical stability.

Film Sub-Model

The film sub-model was developed from the work of Nguyen et al [13] on the drainage of liquid magma from the films around gaseous bubbles at the surface of the melt. Two relationships are given for the drainage time, derived from the equations of fluid flow within the film:

$$\frac{d\delta}{dt} = -\frac{1}{C_g}\frac{\rho g d_b}{\mu}\delta \tag{5a}$$

$$\frac{d\delta}{dt} = -\frac{1}{C_c}\frac{\sigma}{\mu d_b}\delta \tag{5b}$$

(5a) describes film drainage due to gravitational effects, and (5b) describes film drainage as a result of capillary pressure. Here, δ is the film thickness, and C_g and C_c are constants independent of the fluid properties and bubble geometry (C_g/C_c is given as 0.25). Combining (5) and integrating gives a relationship for the evolution of the film thickness used in the sub-model:

$$\frac{\delta}{\delta_0} = \exp\left(-\frac{1}{C_g}\left(\frac{\rho g d_b}{\mu} + \frac{C_g}{C_c}\frac{\sigma}{\mu d_b}\right)t\right) \tag{6}$$

(6) is used to track the film thickness for each bubble phase in the slag foaming model over time after activation of the film sub-model. Once the film thickness drops below a certain critical value, bursting of the bubble phase into the main gas phase is triggered.

Geometry and Boundary Conditions

At present only two-dimensional models are considered – although the solver is capable of working in three dimensions, this is computationally expensive and is expected to be of value only once the model's robustness and quantitative accuracy have been more firmly established.

The model region is a simple rectangular vessel, with one or more injection ports at the base through which the gas to form the bubbles is introduced. A diagram of the region showing the various phase fields and boundaries is shown in Figure 3.

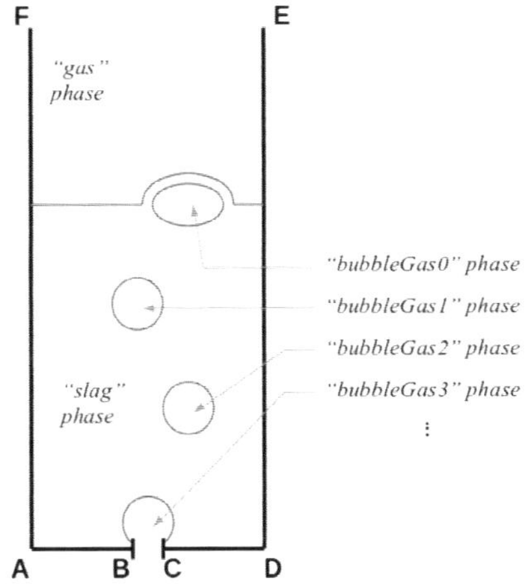

Figure 3. Diagram of model region

The following boundary conditions are applied at the perimeter of the region: FAB and CDE are treated as solid walls, with velocity set to zero and zero pressure gradient. The gradient of all phase fields is also assumed to be zero at these surfaces (equivalent to a wetting angle of 90°). EF is assumed open to atmosphere, at a specified pressure and a velocity calculated from the pressure field. The phase fields are subject to an inlet-outlet condition depending on the direction of flow – zero gradient if outflow, and a gas phase fraction of 1 if inflow. Finally, at boundary BC an intermittent inflow condition is specified, with a fixed velocity for short, regularly-spaced periods to inject bubbles of fixed size at a known average flowrate. During each injection period, the phase fractions of the various bubble phases at this boundary are modulated to ensure that each phase is injected as a separate bubble.

Model performance

Model runs were executed using a combination of local cluster resources (Intel i7 quad-core machines over gigabit ethernet) and nodes of the Lengau machine at the Centre for High Performance Computing. Each model run made use of up to 16 processor cores, however, parallel scaling was generally observed to start falling away from linear at around 8 cores or more due to the relatively small size of the 2D models (this is expected to improve significantly in 3D). Coarse-grained parallelism was employed during the parameter space studies by running multiple model cases simultaneously, with each case using four processors to ensure maximum computational efficiency. Under these conditions, run times of between 16 and 24 hours for a model time of 5 s were typical.

MODEL VALIDATION

The accuracy and validity of the results from the computational model were analysed using an experimental case from slag foaming literature (Ozturk and Fruehan [3]).

The assumed composition of the slag was 48% CaO, 10% FeO, 32% SiO_2 and 10% Al_2O_3 (by mass). Properties were calculated at 1773 K. The effect of a small amount of precipitated solid slag were ignored for all calculations, since the composition of the liquid portion of the slag does not differ significantly from the bulk composition. The viscosity of the slag was estimated using the 'Viscosity' module of FactSage™ (Bale et al [14]), and surface tension and density were calculated based on empirical methods (Mills et al [15]). For the Mills-Keene method, the molar volumes for each of the slag components were taken from Mills et al [16]. Properties of the gas, assumed to be 100% CO at 1773 K, were estimated using ideal gas law relationships. The parameters used in the model are shown in Table 2. Dimensions were chosen to match the size of the crucibles and the gas injection depth used in Ozturk and Fruehan [3].

In the first set of tests, the model's dependence on the resolution of the computational mesh was examined. A value of 20 s was assumed for the constant C_g in the film sub-model. A single gas bubble was injected at the base of the slag layer at the beginning of each simulation, and the upper position of the slag surface on the region center-line was recorded over time. The results are shown in Figure 4 and Figure 5.

Table 2. Model parameters used in validation and testing

ρ	1990 kg/m^3	Gas density, ρ_g	0.192 kg/m^3
μ	0.1407 Pa.s	Gas viscosity, μ_g	5.64 x 10^{-5} Pa.s
σ	0.509 N/m	d_b	0.005 m
Initial film δ	10^{-6} m	Critical film δ	10^{-8} m
Region width	0.020 m	Initial slag height	0.037 m

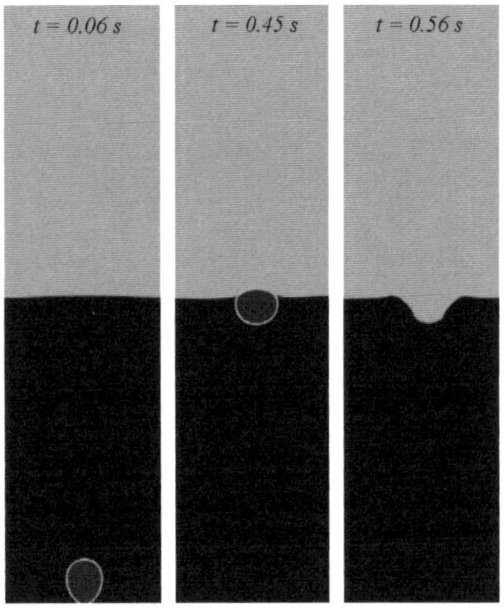

Figure 4. Phase field at various times for mesh resolution 1.67 x 10^{-4} m

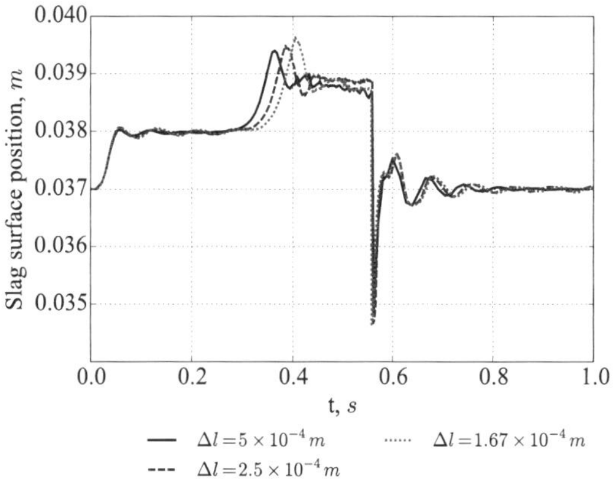

Figure 5. Evolution of slag surface at different mesh resolutions

In general the slag foaming model scales well with mesh resolution, producing relatively similar results even on very coarse meshes. In particular, the time taken for the bubble to burst after it reaches the surface is repeatable across a range of mesh sizes, suggesting that foam behaviour will be captured with reasonable accuracy.

In the second set of tests, a value for C_g was estimated using experimental slag foaming index measurements (Ozturk and Fruehan [3]) from which $\Sigma \approx 0.96$ s at 1773 K. C_g was fitted by comparing the simulated foam height with the experimental data at a fixed superficial gas velocity of 1 cm/s, giving a value of 84.5 s. The gas velocity was then varied to test if the model correctly captures the expected response of the foam height. In all cases 30 different bubble phases were used, and a total model time of 5 s was simulated. Unless otherwise indicated, the model region was meshed at a resolution of 3.3 x 10^{-4} m.

Foam height was calculated from the model data as the difference between the lowest extremity of the "gas" phase and the highest extremity of the "slag" phase along a selection of vertical line samples taken through the domain at various times. The resulting time series of instantaneous foam heights were averaged over the second half of the 5 s time period to give an indication of the steady-state foam height.

Figure 6 and Figure 7 show results from the slag foaming model running at lower and higher gas flowrates. The foam layer that is formed at the slag surface is visibly thicker in the case of higher gas flowrates, as would be expected from the theory of slag foaming. The calculated average foam heights are compared in Table 3.

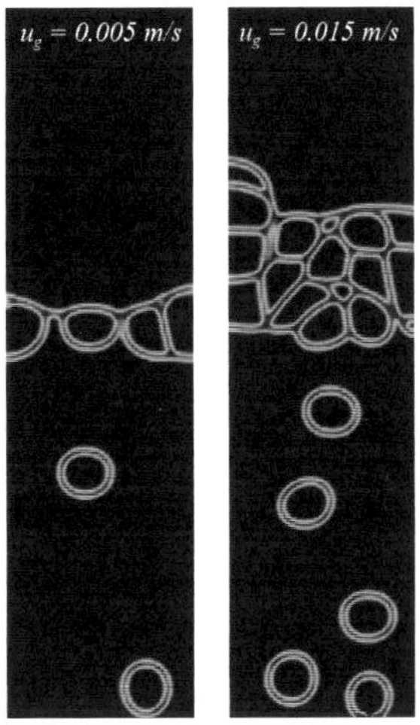

Figure 6. Phase field boundaries for different superficial gas velocities

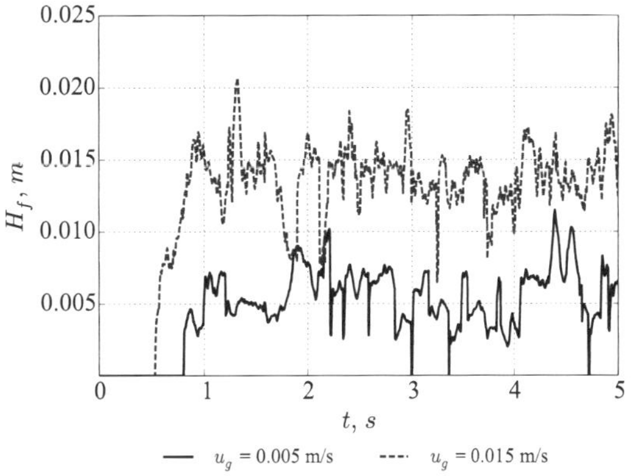

Figure 7. Evolution of foam thickness for different superficial gas velocities

Table 3. Response of computational model to superficial gas velocity

u_g	H_f *(theorerical with* $\Sigma = 0.96$ *s)*	H_f *(computational model)*
0.005 m/s	0.0048 m	0.0055 m
0.015 m/s	0.0144 m	0.0138 m

In both cases, the computational model predicts foam heights within a few percent of the theoretical value assuming a constant value of the slag foaming index. The slight inverse dependence of Σ on u_g observed in the modelling results may also be physically realistic in light of empirical correlations such as (3).

RESULTS AND DISCUSSION

In order to examine the effect of various parameters on the behaviour of the slag foaming model, a base case set of conditions was first identified based on typical average values for a wide range of different slag materials (Mills et al [15], Mills et al [16]). These are shown in Table 4 (other model parameters are as indicated in Table 2).

Table 4. Model parameters for base case

ρ	3000 kg/m^3	σ	0.25 N/m
μ	0.2 Pa.s	d_b	0.005 m
u_g	0.01 m/s	C_g	84.5 s

Results from the base case model are shown in Figure 8 and Figure 9. The slag layer in the model is seen to grow from zero thickness initially while the bubbles rise through the slag to reach the gas interface. Formation of foam is then rapid, and the thickness of the foam layer stabilises at slightly over 10 mm after approximately 1.5 s.

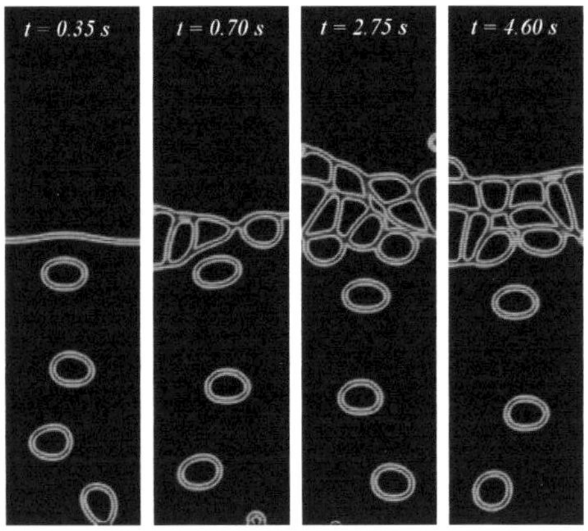

Figure 8. Phase field boundaries at different times, base case

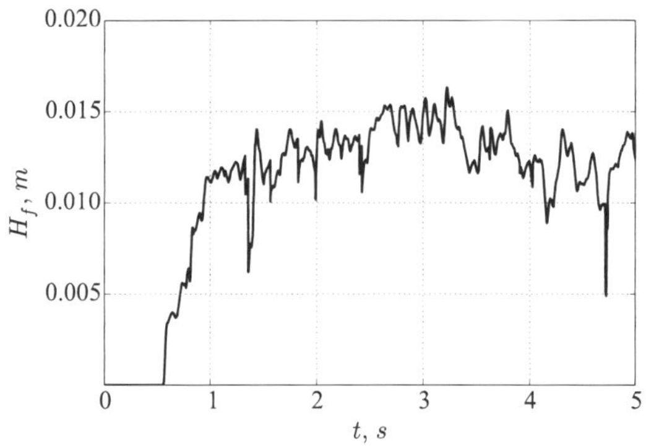

Figure 9. Evolution of foam thickness over time, base case

From the time-averaged value of the foam thickness (0.0130 m), the slag foaming index was determined as 1.30 s for the base case set of parameters.

Response of the model to primitive variables

The response of the computational model's prediction of slag foaming behaviour to changes in primitive variables used to describe the slag properties and operating conditions was tested. Several of the parameters shown in Table 4 were altered one by one while keeping all other variables constant, and the effect on the nature and degree of slag foaming was studied. The results in terms of the slag foaming index are shown in Table 5.

Table 5. Effect of primitive variables on modelled value of Σ

Parameter variation	*Σ (model)*
Base Case	1.30 s
μ = 0.4 Pa.s	2.56 s
σ = 0.125 N/m	1.71 s
ρ = 1500 kg/m^3	1.81 s
d_b = 0.004 m	1.38 s
d_b = 0.006 m	1.17 s

It can be seen that doubling the slag viscosity has the most significant impact on the slag foaming index. Halving the surface tension results in a moderate increase in the index relative to the base case, and halving of slag density has a similar effect. The model is considerably less sensitive to bubble diameter, but a slight trend of foaming index increasing with decreasing bubble diameter is observed. Overall, the functional dependence of the computational model matches well with one of the most common forms of empirical model, that shown in equation (1).

Selected visualisations of the phase field boundaries for two of the cases are shown in Figure 10 and Figure 11. The entrainment of the slag phase into the foam layer at different bubble sizes is shown in Figure 12.

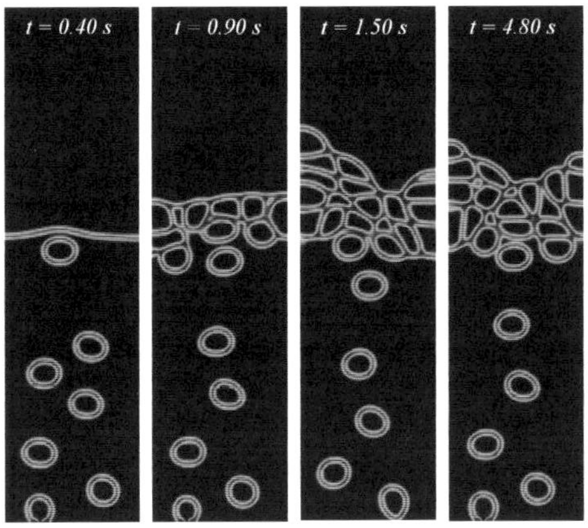

Figure 10. Phase field boundaries at different times for d_b = 0.004 m

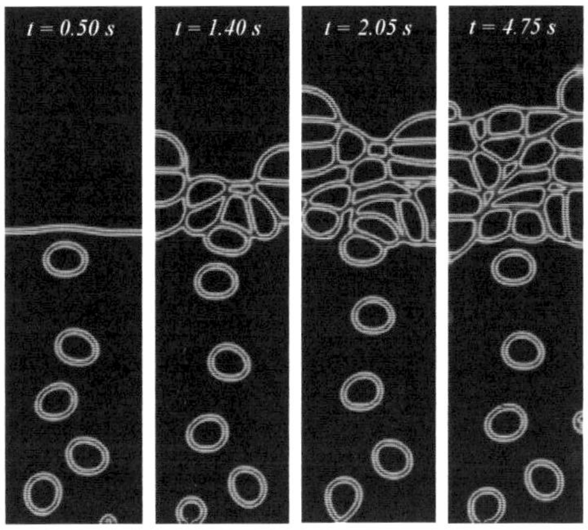

Figure 11. Phase field boundaries at different times for μ = 0.4 Pa.s

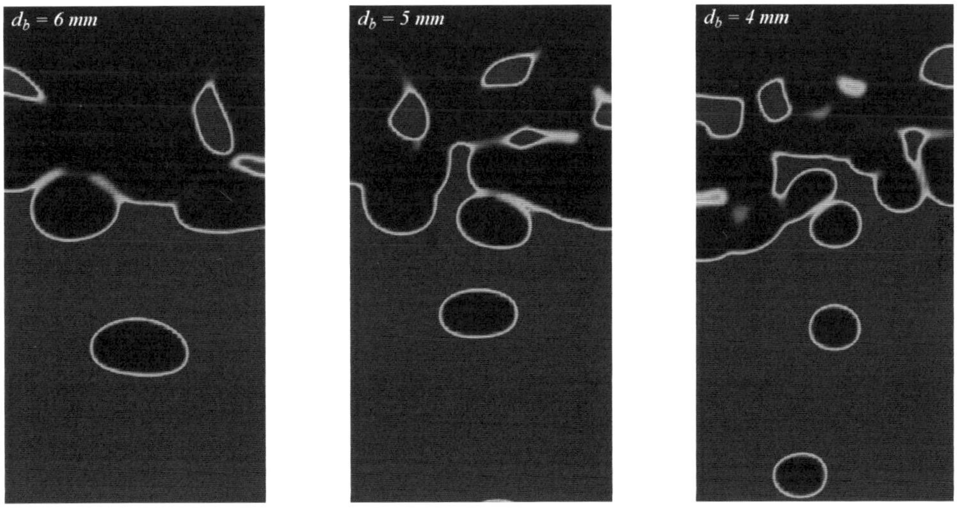

Figure 12. Close-ups of foam layer at end of simulation showing slag phase fraction (blue = 0, red = 1) at different bubble diameters

Smaller bubbles are seen to hold up more of the slag phase in the spaces between them, keeping the film between bubbles "wet" and increasing the lifetime of the foam. Similarly in the case of higher slag viscosity, drainage from the films (controlled predominantly by the film sub-model in use) is substantially slowed when viscosity increases, increasing the time taken before film rupture. As a result, the foam layer thickness at a set gas flowrate is larger, producing a higher slag foaming index.

Response of the model to dimensionless variables

In order to study the dependence of the computational model on the dimensionless groups defined in equation (4), several of the slag parameters were altered simultaneously so as to keep all but one of the dimensionless groups constant. For the base case set of parameters in Table 4, N_{Mo} = 3.35 x 10^{-4}, N_{Ca} = 8.00 x 10^{-3}, and N_{Ar} = 276.

Six additional cases were set up by multiplying or dividing the value of each of the dimensionless groups by two while keeping the others constant. The average steady-state height of the foam layer generated by the computational model was then estimated and used to determine the slag foaming index as before, allowing the value of the foam index group N_Σ to be calculated. Results are shown in Figure 13 and Figure 14.

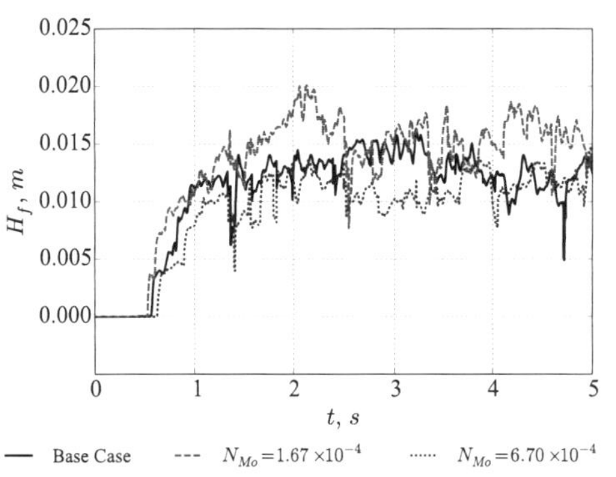

Figure 13. Evolution of slag foam height at different values of N_{Mo}

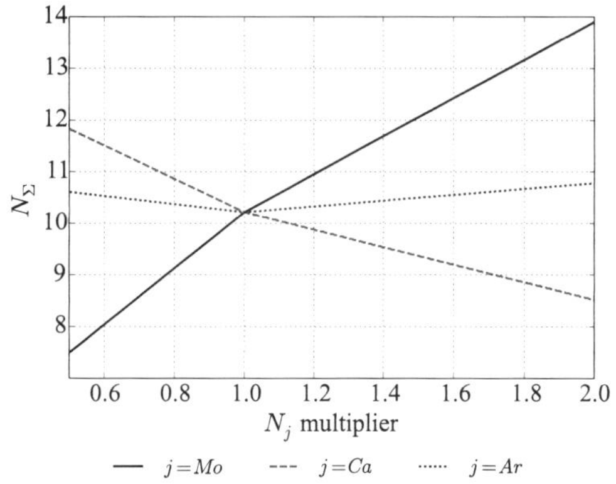

Figure 14. Sensitivity of equation (4) to dimensionless quantities

It was found that N_Σ depends strongly on N_{Mo} (increasing as N_{Mo} increases) and N_{Ca} (decreasing as N_{Ca} increases). Relatively little dependence on N_{Ar} was observed in the range of values tested. This behaviour is in agreement with the results from the primitive variable cases which indicated a strong dependence of Σ on μ, which affects predominantly N_{Mo}, and a much weaker dependence on d_b, which affects predominantly N_{Ar}.

CONCLUSIONS

A computational model was successfully developed to study the problem of slag foaming. Simple extensions of standard volume-of-fluid solvers permitted the inclusion of film drainage models and bubble film rupture dynamics to accurately capture the evolution and behaviour of the foam layer. The model was found to be relatively mesh-independent, and was successfully validated against experimental data.

The sensitivity of the computational model's predictions of slag foaming index to variations in slag physical properties and operating conditions was studied. The results were found to match well with empirical correlations of a form commonly used in the smelting industry, in which foaming index is proportional to slag viscosity and inversely proportional to the square root of slag density and surface tension. In addition, the response of the model to changes in various dimensionless groups used to characterise slag foaming was examined, with the Morton and Capillary numbers found to have the largest effect on foaming behaviour.

Much further work is possible in this area. The computational efficiency of the slag foaming model as currently implemented is rather poor, and scales with the number of bubble phases included – a more elegant reformulation of the multiphase bubble flow algorithm would therefore be of value. In addition, the model suffers from the effects of parasitic currents at the bubble phase interfaces, exacerbated in the case of slag foaming problems by the very high interfacial surface tension and very low gas density. Implementation of improved interface-compression methods such as HiRAC (Oxtoby et al [17]) or similar should be investigated as a means of mitigating this issue.

With the alleviation of numerical performance limitations and further validation against experimental data, it is believed that computational models similar to those presented here could potentially become viable alternatives to laboratory testing for estimation of slag foaming behaviour in pyrometallurgical furnaces.

ACKNOWLEDGEMENTS

This paper is published by permission of Mintek. The authors also gratefully acknowledge support from the Centre for High Performance Computing (CHPC).

REFERENCES

1 Jones, R.T., and Curr, T.R., "Pyrometallurgy at Mintek", SAIMM Southern African Pyrometallurgy 2006, Johannesburg, South Africa, 2006.

2 Seetharaman, S., McLean, A., Guthrie, R., and Sridhar, S., "Treatise on Process Metallurgy Volume 2 (Process Phenomena)", Elsevier, Oxford, United Kindom, 2014.

3 Ozturk, B., and Fruehan, R.J., "Effect of Temperature on Slag Foaming", Metallurgical and Materials Transactions B, Vol. 26B, pp 1086 – 1088, 1995.

4 Lotun, D., and Pilon, L., "Physical Modeling of Slag Foaming for Various Operating Conditions and Slag Compositions", ISIJ International, Vol. 45, No.6, pp. 835 – 840, 2005.

5 Jiang, R., and Fruehan, R.J., "Slag Foaming in Bath Smelting", Metallurgical and Materials Transactions B, Vol. 22B, pp. 481 – 489, 1991.

6 Zhang, Y., and Fruehan, R.J., "Effect of the Bubble Size and Chemical Reactions on Slag Foaming", Metallurgical and Materials Transactions B, Vol. 26B, pp. 803 – 812, 1995.

7 Pilon, L., Fedorov, A.G., and Viskanta, R., "Steady-State Thickness of Liquid-Gas Foams", Journal of Colloid and Interface Science, Vol. 242, No. 2, pp. 425 – 436, 2001.

8 Sattar, M.A., Naser, J., and Brooks, G., "Numerical simulation of slag foaming on bath smelting slag ($CaO–SiO_2–Al_2O_3–FeO$) with population balance modeling", Chemical Engineering Science, Vol. 107, pp. 165 – 180, 2014.

9 Lain, S., Broder, D., Sommerfeld, M., and Goz, M.F., "Modelling hydrodynamics and turbulence in a bubble column using the Euler–Lagrange procedure", International Journal of Multiphase Flow, Vol. 28, pp. 1381 – 1407, 2002.

10 OpenFOAM®, Version 3.0.1, http://www.openfoam.org, accessed 2016.

11 Bazhlekov, I.B., Chesters, A.K., and van de Vosse, F.N., "The effect of the dispersed to continuous-phase viscosity ratio on film drainage between interacting drops", International Journal of Multiphase Flow, Vol. 26, pp. 445 – 466, 2000.

12 Chan, D.Y.C., Klaseboer, E., and Manica, R., "Film drainage and coalescence between deformable drops and bubbles", Soft Matter, Vol. 7, pp. 2235 – 2264, 2011.

13 Nguyen, C.T., Gonnermann, H.M., Chen, Y., Huber, C., Maiorano, A.A., Gouldstone, A., and Dufek, J., "Film drainage and the lifetime of bubbles", Geochemistry Geophysics Geosystems, Vol. 14, No. 9, pp. 3616 – 3631, 2013.

14 Bale, C.W., Chartrand, P., Decterov, S.A., Eriksson, G., Hack, K., Mahfoud, R.B., Melançon, J., Pelton, A.D., and Petersen, S., "FactSage Thermochemical Software and Databases", Calphad, Vol. 26, No. 2, pp. 189 – 228, 2002.

15 Mills, K.C., Yuan, L., Zhang, G.H. and Chou, K.C., "A Review of the Factors Affecting the Thermophysical Properties of Silicate Slags", High Temperature Material Processes, Vol. 31, No. 4-5, pp. 301 – 321, 2012.

16 Mills, K.C., Yuan, L., and Jones, R.T., "Estimating the physical properties of slags", Journal of the Southern African Institute of Mining and Metallurgy, Vol. 111, No. 10, pp. 649 – 658, 2011.

17 Oxtoby, O.F., Malan, A.G., and Heyns, J.A., "A computationally efficient 3D finite-volume scheme for violent liquid-gas sloshing", Numerical Methods in Fluids, Vol. 79, No. 6, pp. 306 – 321, 2015.

10th South African Conference on Computational and Applied Mechanics
Potchefstroom
3 – 5 October 2016

Uncertainties in Modelling an Electromagnetic Levitation Cell (EMLC)

Suzanne Roberts[1]*, Schalk Kok[1], Helen Inglis[1], Johan Zietsman[2]

[1]Department of Mechanical and Aeronautical Engineering, University of Pretoria, South Africa
[2]Department of Materials Science and Metallurgical Engineering, University of Pretoria, South Africa

suzanne.elize.roberts@gmail.com

ABSTRACT

Electromagnetic levitation melting shows some promise as a containerless heating method for measuring material properties. However, the design of the levitation cell influences what temperature ranges are achievable with certain sample materials and sample sizes. It is therefore desirable to have a numerical model available to design levitation cells that allow experiments for measuring material properties of specific materials within specified temperature ranges.

A number of levitation cell models exist in literature but these models contain uncertainties that present challenges when used to design levitation cell experiments. Two of the main causes of uncertainty are the temperature dependent material property values of the levitated sample, and accurate measurement of the large, high frequency current in the levitation cell coil for validation purposes. This work describes these uncertainties and attempts to explore strategies to manage uncertainty in electromagnetic levitation cell modelling.

KEYWORDS

pyrometallurgical modelling, electromagnetic levitation melting (ELM), verification, high temperature material properties, uncertainty

INTRODUCTION

The quality of the results obtained from a numerical model depends on how well the physics is represented in the model, and how well the model is solved. Accurately describing the physics in the model includes providing the model with accurate material property data. This presents a challenge in pyrometallurgical modelling, where processes occur at high temperatures and material properties at these high temperatures are required. High temperature material properties of reactive materials are difficult to find in literature because these properties are difficult to measure experimentally. Few container materials can withstand the temperatures required for experimental property measurements of reactive materials and even then contamination of the sample by the container often occurs [1].

Electromagnetic levitation melting is a containerless method that can be used in measuring high temperature materials properties. Figure 1 shows a schematic cross-section of an arbitrary levitation cell design and a photo of a functional levitation cell coil. The levitation cell functions as follows: a large, high frequency alternating electrical current is sent through an induction coil. Due to the magnetic field generated around a current-carrying conductor, a time-varying magnetic field is generated around the coil. According to Faraday's law of induction, eddy currents are induced inside an electrically conductive sample placed inside the coil. The direction of the eddy currents in the sample is such that an opposing Lorentz force acts between the sample and the coil, and this force levitates the sample inside the coil. The eddy currents are also responsible for resistive heating (Ohmic / Joule heating) of the sample. Containerless levitation melting is achieved in this way, provided that the current is large enough to generate a Lorentz force that is equal to the weight of the sample. The gradient of the force field inside the coil should be negative to allow for stable levitation of the sample.

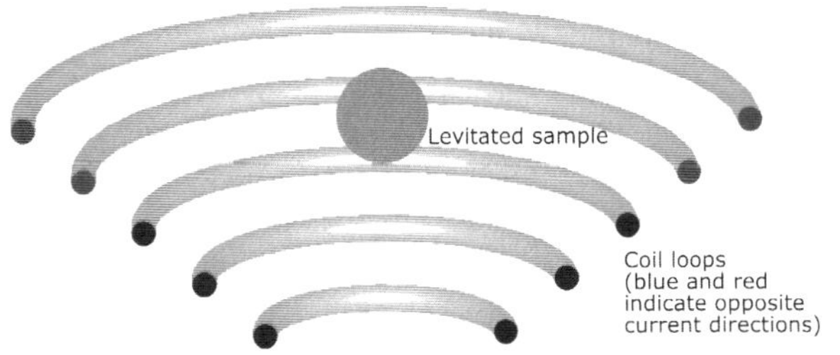

(a) Schematic of a cross-section through a levitation cell with arbitrary coil design. The sample is also shown.

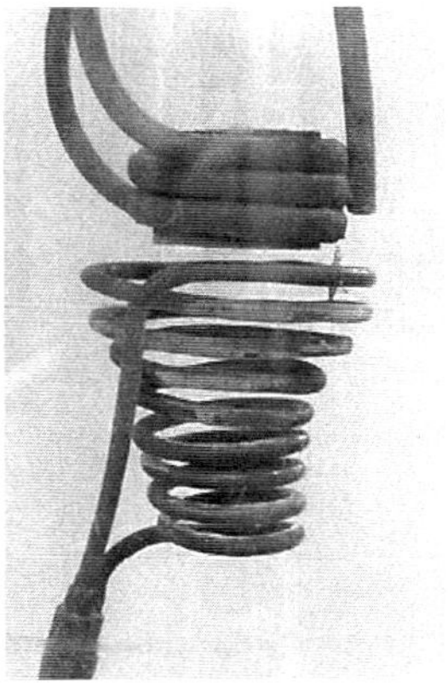

(b) Photo of a functional levitation cell coil.

Figure 1. Electromagnetic levitation cells.

The levitation force and sample temperature depend on the geometry of the induction coil as well as the magnitude, frequency and direction of the current through the coil. It is therefore desirable to have a levitation cell model that can be used as a design tool for levitation cell coils that will levitate a specified sample material of a specified size at a required temperature.

Levitation cell models developed for various purposes are available in literature [2-5]. We followed the finite volume approach proposed by El-Kaddah and Szekely [2] and implemented a model in Python. In this approach the helical coil is modelled as a set of circular loops as shown in Figure 1(a), which allows simplification to a 2-dimensional solution domain due to rotational symmetry. The spherical sample is also discretized into a set of adjacent rings. The induced current density field inside the sample is computed using the principle of mutual inductance between the coil loops and discretized sample loops. With the current density field known, the magnetic field inside the sample can be computed. Finally, the current density field and magnetic field can be used to predict the sample levitation position and temperature.

OBJECTIVES AND PROBLEM DESCRIPTION

For the levitation cell model to be useful in designing levitation cell experiments, it has to be implemented correctly and provide predictions of the actual levitation cell behaviour during experiments that are sufficiently accurate to base a design on. Measuring whether this has been achieved is challenging due to uncertainties on both the modelling and experimental sides of the work.

The three uncertainties we aim to address are:

1. **Uncertainty about the correct implementation of the model.** Literature on levitation cell models generally does not provide methods to verify the correct implementation of the levitation cell physics. How can the model be verified for at least simplified cases and geometries?
2. **Uncertainty about the values of the material properties used in the model.** The purpose of levitation cell experiments is to determine high temperature material properties. However, the model used to design levitation cell experiments already requires these material properties at the high operating temperatures. What is the influence of the uncertainty in the material property values used in the levitation cell model on the parameters used to design levitation cell experiments? Can a

levitation cell model containing uncertain high temperature material property values be used to design levitation cell experiments?

3. **Uncertainty about conditions during experiments.** The available experimental setup is not instrumented to measure the electrical current in the coil. The current in the coil is controlled indirectly by varying the percentage power the power supply supplies to the coil. This is often the case [2], since measurement of the large, high frequency current required for levitation melting can be challenging. Could a levitation cell model still be used to design levitation cell experiments for a experimental setup where the magnitude of the electric current in the coil is unknown?

The rest of this paper deals with finding ways to quantify and manage these uncertainties that will allow the model to be useful as a design tool. This paper only deals with the levitation force calculation, but a similar approach could be followed with the temperature calculation.

VERIFYING THE MODEL IMPLEMENTATION

The implemented model can be used to investigate the effect of the uncertainties around the high temperature material property values and experimental parameters, provided that we can verify that the model is implemented correctly. The first priority is therefore to find a simple case with a known analytical solution that can be used to verify the model against.

Quantitative verification of a single-loop coil model

The Lorentz force acting on one infinitely long straight current-carrying conductor due to the presence of a second parallel infinitely long conductor is given by [6]

$$\frac{F}{\Delta L} = \frac{\mu_0 I_1 I_2}{2\pi d} \qquad \ldots (1)$$

where $F/\Delta L$ is the force per length ΔL of the wire, I_1 and I_2 are the magnitudes of the electrical current in each of the two wires respectively, μ_0 is the magnetic permeability of the air between the two conductors and d is the distance between the two conductors as illustrated in Figure 2(a). Figure 2(b) shows the geometry normally used in the levitation cell model for the simple case of a single loop coil and Figure 3 shows how this geometry can be modified to be more similar to the parallel wire case by replacing the spherical sample with a coaxial ring of the same radius as the coil loop. In this way the levitation cell model implementation can be used to compute the Lorentz force between two coaxial loops. For increasing radii of the two coaxial loops, the force between the loops should tend towards the force between two parallel wires as given in Equation 1 and this can be used to verify the force calculation part of the levitation cell model.

(a) Lorentz force between two parallel wires. **(b)** Single loop levitation cell model geometry.

Figure 2. Using the analytical expression for the Lorentz force between two parallel wires to verify the Lorentz force calculation in the levitation cell model.

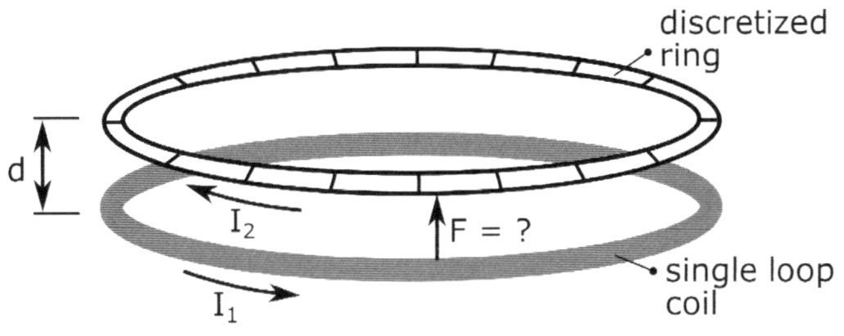

Figure 3. Model geometry for Lorentz force verification.

This test verifies whether the magnetic and force fields are calculated correctly from the induced current density field. It does not verify whether the induced current density inside the discretized sample or ring is computed correctly. However, the set of linear equations derived from Faraday's law of induction that are used by El-Kaddah and Szekely [2] to compute the induced current field reduces to a 2×2 system for the ring geometry given in Figure 2(c). The 2×2 system of linear equations can easily be solved in a hand calculation to verify the induced current density part of the model implementation.

Qualitative verification of a multi-loop coil model

If the model passes the two verification tests above, we know the implementation is at least correct for the simple case of a single loop coil. It is more challenging to verify the model implementation for more complicated coil geometries, but it is possible to ensure that the general behaviour of the model is as expected for a multi-loop coil. The induced current calculation can be monitored by noting that the total current induced in the sample cannot be more than the magnitude of the electrical current in the coil times the number of coil loops. The behaviour of the levitation force calculation can be monitored by modelling a multi-loop symmetrical coil geometry like the one used by El-Kaddah and Szekely [2] given in Figure 4. For a symmetrical coil geometry like the one shown in Figure 4, the magnitude of the levitation force along the central axis of the coil is expected to be symmetric around the symmetry plane of the coil geometry, with the force acting in opposite directions on either side of the symmetry plane.

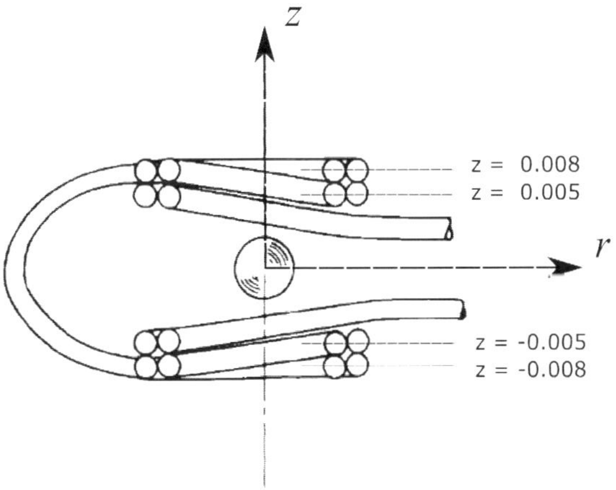

Figure 4. Cross-section through the symmetrical coil geometry used in the investigation by El-Kaddah and Szekely [2]. The sample is also shown. This figure is based on the coil geometry figure by El-Kaddah and Szekely [2].

RESULTS OF THE VERIFICATION STUDIES

Quantitative verification of a single-loop coil model

The parallel wires verification test was implemented for our levitation cell model. For the purposes of the verification test, a cross-sectional area of 0.16 m^2 is used for both the wires and the loops, the distance between the loops is 0.6 m, I_1 has a value of 1 A and I_2 has the value of the induced current as calculated by the model for the specified loop radius. Figure 5 shows the resulting comparison of the modelled force between two coaxial loops with increasing radii and the corresponding analytical calculation for the force between two parallel wires. It can be seen that for very large loop radii the modelled force between the two loops is approximately equal to the force expected between two infinitely long parallel wires. The calculation of the induced current density in the second ring was also confirmed by a hand calculation. The levitation force calculation in our levitation cell model is therefore successfully verified for the simplified case of a single loop coil.

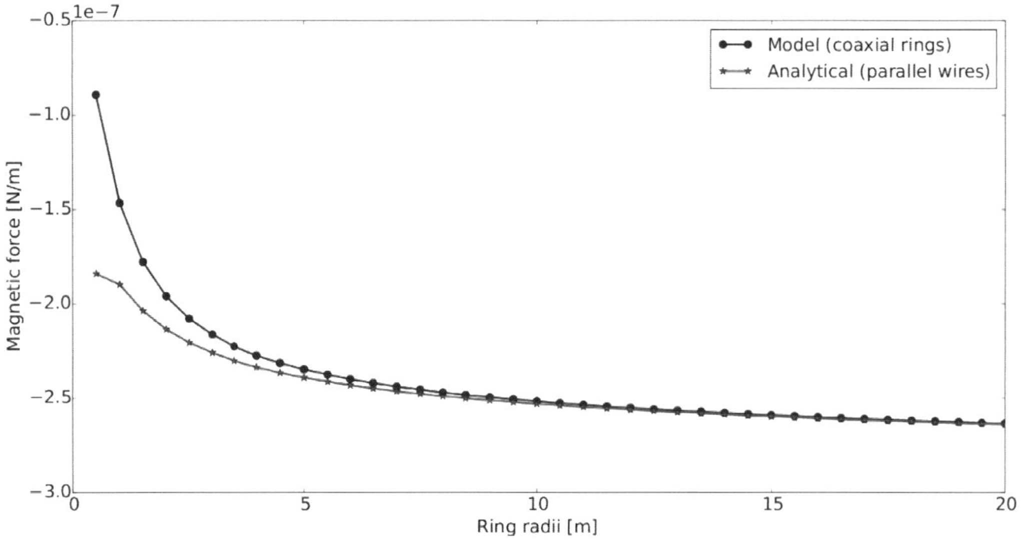

Figure 5. Verifying the modelled force between two parallel loops against the analytical expression for the force between parallel wires.

Qualitative verification of a multi-loop coil model

The behaviour of our model implementation for a multi-loop coil is investigated by modelling the symmetrical coil design (Figure 4) used by El-Kaddah and Szekely [2]. Figure 6 shows the levitation force along the central axis of the coil as predicted by the levitation cell model. It can be seen that the force magnitude is symmetrical around the symmetry plane of the coil geometry but the force is acting in opposite directions, as would be expected for a symmetrical coil design.

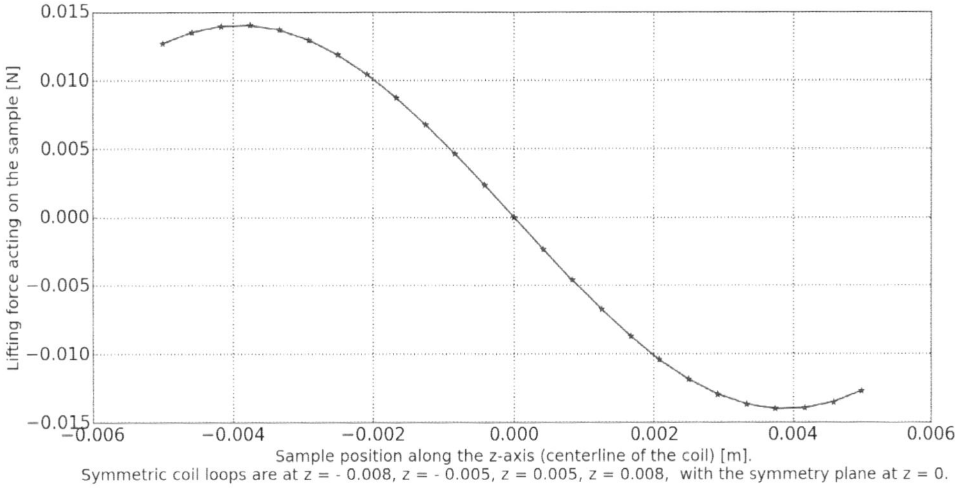

Figure 6. Modelled levitation force along the central axis of a symmetrical, multi-loop coil with the geometry given in Figure 4.

EFFECT OF VARIATION IN MATERIAL AND MODEL PARAMETERS

Once the model is verified, it can be used to investigate the effect of variations in the uncertain material property values and the input coil current on the predicted levitation force. We are specifically interested in finding out which parameters of the levitation cell design are not affected by variations in material properties and input coil current, but can still be used for design purposes.

The high temperature sample material properties required for the force calculation are electrical conductivity, magnetic permeability and density as manifested by changes in the sample radius. Each subfigure in Figure 7 shows model predictions for the effect of a 10% and 50% increase and decrease in each of these material properties as well as the effect of such changes in the input coil current. It can be seen that changes in coil current, magnetic permeability and sample radius have much more significant effects on the levitation force than changes in electrical conductivity.

Changing the material properties and the coil current further only affects the amplitude of the force curve, but not the position of maximum, minimum and zero force along the coil axis. The same effect is also seen on a asymmetric coil design, although it is not shown here. This is fortunate, because one of the important requirements for levitation cell design is that the sample levitation position is stable, i.e. the gradient of the force curve should be negative at the levitation position. A very stable levitation cell design would therefore be a design where the levitation force gradient is negative for a large part of the coil height. None of the uncertain parameters affect the positions along the coil height where the gradient of the force curve changes, and it is therefore possible to use the model to design a levitation cell for stable levitation without precise knowledge of the values of these parameters.

Since changes in coil current magnitude also result only in changes in the amplitude of the force curve, the model can also be used to design levitation cell coils for stable levitation for our available experimental setup where the coil current magnitude is unknown. Levitation cell experiments can be designed for small current magnitudes and experiments can then be performed starting at a low power setting. If sample levitation is not possible at the low power setting, the power supply to the levitation cell can then be increased until the sample is successfully levitated.

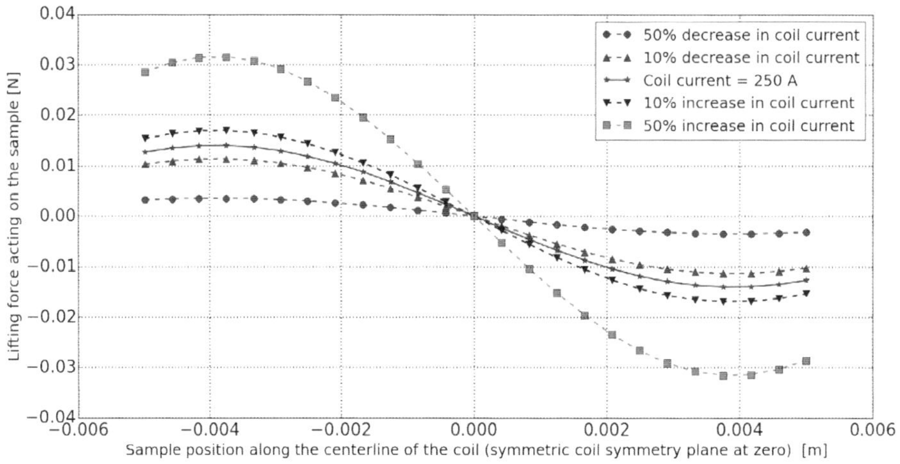

(a). Varying coil current magnitude.

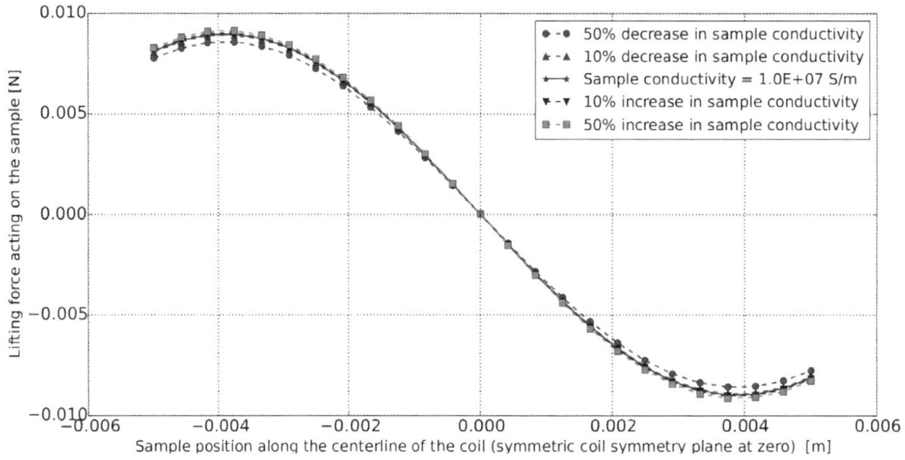

(b). Varying sample conductivity.

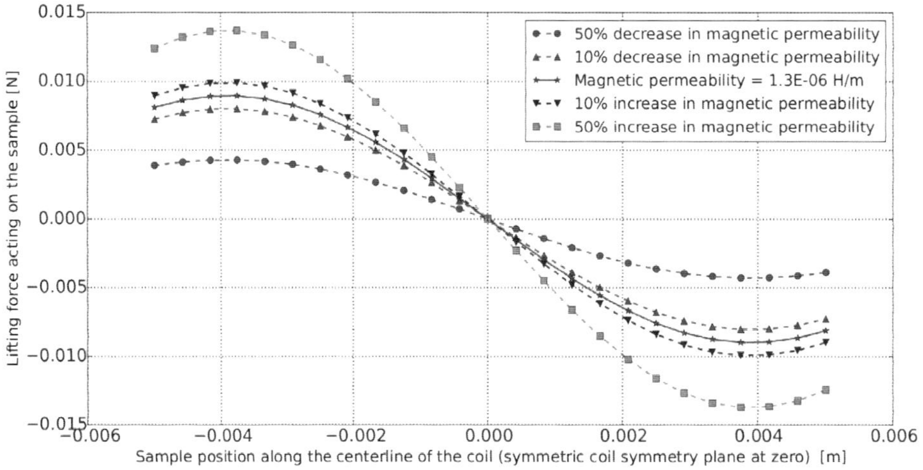

(c). Varying magnetic permeability.

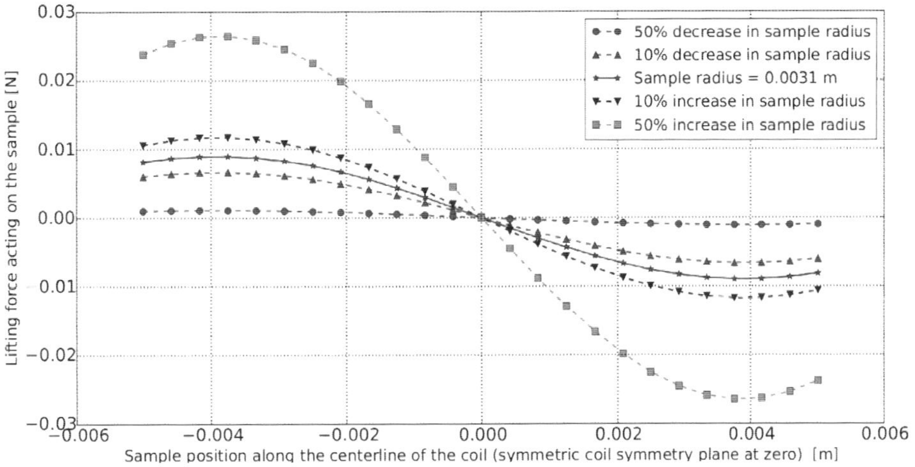

(d). Varying sample radius.

Figure 7. The effect of changes in material property values and coil current on the levitation force along the central axis of the coil.

CONCLUSION

A simple verification case for the force calculation in an electromagnetic levitation cell model is found in comparing the Lorentz force between two coaxial loops of large radii to the known analytical expression for the Lorentz force between two infinitely long wires. This method is used to successfully verify an electromagnetic levitation cell model. The verified levitation cell model is then used to show that variations in the coil current magnitude, sample electrical conductivity, magnetic permeability and sample density do not affect the location and size of the stable levitation region inside a levitation cell coil. The levitation force is found to be especially insensitive to changes in the value of the sample electrical conductivity. It is therefore possible to use the model to design levitation cell experiments for stable levitation in an experimental setup where the coil current magnitude is unknown, and for levitating samples of which the high temperature material properties are not precisely known.

ACKNOWLEDGEMENTS

This research was supported in part by the South African Minerals to Metals Research Institute (SAMMRI, Project no.: S1517).

REFERENCES

1 Wroughton, D.M., Okress, E.C., Brace, P.H., Comenetz, G. and Kelly, J.C.R., "A technique for eliminating crucibles in heating and melting of metals", Journal of the Electrochemical Society, Vol. 99. No. 5, pp. 205 - 211, 1952.

2 El-Kaddah, N. and Szekely, J., "The electromagnetic force field, fluid flow field, and temperature profiles in levitated metal droplets", Metallurgical Transactions B, Vol. 14. No. 3, pp. 401 - 410, 1983.

3 Fromm, E. and Jehn, H. "Electromagnetic forces and power absorption in levitation melting", British Journal of Applied Physics, Vol. 16, pp. 653 - 659, 1965.

4 Moghimi, Z.A., Halali, M. and Nusheh, M. "An investigation on the temperature and stability behavior in the levitation melting of nickel" Metallurgical and Materials Transactions B, Vol. 37. No. 6, pp. 997 - 1005, 2006.

5 Royer, Z.L., Tackes, C., LeSar, R. and Napolitano, R.E., "Coil optimization for electromagnetic levitation using a genetic like algorithm", Journal of Applied Physics, Vol. 113, pp. 214901-1 - 214901-9, 2013.

6 Griffiths, D.J., "Introduction to Electrodynamics", Prentice Hall, 1999.

10th South African Conference on Computational and Applied mechanics

Potchefstroom

3-5 October 2016

The Effects of Grain Size Variants on Fracture Toughness of Nanocrystalline Materials Produced by Accumulative Roll-bonding

P. B. Sob*[1], A. A. Alugongo[2] and T. B. Tengen[3]

[1,2]Departmentof Mechanical Engineering, Faculty of Engineering and Technology, Vaal University of Technology, Vanderbijlpark 1900, Private Bag X021, South Africa

[3]Department of Industrial Engineering and Operations Management, Faculty of Engineering and Technology, Vaal University of Technology, Vanderbijlpark 1900, Private Bag X021, South Africa,[1]baonhe_sob@rocketmail.com, [2]alfayoa@vut.ac.za,[3] thomas@vut.ac.za

Abstract

Although several researchers have investigated fracture behavior of nanocrystalline (nc) materials, difficulties in measuring fracture toughness, since it is as a result of nanoscale-induced crack, have discouraged researchers in gaining insight on the relationships between microstructure and fracture resistance. This current study focuses on crack that occurs during Accumulative Roll-bonding (ARB). The fracture toughness is studied as function of the grain size variants during deformation of conventional materials to nanostructure materials. The results revealed that the fracture toughness increase during grain refinement at different rates when measured as function of different grain size variants. The increase in fracture toughness during grain refinement of nc materials was due to diffusion of materials from the crack region. As materials diffused from the crack region, grain boundary (GB) dislocations took place and more material moved away from the crack section leading to more increase in the crack size.

Keywords: Fracture toughness, grain size variants, nanostructure materials, cracks

Introduction

Nanomaterial is a class of materials that is currently attracting significant attention. This is due to the fact that nanomaterials have more enhanced properties when compared to conventional materials. But the fundamental definition of nanomaterials is that these are materials whose internal constituent structures such as grains, crystallites, pores, vacancies, dislocations, grain boundaries and so on belong to nanometer range of length scale or their sizes are less than 100 nm at least in one direction.

It has been reported by researchers that nanomaterials have superior strength, hardness and good wear resistance but at the expense of low tensile ductility and low fracture toughness [1-10]. The low tensile ductility and fracture toughness have limited the practical applicability of nanomaterials [1-10]. Although most researchers reported on low tensile ductility and fracture toughness, some researchers have equally reported about nanomaterial with enhanced toughness [11-14]. The difficulties in measuring nanomaterials fracture toughness, since it should depend on nanoscale-initiated cracks, have discouraged researchers from gaining more insight into microstructure-fracture resistance relations.

The fracture behaviors of nanomaterials, which is a function of nanoscale-initiated cracks, also greatly depend on the methods of deforming materials from conventional to nanostructure materials [11-18]. Therefore, to understand the toughening mechanisms of nanomaterials, it is important to focus the study on the processing route during grain refinement. Furthermore, the fracture toughness also depends on the type of materials treatment since materials that are not subjected to any heat treatment will have different fracture toughness when compared to materials that are subjected to heat treatment.

Most of the developed theoretical and empirical models of fracture toughness considered the above-mentioned conditions during the modelling processes, but did not consider the varying approaches of measuring grain size. In most of the developed models intergrain sliding; triple junction diffusion creep, Coble creep, rotational deformation and nanoscale deformation twinning have been theoretically described as reasons for nanomaterials toughening [1, 18, 19-29], but yet the effects of the material grain size variants on fracture toughness is still at its premature stage. This might have been due to the fact that most previous studies have been scanning cross-sectional surfaces/planes of the nanomaterials that has the ability of obtaining the grain radius, which depends on the direction of the cutting plane or cross-section. But Transmission Electron Microscopy (TEM) reveals different directions that grain size can be concurrently measured, (herein called grain size variants) during the deformation of conventional material to nanostructured materials by Accumulative Roll bonding (ARB) or Equal Channel Angular Pressing (ECAP) [30-31], with exampled images shown in Fig. 1 (a-b).

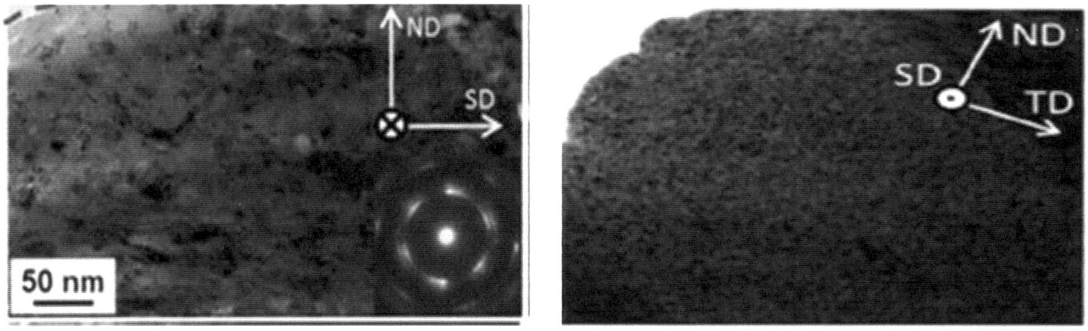

(a) (b)

Figure 1. (a-b) dark-field TEM microstructures and texture evolution revealing that grains in nanomaterials are not spherical and the different directions along which grain size can be measured i.e. along TD, ND and RD or (SD) [30-31].

The TEM revealed that the measurement of grain size during grain refinement can be achieved in three main directions which are the normal direction (ND), rolling direction or sliding direction (RD or SD) and transverse direction (TD). Although the TEM has the ability to measure the nanoscale-initiated cracks and, hence, fracture toughness of nanomaterials for the different directions of measuring grain size, its theoretical study is still at a premature stage Results of the theoretical study should be able to show how fracture toughness affects nc materials properties for the different grain size variants. From the developed models of fracture toughness as function of the grain size variants, other related properties of nc materials can be studied. The models for the evolution of size variants had been derived under a given set of conditions [34] based on the natures of their evolutions during grain refinement; and the fact that the random grain sizes evolve as lognormal distribution is also employed [32].

The reason for incorporating the different grain size variants in most models is that most previous models of nanomaterials dealt only with the equivalent radius r or made the assumptions that the grains in nc materials were spherical in nature [24-29]. Experimentally this spherical shape of all the grains in nanomaterials is not the case. In fact, majority of the grains in nanomaterials are not nearly spherical in nature and thus, these existing models that consider that the grains in nanomaterials are spherical should have limitations in revealing other important (hidden) results or other interesting natures of behavior of the nanomaterials mechanical properties.

In this paper the measurement of grain size along the transverse direction (TD) is termed in the model as the semi major axis length r_1, the normal direction (ND) is represented as the semi minor axis length r_2, and the rolling direction (RD) or sliding direction (SD) is represented as the major axis length r_3 and the equivalent radius r is defined as the radius of an equivalent sphere or circle obtained by the displacement method of volume/area measurement.

Methodology

Manufacturing Process to Produce Nanomaterials

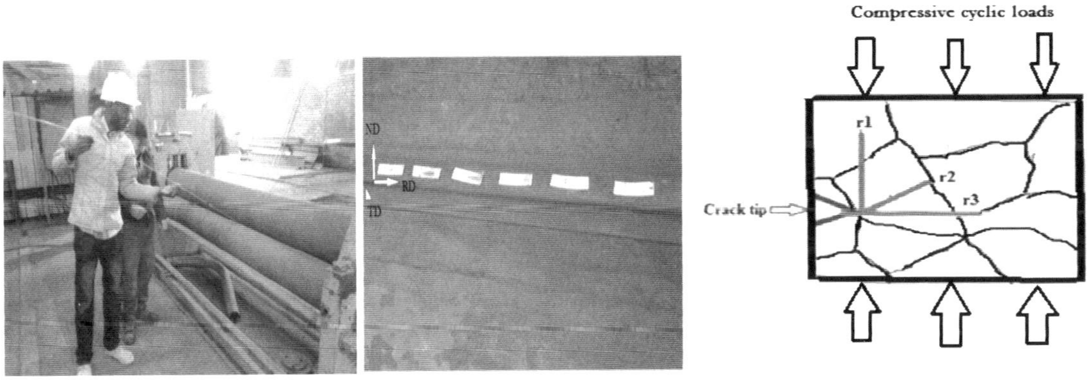

(a) (b) (c)

Figure 2: (a) ARB process (b) deformed materials after ARB cycles (c) schematic of Compressive cyclic loads

The ARB technology using conventional rolling facility as shown in Fig.2 (a) was utilized in this project. Two shafts were placed horizontally so that they were free to rotate by means of a mechanically automated operation as shown in Fig.1 (a). During experimentation the sample was forced or fed through the rotating shafts. The rotating shafts gripped the sample and forced the sample through the rollers in the first pass.

The deformed sample was cut in two pieces and stacked together. Before stacking, the entire surfaces of the strips were wire-brushed (i.e. using stainless steel brush) and degreased with tetrachlorethylene to achieve good bonding. The materials were joined together in the corners using aluminum wires and subsequently rolled. The whole sequence of "rolling, cutting, face-brushing, degreasing and stacking" was repeated again for several "passes" since it is necessary to performed several passes during ARB for conventional materials to become nanomaterials.

The deformed materials and their different directions of measurements are shown in Fig. 2 (b). It must be noted that the nanomaterial was not subjected to any heat treatment during manufacturing. Since it was necessary to perfom several passes during ARB for conventional materials to become nanomaterials, the rate of material straining continuously increased. As the compressive cyclic loads continuously strained the material, the material naturally suffered from fatigue due to crack especially at the fatigue region of the material.

The crack was induced naturally on the material in the transverse direction by the high stress imposed on the material from the normal direction by the roller that resulting in a transverse tensile stress. The induced transverse tensile stress exceeded the strength in the material as shown in Fig. 2 (c). Since several ARB passes were performed for conventional materials to become nanomaterials, the tensile stress increases along the transverse direction (r_2), resulting to fracture and crack growth

Theoretical consideration for model derivation of fracture toughness from crack propagation

The Ilya model [32] which gives the relationship between fracture toughness $\left(K_{IC}^{\sigma}\right)$, shear modulus (G), Possion's ratio (ν) and specific surface energy (γ_e) of nc materials that was employed is given as

$$K_{IC}^{\sigma} = \sqrt{\frac{4G\gamma_e}{(1-\nu)}} \tag{1}$$

The specific surface energy of nc materials is given by [33]

$$\gamma_e = \gamma - T\frac{\Delta\gamma}{\Delta T} \tag{2}$$

Where γ = surface tension, T = temperature, $\Delta\gamma$ = change in surface tension and ΔT = change in temperature. $\Delta T = T_m - T_{m,b}$ where T_m = is the melting temperature of nc material and $T_{m,b}$ = is the bulk (or conventional) material melting temperature, $T_m = T_{m,b}\left(1 - a/r\right)$, a>0, is a dimensional factor which depends on the shape of the material. For a 3-D grain the equivalent radius r was obtained from the equivalent volume of the grain: thus at any instance during grain refinement the equivalent radius r was related to the volume of the grain through $(4/3)\pi r^3=V=(4/3)\pi r_1r_2r_3$, this gave $r^3=V=r_1r_2r_3$ which gives the relationship between r and (r_1, r_2, r_3).

The different models of the evolution of nanomaterial's grain size variants that were derived by [34] are used in this paper stated in equation (3-6) which are

$$dr_1 = M\left(\frac{1}{r_{c1}} - \frac{1}{r_1}\right)dt + r_1^{\frac{1}{2}}DdW(t) - Zr_1V_1d(t) \tag{3}$$

where r_{c1}=local critical grain size, Z and D are Constants, $V_1=\tau_1 r_1^2$ defines rate of grain breakage, $M = M_0\left(1+\frac{CD}{r_1}\right)$ is Grain Boundary (GB) mobility function, $CD= 4(Hm)(h_0)/((k)(T))$, $T_m = T\{ln(m_{01}/m)$ and $M_0= M_{01}exp\{-T_m(inf)/T\}_=$ initial grain boundary mobility constant, $dW(t)$ is change in Wiener process and dN(t) is the number of coalescence events within an infinitesimal time interval.

$$dr_3 = Ratio_1(dr_1) \tag{4}$$

$$dr = -Ordt + IdW(t) \tag{5}$$

where O and I are constants.

$$dr_2 = Ratio_2(dr) \tag{6}$$

The yield stress, σ(r), on nanomaterials grain subjected to plastic deformation is given by Zhao [29]

$$\sigma(r) = \sigma_0' + A\left(r^{-\frac{1}{2}}\right) - B\left(r^{-1}\right) - C\left(r^{-\frac{3}{2}}\right) \tag{7}$$

where $\sigma_0' = \sigma_0 + K_t$ is bulk yield stress, $A = K_d$ is HPR proportionality constant, $B = K_t[2hH_m / RT_r]$, $C = K_d[2hH_m / RT_r]$, K_t is a constant, h is atomic diameter in the case of metal, H_m is the bulk melting enthalpy, R is ideal gas constant, T_r is the room temperature, $K_d > 100K_t$ and $\sigma_0 > 10K_t$.

The stochastic counterparts of expression (1) to (7) are considered, and resulting equations are solved simultaneously using Engineering Equation Solver software (F-Chart Software, Madison, W153744, USA).

Results and Discussion

Observe that the parameters of expressions (1)-(7) are reduced to functions of grain size (either given as equivalent radius or grain size variants). Thus, it is imperative to use previous and newly acquired data of grain size evolution and materials properties to validate the newly proposed models. Thus, the newly proposed models were tested with data from (nanocrystalline) aluminum sample [40], which are $\gamma = 0.98 \pm 0.21 Jm^{-1}$[33], $G=169GPa$, $v=0.23$ [28]

$M_0'=0.01nm^2s^{-1}$, $m=4$, $r_{c1}= 1.95r$, $CC=12$, $a=0.90$, $D=10^{-4}$, $h_0=0.25nm$, $T_m(\infty)=933.47K$, $CV_0=0.3$, $H_m(\infty)=10.71KJMol^1$, $\sigma_0'=16.7MPa$, $K_t=1.3$, $\sigma_0=15.40MPa$, $Kd=1301.77MPa\_nm^{1/2}$, $R=8.31JK^{-1}mol^{-1}$ *and* $T_r=300K$,. The additional data obtained for this work are $O=0.0035$, $I=1.1$, $r_0=100nm$, $Z=0.4$, $Ratio_1=0.81$, $Ratio_2=1.071$, and $\tau_1 = 0.000008$, *which* were obtained through curve fitting of the empirical data from the different measures of the sizes. The obtained results are presented in the plots below.

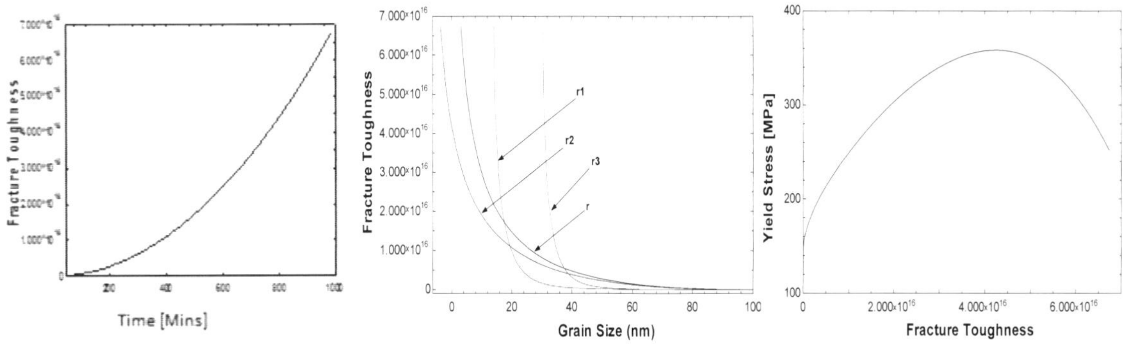

(a) (b) (c)

Figure 2: Fracture Toughness as a Function of (a) Time [Mins] (b) Grain Size [nm] Variants and (c) Yield Stress [MPa] as a Function of Fracture Toughness.

Fig. 2 (a) reveals the evolution of fracture toughness with time during grain refinement of nc materials. From Fig. 2 (b) it can be seen that the fracture toughness increased during grain refinement. Fig 2 (b) shows that the fracture toughness increased at different rates in nc materials for grain size variants during grain refinement. It is demonstrated in Fig. 2 (b) that the fracture toughness was higher for r and r_2 when compared with r_3 and r_1 during the initial process of grain refinement, and as more grain refinement took place the fracture toughness increased more along r_3 and r_1 than along r and r_2. The increase in fracture toughness during grain refinement of nc materials can be explained to be due to the fact that grain boundaries diffusional creep, coble creep, intergrain sliding and rotational deformational of materials took place at the crack region. As more material diffused from the crack region, grain boundary (GB) dislocations took place and more materials move away from the crack section leading to increase crack. The GB dislocation was accompanied by GB sliding that took place at different rates for the different grain size variants, thus, leading to material sliding away from the crack region. Generally the fractures in nc materials are crucially influenced by high amount of grain boundaries. These boundaries defined by the grain size variants experienced different fracture toughness due to their different atomic. It must be noted that diffusion of materials from the crack region, GB sliding and migration mechanism from the crack region are most effective in increasing fracture toughness at smallest grain sizes.

The increase in fracture toughness can also be explained to be due to the fact that the stress concentration at the tip of the crack has been reported to be very high. The stress concentration is usually lower for larger grain size and higher for smaller grain size. As grain refinement took place the material grain size decreased and the stress on the material increased. As this happened the grain boundaries dislocation increased, driven by the stress concentration near the crack tip of the materials, which increased due to fully accommodated grain boundaries sliding. This led to increase in the material crack propagation and fracture toughness as shown in Fig. 2(a-b). The obtained results are similar to those reported by other researcher [35-39]. Fig. 2 (c) shows that the material yield stresses as function of fracture toughness increase and then decrease. The increase and decrease in material yield stress as shown in Fig. 2(c) can be explained due to the fact that during grain refinement properties enhancement takes place and as the material suffers from crack propagation the enhanced material property drops.

Conclusion

The current work was aimed at studying the effects of grain size variants on the fracture toughness of nc materials for cracks induced during deformation of conventional materials to nanostructured materials by ARB. To achieve that the Ilya model [33] which gives the relationship between fracture toughness, shear modulus, Poisson's ratio and specific surface energy of nc materials was modified to be applicable to 3-D grain.

It was observed that the fracture toughness increased in nc materials at different rates for grain size variants during grain refinement. The increase in fracture toughness for grain size variants was due to diffusion of materials from the crack region, GB sliding and migration mechanism from the crack region depended on the material size variants. It was also observed that the material yield stress as function of fracture toughness increased, and then decreased during grain refinement. It can therefore be concluded that fracture toughness have negative impacts in nc material properties during grain refinement.

Acknowledgement

This material is based on the work which is supported financially by the National Research Foundation (NRF) and Vaal University of Technology (VUT).

Reference

1. Meyers, M, A., & Mishra, A and Benson, D. J "Mechanical properties of nanocrystalline materials", Prog. Mater. Sci. 51, 427 (2006).

2. Shen, T. D., & Koch, C. C., & Tsui, T. Y., and Pharr, G. M "On the elastic moduli of nanocrystalline Fe, Cu, Ni, and Cu–Ni alloys prepared by mechanical milling/alloying", J. Mater. Res. 10, 2892 (1995).

3. Kumar, K. S., & Suresh, S., and Van Swygenhoven, H., "Mechanical behavior of nanocrystalline metals and alloys", Acta Mater. 51, 5743 (2003).

4. Koch, C, C., "Structural nanocrystalline materials", An overview. J. Mater. Sci. 42, 1403 (2007).

5. Lu, L., & Wang, L. B., & Ding, B. Z., and Lu, K., "High-tensile ductility in nanocrystalline copper", J. Mater. Res. 15, 270 (2000).

6. Dao, M., & Lu, L., & Asaro, R. J., & De Hosson, J. T. M., and Ma, E., "Toward a quantitative understanding of mechanical behavior of nanocrystalline metals", Acta Mater. 55, 4041 (2007).

7. Liu, Y. G., & Zhou, J. Q., & Shen, T. D., and Hui, D., "Effects of ultrafine nanograins on the fracture toughness of nanocrystalline materials", J. Mater. Res. 26, 1734 (2011).

8. Liu, Y. G., & Zhou, J. Q., and Ling, X., "Impact of grain size distribution on the multiscale mechanical behavior of nanocrystalline materials", .Mater. Sci. Eng., A 527, 1719 (2010).

9. Jiang, B., and Weng, G. J., "A generalized self-consistent polycrystal model for the yield strength of nanocrystalline materials",J. Mech. Phys. Solids 52, 1125 (2004).

10. Ovid'ko, I. A., "Review on the fracture processes in nanocrystalline materials", J. Mater. Sci. 42, 1694 (2007)

11. Figueiredo, R. B., & Kawasaki, M., and Langdon, T. C., "The mechanical properties of ultrafine-grained metals at elevated temperatures". Rev. Adv. Mater. Sci. 19 (1/2), 1–12, 2009.

12. Guo, X., & Weng, G. J., and Soh, A. K., "Ductility enhancement of layered stainless steel with nanograined interface layers" Comput. Mater. Sci. 55, 350–355, 2012.

13. Kaminskii, A. A., & Akchurin, M. Sh., & Gainutdinov, R.V., & Takaichi, K., & Shirakava, A., & Yagi, H., Yanagitani, T., and Ueda, K., "Microhardness and fracture toughness of Y_2O_3 Y and $Y_3AL_5O_{12}$ based nanocrystalline laser ceramics. Crystallogr". Rep. 50, 869–873, 2005

14. Karimpoor, A.A., & Erb, U., & Aust, K.T., and Palumbo, G., "High strength nanocrystalline cobalt with high tensile ductility", Scripta Mater. 49, 651–656, 2003.

15. Morris, D. G., "The origins of strengthening in nanostructured metals and alloys". Revista de Metalurgia,46(2): 173-186, 2010

16. Tengen, T. B., "Designing nanomaterials with desired mechanical properties" .Nanoscale Research,*6(1): 585, 2011*

17. Tengen, T. B., "Stochastic effect of strain-rate sensitivity on nanomaterials mechanical properties". Programme of Eight South African Conferences on Applied Mechanics 2012 (SACAM 2012) 3-5 September 2012. Organised and hosted by Mechanical Engineering Science of the University of Johannesburg South Africa. pp 198 – 202, 2012

18. Feng, H., & Fang, Q. H., & Zhang, L. C., and Liu. Y. W "Special rotational deformation and grain size effect on fracture toughness of nanocrystalline materials" International Journal of Plasticity 42 (2013) 50–64

19. Bobylev, S. V., & Mukherjee, A. K., & Ovid'ko, I.A., and Sheinerman, A.G., "Effects of intergrain sliding on crack growth in nanocrystalline materials" Int. J. Plasticity26, 1629–1644. 2010.

20. Bobylev, S. V., & Morozov, N. F., and Ovid'ko, I. A., "Cooperative grain boundary sliding and nanograin nucleation process in nanocrystalline, ultrafine-grained,and polycrystalline solids" Phys. Rev. B 84, 094103. 2011.

21. Dao, M., & Lu, L., & Asaro, R. J., & De Hosson, J. T. M., and Ma, E., 2007. "Toward a quantitative understanding of mechanical behavior of nanocrystallinemetals" Acta Mater. 55 (12), 4041–4065, 2001.

22. Gutkin, M. Yu., & Ishizaki, T., & Kuramoto, S., & Ovid'ko, I. A., and Skiba, N.V., "Giant faults in deformed gum metal" Int. J. Plasticity 24, 1333–1359, 2008a.

23. Ivanov, K. V., and Naydenkin, E. V., "Grain boundary sliding in ultrafine grained aluminum under tension at room temperature" Scripta Mater. 66, 511–514, 2012.

24. Hall, E. O. "The deformation and ageing of mild steel. III discussion of results" Proceedings of the Physical Society, Section B, 64(9): 747-753, 1951.

25. Petch, N. J., "The cleavage strength of polycrystasl", J. *Iron and Steel Institute, 174: 25-28, 1953.*

26. Hillert, M. "On the theory of normal and abnormal grain growth" *Acta Metal. 13, pp.227, (1965)*

27. Tengen, T. B, & Wejranowski, I. R. and Kurzydlowski, J. K., "The effects of grain size distribution on nanmaterials mechanical properties" , Solid State Phenomena, 140: 185-190, 2008.

28. Tengen, T. B. "Analysis of Characteristic of Random Microstructures of Nanomaterials" PhD. Thesis. Witwatersrand Johannesburg, 2008.

29. Zhao, M. and JIANG, Q. "Reverse hall-petch relationship of metals in nanometer size" Emerging Technologies-Nanoelectronics, IEEE Conference on Vol. pp 472-474, (10-13 Jan. 2006)

30. Hidalgo, P. M. & Cepeda, J. C. M. & Ruano, O. A. and Carreno, R. O. A. "Effect of warm accumulative roll bonding on the evolution of microstructure, texture and creep properties in the 7075 aluminium alloy" Mater Sci Eng A 556 (2012) 287-294 DOI 10.1016/j.msea.2012.06.089

31. Cai, W. and Bellon, P., "Subsurface microstructure evolution and deformation mechanism of Ag-Cu eutectic alloy after dry sliding wear", Wear 303 (2013)602–610

32. Ilya, A. & Ko, O. & Alexander, G. and Sheinerman "Mechanisms of fracture toughness enhancement in nanocrystalline solids" Proceedings of the World Congress on Engineering 2012 Vol III WCE 2012, July 4-6, 2012, London, U.K.

33. Karasevskii "Specific thermodynamic description of nanocrystalls" Journal of Physics: Condensed Matters

34. Sob. P. B. Alugongo, A. A. Tengen, T. B. "Modelling Strain Rate Sensitive Nanomaterials Mechanical Properties: the effect of varying definitions, M. Tech. Dissertation. Vaal University of Technology, 2016

35. Liu, Y. & Zhoua, J. & Shenc, T. D. and Huid, D., "Grain rotation dependent fracture toughness of nanocrystalline materials" Materials Science and Engineering A 528 (2011) 7684– 7687

36. Sob, P. B. & Tengen, T. B. and Alugongo, A. A..,"The effect of deformation activation volume, strain rate sensitivity and processing temperature of grain size variants" International Journal of Mechanical, Aerospace, Industrial, Mechatronics and Manufacturing Engineering Vol:9, No:11, 2015.

37. Sob, P. B. & Alugongo, A. A. & Tengen, T. B., "Determination of strain rate sensitivity (SRS) for grain size variants on nanocrystalline materials produced by ARB and equal channel angular pressing (ECAP)" International Journal of Chemical, Molecular, Nuclear, Materials and Metallurgical Engineering Vol:9, No:12, 2015

38. Liu, Y.G., & Zhou, J.Q., & Shen, T.D., and Hui, D., "Grain rotation dependent fracture toughness of nanocrystalline materials", Mater. Sci. Eng. A 528, 7684–7687, 2011a.

39. Liu, Y. G., & Zhou, J.Q., & Wang, L., & Zhang, S. and Wang, Y., "Grain size dependent fracture toughness of nanocrystalline materials" Mater. Sci. Eng. A 528, 4615–4619. 2011b.

10th South African Conference on Computational and Applied Mechanics
Potchefstroom
3 – 5 October 2016

A Structural-Locking Bicycle Docking Mechanism to Enable Safe and Convenient Bicycle Usage in Urban Environments

Mardu C Swanepoel*(1), Marthinus J Booysen(2), Willie Smit(3).

(1)&(2) Department of Electrical and Electronic Engineering, Stellenbosch University, Stellenbosch, South Africa.

(3) Department of Mechanical and Mechatronic Engineering, Stellenbosch University, Stellenbosch, South Africa.

*email: wjsmit@sun.ac.za

ABSTRACT

Urban cycling as a means of transport is accompanied by a great deal of positive concomitants when executed in an urban environment, but has on a large-scale been embraced by only a few communities. Where actualized, large-scale urban cycling primarily results in improved mobility, reduced congestions, healthier citizens and financial benefits for both the user and the government. Despite these acknowledged benefits, the barriers that hinders the large-scale acceptance of private urban cycling and bicycle-sharing schemes includes bicycle theft, a financial predicament involved in bicycle-related infrastructure, and implementation impediments related to bicycle-sharing platforms. This paper presents the investigation of these barriers, identifies the attitude of recreational cyclists towards urban cycling, and determines the methods and motives involved in urban bicycle theft. It then presents a novel bicycle docking solution that addresses these barriers by means of a structurally-protecting dock design. The docking solution presented here can be further developed to serve as either a bicycle-sharing platform implementation, a private bicycle docking platform, or as a combination of both. The resulting docking solution successfully provides the adequate level of protection to a bicycle docked in a South African urban environment, removes current financial burdens for possible implementable agents, and provides more versatile implementation possibilities.

KEYWORDS

Cycling, bicycle theft, urban cycling, bicycle-sharing, bicycle lock.

INTRODUCTION

Urban cycling as a means of transport is accompanied by a great deal of positive concomitants when actualized in an urban environment. Governments throughout the world are seeking ways to remove cars from congested urban areas, while various are opting for cycling as a substitution. Cities such as Copenhagen, Amsterdam and Utrecht, which were rated the top three most cycle-friendly cities in the world [1], serve as proof of this value that large-scale urban cycling adds to a community. Although urban cycling is extremely beneficial to both the individual as well as the greater community, various barriers exist in the South African context that serves as impediments against the actualization of large-scale urban cycling.

This paper defines "urban cycling" as the regular use (more than twice a week) of a bicycle as a means of transport in an urban area. Urban cycling is further divided into two main categories; *private urban cycling* where cyclists make use of their own personal bicycle to serve as means for their urban transportation needs, and *bicycle-sharing* where an infrastructure is present in the urban environment, which provides individuals with bicycles that can be used on a short-term pay-per-use rental scheme.

Cycling (also known as bicycling or biking) has developed immensely since the first bicycle was invented in 1817 [2], but the core principle on which the bicycle is built has stood the test of time. Today bicycles are used for various means ranging from commuting to recreational use to business use. Although the transport sector also experienced great changes since the introduction of the bicycle, cycling in urban environments is still being used by individuals throughout the world as a means of transport. Unfortunately, large-scale urban cycling has been embraced by only a few communities throughout the world, but as it is shown in this paper, results are overwhelmingly positive concomitant wherever it is actualized. Both the types of urban cycling implementations (as this paper defines them) have several intrinsic advantages and disadvantages when present in an urban environment.

Great effort is being made globally by governments to improve and introduce cycle-enhancing infrastructure, and to also encourage or incentivise society to use cycling as a means of urban transport. Countries such as Italy, Belgium, the Netherlands, the United Kingdom and the United States have all introduced different variations of tax incentives to increase urban cycling frequencies in these countries [3]. The general advantages that any form of cycling introduces to an urban environment includes reduced traffic congestions and therefore shortened travel times, reduced emissions leading to a healthier living environment, flexible and improved user mobility where the available mobility infrastructure is inadequate or overpopulated, and reduced infrastructure costs for the government [4] or party responsible for implementation. An example of these advantages is illustrated by the socio-economic cost analysis published by the City of Copenhagen which states that the city saves DKK 1,22 per kilometre cycled, while losing DKK 0,69 for every kilometre driven by car [5]. Various factors such as transport costs, security, comfort, branding/tourism, transport times and healthcare costs were all taken into consideration.

For the individual, urban cycling presents financial advantages through savings on fuel or transport cost, and in congested areas reduced traveling time [4]. Cycling is classified as a very good form of physical activity, and as urban cycling present a sustainable way of incorporating an individual's required physical activity into a busy lifestyle, this leads to valuable and sustainable health benefits for the cyclist. Physical activity decreases an individual's risk of many adverse health conditions, and as the majority of the world's population are inactive cycling can be very valuable to the individuals and to society when incorporated as form of transport [6].

Additional to the advantages of personal urban cycling, bicycle-sharing platforms in urban environments present an additional mode of public transport that can serve as means to citizens' both habitual or spontaneous transport needs, while also serving as a multi-modal transport connection support system. In areas with inadequate transport infrastructure, bicycle-sharing systems can also serve as the primary transport solution. A big advantage of bicycle-sharing is that it increases bicycle utilisation since the same bike is used by various users. This means that more value is extracted from less infrastructure, reducing the implementation demands and costs.

This paper presents a design solution that can be implemented into urban environments that will aid in increasing urban cycling prevalence. As a means the paper identifies and investigates the three main barriers against urban cycling, which are related to theft, cost and infrastructure implementation. Results from research activities including questionnaires and interviews are presented, followed by the definition of the requirements for a plausible design solution that addresses these barriers. The paper's resulting design solution takes the form of a bicycle docking station that uses an enclosing frame to protect the most important parts of the bicycle (frame, wheels and saddle) from theft, while lowering cost and problems related to the implementation of the infrastructure.

The Introduction section starts by providing a background and definition of cycling, urban cycling and bicycle-sharing. Also provided in this section is the rationale for attempting to increase urban cycling prevalence, the advantages of cycling, and also the status quo of existing cycling communities as corroboration of urban cycling's benefits. The State of the Art section presents the status quo of urban transportation throughout the world, as well as in the South African context. It also explains the status quo of urban cycling in the South African context while presenting the barriers identified against urban cycling. This is followed by the paper's contribution that is made towards these barriers as well as the approach followed. The Research Questionnaires and Research Interviews sections presents the results and conclusions drawn from research questionnaires and interviews, to gain an in-depth understanding of the barriers identified. The Design and Analysis section presents the design and analysis of the solution presented by this paper, followed by a discussion of the results in the last section.

STATE OF THE ART

The benefits of cycling is evident and the realization of the advantages is increasing throughout the world, yet large-scale urban cycling has not actualized in the majority of urban areas throughout the world. Although certain factors such as topography, climate, governmental regulations and infrastructure exclude some urban environments from actualizing large-scale urban cycling, the areas that are suitable experience additional barriers against urban cycling, with these preventing individuals and societies from making large-scale urban cycling a reality. This is especially true in the South African context. This section provides the status quo of urban cycling and bicycle-enhancing infrastructure implementation while elaborating on the three main barriers identified against urban cycling in the South African context.

Related Work

Bicycle theft is one of the biggest challenges for urban cycling, especially in less developed countries where statistics show a higher prevalence of criminal activities in general. According to Midgley [7], bicycle theft is currently one of the greatest challenges in urban cycling implementations taking the form of bicycle-sharing, while Shaheen et al [8] states that in the field of bicycle-sharing platforms one of the main focuses for new bicycle docking stations must be to increase the locking capabilities of the bicycle within the dock. The Nottingham Cycle-friendly Employers Project run by Clearly et al [9] practically evaluated how various implemented measures increase cycling activity among groups of individuals, and found that one of these top three measures is the provision of secure bicycle parking at the cyclist's destination.

Conventional methods used to lock up a bicycle does not measure up to the methods used by bicycle thieves to steal the bicycles. According to individuals actively involved in bicycle theft within the Stellenbosch region, the current methods of locking a bicycle by using conventional bicycle locks are not a barrier to them. They claim that "if [they] want to steel a bicycle secured with a cable lock, chain or even u-lock, [they] can" [10]. The conventional methods of personal bicycle protection is by means of available bicycle locks, with the most common of these locks being U-locks/D-locks, chain locks, folding locks and cable locks. The difference between these locks comes down to variations in their level of security, practicality and cost. U-locks/D-locks are the most secure type of lock but also the most unpractical, and due to its relatively small size all critical parts of the bicycle cannot be locked with a single lock. Cable locks are very practical and more mobile, but offers the worst level of protection. The best locking mechanism is allegedly a chain lock, as it has a good level of security and practicality. Although even they are easy to cut open if it can fit into a bolt cutter, and chain locks capable of resisting a bolt-cutter (larger than diameter 17 mm) becomes heavy and unpractical [11]. All these methods of protecting a bicycle are also limited to only locking the frame and one wheel at the most, as locks that are able to lock both wheels and the frame become unpractical in size and weight.

The *Barclays Cycle Hire* bicycle-share system in London started in 2010 and consisted of 400 docks. The cost of the system were expected to total more than £140 million over the first six years of the project, with this figure excluding the operating costs involved during that duration [12]. The average annual cost per shared bicycle in the US is around $5,900 apiece, including costs related to purchases, maintenance, repairs, software, and redistribution [13]. The US system type is similar to the *Velib'* system implemented in Paris. These figures substantiate the second major barrier which is the financial predicament created when either private bicycle locking infrastructure or a bicycle-sharing platform is to be implemented into an urban environment. The assumption is made that there is usually a strong correlation between a country or area's financial state, and the state of the mobility infrastructure present in that area. This leads to the predicament that the areas with the greatest need for improved mobility also has the least finances available for addressing the problem. Also in a South African context, the weak economy adds to the difficulty of importing possible infrastructure. Also, since the risk of implementation is directly proportional to the cost, the high cost and the nature of permanent-implementation systems increases the risk for any feasible implementation agent.

Figure 1: Velo'v Bicycle-Share System - Lyon

Various bicycle-sharing systems exist globally, varying drastically in size as well as operational principles. Most bicycle-sharing systems are categorized as "fixed-permanent" systems, as implemented in the Velib' (Paris, France), Hangzhou (Hangzhou, China) and Velo'v (Lyon, France) bicycle-share systems. A docking platform from the Velo'v system in Lyon is shown in Figure 1. Fixed-permanent systems consist of bicycle docks that are fixed to the ground, and is dependent on external infrastructure such as an external power supply, networking capabilities, steady mounting platform, etc [14]. Implementation of such a system therefore requires a great amount of resources and additional infrastructure. This adds up to higher implementation cost and thus a greater implementation risk in the case of the system not being successful once it is implemented. Post-installation optimization of the dock placement in a fixed-permanent system is extremely difficult and resource dependant. Installations of such a system therefore require very accurate traffic flow predictions and layout planning to determine the bicycle flow and resulting installation layout. Another drawback of conventional bicycle-sharing systems are the post-installation dependency of the system-specific bicycles that are used with the docks installed. This is caused by a unique design which is specific to the systems docks, bicycles and locking mechanism used. As a result, any existing bicycles that are owned (eg. rental bicycles) by the prospective agent cannot be used in the newly implemented bicycle-sharing platform, and also the servicing and replacement costs of the bicycles after installation is dependent on the manufacturer of the implemented system.

By investigating various existing bicycle-sharing systems, the most promising type of bicycle-sharing system comes in the form of a "fixed-portable" system. This type of system is found in the *BIXI* bicycle-sharing system implemented in Montreal, Canada, and is based on the principle of a "portable modular stations". The service terminal, station control-hardware, power supply and station connectivity hardware are mounted onto a single unit, while the bicycle dock is designed to form a separate modular unit. These are then joined as required to form portable docks where they are needed. Montreal's fixed-portable BIXI system is one of the significant innovators to the usual fixed-permanent system. The BIXI system does not require any anchoring to the ground, allowing stations to be moved around in a time, cost and labour efficient way. The distribution of the system can be adapted to match actual bicycle flow and demand, and the stations can also be moved temporarily to accommodate spikes in demand following events such as festivals or concerts. The BIXI system also allows for constant optimization of the system after installation and during operation, all at very little resource costs. [14]

Barriers & Contribution

Three main barriers towards urban cycling is identified and addressed in this paper. They are problems related to 1) bicycle theft and bicycle protection, 2) a financial predicament involved in bicycle-related infrastructure and 3) implementation impediments related to bicycle-sharing docks.

This paper presents further investigation of these barriers. Insight into the perceptions and attitudes of active recreational cyclists towards urban cycling is presented. The motives and methods behind bicycle theft in the Stellenbosch region is investigated, after which engineering specifications for a solution intended to protect a bicycle in this and similar areas is derived. Then a bicycle docking solution is developed that address the barriers investigated throughout the paper. The ultimate purpose of this docking solution is to improve the cycling frequency of an urban environment. The developed dock (after further development) can be introduced to a targeted urban environment in three possible implementations; the first is for the docks to lock-up personal commuters' bicycles only, the second is to serve as the docking infrastructure for a bicycle-sharing platform by locking the bicycles used in the bicycle-sharing platform, and the third is to create an implementation that consists of a combination of the first two implementations. The solution is therefore able to combine private cycling and bicycle-sharing into a single platform, with each specific implementation adjustable to suit the specific area of implementation. For continuation, the bicycle dock will be further developed and consist of theft detection capabilities that will assist with detecting attempts of theft on the bicycle, audible alarming capabilities, and machine-to-machine communication capabilities that will enable the dock to interact with users via smartphone.

Approach

The approach followed comprises of systematic methods that are used to initially identify and investigate the problems involved in urban cycling, which leads to a better understanding of the problems as well as a design solution that address these problems. The approach starts with an in depth literature study, as presented in the Introduction and State of the Art sections. The literature includes content relevant to existing urban cycling environments, existing bicycle-sharing systems, existing bicycle locking methods and also research on the factors affecting urban cycling prevalence. Research questionnaires identifies the barriers preventing active recreational cyclist from also using their bicycles as a form of urban transport, and also investigates their attitude towards urban cycling and the proposed design solution. As one of the main barriers against urban cycling is identified as bicycle theft, interviews were held with individuals previously involved in bicycle theft to identify requirements for a solution capable of protecting a bicycle in a South African context. The relevant research and requirements identified throughout these activities presents solution criteria that can serve as a basis for an urban cycling solution, and these are used during the design process to guide and improve the design solution. A conceptual design is created, followed by a physical embodiment of the concept design. The embodiment design is then converted to a detailed design which is evaluated relative to the initial design requirements by means of computational mechanics and material analysis. The detailed design is then iteratively improved based on the analysis executed on the frame until the optimal point of protection, cost and implementation characteristics is found.

RESEARCH QUESTIONNAIRES

To obtain supplementary insight into active recreational cyclists' attitudes and perceptions of urban cycling, research questionnaire were handed to 90 cyclists from the Stellenbosch and Cape Town region, South Africa. The questionnaire also investigates the cyclists' attitudes towards this paper's proposed conceptual solution, which helps to determine their personal requirements for such a system. The specific barriers identified from the questionnaire responses are valuable since it identifies the only factors preventing individuals who already agree with the intrinsic value of cycling from adopting urban cycling as a means of urban transport.

Of the 90 questionnaire respondents, 73% use their bicycle three or more times per week. The questionnaire starts by investigating the attitude of the respondents towards urban cycling in their specific environment. The general attitude is predominantly positive and respondents are eager to use cycling as a form of urban transport, with 70% percent of them willing to cycle more than 8km per destination. Next, the questionnaire investigating the cyclists' perceived individual and societal benefits of urban cycling. The responses gathered regarding the perceived benefits of urban cycling correlates substantially to the benefits identified throughout the literature. The most common perceived benefits were those related to health, cost, traffic and environmental impact. The respondents' attitudes towards bicycle-sharing and the proposed conceptual solution is also positive, with 86% of the respondents stating that either bicycle-sharing or the paper's proposed concept will, when implemented, lead to an increase in their frequency of using a bicycle for urban transport. The greatest barriers towards urban cycling that is identified through the questionnaire include bicycle theft, road safety and weather conditions. A total of 46% of the respondents had at least one bicycle stolen from them in their current urban environment, while 19% of all respondents had two or more bicycles stolen. The most important requirements identified for the paper's proposed solution is for it to be able to protect the bicycle during lock-up, with 86% of respondents mentioning this requirement. Quick bicycle lock-up, ease of use and low user cost was other requirements mentioned by a large number of respondents.

The results obtained from the research questionnaires leads to the conclusion that the cyclists' have the correct perception of the various benefits associated with urban cycling, they have an positive and eager attitude towards cycling as form of urban transport, and they are also willing to cycle great distances in an urban environment. Still, the majority of their urban transport is done by modes of transport other than cycling. This suggests that cycling can serve as a sustainable form of urban transport in these urban areas once these factors are addressed.

RESEARCH INTERVIEWS

The most important requirement for the solution to consist of as identified from the research and questionnaires is to protect the bicycle from theft during docking. To gain a better understanding of the motives and especially methods involved in urban bicycle theft, interviews were held with individuals who are or were actively involved in bicycle theft within the Stellenbosch and surrounding region. The interviews were held at the Stellenbosch Parole Observation Centre with interviewees identified by the Observation Centre.

The motivation of individuals involved in bicycle theft in the Stellenbosch region is primarily to acquire money to feed their drug habits. The bicycles are stolen and sold to syndicates operating within ther region. The syndicates then sell the bicycles to clients who intend to use the bicycles or bicycle parts for cycling-related use within the poor surrounding community, or alternatively move the bicycles to other regions to be sold to recreational cyclists unaware of the bicycles background. Therefore there is no specific predefined bicycle specifications for the bicycles that are stolen. The theft of a bicycle is not a spontaneous event. Thieves plan and prepare thoroughly before attempting to steal a bicycle, carefully choosing bicycle locking mechanisms and environments that are the easiest and least risky. If the entire bicycle cannot be stolen, separate bicycle parts will also be stolen and sold. The top priority parts to steal are the frame and the wheels, although they claim that "any other part that can be stolen will also be stolen".

The methods used to steal the bicycles are usually of an impromptu nature, making use of available tools to break conventional bicycle locks. The tools used includes a bolt-cutter, hammer, hacksaw or crow-bar. Allegedly, the easiest way to steal a bicycle is when a bicycle lock or chain can fit into a bolt cutter, as they are easily cut open. More difficult materials such as round bars and steel is not easy to cut open. Desperate measures are sometimes called on to steal a bicycle due to the nature of the motivation for theft, and thieves are prepared to spend time cutting through what is required to steal a bicycle, as long they are not blatantly noticed. The bicycle location and presence of other people are sometimes not a factor, as the thieves "know the fastest escape routes". According to the interviewees, the most effective ways of deterring a thief from stealing a bicycle, in order of descending effectivity, is; a bicycle tracker, visible surveillance cameras, alarms, or other visible security measures. [10]

DESIGN & ANALYSIS

From the conclusions drawn in the interviews, and also from the previous research, safety requirements were derived that are converted into engineering specifications. The engineering specifications is used to direct and evaluate the design. This paper only covers the mechanical design, thus the security-related engineering specifications presented only relates to the structural aspects of protection. The engineering specifications extracted from the interviews are presented in Table 1 found in Design and Analysis section, where they are also compared to the results of the final design solution.

An iterative design approach is used to complete the detailed dock design. The research, requirements and design specifications identified thus far are used as guidelines and evaluation during the design phase. For the detailed design, a three-dimensional frame layout skeleton is created that represents the conceptual design layout, and is constrained according to the embodied dimensions determined. A CAD frame generator is used to generate different frame designs based on the skeleton defined. The various frames created consists of different combinations of materials, structural components and strengthening techniques. The frames are then analysed and iteratively improved by means of computational mechanics and material analysis, using the requirements and engineering specifications as evaluation.

Requirement Definition

The most important requirements that the design solution should adhere to is derived from the research activities covered thus far. These requirements serve as the guidelines during the concept generation and design phase, and is used as the performance measure during the evaluation phase of the final solution. The main requirements that are derived, in order of descending importance, is: to protect the bicycle from theft in the urban areas investigated, to be a low-cost solution for bicycle-share and bicycle docking infrastructure, and to have the characteristics of a fixed-portable bicycle-share system. The dock size should also be universal to fit as many bicycle frames and sizes as possible. The engineering specifications that ensure the dock adequately protects the bicycle from theft is presented and evaluated in Table 1 found at the end of this section.

Embodiment Design

The conceptual solution takes the form of a bicycle docking structure which uses a skeleton-like frame to protect the most important parts of the bicycle, while using a single locking mechanism to enable the protection of these parts. This design concept is then taken and dimensions assigned to the frame. Bicycle geometry is complex, with various variables such as frame angles, wheelbase, wheel trail, tube lengths and the frame type responsible for determining the precise geometry of a bicycle [15]. Defining dock dimensions that fits all bicycle sizes and geometries as specified is the first complex design step. The frame characteristics such as frame geometry, sizing, and materials are then improved by means of rapid iteration using computational mechanics and material analysis as feedback. The technique that the bicycle dock uses to protect the bicycle is presented in Figure 2. The grey shaded area illustrated in Figure 2 is one of the few common features present in the majority of production bicycles, and is referred to as the rear-triangle. It is created by the region bound in by the main frame's rear triangle and the rear wheel's rim. Due to its universality, the rear-triangle is

Figure 2: Bicycle Protection Principle

an important feature that is used to secure the bicycle in the dock. The rest of the frame geometry varies too much between different bicycles to allow for a universal technique to base the locking of the bicycle on. The red shaded areas are barriers that restricts movement through those regions, and can take the form of structural bars or plates. These barriers prevents the frame from moving upwards or forwards. A pin-type locking mechanism consisting of 2 arms and a rod is used to finally secure the bicycle in the position it is docked in. The lock will swivel about the hinge located at the bottom of the arm. A rod is placed through the rear-triangle and arm, thereby allowing only small frame displacements once locked. The combination of the locking mechanism placement and the structure prevents the wheels, frame and saddle from being removed. The locking mechanism used to keep the pin secure is not covered in this paper.

Determining the dock dimensions is an important step since the dimensions ensure that the dock will be able to all 24" to 29" wheeled bicycles as required of it, and also because material costs depend on dock size and therefore increase with an oversized dock. To aid in determining the optimal dock dimensions, the protection technique discussed above is converted into a conceptual design, which is illustrated in Figure 3. This conceptual design is converted into a mathematical model capable of generating various dock dimensions required for a specific bicycle by only using any bicycle's wheel size, wheel base and height as inputs. A dataset of 30 different bicycle geometries including the most common bicycles in South Africa, the most common bicycles on the University of Stellenbosch campus, the *Maties* bicycles used in Stellenbosch, and also the *MTN Qhubeka* bicycle was included in this bicycle geometry dataset. The dataset is used as inputs for the simulation, and the resulting dock geometries required are generated for each bicycle. All the relevant dock dimensions required for the conceptual design's embodiment is illustrated in Figure 3.

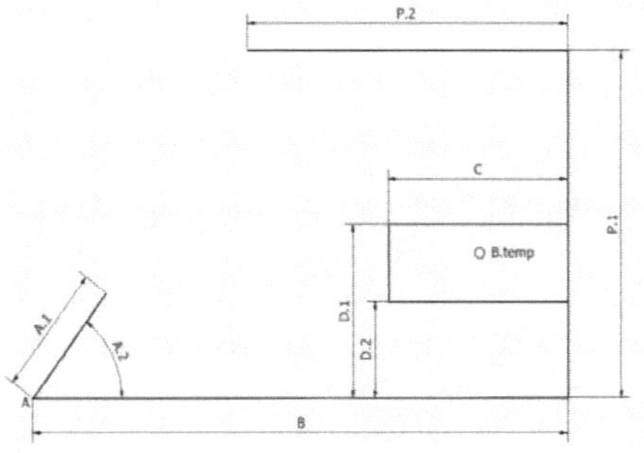

Figure 3: Conceptual Design Geometry

Dimension A.1 is chosen to reach the back triangle of a bicycle (above half diameter) with a 29" wheel (740 mm dia.), while adding a 100mm safety gap to accommodate odd frame geometries. Angle A.2 is limited to 0-90° to ensure that the bicycle cannot be moved backwards and then removed. Dimension P.1 is based on the highest bicycle in the bicycle dimensions data set, while P.2 was chosen to ensure that the seat post is covered for the largest bicycle size within the data. Using the locking mechanism pivot (point A) as datum (null-coordinate), Equation 1 presented below describes the x-coordinate of the front wheel nut (B_{temp}) based on dimension A.1, angle A.2, the bicycle's wheelbase (WB), wheel diameter ($\emptyset_{wheel}$) and pin-spacing-factor (PSF). The pin-spacing-factor (PSF) represents the distance of the locking pin from the rim as a fraction of the total wheel diameter.

$$B_{temp,\ x} = (A.1 \times \cos A.2) - (PSF \times \emptyset_{wheel}) + WB \quad - (1)$$

The height (y-coordinate) of B_{temp} is determined by using half of the wheel diameter. Using the bicycle sizing dataset as input, the simulation result is plotted as illustrated below in Figure 4.

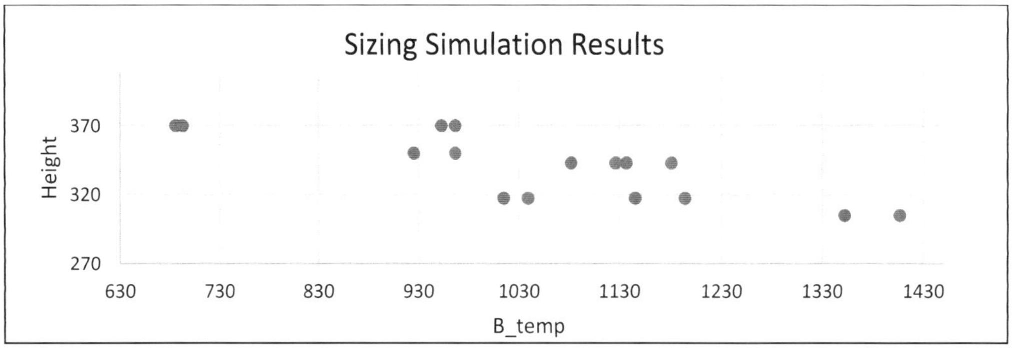

Figure 4: Sizing Simulation Results

Using the outputs of B_{temp} as plotted in Figure 4, dimensions C, D.1 and D.2 required to fit & secure all the bicycles included in the dataset can be calculated and defined.

Detailed Design

The dock's detailed design solution is presented in Figure 5. The frame comprises a combination of 32x32 mm and 25x25mm mild-steel square-tubing, and 1mm steel plates for the front wheel protection. PVC sheets are used for the floor panels on which the wheels rest. The frame and locking mechanism is manufactured as two separate pieces, and they are combined prior to installation. The locking mechanism comprises of a 30x30mm square-tubing bar and a 15mm thick steel bar which makes the fork, and diameter 20mm mild-steel rods that are used for the hinge axle and the locking pin. The user is required to manually dock the bicycle with the front wheel sitting between the two side-plates, and the back wheel axle on the rear wheel plate. After the locking mechanism pin is pushed through the rear-triangle, the rod is locked with a personal lock, or the locking mechanism which will be developed during the continuation of the work.

Figure 5: Dock Detailed Design

Design Analysis

The design analysis evaluates the performance of the paper's design solution as measured relative to the design requirements and specifications initially defined. The factors measured includes the level of protection that the design solution provides the bicycle, the dock's unit cost after production and the resulting implementation characteristics. These are all compared to the initial design requirements and specifications to determine the performance of the solution.

The level of protection that this solution provides is determined through a structural and material analysis of the frame. Computational mechanics is used to simulate the forces involved in bicycle theft (as derived from the corresponding methods) as they are applied to the structure during attempts of theft. The structural deflection, stress and strain distribution within the structure is determined. The resulting deflection simulations are analysed to determine whether the structural protection provided is still adequate in the deflected state. Secondly, the stress distributions are analysed to determine if any structural member will yield or fail due to the applied force. Through this analysis, the materials and structural members' dimensions are iteratively changed to find the optimum point of protection and material cost that still adheres to the initial requirements. The material analysis executed verifies the adequacy of the critical parts' materials in order for it to withstand the methods applied during attempts of theft. The displacements and stress distribution for a force applied by a crowbar on the bottom bar of the dock is illustrated in Figure 6. The various structural analyses performed includes all possible combinations of force and corresponding structural location derived from the methods used in bicycle theft.

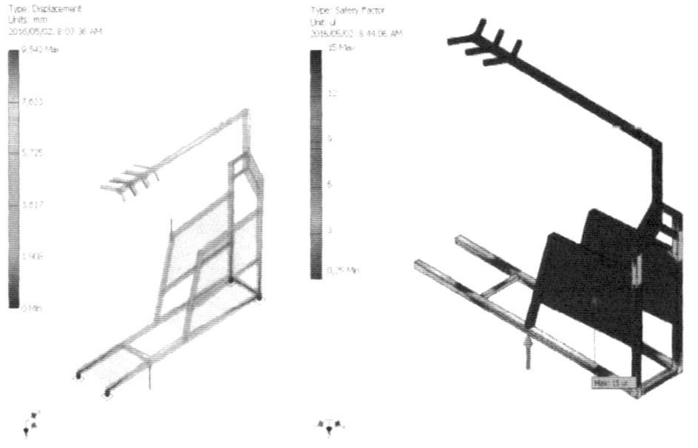

Figure 6: Stress & Deflection Simulation Result – Crowbar

Table 1 presents the engineering specifications identified during the research interviews as well as the results obtained from the structural analysis executed. The methods of bicycle theft are used as motivation for each engineering specification. Table 1 contains the derived engineering specifications, supplemented by the motivation for each specification, and also presented is the resulting performance for the corresponding specification. From the results it is seen that all engineering specifications are met. The smallest safety factor is determined as SF = 3.1, and corresponds to the force applied by the crow bar on the bottom strut.

Table 1: Engineering Specifications

Engineering Specification	Motivation for Specification	Measured Result
Resist **5.4 kN** static force	Resist force applied by a **crowbar**.	**Largest deflection measured =** 3.817mm **Yielding SF** = 3.1
Use materials > **17.5 mm diameter**	To resist bolt cutter, make material dimensions larger than bolt cutter jaws	**Smallest component** = diam. 30mm
Use **mild steel** for critical members	Complicate cutting by **hacksaw**	**All critical components = mild steel**
Resist **1.198 kN** static force	Resist **biomechanical** forces applied	**Yielding SF** = 6.34

As discussed in the State of the Art section, the costs involved to acquire the bicycle docking infrastructure is a big impediment, especially if implemented in a South African context. A direct financial comparison to existing systems is difficult. This is due to the costs available for similar systems which all include various factors such as implementation, labour, operating and added infrastructure costs. An analysis of the resources required to produce the solution is presented in Table 2. Although a direct comparison to existing systems cannot be made, the resource requirements calculated still serve as an illustration of the financial savings that can be accomplished by this system. The system is manufactured locally, creates jobs for unschooled workers, and eliminates the need for importing infrastructure.

Table 2: Resource Requirements

Resource	Total	Comment
Dock Material Weight (Steel)	21 Kg	30x30x2mm square tubing, steel plates.
Manufacturing Time	4.5 hours	Locally produced by hand
Implementation Time	2 hours	Time required to install a dock

The design solution presented is classified as a "portable modular station", as is the case with the fixed-portable bicycle-sharing system present in the *BIXI* bicycle-share. The solution requires minimal anchoring to the ground thereby allowing stations to be moved around in a time, cost, and labour efficient way. The distribution of the station can be adapted to accurately match actual bicycle flow and demand, and the stations can also be moved temporarily to accommodate spikes in demand following outlier events such as festivals or concerts. The solution's semi-permanent installation nature decreases the installation risk of the system because the system can be removed and re-sold to another party in the case of unsuccessful implementation. The resulting design solution's characteristics therefore correlates greatly to the characteristics (and thereby bringing with it the advantages of) a fixed-portable bicycle-sharing system.

DISCUSSION

The paper successfully identifies and elaborates on the three main barriers against urban cycling as they relate to bicycle theft, cost and infrastructure implementation. The research activities in the form of research questionnaires and interviews provides an in-depth understanding of these barriers identified, and also provides a better understanding of the solution requirements to address these barriers. The design solution successfully addresses the barriers identified, and meets all the initial requirements and specifications defined. The resulting dock is a simple design solution that accommodates the majority of urban commuting bicycles. The docking solution presented by this paper is a viable blueprint that can, with continuation of the proposed work, be developed to create a bicycle-sharing implementable platform in the South African context.

ACKNOWLEDGEMENTS

The authors give thanks to the individuals who responded to the research questionnaires distributed among recreational cyclists in the Western Cape. Thank you to the individuals who gave their time and inputs during the interviews that provided clarity on the methods and motives of bicycle theft. The authors also give thanks to the funding received by MTN that contributed to making this research possible.

REFERENCES

[1] Copenhagenize, "Copenhagenize Index 2015," Copenhagenize Design Co., 2 June 2015. [Online]. Available: http://copenhagenize.eu/index/about.html. [Accessed February 2016].

[2] C. Hoefer, "A Short Illustrated History of the Bicycle," April 2009. [Online]. Available: https://www.crazyguyonabike.com/doc/page/?page_id=40616. [Accessed 10 April 2016].

[3] Cycling Embasy of Denmark, "Tax incentives for bike commuting," The Cycling Embassy Of Denmark , Copenhagen, 2010.

[4] Quay Communications Inc., "TransLink Public Bike System Feasibility Study," Quay Communications Inc, Vancouver, 2008.

[5] City of Copenhagen, "Copenhagen City of Cyclists - Bicycle Account 2010," City of Copenhagen - The Technical and Environmental Administration Trafic Department, Copenhagen, 2011.

[6] I. Lee, E. Shiroma, F. Lobelo, P. Puska, S. Blair and P. Katzmarzyk, "Effect of physical inactivity on major non-communicable diseases worldwide: an analysis of burden of disease and life expectancy," *The Lancet,* vol. 380, no. 9838, pp. 219-229, 2012.

[7] P. Midgley, "Bicycle-Sharing Schemes: Enhancing Sustainable Mobility in Urban Areas," New York, 2011.

[8] S. A. Shaheen, S. Guzman and H. Zhang, "Bikesharing in Europe, the Americas, and Asia," *Transportation Research Record: Journal of the Transportation Research Board,* no. 2143, pp. 159-167, 2010.

[9] J. Cleary and H. Mcclintock, "The Nottingham Cycle-friendly Employers Project: Lessons for encouraging cycle commuting," *Local Environment,* vol. 5, no. 2, pp. 217-222, 2010.

[10] M. C. Swanepoel, Interviewee, *Experts in the field of bicycle theft.* [Interview]. March 2016.

[11] The Best Bike Lock, "How to choose the best bike lock in 2016," 2016. [Online]. Available: http://thebestbikelock.com/. [Accessed 10 February 2016].

[12] B. Brady, "Two-thirds of London's Boris bikes need repairs," The Independant, London, 2011.

[13] S. A. Shaheen, E. W. Martin, N. D. Chan, A. P. Cohen and M. Pogodzinski, "Public Bikesharing in North America During a Period of Rapid Expansion: Understanding BusinessModels, Industry Trends and User Impacts," Mineta Transportation Institute, San José, CA, 2014.

[14] Gris Orange Consultant, Bike Sharing Guide, Ottawa: Public Works and Government Services Canada, 2009.

[15] A. Denham, "Understanding Bicycle Frame Geometry," 4 October 2013. [Online]. Available: http://www.cyclingabout.com/understanding-bicycle-frame-geometry/. [Accessed March 2016].

Thermal losses considerations in thermo-acoustic engine design

L.K. Tartibu*

Mechanical Engineering Technology, University of Johannesburg, Doornfontein Campus, Johannesburg 2028, South Africa

ltartibu@uj.ac.za

ABSTRACT

Thermo-acoustic cooling as an environmentally friendly refrigeration system is one of the research areas being pursued. Although not commercially available and simple to fabricate, the designing of thermo-acoustic coolers involves significant technical challenges. Many fundamental issues related to the thermo-acoustic effects and the associated heat transfer must be addressed. The most inhibiting characteristic of current thermo-acoustic cooling devices is the lack of efficiency. The stack has been identified as the heart of the device where the heat transfer takes place. Improving its performance will make thermo-acoustic technology more attractive. Most of the existing efforts have not taken thermal losses to the surroundings into account in the derivation of the models. Five different parameters describing the stack geometry and the angular frequency of the standing wave are considered. This work explores the use of a multi-objective optimization approach to model and to optimize the performance of a simple thermo-acoustic engine. The present study highlights the importance of thermal losses in the modelling of small-scale thermo-acoustic engines using a multi-objective approach. The unique characteristic of this research is the computing of all efficient optimal solutions describing the best geometrical configuration of thermo-acoustic engines.

KEYWORDS

Thermo-acoustic engine, modelling, multi-objective optimization, heat flow, stack

INTRODUCTION

In order to broaden the applications of Micro Electro-Mechanical Systems, sensor networks and small-scale remote systems, the development of miniature power systems is critically important. A lot of efforts are directed toward the development of miniature energy sources [1]. Small-scale power systems are subjected to certain limitations. For instance, the energy densities of batteries are small and scaled-down rotating machinery are challenging to fabricate. Therefore, thermo-acoustic engine is seen as a promising candidate for small-scale electricity generation when coupled with an electroacoustic transformer [2]. Apart from being environmentally friendly, thermo-acoustic systems are potentially highly reliable because of the simplicity of their structures and the limited number of moving parts. Thermo-acoustic processes involve heat and sound interactions and the thermal-to-acoustic energy conversion [3]. A schematic of a simple standing-wave thermo-acoustic engine (or prime mover) is shown in Figure 1. A piece of porous material (called stack) is the heart of the system where acoustic power is generated in the presence of sufficiently large externally maintained temperature gradient. Two heat exchangers located on the sides of the stack supplies and rejects the heat. The acoustic modes are defined by the resonator geometry. Under appropriate conditions, heat is added to gas parcels oscillating inside the stack at the moment of their compression and extracted at the moment of their rarefaction. As a result, an acoustic power is generated and, therefore, acoustic modes can be excited [3]. In order to generate electricity, an electroacoustic transformer installed at the open end of the tube will convert some of the acoustic power into electricity.
The increased roles of thermo-viscous losses, thermal management and fabrication issues, in addition to the difficulty in integrating with heat sources, make the downsizing of thermo-acoustic systems challenging. Several previous studies report the development of miniature thermo-acoustic engines. The construction and

performance of a relatively small 14 cm tube was documented by Hofler and Adeff [4]. Much smaller systems, down to a few centimetres in length, were also built by Symko et al. [5], but their design was not reported in detail sufficient for reproduction

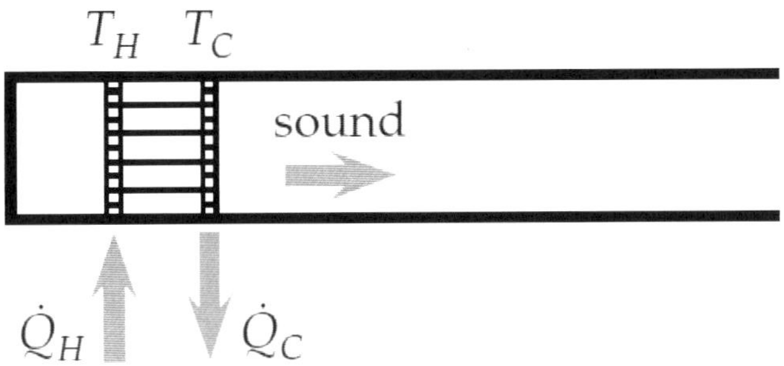

Figure 1: Thermoacoustic engine or prime mover

A study conducted at the Energy research Centre of the Netherlands [6] show that the ratio between thermal losses and acoustic power changes with increasing acoustic power for thermo-acoustic devices. It suggests that heat losses by convection, conduction and radiation need to be adequately covered in the modelling especially with regards to miniaturization of the devices where thermal losses are expected to increase [7]. McLaughlin [8] has thoroughly analysed the heat transfer for a Helmholtz-like resonator, 1.91 cm in diameter and 3.28 cm in length. The loss to conduction has been estimated as 40% of the input power. The losses from convection inside and outside of the device have been estimated as 38%. Radiation accounts for 10% of the input power. This leaves only 12% of input power that can be used to produce acoustic work (Figure 2). Although these losses are approximations not meant to be highly accurate determinations, they suggest that these losses are significant when compared to total heat input and should be considered as design criterion. Therefore, this work aims to highlight one methodology to incorporate thermal losses in the design process.

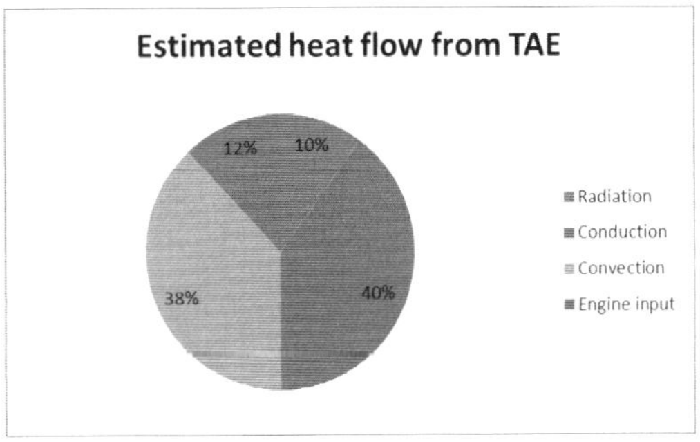

Figure 2: Heat flows from the power supply partitioned by the losses from radiation, conduction, convection and the power input

MATHEMATICAL ANALYSIS AND OPTIMIZATION

Optimization techniques as a design aid for thermo-acoustic engines have been under-utilized. Most previous studies [9, 10] have been limited to parametric studies to estimate the effect of single design parameters on device performance while ignoring thermal losses to the surroundings. These parametric studies are unable to capture the nonlinear interactions inherent in thermo-acoustic models with multiple variables. Therefore, these approaches only guarantee locally optimal solutions. In all likelihood, each optimal design is a local optimum as the solution obtained is optimal (either maximal or minimal) within a neighbouring set of candidate solution. A new approach is proposed in this study to search for global optimum. These solutions will be optimal among all possible solutions in a specific domain, not just those in particular neighbourhood of variables.
This work considers previous optimization efforts by Zink et al. [7] and Trapp et al. [11] in order to illustrate the optimization of thermo-acoustic systems. Thermal losses to the surroundings that are typically disregarded

are taken into account. These losses have been incorporated in the modelling as objectives functions. An effort to effectively implement the Epsilon-constraint method for producing the Pareto optimal solutions of the multi-objective optimization problem is carried out in this work. This has been implemented in the modelling language GAMS (General Algebraic Modelling System, www.gams.com [12]).

MODELLING APPROACH

In this section, the modelling approach for the physical standing wave engine depicted in Figure 1 is presented; the development of the mathematical model equations is included in Tartibu et al. studies [13]. The problem is reduced to a two-dimensional domain because of the symmetry present in the stack. Two constant temperature boundaries are considered; namely, one convective boundary and one adiabatic boundary, as shown in Figure 3. For the model, the stack geometry and the frequency of the sound wave are considered. The model considers variations in operating conditions and the interdependence of stack location and geometry in a quarter-wavelength ($\lambda/4$) resonator tube.
Five different parameters are considered to characterize the stack:

- L: stack length,
- H: stack height,
- Za: stack placement (with Za=0 corresponding to the closed end of the resonator tube),
- dc: channel dimension, and
- N: number of channels.

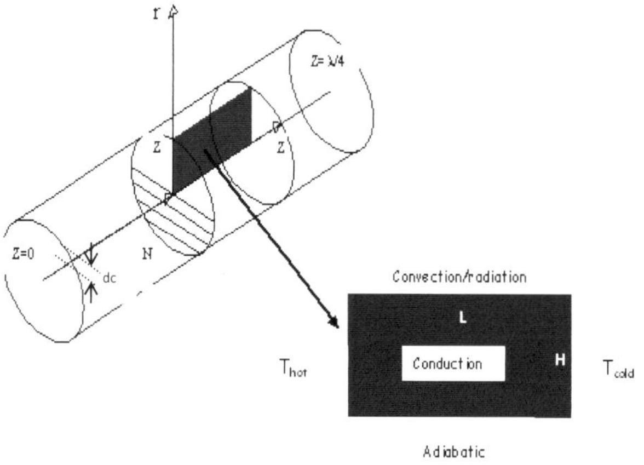

Figure 3: Computational domain

Those parameters have been allowed to vary simultaneously. Five different objectives as described by Trapp et al. [11], namely two acoustic objectives — acoustic work (W) and viscous resistance (R_V) — and three thermal objectives — convective heat flow ($\overset{o}{Q}_{conv}$), radiative heat flow ($\overset{o}{Q}_{rad}$) and conductive heat flow ($\overset{o}{Q}_{cond}$) — are considered to measure the quality of a given set of variable values that satisfy all the constraints. Ultimately, optimizing the resulting problem generates optimal objective function value $G^* = \left[W^*, R_V^*, \overset{o}{Q}{}^*_{conv}, \overset{o}{Q}{}^*_{rad}, \overset{o}{Q}{}^*_{cond} \right]$ and optimal solutions $x^* = [L^*, H^*, dc^*, Za^*, N^*]$. Since the five objectives are conflicting in nature [11, 14], a multi-objective optimization approach has been used.

ILLUSTRATION OF THE OPTIMISATION PROCEDURE OF THE STACK

Boundary conditions

The five variables— L, H, dc, Za, N —may only take values within the certain lower and upper bounds. The feasible domains for a thermo-acoustic stack are defined as follows:

$$
\begin{aligned}
&L_{min} \le L \le L_{max} \\
&H_{min} \le H \le H_{max} \\
&dc_{min} \le dc \le dc_{max} \\
&Za_{min} \le Za \le Za_{max} - L \\
&N_{min} \le N \le N_{max}
\end{aligned}
\tag{1}
$$

$L,\ H,\ dc,\ Za \in \Re+$ and $N \in Z+$

with $dc_{min} = 2\delta_k$ and $dc_{max} = 4\delta_k$ [10]

δ_k is the thermal penetration depth given by:

$$
\delta_k = \sqrt{\frac{2K}{\rho c_p \omega}} \tag{2}
$$

where $\omega=2\pi f$ is the angular frequency of the sound wave, f is the design frequency and K the thermal conductivity. Here, ρ and c_p are the density and isobaric specific heat of the gas, respectively.

$$
N(dc + t_w) \le 2H \tag{3}
$$

where t_w represents the wall thickness around a single channel

The following boundary conditions are defined:

1. constant hot side temperature (T_h) or (T_{hot});
2. constant cold side temperature (T_C) or (T_{Cold});
3. adiabatic boundary, modelling the central axis of the cylindrical stack:

$$
\left.\frac{\partial T}{\partial r}\right|_{r=0} = 0; \tag{4}
$$

4. free convection and radiation to surroundings (at T_∞) with temperature dependent heat transfer coefficient (h), emissivity (ε), Stefan Boltzmann constant (k_B) and thermal conductivity (K):

$$
K\left.\frac{\partial T}{\partial r}\right|_{r=H} = h(T_S - T_\infty) + \varepsilon k_b (T_S^4 - T_\infty^4) \tag{5}
$$

Acoustic power

$$
W = \omega L N \left(\frac{\pi H^2}{2(dc + t_w)}\right)\left[\delta_k \frac{(\gamma-1)p^2}{\rho c^2(1+\varepsilon)}(\Gamma - 1) - \delta_v \rho u^2\right] \tag{6}
$$

where γ, c and Γ are the isentropic coefficient, the speed of sound and the critical temperature gradient respectively.

$$
\Pi = \frac{2A}{dc + t_w} \tag{7}
$$

$$
p = p_{max} \cos\left(\frac{2\pi Za}{\lambda}\right) \tag{8}
$$

$$
u = u_{max} \sin\left(\frac{2\pi Za}{\lambda}\right) \tag{9}
$$

$$
\text{with } u_{max} = \frac{p_{max}}{\rho c} \tag{10}
$$

$$
\varepsilon = \frac{(\rho c_p \delta_k)_g}{(\rho c_p \delta_s)_s} \frac{\tanh((i+1)y_0/\delta_k)}{\tanh((i+1)l/\delta_s)} \tag{11}
$$

Viscous resistance

$$
R_V = \frac{\mu \Pi L}{A_C^2 \delta_V N} = \frac{2\mu}{\delta_V} \frac{L}{(dc + t_w)\pi H^2 N} \tag{12}
$$

where μ, A_C and δ_V (given by $\sqrt{\frac{2\mu}{\rho\omega}}$) are respectively the dynamic viscosity of the gas, the area of the channel and the viscous penetration depth.

Convective heat flow

$$\overset{o}{Q}_{conv} = hA(T_S - T_\infty) \tag{13}$$

$$T_S = T_h e^{\ln\left(\frac{T_C}{T_h}\right)\frac{Za}{L}} \tag{14}$$

$$\overset{o}{Q}_{conv} = H\int_0^{2\pi}\int_0^{L} h(T(z))(T(z) - T_\infty)dzd\phi \tag{15}$$

$$h(Ts) = \frac{k_g}{2H} Nu \tag{16}$$

$$Nu = 0.36 + \frac{0.518 Ra_D^{\frac{1}{4}}}{\left[1 + \left(\frac{0.559}{Pr}\right)^{\frac{9}{16}}\right]^{\frac{4}{9}}} \tag{17}$$

$$Pr = \frac{\nu}{\alpha} \tag{18}$$

$$Ra = \frac{g\beta(T_S - T_\infty)8H^3}{\nu\alpha} \tag{19}$$

where Pr is the Prandtl number; T_S is the surface temperature; T_∞ is the (constant) temperature of the surroundings; ν is the viscosity of the surrounding gas; and α is the thermal diffusivity of the surrounding gas (air).

$$\overset{o}{Q}_{conv} = 2\pi HLh\left[\frac{T_C - T_H}{\ln\left(\frac{T_C}{T_H}\right)} - T_\infty\right] \tag{20}$$

$$Za \geq L\ \log\left(\frac{T_{inf}}{T_C}\right) \tag{21}$$

Radiative heat flow

$$\overset{o}{Q}_{rad} = k_B\varepsilon As(T^4 - T_\infty^4) \tag{22}$$

$$\overset{o}{Q}_{rad} = Hk_B\int_0^{2\pi}\int_0^{L}\varepsilon(T_{(z)}{}^4 - T_\infty^4)dzd\phi \tag{23}$$

$$\overset{o}{Q}_{rad} = 2\pi HLk_B\varepsilon\left[\frac{T_C^4 - T_H^4}{4\ln\left(\frac{T_C}{T_H}\right)} - T_\infty^4\right] \tag{24}$$

Conductive heat flow

$$\frac{\Delta Q}{\Delta t} = -kA\frac{\Delta T}{\Delta x} \tag{25}$$

$$\overset{o}{Q}_{cond} = \int_0^{2\pi}\int_0^{H}\left(k_{zz}\frac{\partial T}{\partial r}\right)drd\phi \tag{26}$$

$$k_{zz} = \frac{k_S t_w + k_g dc}{t_w + dc} \tag{27}$$

$$\text{Therefore: } \left.\frac{\partial T}{\partial z}\right|_{z=L} = \frac{\ln\left(\frac{T_C}{T_H}\right)\left(\frac{T_C}{T_H}\right)}{L} \tag{28}$$

$$\text{And after integration } \overset{o}{Q}_{cond} = \frac{k_{zz}}{L}\pi H^2 T_C \ln\left(\frac{T_H}{T_C}\right) \tag{29}$$

Details description of the derivations of Equations 3 to 29 are available in reference [7], [11] and [13].

SOLUTION METHODOLOGY OF THE MULTI-OBJECTIVE MATHEMATICAL PROGRAMMING PROBLEMS

All the expressions involved in the mathematical programming formulation (MPF) have been presented in the previous section. Together with the following expressions, they represent a mixed-integer nonlinear programming (MINLP) problem:

$$(\text{MPF}) \quad \min_{L,H,Za,dc,N} \xi = w_1(-W) + w_2 R_V + w_3 \overset{o}{Q}_{conv} + w_4 \overset{o}{Q}_{rad} + w_5 \overset{o}{Q}_{cond} \tag{30}$$

This mathematical model characterizes the essential elements of a standing wave thermo-acoustic engine. Restricted cases of objectives functions in order to identify general tendencies of the structural variables to influence individual objective components is described in Tartibu et al. [13] studies . To illustrate the proposed approach, a thermo-acoustic couple (TAC) as described in Atchley et al. [15], which consists of a parallel-plate stack placed in helium-filled resonator is considered. All relevant parameters are given in Table 1 and Table 2.

Table 1: Specifications for thermo-acoustic couple

Parameter	Symbol	Value	Unit
Isentropic coefficient	γ	1.67	
Gas density	ρ	0.16674	kg/m^3
Specific heat capacity of the gas	c_p	5193.1	J/kg.K
Dynamic viscosity of the gas	μ	$1.9561.10^{-5}$	kg/m.s
Maximum velocity	u_{max}	670	m/s
Maximum pressure	p_{max}	114003	Pa
Speed of sound	c	1020	m/s
Thickness plate	t_w	1.91.10-4	m
Frequency	f	600-700-800	Hz
Thermal conductivity Helium	k_g	0.16	W/(m.K)
Thermal conductivity stainless steel	k_S	11.8	W/(m.K)

Table 2: Additional parameters used for programming

Parameter	Symbol	Value	Unit
Temperature of the surrounding	T_∞	298	K
Constant cold side temperature	T_C	300	K
Constant hot side temperature	T_H	700	K
Wavelength	λ	1.466	m
Thermal expansion	β	$1/T_\infty$	1/K
Thermal diffusivity	α	2.1117E-5	m^2s-1
Critical temperature gradient	Γ	3	

All proposed MINLP models are solved by GAMS 23.8.1, using LINDOGLOBAL solver on a personal computer Pentium IV 2.1 GHz with 4 GB RAM.

For the case of multiples objective optimization, all five objective components are considered by regarding acoustic work (W), viscous resistance (R_V), convective heat flow ($\overset{o}{Q}_{conv}$), radiative heat flow ($\overset{o}{Q}_{rad}$) and conductive heat flow ($\overset{o}{Q}_{cond}$) as five distinct objective components. The optimization task is formulated as a five-criterion mixed-integer nonlinear programming problem (MPF) that simultaneously minimizes the negative magnitude of the acoustic work (W) (since it is the only objective to be maximized), the viscous resistance (R_V), the convective heat flow ($\overset{o}{Q}_{conv}$), the radiative heat flow ($\overset{o}{Q}_{rad}$) and the conductive heat flow ($\overset{o}{Q}_{cond}$).

$$(MPF)\ \min_{L,H,Z_a,dc,N} \xi = \left\{ -W_{(L,H,Z_a,dc,N)}, R_{V(L,H,Z_a,dc,N)}, Q_{conv(L,H,Z_a,dc,N)}, Q_{rad(L,H,Z_a,dc,N)}, Q_{cond(L,H,Z_a,dc,N)} \right\} \quad (31)$$

subject to constraints in Equations 3 and 21, and variable restrictions in Equation 33. In this formulation, (L, H, Za, dc, N) denotes the geometric parameters.

There is no single optimal solution that simultaneously optimizes all the five objectives functions. In these cases, the decision makers are looking for the "most preferred" solution. To find the most preferred solution of this multi-objective model, the augmented ε -constraint method (AUGMENCON) as proposed by Mavrotas [16] is applied. The AUGMENCON method has been coded in GAMS. The code is available in the GAMS library (http://www.gams.com/modlib/libhtml/epscm.htm) with an example. While the part of the code that has to do with the example (the specific objective functions and constraints), as well as the parameters of AUGMENCON have been modified in this case, the part of the code that performs the calculation of payoff table with lexicographic optimization and the production of the Pareto optimal solutions is fully parameterized in order to be ready to use.

Practically, the ε -constraint method is applied as follows: from the payoff table the range of each one of the p-1 objective functions that are going to be used as constraints is obtained. Then the range of the ith objective function is divided into qi equal intervals using (q_i-1) intermediate equidistant grid points. Thus in total (q_i + 1) grid points that are used to vary parametrically the right hand side (ε_i) of the ith objective function are obtained (detailed descriptions available in reference [16]). The total number of runs becomecon (q_2 + 1)$\times$ (q_3 + 1) $\times \ldots \times$ (q_p + 1). The augmented ε -constraint method for solving model (Equation 31) can be formulated as:

$$\max\left(W_{(L,H,Za,dc,N)} + dir_1\, r_1 \times \left(\frac{s_2}{r_2} + \frac{s_3}{r_3} + \frac{s_4}{r_4} + \frac{s_5}{r_5} \right) \right)$$

Subject to

$$\begin{aligned} & R_{V(L,H,Za,dc,N)} - dir_2\, s_2 = \varepsilon_2 \\ & \overset{o}{Q}_{conv(L,H,Za,dc,N)} - dir_3\, s_3 = \varepsilon_3 \\ & \overset{o}{Q}_{rad(L,H,Za,dc,N)} - dir_4\, s_4 = \varepsilon_4 \\ & \overset{o}{Q}_{conv(L,H,Za,dc,N)} - dir_5\, s_5 = \varepsilon_5 \end{aligned} \quad (32)$$

$$s_i \in \Re^+$$

where dir_i is the direction of the ith objective function, which is equal to -1 when the ith function should be minimized, and equal to +1, when it should be maximized. Efficient solutions to the problem are obtained by parametrical iterative variations in the ε_i. s_i are the introduced surplus variables for the constraints of the MMP problem. $r_1\, s_i / r_i$ are used in the second term of the objective function, in order to avoid any scaling problem. The formulation of Equation 32 is the augmented ε-constraint method due to the augmentation of the objective function W by the second term. The following constraints (upper and lower bounds) have been enforced on variables in order for the solver to carry out the search of the optimal solutions within each identified ranges:

$$L.lo = 0.005;\ \ L.up = 0.05;$$
$$Za.lo = 0.005; \qquad (33)$$
$$H.lo = 0.005;$$
$$dc.lo > 2.\delta_k;\ \ dc.up < 4.\delta_k$$

We use lexicographic optimization for the payoff table (detailed descriptions available in reference [13]); the application of the model (Equation 32) will provide *only* the Pareto optimal solutions, avoiding the weakly Pareto optimal solutions. Efficient solutions to the proposed model have been found using AUGMENCON method and the LINDOGLOBAL solver. To save computational time, the early exit from the loops as proposed by Mavrotas [16] has been applied. The range of each five objective functions is divided in four intervals (five grid points). The integer variable N has been given values of 20-25-30-35-40-45-50. In addition, the frequency was set (arbitrary) to 600 Hz-700 Hz- 800 Hz. The maximum CPU time taken to complete the results is 2050.212 sec. The following section report only sets of Pareto solutions obtained.

RESULTS AND DISCUSSIONS

Figure 4 represents the Pareto optimal solutions graphically; these results shows that there are numerous optimal solutions that optimizes the geometry of the stack and highlights the fact that the geometrical parameters are interdependent (no clear relation between variables could be derived), supporting the use of a multi-objective approach for optimization of thermo-acoustic engines. To maximize acoustic work W and minimize viscous resistance and thermal losses simultaneously, there is a specific stack length (L) to which correspond a specific stack height (H), a specific stack spacing (dc) and a specific number of channels (N). It should be noted that in all cases, locating the stack closer to the closed end produced the desired effect. All Pareto optimal solutions can be computed in order for the decision maker to select his preferred choice.

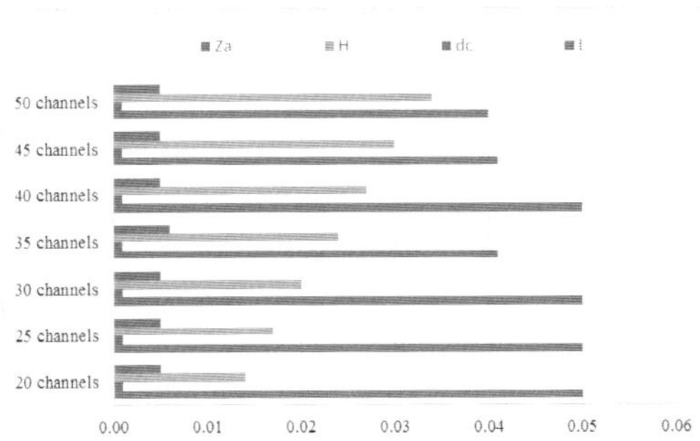

(a) Frequency = 600 Hz

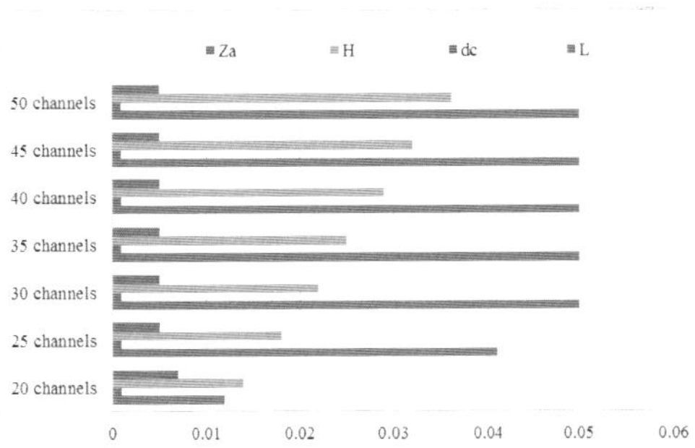

(b) Frequency = 700 Hz

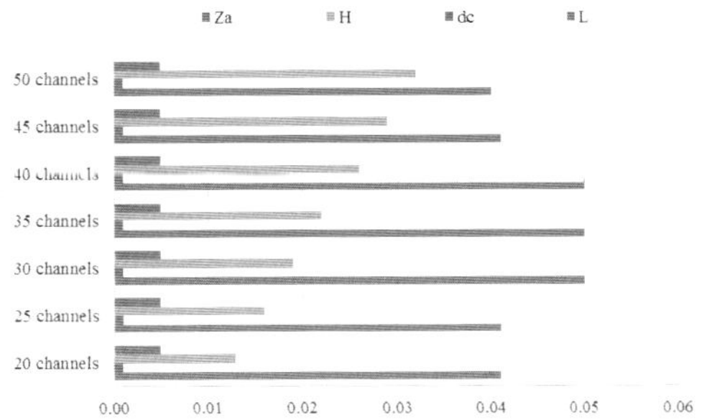

(c) Frequency = 800 Hz

Figure 4: Optimal structural variables

These optimal solutions are then used to construct Figure 5 representing respectively acoustic work, viscous resistance and thermal losses plotted as a function of N, L, dc and H. The unique contribution of this work is the ability to quantify all the thermal losses and select the best geometrical configuration of the stack accordingly. Therefore, the designer can simultaneously maximise acoustic work and minimise losses (viscous resistance as well as heat flows) by considering the thermal efficiency (η) which can be defined as the ratio of the work output over the sum of the work output and losses as follows:

$$\eta = \frac{W}{W + R_v + Q_{conv} + Q_{rad} + Q_{cond}} \tag{34}$$

This ratio can be used to compare the results obtained by the proposed augmented ε-constraint method and identify the preferred solution.

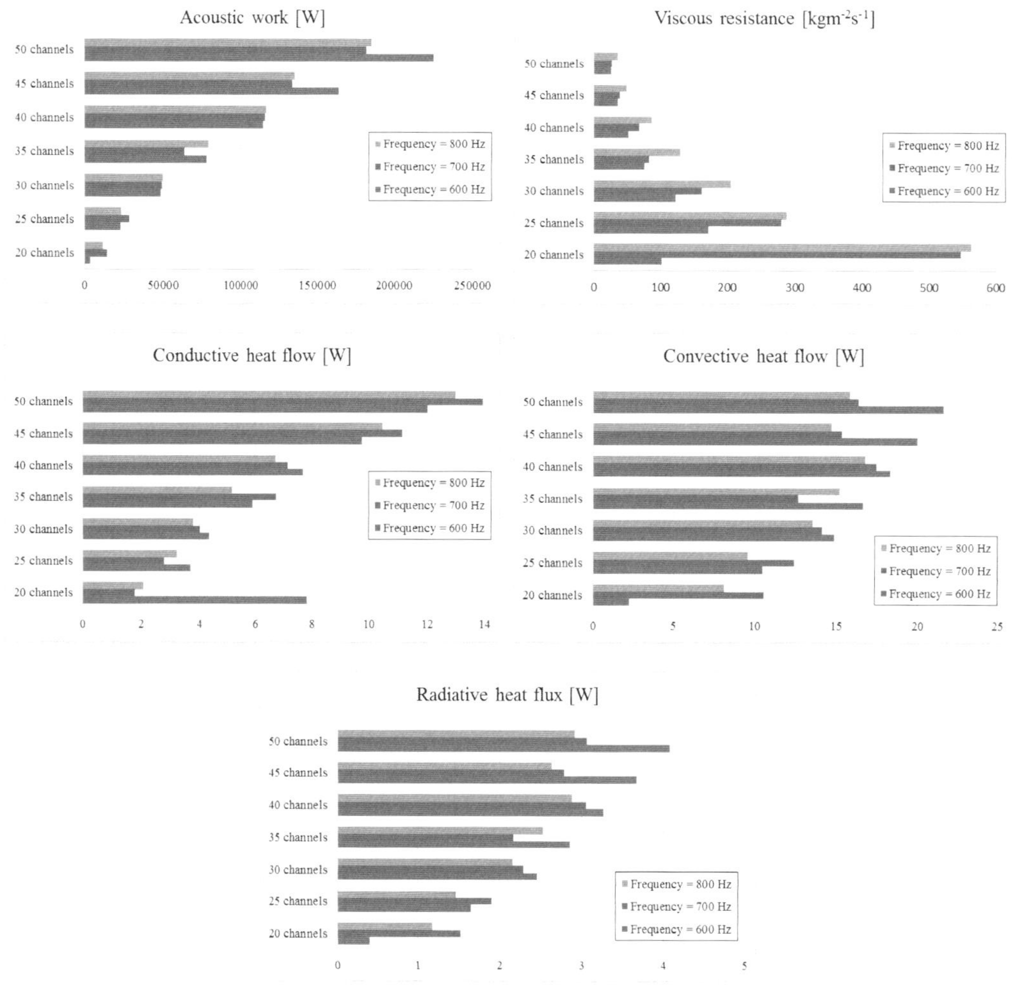

Figure 5: Acoustic work, viscous resistance and heat flows

CONCLUSION

Previous studies reveal that scaling down the device leads to an increase of the ratio between the surface area and the active volume, resulting in higher thermal losses. These thermal losses - the convective, radiative and conductive heat flow - need careful attention since they are not adequately included in current modelling approaches. An estimate of heat losses shows that they are significant relative to the total energy supplied to the thermo-acoustic engine (TAE). This provides a clear motivation to include the aforementioned losses in the modelling approach in an effort to improve the performance of the devices. A new mathematical modelling approach is proposed to model and optimize thermo-acoustic engines while taking into account thermal losses. The idea of incorporating thermal losses in the modelling and optimization of TAE gives the decision maker a clear picture of expected magnitude and the ability to search for the configuration that will simultaneously minimize them. This approach is used to compute the optimal set of parameters describing the geometry of the device: the stack length, stack height, stack position from the closed end of the TAE, stack spacing and the number of channels. These are the variables in the mathematical modelling formulation. The performance of the device is measured through the acoustic (work output and viscous resistance) and the thermal losses (convective, radiative and conductive heat flow) that have been used as objective functions to measure the quality of each set of variable values that satisfies all of the constraints. This problem has been formulated as a five-criterion mixed-integer nonlinear programming problem. This formulation allows for identifying the implication of each objective emphasis on the geometry of the stack. A case study is used for illustration. A set of objective functions and Pareto optimal solutions are computed in this work and guidance for the decision maker's selection of the preferred solution is suggested.

REFERENCES

1 Evans, J. D., "Powering the integrated microsystems", DARPA Microsystems Technology Symposium, San Jose, 2007.

2 Matveev, K. I., Wekin, A., Richards, C. D., Shafeie-Tehrany, N., "On the coupling between standing-wave thermoacoustic engine and piezoelectric transducer", ASME International Mechanical Engineering Congress & Exposition, Seattle, WA, ASME paper IMECE2007-41119, 2007.

3 Swift, G. W., "Thermoacoustics: a unifying perspective for some engines and refrigerators", Acoustical Society of America, Melville NY. 2002.

4 Hofler, T. J. and Adeff, J. A., "An optimised miniature Hofler tube", Acoustics Research Letters Online, Vol. 2, pp. 37-42, 2001.

5 Symko, O. G., Abdel-Rahman, E., Kwon, Y. S., Emmi, M. and Behunin, R., "Design and development of high-frequency thermoacoustic engines for thermal management in microelectronics", Microelectronics Journal, Vol. 35. No. 2, pp. 185-191, 2004.

6 Spoelstra, S., "THermoAcoustic Technology for Energy Applications (THATEA)", Energy research Centre of the Netherlands (ECN), Project reference number: 226415, February 2012.

7 Zink, F., Waterer, H., Archer, R., Schaefer, L., "Geometric optimization of a thermoacoustic regenerator", International Journal of Thermal Sciences, Vol. 48. No. 12, pp. 2309-2322, 2009.

8 McLaughlin B. J., "Study and development of high frequency thermoacoustic prime movers with piezoelectric transducers", Ph.D. thesis, University of Utah, 2008.

9 Wetzel, M., Herman, C., "Design optimization of thermoacoustic refrigerators", International Journal of Refrigeration, Vol. 20. No. 1, pp. 3-21, 1997.

10 Tijani, M. E. H., Zeegers, J. C. H. and De Waele, A. T. A. M., "The optimal stack spacing for thermoacoustic refrigeration", Journal of the Acoustical Society of America, Vol. 112. No. 1, pp. 128-133, 2002.

11 Trapp, A. C., Zink, F., Prokopyev, O. A. and Schaefer, L., "Thermoacoustic heat engine modelling and design optimization". Journal of Applied Thermal Engineering, Vol. 31, pp. 2518-2528, 2011.

12 Generalized Algebraic modelling Systems, (GAMS), [online] available: http://www.gams.com.

13 Tartibu, L. K., Sun, B. and Kaunda, M. A. E., "Multi-objective optimization of the stack of a thermoacoustic engine using GAMS", Applied Soft Computing, Vol. 28, pp. 30-43, 2015.

14 Tartibu L. K., Sun B., Kaunda M. A. E., "Geometric optimisation of micro-thermoacoustic cooler for heat management in electronics", IEEE International Conference on Industrial Technology (ICIT), Cape Town, South Africa, 527 – 532, 2013.

15 Atchley, A. A., Hofler, T., Muzzerall, M. L., Kite, Chianing. A. O., "Acoustically generated temperature gradients in short plates", Journal of Acoustical Society of America, Vol. 88, pp. 251-263, 1990.

16 Mavrotas, G., "Effective implementation of the ε-constraint method in multi-objective mathematical programming problems". Applied Mathematics and Computation, Vol. 213, pp. 455-465, 2009.

Modelling of Thermo-Acoustic Refrigerators Using General Algebraic Modelling System

L.K. Tartibu*

Mechanical Engineering Technology, University of Johannesburg, Doornfontein Campus, Johannesburg, 2028, South Africa

ltartibu@uj.ac.za

ABSTRACT

While thermo-acoustic refrigeration is new among emerging technology, it shows strong potential towards the development of sustainable and renewable energy systems by utilising a sound wave to remove the heat. This work discusses a new mathematical programming approach that provides fast initial engineering estimates to initial design calculations, describing the optimal geometry of thermo-acoustic refrigerators. Three different criteria describing their performances were taken into account: maximum cooling, best coefficient of performance and acoustic power loss. As the stack has been identified as the heart of the device where heat transfer takes place, this new approach aims to optimise its geometrical parameters: namely, the stack position, the stack length, the blockage ratio and the stack pore sizes. Hence, the optimisation task is formulated as a three-criterion nonlinear programming problem with discontinuous derivatives. This approach was implemented in the General Algebraic Modelling Systems. The unique characteristic of this research is the computation of all efficient optimal solutions, allowing the decision maker to identify the most efficient solution. The proposed modelling approach was investigated experimentally to evaluate its ability to predict the best parameters describing the geometry of the stack. Similar trends were obtained to support the use of the proposed approach in the design of thermo-acoustic refrigerators.

KEYWORDS

Thermo-acoustic refrigerator, optimisation, modelling, stack geometry, GAMS, AUGMENCON

INTRODUCTION

In order to address current environmental problems [1], thermo-acoustic refrigerators (TARs) have been proposed as a solution to the current search for alternative refrigerants and alternative technologies for refrigeration (such as absorption refrigeration, thermoelectric refrigeration and pulse-tube refrigeration). Thermo-acoustic refrigerators (see Figure 1) consist mainly of a vibrating diaphragm (a loudspeaker) attached to a resonator filled with gas, a stack usually constructed of thin parallel plates, and two heat exchangers placed at either side of the stack. The heat pumping process occurs within the stack which forms the heart of the refrigerator, and thus is an important element determining the performance of the refrigerator [2].

In a typical standing wave thermo-acoustic refrigerator, an acoustic wave is generated using a sound source such as a loudspeaker to make the gas resonant. The gas within the resonator tube oscillates. This standing sound wave creates a temperature difference along the length of the stack. This temperature change is a result of the compression and expansion the gas undergoes because of the sound pressure and thermal interaction between the surface of the plate and the oscillating gas. Heat is exchanged with the surrounding through the hot and cold heat exchangers located on each side of the stack [3]. The basic mechanics behind thermo-acoustics are already well-understood. Detailed studies on the working of thermo-acoustic coolers have been conducted by Swift [3] and Wheatly *et al.* [4]. Recent research focuses on improving the performance of the designing of the devices to allow thermo-acoustic coolers to compete with commercial refrigerators.

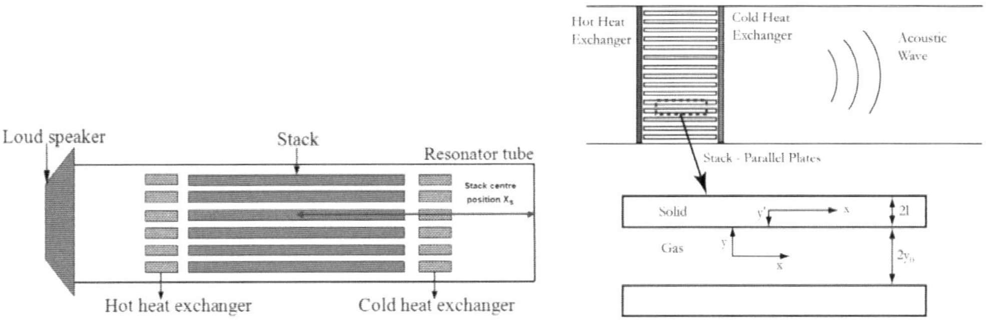

Figure 1: Schematic diagram of a typical thermo-acoustic refrigerator

OPTIMISATION OF THERMO-ACOUSTIC REFRIGERATORS

Mathematical optimisation, defined as the selection of the best element among some set of available alternatives with regard to some criteria, consists of maximising or minimising a function of one or more variables. *Engineering optimisation,* using optimisation techniques to achieve design goals, can be described as the process of searching for the conditions that give maximum or minimum value of a function [5]. The optimisation criterion are normally formulated in mathematical form, as an objective function. This paper addresses the application of mathematical programming techniques suitable for the solution of engineering design problems.

Previous studies have highlighted various parameters affecting the performance of the thermo-acoustic refrigerators (TARs). For example, geometrical parameters and fluid properties are considered in the study reported by Minner *et al.* [6] in order to optimise the coefficient of performance of a TAR. A locally optimal solution is obtained with the Nelder-Mead simplex method. The optimisation of the inertance sections of thermo-acoustic devices is demonstrated by Zoontjens *et al.* [7]. Both Minner *et al.* [6] and Zoontjens *et al.* [7] have in common models that rely extensively upon DELTAE, a black box simulation tool based on linear acoustic theory, initially developed by Swift *et al.* [3]. The evaluation of certain engine parameters affecting the pressures amplitude was offered by Ueda *et al.* [8]. Tijani *et al.* [9] have investigated the influence of the stack spacing with DELTAE. Herman and Travnicek [10] have discussed a systematic design approach that provides fast engineering estimates for initial design calculations of TARs, with results suggesting that sets of parameters leading to two seemingly similar outcomes – maximum efficiency and maximum cooling – are not the same.

More recently, experimental studies conducted by Tartibu [11] have provided evidence of the difference between design for maximum cooling and maximum efficiency for thermo-acoustic refrigerators. With the exception of studies by Minner *et al.* [6], previous works vary no more than a single parameter, holding all others constant. The solution approaches that are applied guarantee only a locally optimal solution, which may potentially be significantly inferior to a globally optimal solution. In addition, parameters like frequency, stack position, stack length, and plate spacing involved in designing TARs were optimised by Hariharan N. *et al.* [12] using Response Surface Methodology (RSM), with results showing that geometrical variables chosen for their investigation are interdependent. While this is by no means a comprehensive list of the 'optimisation' of refrigerator components, it is certainly a thorough overview of optimisation targets.

MOTIVATION

Most of the previous optimisation efforts have in common heavy reliance on the effect of a single design parameter on device performance. In all likelihood, each optimal design is a local optimum as the solution obtained is optimal (either maximal or minimal) within a neighbouring set of candidate solutions. This is in contrast to the global optimum proposed in this particular study, which provides a global optimal solution among *all* possible solutions in a specific domain, not simply those in a particular neighbourhood of variables.

In other words, a novel mathematical programming approach to handling design and choice between maximum cooling and maximum coefficient of performance of thermo-acoustic refrigerators (TARs) is presented in this paper. In addition, the blockage ratio, stack spacing, stack length and position of the stack have been identified as design parameters, so their interdependency is taken into account while computing the optimal set describing optimal performance of TARs, unlike previous studies.

The remainder of this paper is organised in the following fashion: Section 4 presents the model development. The fundamental parameters and equations in our mathematical models characterising the standing wave thermo-acoustic refrigerators are presented. Section 5 describes the proposed optimisation approach using the General Algebraic Modelling System (GAMS). Sections 6 and 7 discuss the valuable contributions of this work.

MODEL DEVELOPMENT

Figure 1 shows the geometry used to derive and discuss the thermo-acoustic equations. The model proposed in this paper does not consider any effect of the stack material and the interdependency between the coefficient of performance of thermo-acoustic core, the effectiveness of heat exchangers and the acoustic power efficiency. In addition, no attempt is made to derive TAR equations, as detailed derivations of the equations are available in Tijani's [13] thesis.

Design parameters of the thermo-acoustic core

The basic design of thermo-acoustic refrigerator requires the following [14]:

(a) to supply the desired cooling load; and
(b) to simultaneously achieve the prescribed cooling temperature.

The coefficient of performance of a thermo-acoustic core COP is dependent on 19 independent design parameters [14]. Herman and Travnicek [10] have collapsed the number of parameters to six normalised parameter spaces. The resultant normalised operation parameters are presented in Table 1 and Table 2. The number of parameters can be reduced by making a choice of some normalised parameters.

Objectives functions

The performance of the thermo-acoustic stack depends on three main stack design parameters: the centre position, the length and the cross-section area of the stack. The normalised heat flow (Φ_H) and acoustic power (Φ_W) neglecting axial conduction in the working fluid as well as in the stack plates are given as follows [15]:

$$\Phi_H = -\left[\frac{\delta_{kn} DR^2 \sin(2X_{Sn})}{8\gamma(1+\sigma)\left(1-\sqrt{\sigma}\delta_{kn}+\frac{1}{2}\sigma\delta_{kn}^2\right)}\right]\times\left[\frac{\Delta T_{mn}\tan(X_{Sn})}{(\gamma-1)BR\ L_{Sn}}\times\frac{\left(1+\sqrt{\sigma}+\sigma\right)}{1+\sqrt{\sigma}}-\left(1+\sqrt{\sigma}-\sqrt{\sigma}\delta_{kn}\right)\right] \tag{1}$$

$$\Phi_W = \left[\frac{\delta_{kn} DR^2 L_{Sn}(\gamma-1)BR\ \cos^2(X_{Sn})}{4\gamma}\right]\times\left[\frac{\Delta T_{mn}\tan(X_{Sn})}{BR\ L_{Sn}(\gamma-1)\left(1+\sqrt{\sigma}\right)\left(1-\sqrt{\sigma}\delta_{kn}+\frac{1}{2}\sigma\delta_{kn}^2\right)}-1\right]$$
$$-\left[\frac{\delta_{kn} L_{Sn} DR^2}{4\gamma}\times\frac{\sqrt{\sigma}\sin^2(X_{Sn})}{BR\left(1-\sqrt{\sigma}\delta_{kn}+\frac{1}{2}\sigma\delta_{kn}^2\right)}\right] \tag{2}$$

Table 1: Normalised cooling load and acoustic power

Operation parameters	
Normalised cooling power	$\Phi_H = \frac{\dot{Q}_c}{p_m aA}$
Normalised acoustic power	$\Phi_W = \frac{\dot{W}}{p_m aA}$

Table 2: TAR parameters

Operation parameters	
Drive Ratio (DR)	$DR = \frac{p_0}{p_m}$ where p_0 and p_m are, respectively, the dynamic and mean pressure
Normalised temperature difference	$\theta = \Delta T_{mn} = \frac{\Delta T_m}{T_m}$ where ΔT_m and T_m are, respectively, the desired temperature span and the mean temperature span
Gas parameters	
Normalised thermal penetration depth	$\delta_{kn} = \frac{\delta_k}{y_0}$ where $2y_0$ is the plate spacing
Stack geometry parameters	
Normalised stack length	$L_{Sn} = \frac{2\pi f}{a} L_S$ where L_S the stack length
Normalised stack position	$X_{Sn} = \frac{2\pi f}{a} X_S$ where f , a and X_S are, respectively, the resonant frequency, the speed of sound and the stack centre position
Blockage ratio or porosity	$BR = \frac{y_0}{(y_0 + l)}$ where 2l is the plate thickness

The normalised cooling load (Φ_C) and the coefficient of performance of the thermo-acoustic core COP and COPR are obtained, respectively, as follows [14]:

$$\Phi_C = \Phi_H - \Phi_W \tag{3}$$

$$COP = \frac{\Phi_H - \Phi_W}{\Phi_W} \tag{4}$$

$$COPR = \frac{COP}{COP_C} = \frac{(|\Phi_H| - |\Phi_W|)/|\Phi_W|}{(2-\theta)/(2\theta)} \tag{5}$$

The cooling load Φ_C is function of eight non-dimensional parameters [10]:

$$\Phi_C = F(\sigma,\ \gamma,\ \varepsilon_S,\ T_{mn},\ L_{Sn},\ X_{Sn},\ BR,\ \delta_{kn}) \tag{6}$$

Where σ , γ , ε_S and T_{mn} represent, respectively, the Prandtl number, the polytropic coefficient, the stack heat capacity correction factor and the normalised temperature difference. The influence of the working fluid on the gas is exerted through parameters σ , γ and ε_S .

In the boundary layer approximation, the acoustic power loss per unit area of the resonator is given as follows [15]:

$$\overset{o}{W}_2 = \frac{dW_2}{dS} = \frac{\delta_{kn} DR^2 L_{Sn}(\gamma - 1)BR\ \cos^2(X_{Sn})}{4\gamma} + \frac{\delta_{kn} L_{Sn} DR^2}{4\gamma} \times \frac{\sqrt{\sigma}\sin^2(X_{Sn})}{BR\ 1\ \sqrt{\sigma}\delta_{kn} + \frac{1}{2}\sigma\delta_{kn}^2} \tag{7}$$

Where the first term on the right-hand side is the kinetic energy dissipated by viscous shear. The second term is the energy dissipated by thermal relaxation. Details description of the model objectives and the constraints used are available in reference [10], [13] and [14].

PROPOSED MMP SOLUTION APPROACH TO EMPHASISE ALL OBJECTIVE COMPONENTS

Most of the equations involved in the mathematical problem formulation (MPF: multi-objective mathematical programming problem) have been presented in the previous section. The three objective components – namely the cooling load (Φ_C), the coefficient of performance (COP) and the acoustic power lost ($\overset{o}{W}_2$) – are the three distinct objective components to optimise simultaneously. Detailed descriptions on the single objective optimisation analysis for each objective function are reported by Tartibu *et al.* [16] revealing that these three objective functions are conflicting and thus suitable for a multi-objective optimisation approach. The optimisation task is formulated as a three-criterion nonlinear programming problem with discontinuous derivatives (DNLP) that simultaneously maximise the magnitude of the cooling load (Φ_C), maximise the coefficient of performance (COP), and minimise acoustic power lost ($\overset{o}{W}_2$).

$$(\text{MPF}) \quad \max_{L_{Sn},X_{Sn},BR,\delta_{kn}} \xi = \Phi_C(L_{Sn},X_{Sn},BR,\delta_{kn}), COP(L_{Sn},X_{Sn},BR,\delta_{kn}), \overset{o}{X}_2(L_{Sn},X_{Sn},BR,\delta_{kn}) \tag{8}$$

Subject to bound limits $\Phi_{C\max}$, $\Phi_{C\min}$, COP_C (details available in studies by Tartibu *et al.* [16]) and the following constraint:

$$\Phi_C = \Phi_H - \Phi_W > 0 \tag{9}$$

The four different parameters describing the geometry of the stack $(L_{Sn}, X_{Sn}, BR, \delta_{kn})$ (Equation 8) are the four variables to optimise. The constraint described by Equation 9 means that a negative cooling load does not have any physical meaning (and thus the solutions for which Equation 9 is not satisfactory have been eliminated). There is no single optimal solution to simultaneously optimise all the three objectives functions. In these cases, the decision makers are interested in the 'preferable' solution.

In order to yield *only* Pareto optimal solutions, avoiding the generation of weak, non-efficient solutions [17], the lexicographic optimisation (Augmented ε-constraint method) for each objective function to construct the payoff table for the multi-objective mathematical programming models (MPF) is proposed and implemented in the software GAMS (Figure 2). The advantages of the augmented ε-constraint method (or AUGMENCON) over traditional weighted methods and ordinary ε-constraint methods can be found in research by Mavrotas [17] and Tartibu *et al.* [18]. A flowchart of AUGMENCON method is presented in Figure 3, with "p" representing the number of objectives function; "q" the number of intervals; and "ni" the index of objective function.

The code describing the AUGMENCON method is available in the GAMS library (http://www.gams.com/modlib/libhtml/epscm.htm) with an example. While the part of the code that has to do with the example (the specific objective functions and constraints), as well as the parameters of AUGMENCON that have been modified in the current modelling of TAR, the part of the code that performs the calculation of payoff table with lexicographic optimisation and the production of the Pareto optimal solutions, is fully parameterised to be ready to use (detailed description of the code is available in Tartibu's doctoral thesis [19]).

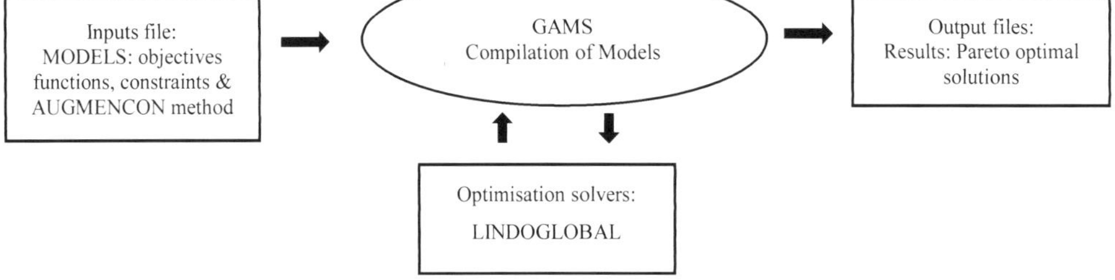

Figure 2: GAMS process illustration

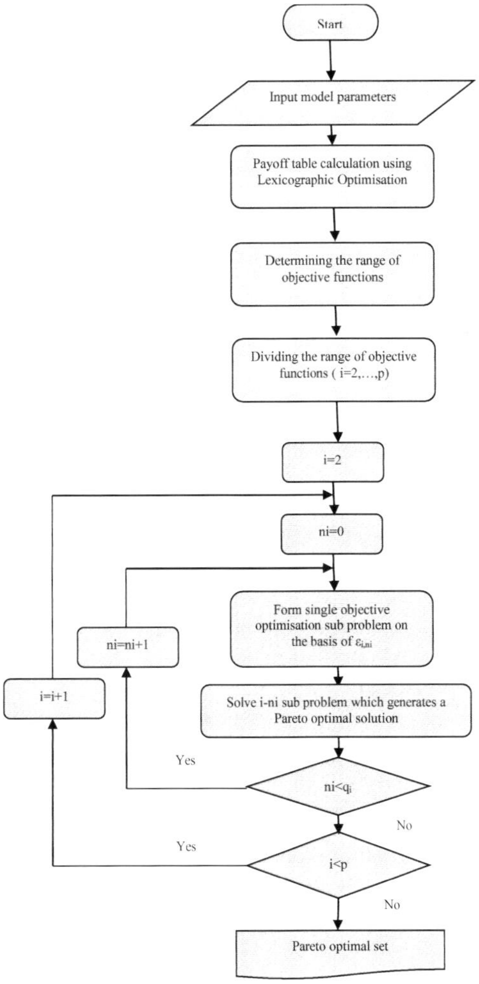

Figure 3: Flowchart of AUGMENCON method for optimisation

Formulation of the problem is a significant aspect of this study. The choice of the most important function or the primary objective function depends on the decision maker. Most of the time, this choice is dependent on problem information and results in partial representation of Pareto optimal sets due to the tendency of the solution to cluster toward the maximum of the primary objective function. Therefore, the preferences and specific limits on objective functions are articulated rather than relying on relative importance of objectives, as suggested by Marler [20], to identify the best problem formulation. Guidance on the best problem formulation is provided by Tartibu *et al.* [16]. Subsequently, the augmented ε -constraint method for solving the model (Equation 8) has been formulated as follows:

$$\max\left(COP_{(L_{Sn},X_{Sn},BR,\delta_{kn})} + dir_1 \; r_1 \times \left(\frac{s_2}{r_2} + \frac{s_3}{r_3} + \frac{s_4}{r_4} + \frac{s_5}{r_5} \right) \right) \quad (10)$$

Subject to

$$\Phi_{C(L_{Sn},X_{Sn},BR,\delta_{kn})} - dir_2 \; s_2 = \varepsilon_2$$

$$\overset{o}{W}_{2(L_{Sn},X_{Sn},BR,\delta_{kn})} - dir_3 \; s_3 = \varepsilon_3$$

$$s_i \in \Re^+$$

Where dir_i is the direction of the ith objective function, which is equal to -1 when the ith function should be minimised, and equal to +1 when it should be maximised. Efficient solutions to the problem are obtained by parametrical iterative variations in the ε_i; s_i represents the introduced surplus variables for the constraints of the MP problem; and r_i s_i / r_i are used in the second term of the objective function to avoid scaling problems.

RESULTS AND DISCUSSIONS

The cordierite honeycomb ceramic stacks (Figure 3) are considered. These stacks have different pore sizes (64 CPSI [Cells per Square Inch], 100 CPSI, 230 CPSI and 300 CPSI) and lengths (100 mm, 70 mm, 48 mm and 26 mm). These stacks were successively positioned at six different locations from the closed end: 100 mm, 200 mm, 300 mm, 400 mm, 500 mm and 600 mm. The experimental results reported by Tartibu [11] are considered for evaluating the ability of the models developed in this research to compute optimal solutions describing the stack geometries.

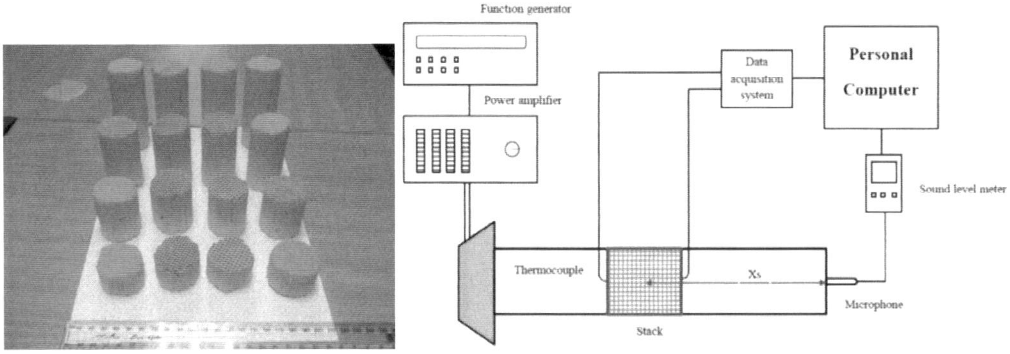

Figure 4: Stack samples used in the experiments and schematic diagram of experimental apparatus

The coefficient of performance (COP) of a thermo-acoustic refrigerator indicates how effective the device is in converting and producing cooling load by absorbing sound energy. The coefficient of performance is given by Equations 4 and 5. For all the stack lengths considered (Figure 4), a study conducted by Tartibu [11] shows that the values of COPR decrease as the distance from the pressure antinode (closed end) increases (Figure 5) (detailed description available in Tartibu's studies [11]). In particular, the shortest stack length shows the highest COPR. Interestingly, this behaviour has been observed by Herman and Travnicek [10] and Tartibu *et al.* [16] using mathematical modelling.

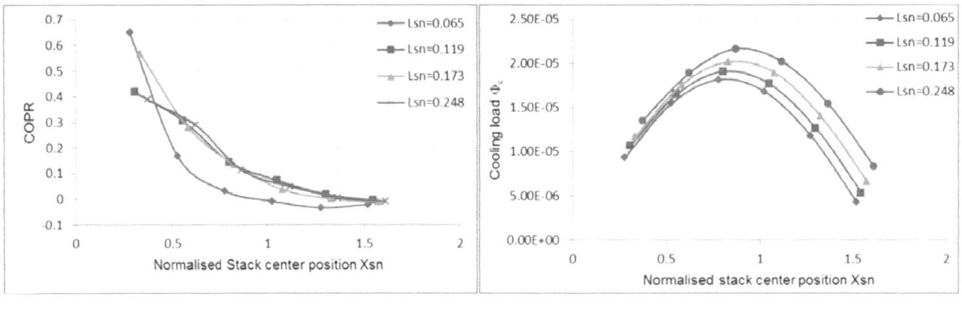

Figure 5: COPR and cooling load for four different honeycomb ceramic stack lengths (detailed description available in Tartibu's study [11])

The cooling load was calculated using Equation 3. Figure 5 shows the cooling load as a function of the normalised stack centre position obtained for four different stack lengths. The results obtained show a maximum cooling load when the stack is moved away from the pressure antinode (Figure 5). The results suggest the cooling load increases with the stack length. Contrary to the maximum COPR, increasing the stack length leads to an increase in cooling load for TAR (Figure 5), a result which concurs with previous studies by Herman and Travnicek [10] and Tartibu *et al.* [16] suggesting that there is a distinct optimum for maximum cooling and maximum coefficient of performance (detailed description available in Tartibu's studies [11]). Table 3 presents the parameters as estimated for the cordierite honeycomb ceramic stack used in this experiment. The normalised values are obtained from Tables 1 and 2.

Table 3: Estimated parameters of TAR

Lsn	Xsn						δkn	F [Hz]	Tm [K]	DR
0.065	0.280	0.528	0.777	1.024	1.272	1.520	0.135	135	250	0.025
0.119	0.307	0.555	0.803	1.051	1.299	1.547	0.168			
0.173	0.334	0.582	0.830	1.078	1.326	1.574	0.255			
0.248	0.372	0.620	0.868	1.116	1.364	1.612	0.291			

The experimental results have been compared with the mathematical programming results of the models. The ε-constraint method as formulated in Equation 10 was applied. For the problem formulation proposed in Equation 10, the following constraints (upper and lower bounds) have been enforced for the solver to carry out the search for the optimal solution in those ranges:

$X_{Sn}.lo = 0.280; \quad X_{Sn}.up = 1.612$

$\delta_{kn}.lo = 0.05; \quad \delta_{kn}.up = 0.1$

The thermal penetration depth range has been shortened to reduce the computational time (as that could take several days) since the result trends were more important. The normalised temperature difference (θ) is set to be equal to 0.030. The blockage ratio has been set to 0.8. More input parameters are shown in Table 3. Efficient solutions for the proposed model have been found using the AUGMENCON method and the LINDOGLOBAL solver. To save computational time, the early exit from the loops as proposed by Mavrotas [17] has been applied. The range of each three objective functions is divided into four intervals (five grid points). The normalised stack length has been arbitrarily given values of 0.065-0.119-0.173-0.248 successively. Table 4 reports the results obtained representing the optimal parameters, namely BR, δ_{kn} and X_{sn}, with the highest performance highlighted in bold. The maximum CPU time taken to complete the results was 324.981 seconds.

For a specific value of normalised temperature difference (θ), the trend of the COPR and the cooling load (Φ_c) is the same as those reported in Figure 5 (the trend lines are shown for visual guidance). These results suggest that the COPR is maximised closer to the closed end, while the cooling (Φ_c) reaches its maximum value away from the closed end.

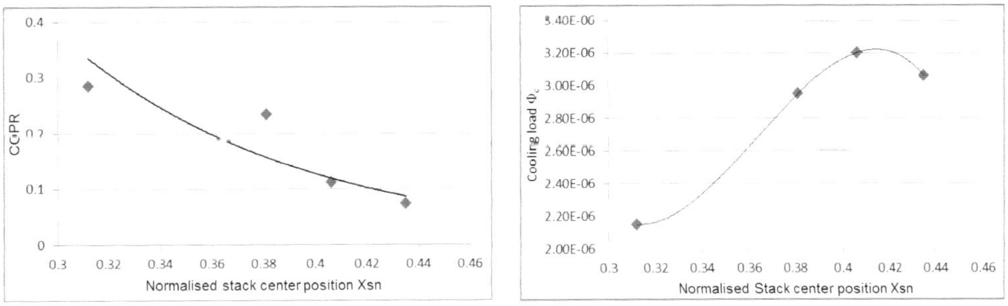

Figure 5: AUGMENCON results representing COPR and Φ_c for cordierite honeycomb ceramic stack

Table 4: Computation results obtained using AUGMENCON

	Results	1	2	3	4	5	6	7	8	9
	BR	**0.800**	0.800	0.800	0.800	0.800	0.800	0.800	0.800	0.800
	δ_{kn}	**0.061**	0.075	0.081	0.100	0.100	0.100	0.100	0.075	0.075
	X_{Sn}	**0.312**	0.312	0.313	0.310	0.311	0.311	0.311	0.314	0.376
$L_{sn} = 0.065$	Φ_C	**2.15E-06**	1.92E-06	1.69E-06	2.07E-06	2.88E-07	2.09E-06	3.04E-08	2.15E-07	1.55E-06
	COP	**9.372**	9.357	8.990	8.992	8.992	8.992	8.993	8.988	8.988
	BR	**0.800**	**0.800**	0.800	0.800	0.800	0.800	0.800	0.800	0.800
	δ_{kn}	**0.050**	**0.100**	0.100	0.084	0.073	0.061	0.050	0.095	0.095
$L_{sn} = 0.119$	X_{Sn}	**0.381**	**0.355**	0.340	0.377	0.379	0.381	0.382	0.376	0.376
	Φ_C	**1.35E-06**	**2.88E-06**	2.57E-06	2.27E-06	1.96E-06	1.66E-06	1.35E-06	2.88E-06	2.57E-06
	COP	**7.686**	**6.484**	6.677	5.361	5.362	5.363	5.364	5.360	5.360
	BR	0.800	0.800	0.800	0.800	0.800	0.800	**0.800**	0.800	0.800
	δ_{kn}	0.050	0.089	0.089	0.085	0.073	0.062	**0.100**	0.097	0.097
$L_{sn} = 0.173$	X_{Sn}	0.413	0.407	0.407	0.408	0.409	0.411	**0.406**	0.406	0.406
	Φ_C	1.44E-06	3.14E-06	2.80E-06	2.46E-06	2.12E-06	1.78E-06	**3.20E-06**	2.80E-06	3.14E-06
	COP	3.651	3.660	3.660	3.647	3.649	3.650	**3.661**	3.645	3.645
	BR	**0.800**	**0.800**	0.800	0.800	0.800	0.800	0.800	0.800	0.800
	δ_{kn}	**0.050**	**0.089**	0.089	0.085	0.074	0.062	0.098	0.050	0.071
$L_{sn} = 0.248$	X_{Sn}	**0.435**	**0.428**	0.428	0.430	0.432	0.433	0.427	0.435	0.400
	Φ_C	**1.43E-06**	**3.12E-06**	2.78E-06	2.44E-06	2.11E-06	1.77E-06	3.12E-06	1.43E-06	2.27E-06
	COP	**2.381**	**2.377**	2.377	2.377	2.379	2.380	2.375	2.381	2.886

CONCLUSION

This paper proposes the use of mathematical analysis and optimisation to model and optimise thermo-acoustic refrigerators (TARs). The General Algebraic Modelling System (GAMS) is used to implement a multi-objective optimisation approach, as this approach provides a fast estimate to initial design calculation of the geometrical configuration describing the stack of thermo-acoustic refrigerators. Four different parameters describing the geometry of the device – stack length, stack centre position, stack spacing and blockage ratio – have been studied. Nonlinear programming models with discontinuous derivatives (DNLPs) have been formulated and implemented in GAMS. An improved version of a multi-objective solution method, the ε-constraint method referred to as *augmented ε-constraint method* (AUGMENCON), has been applied. For different arbitrary values of stack length, this process generates optimal solutions describing geometry of the TAR, solutions which depend on the a priori design goal for maximum cooling or maximum coefficient of performance. This present research reveals the suitability of a multi-objective optimisation approach to model and optimise a TAR using the software GAMS. The test of the proposed models to evaluate their ability to predict the best parameters describing the geometry of the stack has revealed similar trend with experimental results, a finding which reinforces the application of the proposed approach in the design of thermo-acoustic refrigerators.

ACKNOWLEDGMENTS

This research was supported by the Faculty of Engineering at the University of Johannesburg and the Department of Mechanical Engineering at Cape Peninsula University of Technology, South Africa.

REFERENCES

1. Joshi Y. K., Garimella S. V., "Thermal challenges in next generation electronic systems", Microelectron Journal, Vol. 34. No. 3, pp.169, 2003.
2. Swift, G. W., "Thermoacoustic engines", Journal of the Acoustical Society of America, Vol. 84, No. 4, pp. 1145-1180, 1988.
3. Swift, G. W., "Thermoacoustics: a unifying perspective for some engines and refrigerators", Acoustical Society of America, Melville NY. 2002.
4. Wheatley, J. C., Hofler, T., Swift, G. W., Migliori, A., "Understanding some simple phenomena in thermoacoustics with applications to acoustical heat engines", American Journal of Physics, Vol. 53, pp.147–62, 1985.
5. Rao, S. S., "Engineering Optimization: Theory and Practice", 3rd Edition, John Wiley & Sons, 1996.
6. Minner, B. Braun, J. and Mongeau, L., "Theoretical evaluation of the optimal performance of a thermoacoustic refrigerator", ASHRAE Transactions: Symposia, Vol. 103. pp. 873-887, 1997.
7. Zoontjens L., Howard C. Q. and Zander A. C., "Modelling and optimization of acoustic inertance segments for thermoacoustic devices", First Australasian Acoustical Societies Conference: Acoustics: Noise of Progress, Clearwater Resort, Christchurch, New Zealand, pp. 435-441. 2006.
8. Ueda, Y., Biwa, T., Mizutani, U. and Yazaki, T., "Experimental studies of a thermoacoustic stirling prime mover and its application to a cooler", Journal of the Acoustical Society of America, Vol. 72. No. 3, pp. 1134-1141, 2003.
9. Tijani, M. E .H., Zeegers, J. C. H., De Waele, A. T. A. M., "The optimal stack spacing for thermoacoustic refrigeration", Journal of the Acoustical Society of America, Vol. 112. No. 1, pp. 128-133, 2002.
10. Herman, C. and Travnicek, Z., "Cool sound: The future of refrigeration? Thermodynamic and heat transfer issues in thermoacoustic refrigeration", Heat and Mass Transfer, Vol. 42. No. 6, pp. 492-500, 2006.
11. Tartibu, L. K. "Maximum cooling and maximum efficiency of thermoacoustic refrigerators", Heat and Mass Transfer, Vol. 52. No. 1, pp. 95-102, 2016.
12. Hariharan N. M., Sivashanmugam P., "Optimization of thermoacoustic refrigerator using response surface methodology", Journal of hydrodynamics, 2013, Vol. 25. No. 1, pp. 72-82, 2013.
13. Tijani M. E. H., "Loudspeaker-driven thermo-acoustic refrigeration", Ph.D. thesis, Eindhoven University of Technology, Netherlands, 2001.
14. Wetzel, M. and Herman, C., "Design optimization of thermoacoustic refrigerators", International Journal of Refrigeration, Vol. 20. No. 1, pp. 3 21, 1997.
15. Tijani, M. E. H., Zeegers, J. C. H. and De Waele, A. T. A. M., "Design of thermoacoustic refrigerators", Cryogenics, Vol. 42. No. 1, pp. 49-57, 2002.
16. Tartibu, L. K., Sun, B. and Kaunda, M. A. E., "Lexicographic multi-objective optimization of thermoacoustic refrigerator's stack", Heat and Mass Transfer, Vol. 51. No. 5, pp. 649-660, 2015.
17. Mavrotas, G., "Effective implementation of the ε-constraint method in multi-objective mathematical programming problems". Applied Mathematics and Computation, Vol. 213, pp. 455-465, 2009.
18. Tartibu, L. K., Sun, B. and Kaunda, M. A. E., "Multi-objective optimization of the stack of a thermoacoustic engine using GAMS", Applied Soft Computing, Vol. 28, pp. 30-43, 2015.
19. Tartibu, L. K., "A multi-objective optimisation approach for small-scale standing wave thermoacoustic coolers design", Doctoral thesis, Cape Peninsula University of Technology, 2014.
20. Marler, T., "A Study of Multi-Objective Optimization Methods for Engineering Applications", VDM Verlag, Saarbrucken, Germany, 2009.

Modelling of Coupled Vibration Response of a Rotor-Stator System to Multiple Parametric Excitations by Wavelet Transform

B X Tchomeni*, L. M Masu, A A Alugongo, T. Tengen

Support by the Department of Mechanical Engineering, VUT towards this work is highly acknowledged by the authors.

bernardt@vut.ac.za, leonard@vut.ac.za, alfayoa@vut.ac.za , thomas@vut.ac.za

ABSTRACT

A 4-DOF nonlinear model of a cracked de Laval rotor-stator system derived using Energy Principles is presented in this paper. The model has been used to simulate the system response under the parametric excitations; rotor-stator-rub, a breathing transverse crack, eccentric mass and an axial force by Wavelet Transform (WT). CWT and DWT reveal the signal aspects; trends, breakdown points, discontinuities of the non-stationary waveform under the parametric excitations. Presented results provide a useful tool for on-line monitoring of rotor-stator-rub, transverse crack and eccentric mass.

KEYWORDS
Axial force, Crack, Continue Wavelet Transform, Discrete Wavelet Transform, Nonlinear, Rotor-Stator-Rub

INTRODUCTION

THE most commonly and widely used methods to analyse time frequency content of machine vibration signals in condition monitoring are those that are based on the Fast Fourier Transform (FFT). FFT transforms and represents a signal in its frequency analysis domain to facilitate extraction from it and the desired features. Whilst FFT method has proven to be good with the stationary analysis signals [1], it fails to provide information on behaviour of the function in the time domain. This means that although, all the frequencies present in a signal might be determined, to define at what time they occur becomes ambiguous. This is an important problem for signals that harbour faults symptoms, for example, rotor-stator, and a breathing crack with high axial force [2], whose behaviour changes with time. Therefore diagnostic method to identify the above faults and represent them in time and frequency domain may requires use of appropriate tools such as Wavelet Transform (WT).

WT is an adequate method for time varying signal analysis, and it is a description of spectral decomposition via the scaling concept. This method is effective for stationary and non-stationary signal processing [3]. It provides flexibility in describing signals that include regions of varying frequency contents [4]. One of its features is multi-resolution signal analysis with a vigorous function of both time and frequency localization. In comparison to the FFT, WT analysis makes it possible to perform a multiresolution analysis, which makes it an important technique for non-stationary signal analysis. Wavelet analysis can be performed in several ways, as a Continuous Wavelet Transforms (CWT), and as a Discrete Wavelet Transform (DWT).

The current paper focuses on analysis of coupled lateral-torsional vibrations of unbalanced rotor-stator system with parametric excitations; namely, rub-impact, a breathing transverse crack and axial force by WT tools.
A detailed simulation analysis of the signal by CWT and DWT is performed and reported in this paper.

WAVELET TRANSFORM

Signal analysis by the wavelet technique has progressively increased over the last 20 years. As of today, there are many variations of wavelet analysis algorithms which could be deployed to analyse the signals at hand according to several criteria [4], [5]. Therefore, a meaningful comparison of wavelet analysis results is only realizable if the same wavelet analysis algorithm is being used to analyse the signals. The WT uses multi-resolution technique by which different frequencies are analysed with different length windows to provide a best resolution of the signal [6].

Wavelet time-frequency transform is a multiresolution analysis algorithm and is the inner product of the signal and a family of the wavelet. For the mother wavelet or the wavelet prototype $\psi(t)$, there exits a corresponding family of wavelet (son wavelet). The series of the son wavelets are generated by dilation and translation from the mother wavelet, $\psi(t)$ as follows [7]:

$$\psi_{a,b}(t)=\frac{1}{\sqrt{|a|}}\psi(\frac{t-b}{a}) \tag{1}$$

Where, a and b are dilation (scaling) and translation (shift) parameters respectively. The factor $|a|^{-1/2}$ is used to ensure energy preservation. $\psi(t)$ is a square integrable complex function, satisfying the admissibility condition [4].

$$c_{\psi}=\int_{-\infty}^{+\infty}\frac{|\psi(\omega)|^2}{|\omega|}d\omega<\infty \tag{2}$$

$\psi(\omega)$ is the Fourier transform of $\psi(t)$.

The continuous wavelet time-frequency transform of the initial signal $Y(t)$ is expressed as:

$$CWT_s(a,b)=\int Y(t)\cdot\psi_{a,b}^{*}(t)dt \tag{3}$$

Where $\psi_{a,b}^{*}(t)$ is the complex conjugate function of $\psi_{a,b}(t)$.

By varying a and b, the wavelet time-frequency transform coefficients $W_s(a,b)$ can offer the representation of the signal $Y(t)$ at different levels of resolution and time shift. In the process, thus, the wavelet time-frequency transform enables one to extract features of the signal $Y(t)$ WT can be categorized as discrete WT or continuous WT. For vibration-based fault diagnosis, usually CWT and DWT are employed [7].

The DWT is derived from the discretization of CWT. It serves as a tool that separates data into various frequency components, and enable to us to study each component with resolution matched to its scale. DWT is computed with a cascade of filtering followed by a factor 2 subsampling [6], and is given by:

$$DWT_s(j,k)=1/\sqrt{2^j}\int_{-\infty}^{+\infty}Y(t)\psi^{*}((t-2^jk)/2^j)dt \tag{4}$$

where j is an integer, a and b are replaced by 2^j and 2^jk.

By using the wavelets, $Y(t)$ can be analysed at various levels of resolution. The DWT is also invertible and can be orthogonal. The driving force behind wavelet (WT) is to overcome the disadvantages embedded in a Short Time Fourier Transform (STFT), which provides a constant resolution for all frequencies since it uses the same window for the analysis of the inspected signal $Y(t)$. In this work, (CWT) and (DWT) have been used to identify nonlinearities on the rotor system.

MATHEMATICAL MODEL OF THE STUDY AND WAVELET TRANSFORM

A layout of the physical system under consideration is shown in Fig 1.

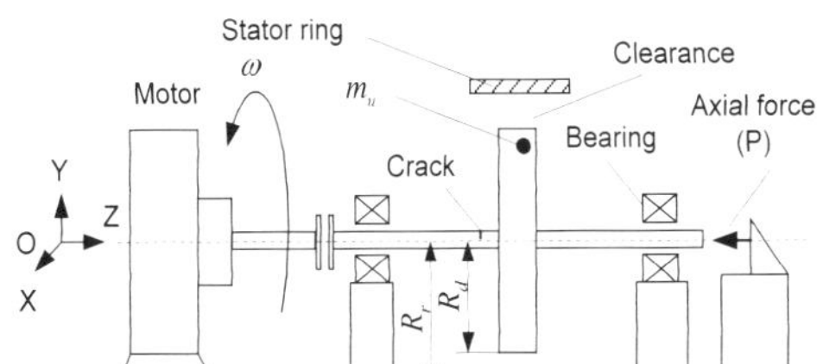

Figure1. Assembly of a Motor-shaft-disc with crack-rub.

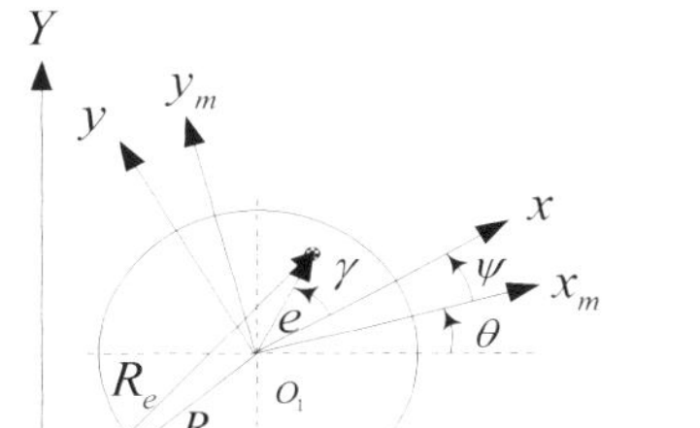

Figure2. Deformed configuration of shaft carrying disk.

Using the Energy Principles, the kinetic energy, the potential energy, the Rayleigh's and the axial force expression of the system in of Fig 1-2 is given as follow:

$$U_D = \frac{1}{2}M\left(\dot{X}^2+\dot{Y}^2\right)+\frac{1}{2}J_D\left(\dot{\theta}+\dot{\psi}\right)^2+\frac{1}{2}J_M\dot{\theta}^2+\frac{1}{2}m_u\dot{R}_e^T\dot{R}_e \tag{5}$$

$$V = \frac{1}{2}K_{XX}X^2+\frac{1}{2}K_{YY}Y^2+\frac{1}{2}K_{\psi\psi}\psi^2 \tag{6}$$

$$D = \frac{1}{2}C_{XX}\dot{X}^2+\frac{1}{2}C_{YY}\dot{Y}^2+\frac{1}{2}C_{\psi\psi}\dot{\psi}^2 \tag{7}$$

$$K_{XX} = K_{YY} = K_b - \frac{P\pi^2}{2L} - \frac{F_\psi\pi^2}{2L^2} \tag{8}$$

$$P = P_0 + P_f\sin\theta, \qquad F_\psi = T_0 + T_f\sin\theta \tag{9}$$

where, K_{XX}, K_{YY}, $K_{\psi\psi}$ are the stiffness coefficients associated with the system degrees of freedom. C_{XX}, C_{YY}, $C_{\psi\psi}$ are the respective damping coefficients. K_b, P and F_ψ correspondingly are, modal stiffness of first bending mode, fluctuating axial force and axial torque. In this paper P and F_ψ are prescribed a priori as in [10]. P_f, T_f and θ are amplitudes and rotation angle respectively. J_D, J_M, θ, ψ and $\dot{R}_e$ respectively are, disc-mass moment of inertia, motor-mass moment of inertia, rigid-body rotation, twist about Z-axis and velocity vector of m_u. Upon differentiating $R_e$with respect to time and making appropriate substitution as in [9], the mathematical model for the non-stationary rotor under the parametric excitations is expressed as:

$$\begin{bmatrix} m_{\theta\theta} & m_{\theta\psi} & m_{\theta X} & m_{\theta Y} \\ m_{\psi\theta} & m_{\psi\psi} & m_{\psi X} & m_{\psi Y} \\ m_{X\theta} & m_{X\psi} & m_{XX} & 0 \\ m_{Y\theta} & m_{Y\psi} & 0 & m_{YY} \end{bmatrix}\begin{Bmatrix} \ddot{\theta} \\ \ddot{\psi} \\ \ddot{X} \\ \ddot{Y} \end{Bmatrix} + \begin{bmatrix} 0 & 0 & 0 & 0 \\ 0 & C_{\psi\psi} & 0 & 0 \\ 0 & 0 & C_{XX} & C_{XY} \\ 0 & 0 & C_{YX} & C_{YY} \end{bmatrix}\begin{Bmatrix} \dot{\theta} \\ \dot{\psi} \\ \dot{X} \\ \dot{Y} \end{Bmatrix} + \begin{bmatrix} 0 & 0 & 0 & 0 \\ 0 & K_{\psi\psi} & 0 & 0 \\ 0 & 0 & K_{XX} & K_{XY} \\ 0 & 0 & K_{YX} & K_{YY} \end{bmatrix}\begin{Bmatrix} \theta \\ \psi \\ X \\ Y \end{Bmatrix}$$

$$+\tfrac{1}{2}f(t)\,\Delta k_{xx} \times \begin{bmatrix} 0 & 0 & 0 & 0 \\ 0 & 0 & 0 & 0 \\ 0 & 0 & 1-cos2\theta & sin2\theta \\ 0 & 0 & sin2\theta & 1+cos2\theta \end{bmatrix}\begin{Bmatrix} 0 \\ 0 \\ 0 \\ Ys \end{Bmatrix} = \begin{Bmatrix} Q_\theta \\ 0 \\ Q_X \\ Q_Y \end{Bmatrix} + \begin{Bmatrix} F_\theta \\ F_\psi \\ F_X \\ F_Y \end{Bmatrix} \tag{10}$$

where, Δk_{xx} ,K_0, and $f(\theta)$ respectively are, change in stiffness in the weaker axis due to the crack, uncracked shaft stiffness and steering function $f(\theta)$ as in [2].

INTERACTION OF ROTOR-STATOR-RUB AND CRACK IN 4 D.O.F

Fully coupled vibrations of unbalanced rotor with rotor-stator rub and a breathing transverse crack and axial force is investigated based on system model equation (10).

Rotor response with torsional lateral rotor model is simulated using CWT under fourth hypothesis. For the first hypothesis, the system response is simulated when the axial force is exerted and without rotor-stator interaction. The torsional and lateral natural frequencies are then equal, $\omega_T = \omega_L$, in the second hypothesis rotor is operated with a crack and an axial force only. The third and fourth hypothesis consider both rub-crack effect for which the torsional resonance behavior takes places when $\omega_T \gg \omega_L$.

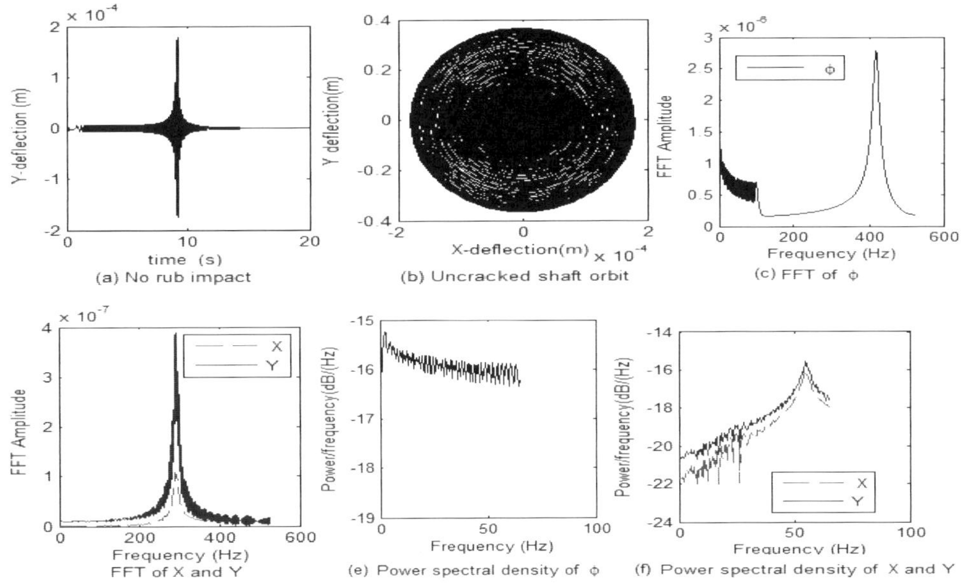

Figure 3 (A). Dynamics system response from lateral-torsional-rotor model, no rub, no crack $\Delta K_\xi/K_b$=0. P=2000N, $\psi \neq 0$, $\omega_T = \omega_L$ (a) Vertical shaft deflection (b) Rotor response orbit (c) Torsional frequency signal (d) Lateral and frequency signal (e) Torsional power spectral density (f) Lateral power spectral density.

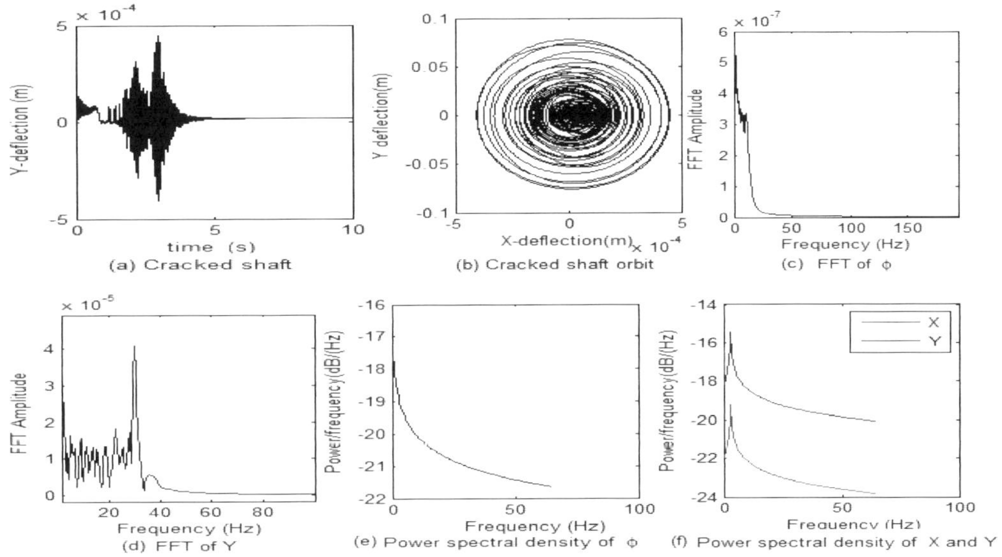

Figure 3 (B). Dynamics system response of the lateral-torsional-rotor, with crack, $\Delta K_\xi/K_b$=0.9542, P=2kN, $\psi \neq 0$; $\omega_T = \omega_L$ (a) Vertical shaft deflection (b) Rotor response orbit (c) Torsional frequency signal (d) Lateral and frequency signal (e) Torsional power spectral vibration. (f) Lateral power spectral density.

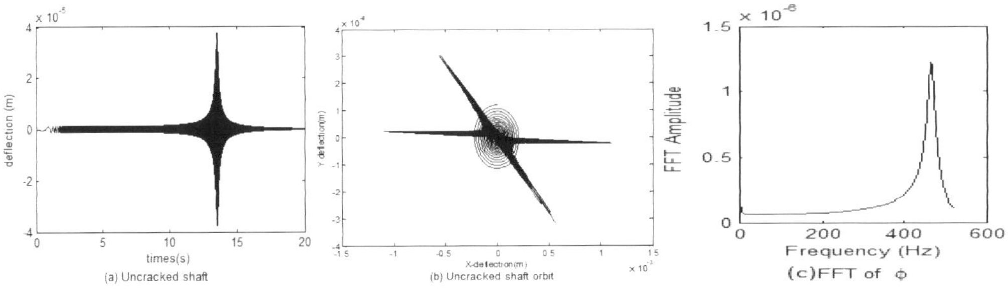

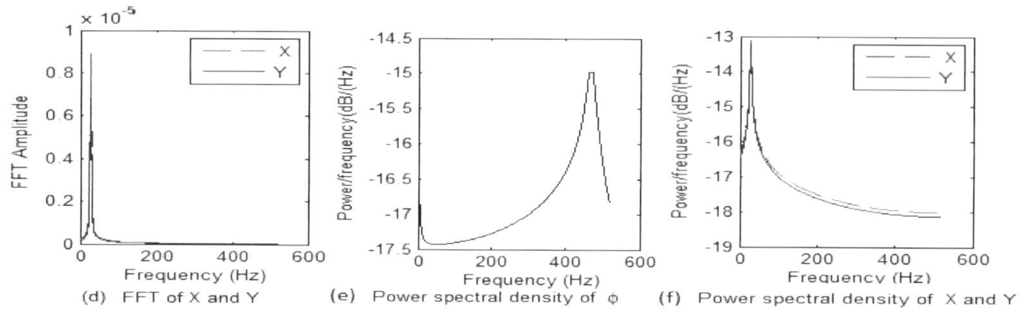

Figure 3 (C). Response of a rotor-to-stator-rub and axial force, $\Delta K_\xi/K_b$=0. P=2000N, $\Delta = 2.65 \times 10^{-7}$m (a) Vertical cracked shaft deflection (b) Rotor response orbit, (c) Torsional frequency signal domain, (d) Lateral frequency signal domain (e) Torsional power spectral vibration (f) Lateral power spectral density.

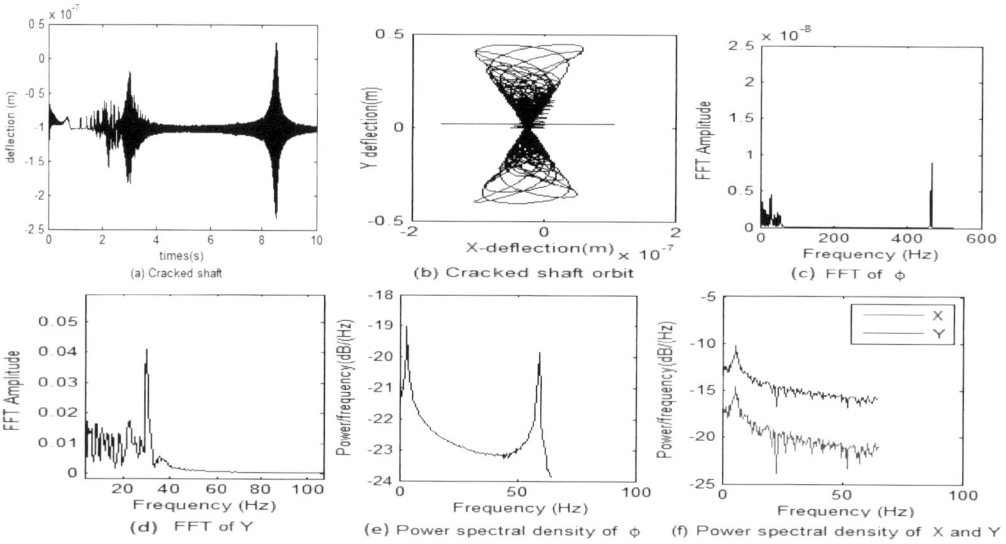

Figure 3 (D). System response with: rub-crack, axial force: $\Delta K_\xi/K_b$=0.9542. P=2kN, $\omega_T \gg \omega_L$, $\Delta = 2.65 \times 10^{-7}$m (a) Vertical cracked shaft deflection (b) Rotor orbit response (c) Torsional frequency signal (d) Lateral frequency signal (e) Torsional power spectral density (f) Lateral power spectral density.

As noted in Fig.3 (A) (c) (d) and Fig.3 (A) (c), (d) once the critical speed is traversed, the amplitude of vibration decreases. This indicates that only unbalance and axial force are the source of the above vibrations. This is verified by the orbit shape depicting a compressed circular response due to the large axial force and the synchronous response shown in the spectral plots in Fig 3(A) (f) and 3(C) (f).

Signal spectrum in Fig 3 (B) (d) and 3 (D) (d) show an increase in the magnitude of frequency components before resonance; while indicating a reduction in magnitude of the second harmonic and a disappearance of higher frequency components. Fig. 3 (D) (b) shows an elliptical orbit due to model anisotropy and shows also the second rigid mode shapes of the modelled rotor. Presence of the second rigid mode can be attributed to multiple factors such as the higher axial force, the transient vibration which follows the removal of the friction torque in the rub free zone and the presence of the flexibility of the cracked shaft.

The orbit presented in Fig. 3 (C) (b) revealed as noted already in [7], that the direction of the two maximum responses changes due to anisotropy model of rotor. Unlike the orbit of the cracked rotor describes the deflection shape of the system due to vibration and to local flexibility of the rotor at the damage location.

Fig. 3(B) (a) and Fig. 3(D) (a) are waveforms registered respectively at a rotational speed of 2309.4 rpm and 4510.21 rpm. Two high resonance points and three resonance points developed from the excitation between the torsional and lateral rotor. The first one is the ordinary resonance and the other two are situated at the median and high resonance in Fig. 3(D) (a).

Two possible factors could lead to the appearance of vibration peak bands. The first is the large axial force and friction which are, of course, a consequence of rub. The axial force contributes to increase bending of shaft effects, which accentuates the rotor-stator contact. However, friction forces cause a frictional torque opposite to rotor motion, resulting in a fluctuation of rotor speed. Therefore, the frequency spectrum (Fig. 3(D) (e)) would contain a vibration component corresponding to these modified processional speeds. From Fig. 3(D), it seems that axial force, crack and rub force have a significant effect on the vibration response in comparison to the case of no rub-crack vibrations in Fig. 3 (A). Examining the maximum power spectral density and shaft trajectory of a cracked rotor, an increase in the spectral power density at half the critical speed can be used to diagnose a transverse breathing crack.

Equation (10) describes rub-impact and a breathing crack whose signals are nonstationary (non-linear). Therefore, localized information is spread out over the whole signal spectrum. FFT is inadequate to determine all their response frequencies and the exact time of their occurrence. A limited performance for machinery diagnostics is offered by FFT. WT technique is applied for extraction and analysis of hidden features of nonlinearity in the response signal in the following section.

The 4-D.O.F dynamic model equation (10) has been numerically integrated to give the responses, X, Y, and ψ, which are analyzed by WT. CWT has been performed to compute the scalogram of wavelet coefficients using the Morlet Wavelet algorithm and a Daubechies Wavelet of order 40, "db40".

The 2-D and 3-D of db40 by CWT under action of a crack, rub, eccentric mass and external axial force at a higher rotation speed is presented in Fig. 4.

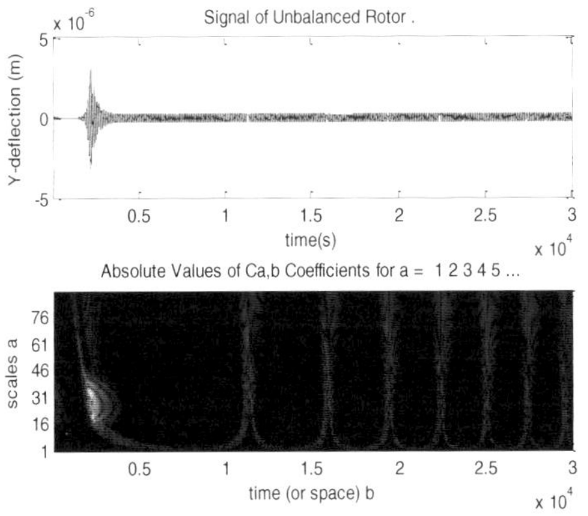

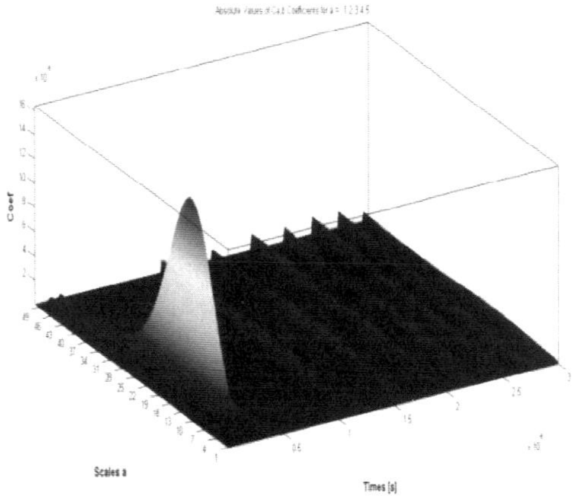

4 (a) 2-D CWT of unbalanced rotor: $\Delta K_\xi/K_b$=0. P=2KN, $\omega_T = \omega_L$ Δ= 0 ; 4 (b) 3-D Morlet Scalogram

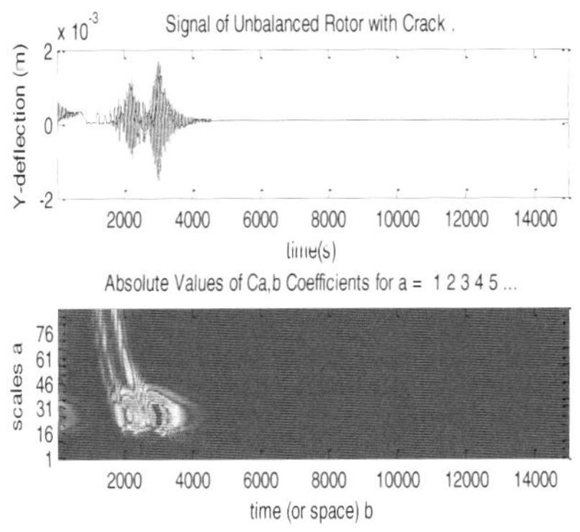

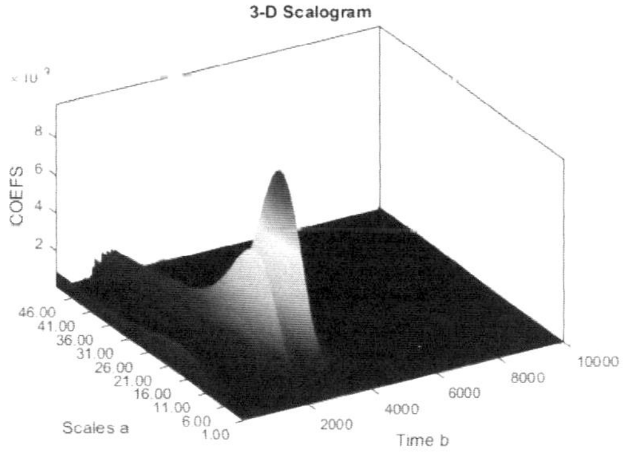

4. (c) 2-D CWT of unbalanced rotor: with crack $\Delta K_\xi/K_b$=0.54, P=2kN, $\omega_T = \omega_L$; 4(d) 3-D Morlet Scalogram

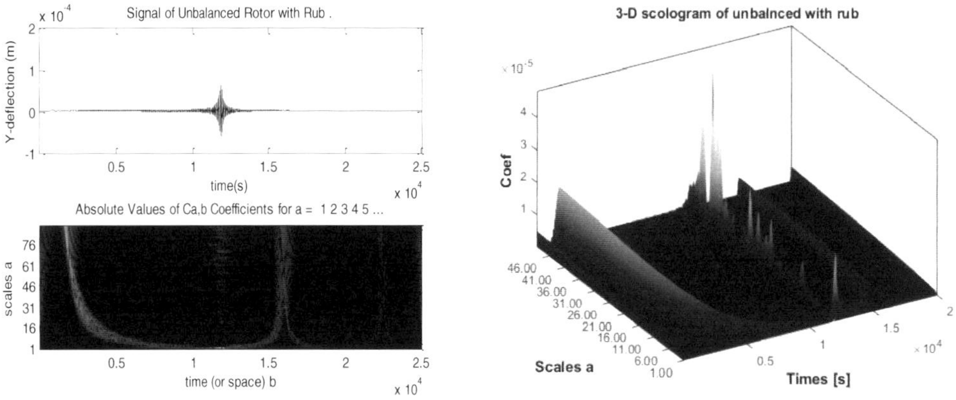

4. (e) 2-D CWT of unbalanced rotor: with rub $\Delta K_\xi/K_b$=0, P=2kN, $\Delta= 2.65 \times 10^{-7}$m ; 4(f) 3-D Scalogram

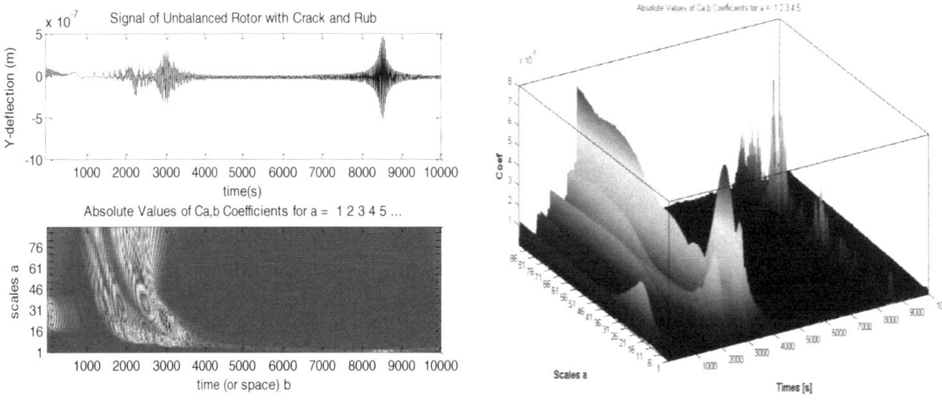

4. (g) 2-D CWT : unbalanced rotor: rub, crack $\Delta K_\xi/K_b$=0.54, $\omega_T \gg \omega_L$ 4(h) 3-D Morlet Scalogram

Figure.4 CWT and power spectrum of the rotor while passing via critical speed, under: unbalance, rub, crack and axial force, $\Delta K_\xi/K_b$=0.542, P=2KN , $\Delta= 2.65 \times 10^{-7}$m.

In the scalogram, maxima of energy are detected at all the simulation at the interval scale [16; 30]. Thorough inspection of fig. 4(a), the first resonance point is characterized by a high concentration of energy represented by a high red peak of resonance at the point *t = 0,428s*. Another similar discontinuity is observed; the breakdown point signal, which is indicated in fig. 4(b) by a small continuous peak after the first resonance. Those similarities are indiscernible in fig. 4(e) and fig. 4(h) because its diffusion power is masked by the strong recurring shocks between rotor-stator, opening and closing of the crack.

The scalogram based on Morlet family in fig. 4 (f) and fig. 4(h), presents the signal corrupted by multiple distributed peaks of variable amplitudes due to rotor-stator rub. In fig. 4(h) the existence of a transverse crack is revealed by the presence of disproportional ridges in the interval of $t = [0, 5000]sec$, and the rub contact at one particular point is manifested by discontinuous peaks appearing between the periods $t = [8000, 9000]sec$ which correspond to the second resonance. The scalogram analysis indicates that the centre of the concentrated energy corresponds to the most irregular regions, alluding to most of the non-linearity. In energy level analysis, an unbalanced rotor dissipates less energy whereas cracked-rotor-rub consumes more energy than the unbalanced rotor.

CWT of the signal, and plots of the scalogram of wavelet coefficients exhibit the distribution of energy of wavelet power spectrum in the direction of scale. From this standpoint, CWT has been useful for vibration analysis of a system with mass unbalance, crack, and rub. Such a technique improves condition monitoring and faults diagnosis in rotating machines.

To examine the constituent components of the original signal at each level, DWT is performed up to decomposition level 7 (D7) and approximation coefficients (A7) using Daubechies (db40) algorithm on Matlab platform. The technique is used in order to study dynamics of each signal separately and enables reconstruction of two different high-level wavelet signals, one corresponding to the approximation signal and the other to the signal details. For a p-level decomposition, the highest frequency limit observed in the approximation wavelet coefficients (A7) has been computed as a function of sample frequency f_S as:

$$f_l = f_s \times 2^{-p-1} \tag{11}$$

The frequency content of the approximation frequency band cA and detail frequency bands cD can be determined as:

$$f_{cA} = [0, f_s \times 2^{-p-1}] \tag{12}$$

$$f_{cD} = [2^{-p-1} f_s, 2^{-p} f_s] \tag{13}$$

The DWT of the signal of equation (10) with a seven level filter bank and a 'db40' wavelet function is shown in figure 5. The DWT result of fig. 5 shows the original signal in the top figure in red colour. The other figures contain the wavelet coefficients of the various sub-bands. The frequency ranges $[f_{Low,} f_{High}]$ of the various sub-bands are given in Table 1.

Table I: Spectral Frequency Band at Different Decomposition Level

Sub-bands	$f_{Low,}(Hz)$	$f_{High}(Hz)$
A7	0	9.76
D7	9.76	19.53
D6	19.53	39.06
D5	39.06	78.125

Figure 5 show the wavelet decomposition details of the reconstructed signals, up to level seven along with final approximation level for the mass-unbalance-rotor and rotor with crack, rub and axial force.

Figure 5(a) shows the upper-level signals A7, D7, D6 and D5 resulting from the wavelet decomposition of the start-up of an unbalance rotor under action of an axial force, rub and crack, obtained from simulation at a sampling frequency f_S =2500 samples/sec. Discontinuities at the start and beginning of the various parts are visible in all frequency bands. The DWT analysis shows that the upper-level signals (A7, D7, D6 and D5) associated do not have any important variation, once the transient vibration of unbalanced rotor ceases. The 7th level approximation signal has a correct shape manifesting rotor unbalanced. Therefore, the 7th level approximation signal is used as a medium of diagnose for the damaged rotor.

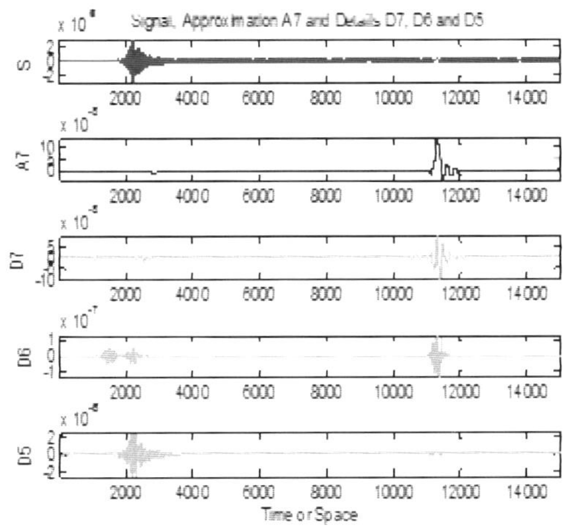

5. (a) D.W.T of unbalanced, rotor with axial force .

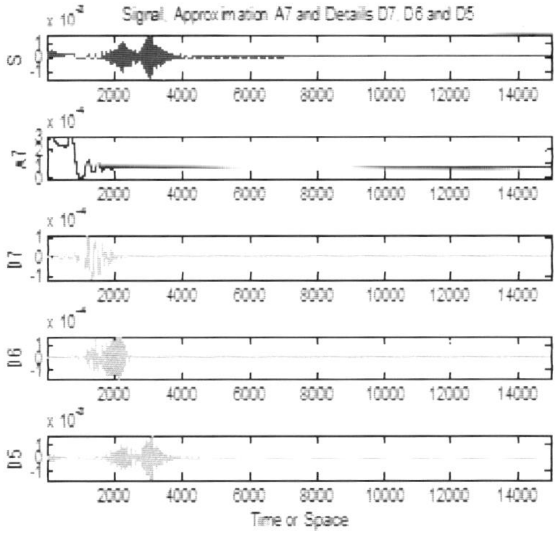

5. (b) D.W.T of unbalanced, cracked rotor

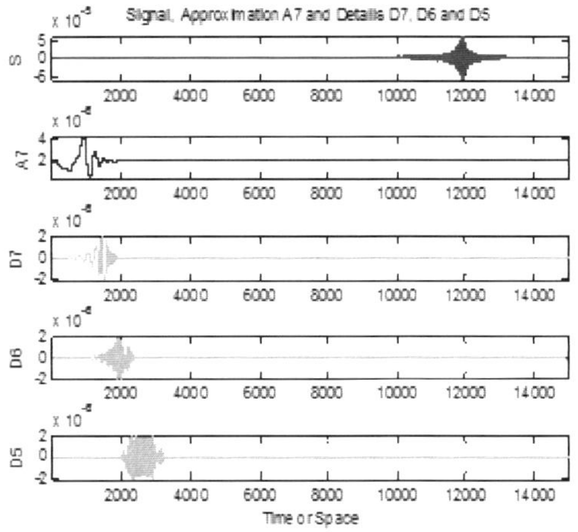

5. (c) D.W.T unbalanced, rotor with rub.

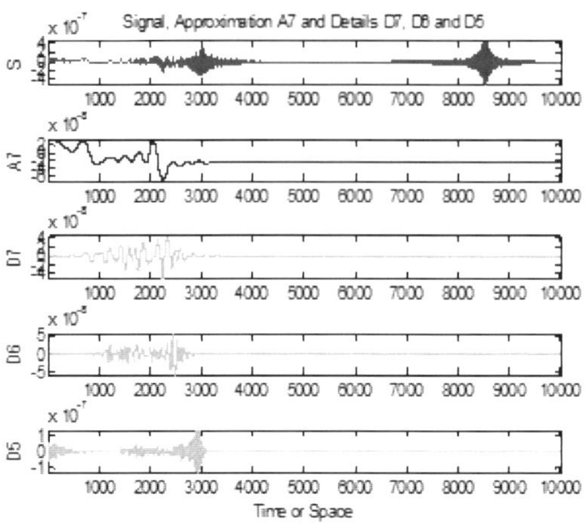

5. (d) D.W.T unbalanced, cracked rotor with rub rotor.

Figure 5. Wavelet decomposition details of the reconstituted signal in 4.D.O.F for mass unbalance (m=6.07 %) with transverse crack $\Delta K_{\xi}/K_b$=0.542, rub and axial force P=2kN.

Fig. 5 (a) shows the approximation 7 (A7) and the corresponding approximation and detail coefficients (obtained by db40) up to seven levels. From *0 s* up to *1000s* in fig 5(a), the signal consists of a simple line which has a pulse at $t = [1000, 1200]s$ with a frequency of *19.53 Hz*. At *0.0 s* up to *2000s* the signal shows a sinusoid with a frequency of *39.06 Hz* in fig.5 (b), (c), (d) which in fig 5(e), changes to *78.125 Hz* at $t = [2000, 4000]s$ when the axial force is increased to *12kN*. The signal is sampled at a frequency of *1 kHz*.
The DWT analysis in fig. 5(a) shows that the upper-level signals (A7, D7, D6 and D5) of the unbalanced rotor do not have any ambiguous variation. The observed oscillations "Perturbations" in fig. 5(a) are due to the unbalance effect at the passage via the critical speed. From this result, it is observed that in the 5th, 6th decomposition levels, in fig. 5(a) that, the unbalanced rotor has large reconstructed detail coefficients, whereas cracked-rotor with rub has less reconstructed detail coefficients in fig. 5 (b), (c), (d).
The comparison between fig. 5(a) of unbalanced rotor and fig. 5 (b), (c), (d) with fault, shows clearly that the signal, which contains a variety of difference according to the parameter of each fault can be identified by means of the discontinuities that appear in high-level D7 and A7. The signal can be stretched due to the transverse crack or totally disturbed due to the rub force.
The results of DWT give the advantageous information to decide the faulty situation. By adjusting the wavelet coefficients the reconstructed signal of the synthesis filter bank can be changed in comparison to the original signal. The wavelet technique presented here is able to extract useful characteristics from unbalanced rotor to rub and cracked rotor by simple analysis of the A7 and D7 signal reconstituted. These variations of statistical parameters of wavelet coefficients at different wavelet levels are useful for fault detection of the rotor.

CONCLUSION

This paper presented the method of analysing and detecting unbalance, transverse crack, and rub of a rotor system under effect of axial force using FFT and CWT and DWT techniques. A combination of CWT and DWT provides a better means to monitor vibration signal induced by a crack, rub, and axial force. Both techniques are capable of extracting the crack features from vibration responses when rotor is operated through the critical speed.

Results of WT scalogram analysis give the advantageous information on the rotor condition, particularly at rotor start-up with torque. This method clearly exhibits time-frequency features of fault signals by increasing the peaks of the time domain waveform, and also increases the magnitude of the perturbed signal. The wavelet technique presented is able to extract useful characteristics the rotor unbalance and nonlinear excitations of rub and crack. The proposed method provides a promising way of mixed-fault diagnosis in a rotor.

REFERENCES

1 Gold, A and Rader, C., ''Digital Processing of Signals''. New York: McGraw-Hill, 1967.

2 Tchomeni, B.X., et al., "In situ Modelling of Lateral-Torsional Vibration of a Rotor-Stator with Multiple Parametric Excitations". World Academy of Science, Engineering and Technology International Journal of Mechanical, Aerospace, Industrial and Mechatronics Engineering Vol: 8 No: 11, 2014.

3 Prabhakar, S., et al., "Detection and monitoring of cracks in a rotor-bearing system using wavelet transforms". *Mechanical Systems and Signal Processing* 15: 447–450, 2001.

4 Mallat, S., "A wavelet Tour of Signal Processing", Academic Press, San Diego 1998.

5 Daubechies, I., "Ten Lectures on Wavelets," CBMS-NSF Lecture Notes nr. 61, SIAM, 1992.

6 Goswami, J. C., and Chan, A. K., "Fundamentals of Wavelets: Theory, Algorithms and Applications," *John Wiley & Sons,* Inc 1999.

7 Daubechies, I., "Orthonormal bases of compactly supported wavelets,"*Communication on Pure and Applied Mathematics*, vol. 41, pp. 909-996, 1988.

8 Merry, J.E., "Wavelet Theory and Applications", A literature study, DCT, 2005.

9 Al-bedoor, B. O., "Transient torsional and lateral vibrations of unbalanced rotors with rotor-to-stator rubbing". Journal of Sound and vibration 229(3), 2000, 627-645.

10 Yigit, A.S., and Christoforou, A.P., "Coupled Torsional and Bending Vibrations of Actively Controlled Drillstings". Journal of Sound and Vibration, 2000, 234:

Analysis of a Cracked Unbalanced Rotor-Stator System with Fluid Film Bearing Forces by Wavelet Transform

B X Tchomeni*, A A Alugongo, T. Tengen

Support by the Department of Mechanical Engineering, VUT towards this work is highly acknowledged by the authors.

bernardt@vut.ac.za, alfayoa@vut.ac.za , thomas@vut.ac.za

ABSTRACT

This paper provides a technique to identify the characteristics of vibration responses of strong nonlinear rotor system, generated by multiple excitations during low and high acceleration by means of the time-frequency plots. A cracked de Laval rotor system characterized by a hinge model for a small crack, Mayes' modified function for a deep crack, and rotor-stator rub-impact are numerically simulated. The effects of the non-linear disturbances due to oil film forces, rotor stator rub forces, and a breathing crack on the rotor shaft is numerically analysed by Fast Fourier Transform (FFT) and Wavelet Transform (WT). An attempt for fault detection based on different factors such as unbalance eccentricity, rub, crack and oil film force with the influence of low and high acceleration vibration are investigated. Simulation results presented show that, the nonlinear phenomena are well observed at low acceleration, through the orbits at the start-up period. However at high acceleration several nonlinear phenomena including chaotic features behaviour are observed and analysis become complicated, and could only be observed through wavelet transform features. The observed frequency spectra and time-frequency features are unique and could serve as identification criteria of rotor faults.

KEYWORDS

Crack, Low and High Acceleration, Nonlinear, Rotor-Stator, Rub, Oil Film Force, Wavelet Transform

INTRODUCTION

IT is conventionally assumed that effect of resonance on vibration of a rotor can be reduced by quick acceleration through the critical speeds. This was analytically and experimentally demonstrated for the case of a Jeffcott rotor on rigid bearings by Hassenpflug et al. [1]. Until the 1980s research on vibration condition monitoring focused on determination of rotor Eigen modes, critical speeds and the prediction of response amplitudes of rotors subjected to an unbalance [2]. Classical analysis of models using a Jeffcott rotor has been proposed by various researchers [2-4]. Dimaragonas et al. derived a rough analytical estimate of a crack compliance based on the energy principle [3], [4]. Sekhar and Prabhu performed a transient analysis on a cracked rotor passing through critical speed at constant acceleration [5]. Dynamic response of a cracked Jeffcott rotor passing through the critical speed with constant acceleration was investigated analytically and numerically by Sawicki et al. [6], using Vibration phase response. At high rotor's acceleration, the vibration responses become typically transient alongside frequency and amplitude modulation [7]. Due to the limitation of FFT some in investigators [8–11] propose various time-scale techniques for analysis.

The present work extends the analytical investigation of crack on rotor at low acceleration in [5], to the case of high rotor acceleration with unbalance transverse crack and rub-impact. It can be assume that, at high speed, centrifugal forces move the rolling elements outward and downward, increase the clearance between the rotor and the stator and contribute to the breathing of the crack, and generate strong vibration of the system.

ESTABLISHING THE SYSTEM MODEL

Consider Fig.1.(a), in which, a de Laval rotor of disc mass m, is supported on a massless elastic shaft of length L with a transverse crack near the disc in the inertial coordinates (X, Y) and rotating coordinates system (ξ,η). Let the disc be located mid-span on the shaft supported on sliding bearings at each end. Given the angular velocity ω and weight dominance, the system equation can be written in the form [12]:

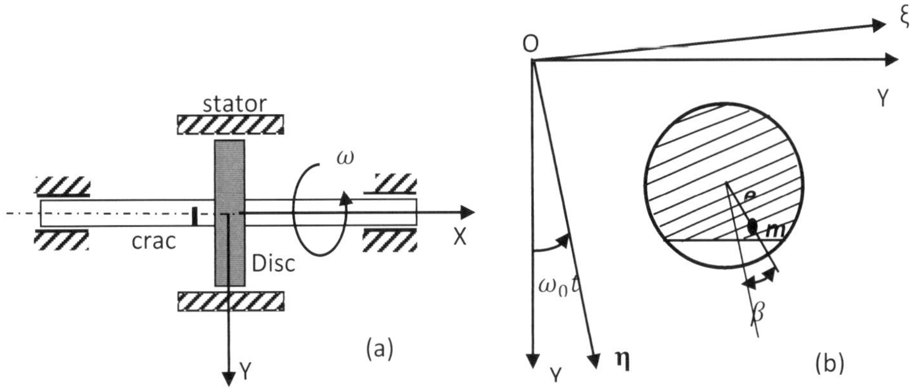

Figure 1. (a) Model of rotor on sliding bearing in rub (b) Coordinate system of the cracked rotor and disc unbalance.

$$\begin{bmatrix} m & 0 \\ 0 & m \end{bmatrix}\begin{pmatrix} \ddot{X} \\ \ddot{Y} \end{pmatrix} + \begin{bmatrix} c & 0 \\ 0 & c \end{bmatrix}\begin{pmatrix} \dot{X} \\ \dot{Y} \end{pmatrix} + \begin{bmatrix} k_X & k_{XY} \\ k_{YX} & k_y \end{bmatrix}\begin{pmatrix} X \\ Y \end{pmatrix} = \begin{pmatrix} f_X \\ f_Y \end{pmatrix} + \begin{pmatrix} mg \\ 0 \end{pmatrix} + \begin{pmatrix} F_X \\ F_Y \end{pmatrix} \tag{1}$$

where c, k, f, F are respectively, damping coefficient, stiffness coefficient, and external disturbance. The bearing is modelled with with zero cross-coupling stiffness coefficients, $k_{XY} = k_{YX} = 0$.

RUB FORCES ACTING ON THE ROTOR'S SHAFT

Let Δ, M, R_d, and ω respectively be, rotor-stator static clearance, mass, radius and angular velocity of the disc about O, as shown in Fig. 2. On touching the stator, the rotor rubs on or bounces on, exciting vibration of the ring and rotor shaft. The shaft deflects due to, torque, crack and unbalance. A friction force F_T, acts against rotation at the contact point. Referring to fig 2, the position of the shaft center and rotor-stator clearance are:

$$\vec{R} = X\vec{\imath} + Y\vec{\jmath}, \qquad \Delta = R_r - R_s \tag{2}$$

A friction force F_T, acts against rotation at the contact point P_c. Impact occurs when,

$$R - \Delta \geq 0 \tag{3}$$

When $R < \Delta$, the contact ceases and the friction force vanishes. The value of R at an arbitrary point P_c, is

$$R = \sqrt{(X - \Delta X)^2 + (Y - \Delta Y)^2} \tag{4}$$

The position of P_c relative to point O in the (X, Y) frame is given by,

$$\vec{R} = (\Delta + \Delta_{XY})(cos\theta\vec{\imath} + sin\theta\vec{\jmath}) \tag{5}$$

where

$$\begin{cases} X = (\Delta_X + \Delta)cos\theta \\ Y = (\Delta_Y + \Delta)sin\theta \end{cases} \tag{6}$$

and

$$tan\emptyset = \frac{Y}{X} \tag{7}$$

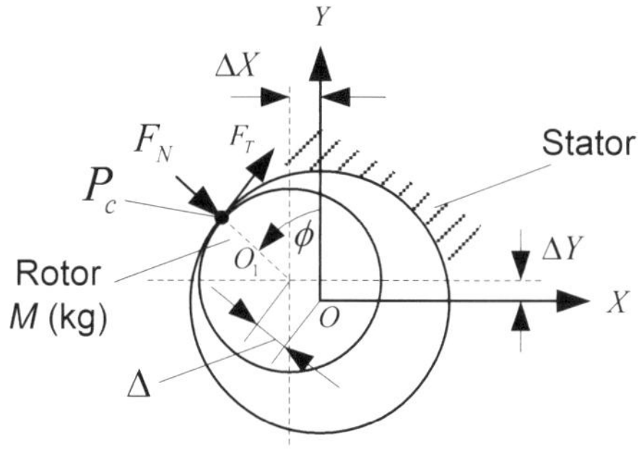

Figure. 2 Geometry of rotor, rubbing impact force and a clearance.

From the geometry of Fig. 2, and the above discussion,

$$\begin{cases} F_N = K_s(R - \Delta) \\ F_T = -\mu K_s(R - \Delta) \end{cases} \quad (8)$$

F_N , μ respectively are, radial component due to impact and friction coefficient. Lateral forces in both X, Y are:

$$F_X = -F_N cos\theta + F_T sin\theta; \qquad F_Y = -F_N sin\theta - F_T cos\theta \quad (9)$$

In Cartesian coordinates, the forces can be expressed as [13],

$$\begin{cases} F_{1X} = -K_s\left(1 - \frac{\Delta}{R}\right)X + K_s\left(1 - \frac{\Delta}{R}\right)\mu Y \\ F_{1Y} = -K_s\left(1 - \frac{\Delta}{R}\right)\mu X - K_s\left(1 - \frac{\Delta}{R}\right)Y \end{cases} \quad (10)$$

FLUID CIRCUMFERENTIAL FORCE

A lumped model of the fluid circumferential force based on the consideration that the journal bearing can generate large reaction forces at high speed has been used. The radius $\boldsymbol{R_o}$ of the bearing and the radius $\boldsymbol{R_i}$ of the rotating shaft differ by a small distance $\boldsymbol{c_r}$, equal to the radial bearing clearance. The shaft rotates inside a cylindrical bearing and a thin clearance between bearing and shaft is surrounded by a thin layer of lubricant as shown in Fig. 3. The fluid circumferential reaction forces $\boldsymbol{F_r}$ (radial) and $\boldsymbol{F_t}$ (tangential), rotating forward at an average speed $\boldsymbol{\lambda\omega}$ are expressed as [13]:

$$-F_{fl} = \begin{bmatrix} -F_r & F_t \\ F_t & -F_r \end{bmatrix} = \begin{bmatrix} K & D\lambda\omega \\ -D\lambda\omega & K \end{bmatrix}\begin{pmatrix} X \\ Y \end{pmatrix} + \begin{bmatrix} D & 2M\lambda\omega \\ -2M\lambda\omega & D \end{bmatrix}\begin{pmatrix} \dot{X} \\ \dot{Y} \end{pmatrix} \quad (11)$$

where F_{fl} is a fluid film reaction forces on the bearing. Assuming isotropy and laminar fluid, the fluid radial stiffness coefficient $K = k_{XX} = k_{YY}$, and fluid radial damping $D = C_{XX} = C_{YY}$, M is fluid inertia, ω is the changing rotor velocity and λ is the fluid circumferential average velocity.

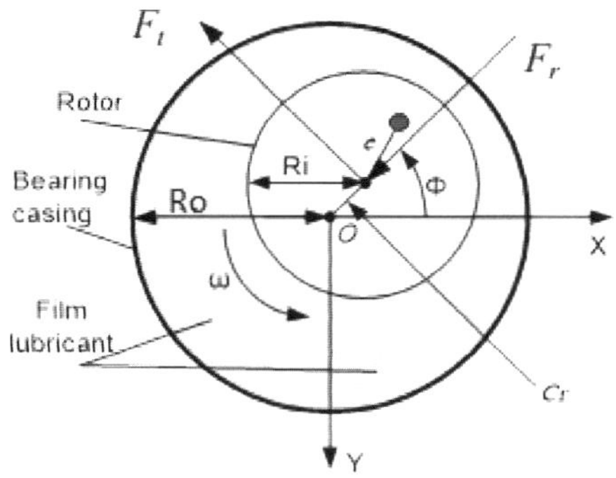

Figure. 3. Simple journal bearing; view in axial direction with the radial bearing clearance c_r

CRACK MODEL AND STIFFNESS MATRIX

The inertial frame may be mapped to the rotational frame by the transformation.

$$\mathbf{T}=\begin{bmatrix}\cos\theta & \sin\theta\\ -\sin\theta & \cos\theta\end{bmatrix} \tag{12}$$

where $\theta=\frac{1}{2}a_1t^2+\omega_0 t$. The stiffness matrix in the rotational frame is given as [7]:

$$\begin{bmatrix}k_\xi & 0\\ 0 & k_\eta\end{bmatrix}=\begin{bmatrix}k_X & 0\\ 0 & k_Y\end{bmatrix}-f(\theta)\begin{bmatrix}\Delta k_\xi & 0\\ 0 & 0\end{bmatrix} \tag{13}$$

where k_ξ, k_η are respectively, the stiffness in the ξ-axis direction, and the stiffness in the η-axis direction. k_X is the stiffness of the non-cracked rotor ($k_X=k_Y=k_0$), Δk_ξ is the stiffness variation due to the crack in the ξ-axis direction. At low speed, the weight dominates the vibrations [12]. The stiffness matrix in the inertial frame assumes the following form:

$$\begin{bmatrix}k_{XX} & k_{XY}\\ k_{YX} & k_{YY}\end{bmatrix}=\mathbf{T}^{-1}\begin{bmatrix}k_\xi & 0\\ 0 & k_\eta\end{bmatrix}\mathbf{T}=\begin{bmatrix}k_0 & 0\\ 0 & k_0\end{bmatrix}-\frac{1}{2}f(\theta)\Delta k_\xi\begin{bmatrix}1+\cos 2\theta & \sin 2\theta\\ \sin 2\theta & 1-\cos 2\theta\end{bmatrix} \tag{15}$$

The unbalance forces f_X, f_Y are

$$\begin{pmatrix}f_X\\ f_Y\end{pmatrix}=me\begin{pmatrix}\ddot{\theta}\sin\theta+\dot{\theta}^2\cos\theta\\ -\ddot{\theta}\cos\theta+\dot{\theta}^2\sin\theta\end{pmatrix} \tag{16}$$

where e is the disc mass eccentricity. Under the stated conditions, the following approximation can be adopted [2, 3]:

$$\begin{pmatrix}X\\ Y\end{pmatrix}\approx\begin{pmatrix}X\\ Y\end{pmatrix}_{st}=\begin{pmatrix}\frac{mg}{k_X}\\ 0\end{pmatrix}=\begin{pmatrix}x_{st}\\ 0\end{pmatrix} \tag{17}$$

Substituting the stiffness matrix from equation (15) into equation (1) and introducing the angular acceleration, the crack excitation, unbalance forces, friction, oil film and gravitational forces including the angular acceleration terms in inertial coordinate frame become:

$$\begin{bmatrix}m & 0\\ 0 & m\end{bmatrix}\begin{pmatrix}\ddot{X}\\ \ddot{Y}\end{pmatrix}+\begin{bmatrix}c+D & 2M\lambda\omega\\ -2M\lambda\omega & c+D\end{bmatrix}\begin{pmatrix}\dot{X}\\ \dot{Y}\end{pmatrix}+\begin{bmatrix}k_0+K & D\lambda\omega\\ -D\lambda\omega & k_0+K\end{bmatrix}\begin{pmatrix}X\\ Y\end{pmatrix}=$$
$$\frac{1}{2}f(\theta)\Delta k_\xi\begin{bmatrix}1+\cos 2\theta & \sin 2\theta\\ \sin 2\theta & 1-\cos 2\theta\end{bmatrix}\begin{pmatrix}x_{stat}\\ 0\end{pmatrix}+me\begin{pmatrix}\ddot{\theta}\sin\theta+\dot{\theta}^2\cos\theta\\ -\ddot{\theta}\cos\theta+\dot{\theta}^2\sin\theta\end{pmatrix}+\begin{pmatrix}F_{1X}\\ F_{2Y}\end{pmatrix} \tag{18}$$

The second terms on the right side of (18) are respectively the crack excitation, unbalance excitation and rubbing forces, and fluid-film bearing cross-coupling terms $k_{XY}=k_{YX}=D\lambda\omega$, and $C_{XY}=C_{YX}=2M\lambda\omega$.

WAVELET TRANSFORM

Wavelet time-frequency transform is a multiresolution analysis algorithm and is the inner product of the signal and a family of the wavelet. For the mother wavelet or the wavelet prototype $\psi(t)$, there exits a corresponding family of wavelet (son wavelet). The series of the son wavelets are generated by dilation and translation from the mother wavelet, $\psi(t)$ as follows [7]:

$$\psi_{a,b}(t) = \frac{1}{\sqrt{|a|}}\psi(\frac{t-b}{a}) \tag{20}$$

where, a and b are dilation (scaling) and translation (shift) parameters respectively. The factor $|a|^{-1/2}$ is used to ensure energy preservation. $\psi(t)$ is a square integrable complex function, satisfying the admissibility condition [4].

$$c_\psi = \int_{-\infty}^{+\infty}\frac{|\psi(\omega)|^2}{|\omega|}d\omega < \infty \tag{21}$$

$\psi(\omega)$ is the Fourier Transform of $\psi(t)$.

The continuous wavelet time-frequency transform of the initial signal $Y(t)$ is expressed as:

$$CWT_s(a,b) = \int Y(t)\cdot\psi^*_{a,b}(t)dt \tag{22}$$

Where $\psi^*_{a,b}(t)$ is the complex conjugate function of $\psi_{a,b}(t)$.

By varying a, b the coefficients $W_s(a,b)$ represent the signal $Y(t)$ at levels of resolution and time shift. $Y(t)$ has been analysed at various resolution levels and the nonlinearities identified by CWT [15].

NUMERICAL SIMULATION

A classical Runge-Kutta algorithm has been used to obtain a dynamic response. To ensure that any and small changes in the system response due to impact-contact and crack condition are captured, a small time step, $\Delta t = 0.001s$ is used. Data of shaft-disc system in [5] is adopted: $k_s = 5\times10^8\ N/m$; $\mu = 0.25$, and ω is varying velocity $x_{stat} = 0.5\ mm$, $k_0 = 9.99\times10^5\ N/m$, $D = d/2m\omega_0 = 0.1$, $\omega_0 = (g/x_{stat})^{0.5}$, $a_1 = 15\ rad/s^2$,$150\ rad/s^2$, $\Delta k_\xi / k_0 = 0.95$, $m = 51\ kg$, $e = 0.01\ mm$, and $\omega = at + (g/x_{stat})^{0.5}$. The values of the fluid average velocity λ and the fluid inertia M are taken as identified in [13] as 0.125 and 4.6 kg respectively.

APPLICATION, RESULTS AND DISCUSSIONS

Initially, the dynamic response of the unbalanced, uncracked system without, rub and oil film is simulated numerically. Further, a crack, rub, unbalance and the effect of a thin film of lubricant are included in the simulation. The results are obtained at $a = 15\ rad/s^2$ and compared. Figure 4 shows deflection in Y direction, orbits patterns and frequency spectrum of the rotor without, a crack and rub while passing the first critical speed. Fig. 4 (d)-(f) show an increase in vibration amplitudes in orbit response of the cracked rotor compared to the uncracked rotor at its first critical speed (at t=10s).

Elliptical orbit observed in Fig.4 (b) changes to an oval curvature due to rub friction and the presence of a crack. Fig. 4 (a), (g) and (j) indicates a large increase in vibrations near the rotational speeds at one-half of the first critical speed (at t=5s). Fig. 4(e) illustrates evolutions of the orbital patterns of the unbalanced rotor with rub occurrence and a breathing crack while passing through one-half of the first critical speed. In figure 4(h) it is noted that, rotor-stator interactions results in numerous irregular impacts or rebounds in the orbit.

A combination of unbalance, breathing crack, and rub in Figs. 4 (j)-(l) manifests a significant domination of crack features with some noise peaks related to the irregular contact of the rotor with the stator. Presence of noise during the regular contact is identified by a multitude of peaks observed in the lateral displacement response in Fig. 4 (g); the peaks are present even in the combination of rub and crack in Fig. 4 (j). Examining Fig. 4 (k), rub contact is not clearly discernible due to nonlinearity of stiffness. Therefore, effects of the crack overshadow those of rotor-stator interaction.

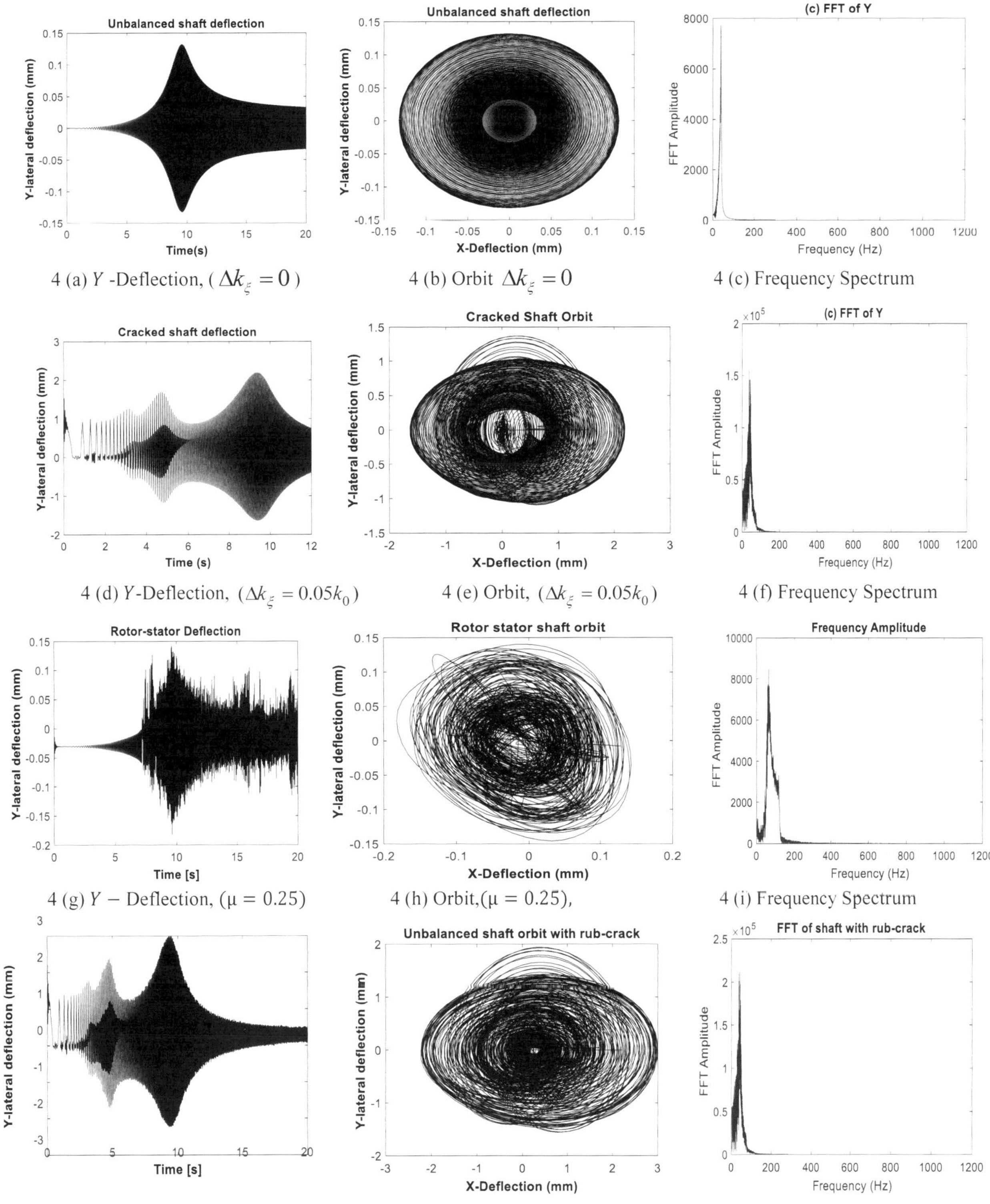

4 (a) Y -Deflection, ($\Delta k_{\xi} = 0$) 4 (b) Orbit $\Delta k_{\xi} = 0$ 4 (c) Frequency Spectrum

4 (d) Y-Deflection, $(\Delta k_{\xi} = 0.05 k_0)$ 4 (e) Orbit, $(\Delta k_{\xi} = 0.05 k_0)$ 4 (f) Frequency Spectrum

4 (g) Y – Deflection, (μ = 0.25) 4 (h) Orbit,(μ = 0.25), 4 (i) Frequency Spectrum

4 (j) Y-Deflection, $(\Delta k_{\xi} = 0.05 k_0)$ 4 (k) Orbit, $(\Delta k_{\xi} = 0.05 k_0, \mu = 0.25)$ 4 (l) Frequency Spectrum

Figure. 4 Dynamic response of rotor at low rotor acceleration $(a = 15 rad / s^2)$ on passing through its first critical speed.

In Figs. 4 (f) and (l) it is noticed that frequency amplitude increases considerably when a breathing crack exists and it decreases considerably during rotor-stator contact as in Fig. 4 (i).

Effects of the oil film on the vibrating system at low accelerations and under multiple parametric excitations are examined and presented in figure 5.

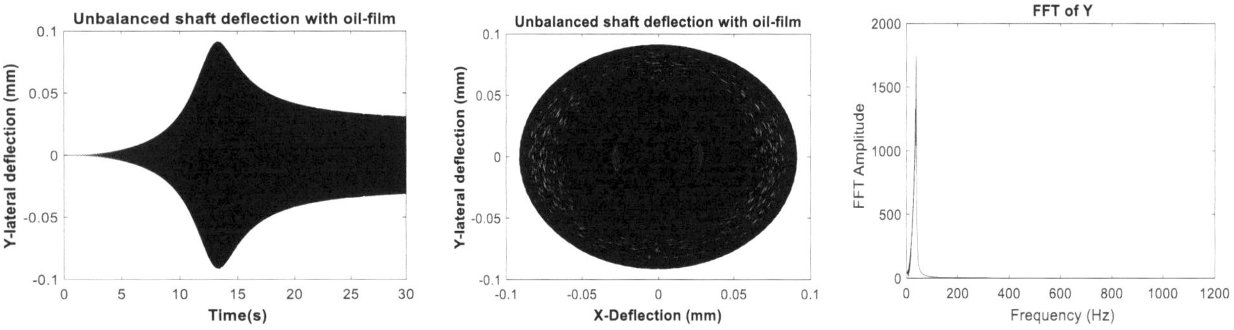

5 (a) Y-Deflection, $(\Delta k_{\xi} = 0.05k_0)$ 5 (b) Orbit 5 (c) Frequency Spectrum

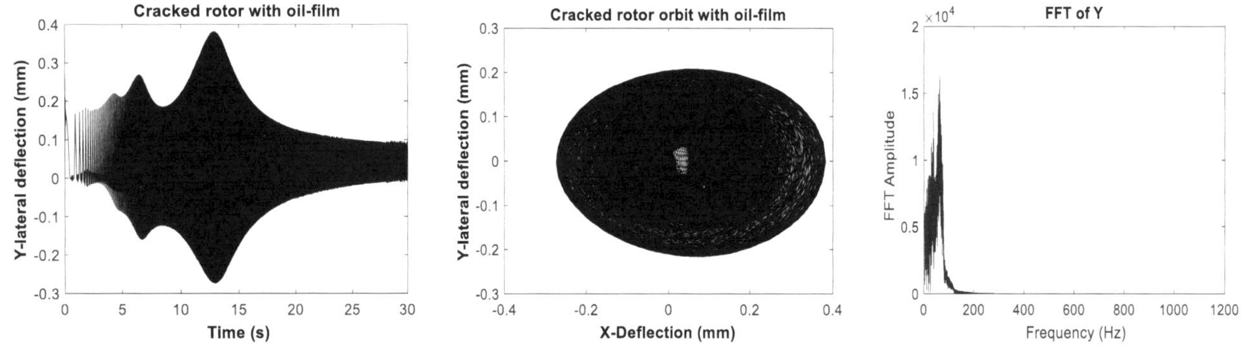

5 (a) Y-Deflection, $(\Delta k_{\xi} = 0.05k_0)$ 5 (b) Orbit 5 (c) Frequency Spectrum

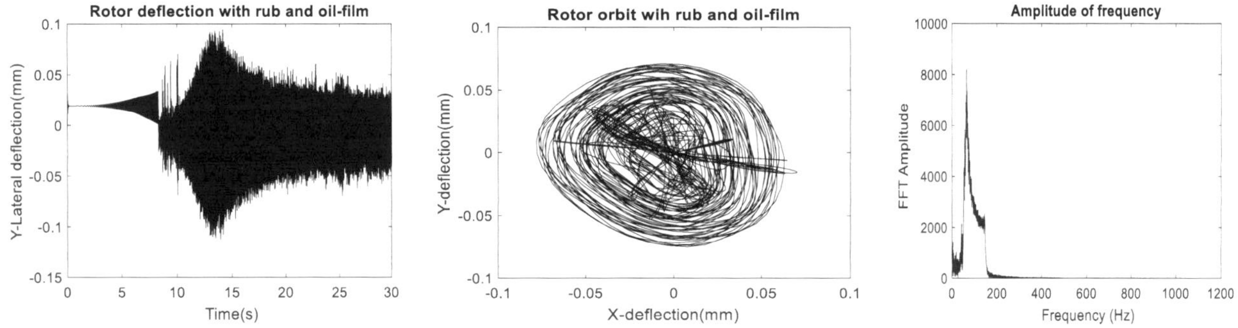

5 (d) Y-Deflection, (μ = 0.25) 5 (e) Orbit,(μ = 0.25), 5 (f) Frequency Spectrum

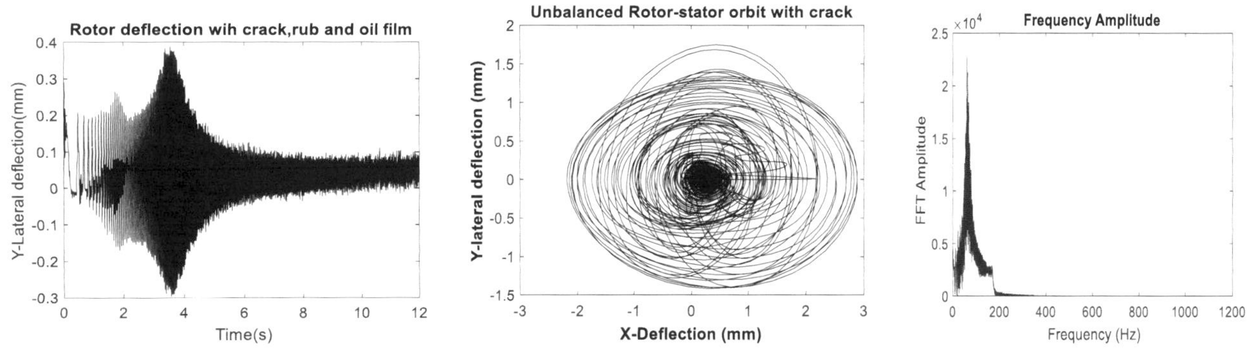

5 (g) Y-deflection, $(\Delta k_{\xi} = 0.05k_0, \mu = 0.25)$ 5 (h) Orbit Patterns 5 (i) Frequency Spectrum

Figure 5. Dynamic response of rotor at low rotor's acceleration $(a = 15\ rad/s^2)$ with unbalance, rubs, crack faults, and oil film passing through its half and first critical speed.

A comparison of the deflection, rotor orbits and frequency amplitude manifested by the system with oil lubricant in Fig. 5 and without oil in Fig 4 reveals a slight difference in the two results. It is observed in time covered region, time frequency resonance of one-half of the critical speed (at t = 4.7s) and the first critical speed (at t = 9,4s) in Fig.4. (a), the time to attain each critical speed has approximately increased to t=10s and t=13s respectively as noted in Fig. 5(a). Comparing amplitudes in Fig. 4 (a), (b), (c) to amplitudes in Fig. 5 (a), (b), (c) the amplitude reduces when the thin oil lubricant is included. The rest of vibration features in the two sets of figures remain unchanged. Inclusion of the thin oil lubricant force in the model particularly minimizes the maximum lateral deflection amplitude, consequently reducing the number of rotor -to-stator contact in the orbit as seen in Fig.5 (e). It appears that, lubrication can be used to influence rotor-stator contact. Observation of the model with and without oil film force (Fig.4 and Fig.5) shows the change both in the orbit and the amplitude of the 1× harmonic components around one-half of the first critical speed.

ANALYSIS BY CONTINUOUS AND DISCRETE WAVELET TRANSFORM

CWT has been performed at a low acceleration ($a = 15\ rad/s^2$) using Daubechies wavelet of order 35, (db35) with filter length of 70 as in [14] and has been employed to extract the approximation coefficient from vibration signals. The CWT result of db35 of the vibration signal of the unbalance mass, crack, and rub sampled at ($a = 15\ rad/s^2$) is given in Fig. 6.

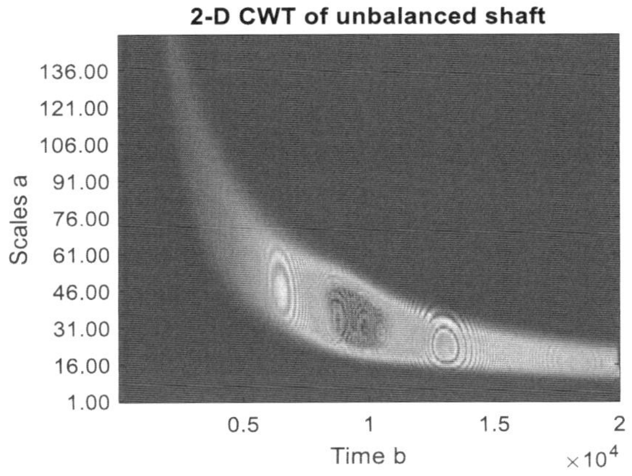

6 (a) CWT of rotor with unbalance fault

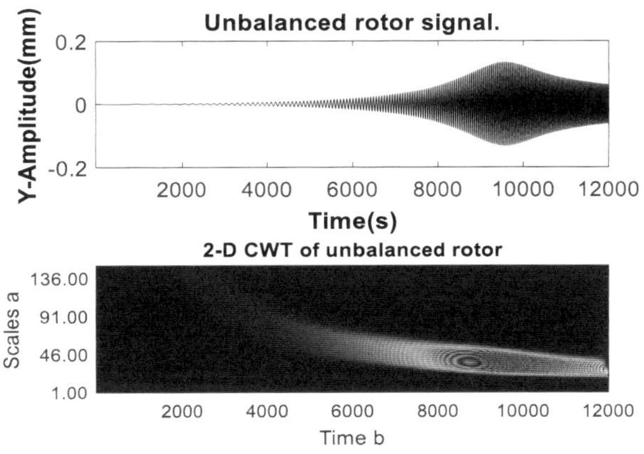

6(b) 2-D Scalogram of unbalanced shaft

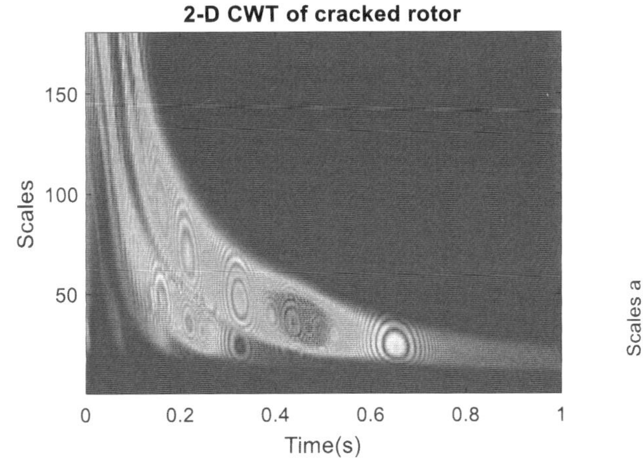

6 (c) CWT of unbalanced rotor with cracked faults

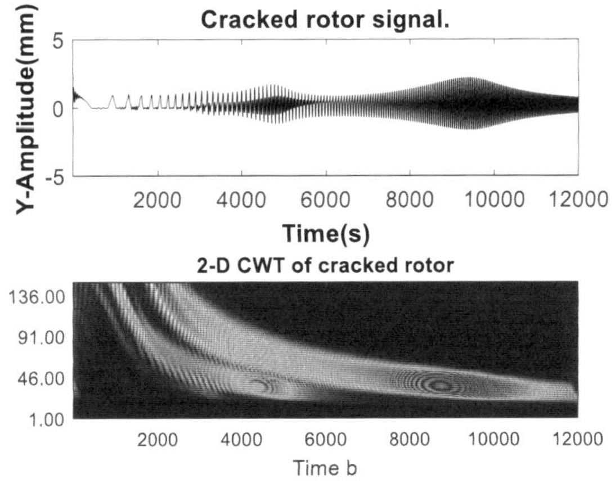

6(d) 2-D Scalogram of cracked shaft

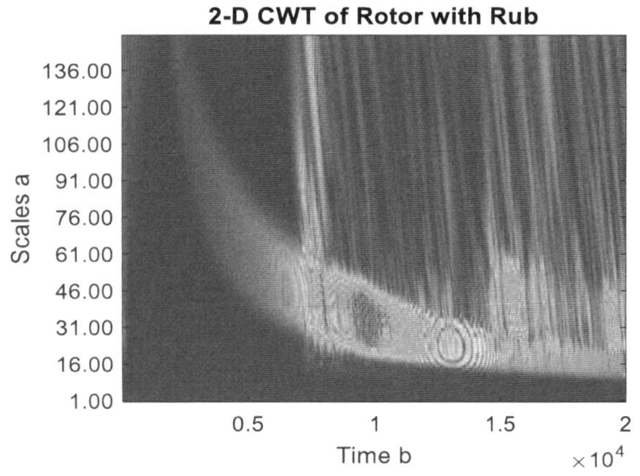

6 (e) CWT of unbalanced rotor with rub faults

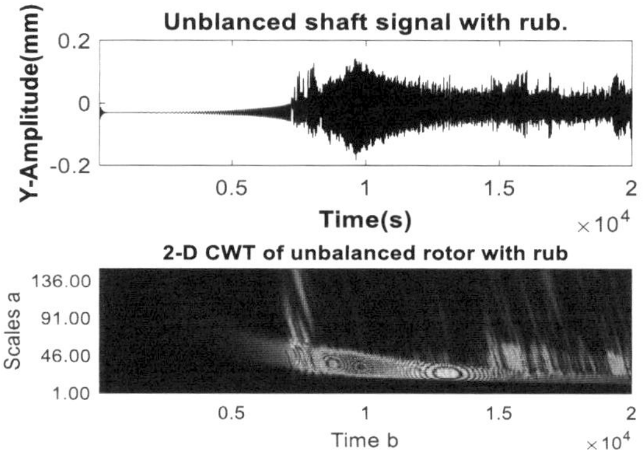

6(f) 2-D Scalogram of rotor with rub (μ = 0.25)

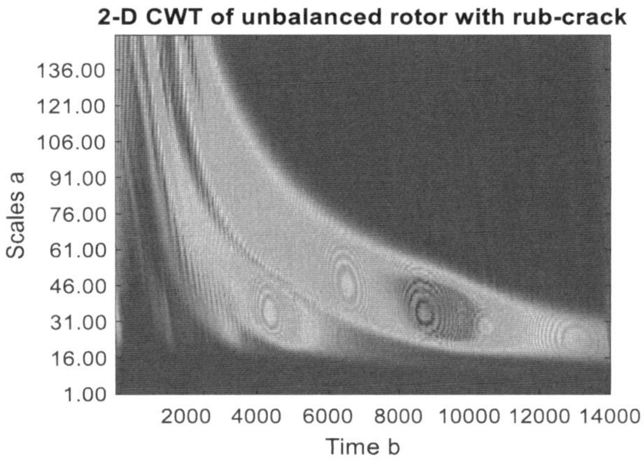

6 (g) CWT of unbalanced rotor with rub-crack faults

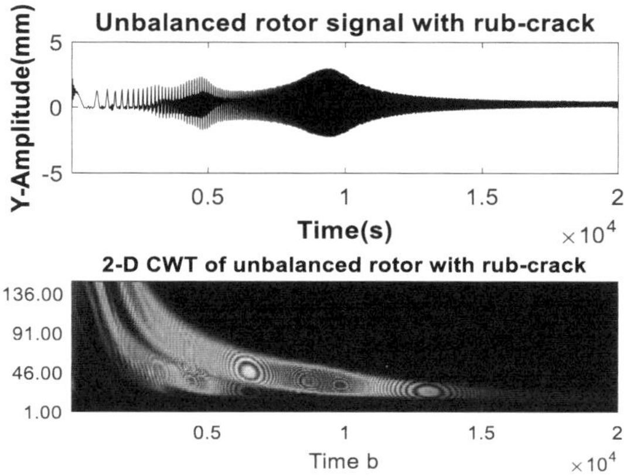

6(h) 2-D Scalogram of rotor with rub-crack (μ = 0.25)

Figure 6. C.W.T and scalogram of unbalanced rotor with crack and rub at $a = 15\ rad/s^2$ acceleration

Figs.6 (a) and (b), show the spectrum variation related to unbalanced rotor during passage via critical speed. The zone of large concentration in the energy distribution is located most of the time around the first critical speed (t=10000s), observed in Fig. 6 (b). Therefore, there is always a fine variation grid around this zone, characterizing specifically the type of fault parameter in the model. Therefore, the CWT at run-up enables by the distribution of energy at the first and second half critical speed, the identification the presence of a transverse crack. Indeed, in Figs. 6 (c) and (d) the distribution of energy of wavelet power spectrum in the direction of scale or frequency shows a high concentration of energy that corresponds to the sub-harmonic resonance of the cracked rotor. In Figs. 6 (c) and (d) it is apparent that, the presence of a transverse crack induces the appearance of a second maximum value of the normalized wavelet power spectrum at the passage via one-half the first critical speed (for t ϵ [0.2–0.7] s or t ϵ [4000–10000] s. The associated normalized wavelet power spectrum is given in Figs. 6 (e) and (f). It clearly appears that, there is significant change in performance of the detection of rub via the CWT. In Figs. 6 (g) and (h) the wavelet power spectrum consists of both rub and crack effects. it appears from these figures that, there is no significant change in performance of the detection of rub in presence of transverse crack via the CWT. The crack effects once again predominate as observed in Fig. 4 (j), (k) and (l). From the above analysis, CWT has been useful on vibration system analysis containing unbalance mass, crack, and rub, and combination of rubs and crack. In Fig. 6 identification of mentioned types of faults gave considerably better results identification. The system response of the signal under influence of fluid-film, unbalance, rub and crack at low acceleration with CWT, are shown in figure 7. Comparison of Fig. 7 with Fig. 6 (without oil-film effects) shows that the signal response to rubbing and crack condition are minimized due to the effect of oil-film. Subsequently, the rotor-to-stator contacts is reduced, in the 2-D scalogram of Fig. 7(c) (e), (g) if compared with Fig. 6(c) (e), (g).

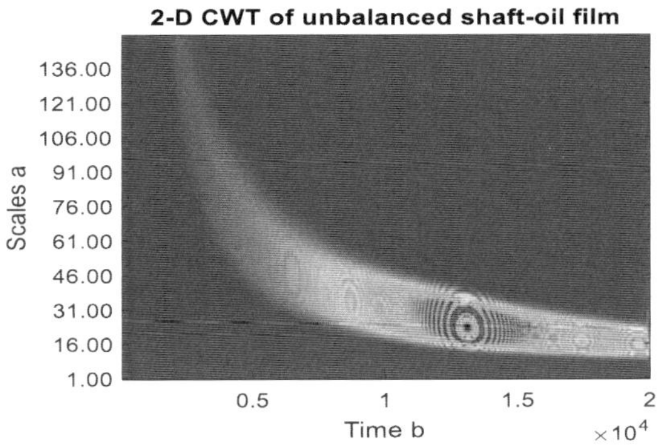

7 (a) C.W.T, unbalanced rotor with oil

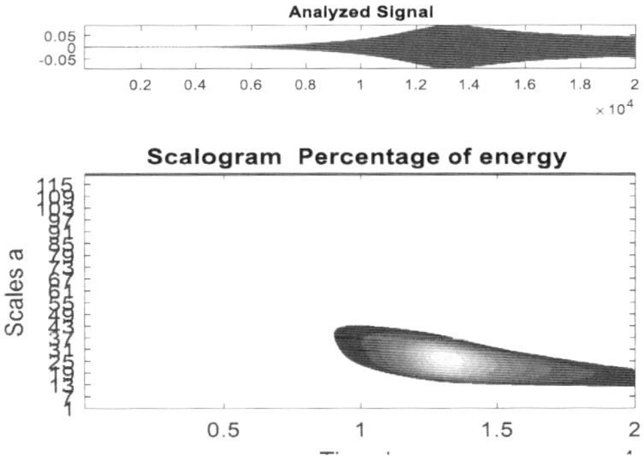

7 (b) Corresponding Scalogram Energy

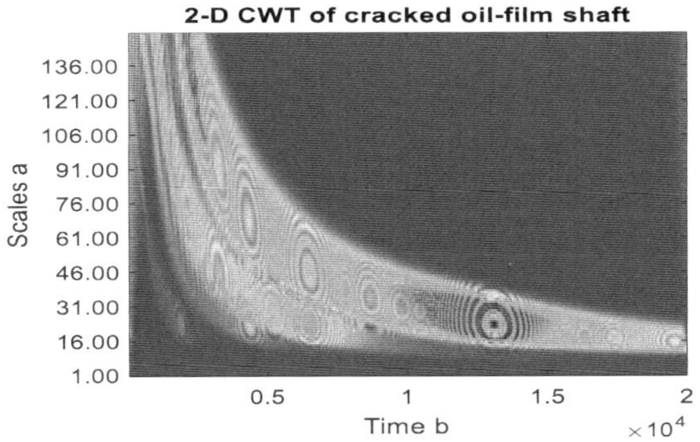

7 (c) C.W.T of unbalanced, cracked rotor with oil

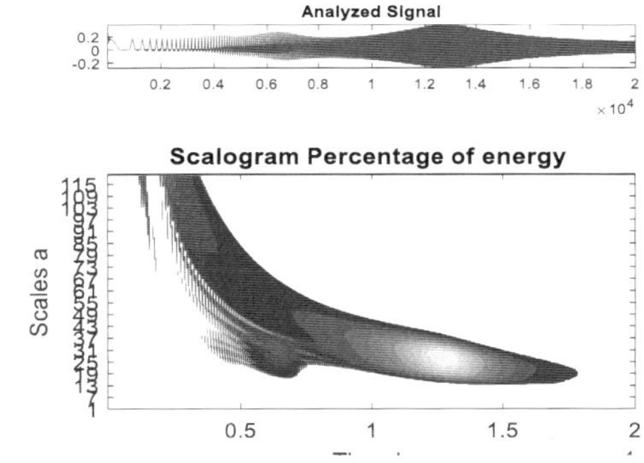

7 (d) Corresponding Scalogram

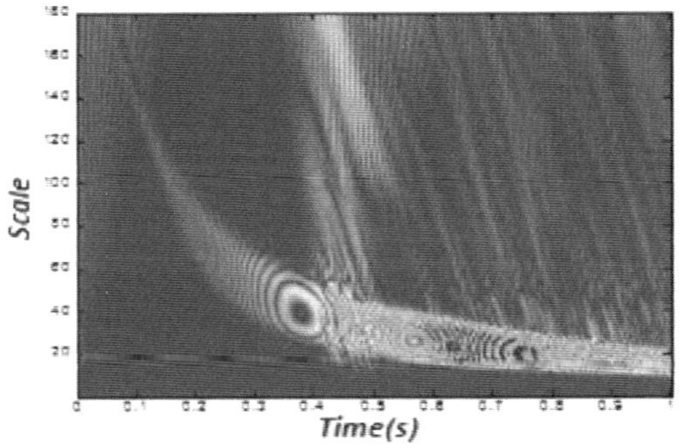

7 (e) C.W.T of unbalanced rotor with rub and oil

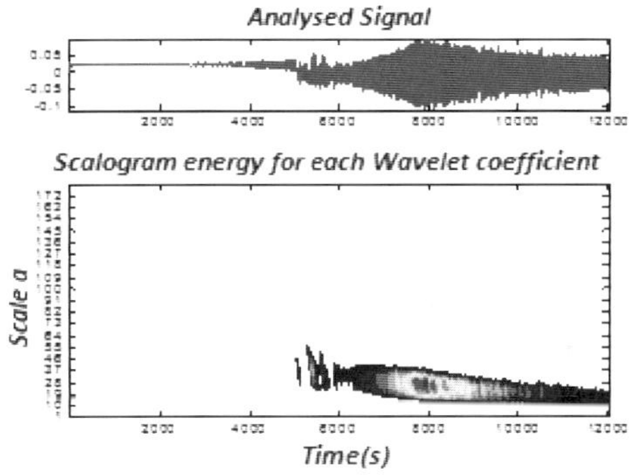

7 (f) Corresponding Scalogram Energy

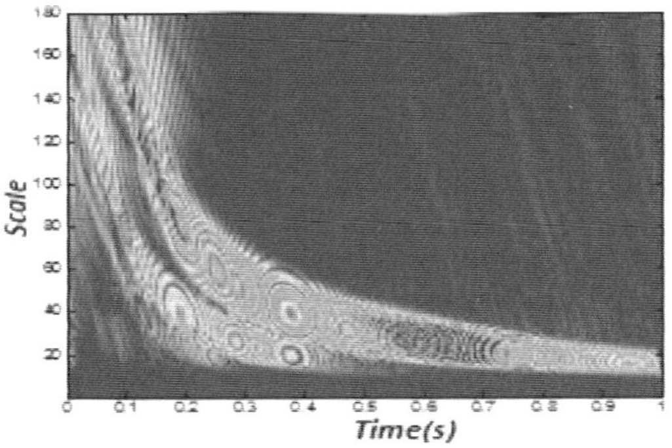

7(g) C.W.T of unbalanced cracked rotor with rub and oil

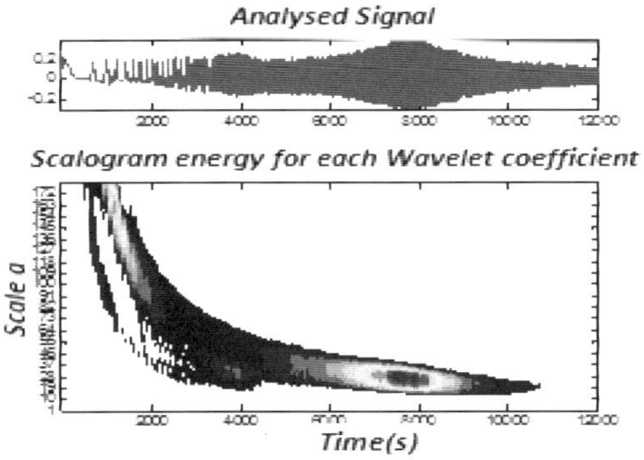

7 (h) Corresponding Scalogram

Figure. 7 Continuous Wavelet, power spectrum of the rotor with crack rub and film oil (with an acceleration of $a=15\ rad/s^2$

CONSLUSION

1- Under unbalance, the oil force is greater, and the reducing vibration effect is better as in Fig.7 (a) (b) compared to Fig.6 (a) (b).
2- The smaller the fluid average velocity λ is, the better is the reducing vibration effect.
3- The rotor-stator-rub system with small fluid-film, under the critical speed, rubbing will generate many vibrations which are considerably reducing with slightly film oil Fig. 7(e) (g).
4- Instability below and above the critical speed generated by unbalanced rotor with crack and rub faults will be greatly reduced with smaller fluid-film diameter Fig.7 compared to Fig.6 with zero film oil lubricant.
5- The presence of oil film lubricant decreases the time and amplitude during passing the resonance region and increases in effective damping.

To examine the constituent components of the original signal, CWT is performed using Daubechies (db35) algorithm in the Matlab toolbox .The selected parameters for calculation are: the Daubechies wavelet with a filter length of 50, and a scaling factor a= 1:1:180. This technique is performed in three steps: (a) selection of the particular wavelet family filters of the signal (b) determine or computing the approximation coefficients vector and detail coefficients vector; and (c) construction in 2-D of the signal.

REFERENCES

[1] H.L. Hassenpflug, R.D, Flack, and E.J, Gunter. Influence of Acceleration on the Critical Speed of a Jeffcott Rotor. J. Engineering for Power, Trans. ASME, vol. 103, no. 1, (1981) 108–113.
[2] R. Gasch, Dynamic behavior of a simple rotor with cross-sectional crack. IMechE Conference Publication c178/76, 1976, 123-128
[3] A. D. Dimaragonas and C. A. Papadopoulos, Vibration of cracked shafts in bending. Journal of Sound and Vibration 91, 1983, 583-593.
[4] A.S. Sekhar, Crack identification in a rotor system: a model- based approach. Journal of sound and vibration 270, 2004, 887-920.
[5] A. S. Sekhar, and B. S. Prabhu, Condition monitoring of cracked rotor through transient response, 1998 Mech. Mach. Theory 33, 1167-1175.
[6] J. T. Sawicki, X. Wu, G. Y. Baaklini, and A. L. Gyekenyesi, "Vibration-based crack diagnosis in rotating shafts during acceleration through resonance," in Nondestructive Evaluation and Health Monitoring of Aerospace Materials and Composites II, vol. 5046 of Proceedings of SPIE, pp. 1–10, San Diego, Calif, USA, March 2003.
[7] J.-J. Sinou, "An experimental investigation of condition monitoring for notched rotors through transient signals and wavelet transform" published in "Experimental Mechanics 49 (2009) 683-695" DOI : 10.1007/s11340-008-9193-6.
[8] A. Sekhar, "Crack detection through wavelet transform for a run-up rotor". Journal of Sound and Vibration, 259, 2003. p. 461-472 .
[9] Z.K, Peng, F.L, Chu, and P.W, Tse. Detection of rubbing-caused impacts for rotorstator fault diagnosis using reasigned scalogram. *Mechanical Systems and Signal Processing*, 2005. Vol. 19, No. 2, 391-409.
[10] L, Xinsheng, and , A.L, Kenneth. Bearing fault diagnosis based on wavelet transform and fuzzy inference. Mechanical Systems and Signal Processing, 2004. Vol. 18, No. 5, 1077- 1095.
[11] F, Wang, Q, Xu, and S, Li. Vibration analysis of cracked rotor sliding bearing systems with rotor-stator rubbing by harmonic wavelet transform. *Journal of Sound and Vibration,* 2004.Vol. 271, No. 3-5, 507-518.
[12] A. A. Alugongo, A dual Time-Frequency-Feature investigation and diagnostics of a cracked de-Laval rotor. *IEEE AFRICON* 2009.
[13] L. T. Tam, A. J. Prezkwas, A. Muszynska, R. C. Hendricks, M. J. Braun and R. L.Mullen. Numerical and analytical study of fluid dynamic forces in seals and bearings *Journal of vibration, Acoustics, Stress, and Reliability in Design* 110,315,325 1988.
[14] I. Daubechies, "Orthonormal bases of compactly supported wavelets,"*Communication on Pure and Applied Mathematics*, vol. 41, pp. 909-996, 1988
[15] B.X. Tchomeni, A.A. Alugongo, L.M. Masu, In situ Modelling of Lateral-Torsional Vibration of a Rotor-Stator with Multiple Parametric Excitations. *World Academy of Science, Engineering and Technology International Journal of Mechanical, Aerospace, Industrial and Mechatronics Engineering* Vol:8 No:11, 2014

Parametric Study of Suitable Orthogonality Conditions for Planar Multilink Flexible Robots

Francis Kunzi Tekweme*, Mechanical Engineering Technology, University of Johannesburg, Doornfontein Campus, Johannesburg 2028, South Africa, ftekweme@uj.ac.za and Andre Nel, Mechanical Engineering Science, University of Johannesburg, Kingsway Campus, Johannesburg 2006, South Africa.

ABSTRACT

Orthogonality conditions and stiffness matrix elements are related. Assumed mode method can be used to obtain a finite model of a flexible linkage. In this paper the normal modes and stiffness matrix elements for multilink flexible manipulators are obtained through a more accurate parametric orthogonality condition equation that accounts for actual mass and moment of inertia at the end of each link including the contribution of masses of distal links. As an alternative to the lengthy orthogonality condition equation, three equivalent and companion orthogonality conditions are proposed. Normal mode shapes are derived from either one of the proposed orthogonality conditions. Lagrangian method is used to derive the closed-form dynamic model. The Jacobian matrices are used to compute the inertia matrix. The centrifugal and Coriolis vector elements are computed from the inertia matrix using the Christoffel symbol. Simulations are conducted to validate the theoretical model.

KEYWORDS

Link, orthogonality, flexible, Jacobian, Coriolis.

NOMENCLATURE LIST

General Symbols

b_i	i^{th} link thickness
$C_{jk}(i)$	Cristoffel symbols
D	Damping matrix
d_i	Distance of center of mass of link i from joint i axis
E_i	Young modulus of link i
$(EI)_i$	Flexural rigidity of link i
f_i	Vector generalized forces
h_i	Coriolis and centrifugal forces vector elements
h_{ri}	Height of link i
$h(q,\dot{q})$	Vector Coriolis and centrifugal forces
I_i	Second moment of area about the neutral axis of cross section of link i
J_{hi}	Moment of inertia of hub i
J_{Li}	Actual moment of inertia at the distal end of link i
J_{oi}	Moment of inertia of link i about joint i axis
J_p	Payload inertia
J_T	Total moment of inertia about the motor armature at hub 1
J_{vi}	The 2 x 6 Jacobian matrix relating the linear velocity of a point located at x_i along link i to the general coordinates
$J_{\omega i}$	The 2 x 6 Jacobian matrix relating the angular velocity of a point located at x_i along link i to the general coordinates
K	Flexible- link system stiffness matrix
K_{ijk}	j^{th} row and k^{th} column element of the stiffness matrix diagonal block relative to link i
L	Lagrangian
l_i	Length of link i
$M(q)$	Inertia matrix

m_{hi}	Hub i mass
m_i	Link i mass
M_{ij}	i^{th} row and j^{th} column inertia matrix element
M_{Li}	Actual mass at the distal end of link i
m_p	Payload mass
$(MD)_i$	Contribution of masses non-collocated at the distal end of link i
n	Number of flexible links
N	Number of selected general coordinates
$P_i(t)$	Position vector of a point along the deflected link with respect to inertial frame
P_i	Scale factor (i.e., normalizing constant)
$\dot{P}_i(t)$	Absolute linear velocity
Q	Input weighting matrix
$\{q(t)\}$	Generalized coordinates vector
t	Time (continuous)
T	Total kinetic energy
u	n- vector of joint torques
V	Total potential energy
x_i	Distance from hub i along link i
$(X_i\ Y_i)$	Rigid body moving frame associated to link i
$(X_{0,}\ Y_0)$	Inertial frame
$y_i(x_i)$	Transversal deflection of link i at abscissa x_i
$Y_i(x_i,t)$	Transversal deflection of link i at abscissa x_i and time t
y_{ie}	Transversal deflection of link i at the distal end of link i
y'_{ie}	Slope of the deflection of neutral axis at the distal end of link i with respect to the abscissa axis of the rigid body moving frame associated to link i

Greek Symbols

$\alpha_i(t)$	Absolute angular rotation of frame (X_i, Y_i)
$\delta_{ij}(t)$	Time varying variable associated with the assumed mode shape $\phi_{ij}(x_i)$ of link i
$\theta(t)$	Rigid motion (joint angle) of link i
$\dot{\theta}(t)$	Joint angular velocity
ρ_i	Mass per unit length of link i

INTRODUCTION

The control of a flexible link mechanism is a challenging task. The efficient control strategies depend on the closed-loop dynamic model being available. Lee & Book [1] have derived the inertia matrix of a flexible link manipulator through the Jacobian matrix. De Luca & Siciliano [2] used the assumed mode method in conjunction with the Lagrange's equation investigate the dynamics of two-link flexible manipulators. They have presented a computationally efficient and explicit dynamic model for a planar multilink flexible robot. Although just the uniform link linear density is used to normalize the mode shapes, they acknowledged the need for a more accurate orthogonality condition. Zhu [3], Zhupping [4] and Ibrahim, Aly & Ismail [5] studied the modelling and dynamic characteristics of single link manipulators using the assumed mode method and the Lagrange's equation. They used orthogonality conditions that include not only the uniform link linear density, but also the actual mass and inertia at the end of the link.

Generalizing the complete orthogonality conditions for single link flexible manipulators as proposed by Zhu [3], Zhupping [4] and Ibrahim, Aly & Ismail [5] to account for interactions between the shoulder and elbow links of the two-link flexible robots yields

$$\int_0^{l_i} \rho_i \phi_{ij}(x_i)\phi_{ik}(x_i)dx_i + H_i(\phi_{ij}(x_i), \phi_{ik}(x_i), M_{Li}, J_{Li}, (MD)_i)\Big|_{x_i=l_i} = \begin{cases} 0 & j \neq k \\ P_i & j = k \end{cases} \quad (1)$$

where H_i is function of actual mass at the end of link i (M_{Li}), actual moment of inertia at the end of link i (J_{Li}) and contribution of non-collocated mass at the end of link i ($(MD)_i$).

The present paper extends the work presented by De Lucas and Siciliano [2]. Specific stress is on deriving complete orthogonality conditions that include the non-zero term $H_i(\phi_{ij}(x_i), \phi_{ik}(x_i), M_{Li}, J_{Li}, (MD)_i)\big|_{x_i=l_i}$ along with three other equivalent expressions. The closed-form dynamic equations arrangements in an explicit form are adapted from De Luca & Siciliano [2] model. In our model the inertia matrix is derived through the Jacobian matrices. Christoffel symbols are used to compute the centrifugal and Coriolis vectors elements from the inertia matrix. Table 1 below shows a summary of some existing model for two-link flexible manipulators.

Table 1: Summary of some orthogonality conditions used for two-flexible link manipulators

Model	$H_i(\phi_{ij}(x_i), \phi_{ik}(x_i), M_{Li}, J_{Li}, (MD)_i)\big\|_{x_i=l_i}$	P_i
1. De Luca & Siciliano [2] 2. Theodore & Ghosal [6] 3. Li, Zhu & Li [7] 4. Chen, Inoue & Shibata [8] 5. Liu, Su & Liu [9] 6. De Bofim, Gripp, Santos, Bernado & Goes [10]	0	m_i
1. Subhudi & Morris [11,12] 2. Morris & Subhudi [13]	0	$m_i\rho_i$
1. Lee & Book [1]	0	1
1. Talebi, Patel & Khorasani [14]	0	ρ_i
1. Henikl, Kemmetmüller, Bader & Kugi [15]	0	-
Current work	$\neq 0$	P_i

KINEMATIC MODELING

Prior to kinematic and dynamic analysis of the flexible link of uniform density, some useful constant and variable parameters that are extensively used need to be defined (De Luca & Siciliano [2]):

$$m_i = \int_0^{l_i} \rho_i dx_i = \rho_i l_i \qquad J_{0i} = \int_0^{l_i} \rho_i x_i^2 dx_i = \frac{1}{3} m_i l_i^2 \qquad z_{ijk} = \int_0^{l_i} \rho_i \phi_{ij}(x_i)\phi_{ik}(x_i) dx_i$$

$$\varpi_{ij} = \int_0^{l_i} \rho_i \phi_{ij}(x_i) x_i dx_i \qquad d_i = \frac{1}{m_i}\int_0^{l_i} \rho_i x_i dx_i = \frac{1}{2} l_i \qquad J_{0i} = \int_0^{l_i} \rho_i x_i^2 dx_i = \frac{1}{3} m_i l_i^2$$

Figures 1 and 2 show the vector positions of a point on deflected link 1 and link 2 respectively.

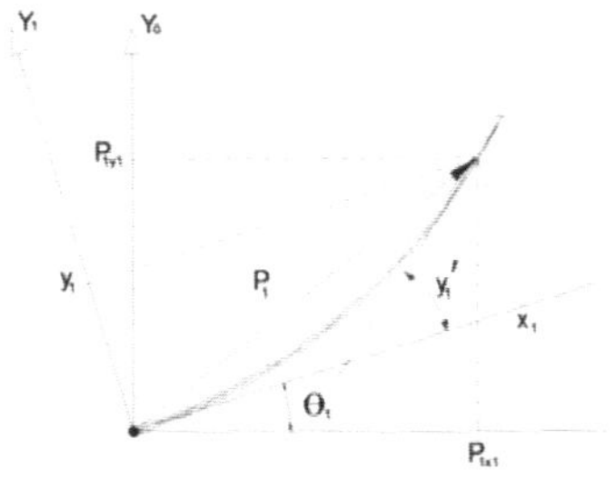

Figure 1: Point on link 1

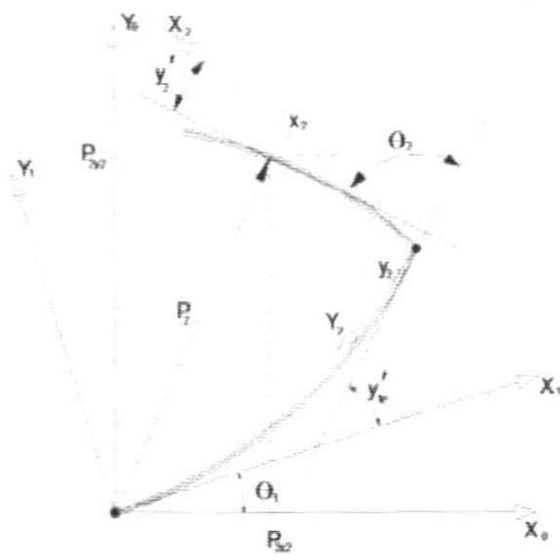

Figure 2: Point on link 2

The position vectors P_1 and P_2 with respect to inertial frame (X_0, Y_0) are expressed as follows :

$$P_1 = \begin{bmatrix} P_{1x1} \\ P_{1y1} \end{bmatrix} \quad \text{and} \quad P_2 = \begin{bmatrix} P_{2x2} \\ P_{2y2} \end{bmatrix} \tag{2}$$

where $P_{1x1} = x_1 \cos\theta_1 - y_1 \sin\theta_1$ and $P_{1y1} = x_1 \sin\theta_1 + y_1 \cos\theta_1$

$$P_{2x2} = l_1 \cos\theta_1 - y_{ie} \sin\theta_1 + x_2 \cos(\theta_1 + \theta_2) - x_2 \sin(\theta_1 + \theta_2) y'_{1e} - y_2 \sin(\theta_1 + \theta_2) - y_2 \cos(\theta_1 + \theta_2) y'_{1e}$$

$$P_{2y2} = l_1 \sin\theta_1 + y_{ie} \cos\theta_1 + x_2 \sin(\theta_1 + \theta_2) + x_2 \cos(\theta_1 + \theta_2) y'_{1e} + y_2 \cos(\theta_1 + \theta_2) - y_2 \sin(\theta_1 + \theta_2) y'_{1e}$$

The scalar absolute angular velocities are given by

$$\alpha_1(t) = \theta_1(t) \quad \text{and} \quad \alpha_2(t) = \theta_1(t) + y'_{1e}(t) + \theta_2(t) \tag{3}$$

To obtain a finite dimensional model the assumed modes method can be used. Therefore the elastic transversal deflections y_1 and y_2 can be evaluated as functions of spatial mode shapes and time dependent general coordinates. If two dominant modes per link are considered, we obtain

$$y_1(x_1,t) = \phi_{11}(x_1)\delta_{11}(t) + \phi_{12}(x_1)\delta_{12}(t) \quad \text{and} \quad y_2(x_2,t) = \phi_{21}(x_2)\delta_{21}(t) + \phi_{22}(x_2)\delta_{22}(t) \tag{4}$$

Six generalized coordinates can be used to completely describe the motion of a two-link flexible robot arm. The vector of generalized coordinates can be expressed as

$$q(t) = [\theta_1(t), \theta_2(t), \delta_{11}(t), \delta_{12}(t), \delta_{21}(t), \delta_{22}(t)]^T \tag{5}$$

MODE SHAPE DERIVATION

The mode shapes can be derived from the solution to the partial differential equation

$$(EI)_i \frac{\partial^4 y_i(x_i,t)}{\partial x_i^4} + \rho_i \frac{\partial^2 y_i(x_i,t)}{\partial t^2} = 0, \quad i=1, 2. \tag{6}$$

and the boundary conditions are

$$y_i(0,t) = 0, \quad y'_i(0,t) = 0, \quad i=1, 2. \tag{7}$$

$$(EI)_i \left.\frac{\partial^2 y_i(x_i,t)}{\partial x_i^2}\right|_{x_i=l_i} = -J_{Li} \frac{d^2}{dt^2}\left(\left.\frac{\partial y_i(x_i,t)}{\partial x_i}\right|_{x_i=l_i}\right) - (MD)_i \frac{d^2}{dt^2}\left(y_i(x_i,t)\big|_{x_i=l_i}\right) \tag{8}$$

$$(EI)_i \left.\frac{\partial^3 y_i(x_i,t)}{\partial x_i^3}\right|_{x_i=l_i} = M_{Li} \frac{d^2}{dt^2}\left(y_i(x_i,t)\big|_{x_i=l_i}\right) + (MD)_i \frac{d^2}{dt^2}\left(\left.\frac{\partial y_i(x_i,t)}{\partial x_i}\right|_{x_i=l_i}\right) \tag{9}$$

where $M_{L1} = m_2 + m_{h2} + m_p$, $M_{L2} = m_p$, $J_{L1} = J_{02} + J_{h2} + J_p + m_p l_2^2$, $J_{L2} = J_p$,

$(MD)_1 = (m_2 d_2 + m_p l_2)\cos\theta_2 - [(v_{21} + m_p\phi_{21e})\delta_{21} + (v_{22} + m_p\phi_{22e})\delta_{22}]\sin\theta_2$, $(MD)_2 = 0$

Solving the differential equation (6) and applying the boundary conditions (7) through (9) yields

$$\begin{bmatrix} F_{11} & F_{12} \\ F_{21} & F_{22} \end{bmatrix} \begin{bmatrix} C_{1i} \\ C_{2i} \end{bmatrix} = 0 \tag{10}$$

where $F_{11} = \sin(\beta_i l_i) + \sinh h(\beta_i l_i) + \frac{\beta_i^3 J_{Li}}{\rho_i}[\cos(\beta_i l_i) - \cosh(\beta_i l_i)] + \frac{\beta_i^2 (MD)_i}{\rho_i}[\sin(\beta_i l_i) - \sinh(\beta_i l_i)]$

$F_{12} = \cos(\beta_i l_i) + \cosh(\beta_i l_i) - \frac{\beta_i^3 J_{Li}}{\rho_i}[\sin(\beta_i l_i) + \sinh(\beta_i l_i)] + \frac{\beta_i^2 (MD)_i}{\rho_i}[\cos(\beta_i l_i) - \cosh(\beta_i l_i)]$

$$F_{21} = -\cos(\beta_i l_i) - \cosh(\beta_i l_i) + \frac{\beta_i M_{Li}}{\rho_i}[\sin(\beta_i l_i) - \sinh(\beta_i l_i)] + \frac{\beta_i^2 (MD)_i}{\rho_i}[\cos(\beta_i l_i) - \cosh(\beta_i l_i)]$$
$$F_{22} = \sin(\beta_i l_i) - \sinh(\beta_i l_i) + \frac{\beta_i M_i}{\rho_i}[\cos(\beta_i l_i) - \cosh(\beta_i l_i)] - \frac{\beta_i^2 (MD)_i}{\rho_i}[\sin(\beta_i l_i) + \sinh(\beta_i l_i)]$$

The nontrivial solutions to the set of equations (10) can be obtained by setting the determinant to zero, therefore

$$\det\begin{bmatrix} F_{11} & F_{12} \\ F_{21} & F_{22} \end{bmatrix} = 0 \tag{11}$$

Computing the determinant (11) yields

$$\begin{aligned} &1 + \cos(\beta_i l_i)\cosh(\beta_i l_i) - \frac{\beta_i M_{Li}}{\rho_i}[\sin(\beta_i l_i)\cosh(\beta_i l_i) - \cos(\beta_i l_i)\sinh(\beta_i l_i)] \\ &- \frac{\beta_i^3 J_{Li}}{\rho_i}[\sin(\beta_i l_i)\cosh(\beta_i l_i) + \cos(\beta_i l_i)\sinh(\beta_i l_i)] + \frac{\beta_i^4 M_{Li} J_{Li}}{\rho_i^2}[1 - \cos(\beta_i l_i)\cosh(\beta_i l_i)] \\ &- \frac{(MD)_i^2 \beta_i^4}{\rho_i^2}[1 - \cos(\beta_i l_i)\cosh(\beta_i l_i)] - \frac{2(MD)_i \beta_i^2}{\rho_i}\sin(\beta_i l_i)\sinh(\beta_i l_i) = 0 \end{aligned} \tag{12}$$

The frequency equation (12) is a transcendental equation with infinite positive solutions, i.e., $\beta_{i1}, \beta_{i2}, \beta_{i3}, \ldots$ which are eigenvalues of the link i a multilink flexible robot arm. The eigenfunction corresponding to the *jth* eigenfunction β_{ij} and its derivatives with respect to x_i can be expressed in forms adapted from (Zhupping [4]):

$$\phi_{ij}(x_i) = C_{1ij}\{\cos(\beta_{ij}x_i) - \cosh(\beta_{ij}x_i) + \xi_{ij}[\sin(\beta_{ij}x_i) - \sinh(\beta_{ij}x_i)]\} = C_{1ij}\bar{\phi}_{ij}(x_i) \tag{13}$$
$$\phi'_{ij}(x_i) = C_{1ij}\beta_{ij}\{-\sin(\beta_{ij}x_i) - \sinh(\beta_{ij}x_i) + \xi_{ij}[\cos(\beta_{ij}x_i) - \cosh(\beta_{ij}x_i)]\} = C_{1ij}\beta_{ij}\bar{\phi}'_{ij}(x_i) \tag{14}$$
$$\phi''_{ij}(x_i) = C_{1ij}\beta_{ij}^2\{-\cos(\beta_{ij}x_i) - \cosh(\beta_{ij}x_i) + \xi_{ij}[-\sin(\beta_{ij}x_i) - \sinh(\beta_{ij}x_i)]\} = -C_{1ij}\beta_{ij}^2\bar{\phi}''_{ij}(x_i) \tag{15}$$
$$\phi'''_{ij}(x_i) = C_{1ij}\beta_{ij}^3\{\sin(\beta_{ij}x_i) - \sinh(\beta_{ij}x_i) + \xi_{ij}[-\cos(\beta_{ij}x_i) - \cosh(\beta_{ij}x_i)]\} = C_{1ij}\beta_{ij}^3\bar{\phi}'''_{ij}(x_i) \tag{16}$$
$$\phi^{iv}_{ij}(x_i) = C_{1ij}\beta_{ij}^4\{\cos(\beta_{ij}x_i) - \cosh(\beta_{ij}x_i) + \xi_{ij}[\sin(\beta_{ij}x_i) - \sinh(\beta_{ij}x_i)]\} = C_{1ij}\beta_{ij}^4\bar{\phi}^{iv}_{ij}(x_i) \tag{17}$$
$$\xi_{ij} = \frac{C_{2ij}}{C_{1ij}} \tag{18}$$

Expression (18) can be derived from (10), i.e.,

$$\xi_{ij} = \frac{C_{2ij}}{C_{1ij}} = -\frac{F_{11j}}{F_{12j}} \quad \text{or} \quad \xi_{ij} = \frac{C_{2ij}}{C_{1ij}} = -\frac{F_{21j}}{F_{22j}} \tag{19}$$

MODE SHAPE NORMALIZATION AND STIFFNESS MATRIX DIAGONALIZATION

The coefficient C_{1ij} of the eigenfunction in formula (13) through (19) is an arbitrary nonzero scale factor that needs to be normalized to obtain unique values (Felippa [16]) which should reduce the amount of nonzero terms in the dynamic equations. Normalized mode shapes and stiffness matrix are tightly connected and both are derived from equations that are equivalent to each other as it will be shown in this section.
Let $\phi_{ij}(x_i)$ and $\phi_{ik}(x_i)$ be two distinct eigenfunctions of link i. Both must satisfy equation (6), i.e.,

$$\frac{d^4\phi_{ij}(x_i)}{dx_i^4} = \frac{\omega_{ij}^2\rho_i}{(EI)_i}\phi_{ij}(x_i) \tag{20}$$
$$\frac{d^4\phi_{ik}(x_i)}{dx_i^4} = \frac{\omega_{ik}^2}{(EI)_i}\phi_{ik}(x_k) \tag{21}$$

Equations (20) and (21) can be solved by generalizing the method used in Erturk & Inman [17] as follows:
Multiplying both sides of (20) by $\phi_{ik}(x_i)$ and integrating over the length of the link i, i.e.,

$$\int_0^{l_i} \phi_{ik}(x_i)\frac{d^4\phi_{ij}(x_i)}{dx_i^4}dx_i = \int_0^{l_i} \frac{\omega_{ij}^2\rho_i}{(EI)_i}\phi_{ik}(x_i)\phi_{ij}(x_i)dx_i \tag{22}$$

Considering link i having a uniform density and integrating the left hand side of equation (22) by parts we get

$$\int_0^{l_i} \frac{d^2\phi_{ij}(x_i)}{dx_i^2}\frac{d^2\phi_{ik}(x_i)}{dx_i^2}dx_i + \left[\frac{d^3\phi_{ij}(x_i)}{dx_i^3}\phi_{ik}(x_i)\right]_0^{l_i} - \left[\frac{d^2\phi_{ij}(x_i)}{dx_i^2}\frac{d\phi_{ik}(x_i)}{dx_i}\right]_0^{l_i} = \frac{\omega_{ij}^2\rho_i}{(EI)_i}\int_0^{l_i}\phi_{ij}(x_i)\phi_{ik}(x_i)dx_i \tag{23}$$

Given that each mode shape satisfies the spatial boundary conditions (7) through (9), when applied to the left hand side of equation (23) we obtain

$$\int_0^{l_i} \frac{d^2\phi_{ij}(x_i)}{dt^2}(EI)_i\frac{d^2\phi_{ik}(x_i)}{dt^2}dx_i = \omega_{ij}^2\left\{\int_0^{l_i}\phi_{ij}(x_i)\rho_i\phi_{ik}(x_i)dx_i + \phi_{ij}(x_i)M_{Li}\phi_{ik}(x_i)\Big|_{x_i=l_i} + \frac{d\phi_{ij}(x_i)}{dx_i}J_{Li}\frac{d\phi_{ik}(x_i)}{dx_i}\Big|_{x_i=l_i} + \frac{d[\phi_{ij}(x_i)(MD)_i\phi_{ik}(x_i)]}{dx_i}\Big|_{x_i=l_i}\right\} \tag{24}$$

Multiplying both sides of equation (21) by $\phi_{ij}(x_i)$ and applying the same procedure we get

$$\int_0^{l_i} \frac{d^2\phi_{ij}(x_i)}{dt^2}(EI)_i\frac{d^2\phi_{ik}(x_i)}{dt^2}dx_i = \omega_{ik}^2\left\{\int_0^{l_i}\phi_{ij}(x_i)\rho_i\phi_{ik}(x_i)dx_i + \phi_{ij}(x_i)M_{Li}\phi_{ik}(x_i)\Big|_{x_i=l_i} + \frac{d\phi_{ij}(x_i)}{dx_i}J_{Li}\frac{d\phi_{ik}(x_i)}{dx_i}\Big|_{x_i=l_i} + \frac{d[\phi_{ij}(x_i)(MD)_i\phi_{ik}(x_i)]}{dx_i}\Big|_{x_i=l_i}\right\} \tag{25}$$

The left hand sides of both equations (24) and (25) being equal, combining both equations we obtain

$$0 = \left(\omega_{ij}^2 - \omega_{ik}^2\right)\left\{\int_0^{l_i}\phi_{ij}(x_i)\rho_i\phi_{ik}(x_i)dx_i + \phi_{ij}(x_i)M_{Li}\phi_{ik}(x_i)\Big|_{x_i=l_i} + \frac{d\phi_{ij}(x_i)}{dx_i}J_{Li}\frac{d\phi_{ik}(x_i)}{dx_i}\Big|_{x_i=l_i} + \frac{d[\phi_{ij}(x_i)(MD)_i\phi_{ik}(x_i)]}{dx_i}\Big|_{x_i=l_i}\right\} \tag{26}$$

From equations (24) and (26) three very important conclusions can be drawn:

1. In (26) if we have two distinct modes $\omega_{ij}^2 \neq \omega_{ik}^2$ the term in the main parenthesis is zero,
2. in (26) for the same mode of vibration $\omega_{ij}^2 = \omega_{ik}^2$ the term in the main parenthesis can be an arbitrary constant P_i playing the role of a scale factor, and
3. when the result of conclusion 2 is applied to equation (24) it reduces to

$$\int_0^{l_i} \frac{d^2\phi_{ij}(x_i)}{dx_i^2}(EI)_i\frac{d^2\phi_{ik}(x_i)}{dx_i^2}dx_i = P_i\omega_{ij}^2 \quad \text{where } j=k \tag{27}$$

The mathematical expressions of the three conclusions are expressions of the orthogonality conditions, i.e.,

$$\int_0^{l_i}\phi_{ij}(x_i)\rho_i\phi_{ik}(x_i)dx_i + \phi_{ij}(x_i)M_{Li}\phi_{ik}(x_i)\Big|_{x_i=l_i} + \frac{d\phi_{ij}(x_i)}{dx_i}J_{Li}\frac{d\phi_{ik}(x_i)}{dx_i}\Big|_{x_i=l_i} + \frac{d[\phi_{ij}(x_i)(MD)_i\phi_{ik}(x_i)]}{dx_i}\Big|_{x_i=l_i} = \begin{cases}0 & j\neq k\\ P_i & j=k\end{cases} \tag{28}$$

$$\int_0^{l_i} \frac{d^2\phi_{ij}(x_i)}{dx_i^2}(EI)_i\frac{d^2\phi_{ik}(x_i)}{dx_i^2}dx_i = \begin{cases}0 & j\neq k\\ P_i\omega_{ij}^2 & j=k\end{cases} \tag{29}$$

Two other equivalent orthogonality conditions can be obtained as follows:
Plugging (29) into (23) yields

$$\omega_{ij}^2\int_0^{l_i}\phi_{ij}(x_i)\rho_i\phi_{ik}(x_i)dx_i + \frac{d^2\phi_{ij}(x_i)}{dx_i^2}(EI)_i\frac{d\phi_{ij}(x_i)}{dx_i}\Big|_{x_i=l_i} - \frac{d^3\phi_{ij}(x_i)}{dx_i^3}(EI)_i\phi_{ij}(x_i)\Big|_{x_i=l_i} = \begin{cases}0 & j\neq k\\ P_i\omega_{ij}^2 & j=k\end{cases} \tag{30}$$

And plugging (22) into (24) yields

$$\int_0^{l_i}\phi_{ik}(x_i)(EI)_i\frac{d^4\phi_{ij}(x_i)}{dx_i^4}dx_i + \frac{d^2\phi_{ij}(x_i)}{dx_i^2}(EI)_i\frac{d\phi_{ij}(x_i)}{dx_i}\Big|_{x_i=l_i} - \frac{d^3\phi_{ij}(x_i)}{dx_i^3}(EI)_i\phi_{ij}(x_i)\Big|_{x_i=l_i} = \begin{cases}0 & j\neq k\\ P_i\omega_{ij}^2 & j=k\end{cases} \tag{31}$$

Equations (28) through (31) are orthogonality condition equations. If one of the four is satisfied, the three others are automatically satisfied. Therefore either one can be used to compute the modal amplitudes that normalize the mode shapes as follows:

plugging (13) through (17) into (28) through (31) and $\beta_{ij}^4 = \frac{\rho_i \omega_{ij}^2}{(EI)_i}$ for $j=k$ yields

$$\int_0^{l_i} \rho_i [\bar{\phi}_{ij}(x_i)]^2 dx_i + M_{Li}[\bar{\phi}_{ij}(l_i)]^2 + J_{Li}\beta_{ij}^2[\bar{\phi}'_{ij}(l_i)]^2 + 2(MD)_i \beta_{ij}\bar{\phi}_{ij}(l_i)\bar{\phi}'_{ij}(l_i) = \frac{P_i}{C_{1ij}^2} \tag{32}$$

$$\int_0^{l_i} \rho_i [\bar{\phi}''_{ij}(x_i)]^2 dx_i = \frac{P_i}{C_{1ij}^2} \tag{33}$$

$$\int_0^{l_i} \rho_i [\bar{\phi}_{ij}(x_i)]^2 dx_1 - \bar{\phi}'_{ij}(l_i)\frac{\rho_i}{\beta_{ij}}\bar{\phi}''_{ij}(l_i) - \bar{\phi}_{ij}(l_i)\frac{\rho_i}{\beta_{ij}}\bar{\phi}'''_{ij}(l_i) = \frac{P_i}{C_{1ij}^2} \tag{34}$$

$$\int_0^{l_i} \rho_i \bar{\phi}_{ij}(x_i)\bar{\phi}_{ij}^{iv}(x_i) dx_1 - \bar{\phi}'_{ij}(l_i)\frac{\rho_i}{\beta_{ij}}\bar{\phi}''_{ij}(l_i) - \bar{\phi}_{ij}(l_i)\frac{\rho_i}{\beta_{ij}}\bar{\phi}'''_{ij}(l_i) = \frac{P_i}{C_{1ij}^2} \tag{35}$$

The modal amplitudes that normalize the mode shapes and diagonalize the stiffness matrix can be derived from any of the four equations (32) through (35). It can be observed that (28) is used to identify the constant elements of the diagonal block of the inertia matrix relative to link deflection while (29) are used to identify the element diagonal stiffness matrix. The stiffness matrix can be derived from the potential energy due to flexible link deflection (Lee & Book [1]):

$$V = \frac{1}{2}\sum_{i=1}^{n}\sum_{j=1}^{m}\sum_{k=1}^{m}\int_0^{l_i}(EI)_i \frac{\partial^2 \phi_{ij}}{\partial x_i^2}\frac{\partial^2 \phi_{ik}}{\partial x_i^2}\delta_{ij}\delta_{ik}dx_i = \frac{1}{2}\sum_{i=1}^{n}\sum_{j=1}^{m}\sum_{k=1}^{m} K_{ijk}\delta_{ij}\delta_{ik} \tag{36}$$

where $$K_{ijk} = \int_0^{l_i} \frac{d^2\phi_{ij}(x_i)}{dx_i^2}(EI)_i \frac{d^2\phi_{ik}(x_i)}{dx_i^2}dx_i \tag{37}$$

Therefore, the stiffness matrix for a two-link flexible robot with respect to the generalised coordinates given by equation (5) can be expressed as

$$K = diag\left(0, 0, P_1\omega_{11}^2, P_1\omega_{12}^2, P_2\omega_{21}^2, P_2\omega_{22}^2\right) \tag{38}$$

It can be observed that the stiffness matrix coefficients are proportional to chosen scale factors. In practice some of the parameters used as scale factors (see Table 1) are: 1, ρ_i, J_{Ti}, J_{Li} and $m_i\rho_i$.

CLOSED-FORM DYNAMIC

The total kinetic energy of the two-link flexible robot can be expressed as

$$T = \frac{1}{2}\sum_{i=1}^{2}\int_0^{l_i}\rho_i \dot{P}_i^T \dot{P}_i dx_i + \frac{1}{2}\sum_{i=1}^{2} J_{hi}\dot{\alpha}_i^T\dot{\alpha}_i + \frac{1}{2}m_{h2}\dot{P}_1^T\dot{P}_1\Big|_{x_1=l_1} + \frac{1}{2}m_p\dot{P}_2^T P_2\Big|_{x_2=l_2} + \frac{1}{2}J_p\dot{\alpha}_p^T\dot{\alpha}_p \tag{39}$$

Expression (39) can be written in terms of Jacobians as follows

$$T = \frac{1}{2}\dot{q}^T\left(\int_0^{l_1}\rho_1 J_{v1}^T J_{v1}dx_1 + \int_0^{l_2}\rho_2 J_{v2}^T J_{v2}dx_2 + J_{h1}J_{\omega1}^T J_{\omega1} + J_{h2}J_{\omega2}^T J_{\omega2} + m_{h2}J_{v1}^T J_{v1}\Big|_{x_1=l_1} + m_p J_{v2}^T J_{v2}\Big|_{x_2=l_2} + J_p J_{\omega p}^T J_{\omega p}\right)\dot{q} = \frac{1}{2}\dot{q}^T M\dot{q} \tag{40}$$

where M is the inertia matrix of the two-link flexible manipulator which elements are expressed as follows

$$M = \begin{bmatrix} M_{11} & M_{12} & M_{13} & M_{14} & M_{15} & M_{16} \\ M_{12} & M_{22} & M_{23} & M_{24} & M_{25} & M_{26} \\ M_{13} & M_{23} & P_1 & 0 & M_{35} & M_{36} \\ M_{14} & M_{24} & 0 & P_1 & M_{45} & M_{46} \\ M_{15} & M_{25} & M_{35} & M_{45} & P_2 & 0 \\ M_{16} & M_{26} & M_{36} & M_{46} & 0 & P_2 \end{bmatrix} \tag{41}$$

$$M_{11} = J_T + 2l_1(MD)_1 + 2(MD)_{01}[(\phi_{11e} - l_1\phi'_{11e})\delta_{11} + (\phi_{12e} - l_1\phi'_{12e})\delta_{12}]S_2 \tag{42}$$

$$M_{12} = J_{L1} + l_1(MD)_1 + (MD)_{01}[(\phi_{11e} - l_1\phi'_{11e})\delta_{11} + (\phi_{12e} - l_1\phi'_{12e})\delta_{12}]S_2 \tag{43}$$

$$M_{13} = \varpi_{11} + \phi'_{11e}J_{L1} + l_1\phi_{11e}M_{L1} + (\phi_{11e} + l_1\phi'_{11e})(MD)_1 - (MD)_{01}[(\phi_{11e}\phi'_{12e} - \phi_{12e}\phi'_{11e})\delta_{12}]S_2 \tag{44}$$

$$M_{14} = \varpi_{12} + \phi'_{12e}J_{L1} + l_1\phi_{12e}M_{L1} + (\phi_{12e} + l_1\phi'_{12e})(MD)_1 - (MD)_{01}[(\phi_{12e}\phi'_{11e} - \phi_{11e}\phi'_{12e})\delta_{11}]S_2 \tag{45}$$

$$M_{15} = \varpi_{21} + \phi'_{21e} J_{L2} + l_2 \phi_{21e} M_{L2} + (v_{21} + M_{L2}\phi_{21e})\{l_1 C_2 + [(\phi_{11e} - l_1\phi'_{11e})\delta_{11} + (\phi_{12e} - l_1\phi'_{12e})\delta_{12}]S_2\} \tag{46}$$

$$M_{16} = \varpi_{22} + \phi'_{22e} J_{L2} + l_2 \phi_{22e} M_{L2} + (v_{22} + M_{L2}\phi_{22e})\{l_1 C_2 + [(\phi_{11e} - l_1\phi'_{11e})\delta_{11} + (\phi_{12e} - l_1\phi'_{12e})\delta_{12}]S_2\} \tag{47}$$

$$M_{22} = J_{L1} \tag{48}$$

$$M_{23} = \phi'_{11e} J_{L1} + \phi_{11e}(MD)_1 - (MD)_{01}\phi_{11e}[\phi'_{11e}\delta_{11} + \phi'_{12e}\delta_{12}]S_2 \tag{49}$$

$$M_{24} = \phi'_{12e} J_{L1} + \phi_{12e}(MD)_1 - (MD)_{01}\phi_{12e}[\phi'_{11e}\delta_{11} + \phi'_{12e}\delta_{12}]S_2 \tag{50}$$

$$M_{25} = \varpi_{21} + \phi'_{21e} J_{L2} + l_2\phi_{21e} M_{L2} \tag{51}$$

$$M_{26} = \varpi_{22} + \phi'_{22e} J_{L2} + l_2\phi_{22e} M_{L2} \tag{52}$$

$$M_{33} = \int_0^{l_1} \rho_1 [\phi_{11}(x_1)]^2 dx_1 + M_{L1}\phi_{11e}^2 + J_{L1}\phi'^2_{11e} + 2(MD)_1\phi_{11e}\phi'_{11e} = P_1 \tag{53}$$

$$M_{34} = M_{43} = \int_0^{l_1} \rho_1 \phi_{11}(x_1)\phi_{12}(x_1) dx_1 + M_{L1}\phi_{11e}\phi_{12e} + J_{L1}\phi'_{11e}\phi'_{12e} + (MD)_1(\phi_{11e}\phi'_{12e} + \phi'_{11e}\phi_{12e}) = 0 \tag{54}$$

$$M_{35} = \phi'_{11e}(\varpi_{21} + \phi'_{21e}J_{L2} + l_2\phi_{21e}M_{L2}) + \phi_{11e}(v_{21} + m_p\phi_{21e})C_2 - \phi_{11e}(v_{21} + m_p\phi_{21e})(\phi'_{11e}\delta_{11} + \phi'_{12e}\delta_{12})S_2 \tag{55}$$

$$M_{36} = \phi'_{11e}(\varpi_{22} + \phi'_{22e}J_{L2} + l_2\phi_{22e}M_{L2}) + \phi_{11e}(v_{22} + m_p\phi_{22e})C_2 - \phi_{11e}(v_{22} + m_p\phi_{22e})(\phi'_{11e}\delta_{11} + \phi'_{12e}\delta_{12})S_2 \tag{56}$$

$$M_{44} = \int_0^{l_1} \rho_1 [\phi_{12}(x_1)]^2 dx_1 + M_{L1}\phi_{12e}^2 + J_{L1}\phi'^2_{12e} + 2(MD)_1\phi_{12e}\phi'_{12e} = P_1 \tag{57}$$

$$M_{45} = \phi'_{12e}(\varpi_{21} + \phi'_{21e}J_{L2} + l_2\phi_{21e}M_{L2}) + \phi_{12e}(v_{21} + m_p\phi_{21e})C_2 - \phi_{12e}(v_{21} + M_{L2}\phi_{21e})(\phi'_{11e}\delta_{11} + \phi'_{12e}\delta_{12})S_2 \tag{58}$$

$$M_{46} = \phi'_{12e}(\varpi_{22} + \phi'_{22e}J_{L2} + l_2\phi_{22e}M_{L2}) + \phi_{12e}(v_{22} + m_p\phi_{22e})C_2 - \phi_{12e}(v_{22} + M_{L2}\phi_{22e})(\phi'_{11e}\delta_{11} + \phi'_{12e}\delta_{12e})S_2 \tag{59}$$

$$M_{55} = \int_0^{l_2} \rho_2 [\phi_{21}(x_2)]^2 dx_2 + M_{L2}\phi_{21e}^2 + J_{L2}\phi'^2_{21e} + 2(MD)_2\phi_{21e}\phi'_{21e} = P_2 \tag{60}$$

$$M_{56} = M_{65} = \int_0^{l_2} \rho_2 \phi_{21}(x_2)\phi_{22}(x_2) dx_2 + M_{L2}\phi_{21e}\phi_{22e} + J_{L2}\phi'_{21e}\phi'_{22e} + (MD)_2(\phi'_{21e}\phi_{22e} + \phi'_{22e}\phi_{21e}) = 0 \tag{61}$$

$$M_{66} = \int_0^{l_1} \rho_2 [\phi_{22}(x_2)]^2 dx_2 + M_{L2}\phi_{22e}^2 + J_{L2}\phi'^2_{22e} + 2(MD)_2\phi_{22e}\phi'_{22e} = P_2 \tag{62}$$

$$(MD)_{01} = (MD)_1\big|_{\theta_2 = 0} \tag{63}$$

The closed-form is obtained through Lagrange-Euler equations expressed as

$$\frac{d}{dt}\frac{\partial L}{\partial \dot{q}_i} - \frac{\partial L}{\partial q_i} = f_i \qquad \text{and} \qquad L = T - V \tag{64}$$

where f_i is the vector of generalized forces.
The closed form dynamic equation can be written in a compact form as [De Lucas & siciliano]

$$M(q)\ddot{q} + h(q,\dot{q}) + D\dot{q} + Kq = Qu \tag{65}$$

$$D_{ij} = 0.1\sqrt{K_{ij}} \qquad Q = [I_{n\times n} O_{n\times(N-n)}]^T \qquad N = n + \sum_i m_i \tag{66}$$

$$h_i = \sum_{j=1}^{N}\sum_{k=1}^{N} C_{jk}(i)\dot{q}_j\dot{q}_k \qquad C_{jk}(i) = \frac{1}{2}\left\{\frac{\partial M_{ij}}{\partial q_k} + \frac{\partial M_{ik}}{\partial q_j} - \frac{\partial M_{jk}}{\partial q_i}\right\} \tag{67}$$

SIMULATION

The dynamic model is tested using a two- link flexible manipulator with the following physical parameters:

$\rho_1 = \rho_2 = 0.1982$ kg/ m	$m_{h2} = 1$ kg	$E_1 = E_2 = 69$ GPa
$P_1 = P_2 = 0.0991$ kg	$l_1 = l_2 = 0.5$ m	$J_{h1} = J_{h2} = 0.1$ kgm^2
$m_1 = m_2 = 0.0991$ kg	$b_1 = b_2 = 0.00153$ m	$J_p = 0.0005$ kgm^2
$m_p = 0.0991$ kg	$h_{t1} = h_{t2} = 0.048$ m	

Solving the frequency equation (12) the following values for the natural frequencies and angular natural frequencies are obtained:

$f_{11} = 0.4391$ Hz, $\omega_{11} = 2.759$ rad/s $f_{21} = 2.1752$ Hz, $\omega_{21} = 13.667$ rad/s

$f_{12} = 1.9748$ Hz, $\omega_{12} = 12.408$ rad/s

$f_{22} = 15.8508$ Hz, $\omega_{22} = 99.5935$ rad/s

The modal amplitudes can be derived using any one of equations (32) and (33). The results are given in Table 2 below.

Table 2: Contribution of different terms in the modal amplitudes equations (32) and (33)

	A_1	B_1	C_1	D_1	E_1	F_1	H_1	C_{1ij}
Mode 11	0.00133	0.07442	0.10292	0.03253	0.21120	0.21120	0.2099	0.6851
Mode 12	0.00527	0.11895	0.09954	-0.02022	0.18331	0.18331	0.1780	0.7354
Mode 21	0.02354	0.09932	0.00448	0	0.12733	0.12733	0.1038	0.8824
Mode 22	0.07215	0.00082	0.10553	0	0.17849	0.17849	0.1063	0.7453

where

$$A_1 = \int_0^{l_i} \rho_i \left[\bar{\phi}_{ij}(x_i)\right]^2 dx_i \qquad C_1 = J_{Li}\beta_i^2\left[\bar{\phi}'(l_i)\right]^2 \qquad E_1 = A_1 + B_1 + C_1 + D_1 \qquad H_1 = B_1 + C_1 + D_1$$

$$B_1 = M_{Li}\left[\bar{\phi}_{ij}(l_i)\right]^2 \qquad D_1 = 2(MD)_i\beta_i\left[\bar{\phi}_{ij}(l_i)\right]\left[\bar{\phi}'_{ij}(l_i)\right] \qquad F_1 = \int_0^{l_i} \rho_i\left[\phi''_{ij}(x_i)\right]^2 dx_i \qquad C_{1ij} = \sqrt{\frac{m_i}{E_1}} = \sqrt{\frac{m_i}{F_1}}$$

From Table 2 it can be observed that $H_i(\phi_{ij}(x_i), \phi_{ik}(x_i), M_{Li}, J_{Li}, (MD)_i)\big|_{x_i = l_i}$ that is missing in most dynamic models for two-link flexible model (see Table 1) is greater than the contribution of the link mass alone (represented by A_1) in the orthogonality condition expression as it can be observed in Table 2 above.

The mode shape functions can then be expressed as follows:

$$\phi_{11}(x_1) = 0.6851\{\cos(1.1116x_1) - \cosh(1.1116x_1) - 1.0444[\sin(1.1116x_1) - \sinh(1.1116x_1)]\}$$

$$\phi_{12}(x_1) = 0.7354\{\cos(2.3574x_1) - \cosh(2.3574x_1) - 1.9771[\sin(2.3574x_1) - \sinh(2.3574x_1)]\}$$

$$\phi_{21}(x_2) = 0.8824\{\cos(2.4741x_2) - \cosh(2.4741x_2) - 0.8520[\sin(2.4741x_2) - \sinh(2.4741x_2)]\}$$

$$\phi_{22}(x_2) = 0.7453\{\cos(6.6787x_2) - \cosh(6.6787x_2) - 1.0638[\sin(6.6787x_2) - \sinh(6.6787x_2)]\}$$

The mode shapes derived from this set of equations are shown in Figure 3.

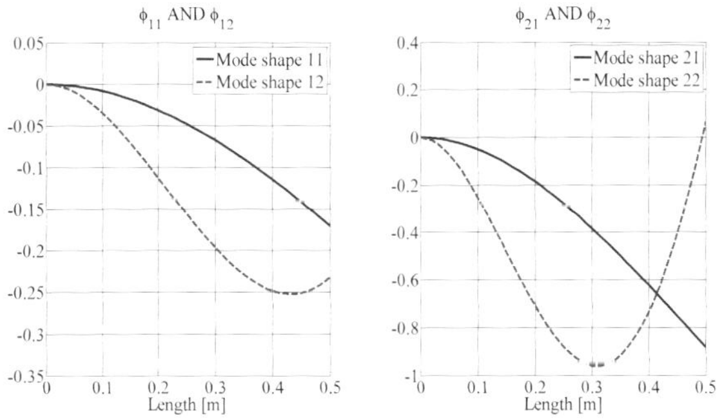

Figure 3: Mode shapes for link 1 and link 2 respectively

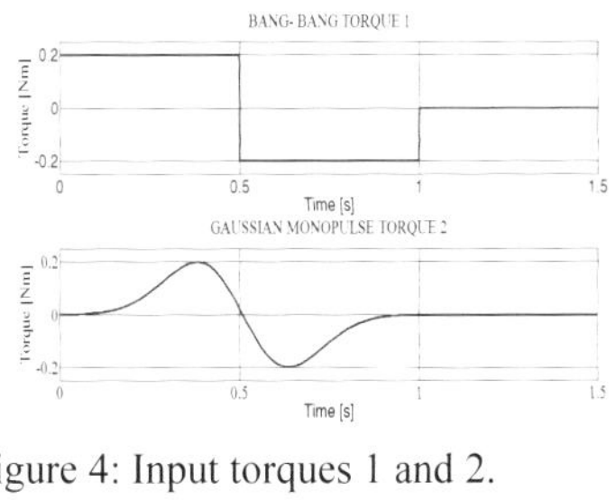

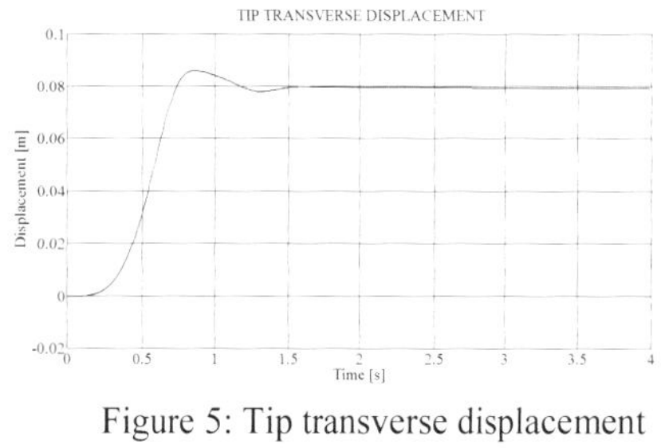

Figure 4: Input torques 1 and 2. Figure 5: Tip transverse displacement

A symmetric bang-bang torque is applied to joint of link 1 while a Gaussian monopulse shaped input is applied to joint of link 2 (Figure 4). The resulting transverse vibration is shown in Figure 5.

CONCLUSION

A parametric study on suitable orthogonality condition for closed-loop dynamic was conducted. Flexible links were modelled as Bernoulli beam. The missing term in most dynamic model for two- link flexible manipulator was derived. Four equivalent companion parametric orthogonality condition equations were derived. The link between orthogonality condition and stiffness matrix diagonalization was established. Simulation was conducted to validate the mathematical model. The two mode shapes selected for each link proved to be orthogonal. The results show that truncated orthogonality condition involving only the cross moment of modes should not result in a diagonal stiffness matrix. Bang-bang and Gaussian monopulse shaped torques applied at hub 1 and hub 2 to actuate the two-link flexible robot. The tip transverse vibration was simulated.

REFERENCES

1 Lee, J.W. and Book, W. J. " Efficient Dynamic Models for Flexible Robots", *Seventh VPI & SV Symposium on Dynamics and Control of Large Structures* (May 8-10, 1989).

2 De Luca, A. and Siciliano, B. " Closed-form dynamic model of planar multilink light weight robots". *IEEE transaction on Systems, man and Cybernetics SMC-21* (4), pp. 826-839, 1991.

3 Zhu, G. " *Modelling and Control of Flexible Robots"*. A Doctoral thesis. National University of Singapore, Singapore, 1997.

4 Zhupping, W. " *Nonlinear Control of Flexible Link Robots and Nonholonomic Robotic Systems"*. A Doctoral thesis. National University of Singapore, Singapore, 2002.

5 Ibrahim, K., Aly, A.A. & Ismael, A.A. " Mode Shape Analysis of a Flexible Robot Arm". *Intrnational Journal of Control, Automation and Systems.* [Online] 1(2). Available from: http://www.researchpub.org/journal/jac/jac.html , 2013. [Accessed: 24 April 2015].

6 Theodore, R.J., & Ghosal, A. " Comparison of the Assumed Modes and Finite Element Models for Flexible Multi-Link Manipulators". *The International Journal of Robotics Research.* 14(2), pp.91-111, 1995).

7 Li, Y., Zhu, M. & Li, Y. " Kinematics of Reconfigurable Flexible-Manipulator Using a Local Product-of-Exponentials Formula". *Proceedings of the 6th World Congress on Intelligent Control and Automation*, June 21-23, 2006, Dalian, China, pp. 9022-9026, 2006.

8 Chen, C.C., Inoue, Y. & Shibata, K. " Impedance Control for a Golf Swing Robot to Emulate Different-Arm-Mass Golfers". *Journal of Dynamic Systems, Measurement, and Control.* 130, 2008.

9 Liu, J., Su, R. & Liu, S. " A Scilab/ Scicos-based Modeling and Simulation Toolbox for Flexible Manipulator". *OSSC-2009 Proceeding of 2009 IEEE International Workshop on Open-source Software for Scientific Computation 5416791*, pp.30-36, 2009.

10 De Bonfim Gripp, J. A., Santos, F. L. M., Bernado, C. R. & Goes, L. C. S. " Modeling and Identification of a Two-Link Flexible Manipulator". *ABCM Symposium Series in Mechatronics.* 5. Section VII-Robotics, pp.1092-1101, 2012.

11 Subhudi, B. & Morris, A.S. " Generalised modelling, simulation and control design for manipulators with flexible links and flexible/ rigid joints". Research report. University of Sheffield, UK., 2001.

12 Subhudi, B. & Morris, A.S. " Dynamic modelling, simulation and control of a manipulator with flexible links and joint". *Robotics and Autonomous Systems.* 41 (2002), pp.257-270, 2002.

13 Morris, A.S. & Subudhi, B. " Use of Singular Perturbation Technique in the Modelling and Control of Flexible Robot Manipulators". In: *Robots Manipulators New Research*: John X. Liu Editor, pp.1-32. New York, USA: Nova Science Publishers, 2005.

14 Talebi, H.A., Patel, R.V. & Khorasani, K. Control and Information Sciences 261, [Lecture Notes] "*Control of Flexible-link Manipulators using Neural Networks"*. Department of Automatic Control and Systems Engineering, University of Sheffield S1 3JD (UK), 2001.

15 Henikl, J., Kemmetmüller, W., Bader & Kugi, A. " Modelling, simulation and identification of a mobile concrete pump", *Mathematical and Computer Modelling of Dynamical Systems*, 21:2, 180-201, DOI:10.1080/13873954.2014.926277 , 2015.

16 Felippa, C. " Introduction to Aerospace Structures" (ASEN3112). Part V: Introduction to Structural Dynamics and Vibrations. Lecture 19 MDOF Dynamic Systems. Lecture 1-slide 1. University of Colorado Boulder. Available from: *www.colorado.edu/engineering/CAS/courses.../IAST.Lect19.Slides.pdf*. [Accessed: 02 October 2015]

17 Erturk, A. & Inman, D. " Piezoelectric Energy Harvesting". John Wiley & Sons, Inc., New York, pp.353-366, 2011.

10th South African Conference on Computational and Applied mechanics

Potchefstroom

3-5 October 2016

The Effects of the Size Variants of Nanocrystalline Materials Produced by Accumulative Roll-Bonding on their Energy, Thermodynamics and Mechanical Properties

P. B. Sob[*1], A. A. Alugongo[2] and T. B. Tengen[3]

[1,2]Departmentof Mechanical Engineering, Faculty of Engineering and Technology, Vaal University of Technology, Vanderbijlpark 1900, Private Bag X021, South Africa

[3]Department of Industrial Engineering and Operations Management, Faculty of Engineering and Technology, Vaal University of Technology, Vanderbijlpark 1900, Private Bag X021, South Africa,[1]baonhe_sob@rocketmail.com, [2]alfayoa@vut.ac.za, [3]thomas@vut.ac.za

Abstract

This paper extends the first law of thermodynamics that deals with the relation between heat, work and energy contents of a system. A general principle or approach is proposed on how to evaluate the internal energy of nanocrystalline materials deformed by Accumulative roll bonding (ARB). The interrelation between thermodynamic, energy and properties of nanomaterials is presented; and the relationship between the variations of these properties and grain size variants of nanomaterials is explained by using 3-D grain deformed by ARB. The results of the current study agree with molecular dynamics (MD) simulation, thermodynamics and experimental results. The results, however, indicate that during grain refinement there exists inverse proportional relationships between grain size variants, internal energy and temperature, which are in good agreement with experimental results. This research is of significance in investigating the effects of size variants of nanomaterials on mechanical property and gives a new approach for studying their energy, thermodynamics and mechanical properties.

Keywords: nanostructure, grain size variants, thermodynamics, grain boundary, internal energy

Introduction

Nanoscience and nanotechnology "span" across (i.e. employ) the whole spectrum of physical sciences to develop the purposeful capabilities of nano-manipulations, nano-structural modifications and miniaturization that were originally proposed by Richard Feynman in his well-known 1959 lecture [1]. For the past years scientists, engineers and technologists have been trying to achieve the objective of Richard Feynman 1959's lecture. Although significant success has been achieved by researchers in what the Nobel Prize winner postulated, the further limitation is that there exists "controversies" when comparing results from different sources, which make the future of nanoscience, nanomaterials and nanotechnology uncertain. One of the breakthroughs in the studies of nanoscience and nanotechnology is that "nanomaterials exhibit novel properties which are different from bulk materials" as reported by different researchers due to the different techniques of producing nanomaterials [1].

The various techniques of producing nanomaterials and their effects on the physical, mechanical and thermodynamic properties have shown that severe plastic deformation (SPD) method is the most widely used method to impose high strains to bulk solids [2]. Temperature has been reported [2] to have significant effects on grain sizes, as well as the strength of the materials during SPD. The phenomenon leading to this conclusion is referred as thermo-mechanical process [2]. Thus, Temperature has a vital role to play during grain refinement since all SPD techniques involve thermo-mechanical processes. Furthermore, Kumar [3] reported that the thermodynamic properties of nanomaterials depend on their vibration energy. Kumar [3] developed a theoretical model to study the effect of temperature on properties.

Kumar [3] compared theoretical results with experimental results. Similarly, Guisbiers [4] also developed a model that predicts the dependence of melting and cohesive energy of nanocrystalline (nc) materials. Guisbiers [4] compared obtained results with different models results. Zhang [5] also demonstrated that, a simple model free from any adjustable parameter can be developed for the melting enthalpy and melting entropy of nanocrystals. The idea was based on Mott's equation for melting enthalpy and melting entropy for non-semiconductor crystals model of melting temperature. The most important characteristic of materials at nanoscales is their high surface-to-volume ratio, which affects their thermodynamic properties and other mechanical properties. It is now well known that the melting temperature of nanoparticles impact nanomaterials thermodynamics and mechanical properties [6] since temperature plays a vital role on material property [2].

It has been reported by several researchers that the thermodynamics behavior of nc materials differs from their corresponding bulk materials [7-15]. The main reason is due to additional interfacial area that causes more free surface energy [7]. It has also been reported that the significant change in thermal stability or phase stability of nc materials is due to the high surface-to-volume ratio of the nc materials, more specifically when the surface possess different interfacial free energies during the deformation of nanostructure materials [8-10].

The deformations of conventional materials to nanostructured materials reduce or refine the sizes of the internal constituent structures such as grains, pores and so on to nanosizes. There are several experimental and theoretical studies that reveal the size dependent properties of nc materials [7]. The results reveal that the thermodynamics and mechanical parameters fall with decreasing size of nanostructure [7]. Most of the findings includes parameters such as melting entropy and melting point, Debye temperature, cohesive energy, diffusion activation energy, amplitude of the thermal vibration, thermal expansion coefficient, specific heat, Young's modulus, and mass density [7]. This behavior of most of these parameters is usually explained to be as a result of high surface to volume ratio of nc materials. The size (or radius) of a 3-D grain can be measured from many different directions based on the orientation of the cutting plane so as to obtain the equivalent radius, which obviously depends on the orientation of the observation plane. This use of equivalent radius assumes that the grains in nanomaterials are perfectly spherical in nature, which is not the case as there are large amount of evidences that show that the grains in nanomaterials are highly elongated. Thus, an approach of measuring the true size of a grain that captures the different shapes of a 3-D grain should concurrently measure the size in, at least, three (3) different directions. These different directions of simultaneously measuring grain size would, definitely, give different values for the radius, and as such are called grain size variants. The high surface/volume ratio of nc materials when obtained from the different grain size variants have not been taken into consideration in most developed models of thermodynamics, energy and mechanical properties [7-15]. The different values of the grain size variants for the same 3-D grain is a clear indicator that the impacts of grain size on thermodynamics, energy and mechanical properties will definitely be different or will produce variable results. These variable results have created a lot of tensions or controversies amongst researchers.

For over the past decades, large volume of experimental data has been collected [7]. However several controversies still exist since the "size effect" is not quite clear because of the variable approaches of measuring grain size [7]. Several models have been developed using thermodynamics and Molecular Dynamics (MD), while considering few parameters and different explanations have been given [7]. This has resulted to more controversies since, for example, there is no common consideration of the "size effects" in nc materials. *The issue "whether it is possible to correlate the variation of properties of nanomaterials" has not received more attention [7].*

The thermodynamics and mechanical properties of nc materials are intrinsically determined by their bonding energy [7]. This is an indication that the change of bonding energy is vital to explain the variation of thermodynamics and mechanical properties of nc materials [7]. It could be observed from the above literature survey that several studies have been done on thermodynamics and mechanical properties of nc materials [2-15], but not much as been done to date on the extension of the first law of thermodynamics which provides us with the relationship between heat, work and energy content of a system to be applicable to this field of materials science. The first law of thermodynamics is the conservation of energy law, which is one of the fundamental laws in physics. The law actually gives the interrelationship between the thermodynamic, energy and properties of nanomaterials and the relationship between the variations of properties. This is an indication that the interrelationship between the thermodynamic, energy and properties of nanomaterials is vital to explain the variation of internal energy, which brings about the instability of nanostructured materials.

Extending the first law of thermodynamics to be applicable to the field of nano-science should give a new approach for studying nanomaterials energy, thermodynamics and mechanical properties which are highly required in the field of nanoscience and nanotechnology. Since this approach would consider the total internal energy involved during the deformation of conventional materials to nanostructure materials, more advanced model of internal energy or nanomaterials stability can be developed.

Theoretical consideration for model derivation

The first law of thermodynamics is a general principle with particular reference to thermal energy (heat) and mechanical energy (work) [1]. When a system completes a thermodynamic cycle the intrinsic energy of the system is the same at the beginning and at the end of the thermodynamic cycle. During the various processes that are involved in the thermodynamic cycle, work is done and heat is supplied (generated). When a system completes a thermodynamics cycle the net heat supplied to the system from its surroundings plus the net-work to the system from its surroundings must be equal zero given as [1]

$$Q_{in} + W_{in} = 0 \quad (1)$$

where W_{in} is the net work input and Q_{in} the net heat supplied. Consider a process in which the intrinsic energy of the system is finally greater than the initial intrinsic energy. The sum of the heat supplied and the net work input increase the intrinsic energy of the system given as [1]

$$dQ_{in} + dW_{in} = dU \tag{2}$$

The first law of thermodynamics is the law on conservation of energy, which is one of the fundamental laws in physics. It, alone, cannot explain everything relating to conversion of thermal to mechanical energy [1]. Another law is required to understand thermal to mechanical energy behavior, which is the second law of thermodynamics, which talks about the differential entropy in a close system [16]. Early in the nineteenth century Lord Kelvin proposed the second law of thermodynamics, which talks about the flow of heat from a hot body to a cold body. This was further investigated by Rudolph Claussius who stated that it is impossible to convert all the energy content of a system to complete work since most of the heat is always released to the surroundings [1]. In fact Lord Kelvin and Rudolph had a microsystems in mind where fluctuations from the average values are insignificant in large time scales [1]. For a materials deformed by ARB cycles, the heat content of the materials continuously increases during monotonic deformation of conventional to nanostructured materials. In fact the high deformation temperature is important for recrystallization [1]. This bring about the definition of differential entropy produced in the material. During the ARB cycles heat flows in the material, and some of the heat is released to the surrounding, therefore the definition of entropy in ARB cycles is given by [1] as

$$dP_S = dS - \frac{dQ_{in}}{T_{ext}} \tag{3}$$

where dS the change of entropy is, dP_s is the differential entropy in a closed system and T_{ext} is the external temperature.

The above expression given in equation (3) involves the change of entropy, the differential entropy in ARB cycles system, the external temperature during ARB and Q_{in} the net heat supplied during ARB cycles. However work is always done during grain refinement. The definition of work done during grain refinement can be defined from first principle. From first principle work is defined as the product of force and the distance $(W = FS)$. The force is the load applied during deformation and the distance is the varying radii (in this case the surface area) during grain refinement. Since, more often, a constant load is applied during deformation of nanostructured materials, as was the situation in the project (so as to achieve constant or variable strain rate depending on the level of material hardness), it follows that $dW=FdS=Fdr^2=2Frdr$, and with $F=\sigma/A=\sigma/S$, it follows that $dW=2\pi\sigma r^3dr$ where $A=S=\pi r^2$. Since the expression for yield stress (σ) contains some constant, it is but logical that those constant will take care of the 2π in the expression for dW. Thus, the dW_{in} the expression given below is obtained

$$dW_{in} = \sigma r^3 dr \tag{4}$$

where dr = change of size, σ = yield stress and r^3 gives the relationship between the different size variants and the equivalent radius during grain refinement by ARB. Substituting equation (4) and equation (3) into equation (2) gives

$$dU = T_{ext}dS - dP_sT_{ext} + \sigma r^3 dr \tag{5}$$

Equation (5) simply gives the interrelation between thermodynamics, energy and mechanical properties of nanocrystalline materials. At any instant during deformation of 3-D grains, the equivalent radius for 3-D grain can be related to the equivalent volume of 3-D grains through $\frac{4}{3}\pi r^3 = V = \frac{4}{3}\pi r_1 r_2 r_3$ this gives

$$r^3 = r_1 r_2 r_3 \tag{6}$$

The different models of nanomaterial's size variants that were derived after experimental observation by [11] based on the different natures of evolutions of the grain size variant during grain refinements are used in this paper given by equation (7-10).

$$dr_1 = M\left(\frac{1}{r_{c1}} - \frac{1}{r_1}\right)dt + r_1^{\frac{1}{2}} D dW(t) - Z r_1 V_1 d(t) \tag{7}$$

where r_{c1}=local critical grain size, Z and D are Constants, $dW(t)$=change of the Wiener process, $V_1 = \tau_1 r_1^2$ defines rate of grain breakage, $M = M_0\left(1 + \frac{CD}{r_1}\right)$ is Grain Boundary (GB) mobility function, $CD = 4(Hm)(h_0)/((k)(T))$, $T_m = T\{ln(m_{01}/m)$ and $M_0 = M_{01}exp\{-T_m(inf)/T\}$= initial grain boundary mobility constant, dW(t) is change in Weiner process.

$$dr_3 = Ratio_1(dr_1) \tag{8}$$

$$dr = -Ordt + IdW(t) \tag{9}$$

where O and I are constants.

$$dr_2 = Ratio_2(dr) \tag{10}$$

Results and discussion

Since most of the parameters presented in the above equations relate to changes in grain sizes (both equivalent radius and grain size variants) and yield stress, thus most of the previous data on the relationship between yield stress and equivalent radius are used and the new data on the evolutions of grain size variant are of presented. Thus the proposed models in this reported are tested with previous data from (nanocrystalline) aluminum sample [16] $dP_s = 1.381023\ln(1000)$, T_{ex}=700°C, $\delta = 96.8$ $M_0' = 0.01 nm^2 s^{-1}$, $m=4$, $r_{c1} = 1.95r$, $CC=12$, $a=0.90$, $D=10^{-4}$, $h_0=0.25nm$, $T_m(\infty)=933.47K$, $CV_0=0.3$, $H_m(\infty)=10.71 KJMol^1$, $\sigma_0'=16.7MPa$, $K_t=1.3$, $\sigma_0=15.40MPa$, $Kd=1301.77 MPa\_nm^{1/2}$, $R=8.31JK^{-1}mol^{-1}$ and $T_r=300K$. The additional data obtained for this work are $O=0.0035$, $I=1.1$, $r_0=100nm$, $Z=0.4$, $Ratio_1=0.81$, $Ratio_2=1.071$, and $\tau_1 = 0.000008$. These data were obtained through curve fitting from the different measures of the sizes. The obtained results are shown in the plots below.

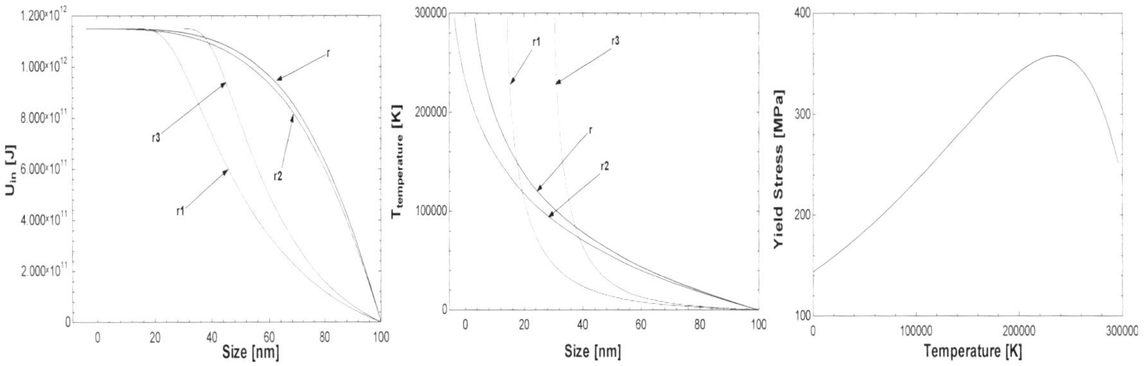

(a) (b) (c)

Figure.1: (a) Internal Energy Versus Grain Size Variants (b) Temperature Versus Grain Size Variants and (c) Yield Stress against Temperature

From most previous experimental or theoretical studies of thermodynamics and mechanical, factors such as structure, shape, processing routes and defects were always considered when deriving the different models to study the thermodynamics and mechanical properties of nanomaterial's. A similar approach has been adopted in this study with the use of size variants and also the first law of thermodynamics. This study has revealed that the conservation of energy in nc materials during grain deformation is not the case as more internal energy is stored in the material as shown in Fig. 1 (a) above and Fig. 2 (a) below. This achievement or revelation should create for more advanced models to be developed from the first law of thermodynamics in the study of nc properties.

It is observed from Fig. 1 (a) that the internal energies as measured from different measures of grain size variants diverge and later converge during the deformation of nanostructure materials. During grain refinement by ARB process dislocation takes place in the grain boundaries of the materials. The dislocation motions that are generated during ARB cycles vary in the different directions of measuring of the materials size variants. And as the ARB cycles increases more dislocation motion are generated in the material which lead to increase in the internal energy of the materials. However the total internal energy in the material diverges and later converged for the different measures of grain sizes or grain size variants.

This is an indication that the total internal energy of the material is not conserved during the deformation of nanostructure materials bringing about nanomaterials instability. This is in line with the first law of conservation of energy which this research has extended to be applicable to the deformation of nanostructure materials. It is therefore shown that the total internal energy is not conserved during deformation of nanostructure materials. It has been revealed as shown in Fig.1 (a) that a decrease in grain size leads to increase in internal energy. It is also observed from Fig.1 (a) that the increase in internal energy varies for the different measurements of the grain size (variants) due to different dislocation motions as functions of the material size variants during materials deformation. From Fig.1 (a) it is observed that the rate of increase of internal energy is higher along r and r_2 when compared to r_1 and r_3 due to higher dislocation and mobility of the grain boundaries along r and r_2. When the materials grain size is around 35nm, the internal energy increases more along r_3 when compared to r, r_2, and r_1 due to higher dislocation and mobility of the grain boundaries along r_3.

The increase in internal energy with decrease in grain size can be explained due to the fact that during grain refinement, the crystalline materials are being "squeezed" due to high compression subjected to the material. The "piezo" (push) effects on the material affect the grain dislocation, grain boundaries and grain mobility. The grain boundaries become larger as the grain size decrease which impacts the internal energy of the material. The mobility of free electrons and atoms also increases which impacts the internal energy of the materials. MD simulations [18] revealed high mobility of atoms when a material is subjected to applied stress when compared with normal bulk material that have not been subjected to any applied stress. The MD simulations results also revealed that the atoms with high mobility when subjected to applied stress are believed to have unique properties due to their small grain sizes. The dislocations that is generated and squeezed closely together result in extremely high internal stress which impacts the internal energy and other mechanical properties of nc materials [18]. A good agreement was obtained between the obtained results in this study when compared to the results of MD simulation and other models of nanomaterials internal energy and size [16-18] presented in the figures below.

It has been revealed as shown in Fig.1 (b) that as the material grain size decreases not only the internal energy in the material increases but also the temperature in the material increases. The increase in temperature during deformation of nanostructure is due to internal heat generated from the material during grain deformation. The internal heats generated from the material tend to increase due to high straining (strain energy) that took place as the material was subjected to several passes by ARB. It must be noted that continuous increase in temperature has a negative impact on nc mechanical properties. It is observed that at extremely high temperature, excessive heating of deformed grain due to adiabatic warming leads to grain growth and nanomaterials start losing its achieved properties as shown in Fig. 1 (c). Several reports on MD simulation have shown the phenomenon of grain growth due to excessive heating at very high temperature [18].

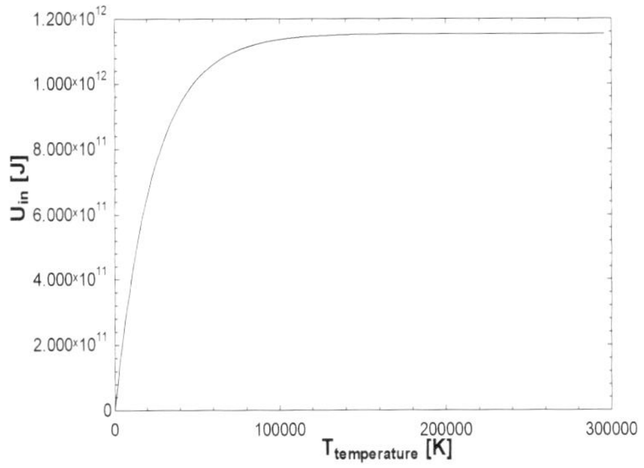

(a)

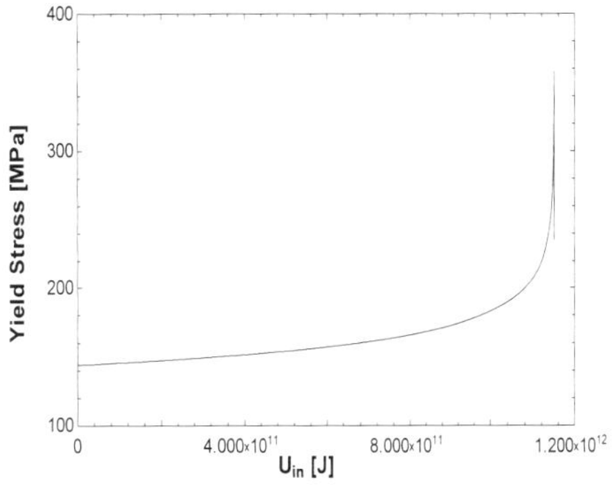

(b)

Figure.2: (a) Internal energy against Temperature and (b) Yield Stress against Internal energy

It is observed from Fig.2 (a) that the increase in temperature lead to increase in internal energy. It is also observed from Fig.2 (b) that during deformation of nc materials, the increase in internal energy is accompanied by increase in material properties (yield stress). Therefore as the internal energy of the material increases the material yield stress also increases. Since it was revealed that a good agreement is obtained between the obtained results in this study when compared to the results of MD simulation and other models of nanomaterial's internal energy and size [18], it is important to present the results from and below are some of the results of MD simulation.

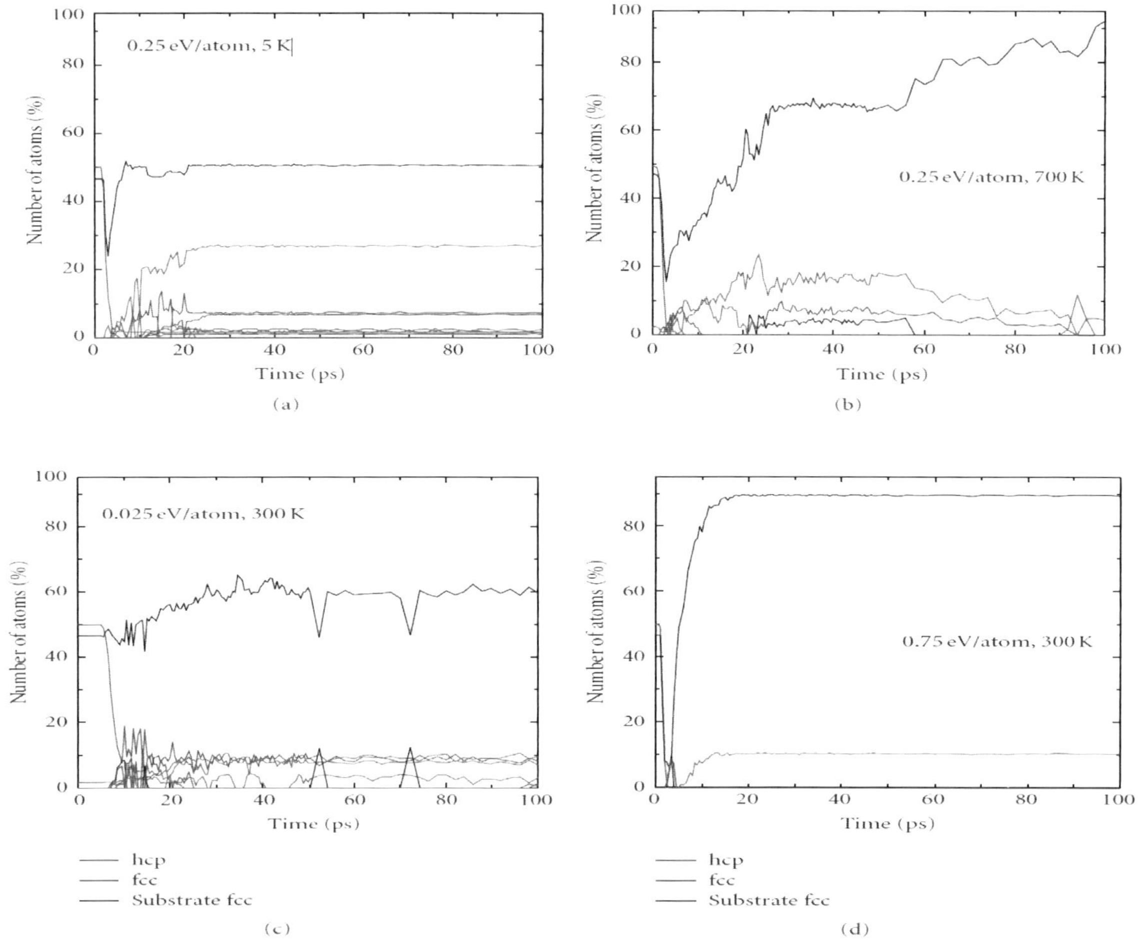

Fig. 3 (a-d) Energy and temperature [16-18]

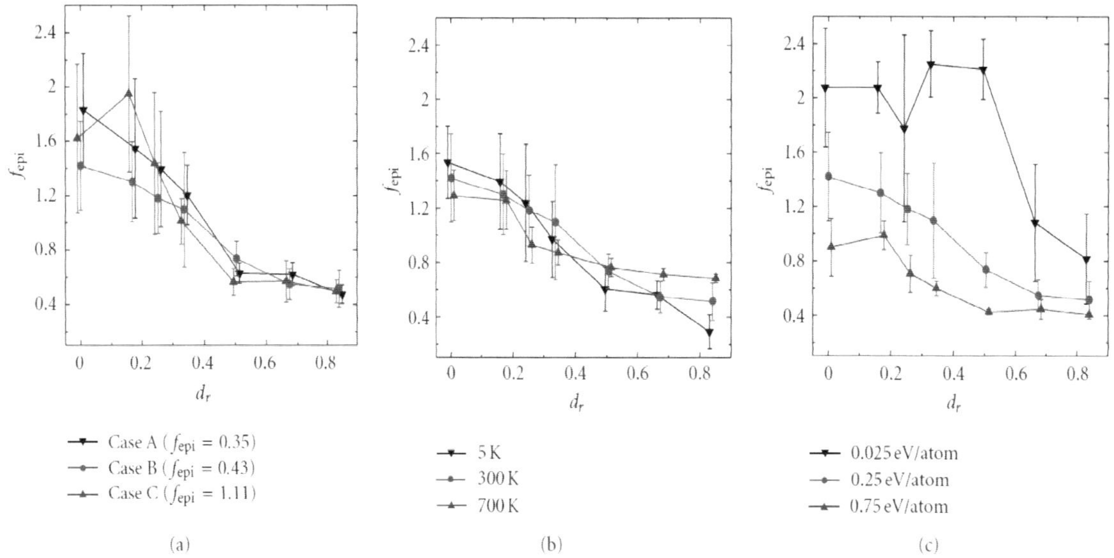

Fig. 4 (a-c) energy and size [16-18]

Conclusions

In summary, we have demonstrated the conservation of energy in nc materials. We have also demonstrated the interrelation between their variables (heat, work and energy) on thermodynamic and mechanical properties. It was observed that the total internal energy is not conserved during grain refinement, which justifies the instabilities of nanomaterials and subsequent high tendencies towards grain growth. It was also revealed that decrease in grain size lead to increase in internal energy. It was however observed that the increase in internal energy varies in different ways to the different measurements of the grain size due to different dislocation activities during deformation.

It was further observed that the increase in temperature led to increase in internal energy. It was however observed that during deformation of nc materials, the increase in internal energy was accompanied by increase in material properties (e.g. yield stress). It was also observed that continuous increase in temperature led to decrease in material properties due adiabatic warming that led to grain growth. It can be concluded that by utilizing the first law of thermodynamics in the study of nc material properties can be vital in future study of nc materials properties.

Acknowledgement

This material is based on the work which is supported financially by the National Research Foundation (NRF) and Vaal University of Technology (VUT).

Reference

1 Ali Mansoori. G., "Principles of nanotechnology: Molecular-Based Study of Condensed Matter in Small Systems", *World Scientific Publishing Co. Pte. Ltd.5 TohTuch Link, Singapore 596224, 2004.*

2 Estrin, Y. & Vinogradov, A., "Extreme grain refinement by severe plastic deformation", A *Wealth of Challenging Science: Act materials, 61: 782-817, 2013.*

3 Kumar, R., and Sharma, G., & Kumar, M., "Effect of size and shape on the vibrational at hermodynamics properties of nanomaterials", *Journal of thermodynamics Vol. pp 5, 2013.*

4 Guisbier, G., "Size dependent materials properties towards a universal equation", *Nanoscale Research Letters, Vol. 5, No.7, pp. 1132-1136, 2010.*

5 Zhang, Z. and Lii, X. X. & Jiang, Q., "Finite size effect on melting enthalpy and melting entropy of nanocrystals", *Physical B Vol. 270, No. 3-4, pp. 249-254, 1999.*

6 Xiong, S., and Qi, W., and Cheng, Y., & Huang, B., & Wang, W., & Li, Y., "Universal relation for size 75 dependent thermodynamic properties of metallic nanoparticles", *Physical Chemistry Chemical Physics, Vol. 13, No. 22, pp. 10652-10660, 2011.*

7 Lu, M. H. & Jiang, Q., "Size-Dependent Surface Energies of Nanocrystals", *J. Phys. Chem. B* 2004, *108,* 5617-5619, 2004.

8 Mchale, J. M. and Auroux, A., and Perotta, A. J., & Navrotsky, *A. Science1997, 277, 788.*

9 *Jacobs, K. and Zaziski, D. and Scher, E. C. and Herhold, A. B. & Alivisatos,A. P., Science 2001, 293, 1803.*

10 Xiaohua, Y. & and Zhaolin, Z., "The effects of the size of nanocrystalline materials on their thermodynamic and mechanical properties" , Nanoscale Research Letters 2014, 9:516

11 34. Sob. P. B. Alugongo, A. A. Tengen, T. B. "Modelling Strain Rate Sensitive Nanomaterials Mechanical Properties: the effect of varying definitions, M. Tech. Dissertation. Vaal University of Technology, 2016

12 Sob, P. B. & Tengen, T. B and Alugongo, A. A., "The effect of deformation activation volume, strain rate sensitivity and processing temperature of grain size variants", *International Journal of Mechanical, Aerospace, Industrial, Mechatronics and Manufacturing Engineering Vol:9, No:11, 2015.*

13 Sob, P. B & Tengen, T. B and Alugongo, A. A., "Determination of strain rate sensitivity (SRS) for grain size variants on nanocrystalline materials produced by ARB and equal channel angular pressing (ECAP)", *International Journal of Chemical, Molecular, Nuclear, Materials and Metallurgical Engineering Vol:9, No:12, 2015.*

14 Tengen, T. B., "*Analysis of Characteristic of Random Microstructures of Nanomaterials*", PhD. Thesis. Witwatersrand Johannesburg. 2008.

15 Yamakov V, Wolf D, Salazar M, Phillpot SR, Gleiter H. Length-scale Dects in the nucleation of extended dislocations in nanocrystalline Al by molecular-dynamics simulation. Acta Mater 2001;49:2713e22.

16 Chang, Q., and Sun, "Size dependence of nanostructures: Impact of bond order deficiency", Progress in Solid State Chemistry 35 (2007).

17 Haslam, A. J. & Moldovan, D. & Phillpot, S. R. & Wolf, D. and Gleiter, "Combined atomistic and mesoscale simulation of grain growth in nanocrystalline thin films", *Computational Material Science 23 (2002) 15-32*

18 Jimenez-Saez, J. C. & Perez-Martın, A. M. C & and Jimenez-Rodrıguez, J. J. "Influence of Energy and Temperature in Cluster Coalescence Induced by Deposition" Advances in Condensed Matter Physics, 2012

10th South African Conference on Computational and Applied Mechanics
Potchefstroom
3 – 5 October 2016

A methodology for the integrated system simulation of the heat transfer and combustion in a coal-fired boiler furnace

W.A. van der Meer[1*], Prof. P.G. Rousseau[2], Prof. L. Jestin[3].

[1] University of Cape Town; Mechanical Engineering; vmrwil011@uct.ac.za

[2] University of Cape Town; Mechanical Engineering; pieter.rousseau@uct.ac.za

[3] University of Cape Town; Eskom Power Plant Engineering Institute; louis.jestin@uct.ac.za

ABSTRACT

Improving availability, efficiency and safety while reducing emissions in coal-fired power plants necessitate an integrated approach. Having an integrated system process model where all of the relevant subsystems are coupled and the interdependencies can be ascertained is therefore invaluable. Such a model can be constructed by making use of a one-dimensional thermal-fluid network approach to model the water to steam cycle and the coal and air to flue gas cycle. Modelling of the furnace, however, requires a three-dimensional approach, especially with regards to modelling the thermal radiation heat transfer. This paper proposes a methodology which combines the zonal method to model the radiation heat transfer with a one-dimensional burnout model for the heat generation, and characteristic flow maps for the advection.

The results obtained for a pilot scale test furnace with a single burner show that a suitable numerical solution can be obtained within reasonable computational time with a discretisation of 400 cells. The predicted temperatures compare well with the expected range for a typical coal burning flame. This demonstrates that combining the multi-dimensional zonal approach, a one-dimensional burnout model and a flow map can be used to model the integrated combustion and heat transfer behaviour in a furnace.

KEYWORDS

Combustion, Zonal method, Radiation heat transfer, One-dimensional network approach.

NOMENCLATURE

A	surface area [m²].
E	blackbody emissive power [W/m²].
du_j	burnout of a parcel of fixed carbon [-].
f_r	fraction radiation heat transfer to particles [-].
$h_{g,e}$	enthalpy of the gas exiting [kJ/kg].
$h_{g,i}$	enthalpy of the gas entering [kJ/kg].
$h_{p,e}$	enthalpy of the particles exiting [kJ/kg].
$h_{p,i}$	enthalpy of the particles entering [kJ/kg].
FC_{in}	fraction fixed carbon in [-].
FC_{out}	fraction fixed carbon out [-].
$\overline{G_iG_j}$	total exchange area from volume *i* to volume *j* [m²].
$\overline{G_iS_j}$	total exchange area from volume *i* to surface *j* [m²].
H	enthalpy [kW].
$K_{diff,j0}$	diffusional reaction rate constant for the *j*th size fraction [g/m²•s•atm].
K_s	surface reaction rate constant [g/m²•s•atm].
$\dot{m}_{in}$	total mass flow rate of gas and particles [kg/s].
$\dot{m}_g$	total mass flow rate of gas [kg/s].
$\dot{m}_p$	total mass flow rate of particles [kg/s].
$P_g(U)$	partial pressure of oxygen in the free stream [atm].
$\dot{Q}_{conv}$	convection heat transfer [kW].
$\dot{Q}_{g,i}$	net radiation heat transfer from volume *i* [kW].
$\dot{Q}_{s,i}$	net radiation heat transfer from surface *i* [kW].
$\dot{Q}_{source,g}$	heat source to the gas [kW].
$\dot{Q}_{source,p}$	heat source to the particles [kW].
$\dot{Q}_{rad,g}$	radiation heat transfer to the gas [kW].
$\overline{S_iG_j}$	total exchange area from surface *i* to volume *j* [m²].
$\overline{S_iS_j}$	total exchange area from surface *i* to surface *j* [m²].
S_{j0}	surface area of the *j*th size fraction immediately after devolatilisation per unit weight of residual char [m²].

T_g temperature of the gas [K].

t time [s].

u_j weight of unburnt fixed carbon in the *j*th size fraction [g].

V volume [m^3].

VM_0 fraction total volatile matter in the coal [-].

VM_x fraction of total volatile matter evolved [-].

w_j weight of raw coal in the *j*th size fraction per unit weight of raw coal [g].

x distance [m].

x_f distance to complete evolution [m].

β extinction coefficient [m^{-1}].

ε wall emissivity [-].

$\emptyset_c$ mechanism factor [-].

INTRODUCTION

The power utility industry is under continuous pressure to lower electricity production costs while at the same time complying with new industry regulations. This requires a balance between the following objectives in order for the power utility industry to be successful and sustainable (EPRI [1]):

- Maintaining capacity.
- Improving efficiency.
- Reducing emissions.
- Preserving reliability and availability.
- Ensuring safety.
- Minimizing production cost.

The following furnace and boiler operating parameters directly influence the performance of a pulverised coal-fired power station and therefore impact the objectives stated above (EPRI [2] and Bhatt [3]):

- Boiler exit temperature.
- Excess air.
- Superheater and Reheater temperature.
- Furnace exit gas temperature (FEGT).
- Coal quality.
- Slagging and fouling characteristics.
- Feedwater temperature.

It can be shown that many of these operating variables are interdependent. Due to the interdependency of these parameters, having access to an integrated model that captures all of the significant phenomena and processes will be invaluable. This can be done with a systems level process model which integrates the coal and air to gas cycle with the water to steam cycle.

The furnace is a very complex zone within a coal-fired boiler because it is where combustion occurs and where radiation heat transfer is the dominant mode of heat transfer to the water walls. The operation in this zone directly influences at least four of the seven operating parameters mentioned above.

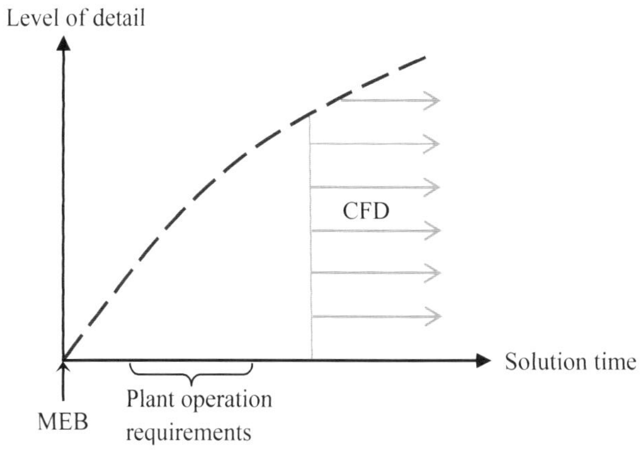

Figure 1. The current technology state paradigm for solution time and level of detail.

A simple overall mass-and-energy balance (MEB) for the furnace will not provide answers regarding the FEGT profile or combustion efficiency (carbon burnout) which are dependent on the coal quality, excess air, mill selection, burner operation or furnace heat absorption rate. A tool that can provide such solutions is computational fluid dynamics (CFD). However, as illustrated in Figure 1, a limitation of CFD is that the solution times are too long for decisions to be made during operation based on on-line process condition monitoring. Currently CFD is more useful in the design phase or in assisting with long-term technical decision making.

The hypothesis for this study is therefore that a suitable systems level process model of a coal-fired boiler furnace can be developed with sufficient detail of the operational parameters and variables, so that it can be used to investigate anomalies and provide guidance in the optimisation of the plant operation. This means that the important parameters that should be taken into consideration are the flame behaviour, the furnace temperature distribution (especially focussing on the FEGT), excess air variations and changes in coal quality.

The specific objective of this paper is to present the first steps in developing a code implementing a zonal approach for a generic furnace enclosure combined with the heat source terms to emulate the flame behaviour. This is combined into a single computer code which integrates the mass flow field, heat source terms, radiation characteristics and exchange areas. The verification was done to test the integrity of the code and the initial results of the predicted gas temperature profiles as well as the numerical performance will be presented.

THEORETICAL BACKGROUND

Hottel and Cohen [4] introduced a method which can be used to model radiation heat transfer in gas-filled enclosures with a non-isothermal temperature distribution. This method was widely applied since the 1960s in studies involving combustion processes in a furnace because it requires less computational resources than the methods traditionally applied in CFD codes. In the zonal method the furnace enclosure is subdivided into a number of uniform property control volumes (also referred to as cells) with distinct surface areas, each containing an isothermal participating medium. The heat transfer between each of the cells (neighbours and non-neighbours) is then accounted for by what is essentially a multi-dimensional network approach.

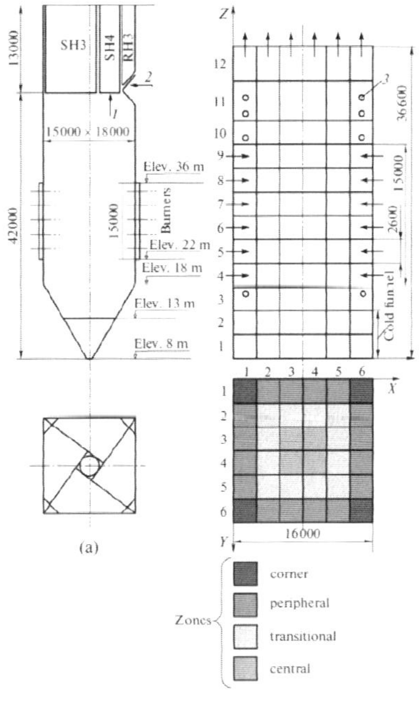

Figure 2. A coal-fired power station and its zonal representation (Karasina et. al. [5]).

An application of the zonal method can be found in the collaboration of researchers at the Israel Electrical Company and the All-Russia Research Institute (VTI) to develop a code named FURNACE (see Figure 2). It was first introduced by Karasina et. al. [6] and further development can be seen in Abryutin et. al. [7] and Karasina et. al. [8]. The limitation of this code is that the temperature distribution is determined based on empirical information. This is because the zonal model calculates only the radiation heat transfer and requires either the temperature or heat sources in the volumes as an input.

In the zonal method where gas and particles are passing through each volume the energy balance for the gas will be:

$$\dot{m}_g\left(h_{g,i} - h_{g,e}\right) + \dot{Q}_{source,g} + \dot{Q}_{conv} - \dot{Q}_{rad,g} = 0 \quad \ldots (1)$$

and for the particles:

$$\dot{m}_p\left(h_{p,i} - h_{p,e}\right) + \dot{Q}_{source,p} - \dot{Q}_{conv} - \dot{Q}_{rad,p} = 0 \quad \ldots (2)$$

$\dot{m}_g$ and $\dot{m}_p$ is the total mass flow rate of gas and particles into the volume, $h_{g,i}$ and $h_{g,e}$ is the enthalpy of the gas entering and exiting the volume, $h_{p,i}$ and $h_{p,e}$ is the enthalpy of the particles entering and exiting the volume, $\dot{Q}_{source,g}$ the source term where heat is added to the gas due to combustion, $\dot{Q}_{conv}$ the heat transfer due to convection from the particles to the gas, $\dot{Q}_{rad,g}$ and $\dot{Q}_{rad,p}$ are the heat transfer due to radiation from the gas and the particles respectively, $\dot{Q}_{source,p}$ the source term where heat is added to the particles due to combustion.

The zonal method requires direct exchange areas as described in Monnaemang [9]. The direct exchange areas are converted to total exchange areas to include the surface and volume properties and then used to solve the radiation heat transfer network (Modest [10]). For the surfaces we have:

$$\dot{Q}_{s,i} = \sum_{j=1}^{N} \overline{S_i S_j} E_{s,j} + \sum_{j=1}^{K} \overline{S_i G_j} E_{g,j} - A_i \varepsilon_i E_{s,j} \quad \ldots (3)$$

and for the volumes:

$$\dot{Q}_{g,i} = \sum_{j=1}^{N} \overline{G_i S_j} E_{s,j} + \sum_{j=1}^{K} \overline{G_i G_j} E_{g,j} - 4 V_i \beta_i E_{g,j} \quad \ldots (4)$$

where $\dot{Q}_{s,i}$ is the net radiation heat transfer from surface i, $\dot{Q}_{g,i}$ the net radiation heat transfer from volume i, $\overline{S_i S_j}$ the total exchange area from surface i to surface j, $\overline{S_i G_j}$ the total exchange area from surface i to volume j, $\overline{G_i S_j}$ the total exchange area from volume i to surface j and $\overline{G_i G_j}$ the total exchange area from volume i to volume j. A is the surface area, ε the wall emissivity, V the volume, E the blackbody emissive power and β the extinction coefficient. All subscripts with s refer to a surface and g to a volume. The radiation heat transfer from the gas in each volume calculated with Equation (4) in the zonal model is divided between the heat transferred from the gas ($\dot{Q}_{rad,g}$) and from the particles ($\dot{Q}_{rad,p}$) with the aid of the factor f_r, i.e.:

$$\dot{Q}_{g,i} = f_r \dot{Q}_{rad,p} + (1 - f_r)\dot{Q}_{rad,g} \quad \ldots (5)$$

Previously empirical data or a characteristic flame model fitted from CFD simulation was used to emulate the heat sources due to combustion of the coal leaving a burner. A more fundamental approach to calculate the burnout was developed by the British Coal Utilization Research Association (BCURA) (Field et. al. [11]) and combined with a radiation heat transfer method used in well-stirred reactors. In this model, the burnout of a parcel of fixed carbon, du_j, when it travels for time dt is given by:

$$\frac{du_j}{dt} = \frac{\left(u_j/w_j\right)^{2/3} S_{j0} P_g(U)}{\frac{\left(u_j/w_j\right)^{1/3}}{K_{diff,j0}} + \frac{1}{K_S}} \quad \ldots (6)$$

where u_j is the weight of unburnt fixed carbon in the jth size fraction, w_j is the weight of raw coal in the jth size fraction per unit weight of raw coal, S_{j0} is the surface area of the jth size fraction immediately after devolatilisation, $P_g(U)$ the partial pressure of oxygen in the free stream, $K_{diff,j0}$ is the diffusional reaction rate constant for the jth size fraction and K_s is the surface reaction rate constant.

The assumption for the rate of devolatilisation used in the original model (see Field et. al. [11]) resulted in an unrealistic temperature spike at the inlet of the burnout model. This was improved by Gill [12] by extending the model with the introduction of the progressive evolution of volatiles model, postulating a constant rate of evolution with respect to the distance from the burner, i.e.:

$$VM_x = VM_0 \frac{x}{x_f} \quad \dots (7)$$

where VM_x is the fraction of total volatile matter evolved (burned) at distance x, x_f is the distance to complete evolution and VM_0 is the total volatile matter in the coal.

An energy source due to fixed carbon burnout based on a mechanism factor ($\emptyset_c$) is given in Gill [12] and can be extended to include the energy source due to volatile burnout. After including the coal to air ratio the combustion heat generation attributed to the gas volume is given by:

$$\dot{Q}_{source,g} = 4.184\dot{m}_{in}(1 - f_r)\left(9120\left(1 - \frac{1}{\emptyset_c}\right)(FC_{in} - FC_{out}) - 7900(VM_{in} - VM_{out})\right) \quad \dots (8)$$

and to the particles by:

$$\dot{Q}_{source,p} = 4.184\dot{m}_{in}f_r\left(\left(\frac{11120}{\emptyset_c} - 3220\right)(FC_{in} - FC_{out}) - 7900(VM_{in} - VM_{out})\right) \quad \dots (9)$$

$\dot{m}_{in}$ is the total mass flow rate of gas and particles into the volume, FC_{in} is the fraction fixed carbon flowing into each volume, FC_{out} is the fraction fixed carbon flowing out of each volume, VM_{in} is the fraction volatile matter flowing into each volume and VM_{out} is the fraction volatile flowing out of each volume.

The burnout model requires as input the mass flow rates of available combustible matter etc. Therefore, a mass flow map must be superimposed for the volumes. The first attempt to combine the zonal method with a mass flow map was by Lowe et. al. [13]. A later attempt in quantifying a mass flow map was proposed by Hesselman [14] when CFD tools became more readily available. The flow paths were characterised with an isothermal simulation of the mass flows in a CFD package using the geometry of an actual wall-fired furnace. Hesselman [14] also used a CFD simulation to characterise the heat sources created by the flames and combined it with the mass flow map to create the furnace code called HotGen (originally known as BWHOT) that uses the zonal method to solve for the radiation heat transfer.

METHODOLOGY

The methodology presented in this paper combines the zonal method with a one-dimensional burnout model together with a superimposed mass flow map using Euler integration with respect to time. The method will follow a network approach similar to that found in one-dimensional network models in order to produce a quick solving furnace model which can be easily integrated into a full plant simulation. The present case study is the rectangular shaped pilot scale test facility (PSCTF) described in Rajoo [15].

The first step is to divide the furnace into smaller rectangular prism shaped finite volumes (also referred to as cells in numerical methods) as shown in Figure 3. The numbering of the cells is similar to the approach used for a structured mesh.

Figure 3. Discretisation of a rectangular shaped furnace.

A mass flow map is required to account for the advection of gas and particles between the cells. This mass flow map consists of a list containing the "From cell number", "To cell number" and "Mass flow rate" values. This can be viewed as a series of one-dimensional mass flow elements, as illustrated in Figure 4a, where each volume is represented by a node and a component represents the cell face. A CFD analysis of the flow pattern within the furnace can be used to pre-characterise the mass flow patterns and then translated into a mass flow map.

It would of course imply significant simplification to assume that a static flow map can be used for a range of different furnace operating conditions. The detail regarding the use of different flow maps for different operating conditions should therefore receive proper attention. However, for purposes of illustrating the integrated modelling methodology here, a static flow map is assumed as a first step in developing the model.

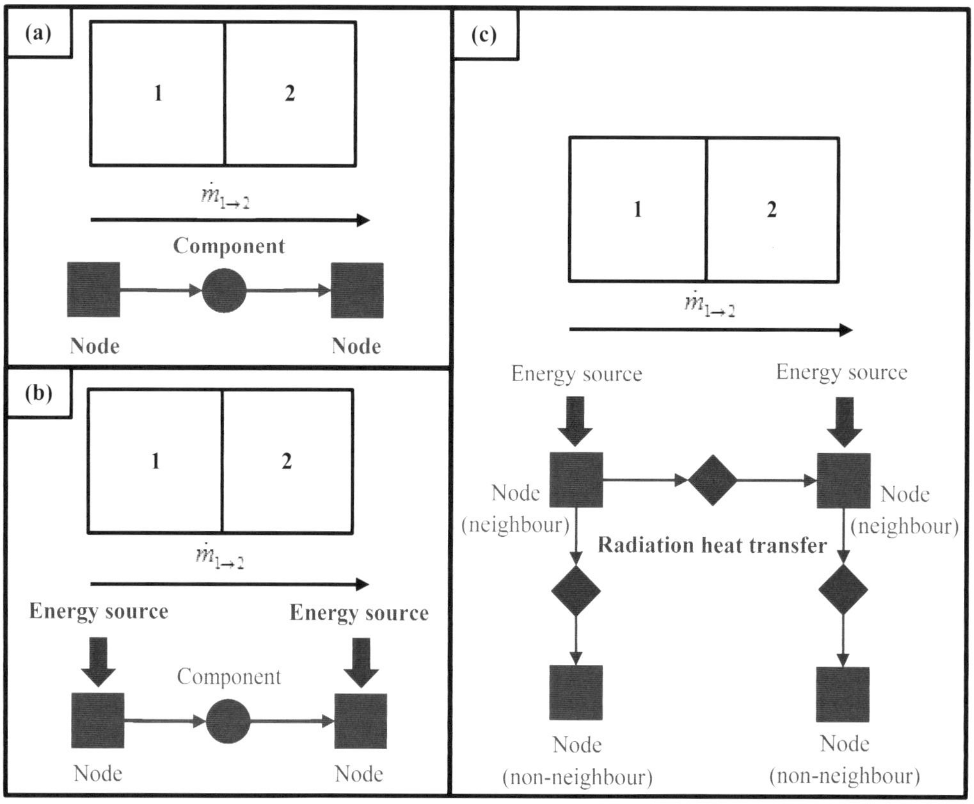

Figure 4. One-dimensional discretised representation of the (a) mass flow between nodes, (b) energy source added to a node, (c) heat transfer between the nodes.

The next step is to calculate the heat sources created due to the burnout of the fixed carbon and volatiles. The heat source is added as an energy increase to each of the nodes as shown in Figure 4b. The model used is the BCURA one-dimensional burnout model. Originally this model was developed to estimate the burnout, oxygen partial pressure, radiation to the walls and temperature along the axial length of a cylindrical combustion chamber (Field et. al. [11]). BCURA applied the model in a single direction but for calculating the heat source in the present code the model is applied for each of the streams flowing out of a node. This effectively extends the application of the BCURA model from calculating the burnout in one dimension to multiple dimensions.

The radiation heat transfer between the volumes and surfaces is a superimposed network of one-dimensional heat transfer elements from each of the nodes to all the other nodes as illustrated in Figure 4c. The radiation heat transfer is calculated based on the zonal method.

MODEL SETUP

The purpose of this paper is to illustrate the viability of solving the radiation heat transfer, combustion and the advection networks in an integrated manner. The purpose here is not to demonstrate the complexities associated with each of the separate models. Therefore, it should be seen as a first step in the development and the following simplifying assumptions were made:

- Gray gas behaviour.
- No recirculation, which results in volumes with no in and out flows (these volumes were assumed to be at the same temperature as the furnace walls).
- Single particle size.
- Constant extinction coefficient throughout the gas volume.
- Gas and wall properties are constant for the current temperature range.
- Only fixed carbon and volatiles participate in the combustion process with 10% O_2 available.
- 90% of the radiation in each volume goes to the gas and the remaining 10% to the particles.

The convergence criteria for the code is set as:

$$\sum_{K}^{k=1}(T^{n}{}_{g,k} - T^{n-1}{}_{g,k}) < 0.0001 \qquad \ldots (10)$$

where $T^{n}{}_{g,k}$ is the temperature for the gas for iteration n and K the total number of volumes.

The mass flow map and inputs for the code is shown in Figure 5 and Table 1. The geometry, coal and gas properties in Table 1 were chosen based on the actual values expected for the PSCTF.

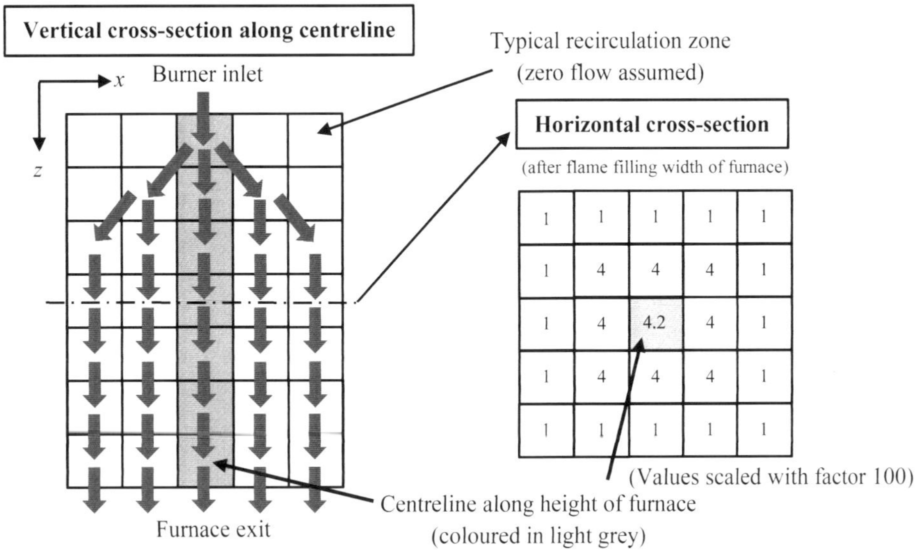

Figure 5. Assumed mass flow map.

Table 1. Model inputs.

Description	Unit	Value	Description	Unit	Value
Length	m	5.76	Coal inlet temperature	K	550
Width	m	1.2	Air inlet temperature	K	550
Height	m	1.2	Coal particle specific heat	kJ/kgK	1.25
Discretised cell size	m	0.24	Gas specific heat	kJ/kgK	1.1
Wall temperature	K	1000	Mass fraction fixed carbon	-	0.3
Wall emissivity	-	0.7	Mass fraction volatiles	-	0.3
Gas extinction coefficient	m^{-1}	0.5	Complete evolution at	m	0.48
Burner mass flow rate	kg/s	0.522	Particle size	μm	50
Coal to air ratio (f_r)	-	0.1			

VERIFICATION

The Flownex® model shown in Figure 6 was created to verify the results of the code. Flownex® uses a one-dimensional flow network approach for thermal-fluid calculations and Rousseau et. al. [16] provide more details about the software as well as how it is used to compare with another code. Script A calculates the energy source to the gas and particles using the same formulas as the code (i.e. Equations (8) and (9)). Script A also uses the radiation heat transfer calculated in the zonal method (amount of exchange in the moving and stagnant volume zones) and divides the heat transfer as energy removed from the two streams based on the coal to air ratio. Script B calculates the enthalpy, H, moving in and out of each node by multiplying the specific enthalpy given by Flownex® at the node with the mass flow rate through the node ($H = \dot{m}_{in}h$).

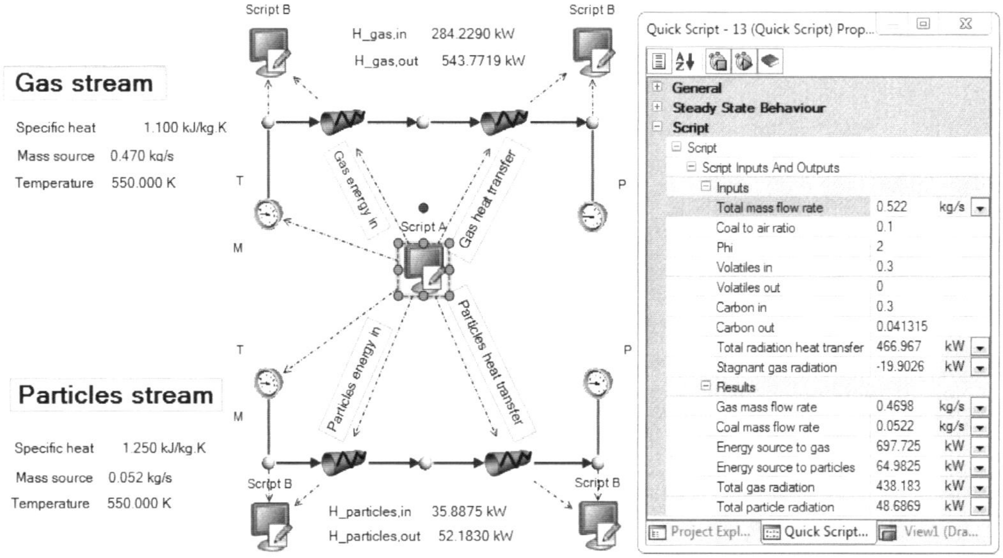

Figure 6. Flownex® model used for mass and energy balance comparison.

Using the radiation heat transfer, fixed carbon and volatile burnout values provided by the new code, the mass and energy balance for the model can be calculated and the results are shown in Table 2. For this verification the convection heat transfer component was neglected (i.e. $\dot{Q}_{conv} = 0$). The values of the codes compare well to the benchmark created in the Flownex® model.

Table 2. Comparison between Flownex® and the code presented in this paper.

		New Code		Flownex		
Description	**Unit**	**Input**	**Result**	**Input**	**Result**	**Difference [%]**
Fraction fixed carbon in	-	0.3000		0.3000		
Fraction fixed carbon out	-		0.0041	0.0041		
Fraction volatile matter in	-	0.3000		0.3000		
Fraction volatile matter out	-		0.0000	0.0000		
Energy transfer in walls	kW		-466.9669			
Energy transfer from volume	kW		466.9669	466.9669		
Energy of inflowing particles	kW		35.8875		35.8875	0.0000
Energy of outflowing particles	kW		52.1832		52.1830	0.0004
Energy source to particles	kW		64.9827		64.9825	0.0003
Energy of inflowing gas	kW		284.2290		284.2290	0.0000
Energy of outflowing gas	kW		543.7736		543.7719	0.0003
Energy source to gas	kW		697.7285		697.7254	0.0004
Energy from stagnant gas	kW		-19.9026	-19.9026		

RESULTS AND DISCUSSION

The predicted temperatures, radiation heat transfer and heat sources along the centre of the furnace are shown in Figure 7a, together with the measured gas temperatures obtained from the PSCTF [15]. Only a limited number of measured temperature points are available, but it shows that the predicted values are of the same order of magnitude. Note that the purpose here is not to make a conclusion about the absolute accuracy, but rather to illustrate the viability of solving the different phenomena models in an integrated manner.

The linear temperature profile at the inlet represents the volatile burnout zone where after the fixed carbon starts to burn. The temperatures is within the expected range of a coal-fired furnace (which is on average a flue gas temperature of 1425K in the PSCTF) and evens out along the length of the flame to a value approaching the wall temperature of 1000K. The heat sources are a maximum at the inlet due to the large amount of heat release due to volatile burnout and then declines as the remaining fixed carbon fraction decreases. The radiation heat transfer follows the same trend as the gas temperature profile because radiation heat transfer is strongly dependent on the temperature differences between the gas volume and the walls. The furnace exit temperature profile in Figure 7b shows that the temperatures are higher in the centre and decrease towards the wall. This is consistent with the temperature profile of a gas heating the walls.

(a)

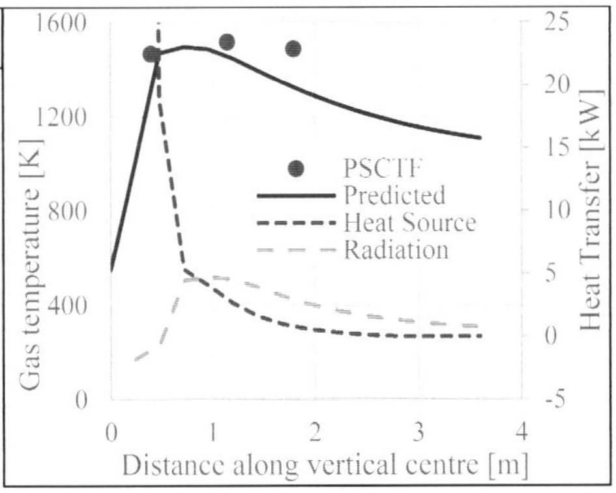

(b)

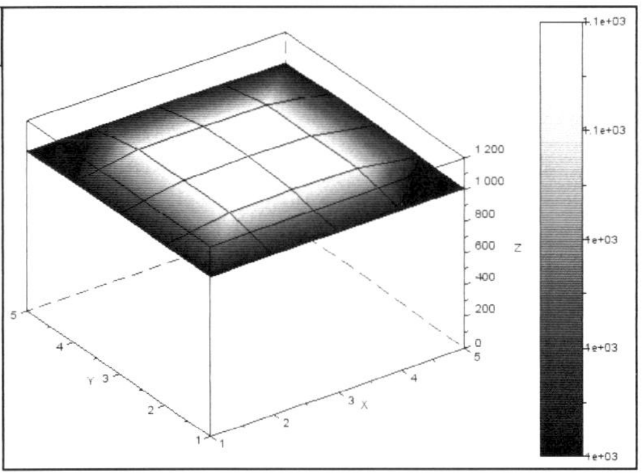

Figure 7. (a) Temperature, radiation heat transfer and heat sources along the vertical centre of the furnace and (b) the furnace exit temperature.

A comparison of the rate of convergence for two different numbers of volume zones is shown in Figure 8a. As expected, the graph shows that the code requires less iterations to converge when the number of volumes is decreased.

(a)

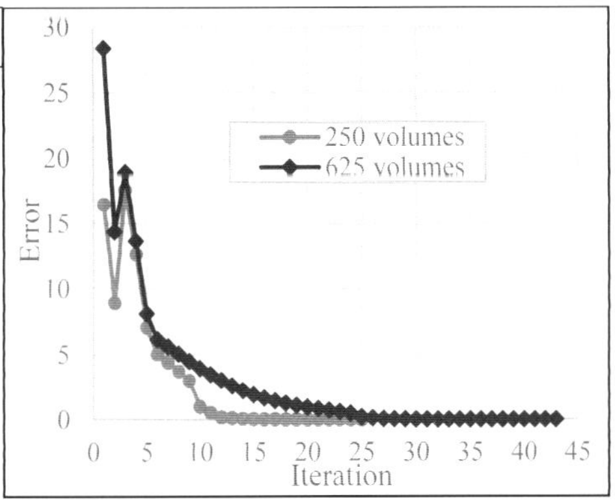

(b)

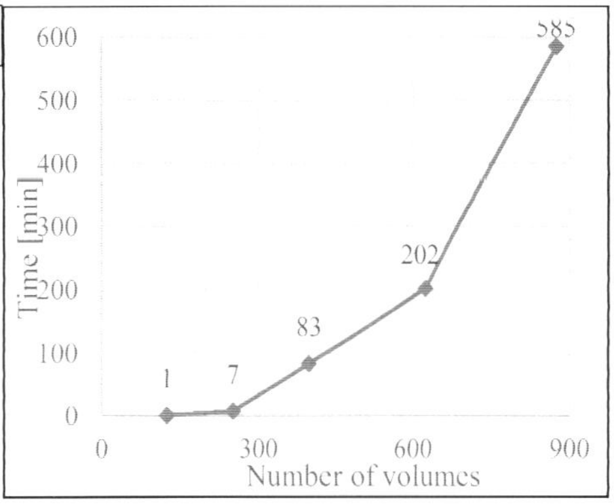

Figure 8. Comparison of the (a) convergence and (b) solution time for increase in number of volumes.

A parametric study of the solution time when the number of volume zones is increased is shown in Figure 8b. This was done by changing the modelled length along the height (i.e. the z-direction as illustrated in Figure 5) between 1.2m and 8.4m. The goal was mainly to investigate solution time but in the future the study will be modified to investigate grid independency. Currently the code provides a solution within a reasonable run time when the furnace is divided into 250-400 volume zones. This discretisation is similar to that of the code FURNACE shown in Figure 2 (also see Karasina et. al. [6], Abryutin et. al. [7] and Karasina et. al. [8]). It is worth noting that all the radiation heat transfer calculations (using the zonal method) use matrices, but during the implementation of the burnout model mainly for/while loops were used. The solution speed can therefore be increased if the implementation of the burnout model solution is optimised using matrix operations.

CONCLUSION

This paper presented a methodology for an integrated coal fired furnace process model which combines the zonal method to model the radiation heat transfer with a one-dimensional burnout model for the heat generation, and characteristic flow maps for the advection. A mass flow map was assumed on which the combustion model was superimposed. The BCURA burnout model was extended from its original one-dimensional application to be applied in a three-dimensional domain. It was combined with the zonal method (which models the radiation heat transfer) to complete the heat balance for each volume in order to predict the flame geometry and position.

The results obtained for a pilot scale test furnace with a single burner show that the predicted temperatures along the vertical centre of the furnace and the furnace exit are consistent with the values expected in actual coal-fired furnaces. The solution time is acceptable when the discretisation of the furnace is 400 volumes or less, which will be important when integrating with a full plant model and quick solution times are required.

The following work is envisaged as a continuation of the current study:

- Optimise the solution algorithm to decrease the computational time required.
- Investigate in more detail the use of characteristic mass flow maps generated via a CFD model.
- Compare the results obtained with the new network code with that of a detailed CFD model for the pilot scale single burner furnace test facility.
- Apply the new methodology to a full scale multi-burner furnace configuration and validate the code with the aid of real plant data.
- Show how the code can be integrated with a system network code such as Flownex® to model the complete power plant.

ACKNOWLEDGEMENTS

The authors acknowledge the financial support provided by the Eskom Power Plant Engineering Institute (EPPEI). The first author also acknowledges the funding provided by THRIP and M-Tech Industrial (Pty) Ltd.

REFERENCES

1 EPRI, 2001, "Statistical analysis methodology for predicting impact of operation factors on boiler availability", EPRI, Palo Alto, CA. TR-1004064.

2 EPRI, 2000, "Impact of Operating Factors on Boiler Availability", EPRI, Palo Alto, CA. TR-1000560.

3 Bhatt, S.M., 2007, "Effect of air ingress on the energy performance of coal fired thermal power plants", Energy Conversion and Management, Volume 48, pp. 2150-2160.

4 Hottel, H.C., Cohen, E.S., 1958, "Radiant heat exchange in a gas-filled enclosure: Allowance for nonuniformity of gas temperature", A.I.Ch.E. Journal, vol. 4, no. 1, pp. 3-14.

5 Karasina, E.S., Livshits, B.N., Chudnovskii, B.R. & Talanker, A.E., 2010, "Experience gained from using a computer program implementing the 3D zone-wise method for calculating heat transfer in the furnace chambers of the coal-fired boilers of the 350- and 575- MW power units", Thermal Engineering, vol. 57, no. 10, pp. 897-900.

6 Karasina, E.S., Z.K., Aleksandrova, T.S. & Borevskaya, S.E., 1982, "An algorithm and computer program for zonal calculation of furnace chambers", Teploenergetika (Abstract available in English), no. 8, pp. 42-50.

7 Abryutin, A.A., Karasina, E.S., Livshits, B.N., Shnirman, A. & Chudnovskii, B., 1998, "Development of a method and a code for the 3D zonal calculation of heat transfer in the furnaces of coal-fired boilers", Thermal Engineering, vol. 45, no. 6, pp. 461-466.

8 Karasina, L.S., Livshits, B.N., Chudnovskii, B., Talanker, A. & Abryutin, A.A., 2000, "Adapting the code Furnace for calculations of heat transfer in the boiler furnace of a 575-MW power-generating unit", Thermal Engineering (English translation of Teploenergetika), vol. 47, no. 11, pp. 1031-1036.

9 Monnaemang, M.O., 2015, *A Zonal Model for Radiation Heat Transfer in Coal-fired Boiler Furnaces*, Master's Dissertation, University of Cape Town, South Africa.

10 Modest, M.F., 2003, "Radiative Heat Transfer", 2nd Edition, Elsevier Science, 200 Wheeler Road, Burlington, Massachusetts, 01803, USA.

11 Field, M.A., Gill, D.W., Morgan, B.B., Hawksley, P.G.W., 1967, "Combustion of pulverised coal", British Coal Utilisation Research Association (BCURA).

12 Gill, D.W., 1969, "Extension to a one-dimensional mathematical model of a pulverised-coal flame", British Coal Utilisation Research Association (BCURA).

13 Lowe, A., Wall T.F., Stewart, I. M., 1975, "A zoned heat transfer model of a large tangentially fired pulverized coal boiler.

14 Hesselmann, G., 1998, "Modelling of pulverised coal fired furnaces with advanced combustion systems by integrating performance programs", Performance Prediction in Advanced Coal Fired Boilers, Contract No JOF3 - CT95 – 0005, Project Area 1 - Project 111, Final Technical Report.

15 Rajoo, P., 2010, "The application of pilot scale coal evaluation to full scale boilers", Master of Science in Engineering dissertation at the University of Witwatersrand, South Africa.

16 Rousseau, P.G., du Toit, C.G., Jun, J.S., Noh, J.M., 2015, "Code-to-code comparison for analysing the steady-state heat transfer and natural circulation in an air-cooled RCCS using GAMMA+ and Flownex", Nuclear Engineering and Design, no. 291, pp. 71–89.

The CFD modelling of the air flow through a finned coil heat exchanger

M Heystek*, M van Eldik, PvZ Venter

Markohey123@gmail.com, Martin.VanEldik@nwu.ac.za, 12330825@nwu.ac.za

School of Mechanical and Nuclear Engineering, North-West University

ABSTRACT

Finned coil heat exchangers are used in a variety of industrial applications for heating air, where warm water flows inside the coils. Fins increase the air side heat transfer area, where the pattern and spacing also have an effect on the air flow through the coil. To accurately simulate the heat transfer on the air side of a finned coil requires an understanding of how air flow is influenced by a specific fin pattern. This paper focuses on the development of a CFD model of a finned coil heat exchanger in Star-CCM+, incorporating a typical fin pattern and spacing found in industry with a staggered tube configuration. The results obtained from this detailed fin pattern are compared to those obtained by using a simplified flat plate type fin approach.

The CFD model was verified by experimental results obtained from a test facility. Four different simulation tests were carried out and the average error calculated for both models. For the flat plate fin approach the model predicted the air temperature change (ΔT) over the coil within an average error of 25.7% and the more complex fin configuration approach yielded an average error of 14.9%. Based on the results obtained, there are advantages to incorporate a more complex fin configuration when modelling a finned coil heat exchanger.

KEYWORDS

CFD, Finned coil heat exchanger, air flow

Nomenclature

Q	vector of conserved variables. [-]	**Subscripts**	
F	vector of fluxes. $[\frac{mol}{m^2 s}]$	air	air
V	volume of control volume. $[m^3]$	in	inlet
A	surface area of control volume. $[m^2]$	out	outlet
ΔT	temperature increment [°C]	$water$	water
ρ	density $[\frac{kg}{m^3}]$		
$\bar{v}$	velocity $[\frac{m}{s}]$		
g	gravitational constant $[\frac{m}{s^2}]$		
z	vertical height $[m]$		
u	internal energy $[J]$		
$\dot{W}$	rate of work done $[\frac{kJ}{kg}]$		
$\dot{Q}$	rate of heat transfer $[\frac{kJ}{kg}]$		
τ	torque $[N.m]$		

1. INTRODUCTION

Finned coiled heat exchangers (FCHXs) are widely used in the heating-ventilation-air conditioning and refrigeration (HVAC&R) industry. For FCHXs a primary fluid flows along the inside of a coil where a secondary fluid, typically air, flows on the outside over the fins and perpendicular to the tubes [1]. Heat exchanger (HX) layout and configuration impact the heat transfer flow characteristics of the fluids. These include characteristics such as the fin pitch, diameters and fin width. The fins can be plain or louvered, the coil circular or elliptical and both affect the HXs performance [2].

The aim of this paper is to develop a CFD model for a FCHX, using Star-CCM+ software, incorporating a typical fin pattern and spacing found in industry and a staggered tube configuration. The use of CFD improves the accuracy of analysing a thermal fluid problem by incorporating fundamental laws. Air flows on the outside over the fins and water inside the coils. Simulated results will be compared with experimental data gathered from a FCHX test bench.

A FCHX consists of parameters including the frontal width and height, fin thickness, fin spacing and number of rows high and deep that define the physical size of the HX. Parameters including the longitudinal and transverse tube spacing, fluid inlets and outlets define the function of the HX. Figure 1 illustrates a typical layout of a FCHX [15].

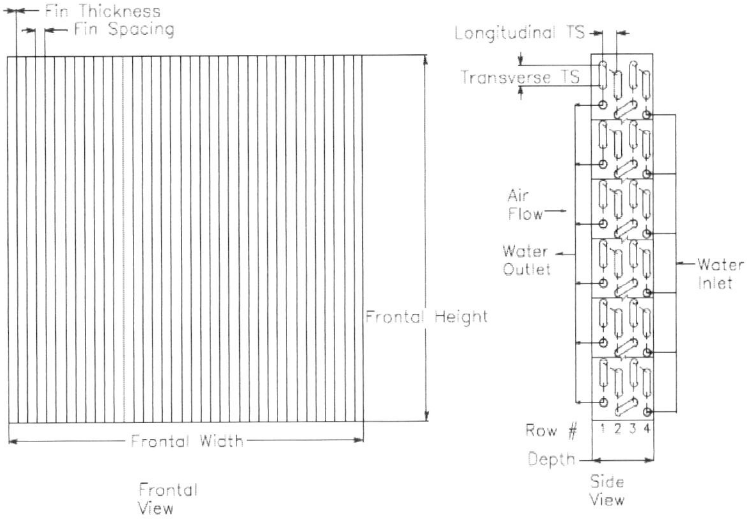

Figure 1 - Finned coil HX design [15]

Figure 2 visually illustrates the HX from the front. The air is rushed through the fins and across the coil as illustrated.

Figure 2 - Photograph of HX air passages

2. SIMULATION MODEL

This section presents the computational domain, governing equations, boundary conditions, and mesh generation that formed part of developing the CFD model in STAR-CCM+. All these fields must be defined to increase the accuracy of the results.

2.1 Computational domain

In this study a plain fin type approach is compared to a more detailed contoured fin pattern of a circular tube FCHX. For the configuration investigated, warm water flows inside the tubes and is cooled down by air flowing on the outside, unmixed across the staggered finned coil. The materials used for construction of the HX include copper for the tubes and aluminium for the fins.

Figure 3 illustrates a side view of the domain that was selected from the HX used for the experimental tests.

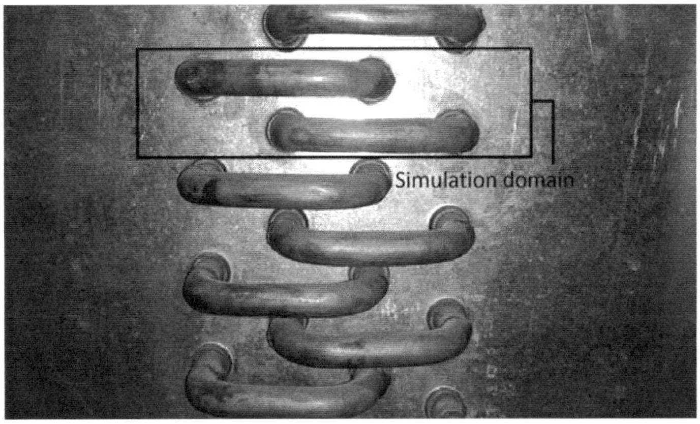

Figure 3 – Side view of the domain selected from the HX

One repeated domain will be used for the simulation process as the HX is symmetric across both sides in an attempt to reduce computational time and memory usage. From literature it follows that this approach has an acceptable accuracy with a dramatic decrease in simulation increment time [1, 6 - 12]. The domain is split into three regions, namely, the main region, the upstream air region and the downstream air region. The main region consists of two half thickness aluminium fins, copper coil rows and a cross flow air region, so the heat transfer between the air and water can be solved. It should be noted that the conductance between the copper tubes and aluminium fins are 100% efficient and the fins fit without weldments on the tube region. The upstream air region consists of a duct section with the assumption that the approaching air is turbulent in nature. The reason for this assumption is due to an air straightener installed upstream from the test HX. Comparisons can be made at the downstream air region of how the flow pattern changed as a result of induced turbulent flow through the HX. The downstream air region stabilizes the flow to avoid possible recirculation at the air outlet, and thus to ensure numerical stability [4].

The dimensions used to set-up the model are as follows:

- Width: 6.65 mm
- Height: 38 mm
- Distance between HX outlet and point probe: 240 mm

The point probe mirrors the values of the closest cell centre. From this, all pressures, velocities and temperatures are calculated and shown as a single point value. To create an assembly with various parts, Solidworks™ software is implemented. The assembly is illustrated in figure 4 that was exported to STAR-CCM+.

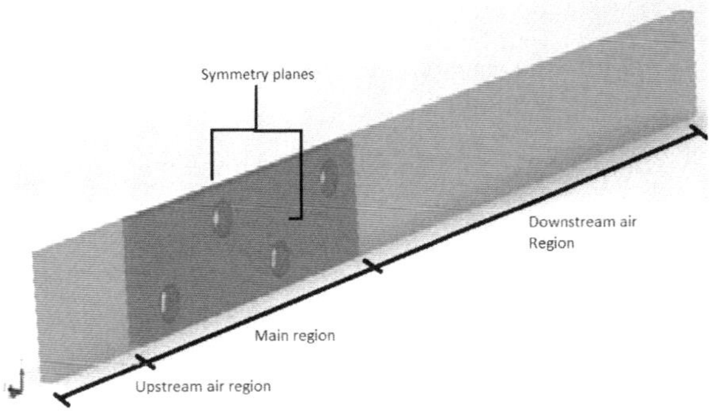

Figure 4 - Domain assembly from Solidworks™

Figure 5 illustrates the difference between the flat plain fin approach compared to the more realistic contoured fin configuration. The added contouring changes the geometrical features of the cells and has a major influence on the flow pattern. The only difference thus being the added riffle effect to the updated version of fins.

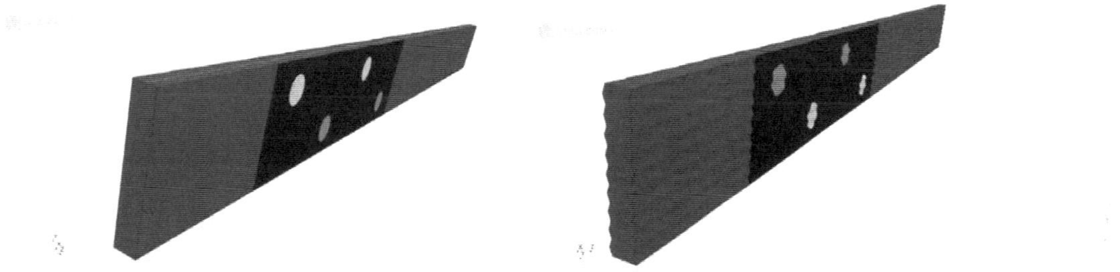

Figure 5 – Plain fin vs. Contoured fin

2.2 Governing equations and Boundary conditions

When importing the assembly file to STAR-CCM+, the user needs to define the geometry and physics of the domain. All solid contacts should be defined as walls and the rest as fluids (water and air). The properties of the air are set as a compressible gas being turbulent, three dimensional and steady state with the use of the K-epsilon turbulence model. The reason for using this specific turbulence model is can be attributed to the good convergence rate and relatively low memory requirements. For water the same conditions were selected, except for the liquid being incompressible and laminar. Kritikos *et al.* studied the effects of the air flow being transient, due to phenomena such as a vortex, and found that it could be neglected as the effect was insignificant towards heat transfer [13].

A common approach in CFD is the finite volume method as the memory used when performing iterations is lower than other methods and as a result shorter solving times are observed. The partial differential equations are recast in a conservative form. It is given by the following equation [14]:

$$\frac{\partial}{\partial t}\iiint \boldsymbol{Q}\, \boldsymbol{dV} + \iint \boldsymbol{F}\, \boldsymbol{dA} = 0 \qquad [1]$$

Star-CCM+ uses the Finite volume method making use of the following conservation equations:

- Conservation of mass.

$$\frac{\partial}{\partial t}\left(\iiint \rho d\, V\right) + \oiint\left(\rho\bar{v}.\overline{dA}\right) = 0 \qquad [2]$$

- Conservation of energy.

$$\dot{Q} + \dot{W} = \frac{\partial}{\partial t}\left(\iiint \left(u + \frac{1}{2}v^2 + gz\right)\rho dV\right) + \oiint\left(h + \frac{1}{2}v^2 + gz\right)\rho\bar{v}.\overline{dA} \qquad [3]$$

- Conservation of momentum.

$$\oiint \tau \overline{dA} + \iiint \bar{B}\,\rho dV = \frac{\partial}{\partial t}\left(\iiint \bar{v}\rho dV\right) + \oiint \bar{v}\left(\rho\bar{v}.\overline{dA}\right) \qquad [4]$$

For comparison, both the plain fin and contoured fin models are set to the same boundary values. The inlet air is set to a velocity of 4.4 [m/s] with a Temperature of 20.7 [°C]. The water is set to 70.0 [litres/min] (complete HX water supply converted to kg/s per tube) with an inlet temperature of 39.7 [°C]. The water temperature at each tube inlet is set to decrease incrementally with a fixed ΔT_{water} from the inlet to the outlet with the flow being in a counterflow configuration with the air. At the air and water outlets, a uniform atmospheric pressure is applied with symmetry conditions on the four lateral boundaries of the domain [4]. The contact between fluid-solid states is defined as a no slip condition. To evaluate the downstream air temperature and velocity, probes are placed at the end of this region. Figure 6 illustrates the position of the point probe.

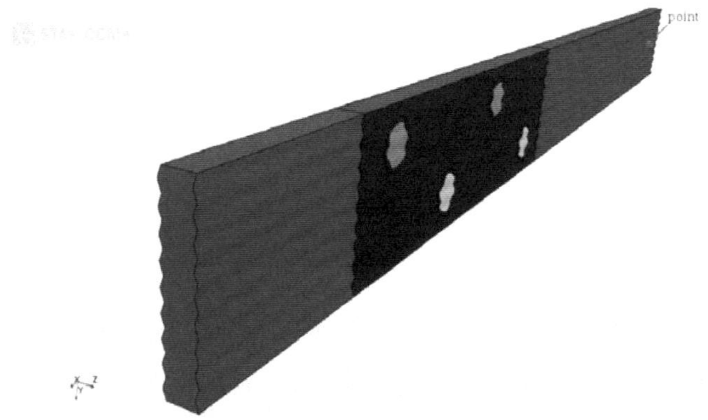

Figure 6 - Probe location

2.3 Mesh generation

When discrete points are placed on a volume taking up a certain space, a grid is generated by the program to define the properties of each point after a given number of iterations have been carried out. The mesh determines to accuracy of the results, as a finer mesh will yield more accurate results for the equivalent body/region size when compared to a coarser mesh. The downside to having a finer mesh is the additional computing time required to complete an iteration.

A sufficiently fine mesh was used for the domain as the domain chosen is relatively small. Compared to the plain flat fin approach the mesh for the contoured fin pattern will be substantially finer to incorporate the bending of the contouring along the adjacent sides of the fins. An unstructured mesh with polyhedral cells was used as this would greatly affect the accuracy of the simulation [1]. Finer meshing was done around points of interest which includes the copper tubes and downstream air region. The mesh properties are as follows:

Mesh 1 (Normal)

- Base size: 0.012 m
- Applied to the general, uninterested parts of the domain such as inlet air and far end of downstream air region.

Mesh 2 (Fine)

- Base size: 0.0027 m
- Applied to the areas of more interest, such as the flow around the tube region, the aluminium fins and the immediate downstream air region.
- Added to 2015 model to be able to implement finer details of a more contoured fin profile.

Figure 7 illustrates the fineness of the mesh on the entire domain, with figure 8 a zoomed in picture of the contoured fin's main section investigated.

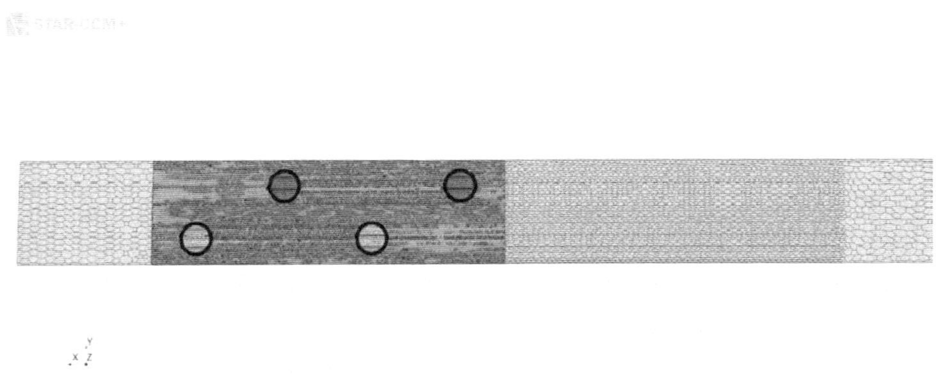

Figure 7 - Domain mesh

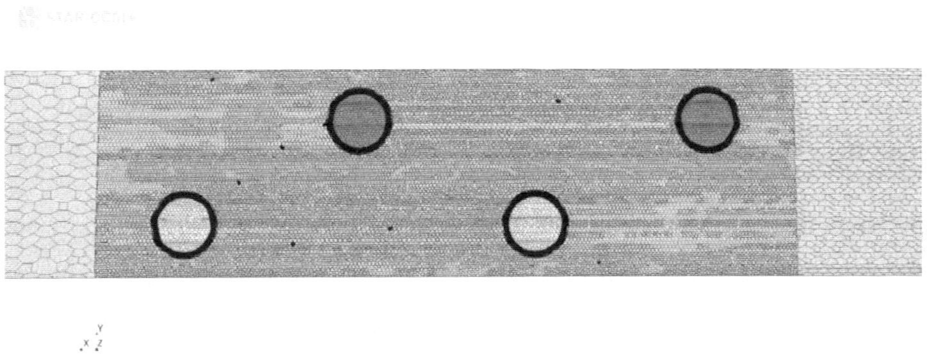

Figure 8 - Zoomed in on domain (Mesh)

3. EXPERIMENTAL SETUP AND TEST PROCEDURE

The test bench is designed for testing HXs where turbulence is minimized with an air straightener implemented before the test section. Using four parallel axial fans the supply air from outside the test bench are induced at the inlet actuators that regulate the volume flow. The louvers controlled by the actuators can be opened and closed to re-circulate a percentage of the air or control the volume flow rate through the tunnel. After being straightened, the air passes the first set of air psychrometers and pressure gauges before reaching the FCHX. A second set of gauges are used to determine the pressure drop over the HX and the air velocity. Air temperatures before and after the test coil are measured by arithmetically averaging the values obtained with nine thermocouples installed in a uniform grid at each location as done by Kim *et al.* [4]. A variable amount of air volume exits the test bench via the outlet louver, with the remainder re-circulated back to the inlet.

Warm water is pumped via a mixing bench to the test coil from pre-heated reservoirs. The mixing bench is used to pre-mix the supply water to the required temperature. The cooled water either returns to the mixing bench to be reused or to the reservoirs based on the inlet temperature required.

To verify the accuracy of the tests an energy balance was implemented. The first law of thermodynamics states that energy cannot be created or destroyed and can for that matter only be transferred from one form/medium to another [5]. Nine tests were done on a total of five different HXs. As the ambient temperature outside the test bench has a marginal effect on the testing data, the tests were done during the same time of the day on consecutive days. The nine tests consisted of three variations in air velocity and three variations in the water flow rate. The air inlet velocity was varied between 2, 4 and 6 [m/s] and the water flow rates for 30, 50 and 70 [litres/min].

After each test run, an energy balance was calculated to ensure the data received via the logging equipment was accurate. When the data satisfies the energy balance the inputs are used for verifying the simulation.

Table 1 illustrates the experimental measured values that were used to create 4 different simulations:

Table 1: Simulation inputs

Test	Water Flow [kg/s]	Air Velocity [m/s]	$T_{air,in}$ [C]	$T_{air,out}$ [C]	$T_{water,in}$ [C]	$T_{water,out}$ [C]
1	1.155	4.387	20.7	27.6	39.7	35.5
2	0.8249	6.02	20.6	26.3	42	34.2
3	0.8249	4.345	20	27.3	40.5	34.2
4	1.155	6.037	20.3	27.3	43.6	37.1

4. RESULTS

The results as generated after 10000 iterations on STAR-CCM+ are illustrated in this section. The stopping criteria for the iterations included the evaluation of the change in temperature at the probe. Creating a plot from the iterations allowed the temperature plateau to become visible meaning the iterations could be stopped. The residuals also converged to below 10^{-4} which is a good indication of convergence for the simulation. Figures 10 to 13 illustrate the velocity profiling and temperature distribution of the domain for both the flat plate fin configuration and the contoured fin approach from the test 1 inputs.

Turbulence is modelled throughout the domain but is induced by a disturbance in the flow direction. The contoured fins approach experienced more turbulence in the downstream air region compared to the flat, plain fins approach.

Plain fins

Velocity profiling

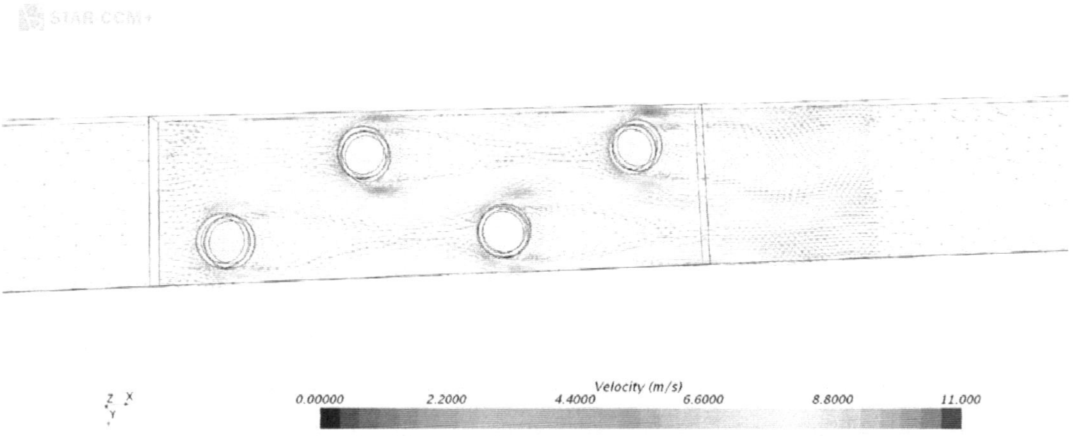

Figure 10 - Plain Velocity profiling

Temperature distribution

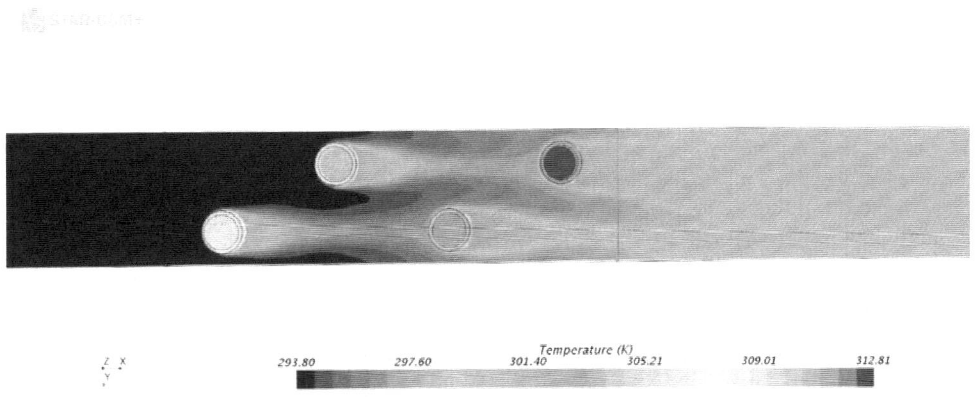

Figure 11 - Plain Temperature distribution

Contoured fins

Velocity profiling

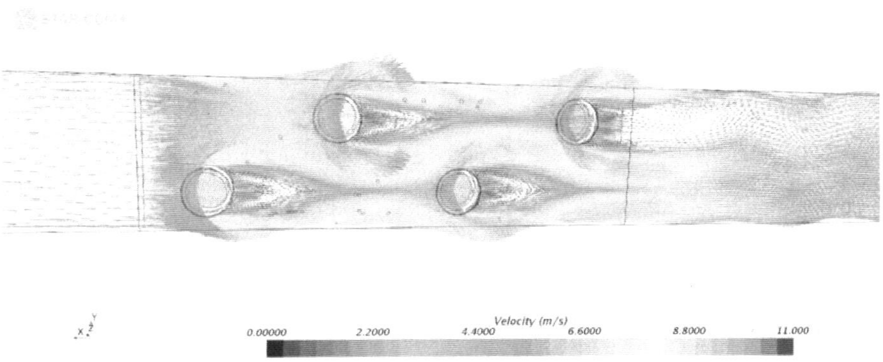

Figure 12 - Contoured Velocity profiling

Temperature distribution

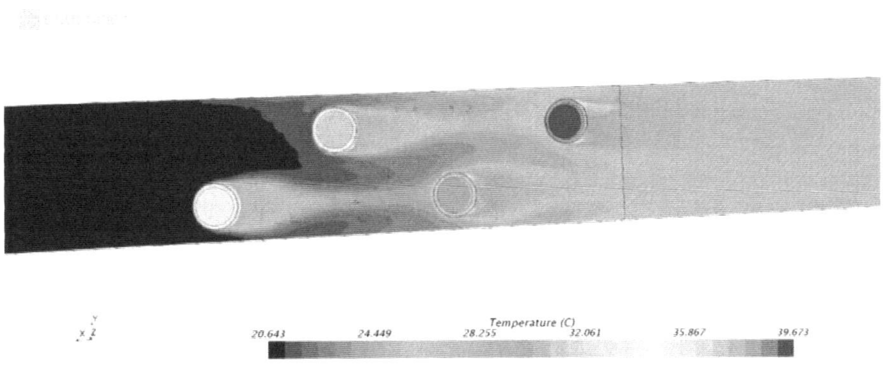

Figure 13 - Contoured Temperature distribution

As can be seen from Figures 10 to 13, the air becomes more turbulent as it passes over the copper tubes and reaches maximum turbulence in the downstream air region. The velocity of the air is at a maximum around the edges of the tubes and a minimum velocity straight behind each tube. This is also the region where a vortex phenomenon occurs. Air spirals in a loop pattern behind each tube creating the perfect environment for heating the air as air closer to stagnation takes up heat more effectively. This region is where the maximum temperature increase of the air takes place and collectively increase the total air in the downstream region.

For the flat plate fin approach the model predicts the downstream air temperatures within an error range of 5.6% compared to the experimental values. The more complex contoured fin configuration approach yielded an error range of 1.6% compared to the experimental values. It can be seen as a substantial improvement in accuracy with a 4.0% decreased error. Based on the results obtained, there are advantages to incorporate a more complex fin configuration when modelling a FCHX. Table 2 summarizes the comparison between models for the air outlet temperatures. Four separate tests are used:

Table 2: Comparison between models for the air outlet temperatures

Test	Experimental $T_{air_{out}}$ [C]	Plain fin model $T_{air_{out}}$		Contoured fin model $T_{air_{out}}$	
		Temp. [C]	Error [%]	Temp. [C]	Error [%]
1	27.55	26	5.6	27.1	1.6
2	26.25	25.1	4.6	25.5	2.9
3	27.34	25.9	5.3	26.4	3.4
4	27.25	30.1	-10.5	29.2	-7.2

To determine the effect of the added contoured profiles on heat transfer, the table below summarize the accuracy in predicting the air temperature change over the coil between the inlet and outlet:

Table 3: Effect on heat transfer

Test	Experimental ΔT_{air} [C]	Plain fin model ΔT_{air}		Contoured fin model ΔT_{air}	
		Temp. [C]	Error [%]	Temp. [C]	Error [%]
1	6.9	5.4	22.5	6.45	6.5
2	5.7	4.5	21.1	4.95	13.2
3	7.8	6.4	18.5	6.85	12
4	7	9.9	-40.7	8.95	-27.9

The results of both approaches are relatively close to the experimental values, but by improving the details of the fin contours the error range should be lowered. By not having contours in the fins, the characteristics of the air flow will also be different than the added contours and may have a noticeable effect on the results, as can be seen from Figure 10 to 13. A more turbulent air flow is formed with the contoured fins. An increase in the fin area due to the contoured shape will create a larger amount of drag that will have a noticeable effect on the air flow pattern. To further this study additional improvements can be made during future research. The differences in heat transfer and effectiveness between staggered and aligned tube arrangements can be incorporated, as the tube arrangement has an effect on the FCHX's performance. It is also recommended that the next step in improving the simulation model will include an even more realistic fin profile as typically found in industry and investigating the least number of cells in order to generate accurate results.

CONCLUSION

This paper focused on the development of a CFD model of a finned coil heat exchanger in Star-CCM+, incorporating a typical fin pattern and spacing found in industry on the air side, and a staggered tube configuration. The results obtained from this detailed fin pattern were compared to those obtained by using a simplified flat plate type fin approach.

A test facility was used to obtain experimental results for a range of air flow rates of 2, 4 and 6 [m/s] and water flow rates for 30, 50 and 70 [litres/min] and temperatures. The experimental results, including the heat transfer rate and downstream air temperature, were used to verify the results obtained by the CFD models. Four different simulation tests were carried out and the average error calculated for both models. For the flat plate fin approach the model predicted the air temperature change (ΔT) over the coil within an average error of 25.7% and the more complex fin configuration approach yielded an average error of 14.9%. Based on the results obtained, there are advantages to incorporate a more complex fin configuration when modelling a finned coil heat exchanger.

REFERENCES

1. Sun et al., "Evaluation of elliptical finned-tube heat exchanger performance using CFD and response surface methodology", International Journal of Thermal Sciences, Vol. 75, pp 45, 2013
2. Kong et al., "Effects of continuous and alternant rectangular slots on thermo-flow performances of plain finned tube bundles in in-line and staggered configurations", International Journal of Heat and Mass Transfer, Vol 93, pp 97 - 107, 2015
3. Aslam Bhutta et al., "CFD applications in various heat exchangers design: a review", Appl. Therm. Eng. Vol 32, 2012
4. Kim et al., "Heat transfer characteristics of flat plate finned-tube heat exchangers with large fin pitch", International Journal of Refrigeration, Vol 28, pp 851 – 858, 2005
5. Incropera et al., "Fundamentals of heat and mass transfer", 6th edition, pp 2, 2013
6. Borrajo-Pelaez et al., "A three-dimensional numerical study and comparison between the air side model and the air/water side model of a plain fin-and-tube heat exchanger", Applied Thermal Engineering, Vol 30, pp 1608 – 1615, 2010
7. Erek et al, "Effect of geometrical parameters on heat transfer and pressure drop characteristics of plate fin and tube heat exchangers", Applied Thermal Engineering, Vol 25, pp 2421 – 2431, 2005
8. Lu et al., "A numerical investigation of the geometric effects on the performance of plate finned-tube heat exchanger", Energy Conversion and Management, Vol 52, pp 1638 – 1643, 2011
9. Simo Tala et al., "Tube pattern effect on thermal hydraulic characteristics in a two-rows finned tube heat exchanger", International Journal of Thermal Sciences, Vol 60, pp 225 – 235, 2012
10. Tian et al., "A comparative study on the air-side performance of wavy fin-and-tube heat exchanger with punched delta winglets in staggered and in-line arrangements", International Journal of Thermal Sciences, Vol 48, pp 1765 – 1776, 2009
11. He et al., "Three-dimensional numerical study of heat transfer characteristics of plain plate fin-and-tube heat exchangers from view point of field synergy principle", International Journal of Heat and Fluid Flow, Vol 26, pp 459 – 473, 2005
12. Tao et al., "Three-dimensional numerical study and field synergy principle analysis of wavy fin heat exchangers with elliptic tubes", International Journal of Heat and Fluid Flow, Vol 28, pp 1531 – 1544, 2007
13. Kritikos et al., "Investigation of the thermal efficiency of a staggered elliptic-tube heat exchanger for aero engine applications", Applied Therm. Engineering, Vol 30, pp 134 – 142, 2010
14. Eymard, R. Gallouët, T. R. Herbin, R. (2000) The finite volume method Handbook of Numerical Analysis, Vol. VII, 2000, p. 713–1020.
15. Taylor Creed, "Measurement of finned-tube heat exchanger performance", magister thesis at Georgia Institute of Technology, 2004

Numerical Replica of a Dunnage Bag Certification Test

Martin Philip Venter* & Gerhardus Venter

Department of Mechanical and Mechatronic Engineering, Stellenbosch University

mpventer@sun.ac.za

ABSTRACT

Dunnage bags are a commodity grade inflatable restraining device used to restrain goods in multi-modal containers. In order to correctly comply with guidelines laid out by insurance underwriters this type of restraining devise is often required to be independently certified as being suitable for a particular application. The long lead-times and high cost of obtaining certification increase the risk of a new product being denied certification.

This research attempts to develop a numerical model in LS-DYNA that can replicate a standard certification test in order to improve product knowledge and predict performance of a new dunnage bag undergoing a certification test. The model accounts for the non-linear material properties of plain woven polypropylene, the inflation of the dunnage bag to working pressure and a single loading cycle.

The model produced was capable of capturing the pressure drop in the bag due to plastic deformation of the material under loading to within 3 % of measured results, operational restraining loads were predicted to within 2 % of measured results, and burst pressures for selected load cases within to 10 %.

KEYWORDS

Dunnage Bag, Finite Element Methods, Inverse Method

INTRODUCTION

Dunnage bags are inflatable dunnage variants used to restrain goods in multi-modal containers or hold of a vehicle while in transit. Dunnage bags are placed between packaged goods and inflated to fill voids and apply a restraining force, Figure 1. Dunnage bags are inexpensive single use products manufactured from commodity grade materials such as Kraft paper, blown linear low density polyethylene and plane woven polypropylene.

For insurance purposes many dunnage bag consumers require new products to be certified through an

Figure 1: Kraft paper dunnage bags used as lateral void fillers in a multi-modal container

appropriate certification body. In each case certification testing is expensive and testing opportunities are limited. The failure of a new product to be certified can severely increase the time to market of new products and reduce the flexibility required to deliver customer specific solutions. With this in mind dunnage bag manufacturers attempt to ensure that new products are of an appropriate standard before sending them for certification.

The conventional process followed is to design new bags making use of extensive prototyping and physical testing which is limited by large minimum material order quantities and a limit to how large a fraction of the production process can be diverted to sustain new product development.

The dunnage bag industry desires an alternative or supplemental design process that can provide more insight into the design of new bags and reduce the number of prototypes required in the design process. This research centres on the development of a numerical model suitable for replication of a standard certification test to provide estimates of product performance before physical prototypes are manufactured.

CERTIFICATION TESTING

A representative certification body is the Association of American Railroads (AAR), which certifies inflatable dunnage bags in North America according to its 'Product Performance Profile for Pneumatic Dunnage' [1]. This is a pass/fail certification test for pneumatic dunnage intended for use on railroads. The AAR performance profile for pneumatic dunnage considers the application and intended performance level of a dunnage bag and specifies its performance measures.

The two intended applications for dunnage bags are for use as either lateral or lengthwise void fillers. As a lengthwise void filler the bag will be subjected to higher loads resulting from braking or shunting operations than is expected for lateral void fillers.

The AAR also certifies by performance level. The performance level of a bag dictates the maximum weight of the goods that can be restrained using a particular bag. Five performance levels are specified, of which levels 2 to 5 are applicable to lengthwise void fillers. They range from restraining goods weighing 34 000 kg for level 2 bags to 93 000 kg for level 5 bags. This research focuses on level 4 bags intended for certification as lengthwise void fillers. This is a general application that is subject to higher performance measures.

Certification is awarded on the performance of a set of ten bags subjected to three sequential tests. The three tests performed are: a leak test, a cyclic load test and a burst test. A level 4 dunnage bag intended for use as a lengthwise void filler will receive certification if:

- All ten samples are inflated into a 305 mm void at 70 kPa and held for 19 days. At the end of the 19 days no individual bag may have lost more than 7 kPa and the standard deviation for all ten samples cannot be more than 1 kPa.
- Five of these samples are then randomly selected for dynamic testing. For the dynamic testing the bag is inflated in a 305 mm at 70 kPa. The bag is then loaded a minimum of 10 times to a maximum load of 265 000 N. The sample should not burst and the final pressure in the bag once returned to the 305 mm void should be within 10 % of the initial pressure.
- At the completion of the dynamic test, each of the five bags tested is then further inflated to 207 kPa while maintaining a 305 mm void. The bag should maintain the pressure for at least one minute without bursting.

NUMERICAL MODEL

A numerical model is required to replicate the key aspects of the certification tests. Previous work by Venter and Venter [2] explored using a numerical model of a Kraft paper dunnage bag as a design tool for new dunnage bags. The research showed that a numerical model was suitable for the prediction of areas where dunnage bags are most likely to fail both due to stresses induced on a large scale due to the geometry of a dunnage bag, and on a small scale by the rigid plastic inflation valve. Presented here is research focused on adding further functionality to the design tool by replicating the response of a dunnage bag to cyclic loading. When a bag undergoes a cyclic overload, the pressure in the bag drops once the bag is returned to the initial state due to material plastic deformation. The presented research explores plane woven polypropylene as the market replacement for Kraft paper.

To add the desired functionality, the new numerical models are required to account for the non-linear plastic response of plane woven polypropylene both in the load and unload phases, the compressible nature of the inflation medium which increases in pressure as the bag is loaded, and the moving boundary conditions dictated by a cyclic void size.

At this stage the model does not attempt to simulate the permeability of the material or any time or rate dependant effects observed in high speed or long duration physical bag testing. These models assume that the material is not permeable, and the tests are assumed to be performed under quasi-static conditions, with temperature being consistent for all tests.

The geometry for the standard dunnage bag investigated is shown in Figure 2. The bags are manufactured from a woven polypropylene tube that is 2590 mm long and 1219 mm wide when folded flat. Two 203 mm regions are shown where the woven polypropylene is folded back into itself to prevent fraying. The general valve position is also shown. The mesh used is a simple orthogonal rectangular shell mesh with the number of elements along each edge also shown in parentheses in the figure. Two flat meshes are generated with the overlap regions modelled simply as double thick areas. The two meshes are then merged along the outside edge of the geometry to form a closed volume, similar to a rectangular pillow with a small internal gap of 10 mm. The mesh was checked for convergence in terms of both inflated shape and strain and found to be suitable.

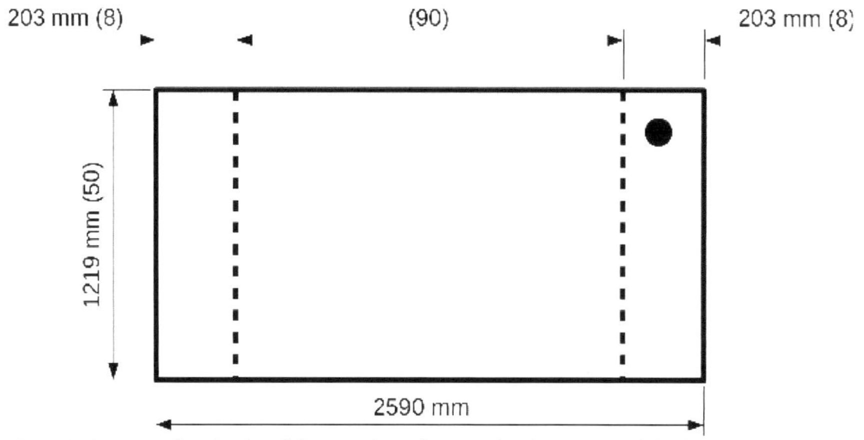

Figure 2: Geometry for simulated dunnage bag, showing the dimensions of the bag, the 203 mm folded overlap regions and the general position of the inflation valve in the top right corner. The number of elements used along each edge is shown in brackets.

LS-DYNA (Hallquist [3]) was chosen because it makes use of an explicit solver that reduces the risk of premature termination of the simulation due to material wrinkling, Jetteur and Bruyneel [4].

The numerical model makes use of analytical rigid walls to account for the void into which the bag is inflated. Two parallel analytical planes are defined within a bounding box that is slightly larger than the deflated dunnage bag geometry. A displacement profile is applied to these two planes to account for the change in void size observed in physical testing. For the replica AAR test the void starts at 305 mm and is reduced to 200 mm. At this point the tested bags reach the prescribed restraining force of 265 000 N. The void is held for a short time and the simulation is allowed to settle then the void is returned to 305 mm. This represents a full load cycle in the AAR test. Since time and rate dependencies are not considered, each additional load cycle will give the same results.

The model is inflated using the hybrid airbag model native to LS-DYNA. This model requires the user to define a control volume representing the inner void of the bag and provide a curve for inflation rate over time. The hybrid airbag model calculates the pressure on the interior surface of the control volume at each time step using the Ideal Gas Law, from the calculated volume and the mass of gas in the bag at that time. To replicate the physical test it is important that the pressure in the bag before starting the load cycle is the prescribed 70 kPa. This is assured by using a numerical optimiser, LS-OPT (Stander et. al. [5]) to scale a generic mass rate over time curve provided until the desired pressure is achieved before the load is cycled. This optimization step takes into account variations in the internal pressure of the bag due to changes in the material properties or geometry. Once the bag has been inflated to the desired pressure, before the load cycle, the inflation is stopped and the mass of gas is kept constant.

The key aspect to be included is the material representation which must capture the non-linear plastic response of the plane woven material in both the load and unload phases to reasonably capture the pressure drop. This is discussed in detail in the next section

MATERIAL HOMOGENISATION

Dunnage bags have a two layer construction with an airtight LLDPE bladder reinforced with a plane woven polypropylene tube stitched on both ends, Figure 3. The bladder is intended to be compliant and is only used to maintain an airtight seal. The structural component is the plane woven polypropylene. This material has a complex material response due to the highly non-linear material properties of polypropylene and the complex inter-tow interactions of plane woven uncoated textiles. Several approaches are available to model this class of material, kinematic and analytical material models (see Boisse et. al. [6], Apedo et. al. [7], Brueggert and Tanov [8], as examples) and macro-, micro-, and meso, scale modelling of weave architecture (see Lomov et.al [9], Durville [10], Nilakantan et. al. [11] as examples). Each of these methods have limitations.

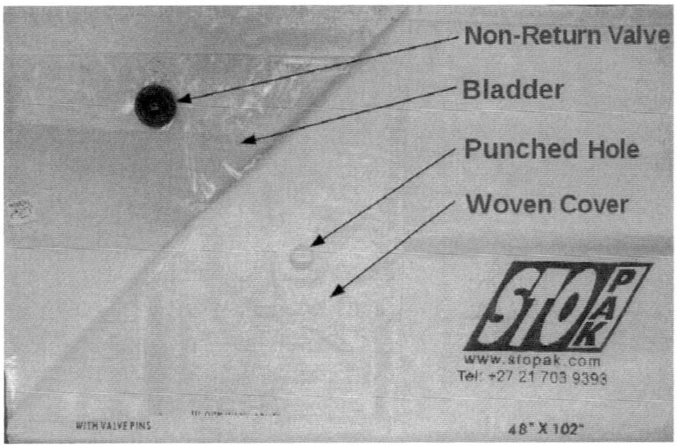

Figure 3: Two layer construction of a woven polypropylene dunnage bag.

The approach taken here is to shift the focus back to the global simulation of the full bag and not fixate on a numerical material model accuracy. The objective was to find a simple homogenised material representation of the overall material response that makes use of only standard material models in an existing commercial FE software.

A preliminary ranging test was performed where a full bag was inflated to near burst pressure. The surface strain was measured using digital image correlation. The results showed that the strain ranged from 0 to 4.9 % in the warp (lengthwise) direction and from 0 to 1.3 % in the weft (hoop) direction. These can be considered to be reasonable order of magnitude quantities because the woven polypropylene material is unlikely to have significantly higher strain as a higher strain would cause the restraining force of the bag to reduce. Based on the results a bi-axial swatch test was chosen as the most representative material test for the woven textile. Cruciform samples were subjected to a displacement up to the maximum strain range in each direction with 1:1 displacement ratio.

An inverse method is used to match the load response of a physical swatch test with the response of a equivalent numerical test. This method produces a homogenised equivalent material representation within the bounds of the desired strain range, Figure 4. Swatch tests and simulations are performed, and force displacement (FD) curves are generated.

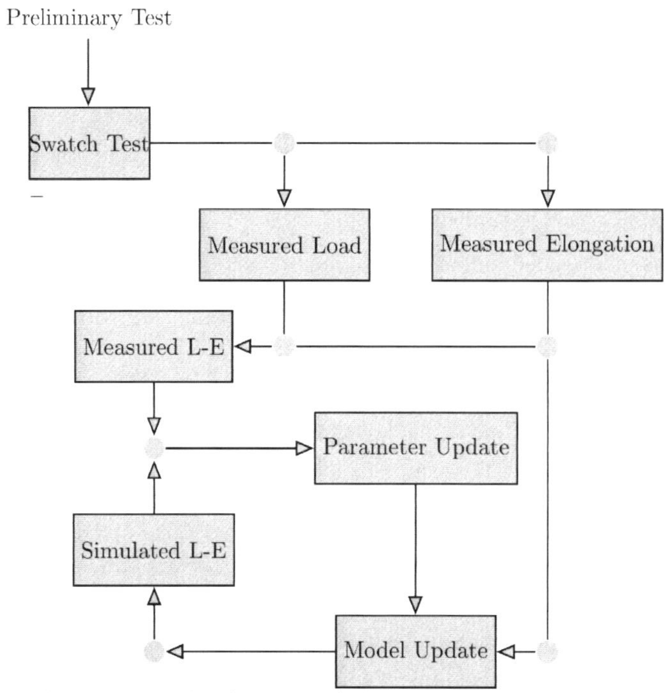

Figure 4: Flow diagram used to characterise the homogenised material representation.

A series of swatch tests were performed on plane woven polypropylene over the strain ranges obtained from the preliminary test. Load and elongation are measured for each swatch in both the warp and weft directions. A load-elongation curve is then generated for the swatches in the warp and weft directions. The measured elongations are then passed as a prescribed displacement boundary conditions to a numerical model of the cruciform sample with comparable geometry. The reaction forces required to enforce the defined displacement boundary condition are recorded and summed to generate a load-elongation curve in the warp and weft directions for the simulated swatch test. LS-OPT is then used to minimise the difference between the two load-elongation curves by manipulating the homogenised material parameters.

Two material models were considered. A non-linear orthotropic elastic plastic material model native to LS-DYNA and a hybrid combination of non-linear springs and membrane elements. It was found that the non-linear orthotropic model did not provide sufficient flexibility to replicate the load-elongation curves observed for woven polypropylene. The orthotropic model was unable to capture significantly different curves for the warp and weft directions simultaneously and was unable to capture the non-linear unload response.

An alternative method using spring and membrane elements was used was to build the overall geometry from smaller unit cells with a higher degree of flexibility. Four non-linear spring elements are combined with one membrane element to create one such unit cell, Figure 5. Two springs are used to represent the warp direction tows and two springs are used to represent the weft direction tows. Each spring can be assigned an arbitrary load elongation curve in both the load and unload directions.

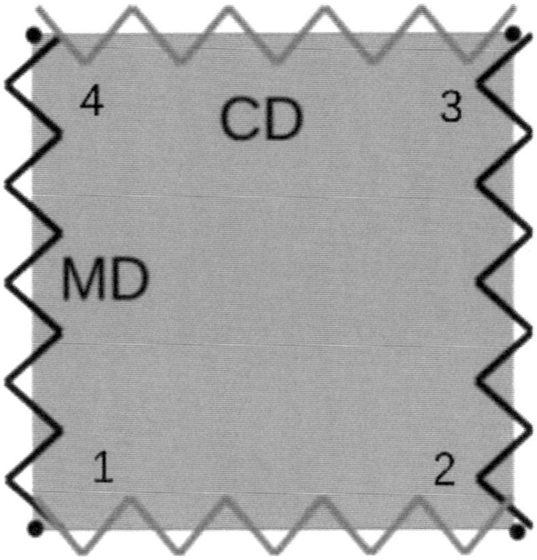

Figure 5: Hybrid spring-membrane unit cell. Two springs are used to represent the response in the warp direction (MD) between nodes 1 & 4 and 2 & 3. Similarly the weft (CD) response is represented between nodes 1 & 2 and 3 & 4.

Using this method the load elongation relationship observed during physical testing could be replicated numerically as seen in Figure 6.

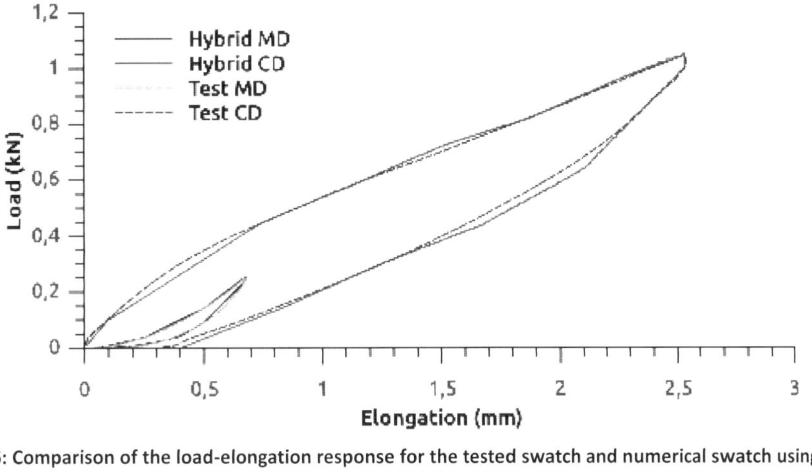

Figure 6: Comparison of the load-elongation response for the tested swatch and numerical swatch using the hybrid element.

RESULTS

The material representation proposed was implemented in a numerical simulation of a full-scale dunnage. Figure 7 shows representative results.

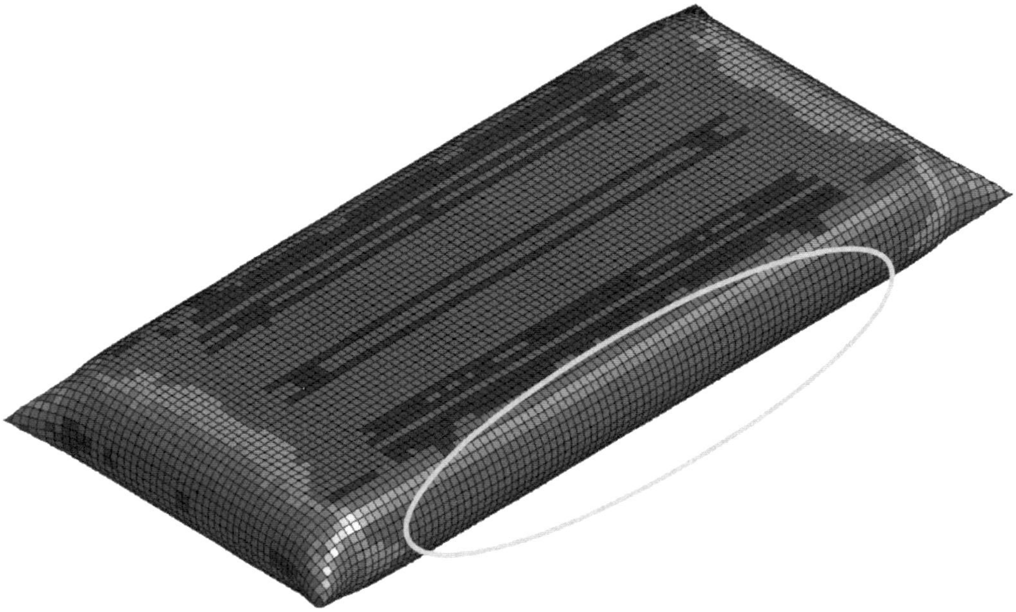

Figure 7: Representative stress results produced by a full scale simulation of a dunnage bag, indicating the critical stress region.

The cycled load portion of the AAR certification test was conducted on ten sample bags. The pressure in the bag, void size and restraining load were measured. The cycle test was then replicated using the numerical model developed here. The numerical model requires that the initial pressure, the void size profile and bag geometry be replicated and predicts the pressure drop in the bag over one cycle as well as the restraining force produced by the bag. Table 1 shows a comparison of the measured and predicted pressure before the load cycle starts, after the cycle and the calculated pressure drop. Table 2 shows similar results for the restraining load

Table 1: Measured and predicted pressure in a dunnage bag before and after a cycled load. The pressure drop is calculated and the standard deviation for the physical tests are shown in brackets.

	Before	After	Drop
Measured Pressure kPa (σ)	70.1 (0.3)	62.2(1.9)	7.9(1.9)
Simulated Pressure kPa	70	64	6

Table 2: Measured and predicted restraining force produced by a dunnage bag before and after a cycled load. The restraining force drop is calculated and the standard deviation for the physical tests are shown in brackets.

	Before	After	Drop
Test Load kN (σ)	116(2.7)	101(4.6)	14.8(3.8)
Simulated Load kN	126	120	6

An additional question was then investigated regarding how well the burst pressure can be predicted for bags deployed in different voids once confidence has been built in the model for the AAR certification test. To this end the burst pressures recorded for the bag inflated in a 305 mm void are used as a benchmark. The maximum stress predicted by the simulation at the average burst pressure observed in physical testing is used as a failure stress. A numerical model is then inflated into a different void and when it reaches this failure stress level the numerical model is considered to have failed.

The results of over-inflation to burst in a 200 and 400 mm void were then compared to the failure pressures predicted by the numerical model in those same voids, the results of which are shown in Table 3.

Table 3: Measured and predicted burst pressures for dunnage bags inflated between 200 and 400 mm static voids. The standard deviation for the physical tests are shown in brackets.

	200 mm Void	400 mm Void
Test Pressure kPa (σ)	320.7 (3.98)	182.7 (13.91)
Simulated Pressure kPa	348	181

CONCLUSION

A numerical replica of the AAR certification test for dunnage was replicated. The model replicates the core characteristics of the inflation of a flat bag into a given void and a cycled load due to a cycled void size. The model takes into account the compressible nature of the inflation medium as well as the non-linear response of the reinforcing material, plane woven polypropylene.

An equivalent material response model making use of non-linear spring elements is used to capture both the load and unload response of a swatch of sample material in both the warp and weft directions simultaneously.

A full simulation of a load-unload cycle was accomplished and the model was capable of capturing the pressure drop in the bag due to plastic deformation of the material under loading within 3 % of measured results and operational restraining loads were predicted to within 2 % of measured results.

The model was then investigated for its ability to predict burst pressures in voids other than that prescribed by the certification test. Two alternative void sizes were tested, a 200 and a 400 mm void. The burst pressures for selected load cases were predicted within 10 %.

REFERENCES

1. "Product Performance Profile for Pneumatic Dunnage", Association of American Railroads, 2012.
2. Venter, M. P. and Venter, G., "Overview of the Development of a Numerical Model for an Inflatable Paper Dunnage Bag", Packaging Technology and Science, Vol. 25. No. 8, pp. 467 – 483, 2012.
3. Hallquist, J. O., "LS-DYNA Theory Manual", Livermore Software Technology Corporation, 2006.

4 Jetteur, P. and Bruyneel, M., "Advanced capabilities for the simulation of membrane and inflatable space structures using SAMCEF", In: Textile Composites and Inflatable Structures II, 2008.

5 Stander, N., Roux, W., Goel, T., Eggleston, T. and Craig, K., "LS-OPT User's Manual - A Design Optimization and Probabalistic Analysis Tool for the Engineering Analyst", Livermore Software Technology Corporation, 2012.

6 Boisse, P., Borr, M., Buet, K. and Cherouat, A., "Finite element simulations of textile composite forming including the biaxial fabric behaviour", Composites:Part B, Vol. 28B., pp. 453 – 464, 1997.

7 Apedo, K.L., Ronel, S., Jacquelin, E., Bennani, A. and Massenzio, M., "Nonlinear finite element analysis of iflatable beams made from orthotropic woven fabric.", International Journal of Solids and Structures, Vol. 47. , pp. 2017 – 2033, 2010.

8 Brueggert, M. and Tanov, R.R., "An LS-DYNA user defined material model for loosely woven fabric with non-orthogonal varying weft and warp angle", In: 7th International LS-DYNA User Conference: Material Technology, 2002.

9 Lomov, S.V., Ivanov, D.S., Verpoest, I., Zako, M., Kurashiki, T., Nakai, H. and Hirosawa, S., "Meso-FE modelling of textile composite: road map, data flow and algorithms", Composites Science and Technology, Vol. 67. , pp. 1870 – 1891, 2007.

10 Durville, D., "Finite element simulation of the mechanical behaviour of textile composites at the mesoscopic scale of individual fibers", In: Textile Composites and Inflatable Structures II, 2008.

11 Nilakantan, G., Keefe, M., Bogetti, T.A., Adkinson, R. and Gillespie, J.W., "On the finite element analysis of a woven fabric impact using multiscale modeling techniques", International Journal of Solids and Structures, Vol. 47., pp. 2300 – 2315, 2010.

10th South African Conference on Computational and Applied Mechanics
Potchefstroom
3–5 October 2016

Gradient-only optimization in review for engineering optimization problems

Daniel N. Wilke

Centre for Asset and Integrity Management
University of Pretoria, South Africa
wilkedn@gmail.com.

Keywords: Gradient-only optimization, non-smooth, discontinuous, differential equation, subgradients, surrogates, complex-step

ABSTRACT

Conventionally optmization problems are formulated either as minimization problems or alternatively as problems that aim to find stationary points (optimality criterion approaches). The former problem being applicable to smooth and non-smooth problems whereas the latter is restricted to smooth problems that are at least twice continuously differentiable. In 2005, gradient-only optimization was introduced as an optimality criterion approach for non-smooth and discontinuous problems. Engineering problems that involve differential or partial differential equation based objective and constraint functions often result unwittingly in discontinuous functions. In particular when temporal or spatial discretizations are not continuously and smoothly varied between designs.

In the last ten years a number of advancements have been made to the field of gradient-only optimization. In particular, the calculation of numerical sensitivities for discontinuous functions, additional classes of algorithms and new fields of application. This paper reflects back on the developments in gradient-only optimization over the last ten years within the context of engineering problems. In this paper, special consideration is given to inverse problems and gradient-only constructed surrogates.

INTRODUCTION

Engineering design optimization problems are usually formulated in the standard form Arora [1] given by

$$\min_{\mathbf{x}\in R^n} f(\mathbf{x}), \tag{1}$$

such that

$$\begin{aligned} \mathbf{g}(\mathbf{x}) &\leq \mathbf{0}, \quad \mathbf{g}\in R^m, &(2)\\ \mathbf{h}(\mathbf{x}) &= \mathbf{0}, \quad \mathbf{h}\in R^k, &(3)\\ \mathbf{l}\leq \mathbf{x} &\leq \mathbf{u}. &(4)\\ & &(5)\end{aligned}$$

The objective function $f(\mathbf{x})$, m inequality constraint functions $\mathbf{g}(\mathbf{x})$, and k equality constraint functions $\mathbf{h}(\mathbf{x})$ in the standard form can be smooth, non-smooth and even discontinuous. Having non-smooth and discontinuous functions in engineering applications is not uncommon. For example a non-smooth function as depicted in Figure 1(a) is often the result when abrupt changes in boundary conditions of a simulation occur as for example in structural contact between deformable bodies. In turn, when differential equation or partial differential equation based functions are evaluated numerically discontinuities result when the temporal or spatial discretizations do not vary continuously and smoothly between designs as depicted in Figure 1(b). Both non-smooth and discontinuous functions poses a problem for computationally efficient gradient based optimization approaches.

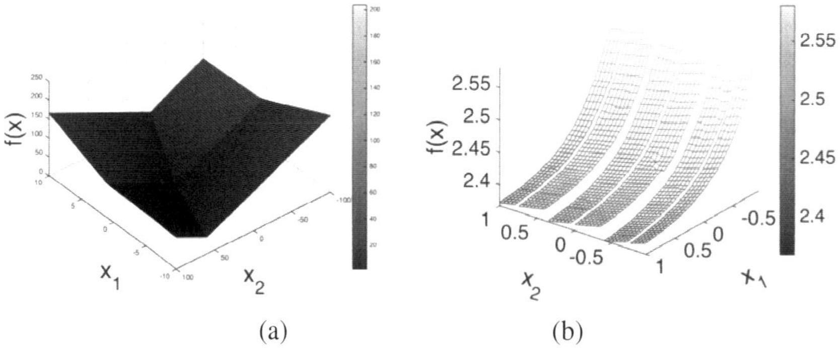

Figure 1: Example of a two dimensional (a) non-smooth function and (b) discontinuous function.

Usually, when the number of design variables are in the low to moderate tens, response surfaces are often employed by evaluating a number of designs and fitting interpolation surfaces (for example radial basis functions Strang [8]) through the data points Khuri and Mukhopadhyay [3]. Alternatively, regression surfaces O'Connell and Wolfinger [6], Nooraziah and Tiagrajah [5] that may alleviate the dominance of discontinuities can be fitted through the data points. Hence, smooth functions are constructed that captures the global response trend so that efficient gradient-based approaches can be employed to optimize the response surfaces. However, larger number of design variables are often solved using inefficient evolutionary approaches combined with conventional gradient based approaches. This however renders many optimization problems impractical to solve unless parallelization is appropriately employed. As a result, the dominant approach to address discretization error based discontinuities for larger number of design variable problems is to explicitely reduce the discretization error by making the time steps smaller for temporal discretizations Strang [8] or by adaptive mesh refinement Schleupen et al. [7] for spatial discretizations. This requires proper error control for each function evaluation. The benefit is that it reduces the size of the discontinuity jumps to the point that the functions can be assumed to be smooth to allow for efficient gradient based approaches to be employed on problems with hundreds to thousands of design variables which is not uncommon when both the material field and geometry are optimized. The only potential limiting factor to employ gradient based strategies is the computation of gradient information. In addition, smooth problems allows for alternative solution approaches, namely finding stationary points or optimality criterion approaches. These approaches however require that all functions are at least twice continuously differentiable. For unconstrained optimization problems stationary points are solved for and the definiteness of the Hessian assessed. In turn, for constrained optimization problems the Lagrangian can be defined. We can then solve for the stationary points of the Lagrangian namely the KKT conditions to solve the problem. In 2006, Wilke et al. [13] introduced gradient-only optimization as an optimality criterion approach for non-smooth and discontinuous problems. The constrained gradient-only optimization problem is defined by finding positive projection points $\mathbf{x}^*$ such that

$$\nabla_A^T f(\mathbf{x}^* + \lambda\mathbf{d})\mathbf{d} \geq \mathbf{0}, \forall\lambda \in (0, r_u], \forall\mathbf{d} \in X : \mathbf{x} + \lambda\mathbf{d}, \tag{6}$$

where X is the convex set of all feasible solutions, r_u a small positive scalar and ∇_A indicates the associated gradient, such that

$$\begin{aligned} \mathbf{g}(\mathbf{x}) &\leq \mathbf{0}, \quad \mathbf{g} \in R^m && (7)\\ \mathbf{h}(\mathbf{x}) &= \mathbf{0}, \quad \mathbf{h} \in R^k && (8)\\ \mathbf{l} \leq \mathbf{x} &\leq \mathbf{u} && (9)\\ & && (10) \end{aligned}$$

To solve the gradient-only optimization problem accurate gradient information is required. Although not apparent at first, to find a positive projection point is easier than the minimum when accurate sensitivities are available. Accurate sensitivities can be computed analytically by differentiating the discretized or continuum equations or numerically via the complex-step method Martins [4], which we will discuss briefly.

ACCURATE SENSITIVITY COMPUTATIONS

Step discontinuous functions require us to consider gradients more carefully, as mathematically they are not everywhere differentiable. The implication is that the gradient field of a scalar step discontinuous function is not defined for all designs in the design domain. However, when we analytically compute the gradient vector for a given temporal or spatial discretization we can compute the sensitivities analytically. A step discontinuity is merely the result of a change in the discretization between designs. Hence, given a discretization we can analytically compute the required gradient vector. In gradient-only optimization we supplement the mathematical gradient field with the associated gradient Wilke et al. [15], which is defined over the entire design domain for step discontinuous functions.

Definition: The **associated gradient** at a point is a vector of partial derivatives where each partial derivative is a one-sided directional derivative when the function is not non-differentiable at that point otherwise it is merely given by the partial derivative at that point when the function is differentiable Wilke et al. [15].

Consider the following simple piece-wise linear step discontinuous function Wilke and Kok [12]:

$$f(x) = \begin{cases} x < 1 : -2x - 0.5 \\ x \geq 1 : -2x \end{cases}, \tag{11}$$

depicted in Figure 2(a). Mathematically, the derivative is not defined at $x = 1$, the derivative is however defined as -2 over the rest of the domain. However, the associated derivative of this function is continuous and defined as -2 over the entire domain, including at $x = 1$.

The univariate function $f(x)$ could be analytically differentiated and evaluated at each point in the domain to obtain the associated derivative. When computing the sensitivities using a finite difference approach we have to acknowledge the fact that the discontinuities may cause complications for conventional finite difference strategies. Fortunately, the complex-step method Martins [4] allows us to accurately compute the associated derivative even at a discontinuity Wilke and Kok [12]. Figure 2(b) depicts the absolute error of the sensitivity at $x = 1$, computed using conventional forward, backward and central difference schemes as well as the complex-step method. The derivative computed using the complex step method yields exactly -2 over the entire domain. Consequently the error is 0 everywhere, including at x = 1, allowing for a full field computation of the derivative of a discontinuous function.

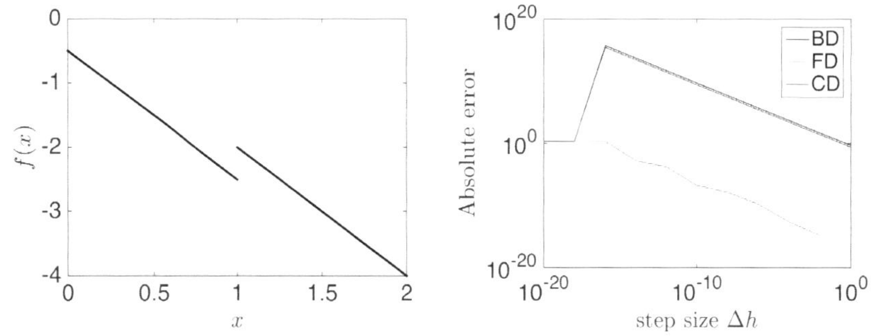

Figure 2: (a) Discontinuous univariate linear function presented in (11) Wilke and Kok [12] and corresponding (b) finite difference sensitivity estimates for the forward (FD), backward (BD) and central difference (CD) schemes.

ALGORITHMS

Conventional gradient based approaches can be classified into three broad categories, namely line-searches, sequential approximations and surrogates.

LINE-SEARCH BASED STRATEGY WITH *POSTERIORI* SELECTED STEP SIZE

A gradient-only line-search approach aims to find a positive projection point along a direction $\mathbf{u}^{\{k\}}$ i.e. to find a sign change in the derivative from negative to positive along the direction $\mathbf{u}^{\{k\}}$ Wilke et al. [14]. Given direction $\mathbf{u}^{\{k\}}$, the aim is to find $\lambda^{\{k\}}$ such that

$$F'_A(\lambda^{\{k\}} + \xi) = \nabla^{\mathrm{T}}_A f\left(x^{\{k-1\}} + \mathbf{u}^{\{k\}}(\lambda^{\{k\}} + \xi)\right)\mathbf{u}^{\{k\}} > 0,$$

and

$$F'_A(\lambda^{\{k\}} - \xi) = \nabla^{\mathrm{T}}_A f\left(x^{\{k-1\}} + \mathbf{u}^{\{k\}}(\lambda^{\{k\}} - \xi)\right)\mathbf{u}^{\{k\}} < 0,$$

where ξ is some small positive value to obtain $\mathbf{x}^{\{k\}} = \mathbf{x}^{\{k-1\}} + \lambda^{\{k\}}\mathbf{u}^{\{k\}}$.

Posteriori selected step sizes resolves $\lambda^{\{k\}}$ within a specified tolerance ξ, using computed sensitivity information.

LINE-SEARCH BASED STRATEGY WITH *A-PRIORI* SELECTED STEP SIZE

Instead of searching for a sign change along a search direction, the step size can be selected *a-priori*. Hence, the user decides on a strategy before the commencement of the optimization run which determines the step size for the entire optimization run. Conventionally, this class of algorithm is referred to as subgradient algorithms and the minimum function value evaluated along the search path is recorded. However, for semi-discontinuous functions subgradient strategies do not necessarily converge to a minimizer. It does however always converge to a positive projection point. The class of gradient-only subgradient approaches merely reports the converged solution as opposed to the minimum function value evaluated along the search path Wilke [10].

SEQUENTIAL APPROXIMATION BASED STRATEGIES

Sequential approximation based strategies usually start with a Taylor series expansion around the current iterate $\mathbf{x}^{\{k-1\}}$,

$$\tilde{f}(\mathbf{x}) = f(\mathbf{x}^{\{k-1\}}) + \nabla_A^{\mathrm{T}} f(\mathbf{x}^{\{k-1\}})(\mathbf{x} - \mathbf{x}^{\{k-1\}}) + \frac{1}{2}(\mathbf{x} - \mathbf{x}^{\{k-1\}})^{\mathrm{T}} \mathbf{H}^{\{k-1\}} (\mathbf{x} - \mathbf{x}^{\{k-1\}}), \quad (12)$$

where $f(\mathbf{x}^{\{k-1\}})$ and $\nabla_A^{\mathrm{T}} f(\mathbf{x}^{\{k-1\}})$ are enforced exactly. Additional information at other points in the design is then used to approximate $\mathbf{H}^{\{k-1\}}$. For spherical approximations the function value at another design usually used to resolve $\mathbf{H}^{\{k-1\}}$.

For gradient-only only approximations we differentiate (12) to obtain

$$\nabla \tilde{f}(\mathbf{x}) = \nabla_A f(\mathbf{x}^{\{k-1\}}) + \mathbf{H}^{\{k-1\}}(\mathbf{x} - \mathbf{x}^{\{k-1\}}), \quad (13)$$

where $\nabla_A f(\mathbf{x}^{\{k-1\}})$ is enforced exactly, while only gradient information at other designs is used to approximate $\mathbf{H}^{\{k-1\}}$ Wilke et al. [14].

SURROGATE BASED STRATEGIES

A number of response surfaces or surrogates strategies approximate some non-linear function $f(\mathbf{x})$ as a linear combination of m chosen non-linear basis functions $\phi_j(\mathbf{x}, \mathbf{x}_j^c)$, $j = 1, \ldots, m$, that are centered around the spatial point $\mathbf{x}_j^c$ given by

$$f(\mathbf{x}) \approx \sum_{j=1}^{m} w_j \phi_j(\mathbf{x}, \mathbf{x}_j^c) = \tilde{f}(\mathbf{x}). \quad (14)$$

The weights are usually computed from the enforcement that the surrogate response should recover the known response at m distinct designs. To construct a gradient-only surrogate we enforce that the gradient of the surrogate response should recover the known gradient at m distinct designs given by

$$\begin{bmatrix} \frac{\partial \phi_1(\mathbf{x}_1, \mathbf{x}_1^c)}{\partial \mathbf{x}} & \frac{\partial \phi_2(\mathbf{x}_1, \mathbf{x}_2^c)}{\partial \mathbf{x}} & \cdots & \frac{\partial \phi_k(\mathbf{x}_1, \mathbf{x}_k^c)}{\partial \mathbf{x}} \\ \vdots & & & \\ \frac{\partial \phi_1(\mathbf{x}_k, \mathbf{x}_1^c)}{\partial \mathbf{x}} & \frac{\partial \phi_2(\mathbf{x}_k, \mathbf{x}_2^c)}{\partial \mathbf{x}} & \cdots & \frac{\partial \phi_k(\mathbf{x}_k, \mathbf{x}_k^c)}{\partial \mathbf{x}} \end{bmatrix} \begin{bmatrix} w_1 \\ w_2 \\ \vdots \\ w_k \end{bmatrix} = \begin{bmatrix} \nabla f(\mathbf{x}_1) \\ \vdots \\ \nabla f(\mathbf{x}_k) \end{bmatrix}. \quad (15)$$

Since any constant added to a surface would not affect the gradient of the surface, the function value at a point is enforced to avoid a singular coefficient matrix Wilke [11]. The overdetermined system is solved using the usual least squares approach.

Consider Figure 3(a) that depict a smooth quadratic function

$$f(\mathbf{x}) = \sum_{i=1}^{n} 10^{i-1} x_{i-1}^2, \tag{16}$$

while Figure 3(b) depicts the same function modified by step discontinuities

$$f(\mathbf{x}) = \sum_{i=2}^{n} 10^{i-2} x_{i-1}^2 + 10^{i-1} x_i^2 + a(\text{sign}(x_{i-1})) - b(\text{sign}(x_i)) + c(\text{sign}(x_{i-1}))(\text{sign}(x_i)), \tag{17}$$

with $n = 2$, $a = 100$, $b = 50$ and $c = 133$ selected Wilke [11].

Depicted in Figure 3(c) is a conventional radial basis surrogate that is constructed by enforcing the function value of the discontinuous test function at nine points in the design domain, while Figure 3(d) depicts the gradient-only constructed radial basis surrogate that enforces the gradient of the discontinuous test function at nine points Wilke [11]. It is clear that the gradient-only constructed radial basis function exactly recovered the smooth quadratic function, while the conventional radial basis surrogate is a poor representation of either the smooth or discontinuous functions.

APPLICATIONS

INVERSE PROBLEMS

Engineering simulations often rely on underlying models that first need to be characterised. These models are mostly systems of non-linear differential equations that describe some constitutive relation, reaction rate or mixing problem Velten [9], Borst et al. [2]. Usually these problems are formulated as weighted least squares problems Strang [8].

Given a known response $\mathbf{d}_e$ and some model $\mathbf{d}_m(\mathbf{x})$ that aims to describe the known response by the following model parameters $\mathbf{x}$. The difference between the known response and the model response is then given by

$$\mathbf{E}(\mathbf{x}) = (\mathbf{d}_m(\mathbf{x}) - \mathbf{d}_e), \tag{18}$$

which can be reduced to a scalar $f(\mathbf{x})$ by the following weighted projections

$$f(\mathbf{x}) = \mathbf{E}(\mathbf{x})^T \mathbf{W} \mathbf{E}(\mathbf{x}), \tag{19}$$

(20)

to obtain the weighted least squares problem Strang [8]. Choosing $\mathbf{W}$ as the identity matrix results in ordinary least squares. The aim is to find $\mathbf{x}$ that minimizes $f(\mathbf{x})$. However, when the underlying system of differential equations are solved using a variable time stepping strategy the resulting sum of errors squared may become discontinuous.

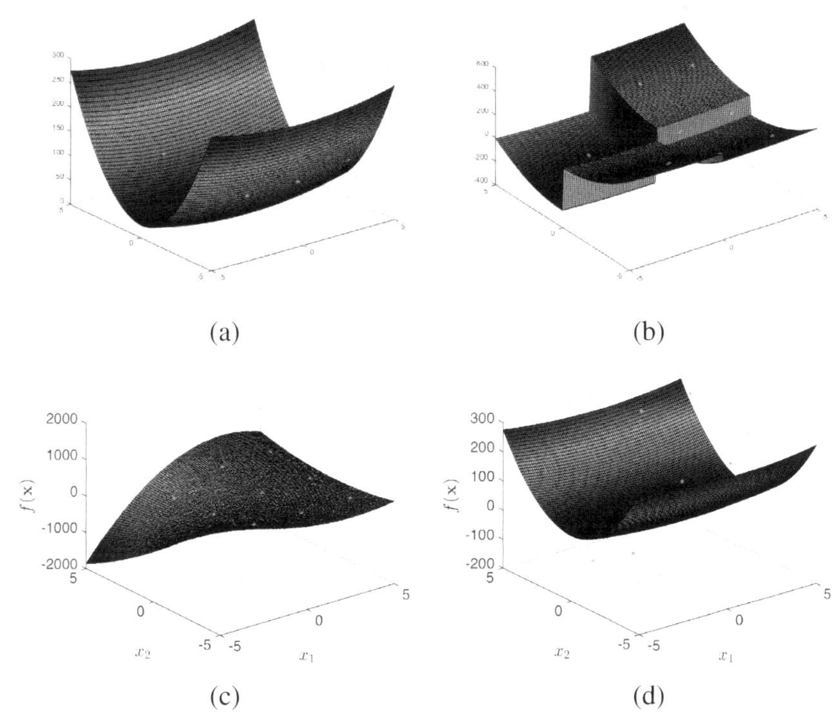

Figure 3: (a) Smooth quadratic function that is modified to be a (b) step wise discontinuous quadratic function that is evaluated at 9 points, from which a (c) conventional and (d) gradient-only radial basis surrogate are constructed Wilke [11].

As an explanatory example, consider the coupled first order non-linear system of differential equations (Lokta-Volterra equations Yorke and W.N. Anderson [16]) given by

$$\frac{dx(t)}{dt} = \alpha x(t) - \beta x(t)y(t) \tag{21}$$

$$\frac{dy(t)}{dt} = \delta x(t)y(t) - \gamma y(t), \tag{22}$$

$$\tag{23}$$

which was first proposed to model autocatalytic chemical reactions. Two initial conditions on $x(0) = x_0$ and $y(0) = y_0$ completes the formulation.

For illustrative purposes we will consider a one dimensional least squares problem that we will illustrate graphically. For our problem we choose $\beta = \frac{1}{3}$, $\delta = \gamma = 1$ and $x_0 = y_0 = 0.9$, while the aim is to find α that best matches $x(t)$ and $y(t)$ at $t = 8$ seconds. Instead of making use of an experimentally measured desired response we compute the desired response for $\alpha = 0.5$ using the forward Euler integration scheme with 1000 equal time steps between 0 and 8 seconds. Hence, the solution is known which allows us to quantify our predicted solutions.

We then solve the least squares problem for α between 0 and 1 using 1000 equally spaced points. For our forward Euler integration scheme we choose three mean μ numbers of equally spaced points, namely 20, 40 and 80. This allows for the accuracy of the solution to be investigated when the number of points to solve the system of differential equations is varied. Note

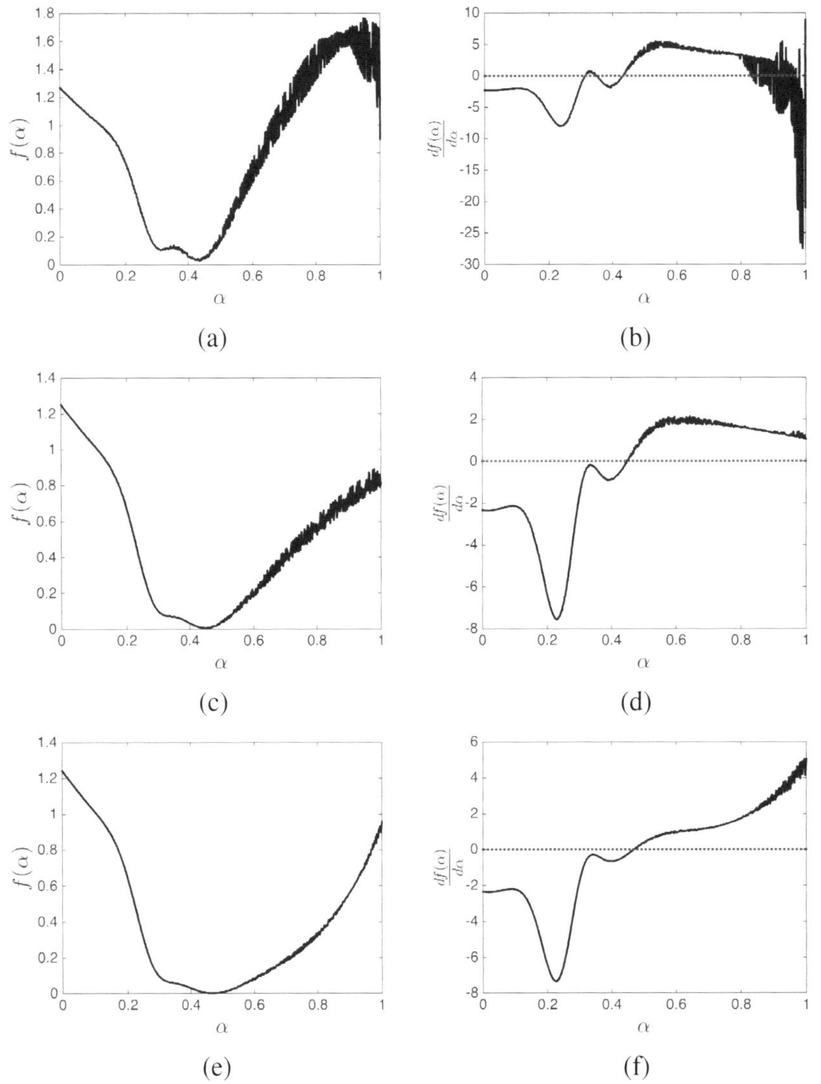

Figure 4: The least squares response for the sum of the error squared $f(\alpha)$ and the derivative of the sum of the error squared $\frac{df(\alpha)}{d\alpha}$ when using (a)-(b) 20, (c)-(d) 40, and (e)-(f) 80 equally spaced points to solve the system of differential equations using the forward Euler integration scheme.

that the number of solution points effectively vary the fidelity of the numerical model. Hence, $\mu = 20$ indicates our lowest fidelity model, whereas $\mu = 80$ represents our highest fidelity model. For a chosen μ, we choose the number of solution points as a uniformly distributed random number between 0.9μ and 1.1μ to demonstrate the effect of small variations in the number of sampling points for low to higher fidelity models. These variations are typical of a variable time stepping approach. Instead of only computing the sum of the errors squared we also compute the derivative of the sum of the errors squared w.t.r. to α to graphically demonstrate the benefits of a gradient-only approach to solve the problem.

The results are depicted in Figures 4(a)-(f) and represent the sum of the errors squared $f(\alpha)$ on Figures 4(a), (c) and (e) as well as the derivatives $\frac{df(\alpha)}{d\alpha}$ of the the sum of the errors squared on Figures 4(b), (d) and (e). On the derivative plot a red dashed line is included to indicate the level around which the sign in the derivative changes.

By considering the least squares response $f(\alpha)$ and $\frac{df(\alpha)}{d\alpha}$ of the lowest fidelity model in Figure 4(a)-(b), it is clear that both function value and derivative information depicts a significant number of local minima for α above 0.8. However, when considering $\alpha < 0.8$, it is clear that based only on derivative information there are only two minima, this is in stark contrast to the function value information that is still hampered by a plethora of local minima for α above 0.45. By increasing the fidelity of the problem, as depicted in Figure 4(c)-(d), it is clear that based on derivative information there is only one minimum, while the function value function value information portrays a highly multimodal function. Finally, by sufficiently increasing the fidelity of the problem, as depicted in Figure 4(e)-(f), the function value and derivative information portrays consistent information, as the discretization error becomes negligible.

SIMULATION BASED DESIGN

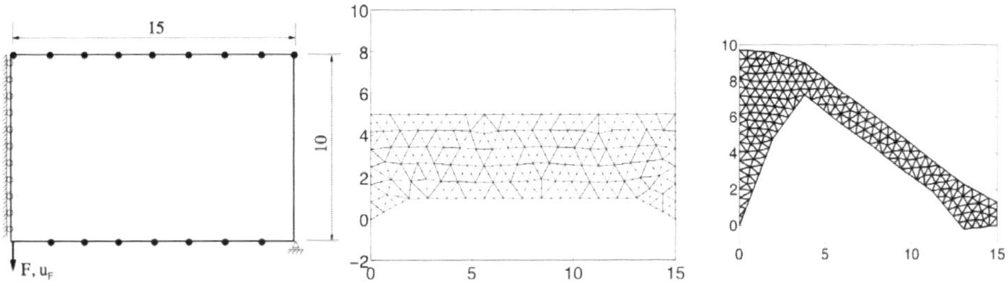

(a) (b) (c)

Figure 5: (a) Half symmetry model for the Michell structure, starting the optimization with an (b) initial coarse mesh that converges to a (c) finer final mesh.

In simulation based design any change in geometry requires the spatial discretization to be adjusted. In some instances the adjustments can be easily affected in a smooth continuous manner e.g. in sizing optimization problems. The general shape optimization problem is stated in Problem 1.

Problem 1: Find the optimal shape Ω^* such that

$$F^*_\Omega = F\Omega(\Omega^*, S^*_\Omega) = min_\Omega F_\Omega(\Omega, S_\Omega) : G(\Omega) \leq V_0, \quad (24)$$

with S_Ω the state solution of the partial differential equations $p_\Omega(\Omega, S_\Omega) = 0$, which characterizes the system [11, 14]. Here, Ω denotes the unknown shape, which is a subset in R^3. Ω is usually defined by a finite set of control variables $x \in R^3$ and $G(\Omega) \leq V_0$ the set nonlinear design constraints, with V_0 given. In general, Problem 1 is intractable. However, we can solve an approximate problem by discretizing the shape Ω with a computational grid Λ. The computational grid Λ can then be used to approximate the state equation $p_\Lambda(\Lambda, S_\Lambda) = 0$ and to define an approximate solution S_Λ to the state equation [11]. The set of nonlinear design constraints $G(\Omega) \leq V_0$ can then also be expressed as m functions of the computational grid Λ as follows

$$g_i(\Lambda(\mathbf{x}), \mathbf{x}) \leq 0, \quad i = 1, \ldots, m \quad (25)$$

Problem 1 is usually cast in a mathematical programming framework, where the aim is to find the unknown shape Ω^* that minimizes the cost function $F_\Omega(\Omega^*, S^*_\Omega)$ subject to $G(\Omega) \leq V_0$. Consider the Michell problem Wilke et al. [14] in Figure 5(a), a significant benefit of gradient-only optimization is that it allows for a continuous grading of model fidelity from low to high as the optimizer progresses as depicted in Figure 5(b)-(c), without hampering the convergence of the problem.

CONCLUSION

In the last ten years a number of advancements have been made to the field of gradient-only optimization. The class of regularly available line search algorithms have been extended to include modified subgradients. Approximation strategies have been extended to include gradient-only constructed surrogates. The complications of computing accurate gradients have been supplemented by convenient complex-step approaches to allow for numerical sensitivities to be computed that are always accurate even when discontinuous functions are considered. As gradient-only optimization has been placed on a solid theoretical foundation it is perceivable that various domains of applications will start to emerge.

References

[1] J. Arora. *Introduction to Optimum.* Academic Press, 3 edition, 2011.

[2] R. De Borst, M.A. Crisfield, J.J.C. Remmers, and C.V. Verhoorsel. *Non-Linear Finite Element Analysis of Solids and Structures.* John Wiley & Sons, Ltd, 2012.

[3] A.I. Khuri and S. Mukhopadhyay. Response surface methodology. *WIREs Computational Statistics*, 2, 2010.

[4] J. Martins. The complex step derivative approximation. *ACM Transactions on Mathematical Software*, 29, 2003.

[5] A. Nooraziah and V. J. Tiagrajah. A study on regression model using response surface methodology. *Applied Mechanics and Materials*, 666, 2014.

[6] M.A. O'Connell and R.D. Wolfinger. Spatial regression models, response surfaces, and process optimization. *Journal of Computational and Graphical Statistics*, 6, 1997.

[7] A. Schleupen, K. Maute, and E. Ramm. Adaptive fe-procedures in shape optimization. *Structural and Multidisciplinary Optimization*, 19, 2000.

[8] G. Strang. *Computational Science and Engineering*. Wellesley-Cambridge Press, 2007.

[9] K. Velten. *Mathematical Modeling and Simulation.* Wiley-VCH, 2009.

[10] D.N. Wilke. Modified subgradient methods for remeshing based structural shape optimization. In *Thirteenth International Conference on Civil, Structural and Environmental Engineering Computing*, 2011.

[11] D.N. Wilke. How to get rid of discontinuities when constructing surrogates from piecewise discontinuous functions. In *The Fifth International Conference on Engineering Optimization*, 2016.

[12] D.N. Wilke and S. Kok. Numerical sensitivity computation for discontinuous gradient-only optimization problems using the complex-step method. In *Tenth World Congress on Computational Mechanics*, 2016.

[13] D.N. Wilke, S. Kok, and A.A. Groenwold. Adaptive unstructured remeshing using gradient-only optimisation algorithms for shape optimisation. In *The Fifth International Conference on Engineering Computational Technology*, number ECT222, 2006.

[14] D.N. Wilke, S. Kok, and A.A. Groenwold. The application of gradient-only optimization methods for problems discretized using non-constant methods. *Structural and Multidisciplinary Optimization*, 40, 2010.

[15] D.N. Wilke, S. Kok, J.A. Snyman, and A.A. Groenwold. Gradient-only approaches to avoid spurious local minima in unconstrained optimization. *Optimization and Engineering*, 14, 2013.

[16] J.A. Yorke and Jr. W.N. Anderson. Predator-prey patterns (volterra-lotka equations). *Proceedings of the National Academy of Sciences*, 70, 1973.

The Sensitivity of Static Tyre Tests and the Effects of Aged Tyres

Kraig RS Wright
University of Pretoria
u11106574@tuks.co.za

Schalk Els
University of Pretoria
schalk.els@up.ac.za

ABSTRACT

With increased dependency on modelling as the preferred method for design and testing, the need for accurate tyre models have become a necessity in vehicle design. The test data gathered from any tyre also needs to be an accurate representation of the tyre to be modelled. This paper analyses the sensitivity of test data and the repeatability of tests as influenced by several factors including tyre age, wear and test rig variations.

Specific conclusions are made on the sensitivity of consecutive cyclic loadings, different waveforms as input loadings and the effect of vertical loading velocity on the resultant force-displacement relationship generated by the tyre. Observable variations were found depending on the loading velocity and input waveform. The repeatability of tests, investigated by comparison between equivalent tests on separate setups, demonstrates significant variations emphasizing the fact that test setup consistency is of considerable importance.

KEYWORDS

Tyre, sensitivity, aging

INTRODUCTION

Modern vehicle design involves the extensive use of computer-aided modelling in place of dynamic tests as a means of lowering costs and improving concept to operation efficiency. Engineering decisions are therefore commonly made based upon results gained from these models. With this in mind it is important to be aware of the validity of the model being used. Whilst the validity of these models is generally acquired through comparison of simulation results to test data, for this paper an investigation was conducted on the repeatability and sensitivity of test data acquired.

Specific to vehicle design there exists a degree of uncertainty in the modelling and testing of tyres. Whilst the purpose of the tyre is simple it is inherently a composite material given its combination of different materials. The interaction between these materials increases the non-linearity of the forces and moments generated by the tyre. Due to the fact that all forces and moments exerted and experienced by all vehicles (except aerodynamic) are transmitted through the tyres an accurate full vehicle model is generally limited by the accuracy of the tyre model.

The field of modelling tyres is a vast and complex one with several different methods and simplifications in attempts to portray the complex non-linearity that is tyres. However, as is the case with most tyre models, test data of the actual tyre requiring a model is a necessity. It is therefore of the utmost importance that the test data acquired is as accurate a representation of the tyre as that of the model to be. With this is mind and depending on the desired accuracy of the tyre model, several factors need to be considered in order to ensure the desired accuracy is achieved.

This article serves to deliver on three of these factors concerning the testing and modelling of tyres, namely the test sensitivity, test repeatability and in future work on the effects of the age of tyres. In order to successfully report on these factors several tests were conducted.

More specifically, tests were conducted on a Michelin LTX AT^2 tyre where the specific intention was to obtain data able to answer questions on the sensitivity of tests in the vertical direction of the tyre and secondly on repeatability of the test setup and procedure. This involved questions on methods for vertically actuating the tyre and how sensitive the tyre is to small changes on these imposed loadings. Furthermore two separate setups using the same equipment were built one after the other to check that the same results in terms of vertical tyre stiffness were achieved.

TEST SETUP

Figure 1 shows the test setup used to statically test the vertical stiffness of a tyre. A 100 *kN* servo-hydraulic actuator is used to load the tyre in its vertical direction via whilst an axial load cell and linear variable displacement transformer (LVDT) measured the vertical load and displacement of the actuator respectively. For redundancy purposes a Wheel Force Transducer (WFT) is used to measure the forces and moments in and about a six component axis at the hub of wheel and two laser displacement transducers to measure the displacement of the plate as well as the deflection of the tyre mounted on the pedestal.

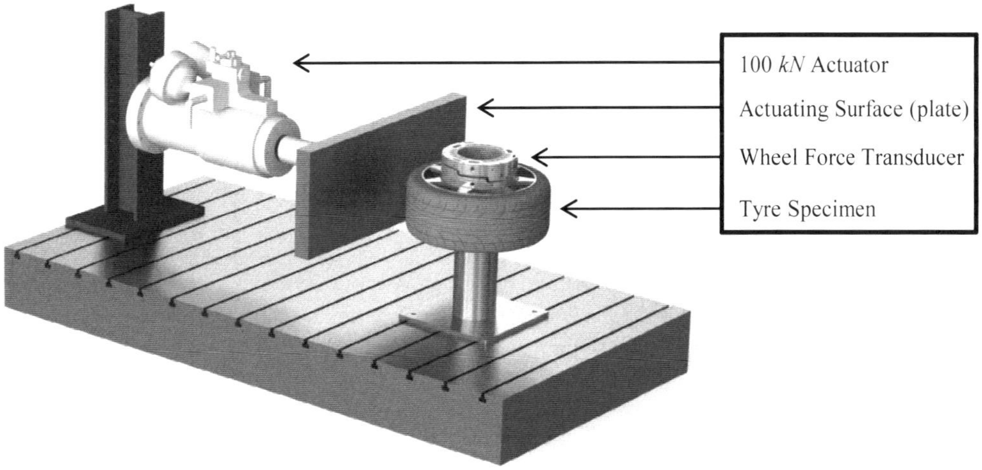

Figure 1. Simplified test setup.

TESTING SENSITIVTY ANALYSIS

The sensitivity of the tests conducted involved the analysis and detailed post-processing of data captured. The intention of such an analysis was to objectively quantify the relative differences between certain parameters and determine which method produces the most consistent results. Thus as seen in Table 1, parameters and their effects on the respective stiffnesses were scrutinised:

Table 1. Sensitivity Parameters on variability of stiffnesses

Parameter	Direction	Variables
Excitation Frequency (Hz)	Vertical	0.1, 0.2, 0.5
Input Function	Vertical	Sine or Triangular
Cycles Sensitivity	Vertical	1 – 4 cycles

In order to scrutinise the sensitivity of the setup a number of variables were held constant. Table 2 shows the two tests that were used for comparison. In both cases the same tyre was used at an inflation pressure of 145 *kPa*, at zero degrees camber on a flat steel surface.

Table 2. Details of the two tests used in the sensitivity analysis

Variables	Test 1	Test 2
Input Waveform	**Sinusoid**	**Triangular**
Excitation Frequency (Hz)	0.1, 0.2, 0.5	0.1, 0.2, 0.5
Analysed Cycles at each Velocity	4, 4, 4	4, 4, 3

CYCLE SENSITIVITY

The cycle sensitivity includes checking the relative difference between vertical stiffnesses attained from consecutive cyclic loads, this was completed for both the sinusoid and triangular waveform input.

Depicted in Figure 2 is the full extent of all the force-displacement relationships (non-linear stiffness) produced by the two tests of interest. From this figure it appears that all force-displacement relationships obtained are exactly the same and some very small variations are observed with the other velocities.

A means of objectively quantifying the relative differences between force-displacement relationships is to plot the change in vertical force exerted by the tyre between the load and unload force-displacement lines. This change in vertical load for the same displacement arises as a result of the hysteresis between rubber particles in the tyre Gillespie [1] amongst other friction mechanisms. Ultimately this loop is generated due to the energy losses in the tyre, as a result it should be a constant value for the tyre which in turn makes it applicable for comparison between different force-displacement lines. Due to the fact that the force-displacement relationships obtained in Figure 2 are so consistent this method will visually amplify their relative error.

These hysteresis difference values are plotted in Figure 3 for each loading cycle. With specific reference to the sinusoid cycles it shows that a variation of approximately 50 *N* difference between cycles is generated. This error is almost completely eliminated with the 0.2 *Hz* cycles, however, due to their magnitude and considering the macro perspective of the tyre data, these differences are very small as they exist in the range of less than 1% of the total load exerted on tyre. These small discrepancies are consistent for the triangular waveform input as well.

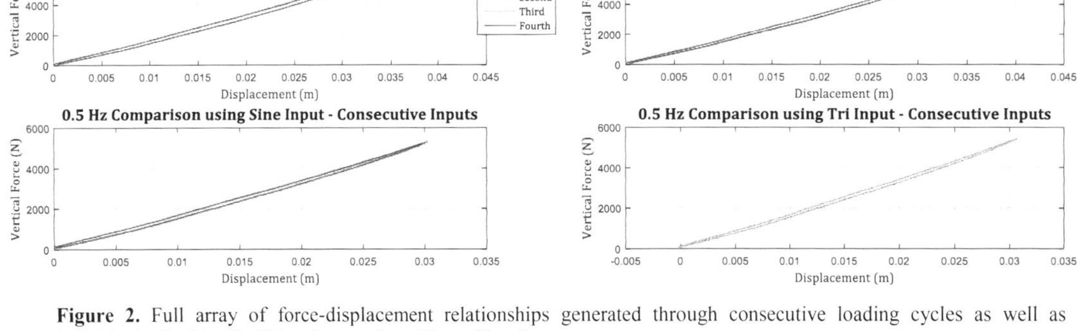

Figure 2. Full array of force-displacement relationships generated through consecutive loading cycles as well as individual velocities for Triangle waveform Sinusoid and.

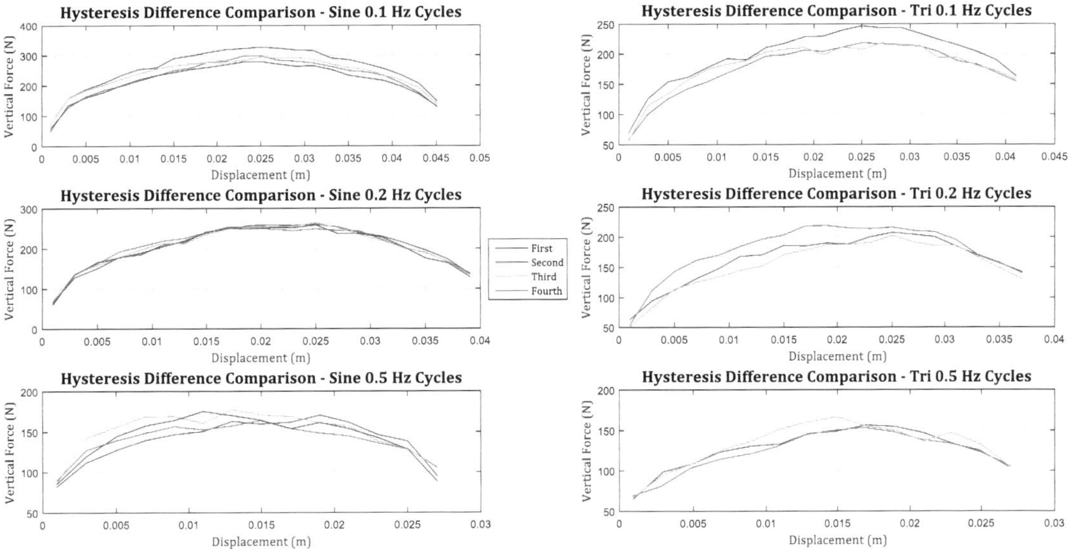

Figure 3. Full array of hysteresis loop differences as a function of displacement for all velocities of Sinusoid and Triangular waveform

Therefore by selecting any single load and unload cycle to generate static force-displacement relationships a maximum potential error of 50 N on the total vertical load is possible relating to less than 1% error. This trait is possibly driven by the variety of friction mechanisms inherent in all tyres, these include the viscoelasticity of the cord-rubber interaction, hysteresis between rubber particles as well as the adhesion and hysteresis at the tyre-road interface. The most influential of which is most likely the tyre contact patch 'stick and slip' action whilst loading and unloading.

VELOCITY SENSITIVITY

Before beginning the analysis of the vertical force-displacement relationships generated by different loading velocities it was important to note the accuracy of the actual input signals compared to that of the desired signal, as shown in Figure 4.

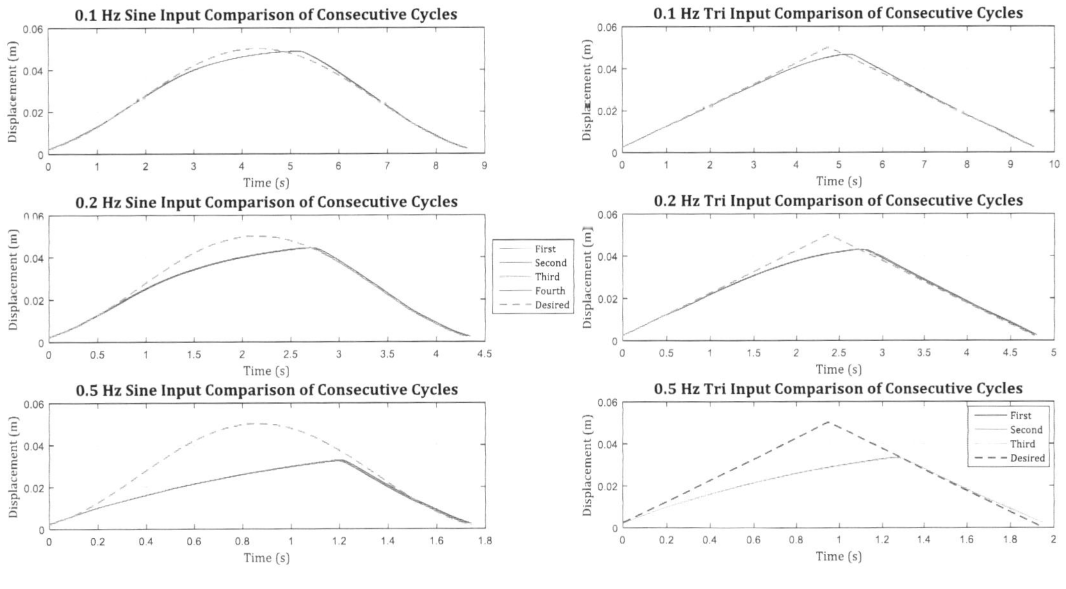

Figure 4. Displacement inputs as a function of time, comparison of desired vs. achieved displacement

Significant discrepancies are noted between the actual and desired signals with a clear dependency on velocity. Importantly the constant velocity expected from the triangular waveform is clearly lost with the two higher velocity inputs. This loss in shape and magnitude is a combination of the actuator reaching its velocity limit as well as possible PID control setting error. As a result it is important to at the very least be aware of these inconsistencies as they are unlikely to have a significant effect on the force-displacement relationships obtained.

The force-displacement relationships generated from the triangular waveform are shown in Figure 5 and whilst similar correlation is observed as with the sinusoidal waveform there is slightly more variation. Nevertheless the upper force-displacement lines at any given displacement only demonstrate variations between velocities of around 50 *N* as was seen with the sinusoidal waveform.

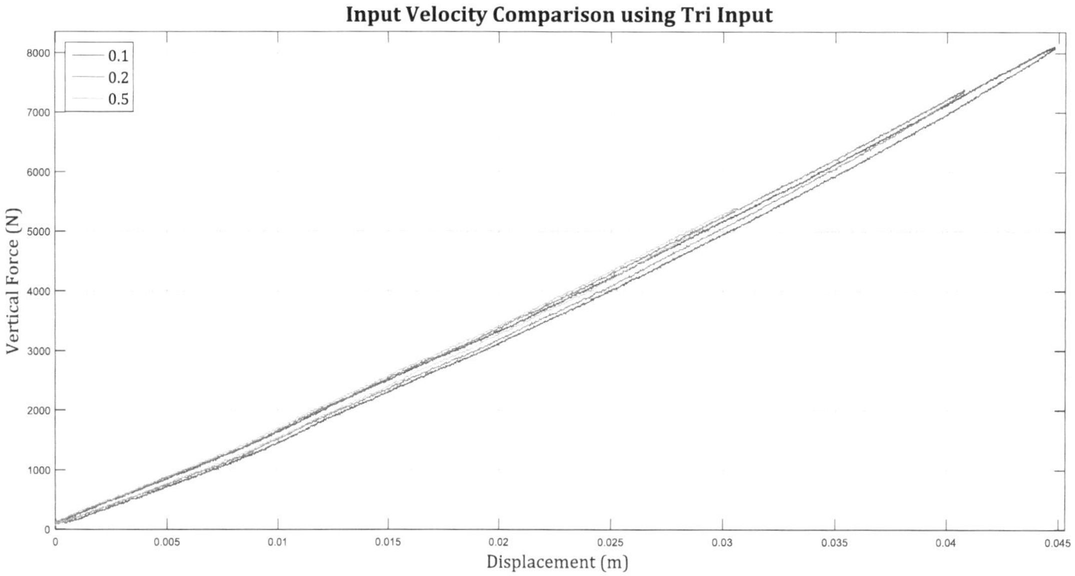

Figure 5. Triangular wave input, comparison of the third cycle at three difference loading velocities (Hz).

HYSTERETIC DAMPING COEFFICIENT COMPARISON

The hysteretic damping coefficient was calculated in an attempt to quantify the variation in both force-displacement relationship and the hysteretic damping loop.

The relevant literature found on this topic was from Section 2.10 of Rao [2] named 'Free Vibration with Hysteretic Damping.' Whilst the formulation of this damping coefficient makes a few crude simplifications of the complexity of the damping in a tyre, the results are still relevant as they will be used by means of comparison. Equation (1) represents the area within the hysteresis loop and can be further simplified to give the hysteretic damping coefficient in *N/m*, h, through Equation (2), where ΔW represents the energy loss due to damping in *Joules*, ω is excitation frequency in *rad/s*, c is the damping constant in *Ns/m* and finally X is the excitation amplitude in metres.

$$\Delta W = \pi \omega c X^2 \tag{1}$$

$$c = \frac{h}{\omega} \tag{2}$$

Equation (2) allows for the independence of the internal friction from frequency in Equation (1) which is subsequently reduced to Equation (3) which can then be used for easy comparison of the hysteretic damping loop generated for each test run.

$$\Delta W = \pi h X^2 \tag{3}$$

Aspects from Table 2 are compared using this coefficient in Figure 6. The red lines for each velocity show the average of the three or four runs depending on the waveform. This method amplifies the errors previously noted on repeatability of consecutive cycles, however, in both cases of sinusoidal and triangular waveforms the largest variation from the mean is 1000 *N/m* which translates to 50 *N* on a stroke of 50 *mm*. This confirms the previously assumed negligibility of these variations in consecutive cycles.

Table 3 illustrates the variation of the hysteretic damping coefficient as a dependence on velocity. Evident is greater consistency than observed between consecutive cycles and thereby demonstrating how the hysteretic damping loop is independent of loading and unloading velocity.

Table 3. Average hysteretic damping coefficient for each velocity on both Sinusoid and Triangle waveform inputs.

Excitation Frequency (Hz)	Sinusoid Coefficients (N/m)	Triangular Coefficients (N/m)
0.1	6 941.9	5 907.4
0.2	6 873.6	5 879.2
0.5	6 839.9	6 054.7
Average	**6 884.4**	**5 947.1**

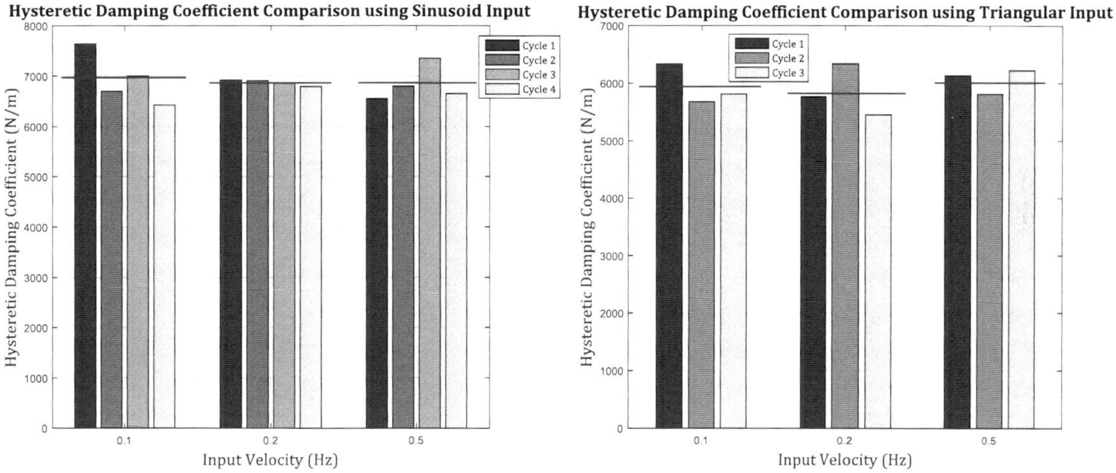

Figure 6. Full array of hysteretic damping coefficients for each cycle at each velocity for both Sinusoid and Triangular waveform

INPUT WAVEFORM COMPARISON

From the set of figures shown thus far as well as Figure 7 it is apparent that the triangular waveform is less repeatable through consecutive cycles as the force-displacement relationship, hysteresis difference and hysteretic damping coefficient demonstrate slightly more inconsistent results compared to the sinusoid waveform. Figure 7, however, shows that the variability that does exist is of negligible magnitude. This is further confirmed in Figure 8 where despite a slightly larger hysteresis difference being observed than previously seen in this article it is still very small compared to the vertical loads applied to the tyre. This variability is more prevalent when comparing the damping coefficients in Table 3. The average value across the input velocities for the sinusoidal input is 6884 *N/m* whereas in the case of the triangular input is 5947 *N/m*. These discrepancies lie in the range of 13 to 15 % and should be accounted for in future tests by using a near frictionless surface to actuate the tyre thereby removing a large amount of tyre contact damping. Alternatively, the surface material, surface finish and roughness should be specified for each test.

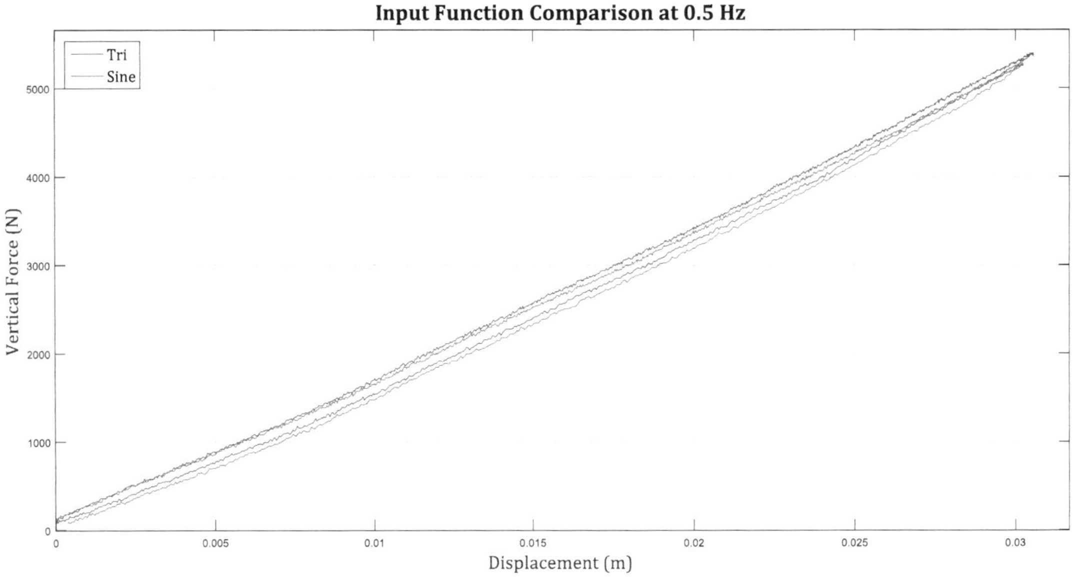

Figure 7. Sinusoid and Triangular waveform as input, force-displacement relationship comparison at 0.5 Hz

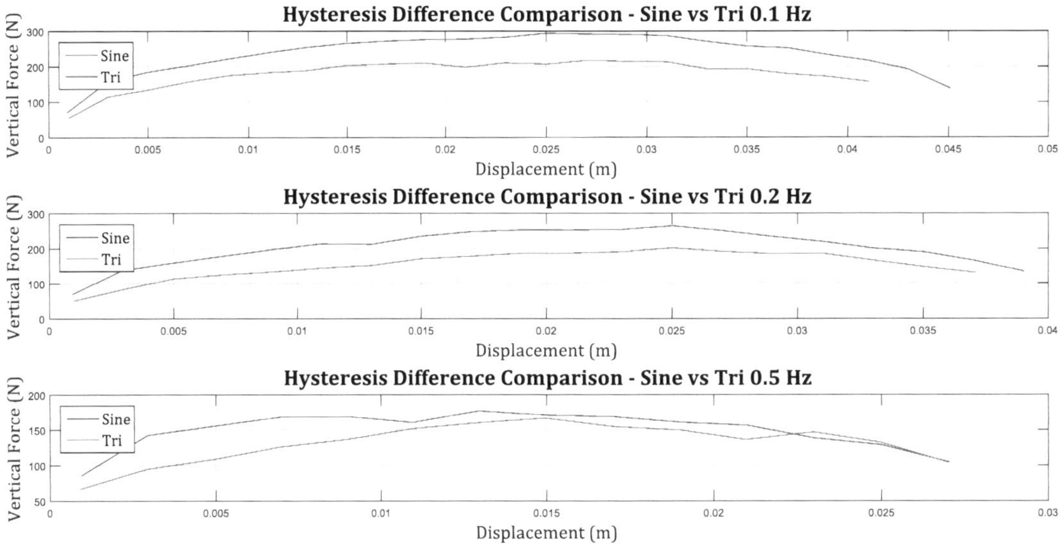

Figure 8. Sinusoid and Triangular waveform as input, comparison of hysteresis loop difference for all velocities

TESTING REPEATABILITY ANALYSIS

The repeatability of the testing procedure was conducted through the repetition of selected tests after the test setup had been broken down completely and then rebuilt. Once again the same tyre was used with the same flat steel surface at zero degrees camber. Table 4 shows the details regarding the two tests.

Table 4. Details of the two tests used in the repeatability analysis

Variables	Test 1	Test 2
Pressure (kPa)	101.3	100.6
Input Waveform	**Sinusoid**	**Load and Hold**
Excitation Frequency (Hz)	0.1, 0.2, 0.5	Constant 0.1
Date of Test	06-08-2015	10-09-2015
Time of Test	07:57	18:45

Two different input waveforms were used for each test, namely a sinusoid in the first and a trapezium in the second. The change in waveform for the second test was an attempt to allow hysteresis in the contact patch to reach a steady state before unloading. As discussed in the testing sensitivity section of this article above, it was fair to assume a negligible change in tyre force-displacement relationship due to the change in these input waveforms.

The inflation pressure in the tyre can have a large influence on the force-displacement relationship characteristics of the tyre, thus making it important to ensure that it remained consistent between and throughout each test. Nonetheless the difference of 0.7 *kPa* between the two tests was confirmed to be negligible based on results from previous tests with inflation pressure differences 4 *kPa*.

Figure 9 depicts the two force-displacement relationships generated by the individual tests. Surprisingly evident from the figure is a significant difference in vertical force-displacement relationship from the two tests. This variation in force-displacement relationship is experienced at the same vertical load and tends to point to an error in the setup of one of the test rigs. This hypothesis can be motivated by the fact that amongst all the sensitivities tested in this report there has never been a difference as large as that demonstrated in Figure 9. Whilst the loading and unloading waveform are very different and the velocities may also differ slightly these are two parameters discussed in detail in the sections above and never produced differences to this extent.

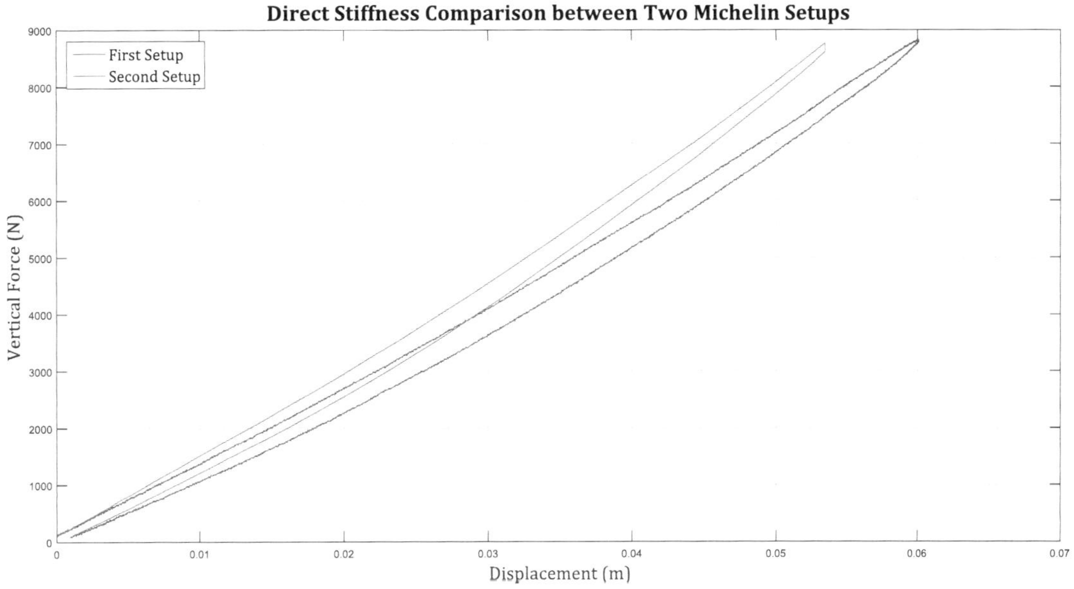

Figure 9. Vertical stiffness comparison of the two separately setup tests

A possible explanation for this apparent change in the force-displacement relationship can be due to a change in the stiffness of the test rig. This implies that a certain assembly error occurred allowing more displacement for the same load on the first setup. It was noticed that the upper bearing on the hub used to mount the tyre was incorrectly assembled and could potentially explain this change in the force-displacement relationship. However, after realising this assembly mistake an additional vertical stiffness test was conducted and it was found that only a smoothing out of the data was noticeable. There were no significant changes in the forces and moments produced by the tyre supporting the error presented in Figure 9.

CONCLUSION

The testing sensitivity analysis revealed interesting however negligible variations between the two input waveforms as well as between consecutive cycles. It is fair to assume less than 1% error is carried if a single loading cycle is used as opposed to the average of multiple. Furthermore it was later discovered that the triangular waveform is more likely able to isolate the influence of the hysteresis present between rubber particles in the tyre from the hysteresis and adhesion present in the tyre contact patch. Although it was noted that a near frictionless contact surface would be needed to accurately prove this hypothesis.

The velocity dependence on the tyre's force-displacement relationship and damping generated was less than anticipated. In this regard the greatest change and possible influence on the results was the discrepancy between the desired signal input and that of the signal delivered to the tyre. Interesting, however, was that a visible but insignificant change in the tyre force-displacement relationship depending on the input velocity was noticed. Due to these small variations in the force-displacement relationship it would be fair to assume a slow and constant velocity would generate the correct force-displacement relationship.

The obvious and most significant observation made in this article was accounted for by the drastic difference in force-displacement relationships obtained for the same tyre on different test setups. This variation in the force-displacement relationship clearly indicates a possible error in setup of one the tests. Nonetheless this demonstrates the necessity for methodical and consistent procedure when setting up a test as significant changes can be easily observed which make all other sensitivities completely insignificant.

Future work will include the testing and validation of tyre stiffnesses as a function of tread wear whilst predicting these potential changes through a validated FTire tyre model.

REFERENCES

1 **Gillespie, Thomas D. 1992.** *Fundamentals of Vehicle Dynamics.* Warrendale : Society of Automotive Engineers, Inc., 1992.

2 **Rao, Singiresu S. 2011.** *Mechanical Vibrations.* Upper Saddle River : Pearson, 2011.

AUTHOR INDEX